AIDS UPDATE 2000

An Annual Overview of Acquired Immune Deficiency Syndrome

GERALD J. STINE, PH.D.

University of North Florida, Jacksonville

PRENTICE HALL, *Upper Saddle River, New Jersey 07458*

> **AIDS IS A WAR THAT NO ONE WANTS TO LOSE**
> **BUT NO ONE YET KNOWS HOW TO WIN**
>
> • • • • • •
>
> *This book, as with my nine other HIV/AIDS college-level textbooks,*
> *is also dedicated to*
> *those who have died of AIDS,*
> *those who have HIV disease,*
> *and to those who must prevent the spread of this plague—*
>
> EVERYONE, EVERYWHERE.

Senior Editor: *Halee Dinsey*
Executive Managing Editor: *Kathleen Schiaparelli*
Assistant Managing Editor: *Lisa Kinne*
Manufacturing Buyer: *Michael Bell*
Director of Marketing, ESM: *John Tweeddale*
Senior Marketing Manager: *Marty McDonald*
Art Director: *Jane Conte*
Cover Designer: *Bruce Kenselaar*
Production Supervision/Composition: *WestWords, Inc.*

ISBN 0-13-082196-9

Prentice-Hall International (UK) Limited, *London*
Prentice-Hall of Australia Pty. Limited, *Sydney*
Prentice-Hall Canada Inc., *Toronto*
Prentice-Hall Hispanoamericana, S.A., *Mexico*
Prentice-Hall of India Private Limited, *New Delhi*
Prentice-Hall of Japan, Inc., *Tokyo*
Pearson Education Asia Pte Ltd
Editora Prentice-Hall do Brasil, Ltda., *Rio de Janeiro*

The war on AIDS is still closer to the **beginning** than it is to the **end** but I have never been more positive about the future for those who are HIV-infected in the United States and other developed nations. The use of anti-HIV drugs in combination has led to a dramatic reduction in HIV-infected babies and AIDS-related deaths, as well as a significant reduction in new AIDS cases and reductions in opportunistic infections. These events, along with new insight into the biology and pathology of HIV may lead to a preventive vaccine and provide a shining light against a stark history of the first 19 years of a pandemic first reported in the United States in 1981.

Unfortunately, the high cost of anti-HIV drugs makes them unavailable for widespread distribution in underdeveloped nations. But hope and help must begin somewhere—someone must first benefit before others can. **Let us pull together so that help, not just hope will be available to all who need it!**

The Author

The information on HIV/AIDS within this text goes where the pandemic has taken us over the last 19 years. This information is not designed to replace the relationship that exists between you and your doctor or other health advisor. For all medical events, consult with your physician.

FOREWORD

Peter Piot

Suzanne Cherney

The fortunate readers of this textbook are about to embark on a journey that may well change their life—and the lives of others.

The journey leads through scientific facts and figures that are anything but dry and boring. What does HIV, the virus that causes AIDS, have in common with other viruses? How do people get infected? Is it true that the virus can be transmitted through oral sex? What makes AIDS similar to and yet different from other epidemics? Why has it been such a biomedical challenge to develop therapy for people with HIV? If vaccines are the traditional way of combating virus disease, why are we still struggling to develop a vaccine against HIV?

The journey takes us through the ethical and moral dilemmas of AIDS. What are the implications of knowing that 95% of the epidemic is now concentrated in the world's developing countries, the ones with the fewest resources for tackling it? Why is it that just about everywhere in the world, the people most vulnerable to HIV are those struggling under the unfair weight of other burdens, for example, being poor or powerless or illiterate, not knowing the language of the country, living separated from one's partner in order to earn a living or serve in the military, or being subject to discrimination on grounds of racial or religious difference or because of sexual orientation? From recent research in two African cities we know that between 15% and 23% of teenage girls were already infected by age 19. These tragically high rates were linked to sex with older partners—men over 25, whose HIV prevalence was as much as ten times higher than the prevalence in teenage boys. What are the ethics of cross-generational sex? What can society do about older men who feel justified in attracting, coercing, or "buying" young girls as sex partners?

Other dilemmas lie at the interface between science and ethics. How can we continue to justify AIDS education programs in school that preach only sexual abstinence? By now research has shown that programs which present the full "menu" of choices for HIV prevention—ranging from abstinence to condom use—are the ones that work best. Comprehensive education programs do not lead to earlier or more sexual activity; instead, they help young people postpone their sexual debut and protect themselves from pregnancy and sexually transmitted disease. Why, for that matter, is there still such resistance to programs that combine AIDS education and drug treatment services with needle-exchange, despite evidence from many countries that comprehensive programs like this reduce HIV transmission and other kinds of harm related to injecting drug use?

Readers of this book are also embarking on a journey of personal empowerment. In the AIDS era, we are not in control unless we master the facts. Some facts are easy to grasp but hard to accept deep down—for instance, that you are genuinely safe when you drink out of a glass shared with an HIV-positive friend, but you're not safe when you have unprotected sex with a partner that you feel comfortable with because you "know" him or her as a student with a 3.50 GPA, or as someone who can sing dozens of Beatles songs from memory, or as a clean person who showers twice a day. The only kind of "knowledge" of relevance to HIV risk is whether your partner is infected—and whether you are. And the only way to know this is through an HIV test.

Even if both partners find out they are uninfected and sincerely intend to remain faithful, they would do well to consider the important benefits of continuing with condom use. The best intentions can fall by the wayside—say, at a party, especially if the alcohol is flowing—and condoms in any case will give protection from unwanted pregnancy and from herpes, chlamydia, or another sexually transmitted infection that one partner may have without knowing it.

So AIDS UPDATE 2000 is valuable reading for anyone. It is twice as valuable for young readers because, as reports from many countries show, the young are those who benefit most from AIDS education. Young people are flexible, open to new ideas, and amenable to new ways of looking at sexual behavior. In countries where the epidemic is turning around, the turnaround is being led by the young, who can boast of the biggest HIV rate decreases of any age group.

Finally, the book is especially valuable for a US-based readership. It is disappointing that barely more than 40% of American college and university students are receiving HIV/AIDS education on campus, as the U.S. Department of Health and Human Services recently reported. Americans have a special responsibility for being knowledgeable about AIDS. The United States is the richest country in the world—and the one with the greatest potential for making a difference to the course of this global epidemic.

Peter Piot is the Executive Director of the Joint United Nations Program on HIV/AIDS (UNAIDS) and Suzanne Cherney is Senior Writer at UNAIDS.

Contents

9 **Preventing the Transmission of HIV** 281

Preface

WHY DO I WRITE ABOUT HIV/AIDS?

I began writing about a new disease, later called AIDS, in 1981 shortly after the Centers for Disease Control and Prevention issued the first of its reports. But these writings were limited in scope because not much was known at the time. My writing then was for classroom use. Little did I know at the time of the passion that I would develop about this disease. As the number of infected people and their deaths continued to rise, fear and discrimination reared its ugly head because in some cases people are what they are, and in other cases, because of the lack of available nonbiased educational material. At the time, in the mid-1980s, I felt a need to write, to help educate people about this disease, hopefully to answer questions and reduce the blatant discrimination occurring against those who already had the overwhelming burden of HIV disease. So, I began—I created a college level HIV/AIDS course, taught it for several years, constantly shaping and reshaping the information necessary to help others learn the facts about this pandemic, and to destroy destructive myths. After I felt I had set the record straight in my classroom, I began writing HIV/AIDS college level textbooks so that information on this pandemic could be shared more broadly.

There are many reasons I have not stopped writing about HIV/AIDS. AIDS summons up the greatest themes in literature, among them sex, faith and death—themes that are universal and unexpectedly permanent. Anyone who had lost a loved one to death, untimely or by nature, can read about AIDS and understand the emotional forces that take place. Anyone who has taken care of someone who has been ill understands the need for compassion. Anyone who has faced death from a prolonged or life-threatening illness should be able to identify with those who suffer with AIDS. Anyone interested in uncovering acts of human kindness or conversely acts of despicable behavior can find them in writings about this disease.

But more than any of these reasons, my writing about AIDS is fueled by a need to do something, anything, to help. For the millions of people who are in pain and dying, I have little to offer except my writing. Although my only known risk factor was a blood transfusion in the mid-1970s, I can not help but feel lucky that so far neither AIDS nor any other serious disease threatens my life. Is there a reason that I have been spared to my present age when so many others have died? I write about AIDS, not just because I live but because it is part of our history. The AIDS pandemic has changed my feelings and attitudes towards people. I find myself tender and more sensitive to others. AIDS has changed the world in some way for most everyone and, at least some of the information about this disease—its the impact and repercussions—needs telling. So, I write about AIDS because someone needs to tell the story.

WHAT IS AIDS?

AIDS is defined primarily by a severe immune deficiency, and is distinguished from virtually every other disease in history by the fact that it has no constant, specific symptoms. Once the immune system has begun to malfunction, a broad spectrum of health complications can set in. AIDS (or Acquired Immune Deficiency Syndrome) is an umbrella term for any or all of 26 known diseases and their symptoms. When a person has any of these 26 microbial- or viral-caused opportunistic infections, and also tests positive for antibodies to HIV, he or she is diagnosed with AIDS. An AIDS diagnosis is also given to HIV-positive people with a T4 cell count of less than 200/μL of blood.

AIDS: A HUMAN AFFAIR

The history of AIDS is a human affair and is part of a cultural process of attempting to come to terms with a new and often terrifying series of events—of young people dying before their time, of the intermingling of sex and death—in a period in which the world itself is changing before our eyes.

The social meaning of the history of AIDS intimately touches upon ideas about sexuality, social responsibility, individual privacy, health, and the prospect of living a normal life span. Understanding how to respond to AIDS and how to think about this pandemic is important not only for what it reveals about the ways in which health policy is created in the United States and elsewhere, but also for what it implies about the human ability to meet the challenge of future emerging diseases and long-standing public health problems. The HIV/AIDS pandemic is a current and long-term public health problem worldwide.

LESSONS LEARNED

From the lessons of history it is difficult to conceptualize how the AIDS epidemic will be halted, let alone reversed, in the absence of a **cheap curative drug or a cheap and effective preventative vaccine.** The syphilis epidemic at the early part of the century displayed a similar kind of epidemiology to the present-day AIDS epidemic. The campaigns which were initiated then closely paralleled those in place at present for AIDS. There were vigorous educational programs to reduce sexual high-risk behavior, which were targeted at brothels and prostitutes as well as at military recruits to the United States Army. **Scare tactics** were spread through the use of posters, pamphlets and radio—today it is television. Serological testing became mandatory before marriages could be licensed in certain states of the USA. But these measures had little appreciable effect on the expansion of the syphilis epidemic. It was only the advent of a cheap, safe, and effective drug, penicillin, which eventually brought the epidemic under control. **BUT** the bacterium (a spirochete) that causes syphilis does not change or mutate as rapidly as the AIDS virus. There are mutant HIV in the human population for drugs that have yet to be tested and a highly effective preventive vaccine may not be a reality for HIV!

One lesson learned from HIV/AIDS is that any disease that is occurring in a distant part of the globe may be in the United States, in your state or in your town, tomorrow. Twenty or even 15 years ago one would not have expected to read that statement.

The advent of miracle drugs and vaccines that conquered the plagues of polio, smallpox, and measles led many people—including scientists—to believe that the age of killer diseases was coming to an end.

The AIDS epidemic changed that perception, and the best-selling nonfiction book *The Hot Zone* graphically illustrates how an exotic killer called the Ebola virus—carried into Virginia in 1989 by 450 imported research monkeys—came close to breaking loose in this country. In the hit film *Outbreak*, Dustin Hoffman plays a doctor battling a deadly virus that neither medicine nor the U.S. government can handle. Today, the Centers for Disease Control and Prevention are rethinking their position on the possibility of new plagues occurring in our lifetimes!

THE AIDS PANDEMIC IN PERSPECTIVE

To place the HIV/AIDS pandemic in perspective, consider Michael Creighton's *The Andromeda Strain*. The essence of this story revolves around the release of an unidentified and untreatable lethal strain of bacteria into the human population. With a few changes—substituting HIV for bacterium, adjusting the time of death after infection, and describing how one dies from the disease—Creighton could have been writing about HIV/AIDS. But with HIV/AIDS, the onset of a real human tragedy, we are learning about our social contradictions, our strengths and our weaknesses, and we are questioning whether this is a morally acceptable disease. How long will it take for this to become a socially acceptable disease?

AIDS: THE WORST-CASE SCENARIO

As presented in this book, the reader will note the ability of HIV to cause a slow, progressive, and permanent disease. Thus, with no recovery, no loss of infectivity, no development of either individual or group immunity, there is at present no known biological mechanism which can stop the continuing expansion of the disease unless an effective vaccine were to come about, and at present there is no feasible design for such an effective vaccine. The progressive increase in the pool of HIV can, in theory, only lead to an exponential increase in the number of individuals who will become infected until eventually the majority of the sexually active population will be infected unless interventions are at least moderately successful.

It is **because** this scenario may become reality and **because** people worldwide are becoming HIV-infected every few seconds and dying of AIDS each minute, and **because** together we must try to prevent the further spread of HIV, that this text continues to be updated.

This text reviews important aspects of HIV infection, HIV disease, and the acquired immune deficiency syndrome. **It presents a balanced review of factual information about the biological, medical, social, economic, and legal aspects of this modern-day pandemic.**

The intricacy of the HIV/AIDS pandemic has unfolded over the last 19 years. Many of the details of basic research, applied biology, medicine, and social unrest are presented in an attempt to convey the few victories and many setbacks within this ongoing saga. Medical and social anecdotes help to convey a sense of the HIV/AIDS tragedy worldwide. The history of the disease describes how the virus is and is not transmitted. Throughout the text there is a special focus on risk behaviors and risk situations involved in HIV transmission and the means to prevent its transmission. The number of social issues raised by this disease are mind-boggling. Many of these issues are presented—some as open-ended questions for class discussions. Such questions help the student reflect on what he or she has read.

PURPOSE

The purpose of this text is to present an understandable scientific explanation of what has been learned about HIV/AIDS over the last 19 years. In addition, it is particularly important to provide students with a conceptual framework of the issues raised by the HIV/AIDS pandemic so that they will be better able to deal with the challenges posed by this disease. Clearly, this pandemic poses new and unforeseen problems with no quick biological solutions. Only through reason can we respond in a socially acceptable fashion.

Because there is a constant stream of new information to be interpreted and shared, a

new edition will be published every year. What has been learned and what must still be learned to bring HIV infection, HIV disease, and AIDS under control is valuable information to most, if not all, of us.

TEXT USE

This text is intended for use in college-level courses on AIDS and as a supplemental HIV/AIDS resource in medical and nursing schools, in colleges of allied health sciences, in psychology department courses on human sexuality and human behavior, in courses on sexually transmitted diseases, in summer teachers programs, in the training of AIDS counselors, in state-mandated AIDS education courses for health care workers, and for physicians who need a ready source of information on how to prevent HIV infection, HIV disease and its progression to AIDS, and types of therapy available. This text is suitable in those cases where information and education about the various aspects of HIV infection, disease, and AIDS are either wanted or required.

During the 19 years since HIV/AIDS was defined as a new disease, more manpower and money has been poured into HIV/AIDS research than into any other disease in history. Information on the AIDS virus has accumulated at an unprecedented pace. Since the discovery of the AIDS virus in 1982–1983, scientists have learned much about how it functions and how it affects the immune system. More was learned about the AIDS virus in the first 6 years after its discovery than had been learned about the polio virus in the first 40 years of the polio epidemic. During the first 6 years of AIDS research, the virus's genetic material was cloned, its structure ascertained, and individual viral genes identified. In fact, so many billions of dollars have been spent on HIV/AIDS that many now believe that HIV/AIDS research has taken away much-needed resources from other diseases that kill many times more people per year than does AIDS.

TEXT OVERVIEW

Because so much of the HIV/AIDS pandemic has been based on the manipulation of dis-

torted scientific fact, it is necessary to counter these distortions and myths by presenting the most consistent, reproducible, and scientifically acceptable facts as possible. Therefore, this text offers answers to questions many people have about the AIDS virus and how their immune systems are affected by it. Covered herein are the activity of the immune system with respect to HIV disease, where the virus might have come from, how it is transmitted, who is most likely to become infected, viral prevalence (geographical distribution of the number of people infected and those expressing HIV disease and AIDS), possible means of preventing infection, signs and symptoms of HIV disease and AIDS, chronological definitions of AIDS, the opportunistic infections most often associated with AIDS, FDA-approved and non-approved drugs and their effectiveness, the potential for a vaccine to the AIDS virus, the tests available to detect HIV infection, the accuracy of these tests, their cost, availability, and confidentiality. The last chapter deals with the fear of HIV infection and the social reaction of the uninfected toward the HIV-infected.

The majority of references herein are dated between 1987 and 1999. Every state in the United States has HIV-infected people and reported people diagnosed with AIDS. The World Health Organization (WHO) reports that of 209 countries reporting to it, 194 have reported diagnosed cases of AIDS through 1999. The virus is truly pandemic and now threatens most of the human population. Yet it is a preventable disorder, and the steps to prevention are presented in Chapter 9.

ORGANIZATION OF THE TEXT

Chapter 1 presents information on the discovery of the disease and naming of the illness. Chapter 2 discusses the cause of AIDS and the possible origin of the AIDS virus. Chapter 3 presents the biological characteristics of the virus. Chapter 4 presents current retroviral therapy and its problems using those drugs to prevent HIV reproduction. Chapter 5 discusses the human immune system as it relates to HIV disease and the loss of the immune system's function. Chapter 6 presents the various opportunistic diseases, how they are affected by the use of retroviral therapy and forms of cancer associated with HIV disease and AIDS. Chapter 7 discusses the clinical profile of biological indicators of HIV disease and AIDS. Chapter 8 presents the means by which HIV is most efficiently transmitted among humans. Chapter 9 details the most effective means to date of preventing HIV infection and the possibilities for a vaccine. Chapter 10 gives the incidence of HIV infections, HIV disease, and AIDS cases among selected behavioral at-risk persons for HIV infection **world wide and in the United States.** Chapter 11 presents the prevalence and problems of HIV disease in women, children, and teenagers. Chapter 12 presents those tests most often used to detect the presence of HIV, biological shortcomings of the tests, the confidentiality or lack of confidentiality before and after test results, and costs. Chapter 13 offers insight to the many social ramifications of having HIV disease or AIDS, human attitudes, and behaviors. There is also a glossary of terms and index and a list of telephone numbers to federal, state, and other groups that offer information and help with all aspects of the HIV/ AIDS pandemic.

SPECIAL FEATURES

Each chapter contains chapter concepts, a summary, review questions, and references. Answers to questions appear at the back of the book. Chapters also contain definitions for new terms as they are introduced, illustrations, photographs, and tables. All 13 chapters contain boxed information. **All of the chapters contain either highlighted points of view, points of information, cases in point, sidebars, Looking Back Features, Sound Off, Points to Ponder or pro and con discussions.** They illustrate and emphasize important events and information about HIV infection, HIV disease, and AIDS. At certain places in the text, there are sets of class discussion statements and class questions. Instructors may wish to challenge the class in these areas.

ACCESS TO INFORMATION

Internet addresses, state HIV/AIDS hotlines, toll-free national HIV/AIDS information on prevention, therapy, health care providers, local and state programs relating to specific areas of research and funding, access to information on women, children and teenage HIV/AIDS programs, and information on where to access materials on what HIV-infected parents may want to tell their children about their condition and the disease in general, can be found within and at the ends of most chapters.

TO THE EDUCATORS

As educators, it is our job to expose our students to the new concepts in biology that are shaping humanity's future. To do that, we must expose them to new vocabulary, new methodologies, new information and new ideas. It is my hope that this text will help in this endeavor.

IN APPRECIATION

The help of the following organizations or people is most deeply appreciated: The Centers for Disease Control and Prevention (CDC) Atlanta, Georgia, for use of slides and literature produced in their *National Surveillance Report* and the *Morbidity and Mortality Weekly Report*; Mara Lavitt and *The New Haven Register* for information on William Bluette; from the CDC, Richard Salik and Patricia Sweeney (Surveillance Branch); Robin Moseley, Patricia Fleming, Todd Webber (Division of HIV/AIDS Prevention), and Eleanor Holland (AIDS Clinical Trial Reference Specialist); E. E. Buff, Biological Administrator, and Barry Bennett, head of Retroviral Testing Services, for their permission and guidance on photographing HIV testing procedures at the Florida Health and Rehabilitation Office of Laboratories Services, Jacksonville, Florida; Denise Reddington, Department of Health and Human Services, Washington D.C., and Regan Alberta of PANOS Institute for their help in obtaining data on the economics of HIV/AIDS worldwide; Russ Havlak of the Infectious Disease Unit, CDC, Atlanta, Georgia; personnel of the National Institute for Allergy and Infectious Diseases, the National Center for Health Statistics, Brookwood Center for Children with AIDS in New York, the George Washington AIDS Policy

Center in Washington, D.C., the National Institutes for Health, Hoffman-LaRoche Co., Abbott Laboratories, the Pharmaceutical Mfg. Association, the National Cancer Institute, Pan American Health Organization, and the Office of Technological Assessment; Teresa M. St. John, University of North Florida for illustrations; the individuals who have contributed photographs; the text reviewers whose work has been greatly appreciated; **a special thank-you to Linda Schreiber, my editor, who has moved on to other endeavors within Pearson Group; to Rhonda Varner and Gilbertine Yadao who word processed the updated material;** Guy Selander, M.D. and Jack Giddings, M.D., who, over the years, have shared their medical journals with me; James Alderman, Eileen Brady, Mary Davis, Signe Evans, Paul Mosley, Ricky Moyer, Sarah Philips, and Barbara Tuck—reference/research librarians at the University of North Florida; **and to my special family, wife Delores and children Sherri and Garrett, who helped with proofreading and demonstrated a great deal of patience and understanding and gave up family weekends so the text could be completed on time.**

This book has benefited from the critical evaluation of the following reviewers:

- Dr. Paul R. Elliott, Florida State University
- Dr. Robert Fulton, University of Minnesota
- Dr. Robert M. Kitchin, University of Wyoming
- Dr. Richard J. McCloskey, Boise State University
- Dr. Wayne B. Merkley, Drake University
- Dr. Linda L. Williford Pifer, University of Tennessee
- Dr. Bernard P. Sagik, Drexel University
- Professor James D. Slack, Cleveland State University
- Dr. Carl F. Ware, University of California, Riverside
- Dr. Phyllis K. Williams, Sinclair Community College
- Dr. Charles Wood, University of Kansas

Gerald J. Stine, Ph.D.
904-641-8979
email: gstine@UNF.edu

"THE FACE OF AIDS"

It is two boys, ages 10 and 12, caring for their mother, who is dying from AIDS. The boys get themselves off to school in the morning, make dinner for their mother after school, carry her to the tub to bathe her, softly towel dry her body, which is covered with sores, then help her into bed. How do you go to school and compete academically when your mother is home dying, your family is broke, and you are afraid to tell anyone about the problem?

It is a young gay man in Orange County, California, who is having difficulty breathing. At the local hospital, he is diagnosed with pneumocystic carinii pneumonia (one of the characteristic infections afflicting people with AIDS) and is told, "We don't take care of this kind of pneumonia here." Packed into a car for a 6-hour drive to San Francisco, he's given an oxygen tank — only the oxygen runs out after 4 hours. Arriving at the emergency room at 2 a.m., in a city with no friends or family, he can barely breathe. Three months later he is dead.

It is a young woman who is knitting a scarf for her 2-year-old daughter. She said "I want to leave some legacy for her." She did not yet know her daughter was HIV-positive and most likely would not live long enough to understand the legacy. Most likely the daughter would die before her mother.

It is a child born to one of the 80,000 or so women in New York City who are injection drug users. The child is delivered in central Harlem, a region where 40% of the women receive late or no prenatal care, where the infant mortality rate is three times the national average, and where the main hospital recently suffered from a shortage of penicillin for a year and a half because it could not pay its bills. Initially the baby undergoes withdrawal from a heroin addiction inherited from her mother. Then 5 months later, the child becomes ill with persistent diarrhea. At the hospital they discover that she's not sick with one of the typical poverty-related infectious diseases that plague malnourished inner-city infants. No, the child has AIDS — and with parents unable to care for her, hospitals too crowded to board her, and foster agencies unable to place her, no one knows where or how this child will endure what promises to be a brief, sickly life.

It is about a 25-year-old male who looks 40 less than a year after his AIDS diagnosis. When he first tested HIV-positive, he decided to kill himself when his T4 cell count dropped to 100. When it did, he prepared his last meal, said goodbye to close friends, and bought a handgun. After dinner he put on his favorite cassette, raised the gun—but could not end his life. What now, he asks, this was my ace in the hole!

It is a young man who lives in Miami, and he's not ready to come home. He wants to remain independent. Eventually he will come home. His family must be prepared to care for him and to deal with all the issues they will face. His mother is a high school teacher, but she has never told anyone about her son. She says when he comes home it will be impossible to hide the fact that he has AIDS. "I am not sure how the community and our friends are going to respond."

The Face of AIDS is Aging. In the United States about 11% of AIDS cases are among people over age 50. Beginning in 1999 that's over 76,000 people—20,000 are between ages 60 and 69 and about 5,000 are age 70 and older.

The Face of Aids is Changing. For 1997 through the year 2000 and beyond, the Face of AIDS, due to the successful use of AIDS drug cocktail therapy, takes on new and different looks—those of rejuvenation and of GUILT. Many who have recovered, even from their death beds, and feel "new" again, feel guilty because many are still dying and many do not have access to these drugs.

It is about establishing new relationships, giving up disability insurance, finding employment, and experiencing a life they thought they had lost.

It is THE FEAR THAT ALL THIS MAY BE JUST SHORT TERM—THAT THE VIRUS WILL WIN IN THE END!—This is a Face of AIDS made possible because science has provided new drug therapies!

The Face of AIDS is surely around us. It is a disease that invites commentary, requires research, and demands intervention. Inevitably linking sex and death, passion and politics, it continues to generate controversy. Appearing as a new and lethal infectious disease, at a time when the biomedical establishment in the United States has shifted its focus to chronic disease, AIDS has called into question many of the premises of the biomedical profession.

Introduction: Histories of Global Pandemics, World AIDS Conferences, Overview of HIV/AIDS, and the AIDS Quilt

TEST OF YOUR CURRENT KNOWLEDGE

Before reading this book, review and answer the following statements about HIV infection and AIDS and answer either True or False to test your current knowledge:

- **Infection by HIV can be prevented.**
- Worldwide, the virus is primarily spread through heterosexual contact.
- Asymptomatic infected persons can infect others.
- High HIV infection risk is associated with having many sexual partners.
- HIV is fragile and easily destroyed by environmental agents when outside the human body.
- **Casual contact does not spread HIV.**

- HIV is not transmitted to humans via insects.
- The current risk of acquiring HIV via blood transfusions in the United States is low.
- Children can safely attend school with a classmate who is HIV-infected or has AIDS.
- Even without using precautions, health care workers are at a low risk for acquiring HIV infection.
- **There is no cure for HIV disease or AIDS.**
- In the United States about 57% of all persons expressing AIDS have died; worldwide it is about 60%.
- AIDS may be close to 100% lethal—sooner or later a person who expresses AIDS dies **prematurely.**
- It is believed that 90% to 95% of those infected with HIV will eventually progress to AIDS.
- **There are people who are genetically resistant to HIV infection.**

All of the preceding statements are **TRUE**. Information concerning these statements and answers to other questions that you may have can be found within.

> It was the best of times, it was the worst of times, it was the age of wisdom, it was the age of foolishness, it was the epoch of belief, it was the epoch of incredulity, it was the season of Light, it was the season of Darkness, it was the spring of hope, it was the winter of despair. . .
>
> Charles Dickens, *A Tale of Two Cities*

"All my friends are dead." This expression is unique in a lifetime and is symbolic of reaching old age—except in a time of war. Too many young people worldwide have said it over the past 19 years because of AIDS.

Nothing in recent history has so challenged our reliance on modern science nor emphasized our vulnerability before nature. We have lived with the acquired immune deficiency syndrome (**AIDS**) epidemic, witnessing its paradoxes every day. People living with the human immunodeficiency virus (**HIV**) live with the fear, pain, and uncertainty of the disease; they also endure prejudice, scorn, rejection, and despair. **This must change!**

HISTORY OF GLOBAL EPIDEMICS

There has never been a time in human history when disease did not exist. The history of epidemics dates at least as far back as 1157 BC to the death of the Egyptian pharaoh, Ramses V, from smallpox. Over the centuries, this extraordinarily contagious virus spread around the world, changing the course of history time and again. It killed 2,000 Romans a day in the 2nd century AD, more than 2 million Aztecs during the 1520 conquest by Cortez, and some 600,000 Europeans a year from the 16th through the 18th centuries. Three out of four

------ POINT OF INFORMATION I.1 ------

INFECTIOUS DISEASE PARADIGM

Most new infections are **not** caused by truly new pathogens. They are caused by microorganisms (bacteria, fungi, protozoa, and helminths) and viruses that have been present for hundreds or thousands of years, but they are newly recognized and named because of recently developed techniques to identify them. Human activities drive the emergence of disease and a variety of social, economic, political, cultural, climatic, technological, and environmental factors can shape the pattern of a disease and influence its emergence into populations. For example, travel can greatly affect the emergence of disease and human migrations have been the main source of epidemics throughout history. Trade caravans, religious pilgrimages, and military campaigns facilitated the speed of plague, smallpox, cholera, and now AIDS. Global travel is a fact of modern life and, equally so, the continued evolution of microorganisms and viruses; therefore, seemingly new infections will continue to emerge, and known infections will change in distribution, frequency, and severity.

DISEASES CAUSED BY VIRUSES AND MICROORGANISMS HAVE HAD 3 BILLION YEARS TO EVOLVE—WE SHOULD NOT EXPECT TO CONQUER THEM EASILY!

------ POINT OF INFORMATION I.2 ------

SMALLPOX—A BIOLOGICAL WEAPON?

Smallpox may have been the most useful weapon of biological warfare in history. European colonialists repeatedly took advantage of the special susceptibility of the Amerindian population, deliberately spreading the deadly virus among the natives who were successfully defending their rights to the lands and resources of the Americas. For example, in 1763 Sir Jeffrey Amherst, commander in chief of all British forces in North America, was having great difficulty controlling the Pontiac Indians in the Western Territories. At Amherst's insistence, blankets inoculated with live smallpox viruses were distributed to the Pontiac, obliterating the tribe. The deliberately induced epidemic quickly spread to the northwest, claiming large numbers of Sioux and Plains Indians, crossed the Rockies and inflicted huge death tolls among Native Americans from south California all the way north to the Arctic Circle tribes of Alaska. This devastation was cited in the official WHO history of smallpox [FENNER, F., et al. (1988). *Smallpox and Its Eradication*. Geneva: World Health Organization].

people who survived the high fever, convulsions, and painful rash were left deeply scarred and sometimes blind. Because victims' skin looked as if it had been scalded, smallpox was known as the **"invisible fire."** Even now, malaria in underdeveloped countries afflicts 300 million people, killing between 2 and 3 million each year. The problem is compounded by development of drug-resistant malarial strains of protozoa. **Thus epidemics are not new to humankind, but the fear they impose on each generation is.**

Major recorded pandemics (global) and epidemics (regional) that have devastated large populations are described in Table I-1.

The 1970s witnessed the emergence of Lyme disease (1975), Legionnaires disease (1976), and toxic shock syndrome (1978). In the 1980s, HIV and three new human herpes viruses (HHV 6, 7, and 8) were identified. In 1994, a physician in Gallup, New Mexico, was called to care for a man who had collapsed at a funeral. Soon, several others in New Mexico had also suffered from sudden respiratory failure.

As victims contracted the disease, they went from seemingly perfect health to flu-like symptoms, to a horrific state where their lungs began to fill with blood. Most of them died.

Within a few weeks, the Centers for Disease Control and Prevention identified this killer: **hantavirus**—a rodent-borne organism that had infected many GIs during the Korean War and killed hundreds. But the Korean type of hantavirus causes death through kidney failure. Somehow the hantavirus had found its way to the United States and mutated(changed) into a virus that invades the lungs.

The outbreak of hantavirus in America occurred first in 1993 in Arizona, Colorado, New Mexico, and Utah. Beginning year 2000, 219 people have been infected in 30 states with the rodent-borne deermouse virus and 96 of these people have died of acute respiratory distress. There is no vaccine for this virus.) In mid-1994 there was an outbreak of necrotizing fasciitis (fash-e-i-tis)—tissue destruction caused by a tissue-invasive strain of Group A streptococcus, a bacterium. This bacterium is a deadly variant strain of the Group A streptococcus that causes "strep throat." The infected may die from bacterial shock, disseminated blood co-

agulation (clotting), or respiratory or renal (kidney) failure. Also, in September of 1994, health authorities in Brisbane, Australia were suddenly confronted with 11 dead horses and the death of their trainer. The killing agent was identified within 2 weeks. It is a new virus called **equine morbillivirus** that causes a fatal respiratory disease. Although nonhuman morbilliviruses have been associated with new and emerging animal diseases, no new human morbillivirus has been reported since the 10th century, when measles was described [Cause of fatal outbreak in horses and humans traced. (1995). *Science*, 268 (32):94–97]. Overall the virus has killed 13 of 25 infected horses and one of the two humans who became infected. This outbreak is an isolated event; no additional infections have been identified. The virus in this case was eliminated by the death of the infected. **But where did the virus come from? Where will it show up next and under what conditions?** The natural host of the virus has not been identified. This virus causes the cells lining the blood vessels to clump, creating holes in the vessel walls that allow fluid leakage into the lungs. The horses and the human drowned in their own fluids.

Although emerging viruses may contain new mutations or represent gradual evolution, more often they emerge because of changes in human behavior or the environment whereby they are introduced to a new host. Viruses that are pathogenic in human and nonhuman hosts include Marburg and Ebola viruses; hantaviruses; human immunodeficiency virus; Lassa virus; dolphin, porpoise, and seal morbilliviruses; feline immunodeficiency virus; and bovine spongiform encephalopathy agent, which in 1996 threatened the possible destruction of the entire cow population of Europe causing economic loss of billions of dollars and a major unemployment problem. And, there are the **rotaviruses.** Rotaviruses are detected in 20% to 70% of fecal specimens from children hospitalized with acute diarrhea, and in developing countries they cause about 870,000 deaths each year. Even in the United States, rotavirus is associated with 3% of all hospitalizations of children younger than 5 years old, which translates to 55,000 to 70,000 hospitalizations per year, with medical and indirect costs in excess of $1 billion. Efforts to prevent disease by improving water or sanitation

TABLE I-1 Plagues in History[1]

Disease	Dates	Place	Number Killed	Causative Organism	Time to Prevention/Cure (in years)
Measles	from 430 BC	Greece/Rome/World	Millions	Paramyxovirus	1,712
Plague	542–1894	Europe/Asia/Africa	7100 million	Yersinia pestis	580
Cholera	1826–1837	New Jersey	900,000	Vibrio cholerae	75
	1849	United Kingdom	53,293		
	1947	Egypt	11,755		
Tuberculosis	1930–1949	United States	1,000,000	Mycobacterium tuberculosis	85
Malaria	1954–1970	Africa/India	150,000	Plasmodia	100
	1847–1875		20 million +		
Scarlet Fever	1861–1870	United Kingdom	972 per million people	Beta-Hemolytic streptococci	45–44
Polio	1921–1970	North America	37,000	Polio viruses Types I, II, III	30–50
Typhus	1917–1921	Russia	2,500,000	Rickettsias	25
Influenza	1918–1919	U.S./Europe	21,640,000	Influenza virus	57
Smallpox	from 1122 BC	Europe (middle ages)	Hundreds of millions	Smallpox virus	3,050
	1926–1930	India	423,000		
Gonorrhea	from 590 BC	United States	57,477	Neisseria gonorrhoeae	1,832
	1921–1992				
Yellow Fever	from AD	Nigeria	10,000	Arbovirus	488
	1986–1988				
AIDS[2]	1981–2000	United States (estimated)	*Cases* 440,000 / *Deaths* 796,000	Human Immunodeficiency Virus	Treatments but no cure
	1981–2000	World (estimated)	18,000,000 / 32,000,000		

1 Historical time line on first suspected cases of the above diseases are: Plague, 11th century B.C.; Cholera, 1781; Tuberculosis, 451 B.C. Malaria, 1748; Scarlet Fever, 1735; Polio, 1894; Typhus, 1083; Influeza, 1580; Smallpox, 429 B.C.; Gonorrhea, 1768; Yellow Fever, 1647; HIV/AIDS, Unites States in the 1970's, in Africa, 1959.

2 Aids data estimated through year 2000.

——— BOX I.1 ———

OTHER LETHAL VIRUSES IN OUR TIME

Richard Preston, in his 1994 book *The Hot Zone* describes the progression of illness as the Ebola virus kills humans:

Ebola...triggers a creeping, spotty necrosis that spreads through all the internal organs. The liver bulges up and turns yellow, begins to liquefy, and then it cracks apart.... The kidneys become jammed with blood clots and dead cells, and cease functioning. As the kidneys fail, the blood becomes toxic with urine. The spleen turns into a single huge, hard blood clot the size of a baseball. The intestines may fill upon completely with blood. The lining of the gut dies and sloughs off into the bowels and is defecated along with large amounts of blood.

The Ebola virus resists all drug therapy to date and kills up to 80% or more of those infected within 1 to 3 weeks after infection. (HIV kills over 90%—it is slower but more lethal)

HISTORY

There have been seven confirmed outbreaks of Ebola. The first was in 1976 when Belgian doctors discovered the virus at a hospital in northern Zaire; it was named Ebola Zaire after the Zairian river flowing through the region. There were two more outbreaks in 1979, one in Zaire and one in southern Sudan. The fourth outbreak occurred once more in Zaire, in a city named Kikwit, a city of 600,000 people, in May of 1995. It struck again in December of 1995 in Gozon, Ivory Coast, again in February 1996 in Mayibout, Gabon, and again in November 1996 in Johannesburg, South Africa. And, each time the virus quickly ran its course of human destruction and disappeared. The result of the new Ebola epidemics was 357 confirmed cases of infection of which 268 have died. In 1989, a strain of Ebola virus appeared in Reston, Virginia, just a short distance from Washington, DC. One hundred monkeys had arrived at the Reston Primate Quarantine Unit from the coastal forests of the Philippines on October 4. Two monkeys were dead on arrival and 10 more died during the next few days. Contrary to all expectations, the illness spread to the unit's non-Philippine monkeys. Samples of monkey tissue were sent to the U.S.

Army Medical Research Institute of Infectious Diseases for examination. Their tests proved positive for Ebola Zaire. The investigation quickly became a potentially volatile political crisis. The quarantine unit was a hot zone containing an organism classified as "biosafety level four": one for which no cure or vaccine existed.

In what has to be considered as an EXTRAORDINARY PIECE OF LUCK, this strain of Ebola virus, even though it appeared identical to the lethal form of Ebola Zaire, was harmless to humans. As it appeared to be spread among the monkeys via the air, if that virus was the lethal form of Ebola, the consequences for this heavily populated area of the Eastern Coast would have been unimaginably horrific.

Class Discussion: In what ways would the entire United States and other countries have been affected?

ORIGIN AND TRANSMISSION

Beginning 2000, a natural host of the Ebola virus has not been identified, though similar viruses are carried by monkeys and rodents. With respect to transmission, Ebola is believed to be spread primarily through an exchange of body fluids and secretions, not casual contact. Most of those infected with Ebola have high fevers and bleed to death from hemorrhaging. There is no treatment or vaccine. Peter Piot said the Zairian hospital where he discovered Ebola in 1976 had 120 beds and only three syringes to inject 300 to 600 patients per day, sometimes for weeks on end. That same year, 274 of 300 infected people died in an Ebola outbreak in one Zairian village. Because Ebola is lethal so quickly, it is ill-suited to sustain an epidemic. **It kills the infected so quickly they don't have an opportunity to transmit the virus to others.** This is why in each of the seven outbreaks in population-dense areas the epidemics were short-lived.

SOME OTHER LETHAL VIRUSES THAT INFECT HUMANS

Ebola is just one of a number of viruses to have emerged from the jungle in the past few decades; others include **Lassa** and **Marburg** viruses in Africa; **Sabía**, **Junin**, and **Machupo** viruses in South America; in 1995 a virus, yet to be named,

——— **BOX I.1** (*continued*) ———

was found, in Brazil, to cause lung destruction and death; and the AIDS virus. HIV probably originated in Africa, but unlike Ebola, it is ideally suited to spread around the globe. The HIV-infected can remain symptom-free for years, which provides the opportunity for them to infect others.

Are there other viruses as dangerous as HIV—or even more dangerous—lurking on the edge of civilization? That's the question that haunts public health officials. Scientists have identified over 100 viruses capable of causing human disease at some level and they believe there may be 5 to 10 times that many still be be identified! In the most recent science fiction-like laboratory scare on August 8, 1994, a scientist at

the Yale Arbovirus Research Unit became infected with Sabia. The newly emerging insect-borne virus had been discovered in Brazil only 2 years earlier. The scientist inhaled virus material from a leaking plastic centrifuge tube. An inquiry by the Centers for Disease Control and Prevention (CDC) charged the researcher with misconduct and the unit with poor emergency response protocols.

[HORTON, RICHARD. (1995). *Infection: The Global Threat.* New York: Reviews April 6, 24–28].

Class Question: Should the public be informed about the dangers of research on viruses lethal to humans before it is allowed to begin?, to continue?

——— **POINT OF INFORMATION I.3** ———

BACTERIAL FOOD POISONING—1997–1998

Between August 1997 and April 1998, 25.5 million pounds of hamburger was recalled from at least 21 states due to possible *Escherichia coli (E. coli) serotype* 0157:H7 (a type of *E. coli* different from other *E. coli* strains as defined by the types of antigens made against it) contamination. This recall caused the financial demise of a giant meat packing company. Type 0157:H7 causes severe cramping, abdominal pain, and bloody diarrhea with little or no fever, and can progress to

life-threatening complications. 0157: H7 produces several toxins that affect the digestive tract. 0157:H7 was first implicated in a food-borne illness in 1982. Over the last 15 years, there have been over 65 0157:H7 outbreaks of food-borne illness in the United States. Most cases of illness have been associated with undercooked beef products in homes or in national chain restaurants. Recovery from the illness takes, on average, 7 days.

seem unlikely to succeed, because all children in developed and developing countries become infected with rotavirus in the first 3 to 5 years of life. A high priority placed on the generation of a safe and effective vaccine resulted in an FDA approved **RotaShield** vaccine against rotavirus in late 1998.

Some 30 newly identified disease-causing agents have emerged over the past 25 years, and 6 old diseases are re-emerging: TB, diphtheria, cholera, dengue fever, yellow fever, and bubonic plague.

Cases of **dengue** or **breakbone** fever have been recognized for over 200 years. Diseases caused by the dengue virus, such as dengue and dengue hemorrhagic fever (DHF), are a greater burden to human health than **any other mosquito-borne viral disease.** About

two-thirds of the world's population live in environments where DHF is either endemic or epidemic, **and an estimated 80 million persons are infected with dengue virus each year.** The nature of dengue epidemics has changed considerably during this century. Where classic dengue was once the norm, we are now witnessing epidemics that affect larger numbers of people and produce more severe symptoms of the disease. This is particularly evident in the Americas, where the disease has made dramatic inroads during the past two decades. In contrast, the virus that causes **hepatitis C** was identified in 1989. An effective blood screen for it was in use by 1992. It is estimated that 4 million Americans, 9 million Europeans, and 170 million people worldwide are infected with hepatitis C, marking an epidemic much larger than HIV. One in 5 will develop "hardening of the liver." There are an estimated 30,000 new infections annually in the United States, with 10,000 **deaths** attributed to the virus. These numbers are expected to triple over the next 20 years. **Hepatitis C is an incurable, bloodborne virus which can cause irreparable liver damage—it is the leading cause for liver transplant in America.** It has an extended latency period, and symptoms are not manifested until the disease has advanced, making treatment difficult. Currently, the only treatment for HVC infection is the use of interferon, which has many side effects, although clinical trials are underway for other possible treatments.

The world is not becoming a safer place in which to live. More people are crowding into cities and humans continue to intrude into once remote areas. Clearly some diseases are more lethal than others and AIDS may be the most lethal of all "new" diseases to strike humans in the 20th century.

SILENT TRAVELERS

In 1962, Sir MacFarlane Burnet, the 1960 Nobel Prize recipient for his work on immunological tolerance, wrote, ". . . one can think of the middle of the 20th century as the end of one of the most important social events in history, **THE VIRTUAL ELIMINATION OF THE INFECTIOUS DISEASE AS A SIGNIFICANT FACTOR IN SOCIAL LIFE.**" This statement represents the complacency that developed in the United States and some other countries in regard to threats to the human race posed by the microbial world. More recently, Joshua Lederberg, president of the Rockefeller University from 1978 to 1990, wrote, "The ravaging epidemic of AIDS has shocked the world . . . **We will face similar catastrophes again** . . . We have too many illusions that we can . . . govern the remaining vital kingdoms, the microbes, that remain our competitors of last resort for domination of the planet." This statement represents a more realistic assessment of the current situation, by recognizing the fact that improvements in sanitation and introduction of antimicrobial agents and vaccines **have failed to eliminate the health risks from exposure to infectious agents.**

In *Harrison's Principles of Internal Medicine*, which was published a little more than a decade ago, Margaret Winker wrote that "infectious diseases are more easily prevented and more easily cured than any other major group of disorders. . . ." A new disease called "acquired deficiency of cell-mediated immunity in young homosexual men" occupied less than a column of text. "Slim's disease," recognized possibly as early as 1952, did not warrant an entry, but the dramatic decline in tuberculosis seen during the previous decade was noted to have "leveled off." This complacency allowed a greater focus on heart disease and cancer. Ten years later, cardiovascular disease mortality has declined, and much of the public knows that high cholesterol and blood pressure should be controlled. **Infectious disease, meanwhile, has climbed to the third leading cause of death in the United States. Worldwide, infectious diseases account for about half of the 52 million global deaths per year and are the leading cause of death among children.**

Since the 1980s, we have come to view infectious diseases with a humbler eye. The victories of a quarter century ago ring hollow as AIDS ravages, disease-causing bacteria become resistant to all standard treatments, and the once easily treated pneumococcus demonstrates antimicrobial drug resistance. Unknown diseases develop with uncomforting frequency, and the Ebola virus has been identified outside the confines of Zaire. Once considered unique and isolated, these events penetrate every corner of the globe. As was recognized in 1892 when the first international sanitary convention on cholera was adopted, infectious diseases cannot be observed, battled, or understood street by street or country by country. A global approach is necessary [WINKER, M.A., et al. (1996). Infectious diseases: a global approach to a global problem. *JAMA*, 275:245–246].

Far from being invulnerable, humans are in fact at the mercy of innumerable microscopic agents that can erupt at any time by mutating (changing) into a virulent form, crossing a species barrier, or escaping from an environmental niche.

There have been some great achievements in the battle against these microbes. Yet the only complete victory over any infectious disease so far has been the eradication of smallpox. However, one of the achievements in the approaching millennium will be the global eradication of poliomyelitis by the end of the year 2000, a target date set by the World Health Organization. During late 1998, a new malaria vaccine shown to be 80% effective in small-scale testing was field tested in Africa. Malaria kills 1 in 5 children in Africa, Southeast Asia, and Central America. Field test results are not yet available.

VIRUS ASSOCIATED WITH CANCERS AND OTHER KNOWN DISEASES

In the mid- to late 1980s, numerous correlations were discovered between viruses and various types of cancers. For example, Epstein-Barr virus was associated with nasopharyngeal carcinoma and B-cell lymphoma, hepatitis B virus with liver cancer, and human papillomavirus with cervical cancer. Now, a decade later, scientists are finding out that viruses may also play a role in an array of other diseases, like atherosclerosis and diabetes, as well as such mental disorders as schizophrenia. Additional connections will be made between viruses and many types of diseases in the future because virologists are turning their attention more and more to diseases to which there is no known cause.

FEAR AND IGNORANCE ABOUT AIDS AND OTHER PANDEMICS

The fear of HIV infection and the ignorance about its causes have created bizarre behavior and at times barbaric practices, strange rituals, and the attempt to isolate those afflicted.

The Black Plague during its most destructive time killed over 500 people a day. Instead of being concerned about providing care to the victims, people spent their time deciding how deep to dig the graves so that none of the horrid fumes would come up and infect others. It was determined that a grave should be six feet deep; and that is exactly how deep it is today. Plague victims were herded together into cathedrals to die or to pray for faith healing to save them. In 14th century Germany and Switzerland, the Christians blamed the Jews for the outbreak of bubonic plague, believing that the Jews were poisoning the water—the very same water that the Jews were drinking. As a result, whole communities of Jews were slaughtered. And in the 1400s and 1500s, when syphilis was spreading across the world killing thousands, the Italians called it the French disease. Of course, the French called it the Italian disease. In the 1930s, cholera was considered a punishment for people unwilling to change their lives—the poor and the immoral. In New York the Irish were blamed. In the early 20th century, polio in America was believed to be caused by Italian immigrants.

Deaths from Disease

Measles, Plague, Smallpox, Polio, Influenza, Tuberculosis, Hepatitis B, and Malaria—These diseases, are responsible for many hundreds of millions of human deaths throughout history. The population of the indigenous Indians of Central and South America dropped from an estimated 130 million to about 1.6 million, more as a result of measles and plague than from war. In the 19th century measles has been held to have been responsible for the total annihilation of the Indian population of the island of Tierra del Fuego. Plague swept through Europe in the 14th century destroying a quarter of Europe's population. Between 1600 and 1650 the population of Italy actually fell from 13.1 million to 11.4 million. In Venice, an average of 600 bodies were collected daily on barges. More than 50,000 Venetians died in the plague of 1630-1631, leaving a population smaller than at any time during the 15th century. Recent outbreaks of bubonic plague occurred in India in August of 1994, in Mozambique in June of 1997 and in Uganda, November 1998.

The smallpox and polio viruses may well go back to the early beginnings of humans but would have mostly been concealed by the enhanced susceptibility of those infected to more common (or more recognizable) infections.

Childhood leukemia, apparently newly emergent in the 1950s, was similarly an artifact of the widespread use of antibiotics for the treatment of the infections from which leukemic people previously died.

Worldwide, smallpox killed over 100 million people by 1725. Smallpox is the only viral disease ever eradicated. The smallpox death of Egyptian pharaoh Ramses V in 1157 BC is the first known. Invading armies then spread smallpox through Africa and Europe. The Spanish brought it to Mexico. Edward Jenner discovered a virus related to smallpox (cowpox) and created a vaccine in the 1870s. Only humans got smallpox. It disappeared after a global vaccination campaign in the 1960s. The last known natural case was in Somalia in 1977.

Influenza— Each year as winter approaches, millions of shots of influenza (flu) vaccine are given, mainly to people who are especially susceptible to the virus (the very young, the elderly, asthmatics, and diabetics). Each year in the United States, from 20 million to 50 million people contract the flu, and between 10,000 and 40,000 (in a bad year) die from it! The vaccine is a cocktail of recent influenza A and B strains and affords some protection against genetic changes in these strains. Nevertheless, in recent history many thousands have died from influenza infection; in the 1957–1958 epidemic, for instance, 1 in 300 over age 65 died. Influenza was first described by Hippocrates in 412 BC. The first well-described flu pandemic occured in 1580. Thirty-one flu pandemics have occured since then. In 1890, in China, the duck-borne flu virus crossed over into swine, then into humans, causing thousands of deaths. This epidemic was followed by **severe flu pandemics** in 1900, 1918 (the Spanish flu), 1957 (the Asian flu), 1968 (the Hong Kong flu), and in 1977 the Russian flu. The flu pandemic of 1918 caused between 20 and 50 million deaths worldwide. This virus was antigenically similar to the swine virus and it appears to have disappeared from the human population following the 1918–1919 pandemic. This virus is, however, still carried in the swine population with an occasional cross-over into humans without person-to-person transmission. Why such widespread transmission of the swine virus among humans has not reoccurred is not known—and that in itself is fearful! It could happen again, even though there is a vaccine. The current theory from the Armed Forces Institute of Pathology on where the 1918 flu virus originated is as follows: The virus most likely traveled **back and forth** between pigs and people until there was a single mutation or a sufficient number of lesser mutations that allowed the virus to cause a serious disease in humans and pigs!

UNITED STATES 1918, America was caught up in the last year of World War I. And, deadly influenza, the so-called "Spanish flu," was sweeping the country, spreading terror everywhere. The first documented deaths were in Boston. Explanations were offered, but in fact no one had an answer. Viruses were still largely unknown. That particular strain of flu virus is still one of the mysteries of 1918 pandemic. Once started, the disease moved west in lethal waves that appeared to follow railroad lines. The speed with which it killed was appalling, the loss of life unimaginable. The pandemic took more than 600,000 lives in just a few months! It would be as if today, with our present population, more than 1,400,000 people were to die in a sudden outbreak for which there was no explanation and no known cure. It was said every family lost someone. **Time wise,** this is the worst epidemic/pandemic America has ever known. It killed more Americans than all the wars of this century combined. There were so many people dying from the flu that communities nationwide were running out of caskets. Some people believe that 1918 pandemic began in the spring of 1918, when soldiers at Fort Riley, Kansas, burned tons of manure. A gale kicked up. A choking dust storm swept out over the land—a stinging, stinking yellow haze. The sun went black in Kansas. Two days later—on March 11, 1918—an Army private reported to the camp hospital before breakfast. He had a fever, sore throat, headache—nothing serious. A minute later, another soldier showed up. By noon, the hospital had over a hundred cases; in a week, 500. That spring, forty-eight soldiers died at Fort Riley. The cause of death was listed as pneumonia. The sickness then seemed to disappear—leaving as quickly as it had come. But that same summer and fall, over one and a half million Americans crossed the Atlantic for war. Some of those soldiers came from Kansas, and it is believed that they carried the virus with them. Almost immediately, the Kansas sickness resurfaced in Europe. American, English, French and German soldiers became ill. As the sickness spread, it appeared to get worse. By the time the disease came back to America, it was a deadly killer.

On a rainy day in September, Dr. Victor Vaughan, acting Surgeon General of the Army, received urgent orders: proceed to a base near Boston called Camp Devens. On the day that Vaughan arrived, 63 men died at Camp Devens. An autopsy revealed lungs that were swollen, filled with fluid, and strangely blue. Doctors were stunned. When the strange new disease was finally identified, it turned out to be a very old and familiar one: influenza, the flu. But it was unlike any flu that anyone had ever seen.

During the epidemic of 1918, in San Francisco, all citizens were required to wear masks. One form of therapy was "cabbage baths." People did not jump into a tub with some cooked cabbage; they ate the cabbage and urinated into the tub. The flu victims then got into the tub. There may have been a positive side to the bath because once you got out, a distance was created between you and other people. In this way, the baths may have helped stop person-to-person transmission. [Researchers seek new weapon against flu. (1997). *Science*, 275:756; Influenza 1994–1995: Now's the time to prepare. (1994). *J. Resp. Dis.*, 15:675].

UPDATE 1998: More than one million chickens, ducks and geese were gassed to death in the first week of january in Hong Kong in an effort to stop the "bird flu" virus before it spread beyond the boundary of Hong Kong. The new strain of influenza type A virus was isolated from a dead three-year-old boy in Hong Kong in May 1997. CDC scientists are investigating how this virus made the huge species jump from chickens into humans. By the end of January 1998, 21 cases were confirmed and 6 people died of this flu.

Tuberculosis, Hepatitis B, and Malaria— In 1937, 112,000 Americans contracted tuberculosis and 70,000 died. About **one billion** people worldwide have died from Mycobacterium tuberculosis (a bacterium 1/25,000th of an inch long) since 1770. **Over ninety million people died of TB from 1990 through the year 2000!** One in three people worldwide carry this organism! Currently, about 10 million people in the United States are infected with TB, but only 5% develop the disease. In 1997, over 2 million adults worldwide died of hepatitis B and of the 44 million children who contracted measles, 1.5 million died. Five to six thousand people die every day from malaria. **Despite over 50 years of global experience in malaria control, more people are dying of malaria now than when malaria control campaigns began! Malaria, like all of the major infectious diseases, is a disease of poverty.** Parents and children quietly cope; the ill were and are served, not shunned—that is, until the HIV/AIDS epidemic. The nine leading causes of death due to bacterial, protozoal or viral infections are given in Table I-2.

AIDS: ITS PLACE IN HISTORY

As the statistics on AIDS cases mounted, its identity as an inescapable plague seemed confirmed. It appeared to mimic the frightening

TABLE I-2 Microbes Causing Most Deaths Worldwide, 1999

Infectious disease	Cause	Annual Deaths
Acute respiratory infections (mostly pneumonia)	Bacterial or viral	4,400,000
Diarrheal diseases	Bacterial or viral	3,200,000
Tuberculosis	Bacterial	1,500,000
Hepatitis B	Viral	2,000,000
Malaria	Protozoan	3,100,000
Measles	Viral	1,500,000
Neonatal tetanus	Bacterial	600,000
AIDS	Viral	2,500,000
Pertussis (whooping cough)	Bacterial	360,000

Sources: World Health Organization; Harvard School of Public Health.

epidemics of the past: cholera, yellow fever, leprosy, syphilis, and the plague or Black Death. The history of AIDS—the history that seemed relevant to understanding the new pandemic—would be the history of the epidemics of the past. Medical history suddenly gained new social relevance; policy analysts, lawyers, and journalists all wanted to know whether past epidemics could provide some clues to the current crisis. How had societies attempted to deal with epidemics in the past? The contemporary meaning of past plagues is read in the face of AIDS.

THE AIDS PANDEMIC IS CERTAINLY ONE OF THE DEFINING EVENTS OF OUR TIME. There are stories to be told from it, stories of the people infected and affected by it—the well, the ill, the dying, and the survivors. And there are the stories of scientific discovery, of HIV and viral mechanisms, and of genetic mysteries being understood. And then there are stories of scientific politics, claims and counterclaims, and the manipulating that goes on in the stratosphere of high-level science.

"A third of the world died," wrote Jean Froissart at the end of the 14th century, when medieval medicine had little to offer against the Black Death. Now, at the end of the 20th century, we see modern science registering its progress about a plague of our own time.

AIDS was identified on planet earth just 20 years ago and it is now the leading killer among all known infectious diseases. AIDS is ranked fourth among **all** causes of death.

AIDS is the most dramatic, pervasive, and tragic pandemic in recent history. So deep an impression has the AIDS epidemic made on public perceptions that according to a Louis Harris poll in 1994, almost 30% of the population believes that the "greatest threat to human life" is AIDS or some other kind of plague. **THE HIV/AIDS PANDEMIC CONSISTS OF TWO PARTS: ONE MEDICAL, THE OTHER SOCIAL.** HIV/AIDS infection has provoked a reassessment of society's approaches to public health strategy, health care resource allocation, medical research, and sexual behavior. Fear and discrimination have affected virtually every aspect of our culture. Both the medical challenge and, in particular,

the social challenge will continue in the foreseeable future. Arthur Ashe, a world-class tennis player, so feared discrimination against himself and family that he lived with AIDS for $3\frac{1}{2}$ years before he was forced to reveal he had the disease (Figure I-1).

OVERVIEW ON HIV/AIDS

What Is AIDS?

AIDS is defined primarily by severe immune deficiency, and is distinguished from virtually every other disease in history by the fact that it has no constant, specific symptoms. Once the immune system has begun to malfunction, a broad spectrum of health complications can set in. AIDS is an umbrella term for any or all of some 27 known diseases and symptoms. When a person has any of these diseases or has a T4 lymphocyte count of less than $200/\mu L$ of blood and also tests positive for antibodies to HIV, an AIDS diagnosis is given.

Who Are the Most Likely to Become HIV-Infected?

When all things are held equal, the most important identifying variable is **income.** Regardless of race, orientation, or language, those in the lower economic brackets are more likely to become HIV-infected. As the pandemic spreads, it is becoming more of a disease that can affect any person regardless of sexual preference or race, but in particular it is becoming more of a disease of the poor, the group within any society that has inadequate access to health care. As the pandemic increases globally, it is becoming more of a disease of the underdeveloped nations. **And where HIV is becoming more a disease of heterosexuals, it is also becoming a disease of women, because transmission to women from heterosexual contact is easier than for men** (see Chapter 11). Also, remember that women are disproportionately represented among those living in poverty anywhere in the world.

The impact of this pandemic is unique. Unlike malaria or polio, previous modern pandemics, it mostly affects young and middle-

aged adults. This is not only the sexually most active years for individuals, but also their prime productive and reproductive years. Thus the impact of HIV/AIDS is demographic, economic, political, and social. HIV/AIDS is a disease of human groups and its demographic and social impacts multiply from the infected individual to the group. In the most affected areas, infant, child, and adult mortality is rising and life expectancy at birth is declining rapidly. The cost of medical care for each infected person overwhelms individuals and households.

Acquired Immune Deficiency Syndrome (AIDS), first identified in 1981, is the final stage of a viral infection caused by the Human Immunodeficiency Virus (HIV). Medical experts recognize two strains: HIV-1, discovered in 1983, which is generally accepted as the cause of most AIDS cases throughout the world; and HIV-2, discovered in West Africa in 1986 and later found in some former Portuguese colonies elsewhere and in Europe. **HIV is a retrovirus: a virus that inserts—probably for life—its genetic material into the genetic material of the host cells at the time of the infection.** Inasmuch as the ability to remove genetic material from cells is beyond the capability of current medical science, the infection may be said to be **incurable**. AIDS is characterized by a profound loss of cell-mediated immune function and the depletion of a special kind of white blood cell called a **CD4 or T4 lymphocyte.**

FIGURE I-1 Arthur Ashe—Winner of Two United States Tennis Championships. On April 8, 1992 Ashe announced that he had AIDS. He died on February 6, 1993 at age 49. He became infected with HIV from a blood transfusion during heart bypass surgery in 1983. In 1988 his right hand suddenly became numb. Brain surgery revealed a brain abscess caused by toxoplasmosis. He asked two questions, **"Why do bad things happen to good people?"** and **"Is the world a friendly place?"** Ashe was the United States Davis Cup Captain for 5 years and supported the NCAA Proposition 42 academic requirements for athletes. In 1975 he became the first black to win a Wimbledon men's title. After learning that he had AIDS, Ashe completed his third book, *A Road to Glory*, a three-volume history of black athletes in the United States. (*Photograph courtesy of AP/Wide World Photos*)

AIDS: A CAUSE OF DEATH

Peter Piot, (Figure 1–2) Executive Director of the United Nations Program on HIV/AIDS, speaking at the 37th Interscience Conference on Antimicrobial Agents and Chemotherapy said that "HIV has transformed the world joining tuberculosis and malaria as a major cause of death worldwide. This epidemic won't be under control in any country until it is brought under control everywhere."

AIDS is now the seventh leading cause of death among 1- to 4-year-olds, sixth among 15- to 24-year-olds, and second among 25- to 44-year-olds in the United States.

WHY ALL THE FUSS ABOUT HIV/AIDS?

So it is now, after 19 years, killing **2.5 million** people a year worldwide. But what of the other diseases, and how many people are they killing each year? Respiratory infections, **over 4 million**; diarrheal diseases, **over 3 million**; TB, **over 3 million**; hepatitis B infection, **over 2 million**; measles, **about 1.5 million**; malaria, **over 2 million**; and there are about **12 to 15 million** who die each year from cardiovascular diseases! **So, why is HIV/AIDS recognized or believed to be a larger threat?**

First, this virus is mainly transmitted during sexual intercourse. Since few human societies talk openly and honestly about sex, this makes this disease difficult to discuss; and since sex is a very private activity, it makes the transmission of HIV very difficult to control.

Second, it has an extraordinary capacity for change and rapid global spread. Through 1998, an estimated 9 million people will have died of AIDS worldwide; by the end of the century this figure is expected to be at least 11 million. Malaria, TB, heart disease, cancer: no other disease is spreading or growing at this rate. And in spite of HIV/AIDS campaigns in most countries, there is little evidence so far of any slowdown in the **spread** of the epidemic. By the end of the century, about 68 million people are expected to have been infected by HIV.

Third, there is a long asymptomatic period between infection and illness. On average, it takes about 10 to 12 years for someone infected with HIV to develop AIDS. During this time, the HIV-infected will show few if any recognizable symptoms, but he or she will be able to infect other people.

This long asymptomatic period is **rare in human infectious diseases**. The long asymptomatic period that occurs between cause and effect has, for years, made it difficult to persuade individuals or governments to take the pandemic seriously. People **dying** today represent those infected 10 to 20 years ago; the results of anything

we do now to reduce transmission will not be apparent until the next century.

Fourth, HIV/AIDS is more serious than many more common diseases **because of the age groups it attacks**. Some diseases—measles and diarrhoea, for example—affect mainly **infants and children**; others, such as heart disease and cancer, **affect mainly the old**. But because HIV is predominantly transmitted sexually, AIDS mainly kills people in **their 20s through their 40s**. A major increase in deaths among these age groups, the most productive section of society, presents a much greater impact socially and economically than deaths that occur among children or old people would have.

Fifth, in the past, plagues were often marked by their lack of discrimination, by the way in which they killed large numbers of people with little regard for race, wealth, sex, or religion. **But AIDS was different from the beginning.** It immediately presented a **political as much as a public-health problem**. Homosexuals, who until this pandemic had been mostly closeted in the United States, were suddenly at the heart of a health crisis as profound as any in modern American history.

Sixth, with respect to therapy, HIV disease/AIDS requires the use of some of the most **expensive** and **toxic drugs** in medical history.

So why all the fuss about HIV/AIDS? It is caused by a **unique virus** that **changes itself faster** than a rumor making the gossip column of a tabloid. It is lethal, it is transmitted, most often sexually, it affects people in their reproductive and most productive years, it is exceptionally expensive to treat, it defies our best scientists who are working to create a vaccine for preventing infection and transmission, and it is a disease that has severely stigmatized those who have it. And lastly, it has awakened our most primal fear, that of dying a horrible death caused by unknown agents—**this new virus means there are others in waiting.**

HIV HAS BUT ONE REQUIREMENT TO CONTINUE ITS PRESENCE ON EARTH: A HUMAN HOST!

There is the expectation that parents will die before their children. Because of the HIV/AIDS epidemic, it is not working out that way for many thousands of parents. They are watching their children die in the prime of life.

The facts on HIV infection, disease, and AIDS that are presented in the following chapters, when understood, clearly place the responsibility for avoiding HIV infection on *YOU*. You must assess your lifestyle; if you choose not to be

FIGURE I-2 Peter Piot Executive Director of the Joint United Nations Program on HIV/AIDS (UNAIDS). He is a Belgian physician, microbiologist and codiscoverer of the Ebola virus in 1976. He is the world's leading advocate for HIV/AIDS control and prevention. *(Photograph courtesy of UNAIDS/Yoshi Shimizu)*

abstinent, you must know about your sexual partner and you must practice safer sex. **NEVER THINK THAT YOU ARE IMMUNE TO HIV INFECTION.**

June 5, 2000 will mark the beginning of the **19th year** of the AIDS pandemic. No cure has been found and, although AIDS is now called a "manageable illness," people who are sick must make endless compromises to this disease. The AIDS pandemic forces people to face their mortality daily, for months and years.

FIRST REPORTS ON AIDS CASES IN THE UNITED STATES

In January 1981, while Ronald Reagan was taking his first oath of office as president, doctors around the country were just discovering the pattern of symptoms and infections in patients that was to become a very new disease.

On June 5, 1981, the first cases of the illness now known as AIDS were reported from Los Angeles in five young homosexual men diagnosed with *Pneumocystis carinii* pneumonia and other opportunistic infections.

Initially, AIDS appeared among homosexual males; most frequently those who had many sex partners. Further study of the gay population led to the conclusion that the agent responsible for AIDS was being transmitted through sexual activities. In July of 1982, cases of AIDS were reported among hemophiliacs, people who had received blood transfusions, and injection drug users. These reports all had one thing in common—**an exchange of body fluids,** in particular blood or semen, was involved. In January 1983, the first cases of AIDS in heterosexuals were **documented.** Two females, both sexual partners of IDUs, became AIDS patients. This was clear evidence that the infectious agent could be transmitted from male to female as well as from male to male. Later in 1983, cases of AIDS were reported in Central Africa, but with a difference. The vast majority of African AIDS cases were among heterosexuals **who did *not* use injection drugs.** These data supported the earlier findings from the American homosexual population: that AIDS is primarily a sexually transmitted disease. And the risk for contracting AIDS increased with the number of sex partners one had and the sexual behaviors of those partners. Thus early empirical observations on which kinds of social behavior placed one at greatest risk of acquiring AIDS were later supported by surveillance surveys, testing, and analysis.

——— **POINT OF VIEW I.2** ———

SILENCE = DEATH: SILENCE AND STIGMA

The silence comes because we are talking about sex, we are talking about needles and injection drug use—things we are not good at talking about as a community. But silence is also perpetuating this pandemic, and it must be talked about everywhere, by everyone! By the time former president Ronald Reagan gave his **first** speech on the AIDS crisis in America, hundreds of thousands of Americans were HIV-infected, 36,000 men, women, and children had been diagnosed with AIDS, and over 25,000 had died of AIDS—the year, 1987!

FIRST REACTION TO AIDS: DENIAL

When the disease that would eventually be called AIDS first emerged in 1981, a few officials within agencies like the Centers for Disease Control and Prevention realized that a new infectious agent was almost certainly at work and that it could well be spreading rapidly. They most cautiously tried to sound the alarm, but the nation was not ready to talk about subjects like **sex, needles** and **condoms.** Among those most heavily in denial were gay men, who were at most risk. They were still enjoying sexual liberation won in the 1970's, and nobody was in a mood to call the party off, even as close friends and sexual partners began dying.

New York playwright Larry Kramer attempted to break through this denial in early 1983 with an article in a widely read gay magazine. Headlined "1,112 and Counting" Kramer warned: "If this article doesn't rouse you to anger, fury, rage and action, gay men have no future on this Earth."

As they turned their fears into political engagement, the activists confronted a Washington that resisted action. Blood banks denied that any extra precautions were needed to prevent transmission. AIDS was buried deep inside newspapers, seldom mentioned on television. The death of movie star Rock Hudson in 1985 finally put AIDS on the front pages. But still, three young hemophiliacs, Ricky, Robert, and Randy Ray, were firebombed out of their Florida home when their neighbors learned they were HIV positive two years later. While there was already convincing evidence that AIDS could not be transmitted by casual contact, people were never the less fired from their jobs across the country because of fears that they posed a threat to coworkers.

THE EDUCATIONAL COMMUNITY

Like most complex problems, the AIDS epidemic poses special problems for educators. One of the most disturbing is discrimination against the HIV-infected. People worldwide have from the very beginning of the AIDS pandemic learned to **categorize, rationalize, stigmatize,** and **"persecutize"** those with HIV disease and

AIDS. Perhaps the worst display of stigmatization and discrimination occurs against children, especially those of school age. Guidelines from the U.S. Surgeon General, federal and state health officials, and the medical community have not calmed the fears of misinformed parents. Many stories have made headlines and television news concerning children who have been barred from attending school. While the courts can order admission, they **cannot** assure peer and adult acceptance. Persuasion must come through a better understanding of the disease. Parents can be reassured through reminders that HIV/AIDS is not transmitted by casual contact. Students must also be educated about HIV/AIDS. Gallup Poll results (Table I-3), although taken in 1988, are about what one would still find true today.

Since the latter half of the 1980s, schools throughout the United States have been offering education programs to teach adolescents about human immunodeficiency virus (HIV) infection. Since 1998, at least 40 states required HIV/AIDS education for adolescents

TABLE I-3 1988 Gallup Poll Results on Age Related HIV/AIDS School Education

	National Totals %	Public School Parents %	Non-Public School Parents %
The Gallup Poll asked those who favored having public schools developed an HIV education program (90% of all respondents) the following questions: At what age should students begin participating in an HIV education program?			
Under 5 years	6	5	11
5–9 years	40	43	42
10–12 years	40	39	32
13–15 years	10	11	13
16 years or older	1	1	1
Don't know	3	1	1
Should public schools teach what is called "safer sex" for HIV prevention? (This was understood to mean teaching about the use of condoms.)			
Should	78	81	72
Should not	16	16	25
Don't know	6	3	3

Source: 20th Annual Gallup Poll on the Public's Attitudes Toward the Public Schools.

(40 states offered some HIV/AIDS education in their elementary schools). However, the effect of such programs on HIV-related knowledge and behavior among adolescents is largely unknown.

The U.S. Department of Health and Human Services report in their "Healthy People 2000 Review—1998–1999" that 41.4% of American's college students received information on HIV/AIDS at their respective college campuses. The year 2000 target is to raise the number of college students receiving HIV/AIDS information on their campuses to 90%.

WORLD INTERNATIONAL AIDS CONFERENCES

The World International AIDS Conference is believed to be the single most important meeting on AIDS.

The International AIDS Conference has been the timepiece of the pandemic. The venue, content, style, and mood of each meeting temporarily freezes in time the state of the worldwide HIV/AIDS pandemic. (Figure I–3)

In 1983, there were relatively so few groups involved with HIV/AIDS research that they could stay in contact by telephone. As the virus spread many countries became involved in HIV/AIDS research and clinical care. International, national and state meetings were formed as a way to exchange new information. The international meetings continue to receive the most press coverage. Since the **first** international AIDS conference in 1985, the conferences have grown in size to the point that scientists questioned their usefulness. As of 1994, the International Conference on AIDS is held every 2 years (Table I-4).

FIGURE I-3   Dr. David Ho, Eleventh International HIV/AIDS Conference, Vancouver, BC, July 11, 1996. Dr. Ho, a leading AIDS researcher from the Diamond AIDS Research Centre in New York is speaking about the recent success in reducing viral load using protease inhibitors and combination drug therapy. His pioneering efforts, using new HIV/AIDS drugs, have helped lift a death sentence—for a few years at least, and perhaps longer—on tens of thousands of AIDS sufferers. He asked the question, **"Can HIV be eradicated from the body?"** With this single question based on his research, he **galvanized** the AIDS community. David Da-i Ho was *Time's* **MAN OF THE YEAR FOR 1996.** (*Photograph courtesy of REUTERS/Jeff Vinnick/Archive Photos*)

TABLE I-4 International AIDS Conferences[1]

Presented is a list of past International AIDS Conferences by month, year and location. The number of **reported** AIDS cases and deaths, are for each year in the United States.

Number	Month	Year	Location	AIDS[2] Cases	Deaths
1st	June	1985	Atlanta, Georgia	17,000	8,000
2nd	June	1986	Paris France	36,000	19,000
3rd	June	1987	Washington, DC	65,000	33,000
4th	June	1988	Strockholm, Sweden	100,000	55,000
5th	June	1989	Montreal, Canada	143,000	83,000
6th	June	1990	San Fransisco, California	191,000	116,000
7th	June	1991	Florence, Italy	249,000	147,000
8th	July	1992	Amsterdam, Netherlands	327,000	192,000
9th	June	1993	Berlin, Germany	432,000	259,000
10th	August	1994	Yokohama, Japan	502,000	286,000
11th	July	1996	Vancouver, Canada	624,000	378,000
12th	June	1998	Geneva, Switzerland	683,000	410,000
13th	July	2000	Durban, South Africa	–	–
14th	July	2002	Barcelona, Spain	–	–
15th	July	2004	Toronto, Canada	–	–

1. Based on UNAIDS data, during the 6 days of the 12th conference, about 100,000 new HIV infections occurred and over 44,000 people died of AIDS worldwide.

2. AIDS cases and deaths are minimum estimates for the United States, based on reported data to the CDC.

3. UNAIDS Executive Director, Peter Piot reported that between the 10th 1996 AIDS Conference and the 12th in 1998, 10 million people were infected with HIV! This, he said "represents a collective failure of the world."

THE AIDS MEMORIAL QUILT

Conception of the Quilt

During the eigth candlelight march in San Francisco, Jones asked fellow marchers to write on placards the names of friends and loved ones who died of AIDS. At the end of the march, Jones and others stood on ladders, above the sea of candlelight, taping these placards to the walls of the San Francisco Federal Building. The walls of names looked to Jones like a patchwork quilt.

The purpose of the quilt is to educate. The "AIDS Quilt" is made up of individual fabric panels, each the size of a grave, measuring three feet by six feet, stitched together into 12 foot × 12 foot sections. In October 1987, the AIDS quilt was first put on display on the mall in Washington, D.C. At that time it contained

--------- BOX I.2 ---------

MARY D. FISHER SPEAKS ABOUT AIDS AT THE 1992 REPUBLICAN NATIONAL CONVENTION (Figure I–3)

I bear a message of challenge, not self-congratulation. I want your attention, not your applause. I would never have asked to be HIV-positive. But I believe that in all things there is a good purpose, and so I stand before you, and before the nation, gladly.

The reality of AIDS is brutally clear. Two hundred thousand Americans are dead or dying; a million more are infected. Worldwide, 40 million, or 60 mil-

lion, or 100 million infections will be counted in the coming few years. But despite science and research, White House meetings and congressional hearings, despite good intentions and bold initiatives, campaign slogans, and hopeful promises—despite it all, it's the epidemic which is winning tonight.

In the context of an election year, I ask you—here, in this great hall, or listening in the quiet of your home—to recognize that the AIDS virus

———— **BOX I.2** (*continued*) ————

is not a political creature. It does not care whether you are Democrat or Republican. It does not ask whether you are black or white, male or female, gay or straight, young or old.

Tonight, I represent an AIDS community whose members have been reluctantly drafted from every segment of American society. Though I am white, and a mother, I am one with a black infant struggling with tubes in a Philadelphia hospital. Though I am female, and contracted this disease in marriage, and enjoy the warm support of my family, I am one with the lonely gay man sheltering a flickering candle from the cold wind of his family's rejection.

This is not a distant threat; it is a present danger. The rate of infection is increasing fastest among women and children. Largely unknown a decade ago, AIDS is the third leading killer of young adult Americans today—but it won't be third for long. Because, unlike other diseases, this one travels. Adolescents don't give each other cancer or heart disease because they believe they are in love. But HIV is different. And we have helped it along—we have killed each other—with our ignorance, our prejudice, and our silence.

We may take refuge in our stereotypes, but we cannot hide there for long. Because HIV asks only one thing of those it attacks: Are you human? And in this is the right question: Are you human?

Because people with HIV have not entered some alien state of being. They are human. They have not earned cruelty and they do not deserve meanness. They don't benefit from being isolated or treated as outcasts. Every one of them is exactly what God made: a person. Not evil, deserving our judgment; not victims, longing for our pity. People. Ready for support and worthy of our compassion.

My call to the nation is a plea for awareness. If you believe you are safe, you are in danger.

FIGURE I-4   Mary Fisher, age 44, addressed the delegates at the 1992 Republican National Convention. Fisher, previously diagnosed with AIDS, called the disease a "present danger" that leaves no one safe. She also called for compassion in place of ignorance and silence. She said, **"The virus asks only one question: Are you human?"** (*Photograph courtesy of REUTERS/ Steven Jaffe/Archive Photos*)

Because I was not hemophiliac, I was not at risk. Because I did not inject drugs, I was not at risk. My father has devoted much of his lifetime to guarding against another holocaust. He is part of the generation who heard Pastor Neimoller come out of the Nazi death camps to say, "They came after the Jews and I was not a Jew, so I did not protest. They came after the Trade Unionists, and I was not a Trade Unionist, so I did not protest. They came after the Roman Catholics, and I was not a Roman Catholic, so I did not protest. Then they came after me, and there was no one left to protest."

The lesson history teaches is this: If you believe you are safe, you are at risk. If you do not see this killer stalking your children, look again. There is no family or community, no race or religion, no place left in America that is safe. Until we genuinely embrace this message, we are a nation at risk.

Tonight, HIV marches resolutely towards AIDS in more than a million American homes, littering its pathway with the bodies of the young. Young men. Young women. Young parents. Young children. One of the families is mine. If it is true that HIV inevitably turns to AIDS, then my children will inevitably turn to orphans.

My family has been a rock of support. My 84-year-old father, who has pursued the healing of the nations, will not accept the premise that he cannot heal his daughter. My mother has refused to be broken; she still calls at midnight to tell wonderful jokes that make us laugh. Sisters and friends, and my brother Philip (whose birthday is today)—all have helped carry me over the hardest places. I am blessed, rich and deeply blessed, to have such a family.

But not all of you have been so blessed. You are HIV seropositive but dare not say it. You have lost loved ones, but you dare not whisper the word AIDS. You weep silently, you grieve alone.

I have a message for you: It is not you who should feel shame. It is we. We who tolerate ignorance and

—————— **BOX I.2** (*continued*) ——————

practice prejudice, we who have taught you to fear. We must lift our shroud of silence, making it safe for you to reach out for compassion. It is our task to seek safety for our children, not in quiet denial, but in effective action.

Some day our children will be grown. My son Max, now four, will take the measure of his mother; my son Zachary, now two, will sort through his memories. I may not be here to hear their judgments, but I know already what I hope they are.

I want my children to know that their mother was not a victim. She was a messenger. I do not want them to think, as I once did, that courage is the absence of fear; I want them to know that courage is the strength to act wisely when most we are afraid. I want them to have the courage to step forward when called by their nation, or their Party, and give leadership, no matter what the personal cost. I ask no more of you than I ask of myself, or my children.

To the millions of you who are grieving, who are frightened, who have suffered the ravages of AIDS firsthand: Have courage and you will find comfort.

To the millions who are strong, I issue the plea: Set aside prejudice and politics to make room for compassion and sound policy.

To my children, I make this pledge:
I will not give in, Zachary, because I draw my courage from you. Your silly giggle gives

me hope. Your gentle prayers give me strength. And you, my child, give me reason to say to America, "You are at risk."

And I will not rest, Max, until I have done all I can to make your world safe. I will seek a place where intimacy is not the prelude to suffering.

I will not hurry to leave you, my children. But when I go, pray that you will not suffer shame on my account.

To all within the sound of my voice, I appeal: Learn with me the lessons of history and grace, so my children will not be afraid to say the word AIDS when I am gone. Then their children, and yours, may not need to whisper it at all.

God bless the children, and bless us all—and good night.

UPDATE 1999—Seven years after her 13-minute speech, Mary Fisher is still introduced as the woman who spoke at the 1992 **Republican National Convention.** But, she would prefer to be remembered for being an ordinary woman who tried to tell everyone that they were at risk for this disease. That ordinary people need extraordinary leadership and that the voices of ordinary people must be heard. Because "we live on a shrinking planet, what is **global today is local tomorrow.**"

1,920 panels and covered an area larger than two football fields and it took less than 2 hours to read all the names. In 1992 it took 60 hours. By mid-1994, the quilt contained over 28,000 panels bearing over 40,000 names, required over 30 acres to lay out and weighed 28 tons. As of mid-1999 the quilt weighed 55 tons with 46,000 panels about 79,000 names. Displayed in its entirety, it will cover over 25 football fields. **It is the largest piece of folk art in the world and continues to increase in size daily.** The quilt was the idea of Cleve Jones, of San Francisco, who in 1985 feared AIDS would become known for the number of people it killed. He wanted a way of remembering the people, who were, in many cases, his friends. There are about 50 miles of seams and 26 miles of canvas edging. There are panels from each of the 50 states. Each day new

panels arrive from across the United States and 29 foreign countries to be added to the quilt (Figures I-5 and I-6). For those left behind, the panels represent an expression of love and a sign of grief—a part of the healing process (Figure I-7, I-8 and I-9). A few of the well-known people who have died of AIDS are seen in Figures I-10 through I-13.

Portions of the quilt tour in major cities. Donations made for viewing the quilt are being used to support local Names Project chapters and their staffs.

Each panel has its own story. The stories are told by those who make the panels for their lost friends, lovers, parents, and children. The complete quilt was displayed in Washington, D.C., October 11–13, 1996 for the last time—it is too large to view again as a whole. It took 10 boxcars

and a freight train to transport this work of art to the nation's capital.

Beginning year 2000, about 450,000 Americans will have died from AIDS. If each of their names made up a separate panel, imagine the size of the quilt. If a name is read every 10 seconds, it would take about seven and a half weeks of calling names, 24 hours a day!

Decade of Remembrance

People nationwide will celebrate the 13th anniversary of the AIDS Quilt on July 21, 2000.

For more information about the Names Project's AIDS Memorial Quilt, call (415) 882-5569, ext. 375, email: chapters@aidsquilt.org or write: The NAMES Project Foundation, 310 Townsend St., Suite 310, San Francisco, CA 94107.

For information about your state chapter, contact the Names Project coordinator at (201) 888-1790.

WORLD AIDS DAY

December 1, 1988 was the first acclaimed *World AIDS Day* (WAD). The theme was "**Join the Worldwide Effort.**" World AIDS Day is a day set aside to pay tribute to those who have AIDS and to those who have died of AIDS. Its purpose is to increase our awareness of AIDS. The first WAD did not attract much attention outside the

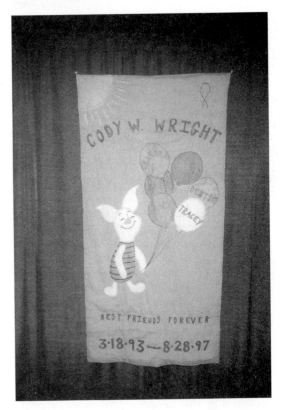

FIGURE I-6 An AIDS panel depicting "Best Friends Forever"—a life time of just over four years. Photo taken July, 1998. (*Photograph courtesy of the author*)

Connie
one year
one week
one day
12·23·86 — 12·30·87

FIGURE I-7 A photograph of a single panel of the quilt taken in July of 1991. (*Photograph courtesy of the author*)

gay community. In 1989, the theme was "**Our Lives, Our World—Lets Take Care of Each Other.**" Artists got together and held the first annual *Day Without Art* to coincide with World AIDS Day. For the second annual Day Without Art/World AIDS Day on December 1, 1990, at least 3,000 art organizations in the United States were involved. They included the Smithsonian Institution as well as small American art galleries and art galleries in Canada, England, France, and The Netherlands.

The second annual Day Without Art/World AIDS Day was commemorated by shrouded sculptures, darkened marquees, and exhibits depicting the loss of life to HIV infection and AIDS. At 8 PM on December 1, 1990, the Manhattan and San Francisco skylines were dimmed for 15 minutes. On Broadway, the marquees were darkened for 1 minute and 23 cable TV stations as well as broadcasts in England, Canada, and Australia were interrupted with a 1-minute announcement about AIDS. (Theme was "**Women and AIDS**")

The third annual Day Without Art/World AIDS Day, 1991, was the largest AIDS event ever. The theme was "**Sharing the Challenge.**" It was intended to underline the global nature of the pandemic and to foster awareness that only by pooling efforts, resources, and imagination can hope prevail against the common threat. Many more people, businesses, and industries became involved for the first time. World AIDS Day 1992 adopted the theme "**A Community Commitment.**" The theme for World AIDS Day 1993 was "**A Time to Act.**" Topics covered were: fighting denial, discrimination, and complacency; bridging the resource gap; reducing women's vulnerability to infection; prevention; and providing humane care. The theme for 1994 was "**The Global Challenge;** AIDS and the Family." World AIDS Day 1995 carried the theme "**Shared Rights, Shares Responsibilities.**" The meaning is that all people—men, women, children, the poor, minorities, migrants, refugees, sex workers, drug injectors, gay men—have the right to be able to avoid infection, the right to health care if sick with AIDS, and the right to be treated with dignity and without discrimination. For 1996, the theme was "**One World. One Hope.**" For 1997, the theme was "**Children Living in a World with AIDS.**" For 1998 the theme was "**Force for Change: World AIDS Campaign with Young People.**" **For 1999 the theme was "Listen, Learn, Live! Challenges for Latin America and the Caribbean."** For information on World AIDS Day 2000, write 108 Leonard Street, 13th Floor, New York, NY 10013 or call (212) 513-0303, or world AIDS Day Public Information Office, WHO-GPA, 1211 Geneva 27, Switzerland.

THE WIDENING GAPS

In the developed countries, AIDS is a "mature epidemic." For example, in the United States, some 40,000 people continue to become

FIGURE I-8 Mark's Panel. He died of AIDS at age 38. Photo taken July, 1998. (*Photograph courtesy of the author*)

FIGURE I-9 Wanda's Panel. She died of AIDS at age 45. Photo taken July, 1998. (*Photograph courtesy of the author*)

infected each year. A million Americans are thought to be infected with HIV, but this estimate has not grown in years. However, a huge, ever-expanding gap exists between what is happening in the US (where less than 1% of the world's HIV population resides) and in other industrialized nations and the developing world. It is easy to make the argument that the bulk of the populations in the developing world are subject to so much poverty, infectious disease, and turmoil that one more disease won't change their lives very much. For example, just substitute AIDS for malaria, TB or measles. They are all major killers in Africa.

People of developed nations most often compare AIDS with plague, syphilis, leprosy and influenza, but they are discussed in terms of European and American events, as if those are the only settings in which such catastrophes matter. Historically, few major diseases have been noted in the developing world, where similar epidemics may be considered just another cycle of nature. Yet every day, 14,000 adults and 2,000 children become infected with HIV in sub-Saharan Africa, India, and the rest of Asia.

FIGURE I-10 Brad Davis, Film Star (center). Voted best actor by the Foreign Press Association's Golden Globes Awards (1979) for his work in *Midnight Express*. The film was the story of an American imprisoned in Turkey. Brad became HIV-infected through drug use, was diagnosed with AIDS in 1985, and died at age 41 in 1991. (*Photograph courtesy of AP/Wide World Photos*)

The gains in childhood disease prevention, vaccination, and care programs developed and implemented at great expense by global health organizations since the late-1970s have been obliterated by the impacted of HIV/AIDS on children's lives and development. But apart from this human tragedy, ignoring AIDS in the world is not in anyone's interest.

THE WORLD HEALTH ORGANIZATION

In May 1998 Gro Harlem Brundtland, a former prime minister of Norway was elected to a 5 year term as director-general of the World Health Organization.

The World Health Organization has established a global program on HIV infection and AIDS. The program has three objectives: to prevent new HIV infections, to provide support and care to those already infected, and to link international efforts in the fight against HIV infection and AIDS.

In 1995 the United Nations Secretary-General chose Peter Piot to begin a joint United Program on HIV/AIDS (UNAIDS). Piot's job is to coordinate actions and reduce duplication among the seven co-sponsors of UNAIDS: United Nations Children's Fund (UNICEF), the United Nations Development Programme (UNDP), the United Nations Educational, Scientific and Cultural Organization (UNESCO), the United Nations Population Fund (UNFPA), the United Nations Drug Control Porgram (UNDCP) and the World Bank.

HIV/AIDS data released by UNAIDS was used to estimate that ending year 2000, 57 million people worldwide will be HIV infected, about 29 million will be living with HIV disease, 12 million are living with AIDS, and 18 million have died of AIDS. And those deaths left over 10 million

FIGURE I-11   Rock Hudson, Movie and Television Star. A Hollywood legend and undisclosed homosexual. He was the first major public figure to reveal he had AIDS. The conventional wisdom, expressed by journalist Randy Shilts in 1987 (Figure I-12) is **"that there were two clear phases to the disease in the United States: there was AIDS before Rock Hudson and AIDS after."** This common observation has been confirmed by others who also note that Hudson's 1985 disclosure that he was suffering from AIDS—rather than a change in the character of the epidemic or the introduction of new, substantive information about AIDS—led to a permanent increase in media attention to the disease. Hudson died in 1985 at age 59. (*Photograph courtesy of AP/Wide World Photos*)

orphans. About 90% of all new HIV infections occur in developing nations, with about 1 in 10 people aware that they are HIV infected.

BREAKTHROUGHS OFFER NEW HOPE!

"And hope is brightest when it dawns from fears."—Sir Walter Scott, 1771–1832

Through 1995, the diagnoses of AIDS was a devastating death sentence. Those seeking to treat it seldom uttered the words "AIDS" and "hope" in the same sentence. **What a difference a year can make!** In 1996 and through 1999, those terms have become inextricably linked in the minds of researchers and patients. And while the new optimism must be tempered with numerous caveats, 1996, 1997, and 1998 ushered in a series of stunning break-

FIGURE I-13 Wladziu Valentino (Lee) Liberace, Internationally Known Pianist and Entertainer. Died of AIDS February 4, 1987 at the age of 67. (*Photograph courtesy of AP/Wide World Photos*)

FIGURE I-12 Randy Shilts, author of "And The Band Played On" (Viking Penguin Press, 1987), died of AIDS February 17, 1994 at the age of 42. (*Photograph courtesy of AP/Wide World Photos*)

throughs, both in AIDS treatment and in basic research on HIV. The newfound optimism continues into 2000 but, the emergence of drug-resistant HIV and serious side effects from the drugs have altered expectations.

THERAPY

On the therapeutic side, new drugs called **protease inhibitors** can now dramatically reduce HIV levels in the blood when taken alone or with other antiviral compounds. At the same time, polypeptide molecules called **chemokines** have been unveiled as potent foes of HIV. To enter cells, HIV must bind to cell surface proteins that normally serve as receptors for the chemokines—and **people born with de-fective receptors are less susceptible to HIV infection.** This work offers new insight into the pathogenesis of the disease and may one day blossom into new treatment or even vaccines. The years from 1996 into year 2000 offer greater hope and success in the fight against HIV/AIDS.

THE FUTURE

The history of HIV/AIDS is one of remarkable scientific achievement. Never in the history of humans has so much been learned about so complex an illness in so short a time. We moved into the 1990s with hope and the determination to find better therapies and a vaccine. The task is formidable but it has to be done and it will be accomplished. The evolving story of the HIV/AIDS epidemic has been one of the major medical news events of the past 19 years. It is getting hard to imagine medicine or society without HIV/AIDS. But we must be careful not to blame this disease on our already failing health care system. As the National Commission on AIDS said:

The AIDS epidemic did not leave 37 million or more Americans without ways to finance medical care, but it did dramatize their plight. This epidemic **did not** cause the problems of homelessness—but it has expanded it and made it more visible. The epidemic **did not** cause the collapse of the health care system—but it has accelerated the disintegration of our public hospitals and intensified their financing problems. The AIDS epidemic **did not** directly augment problems of substance abuse—but it has made the need for drug treatment for all who request it a matter of urgent national priority.

HIV/AIDS is a truly persistent global pandemic and will require a proportionate response to bring it to heel. It is the plague of our lifetimes—and probably that of our children's lives as well. **We already have children age 18 and younger who don't know what an HIV-free world is. They were born into this recognized pandemic(1980).** To survive this pandemic, society must prevent the face of AIDS from becoming faceless. The chapters on HIV infection, HIV disease, and AIDS in this book will help bring widespread information on the virus into focus. The information within these chapters should also help eliminate many of the myths and irrational fears, or FRAID (**fear of AIDS**), generated by this disease. There is much work to be done by both scientists and society.

Perhaps a borrowed anecdote says it best:

As the old man walked the beach at dawn, he noticed a young woman ahead of him picking up starfish and flinging them into the sea. Finally catching up with the youth, he asked her why she was doing this. The answer was that the stranded starfish would die if left to the morning sun.

"But the beach goes on for miles and there are millions of starfish," countered the old man. "How can your effort make any difference?"

The young woman looked at the starfish in her hand and threw it to safety in the waves. "It makes a difference to this one," she said.

(Adapted from *The Unexpected Universe* by Loren Eiseley. Copyright 1969, Harcourt Brace, New York)

Too easily can we become overwhelmed by the enormity of the AIDS pandemic. The numbers of patients and their constant needs have caused many to become paralyzed into inactivity and lulled into indifference. Like the old man, many ask "Why bother?"

For the sake of every individual living with HIV, we must focus on what each one of us can do. Each person can make a difference. Believing this, we are empowered to cope with the larger whole. **WE MUST NOT LET AIDS DICTATE THAT 49 IS OLD AGE AND AGE 39 OUR LIFE EXPECTANCY.**

TOLL-FREE NATIONAL AIDS HOTLINE

- For the English-language service (open 24 hours a day, 7 days a week), call 1-800-342-AIDS (2437).
- The Spanish service (open from 8 AM to 2 AM, 7 days a week) can be reached at 1-800-344-SIDA (7432).
- A TTY service for the hearing impaired is available from 10 AM to 10 PM Monday through Friday at 1-800-243-7889.
- National Prevention Information Network, 1-800-458-5231.
- National HIV Telephone Consultation Service for Health Care Professionals, (800) 933-3413, San Francisco General Hospital, Bldg. 80, Ward 83, Room 314, San Francisco, CA 94110.
- HIV/AIDS Treatment Information Service, (800) HIV-0440 (448-0440), 9 AM- to 7 PM Eastern time, Monday-Friday. (800) 243-7012 Teletype number for the hearing-impaired, 9 AM- to 5 PM Eastern time, Monday-Friday, Box 6303, Rockville, MD 20849-6303.
- AIDS Clinical Trials Information Service, (800) TRIALS-A 874-2572, 9 AM-7 PM Eastern time, Monday-Friday. Information on clinical trials of AIDS therapies.
- National Gay and Lesbian Task Force AIDS Information Hotline (800) 221-7044.
- Gay Men's Health Crisis AIDS Hot Line, (212) 807-6655.
- National HIV/AIDS Education & Training Centers Program, (301) 443-6364, Fax: (301) 443-9887.
- AIDS National Interfaith Network, (202) 546-0807, Fax: (202) 546-5103, 110 Maryland Ave., NE, Room 504, Washington, DC 20002.

- National Hemophilia Foundation, (212) 219-8180, 110 Greene St., Suite 303, New York, NY 10012.
- Pediatric AIDS Foundation, (310) 395-9051.
- National Pediatric HIV Resource Center, 1(800) 362-0071.
- AIDSLINE via the National Library of Medicine. Free access via Grateful Med, obtained from NLM at (888) 346-3656.
- National Institute on Drug Abuse Hotline, (800) 662-HELP (4357).
- National Sexually Transmitted Diseases Hotline (800) 227-8922.
- American Civil Liberties Union Guide to local chapters, 202-544-1076.
- AIDS Policy and Law, (215) 784-0860.
- National Conference of State Legislatures HIV, STD and Adolescent Health Project, (303) 830-2200.
- United States Conference of Mayors, (202)293-2352.
- Centers for Disease Control and Prevention, Public Inquiry, (404) 639-3534.
- Food and Drug Administration, Office of Public Affairs, (301) 443-3285.
- American Red Cross, Office of HIV/AIDS Education, 1(800) 375-2040.
- World Health Organization, (202) 861-4354.

USEFUL INTERNET ADDRESSES

- AIDS Treatment Data Network, (800) 734-7104, 611 Broadway, Suite 613, New York, NY 10012. http://health.nyam.org:8000/public_html/network/index:html, e-mail: AIDSTreatD@AOL.COM. A home page on the internet for people with AIDS and their caregivers, it provides information on approved and experimental treatments for AIDS-related conditions. It also publishes a quarterly directory of clinical trials on HIV and AIDS, in English and Spanish.
- AMA HIV/AIDS Information Center, World Wide Web site (http://www.ama-assn.org) offers clinical updates, news, and information on social and policy questions. Cosponsored by Glaxo Wellcome Inc.
- Gay Men's Health Crisis (GMHC), Site on the World Wide Web (http://www.gmhc.org) provides on-line forums hosted by GMHC representatives.

For additional help you may wish to consult with your college or community library. They may have access to the following AIDS-related data bases:

- Aidsline—90,000 references to journals, books, and audiovisuals
- Aidstrials—information on 500 clinic trials of drugs and vaccines
- Aidsdrugs—a dictionary of drugs and experimental chemicals and biological agents against the virus
- Dirline—lists over 2,300 organizations and services that provide public information on HIV disease and AIDS
- 12th World AIDS conference, http://www.aids98.ch: (good through 1999)
- AEGIS (AIDS Education Global Information System): http://www.aegis.com
- HIV Info Web: http://www.infoweb.org
- Immunet: http://www.immunet.org
- Project Inform: http://www.projinf.org
- The Body: http://www.thebody.com/cgi/treatans.html
- HIVInSite:http://www.hivinsite.ucsf.edu/medical/tx-guidelines
- Search AIDSLINE, MEDLINE: http://www.igm.nim.nih.gov
- Vaccines: http://www.avi.org
- Women, children, health care workers, hemophiliacs, blind, deaf and other affected groups: http://beaconclinic.org/website/groups
- AIDS/HIV statistics: http://www.avert.org/statindx.htm
- http://www.healthcg.com/hiv/links.html (provides linkage to 9 major US Guidelines for HIV Testing, OIs, Treatment etc.)
- The *Centers for Disease Control and Prevention's (CDC) National Prevention Information Network (NPIN) Links:* (http://www.cdcnac.org/hivlink.html and http://www.cdcnac.org/daynews.html
- National Institute of Allergy and Infectious Diseases (NIAID) online at: http://www.niaid.nih.gov.
- *Critical Path AIDS Project* a Philadelphia organization for people with HIV disease, provides another online source for the latest news in HIV disease prevention, research, clinical trials, and treatments. The publication's hot link leads to a directory of AIDS-related publications:(http://www.critpath.org)
- *Managing Desire: HIV Prevention Counseling for the 21st Century* targets the HIV test counseling community as well as the general consumer. The site is

produced by Nicholas Sheon, the Prevention Editor of the *HIV Inside* web site of the UCSF Center for AIDS Prevention Studies (http://hivsinsite.ucsf.edu) and an HIV test counselor at the Berkeley Free Clinic: (http://www.managingdesire.org)

- *AIDS* offers abstracts from recent issues: (http://www.aidsonline.com)
- *AIDS Weekly Plus* contains more than 35,000 articles on health-related topics. Full access is available by subscription: (http://www.newsfile.com)
- *AIDS Treatment News* posts the contents of every issue since the publication began in 1986: (http://www.immunet.org/immunet/atn.nsf/homepage)
- The *Bulletin of Experimental Treatment on AIDS (BETA)* published by the San Francisco AIDS Foundation, is free online: (http://www.sfaf.org/beta.html)
- *Treatment Issues* published by the Gay Men's Health Crisis, provides free access to issues dating back to 1995: (http://www.gmhc.org/aidslib/ti/ti.html)
- *Project Inform* email (INFO@projinf.org), website, established in 1985 as a national, non-profit, community-based HIV/AIDS treatment information and advocacy organization, serves HIV-infected individuals, their care givers, and their health care and service providers through its national, toll-free treatment hotline: (http://www.projinf.org)
- Western Canada's largest AIDS group (in Vancouver, BC) has launched its redesigned website featuring online publications, a map of provincial resources, and links to over 100 AIDS websites. The website, published by the British Columbia Persons with AIDS Society, is one of the most popular sites in Canada and has operated for two years. The site carries information about:
  - *Treatments* http://www.bcpwa.org/treat.htm;
  - *AIDS news* http://www.bcpwa.org/news.htm;
  - *Organizational activities* http://www.bcpwa.org/AboutBCPWA/board.htm; and
  - *Links* http://www.bcpwa.org/Resources/links.htm.
- For more information: Pierre Beaulne, Developer, Communications and Marketing, British Columbia Persons With AIDS Society, mail to: pierreb@parc.org.

The following is a sampling of general internet resources for community research and HIV/AIDS information in Canada:

- The Community-Based HIV/AIDS Research Program National Health Research and Development Program, Health Canada http://www.hc-sc.gr.ca/hppb/nhrdp/cdr.htm
- The HIV/AIDS Aboriginal Research Program National Health Research and Development Program, Health Canada http://www.hc-sc.gr.ca/hppb/nhrdp/abrfp.htm
- Community-University Research Alliances (CURAs), Social Sciences and Humanities Research Council of Canada http://www.sshrc.ca/ english/program-info/grantsguide/cura.html?
- Canadian Strategy on HIV/AIDS http://www.hc-sc.gc.ca/hppb/hiv aids/
- Canadian HIV/AIDS Clearinghouse http://www.cpha.ca/clearinghouse e.htm
- Canadian AIDS Society http://www.cdnaids.ca/
- Canadian Aboriginal AIDS Network http://www.caan.ca/
- Community AIDS Treatment Information Exchange http://www.catie.ca/
- Canadian HIV/AIDS Legal Network http://www.aidslaw.ca/
- Bureau of HIV/AIDS, STD and TB, Health Canada http://www.hc-sc.gc.ca/hpb/lcdc/bah/epi/epi e.html
- Global Network of People Living with HIV/AIDS (GNP+) email: gnp@gn.apc.org.

Global AIDS Directory 2000: It contains a comprehensive listing of 250 agencies—a "who's who" among groups working globally on HIV/AIDS. Copies of Global AIDS Directory 2000 are US$50 each, US$25 for Global Health Council members. For shipping and handling in the United States, please include US$5; for international orders, please include US$15. To order a copy of Global AIDS Directory 2000, send your name and address with a check drawn on a U.S. bank, a U.S. postal money order, U.S. traveler's check or credit card information to: The Global Health Council, c/o First Union National Bank, Dept. 5100, Washington, D.C., 20061-5100, USA. VISA, MasterCard and American Express credit card orders are accepted and should include account numbers, expiration date and signature. If you are unable to afford Global AIDS Directory 2000 or have further questions, please contact: William Craig, Information Officer, Tel: +1-802-649-1340 or email: wcraig@globalhealth.org.

TOLL-FREE STATE AIDS HOTLINES

For information about HIV-specific resources and counseling and testing services, call your state AIDS hotline:

Alabama	800-228-0469	Michigan	800-872-2437
out of state	334-206-5364	out of state	517-335-8371
Alaska	800-478-2437	Minnesota	800-248-2437
out of state	907-276-4880	out of state	612-373-2437
Arizona	800-342-2347	Mississippi	800-826-2961
out of state	602-234-2752	out of state	601-576-7723
Arkansas CDC	800-342-2437	Missouri	800-533-2437
out of state	501-375-0352	out of state	816-561-8784
California	800-367-2437	Montana	800-233-6668
out of state	415-863-2437	out of state	406-444-3565
Colorado	800-252-2437	Nebraska	800-782-2437
out of state	303-692-2700	out of state	402-552-9255
Connecticut	800-203-1234	Nevada USA	800-842-2437
out of state	860-509-7800	out of state	775-684-5941
Delaware	800-422-0429	New Hampshire	800-752-2437
out of state	302-652-6776	out of state	603-271-4502
District of Columbia	202-332-2437	New Jersey	800-624-2377
Spanish	202-328-0697	out of state	973-926-7443
Florida	800-352-2437	New Mexico	800-545-2437
out of state	850-681-9131	out of state	505-476-3614
Spanish	800-545-7432	New York	800-541-2437
Hatian Creole	800-243-7101	out of state	716-845-3380
Georgia	800-551-2728	North Carolina CDC	800-342-2437
out of state	404-876-9944	out of state	919-733-7301
Hawaii	800-321-1555	North Dakota	800-782-2437
out of state	808-922-1313	out of state	701-328-2378
Idaho	800-926-2588	Ohio	800-332-2437
out of state	208-321-2777	out of state	614-299-2437 x106
Illinois	800-243-2437	Oklahoma	800-535-2437
out of state	773-973-2849	out of state	918-834-4190
Indiana CDC	800-342-2437	Oregon	800-777-2437
out of state	317-233-7867	out of state	503-223-2437
Iowa	800-445-2437	Pennsylvania	800-662-6080
out of state	515-244-6700	out of state	717-783-0573
Kansas CDC	800-342-2437	Rhode Island	800-726-3010
out of state	785-232-3100	South Carolina	800-322-2437
Kentucky USA	800-420-7431	out of state	803-898-0749
out of state	502-564-6539	South Dakota	800-592-1861
Louisiana	800-922-4379	out of state	605-773-3737
out of state	504-821-6050	Tennessee	800-525-2437
Maine USA	800-851-2437	out of state	615-741-7500
out of state	207-774-6877	Texas	800-299-2437
Maryland	800-638-6252	out of state	512-490-2500
out of state	410-333-2437	Spanish	800-333-7432
Massachusetts	800-235-2331	Utah	800-366-2437
out of state	617-536-7733	out of state	801-487-2323

Vermont	800-882-2437
out of state	802-863-7245
Virginia	800-533-4148
out of state	804-371-7455
Washington	800-272-2437
out of state	360-236-3434
West Virginia	800-642-8244
out of state	304-558-6335
Wisconsin	800-334-2437
out of state	414-273-2437

Wyoming	800-327-3577
out of state	307-777-5800
Newfoundland Canada	800-563-1575
Ontario Canada	800-668-2437

Compiled April, 1999 by volunteers and R. Hunter Morey, MSW, IHS with Project Inform National HIV Treatment Information Hotline Website: http://www.projinf.org

ACQUIRED IMMUNE DEFICIENCY SYNDROME
AIDS

What do we know about AIDS? The next 13 chapters will present the many faces of the AIDS pandemic in the United States and other countries. Unlike people, the AIDS virus (HIV) does not discriminate; and it appears that most humans are susceptible to HIV infection, its suppression of the human immune system, and the consequences that follow. The viral infection that leads to AIDS is the most lethal, the most feared, and the most socially isolating of all the sexually transmitted diseases. **We must, as a people, fight against AIDS, not against each other** (Figure P-1).

FIGURE P-1 The Loneliness of AIDS. Skip Bluette, diagnosed with AIDS July 1986, died July 1988. He suffered the indignity of having to lie to a dentist to get treatment. He suffered the ignorance of nurses afraid to touch him. He suffered the loss of his greatest pleasures—discos, gourmet meals, movies, and sex with men. Family was vital to Skip. So vital that on July 17, the day he died, their presence was his final wish.

Skip Bluette wanted his story told. Photographer Mara Lavitt interviewed and photographed him during the last 8 months of his life. Portions of Lavitt's article, which appeared in *The New Haven Register*, are presented in the following chapters. (*By permission of Mara Lavitt and* The New Haven Register)

CHAPTER

1

Discovering AIDS,
Naming the Disease

CHAPTER CONCEPTS

♦ AIDS is a syndrome, not a single disease.

♦ The first cases of AIDS-related *Pneumocystis carinii* pneumonia (PCP) were reported by the Centers for Disease Control and Prevention (CDC) in June of 1981; the first case of Kaposi's sarcoma in July 1981.

♦ Luc Montagnier discovered the AIDS virus in 1983.

♦ The first CDC definition of AIDS was presented in 1982 and expanded in 1983, 1985, and 1987, and again on January 1, 1993.

A = acquired = a virus received from someone else

I = immune = an individual's natural protection against disease-causing microorganisms

D = deficiency = a deterioration of the immune system

S = syndrome = a group of signs and symptoms that together define AIDS as a human disease

AIDS: A DISEASE OR A SYNDROME?

AIDS has been presented in journals, nonscience magazines, newspaper articles, and on television as a disease. But a disease is a pathological

The letters A, I, D, S (AIDS) are an acronym for Acquired Immune Deficiency Syndrome.

condition with a single identifiable cause. As we learned from the days of Louis Pasteur and Robert Koch, there is a single identifiable organism or agent for each infectious disease.

AIDS patients may have many diseases. Most AIDS patients have more than one disease at any given time. Each disease produces its own signs and symptoms. Collectively, the diseases that are expressed in an AIDS patient are referred to as a **syndrome.** The number of different diseases an AIDS patient has and the severity of their expression reflects the functioning of that person's immune system.

Early in 1983, the agent that destroys an essential portion of the human immune system was identified by French scientists as a virus. From that point on there was a specific infectious agent associated with the cause of AIDS. The symptoms of viral-induced AIDS can begin **only** after one has been infected with a specific virus. This virus is now called the *H*uman *I*mmunodeficiency *V*irus (HIV). The specific viral induced disease is referred to specifically as HIV/AIDS because there are other reasons for a suppressed immune system, like congenital inherited immune deficiencies, exposure to radiation, alkylating agents, corticosteroids, certain forms of cancer and cancer chemotherapy, that also produce AIDS-like symptoms. (Stadtmauer et al. 1997) *Because individuals can express AIDS for reasons other than becoming HIV-infected, unless stated otherwise all information herein will refer only to AIDS caused by HIV and referred to as HIV/AIDS.*

Over time the AIDS virus depletes a subset of Lymphocytes called T4 helper cells, that are essential in the proliferation of cells necessary to cell-mediated immunity and in the production of antibodies (Figure 1-1). Thus, cell-mediated immunity and antibodies are critical components of the human immune system. Without the ability to produce a sufficient number of immune specific cells and immune specific antibodies, the body is vulnerable to a large variety of infections caused by organisms and viruses that normally do not cause human disease. It is these infections that create the symptoms and progression of illnesses that eventually kill AIDS patients. Thus AIDS begins with HIV infection. Technically it can be called **HIV disease** or **HIV T4 helper cell dis-**

ease, but the popular press, scientists, and others still refer to HIV disease as AIDS. But, AIDS is the end stage of chronic HIV infection. **AIDS IS NOT TRANSMITTED, THE VIRUS IS.** People do not die of AIDS per se. They die of opportunistic infections, cancers, and organ failures brought on by a failed immune system.

It is believed that eventually almost everyone who is **correctly** diagnosed with HIV/AIDS will die. But all who become HIV-infected may not progress to AIDS. Estimates are that some 5% of the HIV-infected population will not progress to AIDS. This implies that there is a percentage of the population that is resistant to HIV-associated immune system suppression. In mid-1996, a gene for HIV resistance was identified! (see Chapter 4).

Naming the Disease

Early in 1981, practically coincident with the report of the first cases of a new disease in the male homosexual community in the United States, there were reports of 34 cases of a new disease among Haitian immigrants to the United States and 12 cases of a disease previously unrecognized in **Haiti**—an aggressive form of **Kaposi's sarcoma.** The earliest known patient, a young male with the new disease, died in Haiti in 1978. Michael Gottlieb, who had identified a new disease that seemed to target gay men, found that although each of the cases was different, **all had one thing in common: whatever was making the men sick had singled out the T lymphocyte cells for destruction.** Eventually the body's battered defenses couldn't shake off even the most harmless microbial intruder. The men were dying from what doctors termed opportunistic infections, such as *Pneumocystis* **pneumonia** which attacks the lungs, and **toxoplasmosis,** which often ravages the brain.

In June of 1981, the Centers for Disease Control and Prevention (CDC) first reported on diseases occurring in gay men that previously had only been found to occur in people whose immune systems were suppressed by drugs or disease [*Morbidity and Mortality Weekly Report (MMWR)*, 1981].

The report stated that five young men in Los Angeles had been diagnosed with *Pneumocystis carinii* pneumonia (PCP) in three different

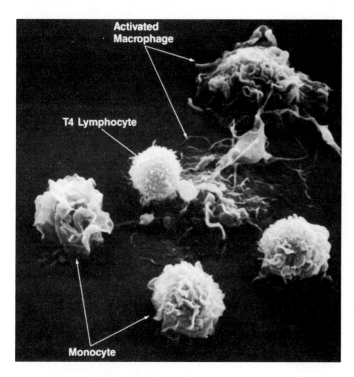

FIGURE 1-1 Normal Human T4 Lymphocytes, Monocytes, and Macrophages. Scanning electron micrograph of monocytes, macrophages, and a T4 lymphocyte, magnified 9,000 times. These white blood cells are the targets of HIV infection. Note that the T4 lymphocyte (round cell, at the center) is adhered to a flattened macrophage. (*Photograph courtesy of Dr. M. A. Gonda.*)

hospitals. Because cases of PCP occurred almost exclusively in immune suppressed patients, five new cases in one city at one time were termed by the report as unusual. The report also suggested "an association between some aspects of homosexual life-style or disease acquired through sexual contact and PCP in this population. Based on clinical exami-

nation of each of these cases, the possibility of a cellular immune dysfunction related to a common exposure might also be involved."

In July 1981, the CDC (*MMWR*, 1981) reported that an uncommon cancer, **Kaposi's sarcoma (KS),** had been diagnosed in 26 gay men who lived in New York City and California. This was also an unusual finding because KS, when it occurred, was usually found in **older** men of Hebrew or Italian ancestry. The sudden and dramatic increase in pneumonia cases, all of which were caused by a widespread but generally harmless fungus, *P. carinii,* and KS cases indicated that an infectious form of immune deficiency was on the increase. This immune deficiency disease, initially called GRID for Gay-Related Immune Deficiency, was noted to be quickly spreading in the homosexual community, among users of injection drugs, and among blood transfusion recipients. This new mysterious and lethal illness appeared to be associated with one's lifestyle. These early cases of immune deficiency heralded the beginning of an epidemic of a previously unknown illness. By 1982 and 1983, the disease was reported in hemophiliacs, adult heterosexuals, and children and was subsequently called AIDS.

——— LOOKING BACK 1.1 ———

The New York Times and A.I.D.S., August 8, 1982

An article, "A Disease's Spread Provokes Anxiety," alerted its readers to the growing health crisis in the homosexual community that was baffling medical science. While the Times had previously reported on a disease causing opportunistic infections in gay men, this was the first time the term acquired immune deficiency syndrome or A.I.D.S. (the punctuation had not been dropped) appeared in the nation's newspaper of record. Later in 1982, the Washington Post joined the Times in reporting on the death of an infant who had received a blood transfusion from an AIDS-afflicted donor. With that, a second major national newspaper was officially in the business of covering the AIDS story.

DISCOVERY OF THE AIDS VIRUS

There was no shortage of ideas on what caused AIDS. It was believed by some to be an act of God, a religious curse or penalty against the homosexual for practicing a biblically unacceptable lifestyle that included drugs, alcohol, and sexual promiscuity. The reverend Billy Graham said "AIDS is a judgement of God." Jerry Falwell stated that AIDS is God's punishment, the scripture is clear, "we do reap it in our flesh when we violate the laws of God." Some believed AIDS was due to sperm exposure to amyl nitrate, a stimulant used by some homosexuals to heighten sexual pleasure (Gallo, 1987). Others believed there was no specific infectious agent. They believed that certain people who *excessively stressed their immune systems* experienced immune system failure, and before it could recover, other infections killed them. But many scientists who had expertise in analyzing the sudden onset of **new** human diseases felt the cause of this form of human immune deficiency was an infectious agent. They believed that the agent was trans-

mitted through sexual intercourse, blood, or blood products and from mother to fetus. They also believed that this agent which led to the loss of T4 cells was smaller than a bacterium or fungus because it passed through a filter normally used to remove those microorganisms. It turned out after further studies that this agent fit the profile of a virus.

In January of 1983, Luc Montagnier (Mont-tan-ya) and colleagues at the Pasteur Institute in Paris isolated the AIDS virus (Hobson et al., 1991). In May of that year, they published the first report on a T cell retrovirus found in a patient with **lymphadenopathy** (lim-fad-eh-nop-ah-thee), or swollen lymph glands. Lymphadenopathy is one of the early signs in patients progressing toward AIDS. The French scientist (Figure 1-2) named this virus **lymphadenopathy-associated virus** (LAV) (Barre-Sinoussi et al., 1983).

Naming the AIDS Viruses: HIV-1, HIV-2

During the search for the AIDS virus, several investigators isolated the virus but gave it different names. For example, Robert Gallo (Figure 1-3) named the virus HTLV III (For the Third Human T Cell Lymphotropic Virus). Because the collection of names given this virus created some confusion, the Human Retrovirus Subcommittee of the Committee on the Taxonomy of Viruses reduced all of the names to one: **Human Immunodeficiency Virus** or **HIV.** This term has now been accepted for use worldwide.

In 1985, a second type of HIV was discovered in West African prostitutes. It was named HIV type 2 or **HIV-2.** The first confirmed case of HIV-2 infection in the United States was reported in late 1987 in a West African woman with AIDS. By December 1990, 16 additional cases of HIV-2 infection were reported to the CDC (*MMWR*, 1990). Beginning year 2000, a total of 88 HIV-2 infections have been reported from 22 states of the United States. Of the 88 infected persons, 66 are black and 51 are male. Sixty-two were born in West Africa, 11 in the United States (including 3 infants born to mothers of unspecified nationality), 2 in India and 3 in Europe. The region of origin was not identified for 10 of the persons, although 4 of

FIGURE 1-2 Luc Montagnier, from the Pasteur Institute in Paris. He and colleagues discovered the human immunodeficiency virus (HIV), the cause of AIDS. In October 1997, he moved to the United States. He accepted an endowed professorship at Queens College, New York to run a new AIDS research institute. *(Photograph courtesy of AP/Wide World Photos)*

FIGURE 1-3 Robert Gallo, Director, Institute of Human Virology. *(Photograph courtesy of AP/Wide World Photos)*

them had a malaria-antibody profile consistent with residence in West Africa. Fifteen have developed AIDS-defining conditions and 8 have died. These case counts represent minimal estimates because completeness of reporting has not been assessed; reporting varies from state to state according to state policy.

The earliest evidence to date of an individual exposed to HIV-2 comes from a case report on an infection most likely occuring in Guinea Bissau in the 1960s. Earlier extensive serosurveys in West Africa are rare or nonexistent. In genetic terms, HIV-2 is much more closely related to SIV, a group of monkey viruses, than to HIV-1. Both HIV-2 and HIV-1 may have been derived from ancestral SIV variants that were from distinct regions and species and do not appear to be direct genetic descendants of each other (Marlink, 1996). Clinically, what has been learned about HIV-1 appears to apply to HIV-2, except that HIV-2 appears to be less harmful (cytopathic) to the cells of the immune system, and it reproduces more slowly than HIV-1.

UNLESS STATED OTHERWISE, ALL REFERENCES TO HIV IN THIS BOOK REFER TO HIV-1.

DEFINING THE ILLNESS: AIDS SURVEILLANCE

The CDC reported that through 1983 there were 3,068 AIDS cases in the United Stated and 1,478 of these had died (48%). All demonstrated a loss of T4 lymphocytes, and all died with severe opportunistic infections. Opportunistic infections are caused by organisms and viruses that are normally present but do **not** cause disease unless the immune system is damaged. Clearly there was an immediate need for a name and definition for this disease so that a rational surveillance program could begin.

Definitions of AIDS for Surveillance Purposes

The initial objective of AIDS surveillance was to describe the epidemic in terms of time, place, and individuals and to recognize immediately changes in rate and pattern in the spread of AIDS. Early surveillance was done to gather data in order to generate information of value in planning control programs. Because a *cellular deficiency of the human immune system* was found in every AIDS patient, along with an assortment of other signs and symptoms of disease, and because the infection was *acquired* from the action of some environmental agent, it was named **AIDS** for **Acquired Immune Deficiency Syndrome.**

In order to establish surveillance, a system for monitoring where and when AIDS cases occurred and a workable definition had to be developed. The definition had to be *sensitive* enough to detect every possible AIDS patient while at the same time *specific* enough to exclude those who may have AIDS-like symptoms, but were not infected by the AIDS virus.

In 1982, there was no *single characteristic* of AIDS that would allow for a useful definition for surveillance purposes. Because immunological testing was essentially unavailable, the first AIDS surveillance definition was based on the **clinical description** of symptoms. The *first* of many criteria for the diagnosis of AIDS were: (1) the presence of a reliably diagnosed disease at least moderately predictive of cellular immune deficiency; and (2) the absence of an underlying cause for the immune deficiency or for reduced resistance to the disease (*MMWR*, 1982). Because the symptoms varied greatly among individuals, **this was a poor first definition.**

Arbitrary AIDS Definitions

The initial definition of AIDS was thus an arbitrary one, reflecting the partial knowledge of the clinical consequences that prevailed at the time.

Various systems for classifying HIV-related illnesses have been devised since 1982 to take into account increasing knowledge about the spectrum of those illnesses. But the definition of AIDS has remained largely unchanged, partly for epidemiological reasons: a standard definition of AIDS makes it easier to monitor the incidence of the disease over time. Had the whole picture of HIV infection and its clinical consequences been known in 1982, the term **"AIDS"** would not have been used. Instead, we would refer to the various stages of **"HIV disease"** (or perhaps, following an older tradition, "Gottlieb's disease," after Mike Gottlieb, who first described it).

The 1982 definition was modified in 1983 to include new diseases then found in AIDS patients. With this modification, AIDS became reportable to the Centers for Disease Control and Prevention (CDC) in every state. In 1985, additional diseases were included in the AIDS case definition. Those diseases were disseminated histoplasmosis, chronic isosporiasis, and some of the non-Hodgkin's lymphomas (a form of cancer).

Part of the 1982 through 1985 AIDS case definitions was the description of patients with AIDS-related complex (ARC). The symptoms of ARC included swollen lymph nodes (lymphadenopathy), weight loss, loss of appetite, fever, rashes, night sweats, and fatigue. But ARC patients did not appear to have opportunistic infections, and because HIV antibody testing was not widely available, their antibody status was unknown. By 1987, the need for a mid-AIDS classification was unnecessary. Antibody tests were available and much more was known about the onset of opportunistic infections. Thus the CDC dropped ARC from their 1987 AIDS definition. Extrapulmonary tuberculosis, HIV encephalopathy (brain disease), HIV wasting syndrome, presumptive *Pneumocystis carinii* pneumonia (PCP), and esophageal candidiasis (both fungal infections) were added to the definition (Table 1-1). For AIDS diagnoses in children, multiple or recurrent bacterial infections were added.

Broadly speaking, the term **AIDS** may be understood as referring to the onset of life-threatening illnesses as a result of HIV disease that results from an HIV infection. AIDS is the end stage of a disease process which may have

TABLE 1-1 1987 CDC Definition of AIDS

ADULT

A. Without laboratory evidence for HIV infection:
  - Lymphoma of brain (<60 years of age)
  - Lymphoid interstitial pneumonitis (<13 years of age)
B. With laboratory evidence for HIV infection:
  - Disseminated coccidioidomycosis
  - HIV encephalopathy
  - Isosporiasis (persisting > 1 month)
  - Lymphoma of brain (any age)
  - Non-Hodgkin's lymphoma (B cell or undifferentiated)
  - Recurrent Salmonella septicemia
  - Extrapulmonary tuberculosis
  - HIV-wasting syndrome
  - Recurrent bacterial infections (<13 years of age)
  - Disseminated histoplasmosis

PEDIATRIC

Children with AIDS may have infectious diseases that are not covered by the adult CDC AIDS definitions. Thus children under 15 months are classified separately from older children because passive maternal antibodies may be present.

To Be HIV-Infected, Children Under 15 Months Must:

1. Show HIV in blood and tissues.
2. Show evidence of humoral and acquired immune deficiency and have one or more opportunistic infections associated with AIDS.
3. Show other symptoms meeting the CDC definition of AIDS.

To be HIV-infected, children between 16 months and 12 years must:

1. Show 1 and 3 above.
2. Show HIV antibodies.

This definition was updated on January 1, 1993 to include all persons with a T4 cell count of less than 200/μL of blood. See text for additional details.

(Adapted from Roger J. Pomerantz, M.D., "The Chameleon Called AIDS," Harvard Medical School, 1988.)

been developing for 5, 10, 15 or more years, for most of which time the infected person will have been well and quite possibly unaware that he or she has been infected.

Thus the number of AIDS cases reported from 1987 through 1992 reflects the revisions of the initial surveillance case definitions. One

major drawback to all of the CDC AIDS definitions is the fact that through 1992, the Social Security Administration (SSA) used the CDC AIDS definition to determine disability. But all the definitions were primarily based on symptoms and opportunistic infections in men. Therefore, about 65% of women with HIV/AIDS symptoms were excluded from SSA benefits. They were excluded because of failure to be diagnosed with AIDS by the CDC AIDS definition (Sprecher, 1991).

Impact of the 1993 Expanded AIDS Case Definition— On January 1, 1993, the newest definition of AIDS was put into the surveillance network. The reason for the new CDC definition was that epidemiologists felt the 1987 definition failed to reflect the true magnitude of the pandemic. In particular, it failed to address AIDS in women. In addition, those with T4 counts under 200/μL of blood are most likely to be severely ill or disabled and in greatest need of medical and social services. With the new AIDS definition, these people are eligible earlier in their illness for federal and state medical and social assistance programs.

CDC revised the classification system for HIV infection to emphasize the clinical importance of the T4 lymphocyte count in the categorization of HIV-related clinical conditions. Consistent with this revised classification system, CDC has expanded the AIDS surveillance case definition to include all HIV-infected persons who have less than 200 T4 lymphocytes/μL of blood, or a T4 lymphocyte percent less than 14% of total lymphocytes (Table 1-2). In addition to retaining the 23 clinical conditions in the 1987 AIDS surveillance case definition, the expanded definition includes (**1**) pulmonary tuberculosis, (**2**) invasive cervical cancer, and (**3**) recurrent pneumonia in persons with documented HIV infection (Table 1-3). This expanded definition requires laboratory evidence for HIV infection in persons with less than 200 T4 lymphocytes/μL or with one of the added clinical conditions. The **objectives** of these changes are to simplify the classification of HIV infection and the AIDS case reporting process, to be consistent with standards of medical care for HIV-infected persons, to better categorize

TABLE 1-2 T4 Lymphocyte Counts Related to Percentages of Total Lymphocytes

T4 Cell Category	T4 Cells/μL	Percent T4 Cells of Total Lymphocytes
(1)	≥ 500	≥ 29
(2)	200–499	14–28
(3)	<200	<14

1. Normal adult T4 cell count ranges from 900 to 1,200 T4 cells/μL of blood.

2. The equivalences of T4 cells to percent of total lymphocytes were derived from analyses of more than 15,500 lymphocyte subset determinations from seven different sources: one multistate study of diseases in HIV-infected adolescents and adults, and six laboratories (two commercial, one research, and three university-based). The six laboratories are involved in proficiency testing programs for lymphocyte subset determinations.

(Adapted from *MMWR*, 1993.)

TABLE 1-3 List of 26 Conditions in the AIDS Surveillance Case Definition

- Candidiasis of bronchi, trachea, or lungs
- Candidiasis, esophageal
- Cervical cancer, invasive[a]
- Coccidioidomycosis, Disseminated or extrapulmonary
- Cryptococcosis, extrapulmonary
- Cryptosporidiosis, chronic intestinal (>1 month duration)
- Cytomegalovirus disease (other than liver, spleen, or nodes)
- Cytomegalovirus retinitis (with loss of vision)
- HIV encephalopathy
- Herpes simplex: chronic ulcer(s) (>1 month duration); or bronchitis, pneumonitis, or esophagitis
- Histoplasmosis, disseminated or extrapulmonary
- Isosporiasis, chronic intestinal (>1 month duration)
- Kaposi's sarcoma
- Lymphoma, Burkitt's (or equivalent term)
- Lymphoma, immunoblastic (or equivalent term)
- Lymphoma, primary in brain
- *Mycobacterium avium complex* or *M. kansasii*, disseminated or extrapulmonary
- *Mycobacterium tuberculosis*, disseminated or extrapulmonary
- *Mycobacterium tuberculosis*, any site (pulmonary[a] or extrapulmonary)
- *Mycobacterium*, other species or unidentified species, disseminated or extrapulmonary
- *Pneumocystis carinii* pneumonia
- Pneumonia, recurrent[a]
- Progressive multifocal leukoencephalopathy
- Salmonella septicemia, recurrent
- Toxoplasmosis of brain
- Wasting syndrome due to HIV

[a]Added in the 1993 expansion of the AIDS surveillance case definition.

(Adapted from the CDC, Atlanta, Georgia.)

HIV related morbidity, and to reflect more accurately the number of persons with severe HIV-related immunosuppression who are at highest risk for severe HIV-related morbidity and most in need of close medical follow-up. The addition of the three clinical conditions reflects their documented or potential importance in the HIV epidemic. Invasive cervical cancer is included on the basis of an epidemiological link between HIV infection and cervical dysplasia, and reports that HIV speeds the progression of both cervical dysplasia and cancer.

The expanded AIDS surveillance case definition has had a substantial impact on the number of reported cases in 1993.

Of the 106,618 Adult/Adolescent AIDS cases reported in 1993, 57,574 (54%) were reported based on conditions added to the definition in 1993; and 49,044 (46%) were reported based on pre-1993 defined conditions (Table 1-3). Of the 57,574 cases reported based on 1993-added conditions, 52,392 persons (91%) had severe HIV-related immunosuppression only; 4,030 (7%), pulmonary tuberculosis (TB); 1,151 (2%), recurrent pneumonia; and 576 (1%), invasive cervical cancer (19 persons were reported with more than one of these opportunistic illnesses). A substantial increase in the number of reported AIDS cases occurred in all regions of the United States. Of areas reporting more than

250 cases, the proportion of cases based on the 1993-added criteria ranged from 35% in North Carolina (*n* = 1,353) to 71% in Colorado (*n* = 1,323).

When compared to 1992 data the increase in reported cases in 1993 was greater among females (151%) than among males (105%). Proportionate increases were greater among blacks and Hispanics than among whites. The largest increases in case reporting occurred among persons aged 13–19 years and 20–24 years; in these age groups, a greater proportion of cases were reported among women (35% and 29%, respectively) and were attributed to heterosexual transmission (22% and 18%, respectively).

Compared with homosexual/bisexual men, proportionate increases in case reporting were greater among heterosexual injecting-drug users (IDUs) and among persons reportedly infected through heterosexual contact.

Women, blacks, heterosexual IDUs, and persons with hemophilia were more likely than others to be reported with 1993-added conditions. Most of these differences were attributable to reports of the three opportunistic illnesses added in 1993.

The pediatric AIDS surveillance case definition was not changed in 1993. During 1993, 990 children aged <13 years were reported with AIDS, an increase of 21% compared with the 783 cases reported in 1992. In 1994, for the first time, pediatric AIDS cases exceeded 1,000 (1,090). For 1995, there were 800 AIDS cases; for 1996, 678; for 1997, 473; and for 1998, 350. Collectively, since 1990, 50% were female, and most were either black (55%) or Hispanic (27%) and were infected through perinatal HIV transmission (93%). New York, Puerto Rico, and Florida reported 51% of the pediatric AIDS cases.

If all of the approximately 1 million persons in the United States with HIV infection were diagnosed and their immune status known, it is estimated that 120,000–190,000 persons who do not have AIDS-indicator diseases would be found to have T4 lymphocyte counts < 200/µL **(in March 1992 the CDC reduced the number of HIV-infected in the United States to between 650,000 and 900,000).** However, since not all of these persons are aware of their HIV infection status, and of those who are, not all have had an immunological evaluation, the immediate impact on the number of AIDS cases will be considerably less. Under the 1987 AIDS surveillance criteria, approximately 49,106 AIDS cases were reported in 1992, but 106,618 were reported for 1993. In 1994, there were 80,691 new AIDS cases. For 1995, the number of new cases was 74,180. For 1996 the number of new AIDS cases was 69,151, 60,634 for 1997, 55,000 for 1998, and an estimated 50,500 for 1999 and 4,800, for the year 2000.

Revised AIDS Definition Creates Backlog

The effects of the expanded surveillance definition was greatest in 1993 because of the backlog in number of persons who fit the new AIDS definition. Once most of these cases were reported through 1998, the yearly incidence dropped off to about 50,000 cases a year into the year 2000. Finally, the expanded HIV classification system and the AIDS surveillance case definition have been developed for use in conducting public health surveillance. They were not developed to determine whether any statutory or other legal requirements for entitlement to federal disability or other benefits are met.

The revised surveillance case definition does not alter the criteria used by the Social Security Administration in evaluating claims based on HIV infection under the Social Security disability insurance and Supplemental Security Income programs. The Social Security Administration has recently proposed a new method for the evaluation of HIV infection and criteria to determine eligibility for disability. Other organizations and agencies providing medical and social services should develop eligibility criteria appropriate with the services provided and local needs (*MMWR*, 1993).

Problems Stemming from Changing the AIDS Definition for Surveillance Purposes

Each time the definition of AIDS has been altered by the CDC, it has led to an increase in the number of AIDS cases. In 1985, the change in definition led to a 2% increase over what would have been diagnosed prior to the change. The 1987 change led to a 35% increase in new AIDS cases per year over that expected using the 1985 definition. The 1993 change resulted in a 52% increase in AIDS cases over that expected for 1993. Such rapid changes alters the baseline from which future predictions are made and makes the interpretations of trends in incidence and characteristic of cases difficult to process. **For the first time because of the 1993 AIDS definition, one could be diagnosed with AIDS and remain symptom-free for years (become HIV positive and have a T4 cell count of less than 200).**

SUMMARY

Much continues to be written about HIV/AIDS. Some of it, especially in lay articles, has been less than accurate and has led to public confusion and fear. **HIV infection is not AIDS. HIV infection is now referred to as HIV disease. AIDS is a syndrome of many diseases,** each resulting from an opportunistic agent or cancer cell that multiplies in humans who are immunosuppressed. The new 1993 CDC AIDS definition will allow, over the long term, earlier access to federal and state medical and social services for HIV-infected individuals.

REVIEW QUESTIONS

(Answers to the Review Questions are on page 487.)

1. The letters A, I, D, and S are an acronym for?

2. Is AIDS a single disease? Explain.

3. When was the AIDS virus discovered and by whom?

4. In what year did CDC first report on a strange new disease which later was named AIDS?

5. Name one acronym for HIV.

6. How many times has CDC changed and expanded the definition of AIDS? In what years?

7. Why did the CDC do away with the ARC definition?

8. What is one major advantage of the new CDC AIDS definition for the HIV-infected?

REFERENCES

BARRE-SINOUSSI, FRANCOISE, et al. (1983). Isolation of a T-lymphocyte retrovirus from a patient at risk for acquired immune deficiency syndrome (AIDS). *Science*, 220:868–871.

GALLO, ROBERT C. (1987). The AIDS virus. *Sci. Am.*, 256:47–56.

HOBSON, SIMON WAIN, et al. (1991). LAV revisited: Origins of the early HIV-1 isolates from Institut Pasteur. *Science*, 252:961–965.

MARLINK, RICHARD, (1996), Lessons from the second AIDS virus HIV-2. *AIDS*, 10:689–699.

Morbidity and Mortality Weekly Report. (1981). Pneumocystis pneumonia—Los Angeles, 30:250–252.

Morbidity and Mortality Weekly Report. (1982). Update on acquired immune deficiency syndrome (AIDS)—United States, 31:507–508, 513–514.

Morbidity and Mortality Weekly Report. (1990). Surveillance for HIV-2 infection in blood donors—United States, 1987–1989, 39:829–831.

Morbidity and Mortality Weekly Report. (1993). 1993 Revised classification system for HIV infection and expanded surveillance case definition for AIDS among adolescents and adults, 41:1–19.

Morbidity and Mortality Weekly Report. (1994). Update: Impact of the expanded AIDS surveillance case definition for adolescents and adults on case reporting—United States, 1993, 43:160–170.

SPRECHER, LORRIE. (1991). Women with AIDS: Dead but not disabled. *The Positive Woman*, 1:4.

STADTMAUER, GARY et al. (1997). Primary Immune Deficiency Disorders that mimic AIDS. *Infections in Medicine* 4: 899–905.

What Causes AIDS:
Origin of the AIDS Virus

CHAPTER CONCEPTS

- AIDS is not caused by HIV infection.
- AIDS is caused by HIV infection.
- An unbroken chain of HIV transmission has been established between those infected and the newly infected.
- HIV is believed to have crossed into humans from animals.
- A third new HIV strain is found.
- The earliest AIDS case to date is reported.

THE CAUSE OF AIDS: THE HUMAN IMMUNODEFICIENCY VIRUS

The unexpected appearance and accelerated spread of an unknown lethal disease soon raised several important questions: **What** is causing the disease? **Where** did it come from?

How does the causative agent function? **What** are the biological characteristics of the agent? These four questions will be answered.

This section has a subtitle which states that AIDS is the result of HIV infection. There are a relatively small number of scientists and non-scientists as of this writing who claim that HIV does **not** cause AIDS. For a balanced HIV/AIDS presentation, this claim will be presented first.

HIV Does Not Cause AIDS: A Minority Point of View

Peter Duesberg is perhaps the most vocal in his concern that the scientific community is investigating the wrong causative agent. Duesberg is a molecular biologist at the University of California at Berkeley and a member of the National Academy of Sciences. Duesberg has advanced his anti-HIV/AIDS hypothesis at

———— BOX 2.1 ————

SEX AND HIV DO NOT CAUSE AIDS!

Medical Doctor Puts His Life on the Line to Prove It.

On October 28, 1993, Robert Willner held a press conference at a North Carolina hotel, during which he jabbed his finger with a bloody needle he had just stuck into a man **who said he was infected with HIV.** Willner was a physician who had his medical license revoked in Florida for, among other infractions, claiming to have cured an AIDS patient with ozone infusions. He is also the author of a book, *Deadly Deception: The Proof that SEX and HIV Absolutely DO NOT CAUSE AIDS.* He insists that jabbing himself with the bloody needle, which he describes as "an act of intelligence," was not meant to sell books. "I'm interested in proving to people that there isn't one shred of scientific evidence that HIV causes any disease".

USA Today, October 3, 1994, carried a full-page advertisement to promote the selling of Willner's book. A full-page ad in this newspaper, regardless of location, if carried nationally costs $57,500. **Question for class discussion:** Did *USA Today* management provide socially responsible advertisement or did money talk and responsibility walk? What is the potential medical downside of this advertisement?

The scientific journal *Science* (1994; 226: 1642–1649) presented a series of six articles by Jon Cohen concerning the *question* of whether HIV causes AIDS. The articles are a balanced review of the scientific facts as they relate to the question.

Robert Willner died April 15, 1995 from an apparent heart attack.

---

great expense to himself. He states that "I have been excommunicated by the retrovirus-AIDS community with noninvitations to meetings, noncitations in the literature and nonrenewals of my research grants, which is the highest price an experimental scientist can pay for his convictions."

Since 1987, Duesberg has **not** been able to obtain new grants to support his laboratory nor attract graduate students. In 1998/1999, with funds low, friends of Duesberg solicited donations by way of the internet and an advertisement in the alumni magazine. The ad brought in a stream of small contributions, which along with $200,000 in foundation money and some other big individual donations amounted to $325,000, enough for another year of operations. With regard to his financial problems, he said, "Most people don't realize how unfree we are to do science in America. They can afford to give millions, but they cannot afford to give me $100,000 or $200,000 to prove them wrong." Duesberg's financial and laboratory problems reflect his lonely battle with HIV/AIDS investigations and drug companies who, he believes, have invested so much in the HIV/AIDS theory that they cannot afford to entertain an alternative theory.

In 1971, Duesberg co-discovered cancer-causing genes in viruses. In the March 1987 issue of *Cancer Research,* he published "Retro-viruses as Carcinogens and Pathogens: Expectations and Reality." The article provoked wide-based scientific discussion and received a lot of popular press coverage. In the article Duesberg argues that there is no evidence that HIV causes AIDS. He has published additional articles in *Science* (1988) and in the *Proceedings of the National Academy of Sciences* (1989) stating that HIV is not the cause of AIDS. In May of 1992, Duesberg was the featured speaker at an alternative AIDS conference, "AIDS: A Different View," promoted by homeopathic physician Martien Brands. The week of meetings was spent giving new interpretations to some of the data used to establish HIV as the causative agent of AIDS. In short, Duesberg suggests that there is no single causative agent, that the disease is due to one's **"lifestyle."** He marshals arguments to support his theory that, in the United States and probably in Europe, AIDS is a collection of noninfectious deficiencies predominantly associated with drug use, malnutrition, parasitic infections, and other specific risks.

Duesberg believes that the tests which detect HIV antibodies are useless. In the June

1988 issue of *Discovery* he said, "If somebody told me today that I was antibody positive, I wouldn't worry one second. I wouldn't take Valium. I wouldn't write my will. All I would say is that my immune system seems to work. I have antibodies to a virus. I am protected."

In June 1990, Weiss and Jaffe wrote a critical refutation of Duesberg's theory that HIV cannot be the cause of AIDS. Duesberg's response suggested that he was unaware of published data that clearly answer the questions he raises concerning HIV involvement in AIDS. For example, one of Duesberg's major points is that no one has yet shown that hemophiliacs infected with HIV progress to AIDS. The data on matched groups of homosexual males and hemophiliacs which show that *only* those infected with HIV develop AIDS have been available for a number of years. (Weiss et al., 1990).

Duesberg's arguments and disagreements with the vast majority of prominent scientists who have researched the causal agent of AIDS are many. But they pale when placed next to the overwhelming evidence which leaves no doubt in the opinion of most scientists that HIV causes AIDS (see Andrews, 1995; Cohen, 1995; Moore, 1996).

Based on an August 1992 report in *Newsweek*, a father discussed his decision, based on Duesberg's claims, to counsel his infected hemophiliac son to avoid zidovudine (ZDV) treatment. This situation is similar to what happened when desperate cancer patients followed the advice of a credentialed academician who recommended vitamin therapy as the cure for cancer. Based on this advice, some people failed to undergo truly effective therapy. **CLASS QUESTIONS: Is Duesberg's opinion on this issue inadvertently harmful to humans? To the scientific process? Will the use of his idea, that HIV does not cause AIDS, provide a course of action that will stop the Acquired Immune Deficiency Syndrome?**

Duesberg's Belief That HIV Does Not Cause AIDS Continues

In 1996, Duesberg's book, *Inventing the AIDS Virus,* was published. Through 1999, Duesberg still insisted that HIV does not cause AIDS. He believes that HIV is just another opportunistic agent like those that cause other opportunistic diseases (Duesberg, 1993, 1995a, 1995b; Moore, 1996). With each new scientific report, it becomes more difficult for Duesberg to maintain his position. However, he remains unconvinced that HIV causes AIDS regardless of the reports that newborn infants with HIV got HIV *only* from HIV-infected mothers and progress to AIDS while noninfected newborns from the same mothers do not progress to AIDS and that some 50% of HIV-infected hemophiliacs have developed AIDS yet **no** HIV-negative hemophiliac has ever developed AIDS (Darby et al., 1995; Levy, 1995; Sullivan et al., 1995). In addition, Duesberg claims that the drug AZT (Zidovudine) causes AIDS — then what does he make of the AIDS Clinical Trials Group (ACTG) Protocol 076 that demonstrated AZT treatment of women during pregnancy and delivery reduced transmission from mother to infant from 25% in the placebo-treated mothers to 8% in those who received AZT (Connor et al., 1994)?

For those who wish to read more on the rebuttal of Duesberg's arguments, read the study reporting on the death rate among HIV-positive and HIV-negative British hemophilia patients (Baum, 1995; Darby et al., 1995; Editorial, 1995). For more by Duesberg see his web page, www.duesberg.com.

In Defense of Duesberg

In defense of Duesberg, Buianoukas (1995) writes that "We at HEAL (Health Education AIDS Liaison) maintain that what is called AIDS is no more complicated than a recreational drug–fear–medical drug disease syndrome which, with the exception of acute medical care, is out of the purview of orthodox medicine." Robert Root-Bernstein authored "Rethinking AIDS" in 1993. This book attempts to support Duesberg's hypothesis that there is no link between HIV and AIDS. His main thesis is that scientists do not have the necessary information to conclude that HIV causes AIDS. He believes, first, that the right studies on the cause of immunosuppression have not been done and, second, that there need not be a common thread (such as the virus) associated with this new disease, no common denominator such as a causative transmissible agent, but on the other hand he does not rule out the case for a single infectious organism that has yet to be identified.

In 1993, Britain's *Sunday Times* took the Duesberg claim that HIV does not cause AIDS to a new level by running a series of articles disclaiming HIV as the cause of AIDS. The *Sunday Times* continued to run these articles throughout 1994 (Dickson, 1994; Karpas, 1994). In mid-1994, Kary Mullis (recipient of the Nobel prize in chemistry for his invention of the polymerase chain reaction—see chapter 12 for explanation) joined Duesberg and the *Sunday Times* by saying, "the idea that HIV was the 'probable cause of AIDS' was not a scientifically proven fact, nor, indeed, was there any clear evidence that AIDS was spread through sexual contact. I think we [spread] retroviruses by our lungs, not by our genitals." Mullis continued by saying, "it is not a scientitic fact, or even a supposed fact, that HIV is the probable cause of AIDS." Mullis is unrepentant in the face of those who complained his ideas were undermining the activities of AIDS health-workers around the world. Critics of Mullis state that he has limited knowledge of HIV/AIDS and no scientific authority to speak on the cause of AIDS (Dickson, 1994).

Mullis himself appeared unfazed by such criticisms. His main concern was that the efforts of "a lot of talented people in the medical profession had been diverted" by misconceptions about HIV. Opinions about Mullis are as mixed as opinions about HEAL. He is famous for taking intellectual risks, sometimes wild ones. His book *Dancing Naked in the Mind Field* (published by Pantheon in August 1998) covers his **disbelief** that HIV causes AIDS and, among other things, his passion for hallucinogenic drugs and his belief in flying saucers.

A Member of the United States Government Enters the Dispute

In March 1995, a letter was sent to eight government scientists and officials who influence AIDS research and policies, in which freshman Representative Gil Gutknecht (R-MN) questions the "HIV = AIDS hypothesis and its inability or effective treatment." His query echoes the arguments of Peter Duesberg. Gutknecht's initial inquiry asks whether recreational drug use and anti-HIV drugs might be the true cause of AIDS and questions whether AIDS is contagious.

Gutknecht's staffers think the federal AIDS effort—based on the conclusion that HIV causes AIDS—"will be seen as the greatest scandal in American history and will make Watergate look like a no-fault divorce" (Stone, 1995).

Talk Show Host Claims AIDS Is a Myth— October 1996, Fort Lauderdale, Florida—A talk show host and two medical professionals have been proclaiming over the airwaves on Miami's WLQY-AM, probably South Florida's most popular Haitian radio station, that AIDS is a myth, and have been urging listeners, an estimated 155,000 Haitians, to discontinue medication and refuse treatment.

The commentators told the *Sun-Sentinel* of Fort Lauderdale that they have ample proof that AIDS does not exist and that the health care industry made it up to increase business. People should stop taking AIDS drugs and wearing condoms, they said in interviews for the article.

"We are not spreading lies. What we are saying are facts," said Henri-Claude Saint-Fleur, a North Miami Beach psychologist who was educated in France. "I'm happy to tell people the truth. I know lots of people who stopped taking their AIDS medication and are living well."

HIV/AIDS counselors say Haitian patients from at least four treatment centers have stopped seeing doctors because of the show.

Class Discussion: Comment on the radio station's right to provide air time to promote this theme "AIDS IS A MYTH." Do the commentators have a right to tell people to stop taking their medications? Are they liable for such advice? Should they be liable?

Evidence That HIV Causes AIDS

It has been firmly established that there is a high correlation between HIV infection and the development of AIDS. With respect to establishing HIV as the causative agent of AIDS, look at some of the evidence:

1. The virus has now been identified in virtually all AIDS patients tested; in over 90% of those with pre-AIDS symptoms (formerly called AIDS related complex or ARC), that is, individuals who are HIV-positive and have symptoms of HIV disease; and in individuals who are HIV-positive but appear to be healthy.

2. The virus has been identified by electron microscopy inside and on the surface of T4 cells in HIV-positive and AIDS patients (Figure 2-1A,B).

3. Recent work of Bruce Patterson and Steven Wolinsky has shown that the genetic material of HIV (HIV-DNA) can be found in as many as 1 in 10 blood lymphocytes of persons with HIV disease (Cohen, 1993).

4. Antibodies against the virus, viral antigens, and HIV RNA have been found in HIV-positive and AIDS patients.

5. There is an absolute chronological association between the emergence of AIDS and the appearance of HIV in humans worldwide.

6. There is a chronological association of HIV-positive individuals who progress to AIDS. People who are truly HIV-negative, and without the need for chemotherapy or radiation treatments, do not demonstrate HIV antibodies and never demonstrate AIDS. For example, people with hemophilia, unlike homosexual men, represent a well-defined group with long-term documentable changes in morbidity and mortality, since they had been well studied as a group before the era of AIDS. This research shows that people with hemophilia began to die of dramatically different things starting about 1982. Looking back shows little evidence of a special incidence of opportunistic diseases in people with hemophilia in the U.S. from the turn of the century up to 1979. Significantly, however, in the years before AIDS, people with hemophilia had never been noted to be particularly susceptible to the more obvious fungal infections, such as candida esophagitis, common to AIDS patients and others with low-lymphocyte type immune deficiency.

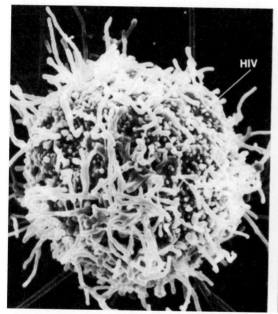
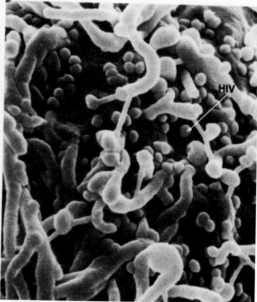

FIGURE 2-1   Viral Replication in Human Lymphocytes. Scanning electron micrograph of HIV-infected human T4 lymphocyte. **A**, A single cell infected with HIV showing virus particles and microvilli on the cell surface (magnified 7,000 times). **B**, Enlargement of a portion of the mountain-like cell surface in (**A**) showing multiple virus particles budding out of the cell surface (magnified 20,000 times). As each HIV exits the cell, it leaves a hole in the cell membrane. (*Photograph courtesy of K. Nagashima,* **Program Resources, Inc.,** *NCI-Frederick Cancer Research and Development Center*)

THE UNITED STATES GOVERNMENT CREATED HIV

At the African-American Summit speech in New Orleans in 1989, Louis Farrakhan told his audience: **"The spread of International AIDS was an attempt by the U.S. government to decimate the population of central Africa."** In 1988 he told Barbara Walters on ABC's 20/20: "Do you know where the AIDS virus was developed? Right outside of Washington. It is my feeling that the U.S. government is deliberately spreading AIDS."

Leonard Horowitz, a past medical advisor to dental and medical supply companies, turned author, claims the United States government is responsible for creating HIV and the Ebola virus! In a speech at the Capital Mall, Washington, D.C., on Labor Day weekend, 1996, Horowitz stated that, "the problem I [Horowitz] had was reconciling the fact that the dentist [David Acer, the Florida dentist who stands accused of infecting his patients with HIV], though a psychopath, was no fool. And he held in his possession one of the most incriminating documents I had ever seen. A 1970 Department of Defense Appropriations request for $10 million for the development of immune system ravaging viruses for germ warfare. In fact, the document, which I lay before you today, reads like this:

> Within the next 5 to 10 years, it would probably be possible to make a new infective microorganism which could differ in certain important aspects from any known disease-causing organisms. Most important of these is that it might be refractory to the immunological and therapeutic processes upon which we depend to maintain our relative freedom from infectious disease. . . . A research program to explore the feasibility of this could be completed in approximately 5 years at a total cost of $10 billion. . . . It is a highly controversial issue and there are many who believe such research should not be undertaken lest it lead to yet another method of massive killing of large populations.

So Dr. Acer created a crime, a mystery, that couldn't be solved, without implicating the government and causing a larger mystery to be investigated. That is, the origin of AIDS and Ebola—the subject of my last three years of research, and why I have come before you today."

Class Discussion: Do you believe the scenarios by Farrakhan and Horowitz? Are they feasible? If Yes, on what factual evidence do you base this conclusion. If No, why not?

CHALLENGING FARRAKHAN, HOROWITZ AND OTHERS ON THE AMERICAN GOVERNMENT'S CREATION OF HIV

To date, the **first** scientific evidence for the presence of HIV in humans came from the detection of HIV antibodies in preserved serum samples collected in central **Africa** in 1959. Although this is for now the earliest evidence, HIV may have entered humans much earlier. Tissue samples are not available for testing. The **first** documented AIDS case in **Europe** occurred in 1976. It was a Danish surgeon who was believed to have contracted HIV in Zaire. In retrospect, the **first** HIV case in the United States occurred in 1968. HIV antibodies were found in a frozen blood sample from a male. He died from a variety of opportunistic infections, wasting, and demonstrated Kaposi's sarcoma. According to the CDC's **first clinical AIDS definition,** at least one case of AIDS occurred in New York City in 1952 and another in 1959. Given the presence of HIV in humans as far back as 1959 it is easy to discount the government conspiracy theorists because the necessary biological techniques for working with or creating HIV, were not known! David Baltimore and Howard Temin did not independently discover **reverse transcriptase,** the marker of a retrovirus, until 1970! And the genetic manipulation of the gene to produce this enzyme would have been crucial to developing an artificial HIV. At this time in history, genetic engineering appeared to be in the distant future!

Conspiracy Theories Continue

Conspiracy theories about HIV still abound despite a wealth of available scientific evidence. The Harlem AIDS Forum featured many outspoken opponents to traditional views on HIV and AIDS. Of 12 speakers, only one believed that HIV is the cause of AIDS, but he also argued that the virus was being spread to people of color throughout the world through World Health Organization via its vaccine programs. Event organizer **Curtis Cost** explained that the objective of the meeting "was to allow people to hear disparate perspectives, and to do their own research." According to a survey conducted by the Institute of Minority Health Research at Emory University's Rollins School of Health, 74 percent of African Americans questioned believed they were likely or somewhat likely to be used as test subjects for studies without their consent. Eighteen percent reported they believed that HIV was an engineered virus and about 10 percent said that AIDS is part of a genocidal plot to kill black people. The AIDS pandemic is particularly prevalent in the African-American community; although African Americans comprise less than 13 percent of the United State's total population, they accounted for 57 percent of new infections in 1998 and 1999. Some AIDS activists among the community disapprove of the conspiracy theories, asserting that they serve to subvert prevention efforts such as testing and safer sex.

After 1984, this type of AIDS-associated opportunistic infection and immune failure rapidly became the single most common cause of death in people with hemophilia in America.

The rise in total mortality in people with hemophilia was sudden: death in this population, which had been stable in 1982 and 1983, suddenly increased by a factor of approximately 900% in the first quarter of 1984. This increase in numbers of deaths was consistent with an epidemic, or some new very toxic contamination of the clotting factor supply. It is not consistent with slower social changes, slower toxin or immune suppression models, multifactorial causation models or the idea that people with hemophilia were actually at no greater risk than before. Mortality figures in hemophilia patients also showed something else important, which was that the new deaths of the late-1980s, by virtue of all being diagnosed with AIDS, demonstrated that most or all of them occurred in people with hemophilia who were HIV-positive.

Since these deaths accounted for almost the entire new increase in mortality, it could be inferred that the mortality rate for HIV-negative people with hemophilia did not increase much in the 1980s, if at all. (Harris, 1998)

7. Hemophiliacs from low- and high-risk behavior groups were equally infected from HIV-contaminated blood factor VIII concentrates.

8. The virus is found in HIV-positive and AIDS patients but not in healthy low behavioral risk individuals.

9. With the exception of persons who had their immune systems suppressed due to genetic causes or by drug therapy, prior to the appearance of the virus, there were no known AIDS-like cases. The virus has been isolated worldwide—but only where there are HIV-positive people and AIDS patients.

10. An HIV-positive identical twin born to an HIV-positive mother developed AIDS, but the HIV-negative twin did not.

POINT OF INFORMATION 2.1

THE RELATIONSHIP BETWEEN THE HUMAN IMMUNODEFICIENCY VIRUS AND THE ACQUIRED IMMUNODEFICIENCY SYNDROME

Each time Peter Duesberg publishes an article claiming that HIV does not cause AIDS, the phone lines at the National Institute of Allergy and Infectious Diseases (NIAID), at the Centers for Disease Control and Prevention, and at AIDS-related research centers across the United States ring for days; nonstop. In order to answer the critics of the HIV-AIDS relationship, the NIAID published a detailed 61-page report ("The Relationship Between the Human Immunodeficiency Virus and the Acquired Immuno-deficiency Syndrome") outlining and discussing the data that leaves no doubt in the HIV/AIDS Scientific Community that HIV is the causative agent (NIAID, 1995). A scientist once said that even if a camel's tail appeared under the tent, he would not believe it was a camel until he could see the hump. The author of this text, looking over volumes of data—for and against HIV being the cause of AIDS—believes he sees the hump!

11. Only HIV-positive mothers transmit HIV into their fetuses and only these HIV-positive newborns progress to AIDS. HIV-negative newborns from HIV-positive mothers **DO NOT** get AIDS!

12. Drugs developed specifically to inhibit the replication and/or maturation of HIV, thereby lowering the level of HIV found in HIV infected people, have delayed the onset of HIV disease and, for HIV infected pregnant women, have **decreased** the birth of HIV infected infants by 66%.

13. **IF HIV does not cause AIDS, how do scientists explain the positive effects of drugs used to affect the early and late stages in the life cycle of HIV that have lowered viral load to unmeasurable levels in the blood of those with HIV disease and AIDS and have virtually restored life for those at death's door and who are now back at work!**

14. Finally, there have been numerous reports in the literature on HIV-infected individuals (homosexual, bisexual, and heterosexual) transmitting the virus to their sexual partners, both eventually dying of AIDS. *The unbroken chain of HIV transmission between prostitutes and their customers, between injection drug users sharing the same syringe, from infected mothers to their unborn fetuses, and so on all lead to the inescapable conclusion that HIV does cause AIDS.*

In short, Koch's Postulates have been satisfied: HIV disease meets all four criteria!

1. The causative agent must be found in all cases of the disease. (**It is.**)

2. It must be isolated from the host and grown in pure culture. (**It was.**)

3. It must reproduce the original disease when introduced into a susceptible host. (**It does.**)

4. It must be found in the experimental host so infected. (**It is.**)

Thus the identification of HIV as the causative agent of AIDS is now firmly accepted by scientists worldwide.

Class Discussion: You have just read some of the evidence for and against HIV being the cause of AIDS. Assuming you agree with the vast majority of HIV/AIDS investigators worldwide, that HIV does cause AIDS, do you think there comes a time at which dissenters forfeit their right to make claims on other people's time and trouble by the poverty of their arguments and by the wasted effort and exasperation they have caused?

NOW discuss the value of the dissenter.

NOW discuss the danger of the dissenter's information or claims.

ORIGIN OF HIV: THE AIDS VIRUS

Clarification ot the term **Origin of HIV.** Scientists are searching for the **source** of HIV or HIV-like ancestor. Finding this source will give us the **Origin of HIV** as it pertains to where and in which animal the virus was housed prior to entering humans. But it does not mean the beginning of the virus per se—that will most likely never be known.

BOX 2.2

THE CAUSE AND EFFECT OF DISEASES: THE UNBROKEN CHAIN

For more than a century, four postulates set down by the German microbiologist Robert Koch have guided the hunt for disease-causing microbes. Koch argued in 1890 that to prove an organism causes a disease, microbiologists must (a) show that the organism occurs in every case of the disease; (b) that it is never found as a harmless parasite associated with another disease; (c) the organism must be isolated from the body and grown in laboratory culture; (d) it must be introduced into a new host and produce the disease again.

Koch's criteria for cause and effect with respect to identifying a disease causing agent (**bacterium**) are now being swept aside by new technology. Many of the microbes isolated in recent years by molecular techniques, that pull segments of their DNA or RNA directly out of infected tissues, **cannot be grown in culture.** That makes it impossible to fulfill all of Koch's postulates. As a result, many biologist are modifying Koch's strict requirements. Researchers can now build a convincing case against a microbe by examining a wide variety of molecular circumstantial evidence, such as how tightly the suspect microbe is associated with infected tissues and how closely the time course of the disease correlates with the amount of microbial genetic material present. But, as with Koch's methodology, the newer methodology for microbe identification with a specific disease also has its pitfalls.

Tracking the origins and early history of a newly-recognized disease is more than just an academic exercise. To appreciate how a disease began can help medical science to combat it. The classic example is **John Snow's** investigation of the cholera epidemic in Golden Square, London, in 1854; his removal of the handle of the Board Street pump contained the outbreak.

More than virtually any other disease, AIDS has generated myths and far-fetched theories about its origin, its causes, and even its very existence. These are probably linked to fear and denial prompted by a virus which is fatal, incurable, and sexually transmitted—and can infect people for years before they show any signs of illness.

Why Do Scientists Want to Know Where HIV Originated?

The object of determining the origin of the AIDS virus is to gain insight into how the virus may have evolved the unique set of characteristics that enable it to destroy the human immune system. Such information will offer valuable clues as to how rapidly the virus is evolving and how to combat it and perhaps help prevent future viral plagues.

If, for example, HIV is a new virus, say less than 30 years old, the many different varieties of HIV now infecting people worldwide probably evolved from a common ancestor sometime after World War II. And new varieties can be expected to continue evolving at what would be a frightening pace for several more decades, possibly producing new strains of the virus that are even more dangerous than those now infecting people. This could mean that vaccines now being developed based on current virus strains may not be useful in 10 to 20 years. But, should the known strains of the virus prove to be hundreds or thousands of years old, it might be possible that the current types of HIVs are in a state of global balance and they would not be expected to offer scientists any shocking evolutionary surprises in the future.

Ideas on the Origin of HIV, UFOs, Biological Warfare, and Cats

Fear stimulates the imagination. Out of human fear have come some rather strange explanations for the origin of the AIDS virus. Early reports had unidentified flying objects (UFOs) crashing to Earth and releasing a "new organism" that would wipe out humanity.

There were frequent reports in the Soviet press linking AIDS with American biological warfare research. The Soviets agreed in August 1987 to stop these reports (Holder, 1988). There are also reports of extremism, as in the case of Illinois State Representative Douglas Huff of Chicago who told the *Los Angeles Times* that he gave over $500 from his office allowance fund to a local official of the Black Hebrew sect to help the group investigate its claim that Israel and South Africa created the AIDS virus in a laboratory in South Africa. Huff said AIDS is **"clearly an ethnic weapon, a biological weapon"** designed specifically to attack nonwhites (*CDC Weekly*, 1988).

Still another myth to surface is that the AIDS virus came from domestic cats. Because of its similarities to human AIDS, feline immunodeficiency virus has been called "feline AIDS." The cat retrovirus may damage cats' immune systems leaving the animals vulnerable to opportunistic infections *or* it may cause feline leukemia. However, the cat virus has never been shown to cause a disease in humans.

The origin of HIV has been attributed to HIV-contaminated polio, smallpox, hepatitis and tetanus vaccines, the African green monkey, African people, their cattle, pigs, and sheep—and the United States CIA. With respect to the use of HIV-contaminated vaccines, a number of recent articles suggest that early monkey kidney cultures used to produce the polio vaccine carried HIV. Review of the literature offers *no* evidence that this occurred. And the argument for the safety of the polio vaccine lies in the absence of any AIDS-related diseases among the hundreds of millions of persons vaccinated worldwide (Koprowski, 1992).

Now 19 years (1981–2000) into the AIDS epidemic, researchers are still baffled by the question: Where did the AIDS virus come from? AIDS is now known to be caused by two human immunodeficiency viruses, HIV-1 and HIV-2. The former, the cause of most AIDS cases worldwide, **appears to have spread from Central Africa; while the latter has so far been confined mainly to West Africa and the islands off its coast.** Several theories have been proposed to explain the origins of the AIDS epidemic. Most have been speculative rather than verifiable; and several have caused offense, particularly those which refer to sexual practices with monkey blood (Gilks, 1991). Charles Gilks (1992) states that HIV may have entered the human population by the direct inoculation of malaria-HIV-infected blood into human prisoner-volunteers. **The problem with this theory is that it is not testable so it, like others, will remain an unproven theory proposed to explain the origin of HIV/AIDS.**

At the Fifth International Conference on AIDS in Montreal (1989), Vanessa Hirsch and colleagues presented evidence that a virus isolated from a species of West African monkey, the Sooty Mangabey (an ash-colored monkey), may have infected humans 20 to 30 years ago. They believe this virus subsequently *evolved into HIV-2*. Hirsch et al. studied a virus known as the simian immunodeficiency virus (SIVsm) that infects both wild and captive sooty mangabey monkeys. They molecularly cloned and sequenced the DNA of the virus and constructed an evolutionary tree of the several known primate immuno-deficiency viruses. This tree showed SIVsm to be more closely related to HIV-2 than to HIV-1.

Gerald Meyers of Los Alamos National Laboratory states that SIVsm and HIV are so closely related that when HIV-2 is found in a human, it may be the sooty mangabey virus. However, HIV-1, that causes AIDS, does not sufficiently resemble HIV-2 or SIVsm, thus HIV-1 probably did not evolve from SIVsm/HIV-2. **Still, the prevailing theory is that humans were first infected through direct contact with HIV-infected primates. The primate to human scenario is easier to accept than humans infecting primates.** Humans have hunted, handled, and even eaten primates for thousands of years. Recent laboratory accidents have shown that SIV can infect humans. Even though at the moment, no identifiable disease has been associated with the SIV/human infections, such accidents have demonstrated the potential for cross-species transmission of HIV-related viruses. Why not believe the same for the origin of HIV-1?

There also remains the possibility that HIV has been present but remained an obscure virus in the human population for a long time before it was recognized as a lethal agent, similar perhaps to the polio virus which surely existed in the human population for years prior to its being discovered as the cause of the polio epidemic of 1894. After all, there is ample evidence that primates, historically, have harbored lentiviruses (the class that includes SIV and HIV). Why should humans be any exception?

Adam Carr (*AIDS in Australia,* 1992) writes that the most widely accepted view on the origin of HIV is that the virus is endemic to a remote part of central Africa, possibly in the mountains of eastern Zaire, and that it began to spread to other parts of Africa only after the area had been penetrated by Europeans in the twentieth century.

In colonial Africa, it is quite possible that a low but persistent level of AIDS cases could have gone unnoticed by the poorly developed health services of the time. In the 1970s, when rapid urbanization and its attendant social changes began in Zaire and neighboring countries, the epidemic began to accelerate and come to the attention of health authorities. By this time it would have begun to spread to other countries. Tourists, soldiers (there were thousands of Cuban troops in neighboring Angola in the mid-1970s), guest workers, and other travelers would have taken the virus to Europe, the Caribbean, and North America.

The first recorded AIDS case in America was that of a 15 year old male prostitute who demonstrated Karposis sarcoma and died in 1969. Frozen tissue samples contained HIV antibodies. These findings were reported at the 11th International Congress of Virology in August 1999.

The first documented case of AIDS in Europe was seen in a Danish surgeon who had worked in Zaire. She died in 1976.

The Current Theory on HIV Origin: A Chimpanzee

Beatrice Hahn and colleagues (1999) from the University of Alabama at Birmingham feel that they have gathered sufficient evidence to believe the origin of HIV-1 in humans to be cross-species transmission from a particular

SCIENTISTS REPORT ON EARLIEST AIDS CASE

Scientists have pinpointed what is believed to be the earliest known case of AIDS—an African man who died in 1959. The scientists looked for signs of HIV in 1,213 blood samples that were gathered in Africa between 1959 and 1982. They found clear signs of the virus in one taken from a Bantu man who lived in Leopoldville, Belgian Congo—what is now Kinshasa, Republic of Congo—in 1959. Scientists compared the genes from the 39-year-old sample of HIV with current versions of HIV. They realized that if they had an old sequence of HIV genes it would serve as a yardstick to measure the evolution of the current HIV. HIV has mutated over the years to form 10 distinct subtypes, lettered A through J. One of these, subtype B, is the dominant strain in the United States and Europe, while subtype D is most common is Africa. The family tree of HIV looks like a bush with the various subtypes forming the limbs. Scientists believe the 1959 HIV is near the trunk, around the point where the subtypes B and D branch off and that this virus is an ancestor to B and D. These data suggest that all HIV subtypes evolved from one introduction of HIV into people, rather than from many crossovers from animals to humans, as some have speculated. Given the steady rate at which HIV mutates, it also means that the virus probably first got into people sometime in the 1940's or early 50s. Simon Wain-Hobson of the Pasteur Institute in Paris believes that this is the oldest, totally unambiguous look at HIV known. This information was presented by Toufu Zhu, University of Washington, Seattle at the 1998, Fifth Conference on Retroviruses and Opportunistic Infections and published in *Nature,* 391:594. Bette Korber and colleagues (1998), reporting on the origin of HIV, state that an analysis of the 1959 HIV DNA sequence suggest that the precursor of today's HIV entered the human population within decades of 1959—1930s or 1940s?

However, the question scientists are trying to answer is why the disease took 20 years to start manifesting itself, ultimately hitting the homosexual and hemophiliac populations in the United States in the late 1970s. Current thinking is that HIV probably first infected rural areas of Africa, slowly moving into the cities and around the world until it hit homosexual communities where conditions were sufficient for rapid transmission of the disease.

subspecies of chimpanzee—meaning that a simian virus **closely related to HIV** jumped from monkeys to humans, and later mutated into its current form, following one of many interspecies transmissions. (Figure 2.2)

Hahn presented three lines of evidence in support of their thesis. **First,** the genes of all four **SIVcpz** isolates cluster on evolutionary trees according to their subspecies or origin, either *Pan troglodytes troglodytes* from West Africa or *Pan troglodytes scheinfurthii* in East Africa. By sequencing parts of the virus' genetic material, the scientists found that more segments of the simian and human virus overlapped than had been identified in three previous simian viruses isolated in recent years from other chimpanzees. The investigation revealed that the other previously studied simian viruses were more closely related to very rare forms of HIV that don't account for most of the infections in the world.

Second, all known HIV-1 strains, including the M group that accounts for over 95% of all HIV-1 infections, as well as the O group and the recently described N group, form a genetic cluster with the West African chimpanzee viruses. This clustering is also geographic and consistent with the likely equatorial West African origin of HIV-1. (Figure 2.3) Hahn said she had initially been equally ready to accept the idea that chimps had gotten HIV from humans, rather than vice versa. However, a **third** line of evidence convinced her that HIV-1 was introduced into the human population from at least three cross-species transmission from chimpanzee—she found evidence of genetic recombination among the SIVcpz strains of the **troglodytes'** lineage. This strongly suggested

FIGURE 2-2 Photograph of *Pan troglodytes troglodytes.* This photograph was taken in the Gabon region of Africa where the sole species of Pan troglodytes is found. Their numbers are declining rapidly due to the "bush meat" trade—the slaughter and selling of their body parts. *(Photograph courtesy of Karl Amman, wildlife photographer.)*

in numerous infection experiments over the years). Most interesting perhaps will be to learn what insight this history of HIV might provide for predicting the ultimate evolutionary accommodation between HIV and humans. For scientists, the new chimpanzee finding is just as much a beginning as an end. Although researchers may have the best evidence so far that HIV came from chimpanzees, no one can yet say how the virus became lethal to humans, and when. Chimpanzees share over 98 percent of the genes that exist in humans, yet they don't get AIDS. So which of the remaining genes protect chimpanzees from an infection that has turned out to be almost universally fatal in people? Also, researchers are still investigating how, after taking root in just a few people, HIV gradually traveled all over the world.

Researchers now believe that AIDS-like viruses jumped from chimp to human more than once, creating different strains of HIV. This means a vaccine against one strain of HIV may not control a new epidemic. As it turned out, **Pan troglodytes troglodytes** lives in the region of Africa where HIV-1 was first recognized. Since there are three separate kinds of HIV-1, the scientists believe that HIV-1 crossed into people from chimps at least three times. Hahn said that SIV appears to have dwelled in primates for hundreds of thousands of years before turning into the deadly human virus, HIV. She believes hunting, which exposes people to excessive amounts of blood during slaughter, allowed precourser HIV to infect humans.

When did HIV Begin to Circulate Among Humans? Scientists do have some idea when the virus began to circulate among people. From looking at samples of HIV taken at different times and different parts of the world, researchers have constructed a type of genetic clock for HIV-1. The speed of the clock is determined by how much the virus changes over time. A key to setting this clock came with the discovery of the oldest known HIV infection, that of a man who lived in what is now Congo in 1959. That year, he had been one of 1,200 Africans who had been given a blood sample as part of a study of the immune system. After the AIDS epidemic began, the frozen samples were

that despite the apparent scarcity of these chimpanzees now, the incidence of SIVcpz in wild chimpanzees must have been high enough, at least at one time, to permit occasional multiple infections of the same chimp from different sources of virus.

It is nice to think that the issue of HIV origin has been solved. However, time will tell if Hahn's current effort will be the final word on the subject. Some questions still remain. Why, for example, does HIV-1 appear to be so benign in chimpanzees (as has been shown

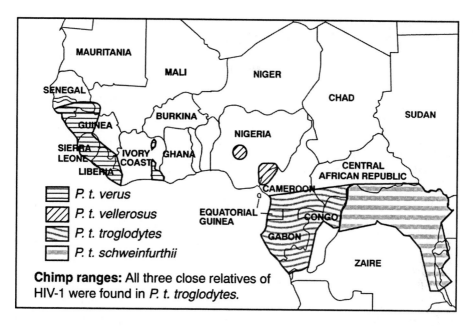

FIGURE 2-3 Geographic Ranges of Four Subspecies of Chimpanzee *Pan troglodytes*. All three close relatives of HIV-1, as determined by genetic sequencing of their genetic material, were found in Pan troglodytes troglodytes *(Adapted from Hahn 1999.)*

thawed and screened. By comparing HIV-1 samples taken over time, Steven Wolinsky and his colleagues determined that HIV began to spread within decades before the 1959 sample was drawn (Balter, 1998). The estimate is based on the assumption that the rate of change for HIV has remained about the same. This would place the cross-over into humans around 1924 to 1946. Studies have also indicated that HIV-2 began to spread about the same time.

Preston Marx of the Aaron Diamond AIDS Research Center said, "Never before in thousands of years of exposure to SIV did these viruses cross over to cause (widespread) disease in people, and yet it happened twice this century.... Not only did this happen in chimpanzees, it happened in sooty mangabeys 1,000 miles away." Many scientists have speculated about why the virus might have become amplified in post-World War II Africa. Among the possibilities: the exploding population of urban centers; social upheaval and commerce that led to migration across the continent; and widespread vaccination campaigns that could have reused needles. In the end, HIV's early accomplices aren't known for sure. Nor is it clear how AIDS came to the United States.

Continued Evolution of HIV in Humans? Possible *good* news (?) is that HIV will, after killing millions of humans, become a harmless passenger as it most likely did thousands of years ago in chimpanzees. The **bad** news would be that HIV is evolving into a virus that is more easily spread and/or that it becomes more lethal. Time will tell.

HIV Is Stigmatizing

Discussion of the African origins of HIV has caused controversy. Some African governments protested at suggestions that Africa was "responsible" for the AIDS epidemic, and even denied that AIDS was a problem in their countries. For most of the 1980s, the military government of Zaire would not allow outsiders to investigate the spread of AIDS and did not report AIDS cases to the World Health Organization. Some Africans denounced "racist" Western media for reporting AIDS as an African plague. Carr states that "by 1990, when it was obvious to all that HIV infection was tragically widespread in central, southern, and eastern Africa, this controversy had largely ended. HIV infection is far too widespread in Africa to be the result of recent importation from outside; the African epidemic clearly precedes the Western one." **Regardless, where HIV/AIDS arose is irrelevant to medical science—when and how HIV entered the human population is relevant.**

SUMMARY ON THE ORIGIN OF HIV

In summary, there are at least three possible origins of the AIDS virus: (1) it is a human-made virus, perhaps from a germ warfare laboratory; (2) it originated in the animal world and crossed over into humans; and (3) HIV has existed in small isolated human populations for a long time and, given the right set of conditions, it escaped into the larger population. Computer modeling of DNA sequence in HIV and SIVsm and the recent work of Beatrice Hahn and colleagues suggests that HIV evolved within the last 100 years. So, for now, the question remains: Is AIDS a new disease or an old disease which was late being recognized—so late that we will never know its true source of origin, the origin of HIV? Continued investigations by Hahn and others may answer these questions.

   An additional question to where HIV came from is whether it has always caused disease. From the study of human history, as it relates to human disease, scientists have numerous examples that show that as human habits change, new diseases emerge. Regardless of whether HIV is old or new, history will show that social changes, however small or sudden, have most likely hastened the spread of HIV. In the 1960s, war, tourism, and commercial trucking forced the outside world on Africa's once isolated villages. At the same time, drought and industrialization prompted mass migrations from the countryside into newly teeming cities. Western monogamy had never been common in Africa, but as the French medical historian Mirko Grmek notes in his book, *History of AIDS*

——— POINT OF INFORMATION 2.3 ———

NEW HUMAN STRAIN OF HIV FOUND

A new, unnamed, strain of HIV was isolated from three people in Cameroon, Africa. Francois Simon, Bichat Hospital, Paris reported, at The 1998 Fifth Conference on Retroviruses and Opportunistic Infections, that this strain of HIV is **closely related to a chimpanzee virus.** It is not unusual for viruses to pass from animals to humans. But in cases like tularemia from rabbits, hantaviruses from mice and Ebola and Lassa virus from unknown carriers, the virus does not spread from human to human very easily. Researchers now know that there were at least 4 times when immunodeficiency viruses passed over from lower primates to humans, once for type M, at least twice for type O and now once for the Cameroon virus. What scientists don't know, is how many other times such cross-species transfer has occured, only to lose out. They also don't know how many times it will happen again.

(1990), urbanization shattered social structures that had long contained sexual behavior. Prostitution exploded, and venereal disease flourished. Hypodermic needles came into wide use during the same period, creating yet another mode of infection. Did these trends actually turn a chronic but relatively benign infection into a killer? The evidence is circumstantial, but it's hard to discount.

   Whatever the forces are that brought us AIDS, they can surely bring other diseases. By encroaching on rain forests and wilderness areas, humans are placing themselves in ever-closer contact with animal species and their deadly parasites. Activities, from irrigation to the construction of dams and cities, can expose humans to new diseases by expanding the range of the rodents or insects that carry them. Stephen Morse, a Rockefeller virologist, studies the movement of microbes among populations and species. He worries that human activities are speeding the flow of viral traffic. More than a dozen new diseases have shown up in humans since the 1960s, nearly all of them the result of once exotic parasites exploiting new opportunities. (Certain of these new disease-causing agents, for example, the Ebola virus, are discussed in Box I.1.)

The Difference Between HIV, HIV Disease, and AIDS

It is important for people to understand the distinction between the terms **HIV, HIV disease, and AIDS.** There is an immense difference between being infected with HIV, expressing various kinds of diseases because of the continued presence of HIV (HIV disease), and being diagnosed as having AIDS.

HIV is the term for the virus that damages the immune system. The continued presence of HIV may eventually cripple the body's ability to fight off opportunistic diseases referred to collectively as HIV disease. AIDS is the end result of HIV disease. People with HIV disease are diagnosed as having AIDS if they are HIV-positive and develop one or more of some 26 diseases or conditions such as *Pneumocystis carinii* pneumonia (PCP), Kaposi's sarcoma, pulmonary tuberculosis, invasive cervical cancer, recurrent pneumonia, or demonstrate a T4 lymphocyte cell count of less than 200/µL of blood.

Many more people, in any country, are infected with HIV or are in some stage of HIV disease than have developed AIDS. Many HIV-infected people can live 10 years or longer without experiencing illness. And using the new anti-HIV drugs, people may live 20 or more years after infection.

SUMMARY

The AIDS virus was discovered and reported on by Luc Montagnier of France in 1983. Identifying the virus that caused the immunosuppression that caused AIDS allowed for AIDS surveillance definitions that began in 1982. The recent recognition of non-HIV AIDS cases is not unexpected and can be explained. **Presently, there is no new threat of another AIDS-causing biological agent.**

The recent work of Beatrice Hahn and colleagues may have pinpointed the reservoir of a precursor, HIV-like virus in the chimpanzee, *Pan troglodytes troglodytes.*

REVIEW QUESTIONS

(*Answers to the Review Questions are on page 487*)

1. What may be the strongest nonlaboratory evidence for saying that AIDS is caused by an environmental agent?
2. Where might HIV have originated and where did the first HIV infections appear?

REFERENCES

ANDREWS, CHARLA. (1995). "The Duesberg Phenomenon." What does it mean? *Science,* 267:157.

BALTER, MICHAEL. (1998). Virus from 1959 sample marks early years of HIV. *SCIENCE,* 279:801.

BAUM, RUDY. (1995). HIV link to AIDS strengthened by epidemiological study. *Chem. Eng. News,* 74:26.

BUIANOUCKAS, FRANCIS. (1995). HIV, an illusion. *Nature,* 375:197.

CDC Weekly. (1988). Extremists seek to blame AIDS on Jews. July: 11.

COHEN, JON. (1993). Keystone's blunt message: 'It's the Virus, Stupid'. *Science,* 260:292–293.

CONNOR, EDWARD. (1994). Reduction of maternal infant transmission of HIV with zidovudine treatment (ACTG 076). *N. Engl. J. Med.,* 331: 1173–1180.

CULLITON, BARBARA J. (1992). The mysterious virus called "Isn't." *Nature,* 358:619.

DARBY, SARAH, et al. (1995). Mortality before and after HIV infection in the complete UK population of haemophiliacs. *Nature,* 377:79–82.

DICKSON, DAVID. (1994). Critic still lays blame for AIDS on lifestyle, not HIV. *Nature,* 369:434.

DUESBERG, PETER H. (1990). Duesberg replies [to the charges of Weiss and Jaffe]. *Nature,* 346:788.

DUESBERG, PETER H. (1993). HIV and AIDS. *Science,* 260:1705–1708.

DUESBERG, PETER H. (1995a). The Duesberg Phenomenon: Duesberg and other voices. *Science,* 267:313.

DUESBERG, PETER H. (1995b). Duesberg on AIDS causation: the culprit is noncontagious risk factors. *The Scientist,* 9:12.

EDITORIAL. (1995). More conviction on HIV and AIDS. *Nature,* 377:1.

GILKS, CHARLES. (1991). AIDS, monkeys and malaria. *Nature,* 354:262.

GILKS, CHARLES. (1992). AIDS and malaria experiments. *Nature,* 355:305.

GRMEK, MIRKO. (1990). *History of AIDS: Emergence and Origin of a Modern Pandemic.* Princeton, NJ: Princeton University Press.

Hahn, Beatrice, et al. (1999). Origin of HIV-1 in the chimpanzee *Pan troglodytes troglodytes*. ***Nature,*** 397:436–441.

Harris, Steven. (1995). The AIDS heresies: A case study in skepticism taken too far. *Skeptic.* 3, no. 2: 42–58.

HOLDER, CONSTANCE. (1988). Curbing Soviet disinformation. *Science*, 242:665.

KARPAS, ABRAHAM. (1994). AIDS plagued by journalists. *Nature*, 368:387.

KOPROWSKI, HILARY. (1992). AIDS and the polio vaccine. *Science*, 257:1024–1026.

KORBER, BETTE, et al. (1998). Limitations of a molecular clock applied to considerations of the origin of HIV-1. *Science* 280:1868–1871

LEVY, JAY. (1995). *HIV and the Pathogenesis of AIDS.* Washington, DC: ASM Press, 359 pp.

MOORE, JOHN. (1996). À Duesberg, adieu! *Nature,* 380:293–294.

NATIONAL INSTITUTE OF ALLERGY AND INFECTIOUS DISEASES. (1995). The relationship between the human immunodeficiency virus and the acquired immunodeficiency syndrome. *National Institutes of Health*, 1–61.

ROOT-BERNSTEIN, ROBERT. (1993). *Rethinking AIDS.* New York: Free Press, 512 pp.

STONE, RICHARD. (1995). Congressman uncovers the HIV conspiracy. *Science*, 268:191.

SULLIVAN, JOHN, et al. (1995). HIV and AIDS. *Nature,* 378:10.

WEISS, ROBIN A., et al. (1990). Duesberg, HIV and AIDS. *Nature*, 345:659–660.

Biological Characteristics of the AIDS Virus

CHAPTER CONCEPTS

♦ Retroviruses are grouped into three families: oncoviruses, lentiviruses, and foamy viruses.

♦ HIV is a lentivirus.

♦ HIV RNA produces HIV DNA which integrates into host cell to become a **proviral** DNA.

♦ HIV contains **nine genes**; its three major structural genes are GAG-POL-ENV.

♦ HIV contains 9,749 nucleotides, its genetic code.

♦ Six HIV genes regulate HIV reproduction and at least one gene directly influences infection.

♦ HIV undergoes rapid genetic changes within infected people.

♦ The reverse transcriptase enzyme is very error prone.

♦ HIV causes immunological suppression by destroying T4 helper cells.

♦ HIV is classified into major (M), outlier (O), and type (N) genetic subtypes. Type M is responsible for 99% of the HIV infections world wide.

Viruses are microscopic particles of biological material, so small that they can be seen only with electron microscopes. A virus consists solely of a strip of genetic material (nucleic acid) within a protein or fatty (lipid) coat.

Viruses are parasitic agents; they live inside the cells of their host animal or plant, and **can reproduce themselves only by forcing the host cell to make viral copies.** The new virus leaves the host cell and infects other similar cells. By damaging or killing these cells, some viruses cause diseases in the host animal or plant. Genetically, viruses are the simplest forms of **"life-like agents"**; the genetic blueprint for the structure of the **human immunodeficiency virus (HIV)** is 100,000 times smaller than that contained in a human cell. The complete sequence of 9,749 nucleotides which form the genetic code for HIV has been identified and their arrangement sequenced (mapped).

Scientists have produced a great deal of information about HIV over a relatively short time. The immediate involvement of so many scientists followed by the rapid identification of the causative agent of AIDS is unequaled in the history of medical science. More is known about HIV than about the viruses that cause such long-standing human diseases as polio, measles, yellow fever, hepatitis, flu, and the common cold. Humankind is very fortunate that HIV entered the human population as a pathogen in the mid to late 1970s. By then scientists had discovered and begun to exploit the molecular aspects of biology. Molecular methodologies necessary to begin the immediate molecular study of HIV were in place to define and refine our knowledge of viruses and, in particular, learn about HIV.

RETROVIRUSES

There are three subfamilies of retroviruses, two of which are associated with human disease: **oncoviruses** (cancer-causing—of which HTLV-I and -II are members) and **lentiviruses** or slow viruses of which HIV-1 and HIV-2 are members. **Spumavirus** is the third group but is not known to be associated with human disease. As a lentivirus, HIV has genetic and morphologic similarities to other animal lentiviruses such as those infecting cats (feline immunodeficiency virus), sheep (visna virus), goats (caprine arthritis-encephalitis virus), and nonhuman primates (simian immunodeficiency virus or SIV).

The retroviruses and in particular the lentiviruses have presumably been present for thousands of years. Modern sociodemographic changes and other human factors have allowed rapid mutating lentiviruses, such as HIV, to find new niches and host populations

---
POINT OF INFORMATION 3.1

VIRAL SPECIFICITY

1. **Viruses are very specific with regard to the types of cells they can enter/reproduce. Not** all viruses can attach to or enter all cells. Humans survive in a world full of viruses that **only** enter or reproduce in a variety of bacteria, protozoa, fungi and higher forms of plant and animal life. It appears that most of these viruses are harmless to humans—if they enter the body they can not reproduce in human cells—they do not cause human damage. For those viruses that do enter human bodies from animals such as pig, chicken, rabbit, mouse, cow, monkey, etc. that cause a human disease, most are very specific as to which cells in a human body they can enter/reproduce and cause damage. For example, the flu virus enters the human respiratory tract cells. Epstein-Barr virus infects cells in the nose and throat. The hepatitis viruses enter liver cells but each of the hepatitis viruses causes a different degree of human cell damage over time. Some cause a more immediate disease—for example, hepatitis A virus—while hepatitis B or hepatitis C may not cause significant cell damage for years. Polio virus enters cells of the human nervous system that are different from those cells of the nervous system that herpes virus invades. HIV enters and reproduces in cells of the human immune system.

2. **Why do different viruses enter specific cell types?** Each of the viruses that cause human disease do so by finding a cell type that carries a receptor molecule (a protein or a protein attached to a sugar molecule) that **fits** with a projection of surface molecule of a given virus—much like the key to a lock imagery. That's why specific viruses are known to be associated with certain types of human tissue. For example, there are viruses that attach only to the receptors of heart, gut, eyes, throat, liver, and other specific human cell type tissues. Find a way to block either the human cell receptors (CD4, CXCKR4 and CCKR5—to which HIV attaches), without harming the cell or block the given viral receptor (in HIV its gp120) and one has a therapy for the given viral disease.

in which to reproduce. Now let us review some of the characteristics of the AIDS virus, HIV.

HUMAN IMMUNODEFICIENCY VIRUS (HIV)

HIV is a retrovirus (Figure 3-1). Retroviruses are so named because they **REVERSE** the usual flow of genetic information within the host cell in order to reproduce themselves. In all living cells, normal gene expression results from the genetic information of DNA being copied into RNA (Figure 3-2). The RNA is translated into a specific cellular protein. In all living cell types, the directions for protein synthesis come from the species' genetic information contained in its DNA:

$$DNA \rightarrow RNA \rightarrow Protein\ Synthesis$$

In brief, retrovirus RNA is copied, using its reverse transcriptase enzyme, into a complementary single strand of DNA (Figure 3-3). The single-strand retroviral DNA is then copied into double-stranded retroviral DNA (this replication occurs in the cell's cytoplasm). At this point the viral DNA has been made according to the instructions in the retroviral RNA. This retroviral DNA migrates into the host cell nucleus and becomes integrated (inserted) into the host cell DNA. It is now a **provirus** (Figure 3-4). From this point on, the infection is irreversible—**the viral genes are now a part of the cells genetic information.** In this respect, HIV can be considered as an acquired dominant genetic disease! A provirus, like the "mole" in a John LeCarre spy novel, may hide for years before doing its specific job. But for HIV, there is evidence that in some human cells the provirus begins to produce new copies of HIV RNA immediately after becoming a provirus or shortly thereafter.

Before the HIV provirus's genes can be expressed, RNA copies of them that can be read by the host cell's protein-making machinery must be produced. This is done by **transcription.** Transcription is accomplished by the cell's own enzymes. But the process cannot start until the cell's RNA polymerase is activated by various molecular switches located in two DNA regions near the ends of the provirus: the **long terminal repeats.** This requirement is reminiscent of the need of many genes in multicellular organisms to be "turned on" or "off" by proteins that bind specifically to controlling sequences.

Production of Viral RNA Strands or RNA Transcripts

Within the host cell nucleus, proviral DNA, when activated, produces new strands of HIV RNA. Some of the RNA strands behave like messenger RNA (mRNA), producing proteins essential for the production of HIV. Other RNA strands become encased within the viral core proteins to become the new viruses. Whether the transcribed RNA strands become mRNA or RNA strands for new viruses depends on whether or not the newly synthesized RNA strands undergo complex processing. RNA processing means that after the RNA is produced, some of it is cut into segments by cellular emzymes and then reassociated or **spliced** into a length of RNA suitable for protein synthesis. The RNA strands that are spliced become the mRNA used in protein synthesis. The unspliced RNA strands serve as new viral strands that are encased in their protein coats (capsids) to become new viruses that bud out of the cell (Figure 3-4).

Two distinct phases of transcription follow the infection of an individual cell by HIV. In the first or early phase, RNA strands or transcripts produced in the cell's nucleus are snipped into multiple copies of shorter sequences by cellular splicing enzymes. When they reach the cytoplasm, they are only about 2,000 nucleotides in length. These early-phase short transcripts encode only the virus's **regulatory proteins;** the structural genes that constitute the rest of the genome are among the parts that are left behind. In the second or late phase, two new size classes of RNA—long (unspliced) transcripts of 9,749 nucleotides making up the new viral genome and medium-length (singly spliced) transcripts of some 4,500 nucleotides—move out of the nucleus and into the cytoplasm. The 4,500 nucleotide transcripts encode HIV's structural and enzymatic proteins (Greene, 1993).

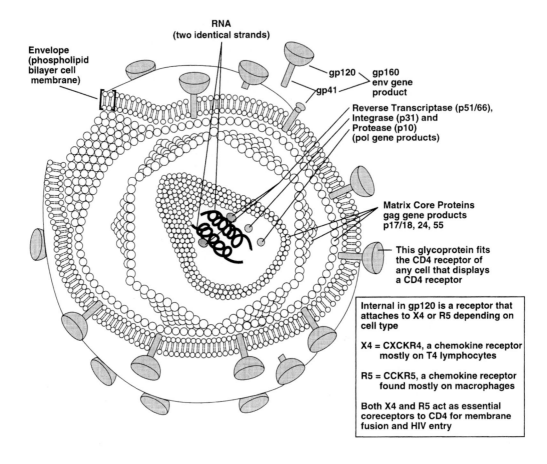

RNA
(two identical strands)

Envelope
(phospholipid
bilayer cell
membrane)

gp120
gp41

gp160
env gene
product

Reverse Transcriptase (p51/66),
Integrase (p31) and
Protease (p10)
(pol gene products)

Matrix Core Proteins
gag gene products
p17/18, 24, 55

This glycoprotein fits
the CD4 receptor of
any cell that displays
a CD4 receptor

Internal in gp120 is a receptor that
attaches to X4 or R5 depending on
cell type

X4 = CXCKR4, a chemokine receptor
mostly on T4 lymphocytes

R5 = CCKR5, a chemokine receptor
found mostly on macrophages

Both X4 and R5 act as essential
coreceptors to CD4 for membrane
fusion and HIV entry

Experimental results reported by Somasundaran et al. (1988) showed that when lyphoid cell lines or peripheral blood lymphocytes were infected with a laboratory strain of HIV, up to 2.5 million copies of the viral RNA were produced by cells; and within 3 days of infection, up to 40% of the total protein synthesized by the cells was viral protein. This is an unprecedented takeover for a retrovirus which typically makes only modest amounts of RNA and protein.

Much of what HIV does after entering the host cell or while integrated as a retroprovirus depends on the activity of its genes.

BASIC GENETIC STRUCTURE OF RETROVIRAL GENOMES

The first retrovirus was isolated from a sarcoma (a cancer) in chickens by Peyton Rous in 1911 and named the Rous sarcoma virus. The basic genetic structure of the Rous sarcoma virus and all animal retroviruses is the same. They all contain retroviral RNA sequences that code for the **same three genes** abbreviated GAG, POL, and ENV. Flanking each end of the retroviral genome is a sequence of similar nucleotides called long terminal repeats or redundancies (LTRs).

$$5' \underset{\text{LTR}}{=} \frac{\text{GAG} \quad \text{POL} \quad \text{ENV}}{} \underset{\text{LTR}}{=} 3'$$

Some of the animal retroviruses such as the Rous sarcoma virus contain an additional *onc* or *oncogene* that, along with its LTRs, causes a rapid form of cancer in chickens which kills them in 1 to 2 weeks after infection. Without the *onc* gene the virus causes a *slow progressive cancer.*

FIGURE 3-1 Human Immunodeficiency Virus. It infects cells by a process of membrane fusion that is mediated by its envelope glycoproteins (gp120-gp41, or Env) and is generally triggered by the interaction of gp120 with 2 cellular components: CD4 and a coreceptor belonging to the chemokine receptor family (CXCKR4 or X4 and CCKR5 or R5). The virus is a sphere measuring 1,000 Å or 1/10,000 mm in diameter. The cone-shaped core in a spherical envelope is the dominant feature. In this diagram the virus has been sectioned to better visualize its internal structure. The membrane of HIV is derived from the host cell. HIV gains the membrane while "budding" out or exiting the cell. Each free HIV leaves a hole in the cell membrane. The membrane, acquired from its host cell, consists of two lipid (fat) layers impregnated with some human proteins, for example Class I and Class II human lymphocyte antigen complexes important for controlling the immune response. The external viral membrane also contains molecules of viral glycoproteins (**gp**)— a sugar chain attached to protein. Each glycoprotein appears as a spike in the membrane. Each spike consists of two parts: **gp41** which contains a **coiled up protein** and extends through the membrane. On interaction between HIV envelope and T4 cell coreceptors (CD4/chemokine receptors X4 and R5—depending on cell type) the gp41 coiled protein is **unsprung** and like a harpoon, pierces the cell membrane initiating the **first step** in HIV replication. **gp120** extends from the end of gp41 to the outside and beyond the membrane (the numbers 41 and 120 represent the mass of the individual gps in thousands of daltons). As a complete unit, gp41 plus gp120 is called gp160. These two membrane or **envelope** proteins play a crucial role in binding HIV to CD4 protein molecules found in the membranes of several types of immune system cells. The gp160 precursor is cleaved into envelope (gp120) and transmembrane (gp41) proteins in the cell's Golgi compartment. The HIV envelope complex is transported via vesicles to patches in the outer cell membrane. Full-length HIV RNA is complexed with capsid proteins and the nucleocapsid is transferred to the cell surface membrane at envelope-containing sites. The binding of gp to CD4 receptors makes such immune system cells vulnerable to infection. Other HIV proteins are located and described in this figure.

Within the cone-shaped core there are two identical strands of viral genomic RNA, each coupled to a molecule of transfer RNA (tRNA) that serves as a primer for reverse transcription of viral RNA into viral DNA. Also present with the RNA are an **integrase**, a **protease**, and a **ribonuclease enzyme**. The released virus is processed internally by HIV protease to form the characteristic dense lentivirus core. Most HIV appear to have initiated DNA synthesis prior to completion of budding and maturation. Actual maturation of HIV takes place after it buds out of the cell (see Figure 4-11).

Retroviral Genome of HIV

What sets the HIV genome apart from all other known retroviruses is the number of genes in HIV and the apparent complexity of their interactions in regulating the expression of the GAG-POL-ENV genes (Figure 3-5).

The Nine Genes of HIV— The HIV genome contains at least nine recognizable genes which produce at least 9 individual proteins. These proteins are divided into 3 classes: 1) Gag, Pol and Env, the 3 major **structural** proteins 2) Tat and Rev, the two **regulatory** proteins and 3) Nef, Vif, Vpu, and Vpr, the 4 **accessory** proteins. As can be seen in Table 3-1, five of the nine genes are involved in regulating the expression of the GAG-POL-ENV genes.

The letters **GAG** stand for group-specific antigens (proteins) that make up the viral nucleocapsid. The GAG gene codes for internal structural proteins, the production of the dense cylindrical core proteins (p**24**, a nucleoid shell protein with a molecular weight of 24,000) and several internal proteins, which have been vi-

Prior to 1970, cell biologists thought that genetic information flowed only in one direction:

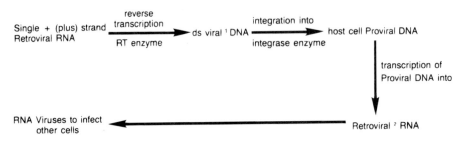

In 1970 the reverse transcriptase enzyme (RT) was found in a virus. These viruses became known as retroviruses.

1 ds = Double stranded DNA

2 If the RNA transcribed is spliced, RNA base sequences are rearranged. This RNA = messenger RNA and is used to make retroviral protein. If RNA transcribed is not spliced, it becomes the RNA genome of the new virus.

FIGURE 3-2 Human and Retroviral Flow of Genetic Information. The general directional flow of genetic information in all living species is from DNA, where the information is stored, into RNA, which serves as a messenger for the construction of proteins which are the cells' functional molecules. This unidirectional flow of genetic information has been referred to as the **"central dogma"** of molecular biology. In the 1960s, Howard Temin and colleagues discovered an enzyme that copied RNA into DNA, a reverse of what was normally expected, thus the name *reverse transcriptase.*

sualized by electron microscopy. The GAG gene has the ability to direct the formation of virus-like particles when all other major genes (POL and ENV) are absent. It is only when the GAG gene is nonfunctional that retroviruses (HIV) lose their capacity to bud out of a host cell. Because of these observations, the GAG protein has been designated the virus particle-making machine (Wills et al., 1991).

The **POL** gene codes for HIV enzymes, protease (p10), the virus-associated polymerase (reverse transcriptase) that is active in two forms, and endonuclease (integrase) enzymes. The integrase enzyme cuts the cell's DNA and inserts the HIV DNA. Evidence from retroviral deletion studies shows that the loss of LTRs on the 3' side of the POL gene stops viral DNA integration into the host genome. However, nonintegrated DNA, without its

LTRs and integrase enzyme, can still produce new viruses. This clearly demonstrates that viral DNA integration is not essential for viral multiplication even though integration is the normal course of events (Dimmrock et al., 1987).

The regulation of HIV transcription appears to be intimately related to the onset of HIV disease and AIDS. Thus interruption or inactivation of the POL gene would appear to have therapeutic effects (Kato et al., 1991).

The ENV gene codes for HIV surface proteins, two major envelope glycoproteins (gp120, located on the external "spikes" of HIV and gp41, the transmembrane protein that attaches gp120 to the surface of HIV) that become embedded throughout the host cell membrane, which ultimately becomes the **envelope** that surrounds the virus as it "buds" out

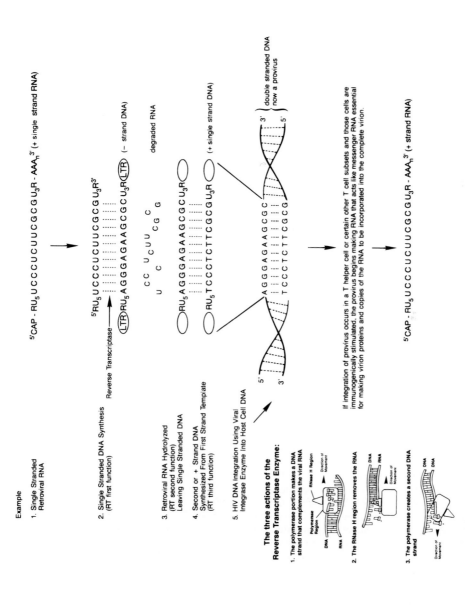

FIGURE 3-3 Proposed Production and Function of Retroviral Provirus. Note the reverse transcriptase enzyme has three functions: (**1**) to act as an RNA-dependent DNA polymerase transcribing single-strand DNA from viral RNA; (**2**) to demonstrate RNase H activity (RNase H is a subunit of the RT enzyme) by hydrolyzing the retroviral RNA off the RNA–DNA complex; and (**3**) to act as a DNA-dependent polymerase and transcribe the second DNA strand complementary to the first DNA strand. The process of viral RNA transcription is complex. When completed, the viral RNA gives rise to the formation of either linear or circular molecules of proviral DNA. Each end of the provirus contains an identical long series of terminal-repeating nucleotides or LTRs. LTRs are not a part of the viral genome. Although retroviral DNA integration is considered to be the normal route for RNA virus reproduction, retroviral reproduction may occur without proviral integration (see Figure 3-4).

63

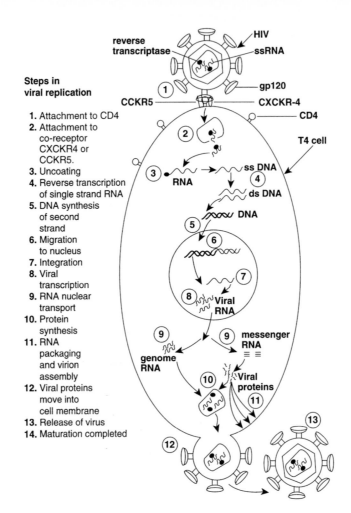

Steps in viral replication

1. Attachment to CD4
2. Attachment to co-receptor CXCKR4 or CCKR5.
3. Uncoating
4. Reverse transcription of single strand RNA
5. DNA synthesis of second strand
6. Migration to nucleus
7. Integration
8. Viral transcription
9. RNA nuclear transport
10. Protein synthesis
11. RNA packaging and virion assembly
12. Viral proteins move into cell membrane
13. Release of virus
14. Maturation completed

FIGURE 3-4   Life Cycle of HIV in a T4 Lymphocyte. On average, the life cycle of HIV in an infected T4 cell is 1.2 days. The lifetime of HIV in the blood is about 7 hours. After fusion of viral and cellular membranes, the inner part of HIV, called the **core**, is delivered into the cell cytoplasm. Uncoating of the core occurs, releasing two identical RNA strands, accompanying structural proteins and enzymes necessary for HIV replication. HIV RNA is then copied into HIV DNA which is then transported into the cell's nucleus for insertion into the host cell's DNA. **This feature of the virus life cycle is essential for the spread of HIV in vivo, because it allows infection of nondividing cells such as monocytes and terminally differentiated macrophages and dendritic cells.** There is still some confusion as to whether HIV becomes a latent infection once HIV DNA becomes inserted or integrated into the host DNA. Evidence from 1994 and 1995 indicates that whether HIV is latent depends on the tissue that one is investigating. For example, in the T4 lymphocytes within the lymph nodes HIV is constantly being replicated, while some T4 lymphocytes in the blood carry HIV in the latent state. Depending on cell type, T4 or macrophage, chemokine coreceptors CXCR4 (FUSIN) and/or CC-CKR-5 are used by HIV to enter T4 and macrophage cells respectively. (See Chapter 5 for details on chemokine receptors and HIV cell entry.)

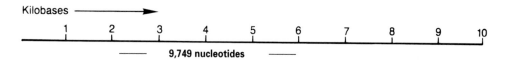

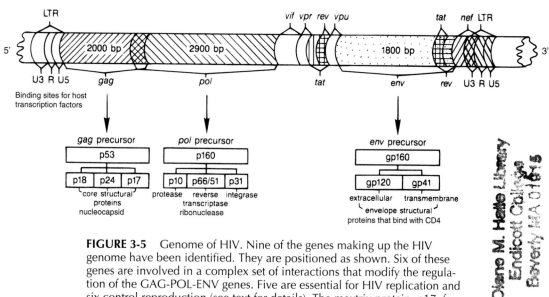

FIGURE 3-5 Genome of HIV. Nine of the genes making up the HIV genome have been identified. They are positioned as shown. Six of these genes are involved in a complex set of interactions that modify the regulation of the GAG-POL-ENV genes. Five are essential for HIV replication and six control reproduction (see text for details). The maxtrix protein, p17, forms the outer shell of the core of HIV, lining the inner surface of the viral membrane. Key functions of p17 protein are: orchestrates HIV assembly, directs gag p55 protein to host cell membrane, interacts with transmembrane protein, gp41, to retain envelope coded proteins within HIV and contains, a nuclear localization signal that directs HIV RNA integration complex to the nucleus of infected cells. **This feature permits HIV to infect nondividing cells, a distinguishing feature of HIV** (Matthews et al., 1994).

(Figure 3-6). Studies on how HIV kills cells have revealed at least one way that the envelope glycoproteins enhance T helper cell death. The envelope glycoproteins cause the formation of **syncytia; that is, healthy T cells fuse to each other forming a group around a single HIV-infected T4 cell.** Individual T cells within these syncytia lose their immune function (Figure 3-7). Starting with a single HIV-infected T4 helper cell, as many as 500 *uninfected* T4 helper cells can fuse into a single syncytium. Continued creation of these syncytia could deplete a T4 cell population.

Several studies have demonstrated that the appearance of syncytium-inducing (SI) HIV strains during the chronic phase of HIV disease heralds an abrupt loss of T4 lymphocytes and a clinical progression of the disease. Although the SI phenotype is detected in many patients with AIDS, it has also been isolated from individuals who have not gone on to an early development of AIDS. However, the marked decrease in the T4 lymphocyte counts after shifting from non-SI (NSI) to SI strains as well as the negative effect SI strains have on primary HIV infection suggests that their appearance may not be just a consequence, but the actual cause of immune system alterations (Torres et al., 1996).

The Six Genes of HIV That Control HIV Reproduction— Collectively, the six additional HIV genes tat and rev (regulatory genes),

TABLE 3-1   Genes of the HIV-1

Name(s)	Molecular Mass (kDa)	Function
GAG (group antigen)	p17	Matrix (protection)
	p24	Capsid (protection)
	p9, p6	Nucleocapsid (protection)
POL (polymerase)	p10	Protease (enzyme)
	p31	Endonuclease (enzyme)
	p50	Reverse transcriptase (enzyme)
ENV (envelope)	gp120	Surface envelope
	gp41	Transmembrane envelope
tat (transactivator protein)[a]	p14	Transactivation of all HIV proteins
rev (differential regulator of expression of virus protein)	p13	Increase production of structural HIV Protein; Transport of spliced/unspliced RNA from nucleus to cytoplasm
vif (virus infectivity factor)	p23	Required for ineffectivity as cell-free virus
nef (negative regulator) factor)	p27	Retards HIV replication
vpr (virus protein R)	?	Triggers cells steroid production to produce HIV
vpu (virus protein U)	p16	Required for efficient viral replication and release-budding

[a] Transactivator gene—the product of their genes influences the function of genes some distance away.

and nef, vif, vpu, vpr (auxiliary genes) working together with the host cell's machinery actually control the reproductive retroviral cycle: adhesion of HIV to a cell, penetration of the cell, uncoating of HIV genome, reverse transcription of the RNA genome producing proviral DNA and immediate production of new viral RNA, or the integration of the provirus and later viral multiplication. The six genes allow for the entire reproductive scenario—from infection to new HIV—to occur in 8 to 16 hours in dividing cells.

Gene Sequence— The HIV proviral genome has been well characterized with regard to gene location and sequence (Figure 3-6), but the function of each gene is not completely understood. The genes for producing regulatory proteins can be grouped into two classes: genes that produce proteins essential for HIV replication (*tat* and *rev*), and genes that produce proteins that perform accessory functions that enhance replication and/or infectivity (**nef, vif, vpu,** and **vpr**) (Rosen, 1991).

Gene Function— Each end of the proviral genome contains an identical long sequence of nucleotides, the long terminal repeats.

Although these LTRs are not considered to be genes of the HIV genome, they do contain regulatory nucleotide sequences that help the five regulatory genes control GAG-POL-ENV gene expression (Figure 3-5). For example, it is known that the **vif** gene is associated with the infectious activity of the virus. Currently, the predominant view is that Vif acts at the late stages of infection to promote HIV processing or assembly (Potash et al, 1998).

The **tat** gene is essential for HIV infection of T4 cells and HIV replication (Parada et al., 1996; Li, 1997; Stevenson, 1998). It is one of the first vital genes to be transcribed. The tat gene produces a transactivator protein, meaning that the gene produces a protein that exerts its effect on viral replication from a distance rather than interacting with genes adjacent to *tat* or their gene product. *Tat* contains two coding regions or exons—areas that contain genetic information for producing a diffusible protein—which, through the help of the LTR sequences, increases the expression of HIV genes thereby increasing the production of new virus particles. The tat protein interacts with a short nucleotide sequence called TAR located within the 5′ LTR region of HIV

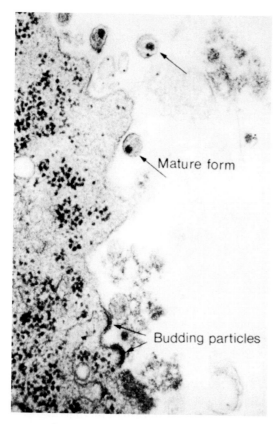

Mature form

Budding particles

FIGURE 3-6 Budding and Mature Retroviruses. This is a photograph of HIV taken by electron microscopy. Note the difference between the free or mature HIV and those that are just budding out through the membrane of a T4 helper cell. This cell came from an HIV-infected hemophiliac. Closely observing the mature HIV, one can make out the core protein area surrounded by the cell's membrane (virus envelope). (*Courtesy CDC, Atlanta*)

messenger RNA (mRNA) transcripts (Matsuya et al., 1990). Once that tat protein binds to the TAR sequence, transcription of the provirus by cellular RNA polymerase II accelerates at least 1,000-fold.

The **rev** gene (**r**egulator of **e**xpression of **v**iral protein) selectively increases the synthesis of HIV structural proteins in the latter stages of HIV disease, thereby maximizing the production of new viruses. It does this by regulating splicing of the HIV RNA transcript (cutting out nucleotide sequences that exist between exons and bringing the exons together) and

transporting spliced and unspliced RNAs from the nucleus to the cytoplasm (Patrusky, 1992; Fritz et al., 1995).

The **nef** gene (negative regulator protein) produces a protein that is maintained in the cell cytoplasm next to the nuclear membrane. It is believed **nef** functions by making the cell more capable of producing HIV. Several antigenic forms of nef protein have been found which suggest multiple activities of nef within HIV-infected cells (Kohleisen et al., 1992; Sagg et al., 1995). Olivier Schwartz and co-workers (Cohen, 1997) showed that **nef** can prompt cells to yank down from their surfaces a molecule known as the major histocompatibility complex (MHC), which displays viral peptides to the immune system. The group predicted that this "down-regulation" of MHC would make HIV-infected cells resistant to cytotoxic T cell killing. (Collins et al., 1998) (See Chapter 5.)

The functions of the **vpr** gene, which codes for a **v**iral **p**rotein **R**, is associated with the transport of cytoplasmic viral DNA into the nucleus. Vpr is also involved in steroid production which in turn helps produce HIV, is required for the efficient assembly or release of new HIV viruses, and stops T4 cell division. **Vif** is necessary for complete reverse transcriptions of viral RNA into viral DNA. **Vpu** codes for a viral protein U that destroys the CD4 protein within the T4 lymphocyte, thus helping HIV to bud out of the cell. Vif, Vpr, and Vpu appear to be necessary for HIV to cause disease.

Of the HIV proteins, Tat, Rev, and Nef are termed **early** proteins, because their production results from the cutting and splicing of full-length HIV mRNA; Vif, Vpu, and Vpr are termed **late** proteins, since their production results from unspliced or single-spliced mRNA.

How Genes Store Genetic Information and the Importance of Mutations or Change Within Genes

Genes store the information necessary for creating living organisms. The information is stored in the form of DNA organized into structures called chromosomes. Apart from sex cells (eggs and sperm) and mature blood cells, every cell in the human body contains 23 pairs of chromosomes. One of each pair is inherited from the mother, the other from the

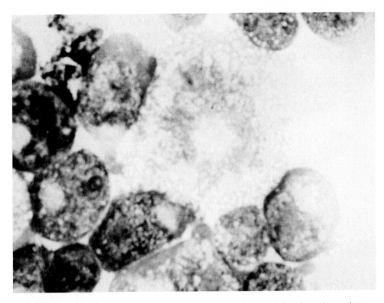

FIGURE 3-7 Formation of T4 Helper Cell Syncytia. A single infected T4 helper cell can fuse with as many as 500 uninfected T4 helper cells. The formation of syncytia lead to a loss or depletion of T4 helper cells from the immune system. (*Courtesy of Tom Folks, National Institute of Allergy and Infectious Diseases*)

father. Each chromosome is a packet of compressed and twisted DNA. Genes are sections of DNA containing the blueprint for the whole body, including such specific details as what kind of receptors cells will have, for example, CD4, CD8, X4, R5 and so on.

DNA is made up of a double-stranded helix held together by hydrogen bonds **between specific pairs of bases. The four bases A, T, G, and C (adenine, thymine, quanine,** and **cytosine)** bond each other in fixed and complimentary patterns that give humans and other species their individuality. A substitution in these letters (bases) replacing, for example, adenine for thymine, or a loss of a base alters the previous DNA base sequence and results in a genetic mutation or change. If a gene is thought of as a sentence, and the nucleotides in DNA as letters, a change or mutation of only one letter can affect the entire sentence or the information the DNA gives the cell. To get an idea of how many mutations can occur in a cell, consider that humans have about 3.2 billion base pair DNA letters which make up

about 50,000 to 100,000 genes or sentences that contain the information that makes a human. There are about a million differences between your 3.2 billion-letter DNA alphabet and that of another person. And the number of differences between the number and sequence of nucleotides (bases) in human DNA and that of other species is much greater. Another way of saying this is that the closer a species resembles a human or vice-versa, the closer will be the DNA base sequences.

Genetic Stability of Species— The individual or collective characteristics (phenotypes) of any virus, cell, or multicelled organism depend on the expression of their genes and the interaction of gene products within a given environment. From a biological point of view, changes in phenotype (observable characteristics) that are inheritable are by definition genetic changes. Such changes occur due to changes in the hereditary material. Changes in hereditary material may occur by nucleotide addition, substitution, and deletion.

These changes are referred to as **genetic mutations.** Genetic mutations provide biological heterogeneity and genetic diversity (similarity as a species but dissimilarity with regard to certain characteristics). Genetic diversity results from slow but ever-present mutations that alter the phenotype. Investigations on the rate at which genetic mutations occur in living species indicate that DNA is a stable molecule with relatively low mutation rates for any given gene. Because of low mutation rates within the DNA of a species' gene pool (all of the genes that can be found in the DNA of a species) and selection pressures by a slowly changing environment, species evolution is constant but very slow.

Genetic mutations in the strain of an organism or virus produce genetic and phenotype **variants** (different members) of that strain. Regardless of the rate at which mutations occur, they are genetic mistakes—they are not intentional, they just happen by chance. Most mutations or genetic mistakes either make no difference to an organism or virus (silent mutations) or they cause a change. **Few genetic mistakes within a stable environment improve the species. After all, the species arrived at this point in time via genetic and environmental selection pressures— those with the best constellation of genes survived to reproduce those genes.** In species that produce large numbers of offspring, genetic mistakes that are lethal or lead to an early death are of little consequence to the species. A genetic mistake that improves the chance of survival and reproduction is retained.

Genetic Stability of HIV— A virus like HIV can produce hundreds of replicas within a single cell. Genetic mistakes during viral replication produce variant HIVs. In biological economic terms, HIV replicas are inexpensive to make. Even if most of these mutant HIV replicas are inactive or throwaway copies, it makes little if any difference to the HIV per se. However, if a few HIV replicas received environmentally advantageous mutant genes, these HIV mutants would survive as well as or better than the parent HIVs. **Both parent HIV and mutant HIV can reproduce in the same cell and exchange genes.** Over time, if only the mutant or variant HIVs were transmitted to other people and undergo still further genetic changes, these variant HIVs could, with sufficient accumulative genetic changes, become a new subtype of HIV—for example, an HIV-3 strain.

Investigations of some of the *RNA* viruses revealed relatively high mutation rates. Thus some of the RNA viruses are our best examples of evolution in **"real"** time. Because of their high error rate during replication they show, as expected, both high genetic diversity and biological heterogeneity in their host, and a rate of evolution about a million times faster than DNA-based organisms (Nowak, 1990). HIV, in particular, fits this category. Heterogeneity of HIV is reflected by: (**1**) the difference in the kinds of cells variant HIV infects; (**2**) the way different HIV mutants replicate; and (**3**) the way different variants of HIV harm infected cells.

It is now known that HIV is capable of enormous genetic flexibility, which allows it to become resistant to drugs, to escape from immune responses, and to be nonsusceptible to an HIV vaccine. What is not known are all the factors contributing to viral diversity in individual infections. Clearly the high error rate of the reverse transcriptase (because this enzme lacks a 3' to 5' exonuclease proofreading ability, i.e., it cannot correct mistakes once made) and the high turnover rate in infected cells generate vast numbers of different virus mutants. The diversity of newly produced variants, however, is shaped by a combination of mutation, recombinations of HIV RNA, and selection forces. The main selective forces that have been proposed to drive HIV diversity are the immune response, cell tropism, (cell types most likely to attract HIV) and random activation of infected cells. At the time of seroconversion a person may carry a homogeneous virus population, but then diversification occurs as HIV infects many different cell types and tissues in the body (Bonhoeffer et al., 1995).

For further information on HIV evolution, diversity, and HIV disease progression, see Chapter 7, Box 7.2.

GENETIC MUTATION

One mechanism for producing HIV variants or HIV mutants results from a **highly error-prone reverse transcriptase** (RT) enzyme of HIV. Preston and colleagues (1988) found that HIV transcriptase makes at least one replication error (mutation) in each HIV genome per round of replication in a cell! Other investigators say that reverse transcriptase, on average, introduces a replication mutation once in every 2,000 nucleotides or about five mutations per round of replication! This unusually high rate of nucleotide misincorporation (substitution, addition, and deletion of nucleotides) as proviral DNA is being made by RT is responsible for generating the genetic diversity and heterogeneity found among the isolates of HIV. In short, the high error rate in producing proviral DNA means that each HIV-infected cell will carry variant or mutant HIV, most of which will be genetically unique (Vartanian et al., 1992). **This high rate of mutation underlies HIV's remarkable ability to become resistant to drug therapies and to hamper the production of an HIV vaccine!**

GENETIC RECOMBINATION IN HIV

HIV is diploid in that it carries two identical RNA molecules that can individually undergo genetic mutation. They are joined together in parallel at one end (5′ end). This RNA strand relationship makes it possible for genetic (RNA) recombination between these two strands and with other RNA strands in the area. Genetic recombination among and between these RNA strands contributes to the remarkable genetic variability of HIV. Thus as new RNA copies accumulate in the host cell cytoplasm, they recombine with each other (exchange parts or participate in RNA strand switching), producing new varieties of HIV (Levy, 1988).

Completing the Picture on Producing HIV Genetic Variability

Both mutation and recombination or RNA strand "mating" contribute, perhaps equally, to the variety of HIV subtypes. Recombination also appears responsible for the rapid emergence of certain subtypes or **clades** within group M. Subtypes A, B, C, and D are distinct throughout their genomes. But type E, prevalent in southeast Asia and India and thought to be particularly capable of transmission heterosexually, appears to be a **recombinant of a primordial E with part of the long terminal repeat (LTR) sequences and the gp120 envelope gene of subtype A.** Identical recombination breakpoints were found in five type E isolates, four from Thailand and one from the Central African Republic.

At least one of every 10 HIV patient isolates are intersubtype recombinants. The true incidence is unknown, as the isolates studied were not obtained at random, nor were genomes completely sequenced. Some strains may be multiple recombinants; the early Zairian isolate HIV-1 MAL is an A/D recombinant. Some HIV recombinants have been spread worldwide as, for example, the recombinant African subtype G was responsible for a large hospital outbreak in Russia following the use of inadequately sterilized needles.

Recombination is probably facilitated by geographic intermixing of subtypes, co-infection of an individual with different subtypes, and co-infection of cells with different susceptibilities for different subtypes (Laurence, 1997).

Michael Saag and colleagues (1988) examined the generation of molecular variation of HIV by sequential HIV isolates from two chronically infected people. They found 39 distinguishable but highly related genomes (HIV variants). **These results indicate that HIV heterogeneity occurs rapidly in infected individuals; and that a large number of genetically distinguishable but related HIVs rapidly evolve in parallel and coexist during chronic infection.** That is, whenever a drug or the immune response successfully attacks one variant, another arises in its place. Pools of genetic distinct variants that evolve from the initial HIV that begin the infection are often referred to as **quasispecies** (Delwart et al., 1993; Diaz et al., 1997).

Additional evidence indicates that some HIV genetic variants demonstrate a preference (tropism) as to the cell type they infect. This means that one genetic change may allow the virus to enter cells that were once immune to the virus. The rapid genetic change which results in altered viral products makes it very difficult to design a vaccine or drug that will be effective against all HIV variants. To date, HIV drug-resistant mutants have been found for all FDA-approved nucleoside and non-nucleoside analogs and protease inhibitors used in the treatment of HIV-infected people and AIDS patients! (Chapter 4 presents a discussion of currently used drug therapies for HIV/AIDS.)

HIV Antigenic Variation— This is the product of the mutations that occur in the GAG-POL-ENV genes. Immunological investigations on GAG and POL gene proteins show that these two genes are relatively stable. That is, although mutant GAG and POL gene proteins have been found in different viral isolates, the amount of variation for these gene products appears to be minimal. Antibodies are made against the GAG-POL-ENV gene proteins, but it is the antibodies made against the ENV gene protein that appear to be the most important. These antibodies neutralize the envelope glycoproteins that seem to be an essential part of HIV's infecting process. **However, it is the ENV gene that is subject to frequent mutations, producing HIV with different envelope glycoproteins within a given individual.** AIDS investigators Mayer-Cheng, E.V. Fenyo, K. Dehurst, and N. Tersmette report that virus isolates obtained from patients with advanced AIDS contain HIV that is more cytopathic than the HIV isolated earlier in the course of the disease and that these more cytopathic strains of HIV have demonstrated host cell tropism. HIV isolated from asymptomatic people normally infects lymphocytes and monocytes that carry the CD4 receptor, but the more cytopathic HIV is capable of infecting and replicating in a variety of **non-CD4 cell types** in the brain, gastrointestinal tract, kidney, and other tissues (Nowak et al., 1991).

It now appears that **HIV can constantly change its surface antigenic composition, thereby allowing it to escape antibody neutralization.** This immune selection of HIV mutants allows the virus to persist in the presence of an immune response. This immune selection viral phenomenon is not new. It is what the influenza virus does yearly so that last year's vaccine will not protect people from this year's variation. However, HIV differs from viruses such as influenza. Influenza and other viruses do not have an RNA to DNA replication step so they are not as mutable as the retroviruses. Because of the error-prone reverse transcriptase enzyme used by the retroviruses, the possibility of genetic change far exceeds that for any other known nonretroviral human pathological virus.

DISTINCT GENOTYPES (SUBTYPES) OF HIV-1 WORLDWIDE BASED ON ENV AND GAG PROTEINS

Because HIV proteins change so rapidly, it is hard to destroy or neutralize them with a single drug or single vaccine. In order to better understand globally circulating strains of HIV, HIV investigators have placed, based on the genetic diversity, the various HIV strains into two major groups, **M,** for the **main** HIV genotypes or **clades** found in different populations and **O** for HIV genotypes or clades that are significantly different from those in the **M** group.

Group M HIV causes over 99% of the world's HIV/AIDS. **Group O** and newly discovered **Group N** (discussed below), less than 1%.

Group M Subtypes

There are 10 subtypes or clades of group M: A, B, C, D, E, F, G, H, I, and J (Table 3-2). The M subtypes have been analyzed with respect to the differences between them based on the variations found in their GAG and ENV proteins (Robertson et al., 1995).

The **10** different M subtypes or clades and O have been globally mapped (geographically located) in Figure 3-8 (Brix et al., 1996; Workshop Report 1997). This figure demonstrates the dissemination of the different subtypes from 1990 through 1997. Clearly, subtype B predominates in North America. The explosive epidemic in Southeast Asia is chiefly attributable to subtype E. There are ample documented introductions of subtype E from endemic areas into the United States and the Western Hemisphere.

TABLE 3-2 Worldwide Distribution of the More Dominant HIV-1 Subtypes (Clades)

Region	Predominant Subtype(s)
U.S.	B
Asia	E, C
South America	B
Africa	A, C, D
Europe	B
Southeast Asia	B
India	C

The classification of HIV into subtypes is provisional and only reflects those isolates that happen to have been collected and characterized. Current classification is primarily based on parts of *env* or *gag*, proteins, the degree to which taxonomies using these areas of the genome accurately reflect viral evolution remains uncertain. Present taxonomic classifications will be modified as more viral isolates are characterized. (Hu et al., 1996). **Also see Figure 3-8.**

International Genetic Subtypes

Numerous groups have shown an increase in intermixing of HIV genetic subtypes in multiple international locales. HIV subtypes A, B, C, D, F, and G have appeared throughout Europe; B, F, and C have been found in Brazil; one-third of new infections in Germany are non-B; 15% of infections in a New York City hospital are non-B; increasingly, E is found among injecting drug users in Thailand, from 24% subtype E in 1991 to 44% in 1995, and among heterosexual risk groups, a change from 86% subtype E in 1991 to 98% in 1995; and group O has been detected in seven African countries, France, Spain, and in the United States, as well as M/O dual infections. In South Africa, there is spreading of subtypes B and D throughout the homosexual population and C/E throughout the heterosexual cohorts studied. The impact of this is an increasing spread of non-B subtypes into previous "B" areas, which has significant implications for vaccine development.

HIV RNA Recombination

Recombination, the intermixing of viral RNA strands, mediated by HIV reverse transcriptase leads to abrupt and extensive genomic shifts, which generates increased diversity and may lead to increased fitness for HIV. Researchers have shown that HIV E and G are not distinct subtypes, but rather represent recombinant viruses between subtype E and G with subtype A. Infection of a chimpanzee with a second HIV strain from a different subtype resulted in recombinants between the two strains. The impact of recombination is highly relevant for vaccine development and demonstrates a need for cross-protection among subtypes and raising the concern of recombination increasing infectivity or enhancing immune escape (Brix et al., 1996).

Robertson and colleagues (1995) also reported that eight of the group M subtypes appear to have been involved in recombination with each other, giving rise to genetic hybrids. **These studies raise questions of whether a vaccine that works against one of the subtypes, will work against a subtype hybrid.** Gerald Meyers has been tracking HIV variants for several years. To do this he has sequenced the nucleotides of the GAG gene. From his analysis he found five major genotypes of HIV subtypes worldwide. Each genotype differs from the other by about 35%. Other families of subtypes are likely to appear over time. Meyers believes that all five families arose from a common viral HIV source and the differentiation of one HIV into another began around 1960 (Goldman, 1992). Two of the five families have been found in Thailand. One family, genotype E, infects heterosexuals almost exclusively. According to Luis Soto-Ramirez and colleagues (1996) subtype E grows more efficiently than other HIV subtypes in Langerhans' cells which are found on the surface of the vagina and represent an initial cell contact for source of HIV infection. Genotype B and E are found in injection drug users (Moore, 1994). Joost Louwagie and colleagues (1993) sequenced the GAG gene from 55 international HIV isolates from people in 12 countries on four continents. Their work resulted in finding seven separate and distinct HIV genotypes A, B, C, D, E, F, and G mentioned above. None of these genotypes were contained within the physical boundaries of one country. Genotype B was found globally. Genotypes A and D were found in a broad east to west belt across sub-Saharan Africa from

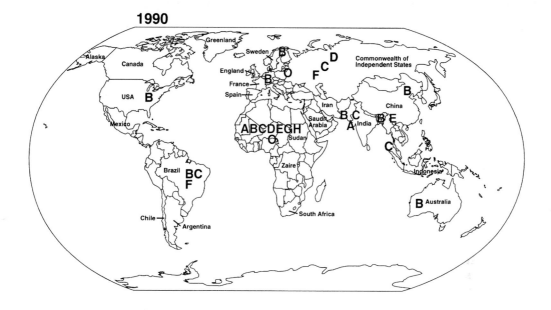

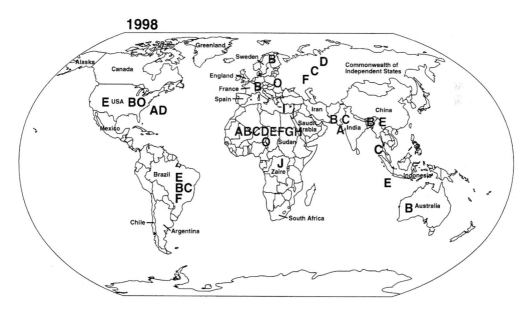

FIGURE 3-8 Global Distribution of the Eight **M** Subtypes and **O** in 1990 and Ten in 1998 in areas of highest prevalence. Clearly this global map shows that HIV subtypes are **no** longer continent based—all subtypes are now present on all continents. No satisfactory explanation exists for the skewed worldwide distribution of HIV-1 subtypes. The worldwide spread of viral subtypes makes it clear that a **world vaccine** will be required, that is, one that is not subtype specific. None appears to be on the horizon, although large-scale trials with existing envelope vaccines are under way in some developing countries (Brodine et al., 1997 updated).

Senegal to Kenya. Genotype C was found in a north-to-south pattern in Africa.

Worldwide, at least these 10 genetic subtypes exist, based on **envelope genetic sequence** (A → J) and **each subtype** differs at the nucleotide level from any other subtype by as much as 35%. Envelope sequences of isolates within a subtype can vary genetically by as much as 15%. Although there are geographic patterns of strain prevalence, multiple strains are found in many countries. A **single sequence envelope subtype** B is dominant in the United States, probably related to a strong **founder effect** (the initial subtype to become established in the United States). Subtype D, found almost exclusively in Africa's Lake Victoria region (Rwanda, Uganda, and Tanzania), appears to be the most rapidly lethal. Subtype E is found almost exclusively among heterosexual HIV-infected persons.

Group O Subtypes

Group O contains at least 30 genetically different subtypes of HIV. Group O subtypes are referred to as "**outlier**" because their RNA base sequence is only 50% similar to the known genotypes of the M group. The O variants have been known since 1987 but have been found mainly in Cameroon, Gabon, and surrounding west African countries that, to date, have only been marginally affected by AIDS. Group O viruses were of concern primarily because their divergence from group M was sufficient to miss their detection by ELISA HIV testing. (The ELISA test is discussed in Chapter 12.) Seven people in France were identified with Group O HIV in 1994. The **first** documented case of Group O HIV infection in the United States was found in April of 1996 in Los Angeles. Beginning year 2000, only two cases

──────── **POINT OF INFORMATION 3.2** ────────

MYSTERIES OF HIV SUBTYPES

Two epidemiologic mysteries have eluded HIV/AIDS researchers for years. The **first** is why heterosexual intercourse has accounted for approximately 10% of HIV transmission in the United States and Western Europe to date, but more than 90% of HIV transmission in Asia and Africa.

The **second** mystery has centered around a subtype B HIV epidemic among injection drug users in Thailand. This epidemic began in the mid-1980s, and plateaued with less than 100,000 people infected. Several years later, however, a subtype E HIV epidemic took hold in Thailand. This epidemic has exploded among heterosexuals, with over a million people already infected. Most surprising, perhaps, is that subtype B—while present in Thailand, India, and several African countries—has not caused heterosexual epidemics in those places.

Although differences in the rate of heterosexual transmission may also be due to factors such as sexual behavior and the presence of other sexually transmitted diseases, none of these considerations has yet adequately accounted for the widespread heterosexual epidemics in Asia and Africa. Investigators have begun to focus on how biologic characteristics of the individual subtypes might also play a role. These findings about subtype cellular affinities do not suggest that subtype B cannot be transmitted heterosexually—it is throughout the United States and elsewhere. The point is that subtype B seems to be transmitted through vaginal intercourse far less efficiently than the subtypes that predominate in Asia and Africa. It has been shown that Langerhans' cells are located in specific areas, such as the vagina, cervix, and penile foreskin. They do not appear to localize in the rectal or colon mucus membranes. Experiments with HIV-infected Langerhans' cells and uninfected T4 cells showed that HIV-infected Langerhans' cells readily formed clusters and syncytia with T4 cells, resulting in efficient cell-to-cell transmission. It was also demonstrated that during the cell-to-cell interaction, Langerhans' cells not only effectively infect T4 cells but also activate them, resulting in enhanced virus replication (UNAIDS, 1997). Thus, the enhanced efficiency of subtype E to attach to Langerhans' cells may help account for the explosive heterosexual spread of the epidemic in countries such as Thailand, while the heterosexual epidemic in the industrialized world has thus far remained at a comparatively low rate (Essex, 1996).

of HIV subtype O infection have been reported in the United States. This subtype was found in a Los Angeles county woman and a Maryland woman; both came into the United States from Africa (*MMWR*, 1996). HIV subtype O is not routinely tested for in the United States.

The rapid evolution of HIV families of variants has researchers abandoning hopes of developing a single vaccine effective against all variants. It may be feasible to concoct multiple-strain cocktail vaccines like those used in the flu vaccinations. (There are three major flu virus families.)

New Group N Subtype?

In August 1998, Francois Simon and colleagues reported the discovery of a **new** HIV-1 isolate that **cannot** be placed in the **M** or **O** subtype category. The authors suggest that the new isolate be classified as group **N** for "**new**" or "**non-M non-O.**" The new isolate, designated **YBF30,** was found in a small number of people in the West African Nation of Cameroon. The first isolate was from a 40 year old woman who died of AIDS. This virus is similar to SIV (cpz-gab) and it either branched with the SIV strain or between it and HIV-1 group M. The authors state that future strains of SIV (cpz) could be found and strains that are closely related to YBF30 might circulate among Cameroonian chimpanzees.

SIV retroviruses are endemic in many species of primates but usually don't cause disease, which makes them natural hosts for what are known as lentiviruses or slow-acting viruses. But when they cross the species barrier into a new host like humans, these viruses become pathogenic, causing illness. Based on evolutionary studies, researchers think SIV has existed in sub-Saharan Africa for thousands of years, adapting to several species including the African green monkey and sooty mangabey. There are six closely linked strains of SIV. Chimpanzees carry a viral predecessor of HIV-1, while sooty mangabeys carry an ancestor of HIV-2. The newly identified group N is likely to have mutated through generations of chimpanzees before crossing the species barrier into humans to cause AIDS. Now the critical questions are how widespread this virus is among humans and whether the virus will get more virulent as it spreads. Simon's discovery

——————— **POINT OF INFORMATION 3.3** ———————

FIRST CONFIRMED CASES OF HIV GROUP O CAUSED AIDS

A Norwegian sailor died of AIDS in 1976, at the age of 29. So did his wife and youngest daughter, born in 1967. The members of this Norwegian family represent the **earliest confirmed cases of AIDS and the first case of HIV type O infection.** The first symptoms appeared in 1966 in the sailor, in 1967 in his wife, and in 1969 in their daughter. Between 1961 and 1965 he traveled the world's oceans, calling at ports in all six inhabited continents. On his first voyage, which began in August 1961 just after his 15th birthday, he worked as a kitchen hand on a Norwegian vessel that sailed down the west African coastline, calling at ports in Senegal, Liberia, Cote d'Ivoire, Ghana, Nigeria and Cameroon. A gonorrheal infection during this trip shows that he was sexually active. He returned home in May 1962, and he never returned to Africa. No known evidence suggests that the sailor was bisexual, which means that sexual contact with a woman is the most straightforward explanation for his infection. This would suggest that HIV group O has been circulating in that part of Africa for at least 35 years.

SECOND CONFIRMED CASES OF HIV GROUP O CAUSED AIDS?

The second case of Group O infection found in the literature is the second child of a French barmaid from Reims, who died in 1981. The child's clinical history is highly suggestive of neonatal AIDS. In 1992 a group O virus was isolated from the mother, who by then had AIDS.

also raises other critical questions and challenges. Looking ahead, how many other SIV strains may have crossed the species barrier? Should we be mass screening for this new virus N or increasing surveillance of the other SIV strains? And what about the human recombinant viruses? How well are we tracking them?

SUMMARY

HIV is a retrovirus. It has RNA for its genetic material and carries reverse transcriptase enzyme for making DNA from its RNA. HIV, using its enzyme, copies its genetic information from RNA into DNA which becomes integrated into host cell DNA and may remain silent for years, or until such time as it is activated into producing new HIV. HIV contains at least nine genes; three of them, GAG-POL-ENV, are basic to all animal retroviruses. The six additional genes are involved in the infection process and regulate the production of products from the three genes. HIV, because of its error-prone reverse transcriptase enzyme, mutates at an unusually high rate. With time, many mutant HIV variants can be found within a single HIV-infected person. A vaccine against one mutant HIV may not work against a second—like the vaccines made yearly against different mutant influenza viruses.
A new group of HIV has been identified, Group N.

REVIEW QUESTIONS

(Answers to the Review Questions are on page 487.)

1. Why is HIV called a retrovirus?
2. What are the three major genes common to all retroviruses? How many additional genes does HIV have?
3. Why are retroviruses, and HIV in particular, believed to be genetically unstable? Give two reasons for your answer.
4. What is believed to be the major reason for the high rate of genetic mutations in HIV production?

REFERENCES

BONHOEFFER, SEBASTIAN, et al. (1995). Causes of HIV diversity. *Nature*, 376:125.

BRIX, DEBORAH, et al. (1996). Summary of track A: Basic science. *AIDS*, 10 (suppl. 3): S85–S106.

BRODINE, STEPHANIE, et al. (1997). Genotypic variation and molecular epidemiology of HIV. *Infect. Med.*, 14:739–748.

COHEN, JON. (1997). Looking for leads in HIV's battle with immune system. *Science*, 276:1196–1197.

COHEN, MITCHELL, et al. (1994). When bugs outsmart drugs. *Patient Care*, 28:135–146.

COLLINS, KATHLEEN, et al. (1998). HIV-1 Nef protein protects infected primary cells against killing by cytotoxic T Lymphocytes. *Nature*, 391:397–401.

DELWART, ERIC, et al. (1993). Genetic relationships determined by a DNA heteroduplex mobility assay: Analysis of HIV envGenes. *Science*, 262:1257–1262.

DIAZ, RICARDO, et al. (1997). Divergence of HIV quasispecies in an epidemiology cluster. *AIDS*, 11:415–422.

DIMMROCK, N.J., and PRIMROSE, S.B. (1987). *Introduction to Modern Virology*, 3rd ed. Oxford: Blackwell Scientific Publications.

ESSEX, MAX. (1996). Deciphering the mysteries of subtypes: communities of color, Spring/Summer: *Harvard AIDS Rev.*, 18–19.

FIELDS, BERNARD. (1994). AIDS: Time to turn to basic science. *Nature*, 369:95–96.

FISCHL, MARGARET. (1984). Combination retroviral therapy for HIV infection. *Hosp. Pract.*, 29:43–48.

FRITZ, CHRISTIAN, et al. (1995). A human nucleoprotein-like protein that specifically interacts with HIV-Rev. *Nature*, 376:530–533.

GOLDMAN, ERIK L. (1992). HIV-1 appears to have at least 5 distinct subtypes. *Fam. Pract. News*, 22:9.

GREENE, WARNER. (1993). AIDS and the immune system. *Sci. Am.*, 269:99–105.

HU, DALE, et al. (1996). The emerging genetic diversity of HIV. *JAMA*, 275:210–216.

KOHLEISEN, MARKUS, et al. (1992). Cellular localization of Nef expressed in persistently HIV-1 infected low-producer astrocytes. *AIDS*, 6:1427–1436.

LAURENCE, JEFFREY. (1997). HIV vaccine conundrum. *Aids Reader*, 7:1–2,27.

LENNOX, JEFFREY, (1995). Approaches to gene therapy. *International AIDS Society—USA*, 3:13–16.

LEVY, JAY A. (1988). Mysteries of HIV: Challenges for therapy and prevention. *Nature*, 333: 519–522.

LI, CHIANG. (1997). Tat protein perpetuates HIV-1 infection. *Proc. Natl. Acad. Sci. USA*, 94:8116–8120.

LOUWAGIE, JOOST, et al. (1993). Phylogenetic analysis of GAG genes from 70 international HIV-1 isolates provides evidence for multiple genotypes. *AIDS,* 7:769–780.

MATSUYA, HIROAKI, et al. (1990). Molecular targets for AIDS therapy. *Science,* 249:1533–1543.

MATTHEWS, STEPHEN, et al. (1994). Structural similarity between p17 matrix protein of HIV and interferon-2. *Nature,* 370:666–668.

MOORE, JOHN, et al. (1994). The who and why of HIV vaccine trials. *Nature,* 372:313–314.

Morbidity and Mortality Weekly Report. (1993). Nosocomial enterococci resistant to vancomycin—United States, 1989–1993, 42:597–599.

Morbidity and Mortality Weekly Report. (1996). Identification of HIV-1 group O infection–Los Angeles County, California. 45:561–565.

NOWAK, MARTIN A. (1990). HIV mutation rate. *Nature,* 347:522.

NOWAK, MARTIN A., et al. (1991). Antigenic diversity thresholds and the development of AIDS. *Science,* 254:963–969.

NOWAK, RACHEL. (1995). How the parasite disguises itself. *Science,* 269:755.

PATRUSKY, BEN. (1992). The Intron story. *Mosaic,* 23:20–33.

POTASH, MARY JANE, et al. (1998). Peptid inhibitors of HIV-1 protease and viral infection of peripheral blood lymphocytes based on HIV-1 ViF. *Proc. National Academy of Science,* 95:13865–13868.

PRESTON, BRADLEY D., et al. (1988). Fidelity of HIV-1 reverse transcriptase. *Science,* 242:1168–1171.

ROBERTSON, DAVID, et al. (1995). Recombination in HIV-1. *Nature,* 374:124–126.

ROSEN, CRAIG A. (1991). Regulation of HIV gene expression by RNA-protein interactions. *Trends Genet.,* 7:9–14.

SAGG, MICHAEL S., et al. (1988). Extensive variation of human immunodeficiency virus Type-1 *in vivo. Nature,* 334:440–444.

SAGG, MICHAEL S., et al. (1995). Improving the management of HIV disease. *Advanced Causes in HIV Pathogenesis,* pp. 1–30. February 25, Swissotel, Atlanta (Michael Sagg Program Chair).

SIMON, FRANCOIS, et al. (1998). Identification of a New Human Immunodeficiency Virus Type I Distinct from Group M and Group O. *Nature Medicine,* 4:1032.

SOMASUNDARAN, M., et al. (1988). Unexpectedly high levels of HIV-1 RNA and protein synthesis in a cytocidal infection. *Science,* 242: 1554–1557.

SOTO-RAMIREZ, LUIS, et al. (1996). HIV-Langerhans' cell tropism associated with heterosexual transmission of HIV. *Science,* 271:1291–1293.

STEVENSON, MARIO. (1998). Basic Science: Highlights of the 5th Retrovirus Conference. *Improving the Management of HIV Disease,* 6:4–10.

TORRES, YOLANDA, et al. (1996). Cytokine network and HIV syncytium-inducing phenotype shift. *AIDS,* 10:1053–1055.

UNAIDS. (1997). Inplications of HIV variability for transmission: Scientific and policy issues. *AIDS,* 11:1–15.

VARTANIAN, JEAN-PIERRE, et al. (1992). High-resolution structure of an HIV-1 quasispecies: Identification of novel coding sequences. *AIDS,* 6:1095–1098.

WILLS, JOHN W., et al. (1991). Form, function and use of retroviral Gag protein. *AIDS,* 5:639–654.

Workshop Report From The European Commission/Joint United Nations Program on HIV/AIDS. (1997). HIV-1 subtypes: Implications for epidemiology, pathogenicity, vaccines, and diagnostics. *AIDS,* 11:17–36.

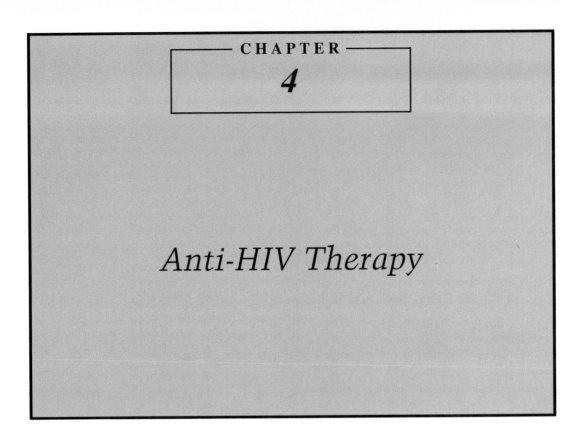

CHAPTER
4

Anti-HIV Therapy

CHAPTER CONCEPTS

♦ There may not be a cure for HIV disease.

♦ From March 1987 through early year 2000, 15 anti-HIV drugs received FDA approval and 2 other drugs have received expanded access approval.

♦ There are no anti-HIV drugs free of clinical side effects.

♦ All nucleoside analogs and non-nucleosides work by inhibiting the HIV reverse transcriptase enzyme which functions **early** in the HIV life cycle.

♦ All protease inhibitors work by inhibiting the HIV protease enzyme from its function **late** in the HIV life cycle.

♦ Viral load is the number of HIV RNA strands found at any one time in human plasma.

♦ Viral load is associated with HIV disease progression.

♦ Viral load is associated with perinatal HIV transmission.

♦ Pediatric anti-HIV therapy is improving.

♦ Anti-HIV drug combinations are extending lives.

♦ AIDS deaths dropping in developed nations due to drug combination therapy.

♦ When to begin anti-HIV drug therapy is the big question.

♦ Strict adherence to drug regimens is essential but is difficult to maintain.

♦ Salvage therapy, the kitchen sink approach—trying to save the patient.

♦ Dispensing current anti-HIV drug therapy requires an HIV/AIDS specialist.

♦ The **best** AIDS drug combinations are still unknown.

♦ Morning-after drug therapy to prevent HIV infection is questionable.

♦ Scientists do not expect the production of an effective anti-HIV vaccine before the year 2005.

We are constantly humbled by the devastation that something so small, HIV, can cast upon something so large, a human (Box 4.1, Figure 4-1). This chapter provides no final answers; there is no curative therapy, no truly

——— BOX 4.1 ———

DRUG THERAPY FOR WILLIAM "SKIP" BLUETTE

William "Skip" Bluette (continued from page 30) 3 weeks before he died. Treatment to maintain Skip was failing—the multiple opportunistic infections were defeating the best medicine had to offer.

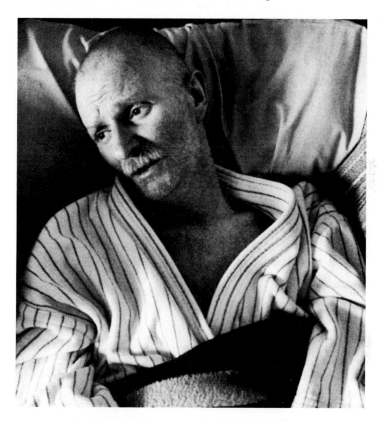

FIGURE 4-1 Drug Therapy for William "Skip" Bluette. Skip's treatment included amphotericin B, an anti-fungal drug, to combat meningitis. Starting in July 1986, he had to visit one of the hospital's clinics at least once a week to receive his treatments, which were given over several hours. Skip started taking zidovudine, a drug known to slow the progress of the disease, during the summer of 1987. Unfortunately, zidovudine also made it difficult for his body to produce blood cells. Skip stopped taking zidovudine when this occurred. In June of 1988, Skip started cleaning out his closets. He said it was part of the dying process to give belongings away "so you know where they're going." Partly, in jest, he said he wanted his ashes scattered on 42nd Street in New York—he thought that's where he "got AIDS."

He was hospitalized July 5, 1988. It was presumed that he had *Pneumocystis carinii* pneumonia, which Skip called "the killer." His breathing and speaking were labored. He had inflammation of the pancreas, for which he received morphine, and his kidneys began to fail. (*By permission of Mara Lavitt and* The New Haven Register)

outstanding therapies (drugs that benefit all HIV-infected without major side effects) against HIV, and, with the limited number of **expensive** anti-HIV drugs that are available, debate continues about the details on which drugs offer the best combination therapy and on the **standard of care** for HIV/AIDS patients. Because some 90% to 95% of HIV-infected people develop AIDS, and because 90% of all **new** HIV infections are occurring in developing nations, an **inexpensive, easily taken,** non-toxic, effective HIV-directed therapy is essential. Physicians need a drug that works against HIV as antibiotics once worked against a large variety of disease-causing bacteria.

GOALS FOR ANTI-HIV THERAPY

The ideal solution would be to prevent HIV from causing an infection. Then anti-HIV therapies would not be necessary. However, there are no means available to stop HIV from entering the body and infecting a limited number of cell types—primarily those cells displaying CD4 antigen receptor sites. Following HIV infection, there is a depletion of cells carrying CD4, especially the **T lymphocyte (T4) cells of the human immune system.** (See Chapter 5 for an explanation of cell types in the human immune system and their function.) With T4 cell loss, over time immunological response is lost. Loss of immunological response leads to a variety of opportunistic infections (OIs). The suppression of the immune system and increasing susceptibility to OIs and cancers give HIV/AIDS a multidimensional pathology. Because HIV/AIDS is a multidimensional syndrome, it is unlikely that a single drug will provide adequate treatment or a cure (Box 4.2, Figure 4-2).

An ideal goal for HIV drug investigators would be to find a drug that excises all HIV proviruses from the cell's DNA. It is very unlikely that this kind of drug will soon, if ever, be available. Alternatively, the **elimination** of all HIV-infected cells might be of comparable benefit, as long as irreplaceable cells are not totally lost through the process. In essence, this is the goal that the human immune system sets for itself, yet falls short of reaching, in the vast majority of HIV-infected individuals.

In the absence of a curative weapon (Box 4.3, Figure 4-3), therapies must be designed to prevent the spread of the virus in the body. Some of these anti-HIV therapies are presented in this chapter.

Can HIV Infection Be Cured?: NEVER SAY NEVER

An effective cure means cleansing the body of the HIV. To find a cure, HIV incorporation into human DNA and subsequent reproduction of new virus must be understood. Without interference, the **HIV provirus** will remain in host-cell DNA for life. The provirus can, at any time, become activated to mass produce HIV which leads to cell death and eventually to the individual's death. As of this writing, there is no way to prevent either the proviral state or proviral activation. However, with the use of combination anti-HIV drug therapy to suppress HIV replication, scientists now believe that if HIV replication can be suppressed long enough, all HIV-infected cells will die off and there will be a cure (see Point of View 4.2, Figure 4.4).

——————— **POINT OF VIEW 4.1** ———————

MEDICAL BENEFIT: A CURE?

Many diseases cause death. "We can cure those diseases," people say, "but we can't cure AIDS." Tell me, how many diseases can we cure? We cannot cure very many viral diseases, nor can we cure many of the diseases that kill people, like heart disease and forms of cancer. The concept of curing people is relatively new. The strength of physicians used to come from their capacity to accompany. Accompanying a person through an illness towards health or towards death was central to the practice of medicine, but it disappeared when the issue became one of curing.

—*Jonathan Mann,* in Thomas A. Bass, *Reinventing the Future: Conversations with the World's Leading Scientists* (Addison-Wesley, 1994)

——— BOX 4.2 ———

VIGIL FOR WILLIAM "SKIP" BLUETTE

The vigil for William "Skip" Bluette (continued from page 79) 48 hours before he died. Time and therapy were running out on Skip. His last wish was to have his sisters Arlene and Nancy near when he died and they were there. On July 15, 1988 the doctors told Nancy and her mother that Skip's kidneys had failed—this time Skip would not recover. He died on July 17. He was 42 years old.

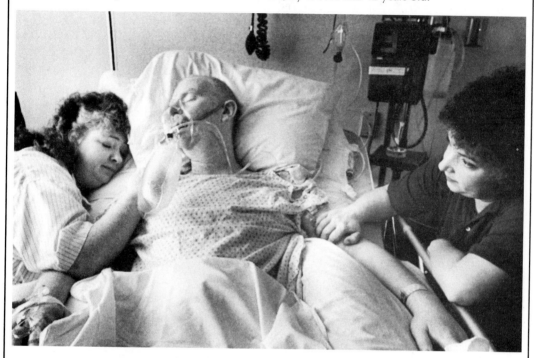

FIGURE 4-2 The Vigil for William "Skip" Bluette. The Bluette family rallied around Skip from the moment they found out he had AIDS. In April 1988, his mother had a heart attack but she survived it.

Skip celebrated his 42nd birthday on May 18; it was his second birthday since his diagnosis. (*By permission of Mara Lavitt and* The New Haven Register)

ANTI-HIV DRUGS HAVING FDA APPROVAL

The **gold standard** for determining the efficacy (effectiveness) of a new treatment is that it alters the disease in a way that is beneficial to the patient. Therefore, the end points most often used in clinical trials of therapies for a chronic disease such as HIV include prolongation of life or the extension of time to a significant disease complication.

But, studies using these end points require large numbers of patients and/or the passage of considerable amounts of time. **Surrogate markers** (i.e., physiological measurements that serve as substitutes for these major clinical events) can eliminate this problem if their validity and correlation with clinical outcome

WILLIAM "SKIP" BLUETTE: MAY 18, 1946–JULY 17, 1988

FIGURE 4-3   Free of Pain—One with the Universe—William "Skip" Bluette. On July 20, 1988 Skip's body was cremated. There was a memorial service. On July 21, Skip's ashes were buried in Evergreen Cemetery. (*By permission of Mara Lavitt and* The New Haven Register)

can be confirmed. The use of surrogates has the potential to shorten the duration of clinical trials and expedite the development of new therapies.

Major Surrogate Markers Used Evaluating Anti-HIV Drug Therapy

The **T4 or CD4+ immune cell number** and viral load are the two best studied and most commonly used surrogates for clinical efficacy of anti-HIV therapies. They are imperfect measurements however, because changes in T4 cell number per microliter of blood and RNA strands per milliter of blood are only partially explained by the therapy. And, T4 cell counts exhibit a high degree of day-to-day variation in individuals, and methods used to count these cells are difficult to standardize.

From March of 1987 through early 2000, 15 anti-HIV drugs have received United States FDA approval for use in persons infected with HIV and two others have been FDA approved for **expanded access use** where standard regimes have failed. (Table 4-1).

Six of the 14 FDA-approved anti-HIV drugs are nucleoside (nuk-lee-o-side) analogs (each drug resembles one of the four nucleosides that are used in making DNA) and three are non-nucleosides. All nine are reverse transcriptase inhibitors. The other six drugs are HIV-protease (pro-tee-ace) inhibitors.

TABLE 4-1 Anti-HIV Therapy: Nucleoside Analogs, Nonnucleosides, and Protease Inhibitors

Name	FDA Approved	Cost/Day[a]
Nucleoside analog (reverse transcriptase inhibitor)		
Adefovir dipivoxil (Preveon)[b]	Gilead Science	Expanded Access[c]
Zidovudine (AZT/ZDV/Retrovir)	March 1987	$7.96
Didanosine (ddI, Videx)	October 1991	$6.12
Zalcitabine (ddC, Hivid)	June 1994	$6.87
Stavudine (d4T, Zerit)	June 1994	$7.76
Lamivudine (3TC, Epivir)	November 1995	$7.68
Abacavir (Ziagen)	December 1998	$9.30

Nucleoside analog reverse transcriptase inhibitors are potent in combination with other drugs; used alone, they lead to HIV resistance. *AZT, d4T, 3TC,* and *abacavir* penetrate the blood-brain barrier. Common side effect: lactic acidosis. **Four** new nucleoside analogues are in some phase of testing in the USA.[b]

Non-nucleoside compounds; (non-nucleoside reverse transcriptase inhibitors)		
Nevrapine (Viramune)	June 1996	$8.26
Delavirdine (Rescriptor)	April 1997	$6.36
Efavirenz (Sustiva)	September 1998	$13.14

Non-nucleoside analog reverse transcriptase inhibitors (NNRTIs, or non-nukes) may interact with other cytochrome *p450-processed drugs:* protease inhibitors, oral contraceptives, etc. NNRTIs have a mixed ability to penetrate the blood-brain barrier. Common side effect: mild rash. Some doctors build up drug doses slowly to avoid rash; the other worry is that dose building increases risk of drug resistance. **Four** new nonnucleoside analogues are in some phase of testing in the USA.

Protease inhibitor drugs		
Saquinavir mesylate (Invirase)[d]	December 1995 (Hoffman-La Roche)	$21.60
Saquinavir (Fortovase)	November 1997	$22.00
Ritonavir (Norvir)	March 1996 (Abbott)	$22.20
Indinavir (Crixivan)	March 1996 (Merck)	$15.00
Nelfinavir (Viracept)	March 1997 (Aguoron)	$15.75
Amprenavir (Agenerase)	April 1999 (Glaxo Wellcome)	$16.80
ABT-378	Expected 1998 (Abbott)	Phase III testing
YX-478	? (Glaxo Wellcome/Vertex)	Phase III testing

Protease inhibitors (PIS) are very potent and may interact with other drugs using *cytochrome p450 metabolic pathways.* Potentially life-threatening if taken with Seldane, Hismanal, Propulsid, Halcion, or Versed. Avoid rifabutin, Nizoral, rifampin. Poor absorption may affect potency. Common side effects: liver toxicity, hypoglycemia, flatulence, bloating, lipodystrophy (fat distribution). **Six** new protease inhibitors are now in some phase of testing in the U.S.A.

Other Anti-HIV Drugs		
Hydroxyurea (Hydrea)		$2.84
Combivir (3TC+ZDU)		$15.64
T2O (Pentafuside)	Timeris	?

[a]Cost is based on wholesale price of 100-pill bottle or largest size available as listed in 1995 Red Book or based on information from the manufacturer, or Hospital Formulatory Pricing Guide, 1998.

[b]Adefovir is technically a *nucleotide* RTI, thus phosphorylation takes two steps.

[c]Expanded Access drugs are for patients failing standard regimes. Beginning year 2000 there are at least 2 new nucleoside and 4 non-nucleoside RTI and 2 PI drugs in early anti-HIV drug trials in the U.S.A. Telephone numbers for information concerning the use of the individual protease inhibitors are:

Saquinavir/Invirase	Hoffman-La Roche In./800-526-6367
Ritonavir/Norvir	Abbot Labs/800-633-9110
Indinavir/Crixivan	Merck & Co./800-672-6372
Nelfinavir/Viracept	Agouron/888-874-2237
Amprenavir/Agenerase	Glaxo Wellcome Inc./800-722-9292
VX-478/–	Glaxo Wellcome Inc./800-722-9292

[d]Saquinavir mesylate was the first nucleotide RTI in Phase III trials. Saquinavir mesylate, ritonavir, and indinavir were approved in 97, 72 and 42 days, respectively. Abacavir is the first 2-deoxyguanosine analog reverse transcriptase inhibitor.

THE END OF AIDS?

This title was placed in 2¾-inch block print on the December 2, 1996 cover of *Newsweek* Magazine (Figure 4-4)! In print ⅜ inch high over to the left side it said, "**not yet—but new drugs offer hope.**" But then, of the many, many thousands of people who walk by the newsstands or read *Newsweek*, how many ever see the small print when such bold print declares "**THE END OF AIDS.**" Clearly, sensationalism sells copy, but what of the **media's moral responsibility?** How many seeing this cover message decided to change their HIV preventive behavior? How many decided that their energies were no longer needed in the fight against AIDS? How many felt their financial, emotional, and physical contributions are no longer necessary? In short, it is **noble** to offer **hope,** but **irresponsible** to offer or suggest myth! Let's **hope** this cover will become a reality soon, but wait until it does before making such questionable promises in a lay publication for millions of readers.

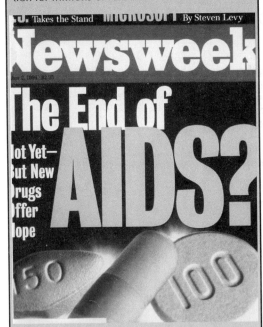

FIGURE 4-4 AIDS CURE? Although this bold suggestion was premature for 1996, the new anti-HIV drugs have given the HIV-infected great hope—they can now contemplate a future. (*Newsweek*-© 1996 John Rizzo, Newsweek, Inc. All rights reserved. Reprinted by permission.)

FDA-APPROVED NUCLEOSIDE ANALOG REVERSE TRANSCRIPTASE INHIBITORS

All of the known mechanisms by which HIV impairs the human immune system depend on HIV reproduction. Therefore, the development of anti-HIV replication drugs would appear to be a positive **first step** in controlling HIV reproduction.

Each of the six nucleoside analogs, sometimes referred to as **nukes** (Figure 4-5), on entering an HIV-infected cell interferes with the virus' ability to replicate itself (Figure 4-6). That is, **when any of the six nucleoside analogs are incorporated into a strand of HIV DNA being newly synthesized, it stops further synthesis of that DNA strand.** The nucleoside analog stops the HIV enzyme, reverse transcriptase, from joining the next nucleoside into position. The six FDA-approved nucleoside analogs are all members of the family of 2′,3′-dideoxynucleoside analogs. By name they are **zidovudine** (ZDV), often **mistakenly called AZT (3′-azido-2′,3′-deoxythymidine**), trade name **Retrovir; didanosine** or **ddI (2′,3′-dideoxyinosine**), trade name **VIDEX; dideoxycytosine,** zalcitabine or **ddC (2′,3′- dideoxycytosine**), trade name **HIVID; stavudine** or **d4T,** trade name **Zerit; lamivudine** or **3TC,** trade name **Epivir,** and Abacavir, trade name Ziagen. (Table 4-1).

Focus on the Six Nucleoside Analogs

Each of the six drugs (Figure 4-5) has limited effectiveness as a **monotherapy.** The principal limitations are: **(1)** they are not 100% effective in stopping HIV reverse transcriptase from making HIV DNA; **(2)** positive clinical effects are short term, they are not sustained; **(3)** each drug has its own set of toxic side effects; **(4)** individually they do not delay the onset of AIDS; and **(5)** HIV rapidly becomes resistant to each of them.

Side Effects

Each of the six reverse transcriptase inhibitors (RTIs) is associated with severe to moderate side effects. For example, zidovudine causes anemia (lowered blood cell count), nausea, headache, and lethargy and

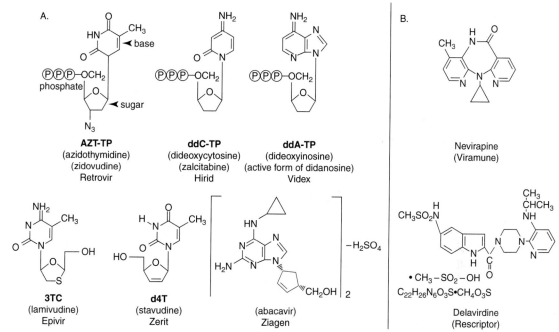

FIGURE 4-5 **A.** Structure of six FDA-Approved Nucleoside Analog Drugs Which Inhibit HIV Replication. These nucleoside analog agents competitively inhibit deoxynucleoside triphosphates (part of the HIV proviral DNA chain) when incorporated by HIV reverse transcriptase, causing DNA chain termination. Azidothymidine triphosphate (AZT-TP) competes with deoxythymidine triphosphate (dTTP), dideoxycytidine triphosphate (ddC-TP) competes with deoxycytidine triphosphate (dCTP), and dideoxyinosine (ddI) is converted to dideoxyadenosine triphosphate (ddA-TP), which competes with deoxyadenosine triphosphate (dATP). Stavudine (d4T) competes with deoxythymidine triphosphate (dTTP). Lamivudine (3TC) competes with deoxycytidine triphosphate (dCTP) and Abacavir competes with deoxyguanosine-5′triphosphate (dGTP). **B.** Structure of the two FDA Approved non-Nucleoside Reverse Transcriptase Inhibitors. Unlike necleoside RTI's, these drugs **do not** have a specific chemical structure that includes a base, sugar, and phosphate group. **These drugs bind directly to the reverse transcriptase.** This blocks DNA polymerase activity and stops HIV DNA replication. The introduction of these drugs brought great promise but that promise did not last long as HIV showed high levels of resistance to the drugs at a very rapid rate. **Cross Resistance:** Note the similarity in base structures among these drugs. Because of such similarities, when HIV becomes resistant to one of these drugs it also, in many cases, becomes cross-resistant to one or more of the other drugs in the nucleoside and non-nucleoside reverse transcriptase inhibitors.

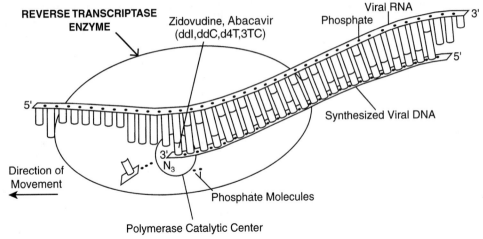

REVERSE TRANSCRIPTASE ENZYME

Zidovudine, Abacavir (ddI,ddC,d4T,3TC)

Viral RNA

Phosphate

Synthesized Viral DNA

Direction of Movement

N₃

Phosphate Molecules

Polymerase Catalytic Center

FIGURE 4-6 Incorporation of FDA Approved Nucleoside Analogs Preventing HIV Replication. Zidovudine (ZDV) is a synthetic thymidine analog that is widely used in the treatment of HIV disease and AIDS. HIV-RT is 100 times more sensitive to ZDV inhibition than is the **human transcriptase.** The incorporation of zidovudine-triphosphate into HIV DNA by the action of HIV reverse transcriptase terminates DNA chain extension because polymerization or the incorporation of the next nucleotide adjacent to the azide (N_3) group cannot occur. Similar blockage of DNA replication occurs when FDA-approved ddI, ddC, d4T, or 3TC are used. Abacavir is the first 2′deoxyguanosine analogue to be used.

inhibits mitochondrial DNA replication in humans (de Martino et al., 1995); didanosine causes peripheral neuropathy (PN) (a burning pain in hands and feet) and pancreatitis (a dangerous swelling of the pancreas) and is very hard to take by people with poor appetites; zalcitabine causes PN and mouth ulcers; stavudine causes PN; and lamivudine causes hair loss and PN.

DEVELOPMENT AND SELECTION OF HIV DRUG-RESISTANT MUTANTS

Development of Drug-Resistant Mutants

HIV reverse transcriptase, the enzyme that copies (transcribes) HIV RNA into HIV DNA, **is unable to edit or eliminate transcription errors during nucleic acid replication** (Roberts et al., 1988). Because there is no repair or correction of mistakes (as occurs in human cells), there are about one to five mutations in each new replicated HIV DNA HIV RNA strand

(Coffin, 1995). This means, that **each new virus is different from all other HIV that is being produced!** Because new virus is being produced at a rate of 1 billion to 10 billion per day! There will be 1 billion to 10 billion new HIV mutants produced each day in one person. Thus, virtually all possible mutations, and perhaps many combinations of mutations, are generated in each patient daily (Ho et al., 1995; Wain-Hobson, 1995; Wei et al., 1995; Hu et al., 1996; Mayers, 1996).

Selection of Drug-Resistant Mutants

Based on the large number of mutant HIV produced in any one person, **it is not surprising that HIV emerges with resistance to drugs used in antiretroviral therapy.** Such strains are referred to as **drug-resistant** mutants. HIV drug-resistant mutants are selected to reproduce most effectively under conditions of selective pressure exerted by the presence of the drugs. Those HIV able to resist the drugs continue to multiply; those that are sensitive to the drugs are destroyed—**selection!**

Problem—Multidrug Cocktails

Multidrug cocktails work well initially because they are attacking the strains of the HIV that are the least resistant to the drugs. But the result of that success is to encourage the rapid spread of HIV strains that are highly resistant to the drugs, which could give rise to a new and even more dangerous AIDS epidemic among people most at risk for the disease.

Several separate research investigations, using two new types of sensitive diagnostic tests, found evidence suggesting that about 30% of individuals newly infected with the HIV are carrying forms of the virus that are already resistant to one of the 15 drugs, and 10% are resistant to two of the drugs used in American combination therapy. Mutations that confer resistance to nucleoside analog reverse transcriptase inhibitors, non-nucleoside reverse transcriptase inhibitors, and protease inhibitors **have all been identified in persons who have never been treated with antiretroviral drugs** (Tables 4-2 and 4-3). The **first** zidovudine-resistant HIV was reported in 1989. Thus, with the wider use of antiretrovirals, pretreatment HIV resistance strains will occur in the population with increasing frequency.

It is clear that the appearance of drug-resistant HIV is a function of the number of HIV replication cycles that take place during infection, and that combination therapy that suppresses HIV replication to undetectable levels can delay or prevent the emergence of resistant strains (Figure 4-7). Abundant evidence has linked the presence of HIV drug resistance to therapy failure (Moyle, 1995).

During 10 years of a person's infection, HIV can undergo as **much genetic change (mutations)** as humans might experience in the course of millions of years, which makes it extremely **difficult to develop treatments** that have **long-term effectiveness** (*Scientific American*, August 1995).

TABLE 4-2 Genetic Alterations of Nucleotides Associated with the Use of Nucleoside and Non-nucleoside Anti-HIV Drugs

Reverse Transcriptase Inhibitor	Codon Mutations[a] in the Reverse Transcriptase Gene[b]
Nucleoside	
Zidovudine (ZDV)	41, 67, 70, 210, 215, 219
Didanosine (ddI)	65, 74, 75, 184
Zalcitabine (ddC)	65, 69, 74, 75, 184
Lamivudine (3TC)	184, 75
Stavudine (d4T)	75 (rare), 184
Abacavir	65, 74, 115, 184
Adefovir dipivoxil	65, 69, 70
Non-nucleoside	
Nevirapine	103, 106, 108, 181, 188, 190
Loviride	103, 181
Delavirdine	103, 181, 236
Efavirenz	100, 103, 108, 179, 188, 190
PNU 142721[c]	

[a]A 24-hour line probe assay that picks up reverse transcriptase mutations. The test helps direct choice of drug to be used.

[b]Cross-resistance is a common feature when using these drugs.

[c]In clinical trials.

(Source: Vella et al., 1996; Vella, 1997 updated)

TABLE 4-3 Genetic Alterations of Nucleotides Associated with the Use of Protease Inhibitor Anti-HIV Drugs

Protease Inhibitor	Codon Mutations in the Protease Gene[a]
Saquinavir mesylate	
Saquinavir[c]	10, 48, 63, 71, 90
Ritonavir	20, 33, 36, 46, 54, 71, 82, 84, 90
Indinavir	10, 20, 24, 46, 54, 63, 64, 82, 84, 90
Nelfinavir	10, 30, 35, 36, 46, 62, 63, 71, 77, 84, 88, 90
Amprenavir	10, 46, 47, 50, 84
ABT 378[b]	
BM 232632[b]	
PNU 140690[b]	

[a]Cross-resistance has been shown in all six FDA protease inhibitors. For example, persons on saquinavir and who developed resistant mutation **48** eventually develop mutation **82** for resistance to indinavir and ritonavir. **It is now clear that in many people combination drug therapy fails not because of the development of drug resistance, as no genetic HIV drug resistance types were found, but because of poor drug absorption by their bodies and/or poor adherence to their drug-taking schedule.**

[b]In clinical trials.

[c]Trade name FORTOVASE

(Source: Vella et al., 1996; Vella, 1997)

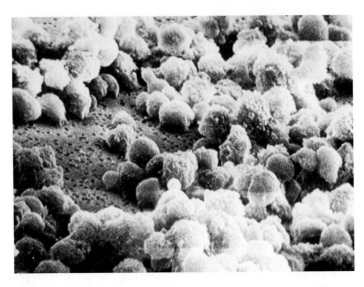

FIGURE 4-7   Production of HIV. Each dark hole in the T4 cell membrane represents the emergence of one new HIV that came off the membrane and is loose in the body. Each T4 cell produces about 250 HIV before it becomes nonfunctional. Each HIV in this photograph is genetically different from any other HIV being released from this T4 cell and most likely genetically different from any other HIV in the body. (*Source: The CDC, Atlanta, GA*)

──────── **POINT OF INFORMATION 4.1** ────────

THE DRUG DEVELOPMENT AND APPROVAL PROCESS

In the United States, the drug discovery and FDA approval currently takes an average of 12 to 15 years and costs about $500 million (from the laboratory to the drugstore). In 1996, Merck & Co. claims to have spent **$1 billion** over 10 years to bring the protease inhibitor **Crixivan** to market! The odds against a new drug making it to the market are about 10,000 to 1. The clinical trial process for a drug usually includes 3½ years of preclinical testing using laboratory and animal studies and 6 years of studies (clinical trials) in humans. **Phase I** includes 20 to 80 healthy volunteers and takes about a year to **test the drug's safety. Phase II** takes about 2 years and involves 100 to 300 persons with the disease to **assess the drug's effectiveness** and to look for side effects. **Phase III** of clinical trials lasts about 3 years and includes 1,000 to 3,000 patients to **verify effectiveness** and **identify adverse reactions of a drug.** Since 1989 the FDA has allowed phases I and II to be combined to shorten the approval process on new medicines for serious and life-threatening diseases. It now takes about 18 months for a drug to go through the review process for approval by the FDA. About **one in five medicines** that begin a clinical trial is **approved** for consumer use.

American pharmaceutical companies have been investing over $1 billion a year since the late 1980s to bring new and improved HIV drugs to market. Since 1987, a new FDA-approved antiviral drug has entered the market, on average, every 9.4 months.

Beginning year 2000, 84 companies had 126 medicines and vaccines for AIDS and AIDS-related conditions in testing. This is in addition to 54 HIV and HIV-related FDA-approved medicines already on the market.

The most difficult part of presenting anti-HIV therapies to the public is the analysis of the incredible amount of new information continuously coming out in the scientific literature and attempting to make sense out of some very contradictory findings!

(*The Drug Development and Approval Process,* Pharmaceutical Manufacturers Association, December 1998)

In December 1997, the United States House of Representatives passed into law the FDA Regulatory Modernization Act of 1997. The force of the law is to make the FDA more responsive to therapies for life-threatening diseases. The bill:

- Enabled people suffering from such afflictions to authorize their doctors to use experimental drugs or devices, not yet approved by the FDA for general use, to treat their ailments.
- Set up a registry of clinical trials to give patients and their doctors ready access to experimental treatments and drugs that might help in ministering to their illness.
- Streamlined the FDA's drug and device-reviewing process to hasten approval of new therapies for treating the severe diseases.

———— **POINT OF VIEW 4.3** ————

DOES SOCIETY WANT MIRACLE DRUGS FOR HIV?: THE NEW TREATMENTS COULD MAKE THE PANDEMIC WORSE!

The drug penicillin cured syphilis and gonorrhea, and many experts confidently predicted that these diseases would soon be eradicated forever.

But both diseases are caused by bacteria and their transmission is caused by human behavior, namely unprotected sex with multiple partners. Scientists at the time (1940s and 1950s) reasoned that if effective treatment was available, people would return to the risky behavior that spread the diseases, leading to unintended consequences. For example, because of the success in curing a variety of sexually transmitted diseases (STDs) using penicillin, the government virtually stopped educational programs on STDs, and the sexual revolution, spurred at least in part by a belief that STDs were now curable, created new opportunities for STD-causing microbes to spread.

By the early 1980s, 2.5 million Americans were contracting gonorrhea every year, and syphilis ranked as the third most common infectious disease in the nation. Things came full circle when the casual use of antibiotics produced drug-resistant strains of gonorrhea that literally destroyed penicillin and rendered other antibiotics useless.

While antibiotics are indeed miracle drugs that have saved millions of lives, in the end these treatments ultimately helped spread and strengthen the bacteria that cause some of the worst STDs.

Relative to treating HIV, studies indicate that when drugs called protease inhibitors are used with other drugs, such as ZDV and 3TC, they can reduce HIV to unmeasurable amounts in the blood of many infected people.

But this news, a triumph for medicine, is a mixed blessing for medical ecology. It could turn out that what happened with penicillin and its use in other STDs could apply equally to HIV. One nightmarish scenario circulating among scientists is that triple-combination therapy–resistant HIV could render the AIDS epidemic more intractable than it already is.

That nightmare is based on three factors. **First,** HIV mutates more quickly than any other known virus, and strains have evolved that evade every drug, including protease inhibitors, and many drug combinations. **Second,** the new combination therapies are extraordinarily expensive and difficult to take. Some drugs must be taken on an empty stomach several times a day with up to a quart of water. Others cause terrible side effects. And, if people do not take the drugs correctly and on time (compliance), the chance of developing resistance to the combination cocktail is greatly enhanced. And if they infect another person, that person may be drug-resistant from the start. **Third,** AIDS prevention efforts have faltered, especially in the most afflicted communities. AIDS is exploding in Third World countries and in the United States among poor and minority people, especially women. Also, the gay male population is undergoing a widely documented "**second wave**" of infections. If the potential for death has not been enough to compel people to practice safer sex, what might happen when that threat lessens? Unfortunately, no one really knows what is going to happen—until it has happened!

Class Discussion: What do you think should be done to eliminate the problem as presented?

In summary, the frequency of genetically variant HIV within an HIV population is influenced by the mutation rate, fitness of the mutant to survive, the size of the available HIV pool for genetic recombination, and the number of HIV replication cycles.

The Demise of Monotherapy

The use of **one** anti-HIV RTI drug at a time (monotherapy) leads to HIV drug-resistant mutants to a clinical end point such that David Cooper, an AIDS-drug therapy researcher declared 1995 the year of the "**demise of monotherapy for HIV and the rise of combination drug therapy**" (Simberkoff, 1996; Stephenson, 1996). The **standard of therapy** now is that **all** anti-HIV drugs must be used in combination and that each combination include two reverse transcriptase inhibitors and a protease inhibitor. The use of this three drug combination is referred to as HAART or Highly Active Antiretro-viral Therapy (Merry et al., 1998).

HOW COMBINATION DRUG THERAPY CAN REDUCE THE CHANCE OF HIV DRUG RESISTANCE

Combination drug therapy works because a single strand of RNA, the genetic material of HIV, must be multiply mutant, that is, it must carry a genetic change to become resistant to **each** new drug used. So, the greater number of drugs used, each capable of stopping HIV replication by interfering with the function of reverse transcriptase, the greater the number of genetic changes that must occur in a single RNA strand of the virus. For reasons of explanation, say that the **chance of change** in one nucleotide of HIV for resistance to a single drug occurs in **ONE** RNA strand during the production of 100,000 (100 thousand) or 1×10^5 such strands. **This is a reasonable figure!** Then to have a second nucleotide mutation occur in that **same** single strand for resistance to a second drug, again at 1×10^5 means that only one RNA strand in 10,000,000,000 (10 billion) or 1×10^{10} would carry both genetic changes. For three separate nucleotide mutations occurring in one RNA strand then, it would be $(1 \times 10^5) \times (1 \times 10^5) \times (1 \times 10^5)$ or **only one RNA strand out of**

1,000,000,000,000,000 (one thousand trillion) would carry a resistance to all three drugs at the same time. But as small as this number is, recall that 1 billion to 10 billion genetically different RNA strands are produced each day in one individual, and there are from 10,000 to 100,000 nucleotide changes possible for each nucleotide of HIV's 9,749 nucleotides (Deeks et al., 1997b)! Also, individual RNA strands can exchange nucleotides a process called recombination, HIV resistant to three drug combinations have appeared! This is why it is so important to **slow** the replication of HIV—fewer rounds of replication mean fewer RNA strands, thus fewer possibilities of producing HIV that are resistant to AIDS drug cocktails! Current AIDS drug cocktails, especially those using one or more protease inhibitors, quickly and significantly reduce HIV replication—this is why people on these combination therapies must take them **on schedule,** in prescribed doses, and maybe for the rest of their lives!

Problems in Drug Selection for Therapy

Recent studies have shown that each HIV-infected T4 cell produces about 250 new HIV. Collectively sufficient numbers of T4 cells and other cells carrying the CD4 protein (macrophage, etc.) are infected to produce say 10 billion HIV per day. Now, if there were a significantly large number of effective drugs available and if their use was carefully scheduled, theoretically HIV disease could become just another chronic illness. **But,** there is a problem of **cross-resistance**—HIV resistant to one drug have been found to be resistant to similar drugs (see Tables 4-2 and 4-3). Because of cross-resistance, new drugs will have to be found to stay ahead of HIV's ability to become resistant to them.

Drug Failure and Salvage Therapy

Drug-resistant HIV strains are a major cause of **treatment failure** (drugs that fail to suppress HIV replication) in the management of HIV disease. Resistant viral strains can evolve whenever the virus is not fully suppressed by a particular drug treatment regimen or via long-term therapy with sub-optimal drug regimens, poor drug absorption or noncompliance with existing regimens. Drug testing is underway to design **salvage therapies** for patients harboring resistant HIV. Salvage therapy is the use of substitute drugs that will continue to suppress viral replication when all else initial therapy fails.

The Kitchen Sink By late 1999 some patients were taking up to 10 of the 15 FDA approved anti HIV drugs in order to suppress HIV replication! Clearly some of the drugs were being recycled in a variety of new combinations. The try anything approach—The kitchen sink!

Genotypic and Phenotypic Drug Susceptibility Profiles

Determining the specific drug resistant and drug susceptibility profiles for a patient strain of HIV can be very useful in order to establish a plausible salvage therapy.

Establishing a list of genetic mutations, as presented in Tables 4-2 and 4-3, related to drug resistance is called **genotypic analysis** and indicates a given HIV isolates' genetic resistance to a given drug. Note that when HIV becomes resistant to one drug, it can at the same time become resistant to others that function in the same manner (cross resistance). The suspected degree of drug resistance as suggested from genotypic analysis can be measured directly by adding the drug in question to an HIV-cell culture and determining the HIV's ability to reproduce. A virus-culture-drug assay is called **phenotypic analysis.** (Table 4-4) A check is made to see if HIV's resistance to a given drug correlates to a given genetic mutation and if other mutations offer a cross resistance to that drug. In this way, drugs that work against the different HIV mutants can be used in **salvage therapy.** Genotype analysis is now available through five commercial laboratories and phenotype testing in two. (Table 4.4).

In general, resistance tests can't predict which drugs will work—only the ones that don't—which is why they're mostly recommended to help people whose regimens are failing. As with viral-load tests, it's critical to be consistent about which tests you use. For now, experts say, the decision to change the regimens should be based primarily on increases in viral load.

VIRAL LOAD: ITS RELATIONSHIP TO HIV DISEASE AND AIDS

Viral load refers to the number of HIV RNA strands in the plasma or serum of HIV-infected persons (a discussion of HIV RNA production is provided in Chapter 3; also see referenced material, Jurriaans et al., 1994, Henrard et al., 1995, Goldschmidt et al., 1998). In general, 2 to 6 weeks after HIV exposure, infected individuals develop a high level of blood plasma HIV-RNA. Methods now exist to quantitate the amount of HIV RNA in the blood plasma or serum of HIV-infected people. (**Plasma** is a transparent yellow fluid that makes up about 55% of blood volume. **Removing** fibrinogen and blood clotting factors from plasma results in **serum.**)

TABLE 4-4   Understanding HIV Drug Resistance: Defining the Terms

Term	Meaning
Resistance (Drug)	A change in HIV that improves its ability to replicate in the **presence** of the drug.
Genotype	The arrangement of 4 nucleotides in making up HIV-DNA (RNA). A change in genotype means a nucleotide has been gained, lost or switched. For example, HIV mutant M184V is a change in genotype at codon 184 of reverse transcriptase from ATG to GTG. This alters the amino acid from methionine to valine abbreviated (M184V), resulting in resistance to lamivudine.
Phenotype	The trait or behavior that results from changes in genotype, e.g., the ability of virus to replicate in presence of drug compared with drug sensitive control virus or wild-type-non mutant HIV.
Cross-resistance	This occurs when **one** mutation occurs that affects HIV's susceptibility to other drugs (generally the drugs are structurally related).

The purpose of HIV Viral Load Testing

Physicians recommend viral load testing for the following reasons:

♦ to help you make decisions about starting or changing drug treatment for HIV;
♦ to find out your risk for disease progression;
♦ to show how well your drug regimen is working;
♦ to help determine your HIV disease stage.

Laboratories report HIV viral load test results as the number of copies of HIV per milliliter of blood plasma (copies/ml). Until recently, HIV viral load tests could only accurately measure HIV levels down to 400 copies. The new AMPLICOR HIV-1 MONITOR UltraSensitive viral load test can measure HIV levels down to 50 copies. **CAUTION: An "undetectable" viral load, however, is not enough. The goal is for people with HIV to live out their normal life span, free of opportunistic infections and with the fewest possible drug-related side effects.**

Pitfalls of Viral Load (HIV-RNA) Testing

Despite the potential advantage of HIV-RNA testing over that of conventional HIV-antibody testing—particularly in a high-risk population with symptoms of acute HIV—**the specificity of all commercially available HIV-RNA technologies is less than 100% and, as a result, likely to yield a significant number of false positives.** According to Chiron Diagnostics, the manufacturer of one such assay, the specificity of Qantiplex branched DNA technology versions 2.0 and 3.0 is approximately 95% to 97% at a copy number of <3,000 and <150 copies/ml, respectively. More conservative published data suggest that the specificity of the three commercially available viral load assays (**bDNA, PCR,** and **NASBA**) may be similar, with a false-positive rate of approximately 5% to 10%. The emotional impact of a **false-positive** screening HIV-RNA test in recently exposed persons can be extraordinarily damaging. The FDA has **not,** at this writing, licensed any viral load tests for diagnostic purposes.

───────── **POINT OF VIEW 4.4** ─────────

WHAT ARE MY OPTIONS?

Paul is a 45-year old HIV-infected man, that had a T4 count of 1,200 cells/mm³ at diagnosis 10 years ago. Two years after viral load testing became available, he measured 45,000 copies/ml with a T4 count of 961 cells/mm³. Pros and cons of medication were explained to him. Paul elected to wait. Since then his T4 counts have ranged from 800 to 900 cells/mm³ with viral loads of 30,000 to 50,000 copies/ml. Six months ago, Paul reported that he had decided to sign up for a research trial. After four months he returned saying that he was still in the trials but had decided **not to take the medications even though "they think I'm taking them."** He was advised to tell the investigators the truth and drop out of the trial. Paul was offered the option of taking a regimen of prescribed HIV medications if he felt that he was ready for them. He said no thank you! Two months ago he reported that he had signed up for another research trial, even though he admitted that the chances that he would adhere to the regimen were small. Paul said, "I just want to see what my options are."

Commentary

Based on his viral load of 30,000 to 50,000 copies/ml, the current U.S. guidelines would identify Paul as a patient who should receive antiretrovirals. Paul's hesitance about antiretroviral medications is understandable. There are still many unknowns about long-term toxicity and efficacy, and the burden of taking 15 to 30 pills a day can make it a difficult decision for anyone. A separate issue is whether antiretrovirals should be recommended for a patient with a T4 count of 991 cells/mm³ and a stable viral load of 30,000 to 50,000 copies/ml, who is clinically well 10 years into his infection. Paul has been fortunate. In an antiretroviral-naïve patient, one would generally anticipate that after five years of HIV infection his T4 count would fall below 500 cells/mm³. Other studies have suggested that older age (greater than 35 years) may increase disease progression and mortality. Paul has clearly beaten these odds, but even at 10 years it is too early in his infection to know if he will remain healthy or for how long (Adapted from Newman, 1999).

When Is Viral RNA Found in the Plasma

HIV RNA strands are present during **all stages** of the disease, and the viral load increases with more advanced disease (Piatak et al., 1993; Saksela et al., 1994). Following infection with HIV, there is usually a rapid, transient increase of HIV proteins and RNA followed by a lengthy period of viral RNA replication at lower but measurable amounts (Figure 4-8). In well-characterized groups with known dates of HIV seroconversion, a high viral load immediately after seroconversion (Mellors et al., 1995) and at 3 years after seroconversion (Jurriaans et al., 1994) appear to be strong predictors of HIV disease progression. However, for most patients with asymptomatic HIV infection, the time of the primary infection is unknown. The levels of viral load prior to the diagnosis are also unknown. In order to determine the predictive value of an isolated viral load measurement under these circumstances, John and colleagues (1996) developed a mathematical model to predict the time to symptomatic disease based on viral load measurements in **untreated patients** with asymptomatic HIV infection and T4 cell counts greater than 500/μL of blood.

Their analysis, based on their mathematical model, shows that in the absence of anti-retroviral treatment, persons with a viral load of 10^5 (100,000) copies/mL serum may begin **progression to AIDS** (PTA) in less than 3 years and people with a viral load half a log higher (at 500,000) may begin PTA in less than 1 year. [See Table 4-5. A **log** just means a 10-fold difference.

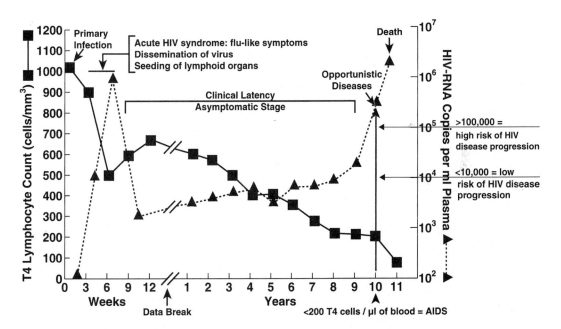

FIGURE 4-8 Clinical Course of HIV Disease as Related to T4 Cell Count and Level (number) of HIV-RNA Copies (viral load) in Plasma. During primary infection (the time period between infection and development of HIV antibody), HIV RNA levels spike (*triangles*) and HIV disseminates (*spreads*) throughout the body. This is followed by an abrupt drop in measurable HIV RNA in the blood (probably due to the production of HIV antibody), followed by a steady rise in HIV RNA until death. Over these same time periods, there is a continuous loss of T4 cells (*squares*). Note that the asymptomatic stage can be quite long—on average about 10 to 12 years, at which time T4 cell counts drop and HIV RNA copies increase to levels where there is high risk of opportunistic infections.

TABLE 4-5 Decimal, Exponent, and Logarithms

Decimal Number	Exponential Form	Log_{10} Value
100,000,000	10^8	8
10,000,000	10^7	7
1,000,000	10^6	6
100,000	10^5	5
10,000	10^4	4
1,000	10^3	3
100	10^2	2

Each number is a 10-fold change from the previous one.

For example, an increase in viral load, say from 1,000 to 10,000 (1×10^3 to 1×10^4), is one log. A decrease from 100,000 to 1,000 is a **two-log reduction.** Thus a **one-log** increase or reduction in viral load means that the viral load has increased or decreased by 90%; a **two-log** increase or reduction is a 99% change; a **three-log** increase or reduction is a 99.9% increase or reduction, and so on.] **In contrast,** people with a viral load of $10^{4.5}$ (50,000) have about 2 years and may have up to 8 years before beginning PTA. People with a viral load of 10^4 (10,000) RNA copies/ mL have about 3 years and may have up to 19 years before beginning PTA. The rate of change of the viral load was also an important predictor of HIV disease progression.

USE OF NON-NUCLEOSIDE ANALOG REVERSE TRANSCRIPTASE INHIBITORS

Non-nucleoside reverse transcriptase inhibitors (NNRTIs), sometimes referred to as **nonnukes, are a structurally and chemically dissimilar group of anti-retrovirals that can be used effectively in triple-therapy regimes.** The

The three tests usually give similar results. However, if the patient's viral load is very low, the bDNA may not detect the HIV RNA. Newer versions of all tests may be more accurate and more sensitive. In general, it is best to stay with one type of test for consistency.

RELEVANCE OF VIRAL LOAD TESTING

The test is new and researchers are still trying to decide on the important values for viral load. In general, though, results can be interpreted as follows:

HIV Viral Load Result	Interpretation
	Death within 6 years
5000 or fewer	6%
10,000 or fewer	18%
10,000 to 30,000	35%
Over 30,000 copies	70%

Viral load numbers will depend on the viral load test used. For example, a 25,000 RNA strand number determined using a branched DNA (bDNA) test might be read as 50,000 RNA strands using Amplicors' FDA-approved PCR-RT (polymerase chain reaction-reverse transcriptase) test and higher or lower numbers using other HIV RNA testing procedures. Overall however, the lower the viral load, the better. But no single test of any type gives a complete picture of health. Trends over time are important, as are nutrition, physical condition, psychological outlook, and other factors. Current viral load tests can now detect as few as 5 to 20 RNA strands/mm^3.

DETECTABLE VS. UNDETECTABLE LEVELS OF HIV RNA

What does an **undetectable** level of HIV RNA mean? Many individuals now have an undetectable level of HIV-RNA in their blood after taking combination therapy. But undetectable RNA levels does not mean that the person is cured or no longer infectious. An undetectable RNA level means that too few RNA strands are present to measure. The number of RNA strands is below the level of tests of sensitivity. Currently, tests measure between 50 and 400 copies/ml. There are experimental assays used in research that are more sensitive and can detect levels of RNA as low as 5 copies/ml.

Achieving **undetectable** viral loads is now the **gold standard** for successful HIV treatment. The more rapidly one's viral load becomes **undetectable** and the longer it remains suppressed, the better the drug and chance for long term survival. A measurable viral load after it was once undetectable is regarded as a sign of drug failure. Never reaching undetectability is also considered a drug failure for that person.

TYPICAL QUESTIONS ABOUT VIRAL LOAD

1. What is the viral **set point?** The viral set point represents an equilibrium between the virus, which wants to replicate, and the host, which tries to control or contain the replication. It appears that each patient establishes his or her own set point within 2 to 6 months after initial infection with HIV. HIV drugs appear to alter the set point for HIV by reducing its ability to replicate. This helps lower the body viral load and gives the body a chance to restore some of the lost immune function.

2. **What is the cost of a viral load test? Does insurance pay for it?** Beginning 1998, a viral load test cost between $100 and $200. Some insurance policies pay for the test.

3. **If a viral load test result was 12,000 and 6 months later it is 18,000, is the change important?** Viral load may rise and fall without any change in health. The rule of thumb is that viral load change is not significant unless it more than doubles.

4. **How often should viral load be measured?** Measuring viral load every 4 months, at the same time T4 cells are measured, has been suggested.

5. **Can viral load be associated with determining anti-HIV therapy?** Measuring viral load about 4 weeks after changing antiretroviral medications can be useful for finding out whether the new treatment is working.

6. **A person's T4 cells had been between 400 and 600 for several years, and suddenly their viral load rose to 50,000 and their physician recommended drug therapy. Why?** Research suggests that people with viral loads over 10,000 are at higher risk for the progression of HIV disease, regardless of T4 cell counts.

mechanism of action of NNRTIs is distinct from that of nucleoside analogs, even though both **prevent** the conversion of HIV RNA into HIV DNA. Nucleoside RT inhibitors constrain, or stop, HIV replication by their incorporation into the elongating strand of viral DNA, causing chain termination. In contrast, NNRTIs **are not incorporated** into viral DNA but inhibit HIV replication directly by binding noncompetitively to RT. NNRTIs at first blush appear to offer hope. All four tested did inhibit the RT-HIV enzyme (atevirdine, delvaridine, loviridine, and nevirapine). But this inhibition was **short-lived** because HIV-resistant mutants for each of the non-nucleoside RTI were found within weeks of their use (Table 4-2). Combination therapies using, for example, zidovudine and nevirapine, have shown some success but it turned out that various drug combinations do not extend survival time. But they do extend the asymptomatic period! The FDA approved nevirapine (Viramune) for market in June of 1996. **The primary advantage of using NNRTIs in therapy is to delay use of protease inhibitors.**

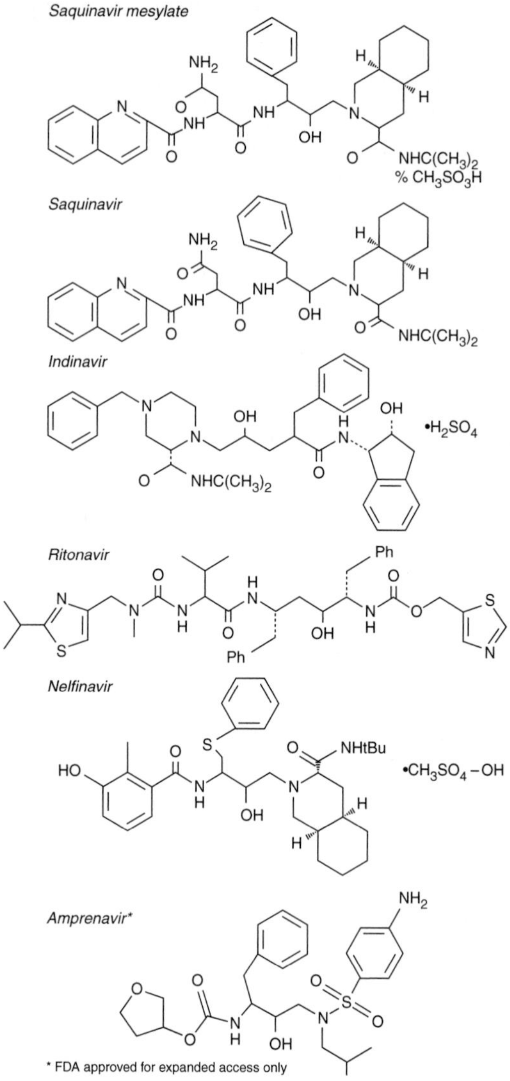

FDA-APPROVED PROTEASE INHIBITORS

Recall from Chapter 3 that some HIV genes code for **reverse transcriptase** (RT) **integrase** and **protease** enzymes. Later on in the reproductive cycle of HIV a specific **protease** is required to process the precursor GAG and POL polyproteins into mature HIV components, GAG proteins and the enzymes integrase and protease (Figure 4-9) (Erickson et al., 1990). **If protease is missing or inactive, noninfectious HIV are produced.** Therefore, inhibitors of protease enzyme function represents an **alternative** strategy to the inhibition of reverse transcriptase in the treatment of HIV infection.

As research progressed, scientists found that HIV protease is distinctly different from human protease enzymes, so a drug that blocks HIV protease should not affect human normal cell function. This means that HIV protease blockers are specific to HIV protease and that reduces dangerous side effects of protease-inhibiting drugs. Beginning 1998, some 20 HIV protease-inhibiting drugs were under study; six

FIGURE 4-9 Chemical Structures of the 6 FDA approved Protease Inhibitors.

are in clinical trials, and six have received FDA drug approval. (Figure 4-9). **PROBLEM: AT LEAST 15% OF PEOPLE MUST GIVE UP PROTEASE INHIBITORS BECAUSE OF SEVERE SIDE EFFECTS, ABOUT 15% OF PEOPLE DO *NOT* RESPOND TO PROTEASE INHIBITORS AND NOW REPORTS INDICATE THAT PROTEASE INHIBITORS ARE FAILING TO SUPPRESS HIV REPLICATION IN ABOUT HALF OF THOSE PEOPLE IN TREATMENT FROM ONE TO THREE**

HIV IS AN IMPRESSIVE ENEMY.

Structure and Function of HIV Protease

Following production of the HIV proteins, the core structure of HIV assembles at the cell membrane. A matrix protein, p17, appears to specify the site of assembly by associating with the cytoplasmic portion of the transmembrane envelope protein gp41. A packaging signal identifies which viral RNA transcripts (messages for making HIV proteins) are to be included into HIV. The assembled HIV core buds through the cell membrane, thus acquiring a lipid envelope complete with envelope proteins. Maturation of the new HIV requires cleavage by HIV protease of the long Gag and Gag-Pol proteins, which occurs during or shortly after HIV budding.

HIV protease is made up of two units, each unit consisting of 99 amino acids, with two aspartyl residues at the base of the active site serving as the target for most of the protease inhibitors under investigation. Protease has two important functions. **First,** it cuts itself loose from other components of the virus by a process known as autocatalysis (self-releasing). **Second,** the aspartyl protease of HIV cleaves the Gag (p55) and Gag-Pol (p160) polyproteins into structural Gag protein subunits and the three viral enzymes—**reverse transcriptase, integrase** and **protease** itself (Figure 4-10). Without protease activity, only immature virions are produced that are incapable of completing a replicative cycle (Laurence, 1996). Thus, protease is absolutely essential for viral replication. HIV protease is also distinct from human proteases, making it an ideal target for antiretroviral therapy. For example, any drug inhibiting protease will also indirectly inhibit the production of **R**everse **T**ranscriptase (**RT**).

Other protease released proteins help organize various HIV components (proteins and nucleic acid) into a mature, infectious HIV (Figure 4-11).

PROTEASE INHIBITORS

In 1988, protease inhibitors with potent and selective antiretroviral activity in cell culture were identified, but the insolubility, poor oral absorption and rapid liver metabolism of candidate drugs delayed the identification of suitable therapeutic agents until 1992. The **protease inhibitors** made up of a small number of amino acids (up to 15) bind to the protease active site and inhibit the activity of the enzyme. This inhibition prevents cleavage of the long HIV proteins, resulting in the formation of immature **noninfectious** viral particles.

The Six Protease Inhibitors in Use

1. Saquinavir mesylate (Invirase)—
Saquinavir mesylate was the first protease inhibitor (PI) that FDA approved for use in combination with nucleoside analog drugs in persons with advanced HIV disease. At the time of approval, there were no clinical trial data on survival or progression of HIV disease using saquinavir (Nightingale, 1996).

Although HIV mutants to saquinavir have been identified, saquinavir mutants appear to develop more slowly and at a lower frequency than formation of HIV mutants to other anti-HIV drugs (Jacobsen et al., 1995). The available oral formulation of saquinavir has **poor absorption (limited bioavailability).**

It is also the least potent of the six FDA-approved protease inhibitors. **The cost is about $648/month.**

2. Saquinavir (Fortovase)— It is a formulation of saquinivar mesylate. It is sold in a softgel capsule and is more easily absorbed by the body making it much more effective than S. mesylate. Fortovase was FDA approved in November of 1997 and is now, according to the latest April 1998 **NIH Guidelines for the Use of Antiretroviral Agents** recommended for use over S. meylate. **The cost is about $660/month.**

3. Ritonavir (Norvir)— Ritonavir is the second PI to be FDA approved **The cost is about $666/month.**

4. Indinavir (Crixivan)— This PI is perhaps the most often used of the six FDA-approved protease inhibitors. **The cost is about $450/month.**

5. Nelfinavir (Viracept)— Nelfinavir has been evaluated at doses similar to the previous

A. Why HIV Protease is Essential for HIV Replication

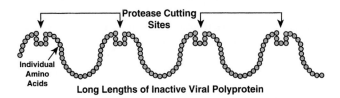

Protease Cutting Sites

Individual Amino Acids

Long Lengths of Inactive Viral Polyprotein

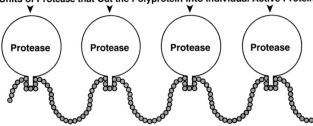

Units of Protease that Cut the Polyprotein into Individual Active Proteins

Protease | Protease | Protease | Protease

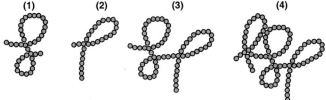

A Different Active Protein From Each Separate Length of Polyprotein

(1) (2) (3) (4)

B. How HIV Protease Inhibits Viral Protease Function

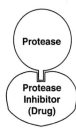

Protease

A Protease Inhibitor Binds to the Cutting Site of the Protease. The Inactivited or Blocked Protease Cannot Cut the Polyprotein.

Protease Inhibitor (Drug)

FIGURE 4-10 Representation of Protease and Protease Inhibitors. **A,** The function of HIV protease is to release the individual replication enzymes, core proteins, and envelope proteins so that HIV can develop into mature, infective HIV. **B,** Blocking the production of these essential HIV proteins using protease inhibitors produces immature, noninfective HIV. The action (cutting of protein lengths into active HIV components) of HIV protease occurs during and just after HIV buds out of the cell!

four protease inhibitors. **The cost is about $473/month.**

6. Amprenavir (Agenerase)—it has been over two years since the last PI received FDA approval. It is approved for use in adults and children. It is also the **first** anti-HIV drug to appear as a result of computer imaging—a structure-based drug design. **The cost is about $504/month.**

Protease Inhibitor Side Effects

Saquinavir mesylate and Saquinavir— Compared to nucleoside analog drugs, **saquinavir mesylate** and **saquinavir** when used alone or with other **non-protease** inhibitor drugs seems to have few side effects. In studies, the most common side effects, occurring in very few people, were rash, diarrhea, stomach discomfort, and nausea.

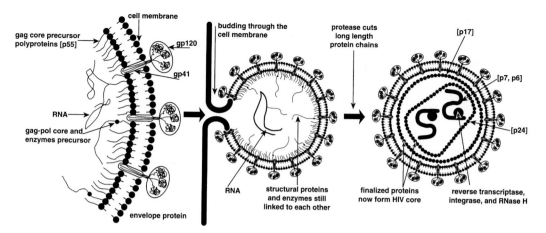

FIGURE 4-11 Representation of HIV Assembly. During the budding process, the viral Gag and Gag-Pol polyproteins assemble at the cell membrane together with viral RNA to form immature HIV. These polyproteins are then cleaved by the HIV-coded protease enzyme to provide the structural and functional (enzyme) proteins essential to form the **mature**, infectious viral core. (*Adapted from Vella, 1995*)

Ritonavir— Used alone, ritonavir blocks a pathway in the liver that is often used to clear drugs from the body. This affects the way other medications are processed in the body, causing drugs to get backed up in the liver. Of the three PIs, ritonavir is the most problematic for patient safety and tolerability, particularly during the first weeks of therapy. Common side effects are diarrhea, nausea, vomiting, anorexia, headaches, fatigue, and taste disturbances (Deeks et al., 1997). Ritonavir, when used with saquinavir, significantly slows saquinavir metabolism resulting in possible overdosing. Other protease inhibitor interactions, when used in combination, must be studied (Merry et al., 1997).

Indinavir— The most common side effects of indinavir are kidney stones and the temporary elevation of a liver enzyme called bilirubin.

Nelfinavir— The most frequent side effect of nelfinavir is diarrhea.

In summary, the protease inhibitors may cause side effects that have not yet been seen because the number of people using them is relatively small at this point in time. For example, it was recently determined that indi-navir and nelfinavir increased plasma concentrations of birth control contraceptives, yet ritonavir **significantly** decreased their levels. All four protease inhibitors interact with and **inhibit** the liver cytochrome P450 enzymes, especially CYP3A. These human cytochrome enzymes are **essential** for metabolizing a large variety of drugs to their nontoxic levels and forms. Because of this effect, many of the medications used in the overall care of HIV-infected people have to be stopped or have their dosage modified, or substitute drugs must be tried.

HIV Resistance to Protease Inhibitors

HIV resistance to all six PIs has been found in clinical trials. The development of HIV resistance to PIs is a major concern (Table 4-3). For example, recent studies (Schmit et al., 1996) found that long-term (1 year) use of **ritonavir** is associated with the production of at least nine different mutations that make HIV resistant to ritonavir. These ritonavir-resistant mutants also demonstrated cross-resistance to **indinavir** and **saquinavir** both. John Mellors (1996) reported that within 12 to 24 weeks after **indinavir** therapy, at least 10 different

———— BOX 4.4 ————

PROTEASE INHIBITORS: EXTENDING LIFE; THE DOWNSIDE OF THIS GIFT—RECOVERY??

THE LAZARUS EFFECT: A RETURN TO FUNCTIONAL STATUS!

In November of 1995, one man in his late forties wrote his obituary—he had been fighting off HIV disease, then AIDS, for over 13 years. In another case, the man's T cell count was zero. He was on oxygen and morphine. Funeral arrangements are made, and his friends and family were on a death vigil. They are but two examples of several thousand men, beginning 1999, all under the age of 50, who had given up hope—they believed they were a short step away from death. Some ran up huge debts—maxed out all their credit cards and gave lavish gifts. Some regretted the way they had lived or not lived to this point, some became very angry, some made peace with themselves and others, but **all** felt death was imminent. **Then it happened**—the first results of combination therapies using nucleoside and non-nucleoside reverse transcriptase inhibitors **dropped viral load counts** and then, a few months later, the protease inhibitors used alone or in combination with two or three of the nucleoside or non-nucleoside reverse transcriptase drugs dropped viral load counts to **unmeasurable** or **undetectable** levels in people with HIV disease and AIDS patients. **And,** those with significantly lowered viral loads demonstrated **surprising recovery— their T4 cells rebounded, in some cases from below 200 back up to, or approaching, normal!** These "**AIDS cocktails**," as the combination therapies were soon called, gave people with AIDS, for the first time since the beginning of the pandemic in the United States in 1981, a new chance at a productive life.

CAN THERE BE A DOWNSIDE TO SUCH A MIRACLE?

Of course the most tragic downside is the fact that not all who would benefit from the **drug cocktail** can afford it—But that is not the issue that is relevant to the question per se. No, this question pertains to a downside, if any, for people who have access to these drugs.

WHAT KIND OF PROBLEMS EXIST?

Guilt

Many who now feel "new" again feel **guilty** because these drugs are not available to all who need them—the poor and the uninsured in the United States and those in underdeveloped nations.

Re-establishing Relationships

Depending on how long and how severe the illness, the affected became more or less isolated— even "best" friends stopped calling or dropping by. Some who also had AIDS died, while some were too sick to care or mourn them; but now that they have recovered? **WHAT NOW? Beginning year 2000, about 300,000 people in the United States were on anti-HIV therapy, which include protease inhibitors. These drugs are not yet available to the other 500,000 to 700,000 HIV-infected in the United States. To date they are unavailable in developing nations.**

Ability to Work Again—Loss of Disability Pay!

Disability insurance has been a **cocoon of safety** for many now experiencing "new" life. Traditionally, people with AIDS received disability checks until death. But suddenly it is not so certain that that is going to be the case. People who were expected to die are going to be coming back. Nobody is prepared for this. Now that drug cocktails have extended life expectancies, it is expected that these people will be re-evaluated and lose their disability payments. **They will have to go back to work.**

Across the United States, AIDS groups are deluged with calls from patients who are excited, confused, or frightened about the prospect of ending disability status and returning to work. In Miami, a psychologist has begun weekly seminars on resumé-writing and job interviews for AIDS patients who have not worked in years. At AIDS Project Los Angeles, counselors field 100 calls each week on return-to-work issues.

AIDS patients face a barrage of challenges as they think about rejoining the work force. Job skills sometimes have atrophied. Old careers do not seem fulfilling anymore after a brush with death. Potential employers may not want to hire someone with a costly disease, even though federal law bars discrimination against people with AIDS. And without clear knowledge of the new drugs' staying power, some patients fear that a return to work could be premature, leaving them in dire shape if protease inhibitors fail them and they lose their jobs once again.

——————— **BOX 4.4** (*continued*) ———————

Dilemmas of Returning to Work: Two Examples

1. A 38-year-old software engineer shudders to think about the myriad new technologies he must master if he is to resume work. He used to be a star computer programmer, deluged with consulting assignments. Then AIDS forced him to go on long-term disability.

"I'm going to need new software, new tools, new products" to be a success once more. He estimates that retraining courses will cost $10,000; where will the money come from?

2. He considered taking a part-time job at a friend's hat-embroidery shop but decided against it, in part for fear of jeopardizing his government benefits, and also because he isn't sure how long his health will hold up. **"The biggest downside of this disease is that you stop being a dependable person,"** he says. "I couldn't be sure that I would be there every day. That's very hard to acknowledge."

PREVENTION DOWNSIDE

From mid-1996 into year 2000, optimism remains high on the medical advances using AIDS cocktail therapy. Though researchers applaud recent drug and immunology findings, many worry about the implications. **Treatment successes may create a false sense of security, and lead to complacency in prevention efforts. If we are close to a cure, people may say, why bother with politically sensitive issues like needle exchange programs, distribution of free condoms, sex education programs in our schools, and so on.** But the therapy is not available to the majority of HIV-infected, even in the developed countries like the United States. What of the other 39 million living HIV-infected worldwide? If too many glowing promises are made too soon— before treatment is available for all, or most of the world—people may revert back to unsafe sex and other behaviors that will increase their risk of becoming HIV-infected. **No, it is not yet time to stop teaching and providing preventive measures!**

POLITICAL DOWNSIDE

As scientists dare ask whether HIV eradication may be possible in some people in the near future (2 to 5 years), the press may be promising more than the scientists can deliver, but the press can and does influence political will. If the political will to fight this disease **weakens** due to the **premature declaration of victory,** this may be the greatest downside of all!

HIV is not yet a chronic disease and has not yet been eradicated from anyone, despite reductions in viral load to **undetectable** levels for over 3 years in some cases.

HIV-resistant indinavir mutations occurred and five HIV saquinavir-resistant mutations occurred with the use of the individual drugs. **Cross-resistance to ritonavir occurred in at least 19 isolates of HIV-resistant indinavir mutants.**

Rapid Production of HIV Protease Inhibitor Mutations

The high rate of viral replication found throughout the course of HIV infection and the high frequency of virus mutations occurring during each replication cycle, due to the lack of RNA proofreading or correction mechanisms, are the basis for the emergence of drug-resistant variants under the selective pressure of antiretroviral drugs. In fact, a **Darwinian** model can be applied to HIV dynamics, with the continuous production of genetically different HIV (variants) and the continuous selection of the **fittest** virus.

Unfortunately, with daily production of billions of HIV, and a mutation rate that produces billions of new mutants each day, **it is likely that any single mutation already exists before any drug is introduced.**

Because resistant variants may exist before treatment is initiated and may evolve (become greater in number) under selective pressure, anti-HIV therapy can address viral resistance in three ways: (**1**) maximize the suppression of viral replication, the fewer rounds of HIV replication, the fewer HIV mutations possible; (**2**) use combination drugs requiring HIV to create multiple drug mutations for resistance; and (**3**) force the emergence of HIV variants with slower replication or decreased virulence.

Protease Inhibitor HIV Resistance Lessons Learned

Data obtained from saquinavir mesylate, saquinavir (Fortovase), indinavir, and nelfinavir studies show a dose-dependent delay in the emergence of resistance. This is the most important property of protease inhibitors that distinguishes them from nucleoside reverse transcriptase inhibitors. Higher doses of nucleosides do not delay resistance or provide a greater effect, but rather increase toxicity. **With increased dose of the protease inhibitors, a more durable reduction in viral RNA response and a delay in resistance are achieved without greater toxicity** (Mellors, 1996).

It appears, based on clinical trial data, that the best way to avoid resistance is to take the drug on time, and not to skip doses. Taking PIs in combination with other anti-HIV drugs has been shown to slow down the development of HIV PI resistance. Researchers have been very clear about one thing: at this point in time, one cannot stop taking a protease inhibitor once started.

Management of HIV/AIDS: Treat Early, Treat Late?

Clinicians face a dilemma of choice when advising asymptomatic patients with established HIV infection on when to begin highly active antiretroviral therapy (HAART) and what drugs to use. **Begin early,** some researchers advise, because later there will be a higher virologic hurdle to overcome. **Begin later,** others recommend, and save potent drugs until the patient's immune system begins to fail.

In order to access the very best time to initiate antiretroviral therapy one needs to know the goal of antiviral therapy! Most HIV/AIDS physicians would agree that the goal is multidimensional: to prolong life while improving the quality of life; to suppress HIV replication to the limits of detection for as long as possible; to select the best possible therapy for the individual; and to minimize all side effects of drug therapy. With these goals in mind, and the drugs available to suppress HIV replication and extend life with quality, the question becomes **"when should one begin antiretroviral therapy?"**

The Decision

There is much debate about when to start anti-HIV therapy, which therapies to start, and in what combinations. Should treatment be used immediately when people first learn they are infected, or should it wait until there are measurable body changes or until noticeable symptoms of HIV disease develop? Such questions have to be considered when deciding when and which drug combinations to use and when to begin using them. When deciding when to start, switch, or change antiviral regimens there are generally three medical or biological factors to consider: **viral load; T4cell counts** and the **patient's medical condition** (with or without therapy). Following these assessments, additional factors to be considered are: the person's psychological readiness to commit to long-term therapy and his or her acceptance of risk to disease progression regardless of therapy. Together the above parameters help the physicians deal with the individual patient—sort of knowing how he/she wants to proceed, but there is **no single absolute signal that tells the physician to START THERAPY NOW!**

One note of agreement among researchers and physicians is that the decision to start anti-HIV therapy should be guided by looking at clinical health and measures of both T4 cell counts and the viral load. Increasingly, information suggests that viral load tests coupled with T4 cell counts provide the most accurate tool to monitor the risk of HIV disease progression.

In 1998 the US Department of Health and Human Services released and has since upgraded a set of Guidelines for the Use of Antiretroviral Agents in the Treatment of HIV-infected Adults and Adoles-cents. In short, the Department suggests that most HIV-infected people should be treated. The one general exception is regarding people with the combination of high T4 counts (above 500) and low viral load (below 20,000 copies of virus). In this population, the Federal Guide-lines recommend **"observing"** the person and continuing diagnostic testing. However, the Guidelines acknowledge that this is just one approach and that some will still prefer to offer treatment, even in this population. The new Federal Guidelines do not argue that all patients must be tested.

Reality of the Decision

The combination of the release of many anti-HIV medications and a lack of long-term research on their effects has resulted in confusion concerning the best course of treatment using the drugs. Some physicians and researchers subscribe to an aggressive regimen, while others adhere to a reactive regimen that initiates medication in response to condition. There is still debate over which strategy provides the best chances for survival, although some evidence presented at the 12th International AIDS Conference in Geneva, 1998 suggests that an "easy does it" plan for long-term HIV control instead of outright eradication of HIV may be more attainable.

Studies have shown that stopping treatment has led to very rapid gains in viral replication, essentially eliminating the gains of therapy. But, longer therapy may lead to the develop-

ment of drug resistance, especially if the therapy proves to be suboptimal. Furthermore, there are concerns about the toxicity of the regimens, as well as adherence problems. Some studies have shown that HIV quickly eliminates the T4 cells that coordinate the immune attack against the virus and that these cells do not appear to regrow, even after the reduction of viral load through treatment. This evidence indicates that it may be best to try and reduce viral levels as soon as possible to save these cells so the body can mount an effective attack against the virus. In short there are about as many physicians for early therapy as physicians and their reasons for the use of therapy later in the disease.

Readers may further wish to confuse themselves by reading the **Pro and Con** on this issue as presented by Bruce Walker and colleagues **"Treat HIV Infection Like Other Infections— Treat It"** (1998—Pro) and William Burman and colleagues, **"The Case For Conservative Management of Early HIV Disease"** (1998—Con).

——————— SIDEBAR 4.1 ———————

GUIDELINES AND RECOMMENDATIONS FOR THE USE OF ANTIRETROVIRAL AGENTS IN HIV-INFECTED ADULTS AND ADOLESCENTS: PRINCIPLES OF THERAPY

The Department of Human Health Services published a draft guideline from a 32-member panel for the use of antiretroviral agents on June 19, 1997. On June 25, a week later, members of the International AIDS Society—USA Panel published, *Antiretroviral Therapy for HIV Infection in 1997* (Carpenter et al., 1997; Carpenter et al., 1998). And in April 1998 NIH also published a guideline on retroviral therapy. On May 5, 1999, a panel on Clinical Practices for Treatment of HIV Infection convened by the department of Human Health Services and the Henry J. Kaiser Family Foundation created a **Living Document** (one that can be updated quickly and routinely): Guide-lines for the Use of Antiretroviral Agents in HIV Infected Adults and Adolescents. (e-mail: atis@hivastis.org or call 800-448-0440) All publications advise the earlier, more aggressive use of antiretroviral therapy, offering therapy to all persons with HIV RNA levels greater than 5,000 to 10,000 copies/mL of plasma regardless of their T4 cell count, discourage the use of a fourth drug to the current three drug combinations, and say

that treatment with only two drug combinations is **less than optimal.** The British HIV Association also published a set of similar guidelines (British Guidelines Coordinating Committee, 1997).

The reason for updated antiretroviral therapy is because of newly acquired knowledge about HIV pathogenesis (how HIV causes disease) and the results of the effective use of combination drug therapies from 1996, into year 2000. It is hoped that these guidelines, which now set a **standard of care,** will move insurance companies to pay for treatments.

NOTE: Although the guidelines have noble intentions, much of the popularity of the recommended regimens is based on expectation raised by limited clinical experience. The guidelines, including those of NIH, have been revised for 1998 and will be revised again mid 1999. **Reason** to provide the best possible combination therapies. Updated guidelines can be found on the Internet at www.hivatis.org.

CURRENT PROBLEMS USING COMBINATION THERAPY

First, DURATION: Many patients ask how long they will have to continue combination therapy? But why would they ask this—it's keeping them alive. But the combinations often involve taking 20 to 70 or more pills a day that must be swallowed according to a rigid schedule that would tax even the most compulsive individual. The timing centers around an empty stomach, and the drugs often cause severe irritation of the stomach. In one case, a patient had to discontinue therapy after 26 weeks because of gastrointestinal intolerance. The viral load went from **undetectable to 98,600 copies/mL** and the T4 cell count fell from 307 to 227/mm³ within 2 weeks! **Pills rule life!** (See Figure 4-12.)

Second, ADHERENCE or COMPLIANCE—Health care providers have been dealing with the issue of client **adherence** or **compliance** for centuries. The medical literature shows that it is difficult for patients to adhere to even the simplest treatment regimens. In the hypertension literature, one-third of patients take medications as directed, one-third take little or no medication, and one-third are intermittently adherent. Age, race, sex, educational level, socioeconomic status, and a past history of alcoholism or drug use are not reliable predictors of poor adherence. Regimen simplicity, however, does affect adherence. For example, in hypertension; once-daily therapy is associated with rates of 90% adherence, twice-daily with 80%, and three or more times daily with 65%. Other factors shown to be associated with better adherence to medications include a severe illness and belief in the efficacy of treatment. Factors associated with poor adherence include unstable housing, mental illness, and major life crises. **Adherence** to a drug regimen means taking all of the prescribed anti-HIV drugs at the scheduled times and not missing any doses. Anytime people are asked to change and/or maintain new behaviors to treat an existing condition or to prevent a threatened one, there is a good chance that they will not comply, consistently and correctly, to the prescribed activities. Combination drug therapy for HIV infection does not cause new compliance issues; it just highlights the known difficulties of an existing problem. **The Achilles Heel of Anti-HIV Therapy:** Skipping only a few pills can trigger the emergence of drug-resistant strains of HIV. Such a development could create a worse problem than the initial infection because the resistant virus could overwhelm the individ-

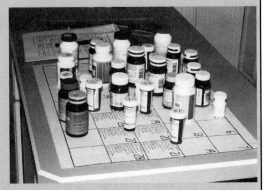

FIGURE 4-12 Patient Compliance. People with HIV disease and AIDS take many different drugs daily, depending on their prescribed medications for the various opportunistic infections and combinations of drugs necessary for anti-HIV therapy. The patient's drug intake, when this photograph was taken, numbered 28 different pills daily. If this patient took 27 pills each day for 10 years, this person would ingest **98,550 pills. Is it any wonder that people fail to take all of their medications!** (*Photograph courtesy of the author*)

ual taking the drugs and anyone else to whom the individual transmitted the virus.

There is a present danger that the behavior of underdosing (not taking enough) or partial compliance (taking the protease inhibitor when they feel like it) or patients who modify their dosage regimens—**to extend their prescription**—may create HIV strains resistant to all currently available drugs. This would lead to an even more devastating AIDS pandemic! The major reason given by 202 HIV-positive people on combination anti-HIV therapy was, **I FORGOT!** (33%). Clearly, the degree of compliance to therapy affects treatment outcome. Andrea Barthwell (1997) reported that from 12 years of research, the clearest finding is that nonadherence is the norm in medical care.

Third, COSTS: AIDS is a disease of poverty. The majority of HIV-infected in the United States and worldwide will **never** get a first dose of a protease inhibitor or an AIDS cocktail. The cost of **just** protease inhibitors is between $12,000 and $20,000 per year! The cost to treat the 335,000 Americans, beginning year 2000, who are receiving HIV treatment (about 80% receive the 3-drug cocktail) is about $7 billion a year. To treat the estimated

———— BOX 4.5 ————

THE NEW AIDS DRUGS: TIMING* IS CRUCIAL

Doctors are deciding that some people should not have access to new anti-HIV drugs because of the complex drug regimen to which HIV and AIDS patients must **adhere.** Failure to stick to the strict schedule could make the virus resistant to treatment and cause a potential health risk by creating drug-resistant strains of the disease. Here is how one patient described his daily medication schedule:

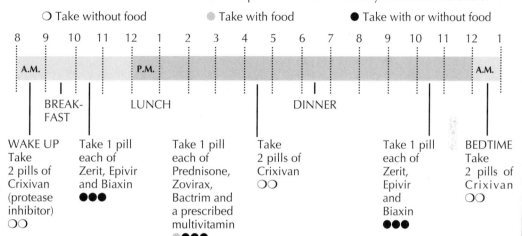

| ○ Take without food | ◓ Take with food | ● Take with or without food |

| | | | |

WAKE UP
Take
2 pills of
Crixivan
(protease
inhibitor)
○○

Take 1 pill
each of
Zerit, Epivir
and Biaxin
●●●

Take 1 pill
each of
Prednisone,
Zovirax,
Bactrim and
a prescribed
multivitamin
◓●●●

Take
2 pills of
Crixivan
○○

Take 1 pill
each of
Zerit,
Epivir
and
Biaxin
●●●

BEDTIME
Take
2 pills of
Crixivan
○○

*AIDS combination therapy regimens are arguably the most demanding of any in medicine. Every day must be choreographed to coordinate such basic activities as working, eating, and sleeping with the consumption of a score of pills and liters of water. **One patient takes 62 pills daily, another takes 72 daily.**

Some drugs require fasting; others must be taken with high-fat meals. Some are taken once a day, some twice, and some three times. It is a rigorous treatment routine that wreaks havoc on already difficult lives. **One patient, who swallows 16 pills a day, said it took him months to devise a schedule that did not leave him hungry.** Now he starts his day with a protease inhibitor, washed down with several glasses of water. Then he has to wait an hour for breakfast.

Between breakfast and lunch, he takes three more pills, and he must eat lunch by 2 P.M. so his stomach can be empty enough for his next protease inhibitor dose at 4:30 P.M. Then he must eat dinner by 6 P.M. to have a full stomach for the next round of drugs. His final challenge is to stay awake to take his last protease inhibitor at 12:30 A.M.

This patient said, "It's a good thing I can actually feel it prolonging my life, because the whole thing is so hard, especially with HIV, which makes you really forgetful."

For most patients, the drugs are also an unwelcome, unending remainder—with side effects like nausea, diarrhea, and kidney stones—that they have HIV (Source: *New York Times,* 3/2/97).

In May 1999, an HIV retroviral drug scheduler, developed by Stephen Piscitelli and Charles Flexner was put on the Internet by Medscape, a free site (http://www.medscape.com). The interactive **Drug-Drug Interactions and Daily Medication Daily Scheduler** tailors patient dosing recommendations to patient preferences for waking, sleeping and meal times.

QUESTIONS: How can people high on drugs or those with AIDS dementia or those just too ill to function remember to take their drugs on time? in the proper sequence? or with the proper diet? How many have access to the proper diet, fat vs. non-fatty foods at given times?

STATEMENT: In some large cities like New York, some 75% of HIV/AIDS patients are on **protease inhibitors.** Not to be compliant (that is, taking the medication as prescribed, without failure) is to allow HIV to replicate and give rise to protease inhibitor mutants that may otherwise not occur! Such mutants will then be transmitted to others so, many doctors say they have an ethical responsibility to withhold drug cocktails from some patients after judging them, not as people, but rather as compliant or noncompliant patients.

Class Discussion: Is this a moral decision? Why? Why not?

800,000 HIV-infected at the same rate (80%) would cost about $17 billion per year. Over **half the costs** are for antiretroviral medication—the pills can cost between $5 and $16 each. At 10 to 30 pills a day the cost per person can run up to about $500 a day or over $180,000 a year. Worldwide, **if** people received equal treatment at the above rate of 80%, the bill for all people living with HIV/AIDS outside the United States beginning year 2000 would be about $17.2 billion per year ($500/day × 34,400,000 infected people).

Larry Kramer, a co-founder of Gay Men's Health Crises, states that the cost of his drugs to combat AIDS, **which do not include a protease,** amounts to about $19,000 a year; this does not include visits to his doctor or the batteries of blood tests that he routinely requires. And he is asymptomatic. A *New York Times* article in 1997 estimated that drugs for someone with symptomatic AIDS cost about $70,000 a year; in response, a New York University adjunct law professor and gay-rights advocate, wrote a letter to the editor saying that his drugs cost $84,000 a year using protease inhibitors; the annual drug cost can exceed $150,000. At these prices, how many of the nation's 650,000 to 900,000 HIV-infected will be able to afford proper HIV therapy? **AT THESE PRICES PEOPLE HAVE TO CHOOSE WHETHER TO PAY RENT, BUY FOOD, OR PAY FOR THEIR MEDICINE—SOME CHOICE!**

Fourth, SIDE EFFECTS: Because of the short time period from development to approval, many of the drug interaction studies with the protease inhibitors are unpublished or have yet to be performed; therefore, it is difficult to make rational decisions about how to use many of those drugs when using protease inhibitors. To date, all FDA approved protease inhibitors (PIs) have important side effects in a large number of PI users (25% to 65% in different surveys). To name a few, PIs have been associated with: dry skin, cracked lips, loss of body hair, sexual dysfunction in men, oral warts, loss of bone mineral density, gastrointestinal problems (diarrhea etc.); inducing high triglycerides and cholesterol levels associated with coronary artery disease; increased blood sugar levels and diabetes; and abnormal body fat metabolism and distribution or **lipodystrophy** in particular on the abdomen, referred to as the **protease paunch** (Figure 4-13) and on the back between the shoulder blades, called "**buffalo hump**" (Figure 4-14) and a loss of fatty tissue in the arms and face. Women may experience narrowing of the hips and breast reduction or enlargement (Lo et al., 1998; Miller et al., 1998; Lipsky, 1998; Carr, et al., 1998; A-E and Gervasoni, et al., 1999). Although abnormal fat distribution may not be considered serious in life and death situations, it probably does matter to the asymptomatic person who cares about body image and who cares about cardiovascular problems in the years ahead.

Some of the newer, not yet FDA-approved, PIs may cause kidney failure and liver problems. PIs also appear to increase blood levels of **VIAGRA**, the new male impotence drug by inhibiting liver enzymes that would normally eliminate Viagra.

Fifth, WEIGHING THE UNKNOWN: Without taking any drugs, about 50% of the HIV infected will be well some 10 years after they have become infected!

Help: for those experiencing drug side effects or who want to know more about the drugs call:

AIDS Treatment News: 800-341-2437
Project Inform: 800-822-7422
Women Alive: 800-554-4876
World: 510-658-6930

Conclusion

Each group, those for and those against early anti-HIV drug therapy, has its own reasons that sound logical, but only one group can be right. Scientific rationale for advocating both very early treatment and treatment after symptoms begin are believable. Which to choose? That is the question HIV-infected persons and their physicians face. Beginning the year 2000, the data for and against early aggressive treatment continues to foster disagreement. For additional ideas on this issue read the article by James Kahn and colleague (1998).

As more drugs arrive to treat HIV, it's very hard to keep up with the latest strategies and guidelines for care. HIV is a complex disease that affects every individual differently. That's why most experts stress that the best approach to HIV treatment is one designed to fit each individual's needs. An integrated, quality-of-life

TESTIMONY ON SIDE EFFECTS: ONE WOMAN'S EXPERIENCE

(This anecdotal account was sent to the author from Ms. Colleen Perez and is used with her permission and the author's appreciation.)

She wrote: I am a single parent with three children—they are the medicine that lights my heart afire and keeps me going! The side effects I lived with from taking Saquinavir/ Fortovase (not simultaneously), 3TC, and AZT were a nightmare. Firstly, I suffered from a severe peripheral neuropathy in my arms. I'd awaken drenched with sweat, unable to feel my arms, that is until the shooting pains hit! Then came severe edema, I went from 145 lbs. to 180 and still climbing in less than one year. I looked full term 9 months pregnant. In fact, not a day went by that someone didn't ask me when the baby was due? You may initially think me vain, but let me tell you that I labored to breathe, and had great difficulty making the smallest movement. I also suffered from acid reflux disease that was so bad I was nauseous, burping, and farting uncontrollably, and constantly! Having always been fairly fit and athletic in the past, those drugs made my life little more than a living hell. The cost of the drugs around $1,200 a month. The tab is picked up by our local MIP (Medically indigenous person) and somewhat federally funded health insurance plan. The pharmacist here in Guam tells me health care is going to collapse.

It was at the 12th World AIDS Conference in Geneva that I became incensed over the unfairness of North/South, bridging the gap! The gap is just getting wider. Thanks to the giant pharmaceutical companies and their medical marketers. I had never seen such bullshit, it made me ashamed to be human. The drug companies all had their stuff strutting displays, talking to me because I'm kind of white, and ignoring my Asian companions. It made me sick and I flushed all my Fortovase down a Swiss toiletet. Why the organizers chose to hold a global conference in one of the world's most expensive cities is beyond me. By the way, I had a breakdown there, right at the Geneva airport, I read my ticket wrong and missed my flight by 12 hours. I spent the night in a Swiss hospital, well at least that was free! I have stopped taking the drugs and started vitamins and mineral supplements, lost the "baby" (fat belly induced from the drugs) and at last I feel OK. My viral load is high at 532,000, my T cells low at 193, my doctor is a bitch and our ONLY HIV/AIDS specialist here. But the worst thing that has happened to me is a whopping coat of white esophagal thrush! I would like to try something that will bring my viral load down. I really don't want to take protease again. Please don't ask me to consult my doctor because she told me to find my own treatment. I'm tough and ornery and you'll continue to come across me and my philosophies like, "no pain, no pain!", far into the next millennium.

approach to HIV care is one that focuses on preventing illness, stopping HIV, and keeping one healthy every day.

Although many experts advocate the initiation of antiretroviral therapy for **all** HIV-infected persons, the decision is usually based on markers of disease severity. Consideration of T4 cell count, viral load, clinical features, and the rate of change in each person provides a basis for initiating treatment. Recent guidelines from NIH and the International AIDS Society-USA panel suggest that HIV plasma RNA levels greater than 5,000 to 10,000 copies/mL and T4 cell counts less than $500/\mu L$ may be used as benchmarks. Theoretically, aggressive treatment early in the

course of HIV infection reduces viral load dramatically and limits the potential for disease progression. Whether very early treatment can prevent the long-term consequences of HIV infection remains to be determined.

Anti-HIV therapy becomes more complicated with the approval of each new anti-HIV drug (Figure 4-13). Just a few years ago (1990/1991) the only anti-HIV therapy was ZDV or nothing, depending on the patient's disease stage. At that time, the ZDV/ddC and ZDV/ddI combinations were considered alternatives for initial therapy, but there were no data to support their use. By late 1998, six nucleoside analogs, six protease inhibitors, and three non-nucleoside RTI had received FDA approval. Still, the end point of the entire arsenal of anti-HIV therapy is premature death. But, for the first time there are some drugs that offer considerable promise for the clinical management of HIV infection. Unlike nucleoside analogs, protease inhibitors have no requirement for metabolic activation and are active in a wider range of cell types. In addition, protease inhibitors are well tolerated and are suitable for combination therapy with all current nucleoside analog reverse transcriptase inhibitors.

Over the past years many billions of dollars have been used to help those with HIV/AIDS die. Now there is the option to let them continue living with quality. However, the costs for anti-HIV combination therapy using the protease inhibitors is expected to be about $12,000 to $18,000/year for life. **Who will be able to afford the therapy? (Class Discussion.)**

RATIONING HIV DRUGS IN THE UNITED STATES

Because of cost, availability and the possibility of producing drug-resistant HIV strains, the International Association of Physicians in AIDS Care (IAPAC) in Chicago has announced its support for a policy to ration HIV/AIDS drugs in the United States. The group has proposed that a team of experts—composed of bioethicists, physicians, community leaders, health care officials, and people with HIV—work together on a plan for explicit rationing.

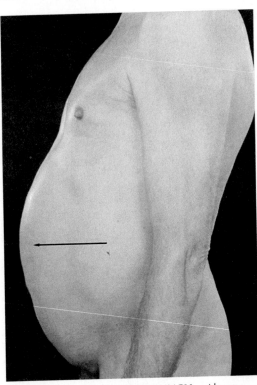

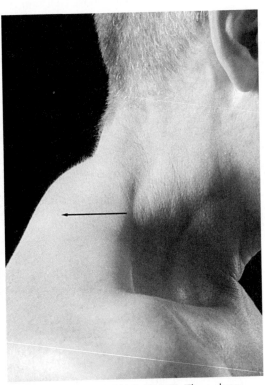

FIGURE 4-13   PROTEASE PAUNCH—Also referred to as Crix Belly. This lipid disorder (lipodystrophy) occurs in many patients using protease inhibitors for anti-HIV therapy. In addition to the visual effects of antiretroviral therapy, there are a number of dangerous side effects—see text for details (Point of Information 4.3). *(Photograph courtesy of Dr. David Cooper)*

FIGURE 4-14   BUFFALO HUMP. The enlargement of a cervicodorsal fat pad. Buffalo hump develops after use of protease inhibitors for anti-HIV therapy. In addition to the visual effects of antiretroviral therapy, there are a number of other dangerous side effects—see text for details (Point of Information 4.3). *(Photograph courtesy of Dr. David Cooper)*

CURRENT DRUG/PATIENT REGIMENS MAY CALL FOR A SPECIALIST

Early year 2000, there were 15 FDA-approved anti-HIV drugs for use in combination drug AIDS cocktails. In addition to the 15 FDA approved drugs (two of them are saquinavirs but have separate FDA approvals) there are at least 5 additional non-FDA approved anti-HIV drugs now available for physicians/patient use. Three of these 5 drugs are expected to receive FDA approval by the end of 1999. As the number of drugs continue to grow, therapy becomes even more complex. (See Tables 4-2; 4-3) Using these drugs, in three-at-a-time combinations, a physician has 8000 (20^3) prescription options available for any one person and more drugs are on the way to physicians! Some physicians are now using 4 drugs at a time—that's 160,000 options to choose from! Some have tried 5, 6, and 7 drug combinations!

WHAT TO CHOOSE?—THAT IS THE QUESTION

The complexity of today's drug regimens may call for specialist care, analogous to the need for an oncologist to manage complex cancer chemotherapy. In the past it was felt HIV patients could be managed by primary care providers, but today these patients may be better served in the care of specialized physicians. Their patients have better outcomes (Hicks, 1997; Lewis, 1997; Zugert et al., 1997).

In fact, **NIH 1998 Guidelines for the Use of Antiretroviral Agents** state, "these recommendations [on how to use the drugs] are not intended to substitute for the judgment of a physician who is **expert** in the care of HIV-infected individuals." A survey reported in the Medical Tribune, July 16, 1998, stated that even among the most experience doctors treating HIV/AIDS patients, 13% did not follow established treatment guidelines. Less experienced physicians waited longer to treat and prescribe fewer drugs than recommended in the guidelines.

Anti-HIV Patient Medication Dilemmas

To quote William Faulkner, **"if it ain't complicated it don't matter whether it works or not because if it ain't complicated up enough it ain't right."**

How well are doctors managing the complexities of treatment? Not very well, suggest the experts, but doctors are not to blame, rather it is the difficulties posed by Highly Active Antiretroviral Therapy (HAART). Even current federal guidelines for HIV treatment serve only as a good, but rough guide for managing patients. In reality, there are so many new drugs and patient factors to consider that many doctors are struggling, and failing, to make sound treatment decisions on issues such as dosing, toxicity, drug tolerance, and potential drug interactions.

Medication Errors

A study conducted by pharmacists at Albany Medical Center Hospital in New York in 1998 highlights the problem. Over 31 months, hospital physicians prescribing HIV antiretroviral drugs made 100 errors in 81 patients. Seventy three percent of the errors were ranked as serious to severe, and 21% were considered clinically significant. The most common prescribing errors were underdosing, 46% and overdosing, 34%. Half the errors involved patients taking **protease inhibitors,** and most of the rest involved the better-known nucleoside analog drugs. These errors increased from two percent in 1996 to 14 percent in 1998-paralleling the arrival of protease drugs.

Some patients infected with HIV are receiving the wrong medication because of confusion over the names of drugs used to treat infection. The **Medication Errors Reporting** program has received over a dozen reports of name confusion including errors in reading and transcribing orders and in filling prescriptions. Mix-ups occur between the antiviral drug Retrovir (the brand name for zidovudine or AZT) and the protease inhibitor ritonavir, between lamotrigine (an anticonvulsant) and lamivudine (an antiviral), between Sinequan (the brand name for doxepin hydrochloride, an antidepressant) and the protease inhibitor saquinavir, and between **Nevirapine or Viramune** (a non-nucleoside reverse transcriptase inhibitor) and **Nelfinavir or Viracept** (a protease inhibitor). Because each drug has its own set of dose/ restrictions, errors in filling/taking the wrong medications can be toxic to the individual. **Adverse drug events are common, preventable and costly!**

Questions People Ask About Drug Interactions

WHAT ARE DRUG INTERACTIONS? The amount of a prescription medication is supposed to be **high enough** to help fight a specific disease, and **low enough** to avoid causing too many **side effects.** Other medications, non-prescription (over-the-counter) drugs or recreational drugs can sometimes cause large changes in the amount of a medication in the bloodstream. This can result in an overdose, leading to serious side effects, or to an underdose, meaning the medication will not be effective.

HOW DOES THE BODY PROCESS DRUGS? The body removes drugs, usually in urine or in bowel movements. Many drugs are taken out of the blood unchanged by the kidneys, and leave the body in urine. Other drugs have to be processed by the liver. Enzymes in the liver change or alter drug molecules such that they can then be eliminated in urine or in bowel movements. For example, after taking a drug by mouth, it goes from the stomach into the intestine, and then to the liver before it is circulated to the rest of the body. If the drug is easily broken down or inactivated by the liver (such as the protease inhibitor Saquinavir), then very little useful drug reaches the body.

HOW DO DRUGS INTERACT? A few drugs slow down the kidneys. This increases the blood levels of substances that are normally removed from the body by the kidneys. The most common drug interactions involve the liver. Several drugs can slow down or speed up the action of liver enzymes. This can cause large increases or decreases in the blood levels of other drugs that are broken down by the same enzyme.

WHY DOES FOOD MATTER? Taken by mouth, drugs go through the stomach. Most drugs are absorbed faster if the stomach is empty. For some medications, this is a good thing, but it can also cause more side effects. Some medications need to be taken with food so that they are broken down more slowly or to reduce their side effects. Other medications are taken with fatty foods because they dissolve in fat and are absorbed better. Stomach acid can also make a difference. For example, it breaks down the drug ddI (didanosine, Videx). But ddI tablets include a buffer that protects the drug from stomach acid. The buffer, however, interferes with the absorption of indinavir (Crixivan) so the drugs should not be taken at the same time.

WHAT ANTI-HIV DRUGS CAUSE THE MOST INTERACTIONS? Protease inhibitors and nonnucleoside reverse transcriptase inhibitors are processed by the liver and cause many drug interactions. Everyone taking anti-HIV drugs needs to be very careful about drug interactions.

NOTE: This is but a small sample of questions people ask about drug interactions. And there are many drugs that cause interactions so, you and your physician should carefully review the information that comes with each medication (the "package insert"). Ask for this information for each drug that you are taking. Also, be sure that a doctor reviews ALL medications or drugs you are taking whenever you make a change.

Among the issues the panel would need to consider are how a patient's age, health, financial status, and ability to comply with drug regimens should factor into eligibility. Public health officials say that without rationing or a national mandate to increase HIV treatment funding dramatically, the cost of new therapies will remain prohibitive for a number of patients. Georgetown University law professor Lawrence Gostin notes that lotteries have been used to ration drugs in the past, citing instances when new therapies for the treatment of multiple sclerosis and Lou Gehrig's disease were first released (AIDS Alert, 1998).

Class Discussion: Do you support a policy on HIV/AIDS drug rationing? Present pros/cons.

Future Drug Therapy

Zidovudine, referred to most often as AZT, was FDA-approved in 1987; today, 13 years later, there are no less than 60 drugs available for HIV disease or its associated conditions. Under development are some 40 anti-HIV drugs, 23 drugs against AIDS-related cancers; 11 to fight opportunistic disease; 5 gene therapies to create resistance to HIV infection and 12 anti-HIV vaccines. There are a number of nucleoside RTIs and nonnucleoside RTIs and protease inhibitors now in clinical trials. **T-20 fusion inhibitor,** is a new anti-viral compound designed to block HIV entry into cells by interfering with gp41, HIV's protein used to enter T4 cells. T-20 completed Phases I/II trials, demonstrating potent anti-HIV activity. No drug-related adverse effects were seen in any of the 89 patients

——— BOX 4.6 ———

CAN COMBINATION THERAPY USING NUCLEOSIDE ANALOGS WITH PROTEASE INHIBITORS CURE HIV DISEASE? AN ALTERNATIVE—REMISSION?

Using a protease inhibitor along with two other nucleoside analogs has created a guarded sense of optimism that a **cure** just might be possible in the near future. There are, however, many, many more skeptics of this belief than believers. To date, physicians marvel at the drop in viral load, to unmeasurable levels, in people who were very sick and their return to a more reasonable quality of life. **The question is, can the viral load be kept that low, and what will happen to the body with that kind of treatment?** There is a real possibility that the therapy will prove too toxic to continue for very long. Or that HIV may find hiding places in the body from which it can eventually reproduce.

But if combination therapy can help people with long-standing HIV infections, it could do even more for those who have just contracted the virus. But, finding people with new HIV infections is not easy; most do not get tested that quickly. But in July 1996, David Ho and colleagues of Aaron Diamond AIDS Research Center found 12 men who had been infected with HIV for no more than 90 days. The men were put on combination therapy and HIV became unmeasurable in their blood's lymph tissues and semen! The ultimate test is to take some of the patients off their medication and see whether HIV increases in their blood. In late 1996, one recovering AIDS patient in San Francisco had no measurable HIV in his blood for the 89 weeks he was on the protease inhibitor-AIDS cocktail. **He decided to stop the drug therapy; the virus became measurable within 1 week**.

Also in 1996, Katherine Luzuriaga from the University of Massachusetts, Worcester, described how she and her co-workers treated a twin baby boy and girl with three anti-HIV drugs after they showed signs of infection at 10 weeks of age. The drug combination soon drove HIV RNA in the blood down to undetectable levels. Their levels of antibodies against the virus also steadily declined, another signal that the drugs worked.

But after 16 months of treatment, one child's blood suddenly tested positive for HIV RNA. Since then, the researchers have used a different drug combination to reduce viral load. But these two cases highlight the fact that the new drugs have yet to cure anyone of HIV infection—and it is still unclear whether they will.

The work of Ashley Haase and colleagues (1997) has shown that a 6-month, three-drug com-

bination, eliminated 99% of HIV in tonsils and lymph nodes of 10 HIV-infected people. Their fear, however, is that HIV may be hiding out in brain and other sheltered body tissues. Perhaps a larger problem in eliminating HIV from the body is that of eliminating slow dividing or resting cells that carry HIV in the **proviral** states, that is, HIV DNA integrated within the cells DNA. Such cells may be important reserves of nondeletable HIV that can at some given time become activated and produce HIV (Chun et al., 1997).

Through 1999, studies showed the HIV-infected people have **compartments or sanctuaries** of replication-competent HIV residing in resting (non-dividing) T4 and T4 memory cells, macrophage, and other cell types and tissues (e.g. brain, nervous system and testes). HIV recovered from the pool of cells/tissues, although they did not show resistance to anti-HIV drugs, were still present. That is, after three years on combination drug therapy, HIV were still present in persons that had **no** measurable HIV-RNA in their blood. Such data clearly indicates that the body's reservoir of HIV has a very slow rate of cleansing itself of HIV. The worst case scenario; it may take between 5 and 20 years of uninterrupted aggressive anti-HIV therapy before the pool of latent HIV is cleared from the body. (AIDS Clinical Care 1998; Schrager et al., 1998; Balter, 1997; Wong et al., 1997) Anthony Fauci (Figure 5-8) and other HIV/AIDS scientists told those at the 12th International AIDS Conference (June–July 1998) that every patient, even those in successful therapy for three or more years, have latent HIV pool which is formed early in infection and remains despite the most effective treatment.

People attending the conference were also told that

- HIV/AIDS vaccines are at least 15 years away at best,

- drug resistance and often unbearable side effects from the most powerful of the new drugs are causing many patients to discard them,

- although prevention campaigns are succeeding in a few countries, those successes are spotty even in the developed nations, and

- the first documented cases of HIV resistant to all **protease inhibitors** and **reverse transcriptase inhibitors** were transmitted from one person to another in the United States and in Switzerland.

————— BOX 4.6 (*continued*) —————

In mid-1999, Diana Finzi and colleagues reported that HIV can remain dormant in resting T4 cells. They described the decay rate of the inactive reservoir of HIV in 34 adults who are taking combination therapy and whose plasma levels had fallen to below-detectable level. On average, the half-life of the latent HIV was 43.9 months, which could mean **eradication could take as long as 60 years, a lifetime if the source of the latent virus holds only 1×10(5) or 100,000 cells.** As a result, the researchers say, latent infection of dormant T4 cells provides an opportunity for HIV to remain in all patients, including those who consistently take antiretroviral therapy. This work implies that most HIV-infected people will not be able to stop their anti-HIV medications. Yet, **current drug therapy** most likely **cannot** be tolerated for a lifetime. To date, most people taking anti-HIV drugs have been on them for 2 to 3 years. The longest experience, for relatively few who participated in the early clinical trials for testing of drug cocktails has been about 5 years

AN ALTERNATIVE—REMISSION

HIV/AIDS investigators remain reluctant to use the "Cure" word but they are beginning, in increasing numbers, to believe in **long-term remission** for most HIV infected people. Cecil Fox, one of the first HIV/AIDS investigators to search for HIV in human lymphoid tissue, believes that the HIV/AIDS scientific community should be relating the treatment and results of treating HIV/AIDS patients much the same as people undergoing cancer therapy with the goal of getting the patient into remission.

To this point HIV/AIDS therapists have been treating HIV/AIDS patients for an **infectious disease.** They are looking for a cure: get rid of every last microbe—in this case HIV. But remission, keeping HIV under control is much more reasonable. Make HIV/AIDS into a long term, manageable disease with medications that are tolerable—fewer side effects. The model for this alternative outlook on anti-HIV therapy can be seen in long term, nonprogressors. These people make up a small percent of the HIV/AIDS population, **but they are infected and remain healthy.**

In contrast, Robert Siliciano of John Hopkins Medical Institute calculated that it would take up to 12 years to eradicate from the body those latent HIV-infected cells that are somehow protected from the drugs. Other scientists estimate that it would take 60 years on Highly Active Antiretroviral Therapy (HAART) for the last of the HIV-infected cells to die off! How does one adhere to the demanding drug regimens for all that time? And adherence is a must because any failure could be enough to restock the body with newly protected HIV infected cells and then the whole drug treatment process would have to begin again!

The AIDS cocktail may not provide the light at the end of the tunnel, but for now, at least, it has lengthened the tunnel for those with AIDS. **THE AIDS COCKTAIL IS NOT A CURE, A MORNING-AFTER-A-CARELESS-NIGHT PILL, AND IT IS NOT A REASON TO ABANDON SAFER SEX PRACTICES!**

---

tested. **Drawback:** T-20 is administered by twice daily injections. About 400 more people will have to receive T20 before the drug can receive FDA approval in mid-2000. This is a new and very different class of drugs for anti-HIV therapy. Also in 1998, volunteers began registering for participation to determine the safety and feasibly of anti-HIV gene therapy. Such trials are to begin in mid 1999.

PEDIATRIC ANTI-HIV THERAPY

Most of the questions in the 1998 AIDS Update concerning **when** and **what** should be used in treating pediatric patients have been somewhat resolved in that the CDC issued new pediatric guidelines that advise aggressive antiretroviral therapy for all HIV-infected children younger than 12 months of age "regardless of clinical, immunologic, or virologic status" (Guidelines, Pediatric, 1998; see Chapter 11). In 1999, Sylvia Lee-Huang and colleagues of New York University, Harvard Medical School and the National Institutes of Health reported on the discovery of three enzymes in urine of pregnant women with anti-HIV activity. The enzymes are **lysozyme** and two separate **Ribonucleases.** Lysome is believe to break down the outer membrane of HIV while the ribonucleases degrade HIV's genetic material. Because these are proteins already present in the body, their

———— BOX 4.7 ————

TESTING OF ANTI-HIV DRUGS FOR POST EXPOSURE PREVENTION

The Forum— A woman has sex with her HIV-positive husband and the condom breaks. A woman is raped by a man who is HIV-positive. A child is sodomized by an HIV-positive male. A prison guard is bitten by an HIV-positive inmate. A couple has unprotected sex—a one-night stand. These were some of the cases brought up by experts as they debated whether doctors should be prescribing AIDS drugs as a **morning-after** or post exposure prophylaxis or prevention (PEP) treatment for those exposed to the HIV.

In mid-October 1997, San Francisco became the first city in the United States to offer new PEP drugs to individuals trying to prevent HIV infection. During this study, researchers hope to find out whether individuals will complete the treatment, which is actually a weeks-long combination therapy; whether public knowledge of successful treatment will cause a reversion to high-risk sexual behavior or intravenous drug use; and whether the drugs will produce toxic side effects in people who are otherwise healthy. "**Post-Exposure Prevention,**" or PEP, is a 3-year research study that will enroll a total of 500 people.

72-Hour Window— The **standard of practice** for about two years at most medical centers is to offer anti-viral drugs promptly when health care workers of AIDS patients are stuck with needles or come into contact with body fluids from infected patients. Many researchers believe the drugs must be given within 72 hours of exposure–the time it takes for HIV to integrate into human cells' DNA after infection. The treatments are given for four weeks. In the San Francisco study, individuals who fear they have been exposed to the HIV in the previous 72 hours can obtain treatment and counseling at San Francisco General Hospital's Ward 4C and at the San Francisco City Clinic at 356 Seventh Street.

Beginning 1999 about 165 health care workers have received PEP care—no one has become HIV positive. However, since the risk of HIV infection from a single exposure is small, these results could be explained by chance rather than the effects of PEP treatment. (**To date PEP is not available for general use.**)

Of interest, is the fact that—like sexual behavior and injection drug use—not all potential exposures are equal. When thinking about sex and drug exposures every sexual exposure is not associated with the same risk of becoming infected; neither is every type of injection drug use exposure. In the occupational world, it is quite obvious that deep needle sticks with hollow bore needles that have visible blood and were used in an artery or vein of a terminal AIDS patient are associated with a higher risk of becoming infected. How are these differences taken into account when offering PEP? For those who think PEP is or will be a magic bullet— **no way**—better to focus on practicing medical procedures correctly or being safe the night before.

More detailed information on the research project is available by calling (800) 367-2437 [(800)-FOR-AIDS]. Information is also available on the **Internet** at hivinsite.ucsf.edu by clicking on "Key Topics." As yet, there is no national list of PEP programs. Your local AIDS service organization may be able to help. The CDC published **"Management of Possible Sexual, Injecting-Drug-Use, or Other Nonoccupational Exposure to HIV, Including Considerations Related to Antiretroviral Therapy"** in September 1998. A must read for those with questions on PEP.

CLASS DISCUSSION: With your current knowledge about HIV/AIDS, if you knew that you had just been exposed to HIV would you ask for, immediate therapy? Why?

use as antiviral agents should be well tolerated. Investigations continue.

PROTEASE INHIBITORS AND PREGNANCY

The use of zidovudine (often referred to as AZT) during pregnancy has cut the transmission rate of HIV into children to 8%. In mid-1997, researchers began putting HIV-infected mothers on combination therapy with protease inhibitors to see if they can cut the rate to lower levels. But protease inhibitors are more powerful and potentially more toxic than zidovudine and no one knows what harm it might do to the developing fetus.

For additional considerations for antiretroviral therapy in the HIV-infected pregnant woman, see "Guidelines for the Use of

AIDS DEATHS DECREASING IN THE UNITED STATES

Although there is **no end** of the HIV/AIDS pandemic in sight with the dramatic drop in **deaths** from AIDS in those nations whose people can afford the new anti-HIV drugs, a beginning of a new era is in view! In the United States, the number of deaths from AIDS peaked in 1994 in adults/adolescents and children. AIDS deaths began to drop marginally in 1994, but, there were **12,911 fewer** deaths in 1996 than in 1995 and **16,685 or 47% fewer** deaths in 1997 than in 1996. It is estimated that in 1998 just 11,302 will die and 11,000 in 1999. A drop in AIDS deaths has occurred in almost every state. For example, New York, with 16% of the nations AIDS cases, reported a 48% drop in AIDS deaths for 1997 and a 65% drop over the past two years. In 1995 there were 19 AIDS deaths per day in New York City; in 1997 there were 7! Data from Canada, France, Germany, and Switzerland show a 60% to 80% drop in their death rates (Palella et al., 1998; Hirschel et al., 1998). The number of AIDS deaths dropped in 1997 by at least 29% in each of the four regions of the United States (Northeast, South, Midwest, and West).

The number of AIDS deaths also declined among all racial/ethnic groups (non-Hispanic whites, 28%; non-Hispanic blacks, 10%; Hispanics, 16%; Asians/Pacific Islanders, 6%; and American Indians/Alaskan Natives, 32%) and among men (22%) and women (7%). During 1996, HIV infection dropped to the **second** leading cause of death among persons aged 25 to 44 years, and should soon drop to third place! The drop in AIDS deaths is a direct reflection of better overall access to new drugs and overall beneficial health care for those who have HIV disease and those with AIDS.

CAUTION—The use of anti-HIV combination drug therapy from 1996 into 1999 has given thousands of very ill people an unexpected, much improved **quality** of life. However, while powerful new drug combinations are delaying disease and death, they have serious limitations, and clinicians and patients who ignore these shortcomings do so at their peril. For example, the new drug therapies **do not** allow the immune system to recover, in most people, even when the virus can be suppressed. Researchers have spelled out just how distant the goal is of completely rebuilding a full range of immune responses in an HIV-damaged person. Late in 1997, three disconcerting facts came to light. **FIRST**, using new more sensitive RNA tests, 133 people of 151 people (88%) who previously had **no measurable HIV, had measurable HIV** using the new test procedures. **SECOND, the increase in T4 cells** that usually occurs after using combination drug therapy results from a **redistribution** rather than a **regeneration** of new T4 cells. That is, T4 cells that are being held in the lymph nodes and other tissues are released into the bloodstream, increasing the number present in the blood (Autran et al., 1997). **THIRD**, in September 1997 at the 37th Interscience Infectious Disease Conference, Steven Deeks (1997a) reported **anti-HIV combination drug therapy failed to significantly reduce viral load in 53% of people taking the drugs outside of a clinical trial setting.**

In 1998, a Swiss study found that 49% of people on the latest combination therapy failed to reach undetectable viral loads (less than 500 RNA copies per ml). (Kaufmann et al., 1998).

CLASS DISCUSSION: No one knows just how long the life-saving benefits of the current anti-HIV drugs will last. Will even those who faithfully take their pills on time eventually regress and lose their struggle to a mutant HIV? Support your conclusion with available scientific data.

Antiretroviral Agents in HIV-Infected Adults and Adolescents" (1999).

DISCLAIMER

This chapter is designed to present information about certain aspects of HIV/AIDS therapy. This chapter does **NOT** provide medical advice or replace the advice or care of a medical professional. **ALWAYS** consult a trained health care provider for medical advice and treatment options.

SUMMARY

New results in the field of HIV therapy have given HIV-infected persons new hope in their battle against HIV disease and AIDS. First, com-

DRAMATIC HEALTH IMPROVEMENTS AMONG AIDS PATIENTS CAUSE A SHIFT IN UNITED STATES HEALTH CARE POLICIES

Powerful new AIDS cocktail (combination therapy) treatments have had dramatic results for thousands of patients, reducing the need for hospital and hospice care. This, in turn, had led some hospitals to consider downsizing their AIDS wards. At the same time, doctors and outpatient clinics are reporting increased demand for HIV-related services. Many physicians now say they are taking care of problems in the office that they used to have to treat in the hospital. Hospices have also reported that fewer AIDS patients are seeking end-of-life care, and patients who do choose hospices are waiting longer to enter them, so stays are shorter. Overall, AIDS health care costs are going down nationwide, and the patients are living longer.

THE PRICE FOR SUCCESS: AIDS SOCIAL SERVICE PROVIDERS

Over the last 14 years, an army of AIDS social service providers have been dedicating their lives to caring for people with HIV disease and AIDS. But now their world has changed. With new med-

ications restoring some people to robust health, pressure is building to shift funds away from social services like transportation, housing, and counseling and into drugs and treatment.

In December 1996, Laurie McGinley (*Wall Street Journal*) wrote that Cleve Jones, founder of the AIDS quilt, got a warm reception in October as he walked to the podium of a Washington, D.C. hotel to speak to AIDS social service providers. But his message sent a chill through the audience.

Mr. Jones, who had almost died of AIDS, was now strong thanks to advanced anti-HIV medication. His message was that drugs, research, and treatment—not social services—should be the priority for public and private funds.

"My point was, 'AIDS Inc., I want you out of business,'" he says. "Some people wanted to string me up."

At press time, the gap between the AIDS social service providers and those favoring increased funding for therapy at the expense of social services continues to grow. There may not be an **acceptable** resolution by the opposing forces to this problem.

---

binations of nucleoside analogs compared to zidovudine alone were shown to have a clinical benefit (prolonged survival and fewer AIDS-defining events) when given to asymptomatic individuals with relatively early-stage disease. This was the first demonstration that an intervention regimen used in patients with early-stage HIV disease could actually be clinically beneficial. Next came the demonstration of nucleoside combination therapy, then the extraordinary capability of protease inhibitors. The introduction of the protease inhibitors is the most powerful intervention against HIV to date. Marked reductions in viral load, striking clinical improvement, and reduction in mortality have been observed among patients able to take these medications **properly**. Whether to initiate protease inhibitor therapy in the very early stages of HIV disease is controversial.

Protease inhibitors do not work for all HIV-infected persons; the reasons for this lack of efficacy are not completely understood. Long-

term adverse effects of protease inhibitors are not known at this time. It is hoped that cumulative toxicities from years of protease inhibitor therapy will not offset the benefits. Drug resistance to protease inhibitor therapy can occur during regular use and especially when the drugs are not taken as scheduled. Resistance to one protease inhibitor can also confer cross-resistance to other protease inhibitors, although cross-resistance has not yet been established for the new protease inhibitors currently under investigation. These factors will need to be considered when discussing antiretroviral strategies with patients and their families.

The use of protease inhibitors in combination with nucleoside and non-nucleoside reverse transcriptase inhibitors holds great promise for a level of control of established HIV disease that has, until now, eluded patients and physicians. Furthermore, the use of these drugs in primary HIV infection holds promise for interfering with the establishment of chronic persistent infection

Targets of Anti-HIV Drugs: Generalized Scheme

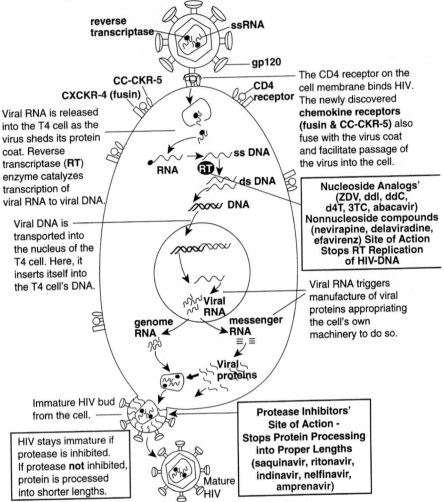

reverse transcriptase

ssRNA

gp120

CC-CKR-5

CXCKR-4 (fusin)

CD4 receptor

The CD4 receptor on the cell membrane binds HIV. The newly discovered **chemokine receptors (fusin & CC-CKR-5)** also fuse with the virus coat and facilitate passage of the virus into the cell.

Viral RNA is released into the T4 cell as the virus sheds its protein coat. Reverse transcriptase (**RT**) enzyme catalyzes transcription of viral RNA to viral DNA.

ss DNA

RNA **RT**

ds DNA

DNA

Nucleoside Analogs' (ZDV, ddl, ddC, d4T, 3TC, abacavir) Nonnucleoside compounds (nevirapine, delaviradine, efavirenz) Site of Action Stops RT Replication of HIV-DNA

Viral DNA is transported into the nucleus of the T4 cell. Here, it inserts itself into the T4 cell's DNA.

Viral RNA

genome RNA **messenger RNA**

Viral RNA triggers manufacture of viral proteins appropriating the cell's own machinery to do so.

Viral proteins

Immature HIV bud from the cell.

Protease Inhibitors' Site of Action - Stops Protein Processing into Proper Lengths (saquinavir, ritonavir, indinavir, nelfinavir, amprenavir)

HIV stays immature if protease is inhibited. If protease **not** inhibited, protein is processed into shorter lengths.

Mature HIV

FIGURE 4-15 Diagram of Anti-HIV Therapy. The six FDA-approved nucleoside analogs, three nucleoside compounds, and six protease inhibitors are represented with respect to their anti-HIV activity. The nucleoside analogs act early after infection, while the protease inhibitors act later in the HIV life cycle, after viral proteins have been synthesized into long strands. Those strands of amino acids contain the individual HIV proteins that become functional after they are cut into their appropriate amino acid sequence lengths. **NOTE: The enzyme integrase is required for HIV DNA to enter human DNA. Drugs called INTEGRASE INHIBITORS are in development.**

and dramatically altering the subsequent course of HIV disease.

The overall goal of antiretroviral therapy will remain largely unchanged in year 2000 from that in 1999. Potent antiretroviral regimens are used to effect sustained maximal suppression of HIV replication. Such suppression is considered important to prevent the emergence

THE COST OF STAYING ALIVE INCREASES WITH EACH NEW THERAPY

Even if protease inhibitors live up to their potential, it is not clear who will be able to afford them. By some estimates, triple drug HIV/AIDS therapy costs between $850 and $1500 a month for life. Hospitalization and other medical care in the final stages of the disease can add $150,000. Future treatments could dwarf even that. Where is this going to stop?

Where will all this money come from? In an era of managed care, insurers do not wish to take on additional liability. And, funds are beginning to dry up. The Federal and State **AIDS Drug Assistant Programs** (ADAPs) were set up in 1987 as a safety net to help provide low-income uninsured and underinsured persons with HIV access to treatment. Beginning year 2000, about 200,000 clients are being served by these programs but experts estimate that anywhere from 140,000 to 280,000 persons may be eligible. This number does not include those persons who may not know their HIV status and who are not in a system of care.

The national ADAP budget for 1998 was $510.2 million (states share was $119.4 million or 23.4%), a 38% increase over 1997. If this rate of increase continues the 1999 budget will be $704 million; for year 2000, it will be $972 million. Regardless of current funding, the **Kaiser Family Foundation** (1999) reported that ADAP programs are spending an average $747 per month per client; that 11 states are restricting access to anti-HIV drugs because they are running out of money; and 14 states expect to run out of money before the end of the fiscal year. Over all, beginning 1999, 98.9% of ADAP clients were age 20 and older, 80% male and 20% female; 40% white, 30% black and 26% Hispanic.

Beginning 1999 46 state ADAPs offered at least one protease inhibitor drug to those with HIV disease. **For fiscal year 1998, ADAPs spent over $200 million buying drugs for 95,000 Americans who are not covered by Medicaid or private insurance.** For 1999 ADAPs will spend about $230 million.

Washington became the latest state, following Illinois, Kansas, Florida, Mississippi, and South Dakota, to dramatically cut its ADAPs as it tried to avert almost certain bankruptcy. Nine other states reported drug assistance shortfalls in 1998.

More cutbacks are pending as states scramble to cover unexpected bills for today's patients, not counting the thousands suddenly demanding protease inhibitor drugs because of headlines promising unprecedented new hope. It may well be that the more sophisticated and effective HIV treatments become in the developed nations, the fewer people will have access to them. What chance does someone with AIDS have to new therapies if they live in an undeveloped nation? Moving into year 2000, **less than 3%** of 43 million people living with HIV infections in developing nations have access to drugs presented in this chapter. People in underdeveloped countries have neither the money to buy the drugs, sufficient people to dispense them if they had them, nor the means by which to keep patients in dosing compliance! And these are just a few of the problems with making medicines available in these nations. Robert Horton, editor of the British medical journal, **The Lancet,** noted that the chance of over 800 million people in the developing world who lack basic health services getting antiretroviral drug therapy is about zero. Calculations on the cost for triple drug therapy world-wide would be about $36 billion and two thirds of that ($24 billion) would have to be spent in Africa.

Antiretroviral Drug Lottery: Guatemala

The time is early 1999. The place—the Luis Angel Garcia AIDS clinic. You can read the torn labels on some of the medicine bottles: Dallas, Los Angeles, the Bronx. The patients they were prescribed for have either died or changed their medications. And so the medicines, $50,000 worth, were donated and have ended up on the shelf in this public AIDS clinic in Guatemala City. The drugs are components of the triple cocktail therapy now staving off death for hundreds of thousands of AIDS patients in the United States and Europe. More than 100 patients at this clinic would like to take the drugs, but there are only enough for six people for six months. This week the clinic will hold a lottery to decide which patients will have the chance to extend their lives.

Who are the contestants? One is a 48-year-old who says she contracted the virus from her husband. The couple and their 17-year-old daughter have since sold most of their possessions and moved into a rent-free shanty on the edge of a ravine. She will have to compete against a 22-year-old male who got the virus while he was living on the streets as a teen-ager. A third competitor is a 45-year-old hairstylist who was infected by a gay lover. But this is all there is—old unused sometimes out-of-date drugs. There is no other way here because there is no money. This is but one example of the global problem in obtaining antiretroviral drugs.

Based just on access to affordable HIV/AIDS treatment, it appears that **AIDS is a two-world dis-** ease, the wealthy versus the indigent. (Class Discussion) Read, "The Cost of Triple-Drug Anti-HIV therapy for Adults in the Americas" JAMA 1997, 279: 1263–1264 and The Kaiser Family Foundation Reported, 1999 to help in your discussion.

of drug-resistant viral variants and to allow the recovery of immune function. While the goal of antiretroviral therapy has remained straightforward, the clinical management of HIV-infected patients has become more complex as an increasing number of drug combinations are evaluated and found potent. In addition, identifying practical strategies for attaining durable viral responses and defining and managing treatment failure remain significant challenges.

SOME AIDS THERAPY INFORMATION HOTLINES

For HIV/AIDS treatment information, call:

The American Foundation for AIDS Research: 800-39AMFAR (392-6237)

AIDS Treatment Data Network: 212-268-4196

AIDS Treatment News: 800-TREAT 1-2 (873-2812)

For information about AIDS/HIV clinical trials conducted by National Institutes of Health and Food and Drug Administration-approved efficacy trials, call:

AIDS Clinical Trials Information Service (ACTIS): 800-TRIALS-A (874-2572)

For more information about HIV infection, call:

Drug Abuse Hotline: 800-662-HELP (4357)

Pediatric and Pregnancy AIDS Hotline: 212-430-3333

National Hemophilia Foundation: 212-219-8180

Hemophilia and AIDS/HIV Network for Dissemination of Information (HANDI): 800-42-HANDI (424-2634)

National Pediatric HIV Resource Center: 800-362-0071

National Association of People with AIDS: 202-898-0414

Teens Teaching AIDS Prevention Program (TTAPP) National Hotline: 800-234-TEEN (8336)

General information:

English: 800-342-AIDS (2437)

Spanish: 800-344-7432

TDD Service for the Deaf: 800-243-7889

General information for health care providers:

National Clinician's Post Exposure Treatment: 800-448-4911

HIV Telephone Consultation Service: 800-933-3413

OTHER ORGANIZATIONS AND WEBSITES

AIDS Action Council

Advocacy; info on public policy and treatment access. 1875 Connecticut Ave. NW, Ste. 700, Washington, DC 20009; 202-986-1300

AIDS Clinical Trials Information Service (ACTIS)

Info in Spanish and English on federal and other clinical drug trials. 800-TRIALS-A; TTY/TDD 800-243-7012; www.actis.org

AIDS Education Global Information System (AEGIS)

Daily news updates, newsletters, trials, search engines; run by Sister Mary Elizabeth, www.aegis.com.

aidsinfonyc.org

Cooperative site run by ATDN, PWA Health Group, Treatment Action Group (TAG), and other organizations. www.aidsinfonyc.org

AIDS Project Los Angeles (APLA)

1313 North Vine St., L.A., CA 90028; 213-993-1600; 204.179.124.76/apla/; Nutrition and Protease Inhibitors

AIDS Treatment Data Network (ATDN)

Clinical trials listing, many glossaries, ADAP, alternative treatments; also in Spanish. 611 Broadway, Ste. 613, NYC, NY 10012; 800-734-7104; aidsinfonyc.org/network

Critical Path AIDS Project

Treatment info, clinical trials, newsletter library. www.critpath.org

Gay Men's Health Crisis (GMHC)

Legal issues, prevention, treatment, counseling and support groups. 129 W. 20th St., NYC, NY 10011; 212-367-1000; Hotline 212-807-6655; www.gmhc.org

Gay and Lesbian Medical Association

459 Fulton St., Ste. 107, San Francisco, CA 94102; 415-255-4547

Health Care Communications

A commercial website with updates on conferences and physicians' and activist reports; www.healthcg.com

HIV/AIDS Treatment Information Services (ATIS)

Info in English and Spanish on FDA-approved treatments. www.hivatis.org

HIV Law Project

Legal advocacy; 212-674-7590

HIV Insite

A top-notch website run by UCSF; hivinsite.ucsf.edu

HIVNet-Information Server. www.hivnet.org

Housing Works

Info, advocacy for HIV-positive homeless. 594 Broadway, NYC, NY 10012; 212-966-0466

Lambda Legal Defense and Education Fund

120 Wall St., 15th Fl., NYC, NY 10005; 212-809-8585

National Association of People with AIDS

Info, advocacy, referrals. 1413 K St. NW, 7th Fl., Washington, DC 20005; 202-898-0414

Project Inform

1965 Market St., Ste. 220, San Francisco, CA 94103; Treatment hot line 800-822-7422; 415-558-8669; www.projinf.org

PWA Health Group

Buyers' club that provides treatment info. 150 W. 26th St., Ste. 201, NYC, NY 10001; 212-255-0520; aidsinfonyc.org/pwahg

San Francisco AIDS Foundation

HIV info. treatment, newsletter, referrals. P.O. Box 426182, San Francisco, CA 94142; 415-487-3000; www.sfaf.org

Internet Addresses

1998 NIH Guideline—Anit-HIV Therapy for Adults/Adolescents **or** for Pediatrics: http://www.hivatis.org

To locate a physician, call your local or state Medical Society.

Note: This is not an all-inclusive list. For other sources of information, contact your state HIV hotline. State HIV hotline numbers are located at the end of the "Introduction."

REVIEW QUESTIONS

(Answers to the Review Questions are on page 487.)

1. Can HIV infection be cured?

2. What is a surrogate marker?

3. From ___ of 1987 through ___ of 1997 ___ anti-HIV drugs were FDA approved.

4. How many FDA-approved anti-HIV drugs are nucleoside analogs?

5. What is the proper drug name for the following drug acronyms: ZDV, ddI, ddC, d4T, and 3TC?

6. How do nucleoside analogs inhibit HIV replication?

7. What are the two major problems in the use of nucleoside and non-nucleoside analogs in HIV therapy?

8. What is HIV viral load? What can its quantitative measurement reveal?

9. Name the six protease inhibitors that have FDA approval for use in the United States.

10. Which of the six FDA-approved protease inhibitors appears to be the most effective and why?

11. How do non-nucleoside drugs inhibit HIV replication?

12. Briefly describe the focus of HIV combination drug therapy.

13. After starting antiretroviral therapy, what is an acceptable target for viral load that indicates the therapy is effective?
    A. 5,000 copies/mL
    B. 5,000–10,000 copies/mL
    C. 10,000–15,000 copies/mL
    D. No acceptable target level has been set.

14. Which of the following statements is correct regarding HIV pathogenesis?

    A. HIV contains two copies of the viral DNA genome.

    B. Viral polyproteins are cleaved by protease.

    C. The RNA copy is integrated in the host-cell chromosome.

    D. gp120 envelope glycoprotein facilitates entry of HIV core particle into the cell's cytoplasm.

15. Go to the library/internet and write a report on the most recent advance in antiretroviral therapy.

16. Compared with the money the federal government spends on research and treatment to combat other health and medical problems such as heart disease and cancer, **do you think federal spending on AIDS research and treatment is too high, too low or about right? Support your choice with credible evidence.**

REFERENCES

AIDS ALERT. (1998). IAPAC Proposes Rationing HIV Drugs in the U.S. 3:9.

AIDS CLINICAL CARE. (1998). Report on the Fifth Conference on Retroviruses and Opportunistic Infections. 10:27–29.

AUTRAN, BRIGITTE, et al. (1997). Positive effects of combined antiretroviral therapy on CD4+ T cell homeostasis and function in advanced HIV disease. *Science,* 277:112–116.

BALTER, MICHAEL. (1997). HIV Survives Drug Onslaught by Hiding Out in T Cells. *Science* 278:1227.

BARTHWELL, ANDREA. (1997). Substance use and the puzzle of adherence. *FOCUS,* 12:1–4.

British Guidelines Coordinating Committee. (1997). British HIV Association guidelines for antiretroviral treatment of HIV seropositive individuals. *Lancet,* 349:1086–1091.

BURMAN, WILLIAM, et al. (1998). The Case for Conservative Management of Early HIV Disease. *JAMA,* 280:93–95.

CARPENTER, CHARLES, et al. (1997). Antiviral therapy for HIV infection in 1997. *JAMA,* 277:1962–1969.

CARPENTER, CHARLES, et al. (1998). Antiretroviral therapy for HIV Infection in 1998: Updated recommendations of the International AIDS Society—USA Panel. *JAMA,* 280:78–86.

CARR, ANDREW, et al. (1998). Pathogenesis of HIV protease inhibitor-associated peripheral lipodystrophy, hyperlipidemia and insulin resistance. *Lancet,* 351:1881–1883.

CARR, ANDREW, et al. (1998a). A syndrome of peripheral lipodystrophy, hyperlipidaemia and insulin resistance in patients receiving HIV protease inhibitors. *AIDS,* 12:F51–58.

CARR, ANDREW, et al. (1998b). Lipodystrophy associated with an HIV protease inhibitor. (Images in Clinical Medicine) *N Engl J Med,* 339:1296.

CARR, ANDREW, et al. (1998c). Abnormal fat distribution and use of protease inhibitors. *Lancet,* 351:1736 (letter).

CARR, ANDREW, et al. (1998d). Pathogenesis of HIV-1 protease inhibitor-associated peripheral lipodystrophy, hyperlipidaemia and insulin resistance. *Lancet,* 351:1881–1883.

CHUN, TAE-WOOK, et al. (1997). Quantification of latent tissue reservoirs and total body viral load in HIV infection. *Nature,* 387:183–188.

CLOUGH, LISA, et al. (1999). Factors that predict incomplete virological response to protease inhibitors-based antiretroviral therapy. *Clinical Infections Diseases.* 29:75–81.

CLUMECK, NATHAN. (1995). Summary to the use of saquinavir for HIV therapy. *AIDS,* 9(Suppl. 2):533–534.

COFFIN, JOHN. (1995). HIV population dynamics *in vivo*: Implications for genetic variation, pathogenesis and therapy. *Science,* 267:483–489.

DANNER, SVEN, et al. (1995). Short term study of the safety, pharmacokinetics and efficacy of ritonavir. *N. Engl. J. Med.,* 333:1528–1533.

DEEKS, STEVEN, et al. (1997a). HIV protease inhibitors. *JAMA,* 277:145–153.

DEEKS, STEVEN, et al. (1997b). Genotypic-resistance assays and anti-retroviral therapy. *Lancet,* 349: 1489–1490.

DE MARTINO, MAURIZIO. (1995). Redox potential status in children with perinatal HIV-1 infection treated with zidovudine. *AIDS,* 9:1381–1383.

DICKOVER, RUTH, et al. (1996). Identification of levels of maternal HIV-RNA associated with risk of perinatal transmission. *JAMA,* 275:599–605.

DOBKIN, JAY. (1997). Fortovase: Son of Invirase *Infections in Medicine.* 14: 926, 934.

DUBÉ, MICHAEL, et al. (1997). Protease associated hyperglycaemia. *Lancet,* 350:713–714.

ERICKSON, JOHN, et al. (1990). Design, activity, and 2.8 angstrom crystal structure of a C_2 symmetric inhibitor complexed to HIV protease. *Science,* 249:527–533.

FINZI, DIANA, et al. (1999). Latent infection of CD4(+) Tcells provides a mechanism for life-long persistence of HIV-1, even in patients on effective combination therapy. *Nature Medicine,* 5:512–517.

FLEXNER, CHARLES. (1996). Pharmacokinetics and pharmacodynamics of HIV protease inhibitors. *Infect. Med.,* 13:16–23.

GERBER, JOHN. (1996). Drug interactions with HIV protease inhibitors. *Improv. Manage. HIV Dis.*, 4:20–23.

GERVASONI, CRISTINA, et al. (1999). Redistribution of body fat in HIV-infected women undergoing combined antiretroviral therapy. *AIDS* 13:465–472.

GOLDSCHMIDT, RONALD, et al. (1995). Antiretroviral strategies revisited. *J. Am. Board Fam. Pract.*, 8:62–69.

GOLDSCHMIDT, RONALD, et al. (1998) Individualized strategies in the era of combination antiretroviral therapy. *J.A.M. Board Family Practice*, 11:158–164.

Guidelines for the Use of Antiretroviral Agents in HIV-Infected Adults and Adolescents. (1999). *MMWR* 47:1–63 No. RR–5.

Guidelines for the Use of Antiretroviral Agents in Pediatric HIV Infection (1998). *MMWR* 47:1–38 No. RR4.

HAASE, ASHLEY, et al. (1997). Kinetics of response in lymphoid tissues to antiretroviral therapy of HIV infection. *Science*, 276:960–964.

HENRARD, DENIS, et al. (1995). Natural history of HIV cell-free viremia. *JAMA*, 274:554–558.

HICKS, CHARLES. (1997). A specialist's view on the case of HIV-infected patients. *J. Clin. Outcomes Manage.*, 4:65–66.

HIRSCHEL, BERNARD et al. (1998). Progress and Problems in the Fight Against AIDS. *N. Engl. J. Med.*, 338:906–908.

HO, DAVID, et al. (1995). Rapid turnover of plasma virions and CD4 lymphocytes in HIV-1 infection. *Nature*, 373:123–126.

HU, DALE, et al. (1996). The emerging genetic diversity of HIV. *JAMA*, 275:210–216.

JACOBSEN, HELMUT, et al. (1995). Reduced sensitivity to saquinavir: An update from genotyping from phase I/II trials. 4th International Workshop on HIV Drug Resistance, Sardinia.

JOHN, PATRICIA, et al. (1996). Predictive value of viral load measurements in asymptomatic untreated HIV infection: A mathematical model. *AIDS*, 10:255–262.

JURRIAANS, SUZANNE, et al. (1994). The natural history of HIV infection: Virus load and virus phenotype independent determinants of viral course? *Virology*, 204:223–233.

KAHN, JAMES, et al. (1998). Acute Human Immunodeficiency virus Type I Infection. *N. Engl. J. Med.*, 339:33–39.

KAISER FAMILY FOUNDATION. (1999). National monitoring project: *Annual Report* 1–45 and Appendix 1–14.

KAUFMANN, DANIEL, et al. (1998). CD4-cell count in HIV-1 infected individuals remaining viraemic with HAART. (Swiss HW Cohort Study) *Lancet*, 351:723–724.

LAURENCE, JEFFREY. (1996). The clinical promise of HIV protease inhibitors. *AIDS Reader*, 6:39–41, 71.

LEE HUANG, SYLVIA, et al. (1999). Lysozyme and RNases as anti-HIV components in beta-core preparations of human chorionic gonadotropin. Proceedings of the National Academy of Sciences. 96:2678–2681.

LEWIS, CHARLES. (1997). Management of patients with HIV/AIDS: Who Should Care? *JAMA*, 278:1133–1134.

LIPSKY, JAMES. (1998). Abnormal Fat Accumulation in Patients with HIV-I Infection. *Lancet*, 351:847–848.

LO, JOAN, et al. (1998). "Buffalo Hump" in Men with HIV-1 Infection. *Lancet* 351:867–870.

MARKOWITZ, MARTIN, et al. (1995). A preliminary study of ritonavir, an inhibitor of HIV protease. *N. Engl. J. Med.*, 333:1534–1539.

MAYERS, DOUGLAS. (1996). Rational approaches to resistance: Nucleoside analogues. *AIDS*, 10 (Suppl. 1):S9–S13.

MELLORS, JOHN, et al. (1995). Quantitation of HIV-1 RNA in plasma predict outcome after seroconversion. *Ann. Intern. Med.*, 122:573–579.

MELLORS, JOHN. (1996). Clinical implications of resistance and cross-resistance to HIV protease inhibitors. *Infect. Med.*, 13:32–38.

MERRICK, SAMUEL. (1997). Managing antiretrovirals in HIV-infected patients. *AIDS Reader* 7:16–27.

MERRY, CONCEPTA, et al. (1997). Saquinavir Pharmacokinetics alone and in combination with nelfinavir in HIV infected patients. *AIDS* 11:F117–F120.

MERRY, CONCEPTA, et al. (1997). Saquinavir pharmacokinetics alone and in combination with ritonavir in HIV-infected patients. *AIDS*, 11:F29–F33.

MILLER, KIRK, et al. (1998). Visceral abdominal-fat accumulation associated with use of indinavir. *Lancet* 351:871–875.

Morbidity and Mortality Weekly Report. (1994). Zidovudine for the prevention of HIV transmission from mother to infant. 43:285–287.

MOYLE, GRACME. (1995). Resistance to antiretroviral compounds: Implications for the clinical management of HIV infection. *Immunol. Infect. Dis.*, 5:170–182.

NEWMAN, MEG. (1999). Trials, Lies, and Options. *AIDS Clinical Care*, 11:12-13.

NIGHTINGALE, STUART. (1996). From the Food and Drug Administration: First protease inhibitor approved. *JAMA*, 275:273.

NOWAK, MARTIN. (1995). How HIV defeats the immune system. *Sci. Am.*, 273:58–64.

PALELLA, FRANK, et al. (1998). Declining morbidity and mortality among patients with advanced HIV infection. *N. Engl. J. Med.*, 338:853–860.

PIATAK, MICHAEL, et al. (1993). High levels of HIV-1 in plasma during all stages of infection determined by competitive PCR. *Science*, 259: 1749–1754.

RICH, JOSIAH, et al. (1999). Misdiagnosis of HIV infection by HIV plasma viral load testing: A case series. *Annals of Internal Medicine Online* (1/5/99) 130:37.

ROBERTS, JOHN, et al. (1988). The accuracy of reverse transcriptase from HIV-1. *Science*, 242: 1171–1173.

SAKSELA, KALLE, et al. (1994). Human immunodeficiency virus type 1 mRNA expression in peripheral blood cells predicts disease progression independently of the number of CF4+ lymphocytes. *Proc. Natl. Acad. Sci. USA*, 91:1104–1108.

SCHMIT, JEAN-CLAUDE, et al. (1996). Resistance-related mutations in the HIV protease gene of patients treated for 1-year with protease inhibitor ritonavir. *AIDS*, 10:995–999.

SCHRAGER, LEWIS, et al. (1998). Cellular and anatomical reservoirs of HIV in patients receiving potent antiretroviral combination therapy. *JAMA*, 280:67–71.

SIMBERKOFF, MICHAEL. (1996). Long-term follow-up of symptomatic HIV-infected patients originally randomized to early vs. later zidovudine treatment: Report of a Veterans Affairs cooperative study. *AIDS*, 11:142–150.

STEPHENSON, JOAN. (1996). New anti-HIV drugs and treatment strategies buoy AIDS researchers. *JAMA*, 275:579–580.

VALDEZ, HERNAN, et al. (1999). Human immunodeficiency vivas 1 protease inhibitors in clinical practice. *Archives of Internal Medicine*. 159:1771–1776.

VELLA, STEFANO. (1995). Clinical experience with saquinavir. *AIDS*, 9(Suppl 2):S21–S25.

VELLA, STEFANO. (1997). Clinical implications of resistance to antiretroviral drugs. *AIDS Clin. Care*, 9:45–47, 49.

VELLA, STEFANO, et al. (1996). HIV resistance to antiretroviral drugs. *Improv. Manage. HIV Dis.*, 4:15–18.

WAIN-HOBSON, SIMON. (1995). Virologies mayhem [editorial]. *Nature*, 373:102.

WALKER, BRUCE, et al. (1998). Treat HIV infection like other infections—Treat it. *JAMA*, 208:91–93.

WEI, XIPING, et al. (1995). Viral dynamics in human immunodeficiency virus type 1 infection. *Nature*, 373:117–122.

WONG, JOSEPH, et al. (1997). Recovery of replication-competent HIV despite prolonged suppression of plasma viremia. *Science* 278:1291–1294.

ZUGER, ABIGAIL, et al. (1997). HIV specialists: The time has come. *JAMA*, 278: 1131–1132.

The Immunology of HIV Disease/AIDS

CHAPTER CONCEPTS

- HIV **attaches** to CD4 receptor sites on T4 lymphocyte, monocyte, and macrophages cells.
- CC-CKR-2, CC-CKR-3, CXCKR4 (FUSIN), and CC-CKR-5 are coreceptors that allow HIV to **enter** cells.
- Monocytes, macrophages follicular dendritic and T4 cells serve as HIV reservoirs in the body.
- Basic immune system terminology is defined.
- T4, T8, and B cell function is related to HIV disease.
- Apoptosis is a normal mechanism of cell death.
- Cofactors may enhance HIV infection.
- Impact of T4 cell depletion is immune suppression.
- Latency refers to inactive proviral HIV DNA, but true latency may not exist for HIV.
- Clinical latency refers to infection with low-level HIV production over time.

THE IMMUNE SYSTEM

All living organisms are continually exposed to substances that are capable of causing harm. Most organisms protect themselves against such substances in more than one way—with physical barriers, for example, or with chemicals that repel or kill invaders. Animals with backbones, called **vertebrates,** have these types of general protective mechanisms, but they also have a more advanced protective system called an **immune system.** The immune system is a complex network of organs containing cells that recognize foreign substances in the body and destroy them. It protects vertebrates against pathogens, or infectious agents, such as viruses, bacteria, fungi, and other parasites. The human body immune system is the most complex.

Although there are many potentially harmful pathogens (agents that cause diseases), no pathogen can invade or attack all organisms because a pathogen's ability to cause harm requires a susceptible victim, and not all organisms are susceptible to the same pathogens. For instance, the virus that causes AIDS in human does not cause a disease in animals such as dogs, cats, and mice. Similarly, humans are not susceptible to the viruses that cause canine distemper or feline (cat) leukemia.

DEVELOPMENT OF HUMAN IMMUNE SYSTEM CELLS

The beginning of the immune system begins during the **first month** of fetal development with occurrence of blood stem cells in the yolk sac. By the third month of development, blood cell development occurs predominantly in the liver until the skeletal elements of the fetus are formed. Thereafter, bone marrow is the major site of blood cell formation. Blood stem cells differentiate into granulocytes, monocytes, and lymphocytes, as well as megakaryocytes and erythrocytes. Beginning 2 months into development, lymphocytes destined to become T cells move from the bone marrow into the developing thymus to complete their maturation. The maturation of B lymphocytes occurs in bone marrow under the influence of the stromal reticular cells. By birth, the immune system has developed into a sophisticated network, connecting central locations of immune cell production with peripheral tissues for immune surveillance. These peripheral components of the immune system include the blood, thymus, lymphatic system, spleen, skin, and mucosa. **Host defense is the culmination of a carefully orchestrated cellular and molecular interactions within the immune system and between the immune system and the rest of the body.**

Function of the Immune System

The immune system filters out foreign substances, removes damaged and dead cells and acts as a security system to destroy mutant and cancer cells. It is composed of a number of specialized cells, several organs, and a group of biologically active chemicals. The human immune system is like a jigsaw puzzle—many parts come together to form an overall defense against disease-causing agents. If parts of the immune system are missing or damaged, illness may occur due to an immune deficiency.

Separating Friend From Foe: How The Immune System Decides

Over most of the body, the skin prevents disease-causing agents from entering the body. But they can enter through body openings, cuts, or wounds. Whether the invader is a life-threatening bacterium or a relatively harmless cold virus, your immune system must control it. A single infectious microorganism that survives and multiplies causes illness. Common infectious agents include bacteria, viruses, and parasites such as fungi, worms and single-cell protozoa. Some of these infectious agents resemble the body's own cells. **How do immune cells know which to attack and which to ignore?** The answer is, that once an agent enters the body, it triggers an **immune response** if the body does not recognize the substance or agent as a part of itself or **"self."** All body cells have special molecules, called **class I proteins,** on their membranes that are like flags or barcodes with the word **"self"** on them. The cells of the immune system try to destroy anything present in the body that is not carrying the self molecules, anything that is **"nonself."** Nonself is any substance or object that triggers the creation of antibodies. Such substances are called **antigens.** Antigens may be whole virus or organisms or parts of virus, organisms, or their products.

In general, most organisms that damage cells do so from the outside by producing toxic chemicals or in some way externally interfering with the cell's metabolism. But viruses generally invade or **enter** different cell types forcing them to produce viral replicas at the expense of the cells' own essential metabolic functions. Gradually, like a machine wearing out, host cells start to malfunction and die. The best thing a virus can do is find a host cell that does not die and that can produce replicas indefinitely. In a biological time frame, new disease-causing viruses are often very deadly to new hosts. If a new virus strain is too deadly, it kills its host before other hosts are infected, and the ensuing epidemic dies out. For example, the **Ebola virus,** which makes its victims very weak shortly after infection and kills them in 7 to 14 days, has had seven brief outbreaks since its first appearance in 1976, the latest in November 1996. This virus kills quickly and vanishes. Its origin is still unknown. Over bio-logical time, successful viruses and their new hosts learn to accommodate each other. This will most likely happen with human-HIV associations, but how many people over how many years will have to die before human cells learn to accommodate HIV is unknown. Perhaps it will never happen. Smallpox virus has been infecting humans for thousands of years and has never been accommodated by humans.

--------- BOX 5.1 ---------

HOW LARGE IS A TRILLION?

The human body is made up of many trillions of cells. For example, the human immune system contains at least a trillion lymphocytes dedicated to destroying foreign substances that endanger health. But, a trillion of anything is a very large number. Can we **really appreciate just how large a trillion of something is?** Perhaps the following will help.

1. A stack of 1 trillion dollar bills would reach a height of 69,000 miles.

2. It would take a person 11.5 days to count to 1 million and 31,688 years to count to 1 trillion—1, 2, 3, 4, 5. . . . !

3. A stack of 1 trillion HIV (Figure 5-1) would be **over** 62 miles high (diameter of HIV = 1,000 angstroms: (1×10^{12}) $(1,000)$ $(1 \times 10^{-10}) = 1 \times 10^5$ meters = 62.15 miles)

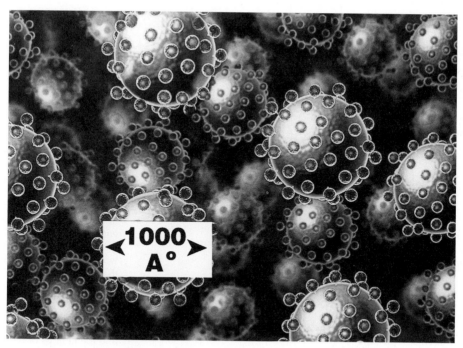

FIGURE 5-1   Dimensions of HIV in Angstroms. HIV has a diameter of 1,000 angstroms (Å). It would take 254,000 HIV laid side by side to equal 1 inch in length. (*Photograph courtesy of the author*)

HUMAN LYMPHOCYTES:
T CELLS AND B CELLS

The hallmark of the human immune system is its ability to mount a highly specific response against virtually any foreign entity, even those never seen before in the course of evolution. It is able to do this because of the number of different kinds of cells called lymphocytes. The human immune system contains about 2 trillion (2×10^{12}) lymphocytes, a relatively small number when compared to the trillions upon trillions of other types of cells in the body (see Box 5.1, Figure 5-1). Most mature lymphocytes recirculate continuously, going from blood to tissue and back to blood again as often as one to two times per day. They travel among most other cells and are present in large numbers in the thymus, bone marrow, lymph nodes, spleen, and appendix (Figure 5-2). By 1968,

lymphocytes had been divided into two classes: lymphocytes called **B cells** that are derived from and mature in bone marrow, and lymphocytes called **T cells** that are derived from bone marrow but travel to and mature in the thymus gland (Figure 5-3). T cells make up 70% to 80% of the lymphocytes **circulating** in the body. Circulating T cells are a heterogeneous group of cells with a wide range of different functions. Each individual T cell expresses a receptor **(T cell antigen receptor, TCR)** which recognizes a ligand (a compound that fits a particular receptor) composed of an antigenic peptide, 8–15-amino-acid-long, bound to a **self-major-histocompatibility-complex (MHC) molecule** (also referred to as HLA-human leukocyte antigen). Thus, a T cell does not directly recognize a soluble antigen, but rather recognizes an antigen displayed on the surface of an **antigen-presenting cell (APC)** like a B cell

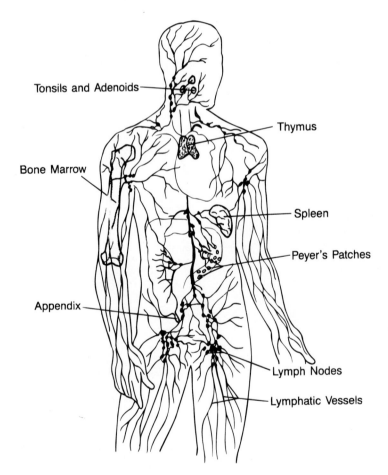

Tonsils and Adenoids

Thymus

Bone Marrow

Spleen

Peyer's Patches

Appendix

Lymph Nodes

Lymphatic Vessels

FIGURE 5-2 Organs of the Human Immune System. The organs of the immune system are positioned throughout the body. They are generally referred to as lymphoid organs because they are concerned with the development, growth, and dissemination of lymphocytes, or white cells, that populate the immune system. Lymphoid organs include the bone marrow, thymus, lymph nodes, spleen, tonsils, adenoids, appendix, and the clumps of lymphoid tissue in the small intestine called Peyer's patches. The blood and lymphatic fluids transport lymphocytes to and from all of the immune system organs.

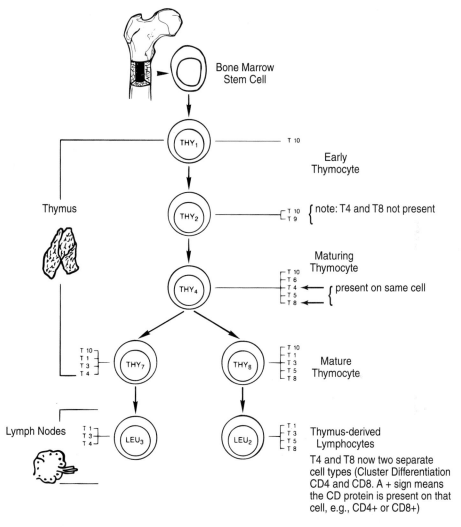

FIGURE 5-3 T lymphocyte Cluster Differentiation (CD) Antigens in Humans. Stages of thymic differentiation (i.e., the presence of the different antigens on their membrane surfaces) are defined on the basis of reactivity to monoclonal antibodies. Schematic pictures of cells represent thymocytes within specific stages of a defined phenotype: T1–T10 and membrane glycoproteins of T cells.

or macrophage (Table 5-1). There are about 10 billion APCs located in the lymphoid organs. The receptors of T cells are different from those of B cells because they are "trained" to recognized fragments of antigens that have been combined or complexed with MHC molecules. As T cells circulate through the body, they scan the surfaces of body cells for the presence of foreign antigens that have been picked up by the MHC molecules. This function is sometimes called immune surveillance.

Three major kinds of T cells are **cytotoxic or killer** T cells, **suppressor** T cells and helper T cells. Both killer and suppressor T cells carry the **CD8 antigen.** T suppressor cells are referred to as **T8 cells.** Helper T cells carry the **CD4 antigen** and are called **T4 cells** (for an explanation of CD8 and CD4 antigens, see Box 5.2, T lymphocytes). Killer T cells bind to cells carrying a foreign antigen and destroy them. **T4 cells do not kill cells; they interact with B cells and killer T cells and help them respond to foreign antigens.**

TABLE 5-1   Cells of the Human Immune System

Cell Type	Function
Stem cells	Self perpetuating cells that give rise to lymphocytes, macrophages, and other hematological cells
T helper cells (T_H), also called T4 or CD4 cells	Cells that interact with an antigen before the B cell interacts with the antigen
Cytotoxic cells (Tc)	Recruited by T helper cells to destroy antigens
Suppressor T cells (Ts), also called T8 or CD8 cells	Cells that dampen the activity of T and B cells; they somehow inhibit the immune response
B cells	Bone marrow stem cells that differentiate into plasma cells that secrete antibodies
Plasma cells	Cells devoted to the production of antibody directed against a particular foreign antigen
Monocytes	Precursors of macrophages
Macrophages	Differentiated monocytes that serve as antigen-presenting cells to T cells; they can also engulf a variety of antigens and antibody-covered cells
Killer cells (K)	Lymphocytes that recognize and kill any cell that is coated with antibody
Natural killer cells (NK or null cells)	Lymphocyte cells that detect and kill tumor cells and a broad range of foreign cells

--------- **BOX 5.2** ---------

TO UNDERSTAND THE DIFFERENCE BETWEEN HEALTH AND DISEASE, IMMUNOLOGISTS NEED TO UNDERSTAND THE IMMUNE SYSTEM!

Basic Immune System Terminology

Of the many mysteries of modern science, the mechanism of **self** versus **nonself recognition** in the immune system must rank near the top. The immune system is designed to recognize foreign invaders. To do so it generates on the order of 10^{12} or one trillion different kinds of immunological receptors so that no matter what the shape or form of the foreign invader there will be some complementary receptor to recognize it and effect its elimination. Understanding how the immune system responds to any foreign substance is puzzle enough, but the added mystery is that the immune system can distinguish foreign carbohydrates, nucleic acids, and proteins from those that exist within us, often in shapes barely distinguishable from the invaders. When the immune system is working well it never gets activated by self substances, but unerringly responds to the nonself substances. When the system is not working well this distinction gets blurred and **diseases of autoimmunity occur—our immune cells attack our own tissue!**

To understand the human immune system certain basic terminology is reviewed:

Antibody—an immunoglobulin (a protein) which is produced and secreted by B lymphocytes in response to an antigen. Antibodies are able to bind to and destroy specific antigens. Antibodies contribute to the destruction of antigen by their interaction with other components of the immune system.

Antigen—a substance that is recognized as foreign by the immune system when introduced into the body. An antigen can be a whole microorganism (such as a bacterium or whole virus) or a portion or a product of an organism or virus.

Leukocytes—all white blood cells (WBCs) including neutrophils, lymphocytes, and monocytes (phagocytes).

Lymphocytes—mononuclear WBCs which are critical in immune defense because they provide the specificity and memory needed for immune function and long-term or even life-long immunity. The two major classes of lymphocytes are B cells and T cells; both recognize specific antigens. (For an excellent review of lymphocyte life span and memory, see Sprent et al., 1994.)

Lymphocyte Surface Receptors—Proteins are located on cell surfaces to serve a physiologic function. Some are enzymes, some are transport proteins, and many are **receptors.** All cells use specific receptors to communicate with their environment or with other cells. The receptors found on lymphocytes may be loosely classified according to their function. Thus lymphocytes require receptors that recognize antigen. They need receptors for antigen-presenting cells and receptors for the many factors that regulate lymphocyte responses.

Chemokines—also called cytokines, activate and **DIRECT** the migration of leukocytes. There are 28 known chemokines separated into two sub-families, those whose protein amino acid structure begins with cytosine–amino acid–cytosine, called the CXC chemokines, and those whose protein amino acid structure begins with cytosine–cytosine, called the CC chemokines.

Cytokines—soluble factors secreted by T cells, B cells, and monocytes which mediate complex immune interactions by acting as messengers. Subcategories of cytokines include: (1) lymphokines (secreted by lymphocytes), and (2) monokines (secreted by monocytes and macrophages). Specific cytokines include interleukin-1 (IL-1), interleukin-2 (IL-2), gamma interferons (γINF), B cell stimulating factor (BCSF), and B cell differentiating factor (BCDF).

Two Branches of the Immune System—The two branches of the human immune system are: Cellular Immunity and Humoral Immunity. Basic terms for each are provided, but as the terms will show, the two branches overlap to provide our immunity. (Adapted from *Mountain-Plains Regional HIV/AIDS Curriculum*, 1992.)

I. *Cellular Immunity*—immune protection resulting from direct action of cells of the immune system. Key cell types offering Cellular Immunity are:

A. *Phagocytes*
   1. Phagocytes are leukocytes which are specialized for ingesting particles and molecules.
   2. The most active phagocytic cells are monocytes and macrophages. Monocytes circulate in the blood but eventually move into body tissues (brain, muscle, etc.) where they mature into macrophages.
   3. Phagocytes initiate the cellular immune response. When an invader (e.g., a virus or bacteria) enters the body, it will be trapped and digested by a phagocyte. This attack against invaders occurs in

various places in the body: the lining of the gut, throat, skin, bloodstream, or in organized lymphoid tissue such as the tonsils, other lymph nodes, or the spleen.

   4. The phagocyte ingests the foreign invader and partially digests it. Pieces of the invader (antigens) can then be displayed on the phagocyte's surface, making the phagocyte an **antigen-presenting cell.** This process alerts other cells in the immune system that a foreign substance is present.

B. *Lymphocytes*
   1. Ninety-eight percent of lymphocytes reside in lymphoid tissue (Pantaleo et al., 1993). Lymphocytes are uniquely specialized. Each lymphocyte has receptors on its surface for one, and only one, of the many millions of possible antigens that can invade the body. When the receptors lock with their matched antigen, the process of neutralizing, inactivating, or destroying the foreign particle begins.

   2. This requires that the body have an immense variety of receptors for an immense number of antigens.

   3. Humans usually have about a trillion lymphocytes. Several types of these lymphocytes play major roles in the immune response.

   a. *T Lymphocytes* mature in the thymus and play a central role in the immune response by destroying infected cells, controlling inflammatory responses, and helping B cells make antibodies. There are several subsets of T lymphocytes (*Subsets of T cells display different proteins in groups (antigens) on their cell membranes. For this reason they are referred to as cluster differentiation (CD) antigens. At least 130 different cluster differentiating antigens are known to reside on different human cell types. The different CD antigens located on T lymphocyte subsets allow investigators to distinguish between the different T cell lines. For example, in physical appearance it is nearly impossible to distinguish T cells from B cells. But each carries a different CD antigen. T lymphocytes attached by HIV carry type 4 antigen and are referred to as CD4 + T lymphocytes or T4 cells, likewise for T8 cells. B lymphocytes do not carry the T4 or T8 antigen.*)*

———————— **BOX 5.2** (*continued*) ————————

1) **Helper T cells** (T_H) alert the immune system to the presence of an antigen and activate other cells in the system. Helper T cells carry about 10,000 copies of the CD4 protein molecule on the surface of each helper T cell so these **CD4+** cells are most often referred to as **T4** cells.

   a) T4 cells have receptors (TCR) on their surface which are specialized for the recognition of antigens found on the surface of antigen-presenting cells. It has been estimated that 10^{11} to 10^{15} differenct TCRs can be generated in humans. The total of the different TCR specificities of a given individual constitutes their "T-cell repertoire." This is a rough measure of the capacity of a given individual to respond to the myriad of foreign microbes, antigens, and so forth, that he or she will be confronted within his or her lifetime.

   b) When the T4 cell randomly contacts the surface of a phagocyte, its receptor binds to antigen on the phagocyte and the T4 cell becomes activated.

   c) The T4 cell then begins to secrete a variety of stimulatory factors (*lymphokines*) into the space around it.

   d) The T4 cell will eventually divide into two cells of exactly the same specificity. If the foreign substance persists, the two daughter cells will divide, and so on. Thus, the number of cells specific for that foreign antigen is greatly expanded.

   e) The T4 cell probably does little to repel the intruding substance on its own, but it is vital for activating the other main classes of lymphocytes (e.g., B cells, natural killer cells and phagocytes). It does this by means of the lymphokines it secretes when activated. The most familiar of these are interleukin-2 (IL-2) and gamma interferon (γIFN).

   f) As a result of activation by T4 cells, B lymphocytes multiply and produce specific antibodies. The antibodies attach to the antigens on free organisms and infected cells, leading to inactivation and destruction of organisms and cells.

2) **Inducer T cells** (T_i) were previously known as the delayed sensitivity T cells. Inducer T cells also carry the CD4 molecule.

   a) When an inducer T cell recognizes antigen on the surface of a phagocyte and is helped by factors from a T4 cell, it also becomes activated and secretes its own family of lymphokines, IL4, IL5, and IL10. These factors attract many more phagocytes from around the body which accumulate in the area where antigen is being recognized.

   b) Phagocytes are specialized for ingesting and destroying the foreign invader.

3) **Cytotoxic T** lymphocytes (CTL) are also known as killer T cells. CTLs express the CD8 molecule and are most often referred to as **T8** cells. They are crucial for the immune response to viral infections. Although they cannot neutralize free virus, by eliminating virus infected-cells they are largely responsible for recovery from a viral infection.

   a) The recognition of invading microbes by CTLs is a complex business. An infected cell must first chemically chop up the microbe's protein into small fragments, or peptides. These peptides are transported to the cell surface, where they become bound to specialized molecules called human leukocyte antigens (HLAs). Special receptors on the CTL recognize the HLA-peptide complex, and the CTL then kills the infected cell by unloading a cocktail of cytotoxic chemicals into it. But even a small mutation can change the microbes peptide's structure enough so that it will no longer bind to HLA—or, alternatively, so that the peptide is no longer transported to the cell surface—and the infected cell then becomes "invisible" to the CTL.

   b) As the infected cell breaks apart, infectious particles can be released from the cell. The released infectious particles may be phagocytized, neutralized by antibody, or may infect new cells.

c) Cytotoxic T8 cells have antigen specificity; this specificity is determined when the cytotoxic T8 cell is exposed to the antigen.

b. **B Lymphocytes** or **B cells** arise principally in the bone marrow and are responsible for making antibodies. The B cell response is referred to as humoral immunity.

1) B lymphocytes have receptors for antigen and also require help from T4 cells to become activated.

2) When activated, they release large amounts of immunoglobulins, called *antibodies*, (about 1000 per second) into the bloodstream and mucosal surfaces.

3) Antibodies circulate in body fluids. When they encounter the specific antigen with which they can interact, they bind to it.

4) Antigen–antibody interactions block the antigen's harmful potential in a variety of ways. For example, if the antigen is a toxin it may be neutralized or if it is a virus, it may lose its ability to bind to its target tissue.

5) In addition, the complex of antigen and antibody activates a group of proteins in the blood called *complement*, which then facilitates the removal of the antigen by calling in a large number of phagocytes.

c. **Memory cells**

1) Two groups of T and B lymphocytes become separate T and B memory cells for when the same antigen invades the body again.

2) Memory cells initiate immune response upon re-exposure to an antigen.

3) Memory cells remember, recognize, and induce a more rapid immune response against the antigen.

4) Memory cells are responsible for disease resistance after immunization and are also responsible for natural immunity.

d. **Suppressor T cells** also carry the CD8 molecule.

1) The function of suppressor T8 lymphocytes in the immune system is not well understood.

2) Suppressor T8 cells turn off antibody production and other immune responses after an invader has been destroyed or eliminated from the body. This provides a balance in the immune system and allows it to rest when its functions are not required.

e. **Natural killer (NK) cells**

1) NK cells are antigen nonspecific lymphocytes which recognize foreign cells of many different antigenic types. That is, one NK cell can recognize many different types of invaders. Therefore, NK cells can attack without first having to recognize specific antigens.

2) NK cells are important in fighting viral infections.

II. **Humoral Immunity**—immune protection by the circulating antibodies which are produced by B cells. (See preceding section on **B Lymphocytes**.)

It is believed that T4 cells recognize only those antigens of viruses, fungi, and other parasites; and trigger only those parts of the immune system necessary to act against these agents. Indeed, it is the viruses, fungi, and other parasites that produce the majority of opportunistic infections when the T4 cells have been depleted by the AIDS virus.

In Summary

The two types of lymphocytes, B cells and T cells, play different roles in the immune response, though they may act together and influence one another's functions. The part of the immune response that involves B cells is often called **humoral immunity because it takes place in the body fluids.** The part involving T cells is called cellular immunity because it takes place directly between the T cells and the antigens. This distinction is misleading, however, because, strictly speaking, all adaptive immune responses are cellular—that is, they are all initiated by cells (the lymphocytes) reacting to antigens. **B cells** may initiate an immune response, but the triggering antigens are actually eliminated by soluble products that the B cells release into the blood and other body fluids. These products are called antibodies and belong to a special group of blood proteins called immunoglobins. When a B cell is stimulated by an antigen that it encounters in the body fluids, it transforms, with the aid of a type of T cell called a helper T cell, into a larger cell called a blast cell. The blast cell begins to divide rapidly, forming a clone of identical cells. Some of these transform further into plasma cells—in essence, antibody-producing factorys. These **plasma cells produce a single type of antigen-specific antibody of about 2,000 antibodies per second.** The antibodies then circulate through the body fluids, attacking the triggering antigen.

The Microbial World Responds to the Human Immune System

Large DNA viruses such as adenovirus, herpes simplex viruses 1 and 2, Epstein-Barr virus, and cytomegalovirus have evolved sophisticated mechanisms to avoid detection by the human immune system. Most of these mechanisms are geared toward preventing the expression of antigenic viral peptides that are bound to major histocompatibility complex (MHC) class I molecules, on the surface of the infected cells (antigen presentation). In this way, the viruses turn off the signal that would alert the immune system to the presence of an intracellular pathogen and, undetected by antigen-specific T lymphocytes, the infected cells continue to provide a safe haven for the viruses.

T4 Cell Function and HIV Disease

Humans have about 1 trillion T cells. In general, individuals with partial or absolute defects in T cell function have infections or clinical problems for which there is no effective treatment. **The T cell disorders are more severe than the B cell disorders.**

The discovery of the relationship between T4 lymphocyte count and HIV disease progression is based on the observation that losses in T4 counts leads to the development of **opportunistic diseases,** the diagnoses of AIDS, and death. T4 cell counts reliably track the natural history of HIV infection, particularly after an AIDS-defining illness is diagnosed. T4 cells are selectively deleted as a result of a series of events initiated by the binding of HIV to the CD4 molecule (located on the T4 cell membrane) by means of the viral envelope protein gp120. Once HIV enters a T4 cell, the cell begins to lose normal function. This change, however, is not

immediately apparent. HIV's takeover is a quiet event. The virus joins the host cell's DNA, then, instead of functioning normally, the T4 cell first manufactures new HIV and then dies. The loss of T4 cells severely reduces cell-mediated immunity, eliminates the T4 cell dependent production of antibodies by B cells, and eventually makes HIV-infected people susceptible to opportunistic infection and subsequent death. **AIDS is diagnosed when the T4 cell count drops to less than 200/µL of blood.** (T4 counts from 800 to 1,200 are considered normal in adults.) T4 cell counts of less than 100 are associated with **profound** immunodeficiency and multiple or disseminated opportunistic infections (Crowe et al., 1991). (See Chapters 6 and 7.)

One of the most interesting findings has been the discovery of a subset of healthy, **long-term survivors** who have lived for years with T4 counts of less than 200 (see Chapter 7, Box 7.4). For the most part, they do not develop the secondary infections that are associated with AIDS, or if they do, they tend to recover. This means that there is a lot about the immune system that immunologists still do not understand. Some researchers believe these men and women managed to press other white blood cells into service to make up for the T4 deficiency. It is also possible that persons with low T4 cell counts actually have more T4 cells than those detectable in peripheral blood measurements. In fact it is now known that most T4 cells are sequestered in lymphoid tissue. Thus the true number of T4 cells in the body can only be established by measuring their presence in tissues other than blood. Thus long-term survivors may have less damage to their lymphoid tissues than do those with less survival time.

B Cell Function and HIV Disease

Humans have about 1 trillion B cells. The ultimate product of the B cells is **immunoglobulin.** Any abnormalities in the transition of the B lymphocyte from the stem cell level to the plasma cell may affect imunoglobulin synthesis. The earlier the rupture in the developmental pathway, the more extensive the loss in antibody-forming capacity.

The B lymphocytes can produce and secrete soluble antibodies in response to direct contact with an antigen (T cell-independent B cell

MEASURES AND COUNTERMEASURES: THE HUMAN IMMUNE SYSTEM VS. VIRUSES

A war has existed between humans and viruses that invade human cells and cause disease. Probably from the first encounter, pre-human to first humans, viruses became able to **hide** inside human cells. The immune system in turn adapted by developing a means of surveying and identifying cells containing hidden virus. Virus then adapted by deceiving the immune surveillance system; the immune system then evolved a better means to **mark** cells carrying the virus. This game of hide-and-seek between the virus and the immune system continues. Learning the means by which the virus attempts to trick the immune system carries over into the study of **tumor cells.** They must also escape the immune system's surveillance system and do so by using several of the same ploys used by the virus. From what microbiologists have learned so far, different viruses have evolved different ways of getting rid of expression of class 1 proteins, the flagpole of self proteins. But the mechanisms are amazingly different in different viruses. Examples are as follows:

Cytomegalovirus (CMV)—A virus that may cause birth defects in the fetus and the rapid onset of blindness in people with AIDS. On entering a cell, CMV stops self identity proteins from reaching the cell membrane. If the class 1 proteins can't reach the membrane neither will pieces of CMV. The immune system is blinded to the fact the CMV is inside the cells. The immune system counters with a natural killer cell that destroys all cells that do not contain a self protein on their surface! But CMV over time evolved a means to produce a **fake** self protein that passes for the real one The natural killer cells are successfully fooled.

Adenovirus—The virus causing common colds. This virus can also stop self proteins from reaching the membrane carrying identifying pieces of virus. Further, this virus is able to make the **cell divide** so it can replicate itself. The cell then becomes cancerous (replication out of control), begins to destroy itself—but the virus has evolved a way to stop the cell's self-destruction, so the cell continues to divide. The process is similar to what happens in tumor cells.

HIV—This virus has the most deadly scheme of all. This virus infects cells of the immune system—most often the T4 cell that is essential to the initiation of the immune response. Similar to other virus, HIV has to disable the "self" alerting system. One of HIV's genes, call **nef,** makes a protein that attaches to the **self proteins,** just inside the cell's membrane. The other end of the nef protein carries an address label, readable by the cell's internal sorting system that directs proteins to their proper place. The message carried by the nef protein says, in the cell's sorting code, "Haul to garbage dump and recycle," tricking the infected cell into pulling down its self proteins and destroying them. The nef protein also tags the CD4 proteins for destruction in the same way. Like the self proteins, the CD4's stick up through the cell membrane. What does HIV gain from having the cell destroy CD4 proteins? That's the million dollar question investigators are working on. The reason may be to prevent other viruses from entering the cell or because in latching onto the CD4 proteins, the nef protein dislodges and energizes another protein that is known to activate the T cells. The cell's activation to make more HIV may be the answer.

response), or the **B cell with the help of T4 cells can produce antibodies specific for a given antigen** (T4 cell-dependent B cell response). The antibodies enter the circulatory system and are carried throughout the body.

T8 Cell Function and HIV Disease

Together, T4 cells and T8 cells regulate the body's immune response to invaders. Also, T8 cells help the immune system recognize (and not attack) the cells of its own body. But some T8 cells are lost in HIV-infected persons. Depending on how many are lost, the immune system may not be properly suppressed when necessary and an autoimmune response (the immune system attacks self) may occur.

Summary: Immune Dysfunction Caused by HIV/AIDS

After some 18 years of intense research and the expenditures of billions of dollars, HIV/AIDS investigators have yet to unravel the mystery of how HIV invades the human immune system, alters its function and eventually stops its

CLASS I AND CLASS II PROTEINS

There is a series of genes located on human chromosome 6 that is referred to as the **major histocompatibility** (tissue type) **complex (MHC).** The genes in this region produce a series of proteins that are **almost** unique to the individual. Because these proteins belong to the individual, they would be recognized as foreign or as an **antigen** when placed in another human, for example, as in a transplanted heart (**transplantation antigens**). **Almost** all cells in an individual have a sample of their own MHC proteins or transplantation antigens located on the surface of each cell's membrane. Thus, a person's transplantation antigens are recognized as **self-proteins** by their immune system. Such MHC proteins are referred to as **class I** (type 1) tissue compatibility proteins. So, every individual will have their own set of class I proteins on their cell membranes. It is these proteins that cause **transplant graft rejections.** So, with minor exceptions, there is not much of your body tissue that would not be rejected if placed in another human body because your class I proteins would be recognized as foreign by the recipient body's immune system. **But how does the recipient's immune system recognize these class I proteins as foreign? ANSWER:** By using a second class of MHC proteins, **class II,** that are **immune response** gene products. Relatively few human cell types carry this **protein on their cell membranes.** All such cells carrying the class II protein are derived from a monocyte-macrophage cell series which includes macrophages located in the lung, liver and spleen, dendritic cells of the gut, skin, spleen and lymph nodes, Langerhans' cells of the skin, and microglial cells of the nervous system.

This class II protein, located on the membrane of these cell types, allows these cell types to act as the body's police force. These cells **digest** foreign protein into smaller products. Some of these products, a small series of amino acids or **peptides,** are then escorted to the cell surface alongside a molecule of the class II protein. In this way, the cells carrying a class II protein **serve up** or **present** the foreign substance or antigen alongside or adjacent to the class II protein or "self molecule" so that immune system cells, **notably** the T4 cells and B cells, can sense the presence of the foreign protein and initiate an immune response against it—the activation of a variety of immune system attack cells (cytotoxic T cells, killer T cells) and the production of very specific antibodies.

In medically important viral infections, neutralizing antibodies are generated within 6 to 14 days. In contrast, such protective antibodies generally appear 50 to 150 days after infection with HIV and the hepatitis B virus (HBV) in humans (Pianz et al., 1996).

JEFFREY GETTY'S HIV-RESISTANT BABOON CELL TRANSPLANT FAILS

In December of 1995 Jeff Getty, in a first of a kind experiment received a transfusion of prepared HIV-resistant baboon bone marrow immune stem cells. After 2 months there were no signs of remaining immune baboon cells. Investigators believe the experiment failed because the low doses of radiation and chemotherapy failed to suppress his immune system. His immune system, sensing the foreign protein, rejected (destroyed) the baboon stem cells. Getty has now been HIV positive for 19 years, 1980–1999.

function. Recent evidence suggests that HIV in some fashion controls and/or inhibits T4 cell production—but how, that is the important question. Meanwhile, lets review what scientists believe is happening to the immune system, to some degree, after HIV infection.

HIV enters the body via infected body fluids: blood, semen, and vaginal secretions. Once inside, HIV specifically infects a type of white blood cell called the T4 cell. These cells direct the body's immune response against infection. As HIV takes over the T4 cells, it alters their growth and reproduction through a complicated process that leads to the T4 cells' destruction. The ratio of T4 to other cells called T8 then changes. In healthy people, the number of T4 cells is greater than the number of T8 cells. In HIV-infected individuals, a decline in the T4 count signals the progress of immune system deterioration. The results are debilitating: the T4 cells are not as responsive to antigen identification, macrophages become less responsive, and B cells produce fewer specific antibodies and lose their normal responsiveness. At this point, the immune system becomes dysfunctional and the host becomes vulnerable to attack from opportunistic infections.

ANTIBODIES AND VIRUSES

Twenty-four hours a day, 365 days a year, the body's lymphocytes must be on the lookout for harmful bacteria, viruses, and other pathogens. And these agents of disease don't look very different, biochemically, from most of the regular molecules lymphocytes normally encounter in the blood. Yet a mistake by these immune cells can leave the body open to infection.

But, the immune system has evolved antibodies for distinguishing the body's own molecular debris from potentially harmful material.

The Antibody

Antibodies are Y-shaped molecules that bind to specific foreign proteins, or pieces of protein, called antigens.

When antibodies on the surface of a B cell snag an undesirable or foreign protein, **both disappear inside the cell.** Eventually, a bit of the foreign protein may reemerge, attached to a "self" recognizable protein molecule called a **class II protein** (Figure 5-4). The pair, the small piece of foreign protein attached to a "self" **class II protein,** acts as a red flag to T4 cells, which set off an aggressive immune response.

The self or class II protein molecules play a critical part in the immune defensive response because they determine which viral peptides will be displayed on certain cells, antigen-presenting cells (APC), and how effectively they are showcased. Any two individuals are likely to differ in the precise structure of these class II molecules they possess. Thus, they will also differ in the peptides their cells exhibit and in the ability of the class II-peptide units to attract the attention of the immune system. Most people infected with HIV seem to recognize just a few of the many potential small peptide units generated from the virus's proteins, usually between one and 10.

Until now, scientists did not know where the cell degraded this undesirable protein and attached it to a "self" recognizable protein.

In mid-1994, four research teams announced the discovery of a special compartment, or organelle, inside cells where this processing occurs. Using sophisticated biochemical and immunological techniques, they independently determined that both the "self" recognizable protein and the antibody-antigen complex wind up in this compartment called an **endosome** (Amigorena et al., 1994; Pennisi, 1994; Schmid et al., 1994; Tulp et al., 1994).

The characterization of this specialized vesicle is an important step. It permits an understanding of the trafficking events and defines the intracellular event of antigen processing. (For an excellent review of the concept of antigen processing and presentation, see Unanue, 1995; for class I and class II proteins, Strominger et al., 1995; for the concept of self, Zinkernagel, 1995; for cell-mediated immunity, Doherty, 1995).

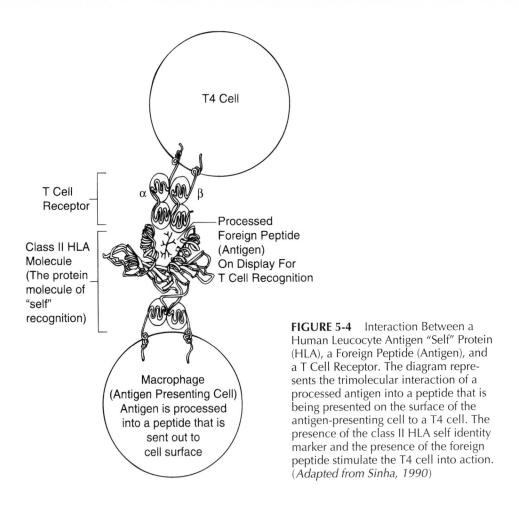

T4 Cell

T Cell
Receptor

α   β

Processed
Foreign Peptide
(Antigen)
On Display For
T Cell Recognition

Class II HLA
Molecule
(The protein
molecule of
"self"
recognition)

Macrophage
(Antigen Presenting Cell)
Antigen is processed
into a peptide that is
sent out to
cell surface

FIGURE 5-4   Interaction Between a
Human Leucocyte Antigen "Self" Protein
(HLA), a Foreign Peptide (Antigen), and
a T Cell Receptor. The diagram represents the trimolecular interaction of a
processed antigen into a peptide that is
being presented on the surface of the
antigen-presenting cell to a T4 cell. The
presence of the class II HLA self identity
marker and the presence of the foreign
peptide stimulate the T4 cell into action.
(*Adapted from Sinha, 1990*)

B cells make antibodies and release them into the bloodstream

After an antibody and virus join, they are digested by macrophages or cleared from the
blood by the liver and spleen. Some B cells and
T cells become **memory cells** which are stored
by the immune system. Memory cells can
remember the antigen they have previously
encountered. If the same virus ever gets into
the bloodstream again, these cells rapidly begin
antibody production. However, if the virus has
mutated (changed), as the flu virus does yearly,
previous antibodies will not affect it. New antibodies must be created to neutralize the new
mutant virus. While this antibody production
is taking place, the viral invader has time to
multiply and infect new cells, and the infected
person suffers the symptoms of the flu.

ANTIBODIES AND HIV DISEASE

Resistance to HIV does not seem to be the same
as more common examples of immunity. The
body's protective countermeasures against
measles and mumps are absolute. Years after
exposure, there is no hint within the body of
the foreign agents that cause those diseases.
After children become immune to mumps,
they can no longer infect other people.

The human body apparently makes effective antibodies against HIV, but over time, the
ratio of antibody production to new virus production becomes disproportionate, that is,
more virus than antibody is made and at this
time the level of T4 cells begin to drop. T4 cells
drop because antibodies to HIV reduce circulating HIV in the plasma (viral load) without
affecting HIV replication or cell-to-cell spread

of HIV. This means that the production of mutant HIV to existing antibody never ceases, resulting in continued T4 cell infection and loss. Eventually there are too few T4 cells to ward off opportunistic infections (see Chapter 7 for additional information on T4 cell replacement and HIV production).

The rapid production of HIV mutants without the same rapid production of neutralizing antibody against each mutant means that sometime after infection, many if not most of the antibody in some people may be nothing more than useless antibody copy (antibody to the initial strain or strains of HIV). In these people, HIV disease would most likely progress more swiftly than in persons whose immune system can keep up with the production of new antibody to match the formation of HIV mutants. This may be one important reason why some people progress to AIDS and death so rapidly when compared to other HIV-infected people.

HIV Protected from Human Antibodies

Humans create antibodies against a number of HIV proteins, namely the envelope proteins (gp120), the transmembrane protein (gp41), and the proteins of HIV's core (gp24). **But, antibodies cannot enter cells.** The antibody can only attack HIV in the plasma. Plasma is the fluid part of the blood and does not include the blood cells. **Once inside a host cell, HIV is protected from antibody.** Cells can generate antiviral chemicals within themselves, but these too seem ineffective against the hidden HIV.

Once HIV gets inside T4 or other cell types as a **provirus,** it is likely to remain there for the rest of the cells life (person's life?) unless some other antiviral mechanism within the body or some chemical agent is able to destroy the provirus. To date, no such drug has been found to be effective against the HIV **provirus.**

How HIV Attaches to the T4 Cell

The envelope of HIV is studded with proteins made by the virus. The major envelope protein is called gp120:

1. gp120 proteins protrude from the viral envelope. The gp120 region contains a conserved or **constant region** (in this region the sequence of amino acids remains constant from virus to virus) and a **variable region** (among HIV, the amino acid sequence is often different in this region). With respect to building a vaccine, the variable or **V** region, made up of five domains, each containing a separate sequence of amino acids, is the more important region of the envelope proteins. The region of the five domains is referred to as the

--------- POINT OF INFORMATION 5.4 ---------

WHAT ARE CELL RECEPTORS

In the study of cells, the term **receptor** is used to describe any molecule which interacts with and subsequently holds onto some other molecule. The receptor is like the **hand** and the object held by the hand is commonly called the **ligand.** The interaction between the receptor and ligand implies specificity; a receptor known to bind with substance X would not normally bind with a different substance. For example, a two-slotted electrical wall outlet is a receptacle (receptor) for a two-pronged plug (ligand). A three-pronged plug will not fit into this receptacle. And, depending on the type of two-slotted receptacle, even some two-pronged plugs may fit.

Where are cell receptors located?

Receptors can be found inside a cell, and especially embedded within and an integral part of all the membranes that a given cell may have. In humans there are organ systems present, and a given receptor may be found associated only with a particular type of cell which comprises a particular type of tissue which makes up a particular organ.

What do receptors do?

Receptors are critical to the life of all cells, whether or not the cells represent an animal, a plant, a fungus, or a bacterium. Every function, response, interaction, pathway, process, and any other term you might think of that concerns the moment-to-moment existence of a cell, is controlled by various receptor/ligand-induced systems.

hypervariable region of gp120 because of the extensive number of amino acid substitutions (new amino acids) that occur within the five areas. Each of the five domains, because of its amino acid arrangement, contains a loop-like structure. These are referred to as the V1 through V5 loops. The V3 loop is the most hypervariable (changeable). The V3 loop is located near the midportion of gp120 and appears to be involved in helping HIV lock into the CD4 receptor of a lymphocyte. A number of investigators are currently attempting to produce V3 loop antibody, believing that if the V3 loop is neutralized, HIV could not attach to the cell. **In summary,** each V-loop undergoes antigen change as its amino acids are replaced by different amino acids as genetic mutations occur in sections of HIV-RNA that code for the amino acid arrangement in the loop regions. Each loop acts as a separate antigen in that each stimulates the production of a loop-specific antibody. The V3 loop, the area that binds to the T4 lymphocyte (CD4 binding domain) is the principal site in the viral envelope for antibody neutralization (antibodies that inactivate HIV) and is believed to be the most important site responsible for HIV's-cell attachment (Belshe et al., 1994; Cohen, 1994; Ghiara et al.,

1994). The V3 loop also contains the determinants responsible for the formation of T4 cell syncytium (fusion of uninfected T4 cells with HIV-infected T4 cells) (Bobkov et al., 1994). A more complete knowledge of the V3 loop should facilitate the design of new drugs and vaccines to inhibit HIV infection.

2. When HIV enters the body, it circulates in the blood until by chance it bumps into a cell with the CD4 receptor on its surface.

3. The viral protein gp120 then binds specifically and tightly to a CD4 receptor site, as well as to other receptors described below.

How HIV Enters CD4+ Cells: T4 cells and Macrophage

HIV researchers have known since 1984 that human CD4 cell membrane receptors alone are sufficient for **binding** HIV to the T4 lymphocyte membrane but CD4 receptors are **not** sufficient for **HIV envelope fusion** with the T4 cell membrane or for **HIV penetration** or entry into the cell's interior. This knowledge has enticed many groups of AIDS researchers to search for addi-

─────── **POINT OF INFORMATION 5.5** ───────

THE FIRST LOOK AT HOW HIV'S gp120 ENVELOPE GLYCOPROTEIN CHANGES SHAPE TO ATTACH TO CD4 AND CHEMOKINE RECEPTORS

In three **landmark** studies, published in two scientific journals, Carlo Rizzuto and colleagues (1998) Richard Wyatt and colleagues (1998) and Peter Kwong and colleagues (1998) revealed and discussed the crystal structures of gp120, HIV's surface (envelope) glycoprotein. These reports have **very important** implications for the virology, immunology and vaccine development against HIV and most other viruses that infect humans. These papers came after 10 years of frustrating efforts to crystallize gp120—a prerequisite to determining its structure by x-ray crystallography. David Baltimore, noted laureate and head of a U.S. government advisory panel on AIDS vaccines said, "This is a big deal." Other notable HIV researchers believe these findings are a major advance in the collective knowledge of the virology and immunology of HIV infections. Revealing the structure of gp120 was to reveal the **passkey** on how HIV escapes antibody attack and invades T lymphocytes. First, **gp120** attaches to the T cell receptor, CD4; then gp120 changes shape, unveiling a binding site that allows HIV to attach to a sec-

ond T cell receptor, one of the chemokines, with both attachments secured, HIV enters the cell. A computer-generated snapshot of the HIV infection process is very informative (Figure 5-5). The picture shows, for example, that some of the virus' most stable—and therefore vulnerable—structures are either located at the bottom of crevices, where the relatively bulky antibodies of the immune system can't reach them, or obscured by great forests of sugar molecules. It appears that the immune vulnerable structures are covered by a top similar to a moveable roof on a stadium. One possible antibody target comes out of hiding but only in that brief moment after gp120 latches onto the CD4 receptor and before it attaches to the chemokine receptor—much too briefly for the immune system to react. Because HIV keeps its' immune-vulnerable structures hidden until the last possible moment, HIV is now called the **"Houdini of viruses."** However, most drugs are very small molecules—much smaller than antibodies. Properly designed drugs might be able to infiltrate some of those crevices in HIV's outer walls.

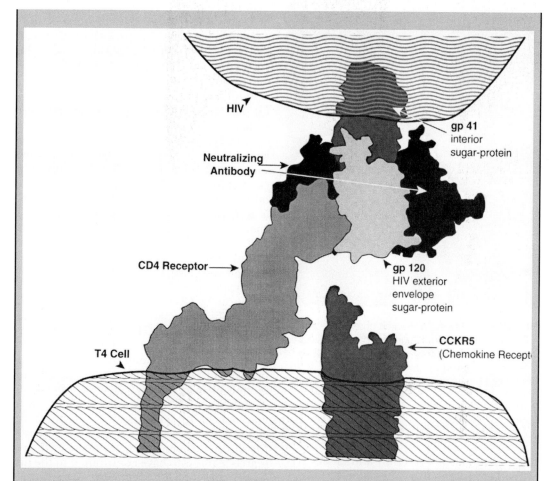

FIGURE 5-5 Unraveling the Secret to HIV Cell Entry. An artist's illustration of HIV attaching to a T4 lymphocyte. **First** gp120 attaches to the T4 cells CD4 receptor. On docking with CD4, gp120 undergoes a configurational change that momentarily exposes a hidden docking site for CCKR5 within the gp120 molecule. Uncovering of gp120's docking site for and attachment to CCKR5 appears to occur rapidly enough so that antibody to the covered site is not made or if made it cannot enter the hidden gp120 domain and neutralize it before attachment to CCKR5. Once both of the T4 cell receptors are locked on by HIV, the virus enters the cell. (Illustration adapted from the cover of *Science,* June 19, 1998, vol. 280, with permission.)

tional receptors, **coreceptors** to CD4, that HIV uses to enter a cell after binding to it. Various candidate receptor molecules were put forward, sometimes with more fanfare than fact, but have not stood the test of critical examination.

In May of 1996, Ed Berger (Figure 5-6) and colleagues reported on finding a receptor that allowed syncytium-inducing (SI) strains of HIV (HIV that causes T4 cells to form clusters—they attach to each other) to enter T4 cells. Strains of HIV that do not induce T4 cell syncytium formation (NSI) could not enter T4 cells. Berger and colleagues named this T4 cell receptor **FUSIN.** In August of 1996 Conrad Bleul and colleagues identified a chemokine, **CXC stromal cell-derived factor-1 (SDF-1)** that

FIGURE 5-6 Edward A. Berger, chief of the molecular structure section, Laboratory of Viral Diseases at the National Institutes of Allergy and Infectious Diseases. He and colleagues, in the spring of 1996, discovered the **first** co-receptor **CXCKR4** or **FUSIN** that HIV needs to complete its attachment and entry into T4 lymphocytes. (*Photograph courtesy of Edward A. Berger*)

binds to the FUSIN receptor and blocks HIV entry. They named this chemokine **CXCKR-4.**

What Are β-Chemokine Receptors and β-Chemokines— β-Chemokine receptors are cell-surface proteins that bind small peptides (a short chain of amino acids) called β-chemokines to their surface. Chemokines are classified into three groups depending on the location of the amino acid cysteine (**C**) in the peptide. These are C_ _, C-C_ _ _ _, and CXC_ _ _chemokines. The X in CXC represents any other amino acid. The chemokine receptors are identified by the individual chemokine(s) that binds to them. So, in that sense, **a reference to a specific chemokine(s) also identifies its receptor.**

FUSIN or CXCKR-4 (X4)

The first chemokine receptor was identified by Edward Berger and colleagues. They named

the coreceptor "FUSIN", and it was subsequently renamed CXCKR4 upon the discovery by others that it is receptor for the CXCKR4 chemokine SDF-1 (**S**tromal-**D**erived **F**actor that attaches to the CXCKR4 receptor). **This coreceptor functions preferentially for T cell line-tropic HIV strains.**

Within two months after the FUSIN receptor data was reported, **five separate research teams** reported on an additional coreceptor called **CC-CKR-5** (R-5). R-5 is the fifth of seven known human chemokine receptors that respond to β-chemokines. The function of β-chemokines is to attract macrophages and other immune cells to sites of inflammation. R-5 is known to be a receptor for three β-chemokine proteins called RANTES, MIP-1α, and MIP-1β. The β-chemokine RANTES (**r**egulated-upon-**a**ctivation **n**ormal **T** expressed and **s**ecreted), chemokine MIP-1α and 1β (macrophage inflammatory proteins) are all produced by T8 cells and are known to inhibit HIV replication, particularly in **macrophage-tropic HIV strains** (HIV strains that infect macrophage over T4 cells) (Deng et al., 1996; Dragic et al., 1996) **and are powerful suppressors of HIV infection, especially HIV infection of macrophage!** That is, these three chemokine proteins suppress HIV entry into macrophage by somehow blocking the coreceptor R-5. R-5 serves as a membrane receptor for all three chemokines. Members of the research teams who have contributed to the discovery of the R-5 receptor believe that **macrophage-tropic or M-tropic HIV strains occur in greatest number early on after HIV infection and then, sometime later during HIV disease the predominant HIV strain shifts to HIV strains that use the FUSIN receptor (X-4) on T4 cells.** HIV, by shifting receptors, may be avoiding the suppressive activity of the chemokines that block the R-5 receptor. Figure 5-7 shows a diagram of this suggested **receptor swap** that occurs sometime during HIV disease progression. One of the great unsolved puzzles of HIV disease is **why,** during disease progression, does HIV lose its ability to infect macrophage and become T-cell tropic? **Why does HIV switch to other cell receptors?**

Recent studies have shown that at least two additional chemokine receptors participate in allowing HIV cell entry. HIV can use both

FIGURE 5-7 Coreceptors Required for HIV to Enter Human Cells. HIV can broadly be divided into two classes: those more suitable (tropic) to infecting macrophage and those which infect T4 cells in the lymph nodes or other tissues. Macrophage-tropic nonsyncytium-inducing (NSI) HIV isolates infect macrophages but fail to infect HIV T cells lines, while T4 HIV syncytium-inducing (SI) strains fail to infect macrophage. But HIV of both classes efficiently infect T4 cells isolated from peripheral blood mononuclear cells (PBMC). **Macrophage-tropic NSI HIV appear to be preferentially transmitted by sexual contact and constitute the vast majority of HIV present in newly infected individuals (Zhu, 1993). The T-tropic SI viruses generally appear late in the course of infection during the so-called "phenotypic switch" that often precedes the onset of AIDS symptoms (Conner, et al., 1994).** The molecular basis of HIV-1 tropism appears to lie in the ability of envelopes of macrophage-tropic and T-tropic viruses to interact with different coreceptors located on macrophage or T4 cells. Macrophage-tropic viruses primarily use R-5 and less often R-3 and R-2, newly described seven-transmembrane domain chemokine receptors while T-tropic HIV tend to use FUSIN (X-4), a previously identified seven-transmembrane protein (Hill et al., 1996; Moore, 1997).

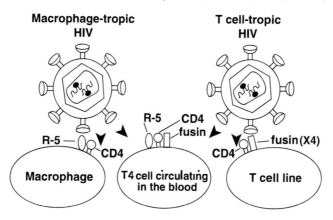

CCKR-2 (R-2) and CCKR-3 (R-3) to help its invasion of cells. For example, dual tropic HIVs use R-2 and R-3 to enter cells and studies with microglial cells (**non-CD4 type cells**) show that both R-5 and R-3 permit HIV to enter these **brain** cells (McNicholl et al., 1997). Current research involves finding ways to block or fill the chemokine receptors with a harmless molecule, thus blocking HIV's binding site on the cell (Chen et al., 1997).

GENETIC RESISTANCE TO HIV INFECTION

At this point in the HIV/AIDS pandemic it is believed that about 95% of HIV-exposed people are susceptible to HIV infection and HIV disease progression. This statement is made because it has long been known that persons who have deliberately avoided safer sex practice and who have had unprotected sex with HIV-infected persons failed to become HIV-infected! The question that has continued to puzzle HIV

investigators is, how can multiple HIV-exposed persons remain uninfected. Pieces of that puzzle, why some frequently HIV-exposed people remain uninfected, began to fall into place when two gay males who, despite repeated unprotected sex with companions who died from the disease, remain HIV-negative. Neither quite understood why he was spared but they were public-spirited enough to press scientists to come up with the answer. HIV investigators Rong Liu and co-workers (1996) and Michel Samson and co-workers (1996) reported that repeated HIV-exposed but uninfected people have a **32-nucleotide deletion in the gene that produces the R-5 receptors on macrophage.** The protein produced by this gene is severely damaged and does **not** reach the cell surface to act as a chemokine receptor. Without these receptors, the envelope of HIV cannot fuse with the envelope of macrophage to gain entrance into the cell. Thus, people who carry both defective R-5 genes' **homozygotes** (they received one defective gene from each parent) are resistant to HIV infection. If one is **heterozygous** (that is, one carries one defective gene and one normal gene), one will produce the R-5 receptor that HIV needs to penetrate macrophage, but they may be fewer in number, thus, chemo-kines (the chemicals that the receptors were made for) would have fewer receptors to fill so there is a greater chance that fewer R-5 receptors are open or available for HIV attachment. As a result, heterozygous people may be **less** resistant to HIV infection than the homozygous mutant (receptorless) people, but may be **more** resistant then people who have two normal genes (homozygous normal) who are the most susceptable to HIV infection. (O'Brien 1998)

Michael Dean and colleagues (1996) present evidence that the frequency of R-5 deletion heterozygotes (those with one copy of the gene) was significantly elevated in groups of individuals who had survived HIV infection for more than 10 years, and, in some risk groups, twice as frequent as their occurrence in **rapid progressors to AIDS.** Survival analysis clearly shows that disease progression is slower in R-5 deletion heterozygotes than in individuals with the normal R-5 gene. Jesper Eugene-Olsen and colleagues (1997) reported that individuals who are heterozygous for the 32-base-pair deletion in the R-5 gene have a slower decrease in their T4-cell count and a longer AIDS-free survival than individuals with the wild-type gene for up to 11 years of follow-up.

Will the Mutant R-5 Gene Protect People from All Subtypes of HIV?

This genetic defect, the 32-nucleotide deletion, prevents infection only with the strain of HIV (Subtype B) that is transmitted sexually and is prevalent in the United States and Europe. It does not necessarily protect against other strains of HIV transmitted through intravenous drug use or blood transfusions, or strains prevalent in Africa.

Should I Be Tested for the Presence of the R-5 Gene?

Researchers agreed that getting tested for the gene would not be difficult, but it would not be of great value because the tested person could still be infected by other strains of HIV. Besides, it must be assumed that most people

do not carry a pair of defective R-5 genes because 95% or more of HIV-exposed people become HIV-infected!

Who Carries the Defective R-5 Gene?

Perhaps most surprising, HIV investigators found that the homozygous genetic defect (a person having two defective genes) is common: **It is present in 1% of whites of European descent.** But it appears to be absent in people from Japan and Central Africa; about 20% of whites are heterozygous for this gene.

Why Does This Gene Exist?

Scientists speculate that the mutant form of R-5 protected against some disease that afflicted Europeans but not Africans. The obvious candidate would be the Black Death of 1346, the plague. If the mutant R-5 blocked plague bacteria, the survivors would be more likely to carry the trait and pass it to the next generations. Stephen O'Brien feels that **chance** of this gene randomly reaching its current frequency in the white population is about **zero.** The idea that a mutant gene can confer protection against a specific infection is not new. The mutation that causes sickle cell anemia provides people carrying one copy of it with resistance (but not immunity) to malaria. There is some evidence that the cystic fibrosis mutation may protect against typhoid fever.

T4 CELL DEPLETION AND IMMUNE SUPPRESSION

An understanding of the mechanisms by which HIV reduces the number of CD4 cells is important. However, it is not yet understood just how the T4 helper cell is killed or impaired by HIV infection. Recent research using the **polymerase chain reaction** (a method of amplifying present but unmeasurable quantities of HIV DNA in T4 cells into measurable quantities) has revealed that about *one in every 10 to 100* T4 cells is HIV-infected in an AIDS patient. Thus T4 cells serve as reservoirs of HIV in the body (Schnittman et al., 1989; Cohen, 1993). Collectively, T4 cell depletion leads to cumulative and devastating effects on the cell-mediated and humoral parts of the immune system.

———— BOX 5.3 ————

COFACTORS EXPEDITE HIV INFECTION

(An excerpt from Michael Callen's article, "Everything Must Be Doubted," *Newsline,* July/August 1988, pp. 44–5.)
(Michael Callen died in 1993.)

By the age of 27 (when I was diagnosed with crypto [cryptococcus infection] and AIDS) I had had over 3,000 different sex partners. Not coincidentally, I'd also had: hepatitis A, hepatitis B, hepatitis non-A, non-B; herpes simplex types I & II; syphilis; gonorrhea; non-specific urethritis; shigella; entamoeba histolitica; chlamydia; fungal infections; venereal warts; cytomegalovirus infections; EBV [Epstein-Barr virus] reactivations; cryptosporidium and therefore, finally, AIDS. For me, the question wasn't why I got AIDS, but how I had been able to remain standing on two feet for so long . . . If you blanked out my name and my age on my pre-AIDS medical chart and showed it to a doctor and asked her to guess who I was, she might reasonably have guessed, based on my disease history, that I was a 65-year-old malnourished equatorial African living in squalor . . . I believe that a small subset of urban gay men unwittingly managed to recreate disease settings equivalent to those of poor Third World nations and junkies.

This excerpt details many of the cofactors which play a role in the development of immune dysfunctions; while any one infection might not cause a problem for a healthy individual, repeated infection and exposure to foreign body fluids takes a cumulative toll. And certain infections and unquestionably more immune-suppressive than others—in particular, the sexually transmitted diseases.

Means by Which T4 Cells May Be Lost

1. Filling CD4 Receptor Sites— There is evidence from in vitro studies that HIV can attack CD4 receptor sites in at least two ways. First, HIV can attach, via its gp160 "spikes," to CD4 receptor sites. Second, HIV is capable of releasing or freeing its exterior gp120 envelope glycoprotein, thereby generating a molecule that can actively bind to CD4-bearing cells (Gelderblom et al., 1985). As a result of filling the receptor sites on the T4 cells, the T4 cells lose their immune functions; that is, the T4 cell does not have to be infected with HIV to lose immune function. In addition, CD4-bearing cells that attach the free gp120 molecule then become targets for immune attack by antibody-mediated antibody-dependent cell cytotoxicity (ADCC) and nonantibody-mediated cytotoxic T cells. Both events can result in the destruction of uninfected CD4 bearing cells. The extent to which this occurs in vivo depends on the level of gp120 synthesis, secretion, and shedding. Free gp120 has not yet been measured in the circulation, but this is not surprising given its powerful affinity for CD4 (Bolognesi, 1989).

2. Syncytia Formation— The formation of syncytia involves fusion of the cell membrane of an infected cell with the cell membranes of uninfected CD4 cells, which results in giant multinucleated cells (Figure 3.7). A direct relation between the presence of syncytia and the degree of the cytopathic effect of the virus in individual cells has been demonstrated in vitro, and HIV isolated during the accelerated phase of infection in vivo has a greater capacity to induce syncytia in vitro. Syncytia have rarely been seen in vivo.

In the asymptomatic phase of HIV infection, predominantly nonsyncytium inducing (NSI), HIV variants can be detected. In about 50% of the cases SI HIV variants emerge in the course of infection, preceding rapid T4 cell depletion and progression to AIDS (Groenink et al., 1993).

3. Superantigens— Superantigens are bacterial or viral antigens that are capable of interacting with a very large number of T4 cells. Unlike a conventional antigen, which usually evokes a response from less than one in a million (0.01%) T4 cells, a superantigen can interact with 5% to 30% of T4 cells. They directly bind to the exposed surfaces of MHC class II molecules on **antigen-presenting cells (APCs)** and to the variable region of the T cell receptors (TCR) β chain (V_β) on the responding T cells. Therefore, superantigens can interact with a large fraction of T cells, resulting in cellular activation, proliferation, anergy (absence of reaction to an antigen), or deletion of specific T cell subsets. Bacterial superantigens include staphylococcal enterotoxins and the toxic shock syndrome toxin-1, which are the causative agents for several human diseases such as food poisoning and toxic shock syndrome (Phillips et al., 1995). The superantigen hypothesis regarding HIV infection stems from the observations that endogenous or exogenous retroviral-encoded super-antigens stimulate murine T4 cells in vivo, leading to the anergy or deletion of a substantial percentage of T4 cells that have the specific variable β regions (Pantaleo et al., 1993).

4. Apoptosis— Programmed cell death, or apoptosis (a-po-toe-sis), is a normal mechanism of cell death that was originally described in the context of the response of immature thymocytes to cellular activation. It is a mechanism whereby the body eliminates autoreactive clones of T cells. It has recently been suggested that both qualitative and quantitative defects in T4 cells in patients with HIV infection may be the result of activation-induced cell death or apoptosis. Since apoptosis can be induced in mature murine T4 cells after cross-linking CD4 molecules to one another and triggering the T4 cell antigen receptor, there has been speculation that cross-linking of the CD4 molecule by HIV gp120 or gp120-anti-gp120 immune complexes prepares the cell for the programmed death that occurs when a MHC class II molecule in complex with an antigen binds to the T4 cell antigen receptor. Thus activation of a prepared cell by a specific antigen or superantigen could lead to the death of the cell, without direct infection by HIV (Pantaleo et al., 1993). For an excellent review of Apoptosis and its role in human disease, see Barr et al., 1994, Hengartner, 1995, and Yeh, 1998.

5. Cellular Transfer of HIV— Infected macrophages, or antigen-presenting cells, which normally interact with the T4 cells to stimulate the immune function, can transfer HIV into uninfected T4 lymphocytes. **In any case, immediate viral or proviral replication, kills the T4 cell.**

6. Autoimmune Mechanisms— One of the older theories, namely that HIV tricks the immune system into attacking itself is back because of the recent work of Tracy Kion and Geoffrey Hoffmann (1991). Their work showed that mice immunized with lymphocytes from another mouse strain make antibodies to the HIV envelope protein gp120, as do autoimmune strains of mice, even though **none of the animals had ever been exposed to the AIDS virus.** One implication of these results is that some component on the lymphocytes resembles gp120 closely enough so that antibodies directed against it can recognize gp120 as well. The converse is that antibodies to gp120 should also recognize the lymphocyte component so that an immune response directed against HIV might also interfere with normal lymphocyte function. The autoimmune theory is consistent with results that show there is a selective loss of particular subsets of T cells in AIDS patients.

7. Cofactors May Help Deplete T4 Cells— HIV-infected people who are asymptomatic show a wide variation in HIV disease time and progression to AIDS. It is believed that cofactors may be responsible for some of the time variation with regard to disease progression.

Many agents may act as cofactors to activate or increase HIV production. Although, in general, it is not believed that any cofactor is necessary for HIV infection, cofactors such as nutrition, stress, and infectious organisms have been considered as agents that might accelerate HIV expression after infection. Three new human herpes virus (HHV-6, 7, and 8) may be cofactors and play a role in causing immune deficiency. They have been shown to infect HIV-infected T4 cells and activate the HIV provirus to increase HIV replication (Laurence, 1996). Cytomegalovirus and the Epstein-Barr virus have also been associated with increased HIV expression. Over time, investigators expect to find other sexually transmitted diseases that behave as cofactors associated with HIV infection and expression.

Drugs may also be cofactors in infection. Used by injection drug users (IDUs), heroin and other morphine-based derivatives are known to reduce human resistance to infection and produce immunological suppression. *Pneumocystis carinii* pneumonia is about twice as frequent in heroin users as in homosexuals. It is believed that the heroin has an immunosuppressive effect within the lungs (Brown, 1987).

Blood and blood products may also act as cofactors in infection because they are immunosuppressive. Because blood transfusions save lives, their long-range effects are generally overlooked. Transfusions in hemophiliacs, for example, result in lowered resistance to viruses such as cytomegalovirus (CMV), Epstein-Barr (Berkman, 1984; Blumberg et al., 1985; Foster et al., 1985), and perhaps HIV. Seminal fluid (fluid bathing the sperm) may also act as a cofactor in infection because it also causes immuno-suppression. One of its physiological functions is to immunosuppress the female genital tract so that the sperm is not immunologically rejected (Witkin et al., 1983; Baxena et al., 1985).

Epidemiologically, homosexuals at greatest risk of AIDS are those who practice passive anal sex, that is, **anal recipients** (Kingsley et al., 1987). It is generally considered that this sexual behavior is a cofactor that enables the AIDS virus to enter the bloodstream by means of traumatic lacerations of the rectal mucosa.

Last but not least of the agents that can suppress the immune system and thereby act as a cofactor in HIV infection is **stress.** Stress can be mental or physical; but it is easier to measure the effects of physical stress.

Although moderate exercise appears to stimulate the immune system, there is good evidence that intensive exercise can suppress the immune system. We still do not really know why (Fitzgerald, 1988; LaPierre et al., 1992). The effect of stress on the immune system is one of the reasons why physicians did not want Earvin "Magic" Johnson to play basketball (see Box 8.6 in Chapter 8).

IMPACT OF T4 CELL DEPLETION

The overall impact of T4 cell depletion is multifaceted. HIV-induced T4 cell abnormalities alter the T4 cells' ability to produce a variety of inducer chemical stimulants such as the **interleukins** that are necessary for the proper maturation of B cells into plasma cells and the maturation of a subset of T cells into cytotoxic cells (Figure 5-8). Thus the critical basis for the immunopathogenesis of HIV infection is the depletion of the *T4 lymphocytes* which results in profound immunosuppression.

Presumably, with time, the number of HIV-infected T4 cells increases to a point where, in terminal AIDS patients, **few normal T4 cells exist.**

ROLE OF MONOCYTES AND MACROPHAGES IN HIV INFECTION

Some scientists now believe that T4 cell infection alone does not cause AIDS because not enough T4 cells are destroyed. They believe that equally important to T4 cell infection is **monocyte** and **macrophage** infection (Bakker et. al., 1992). Monocytes change into various types of macrophages (given different names) in order to search and destroy foreign agents within tissues of the lungs, brain, and interstitial tissues, tissues that connect organs. Despite the name changes, all forms of macrophages basically work the same way: they ingest things. Some macrophages travel around within the body, others become attached to one spot, ingesting.

Macrophages are often the first cells of the immune system to encounter invaders, particularly in the area of a cut or wound. After engulfing the invader, the macrophage makes copies of the invader's antigens and **displays** them on its own cell membrane. These **copies of the invader antigens sit right next to the self molecules. In effect, the macrophage makes a "wanted poster" of this new invader.** The macrophage then travels about showing the wanted poster to T4 cells, which triggers the T4 cells into action. Macrophages also release chemicals which stimulate both T4 cell and macrophage production and draw macrophages and lymphocytes to the site of infection.

Macrophages may play an important role in spreading HIV infection in the body, both to other cells and to HIV's target organs. First, HIV enters macrophage and spreads from macrophage to macrophage before the immune system is alerted. **Second,** macrophages, in their different forms, travel to the brain, the lungs, the bone marrow, and to various immune organs, carrying HIV to these organs.

HIV's ability to infect brain tissue is particularly important. The brain and cerebral spi-

FIGURE 5-8 The T4 Cell Role in the Immune Response. T4 lymphocytes are responsible directly or indirectly for inducing a wide array of functions in cells that produce the immune response. They also induce nonlymphoid cell functions. T cell involvement is effected for the most part by the secretion of soluble factors or **lymphokines** that have trophic or inductive effects on the cells presented in the figure. Lymphokines serve to transfer control of the immune response from the external environmental proteins to the internal regulatory system consisting of ligands and receptors. (*Adapted from Fauci, 1991*)

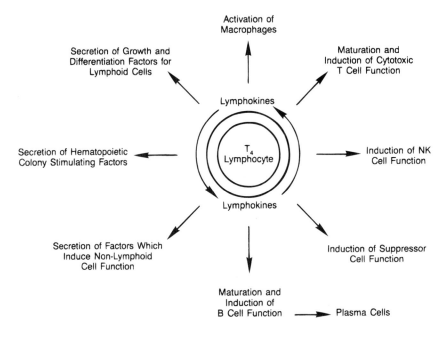

nal fluid (CSF) are specially protected sites. CSF cushions the brain and the spinal cord from sudden and jarring movements. The brainblood barrier, a chemical phenomenon, normally stops foreign substances from entering the brain and the CSF. But HIV-infected monocytes can pass through this barrier. **For HIV, monocyte-macrophages are Trojan horses, enabling HIV to enter the immune-protected domain of the central nervous system—the brain, the spine, the rest of the nervous system.**

HIV isolates taken from macrophages appear to grow better in macrophages than in lymphocytes. In human cell culture experiments, HIV isolates taken from lymphocytes appear to grow better in lymphocytes than in macrophages. These peculiarities of replication rates may be evidence of separate tissue-oriented HIV strains.

HIV-infected macrophages have proven to be a major problem in efforts to control and stop HIV infection.

Clinical Latency: Where Have All the Viruses Been Hiding?

The immunodeficiency syndrome caused by HIV is characterized by **profound T4 cell depletion.** Prior to the progressive decline of T4 cells and the development of AIDS, there is a symptom-free period which may last for 10 or more years, during which the infection is thought to be clinically latent. Scientists thought little if any viral replication occurred during this period. In recent years, a number of studies have offered an alternative view. Some investigations suggest that there is no real latency period in HIV infection. Instead, **starting from the moment of infection, there appears to be a continuous**

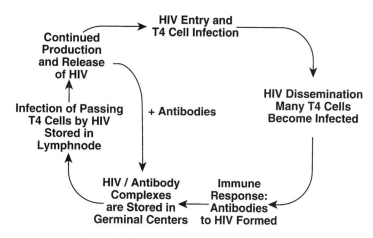

FIGURE 5-9 The Role of HIV Antibody in HIV Seeding of the Lymph Glands and Infection of New T4 Cells. To rid the body of HIV requires a way to cleanse the lymph nodes of HIV. (*Adapted from Fox, 1996*)

struggle between HIV and the immune system, the balance of which slowly shifts in favor of the virus. Cecil H. Fox of the Yale School of Medicine reported in December 1991 that HIV is not lying dormant at all but is continually infecting immune cells in the **lymph nodes** (Figure 5-9).

The **lymph nodes,** which are pea-sized capsules that trap foreign invaders and produce immune cells, are all over the body and are connected by vessels much like those that transport blood. Deep within the lymph nodes of 18 HIV-infected patients, researchers found millions of viruses. Based on those data, Fox suggested that HIV uses the lymph nodes as places to meet up with most immune cells. Fox believes that HIV begins replication in T4 cells in the lymph nodes soon after it enters the body. Infected T4 cells leave the lymph nodes and new uninfected cells arrive to become infected. Years later, when enough immune cells have been killed, the patient's defenses become so impaired that he or she is vulnerable to any one of a wide array of opportunistic infections— AIDS has arrived.

Anthony Fauci (Figure 5-10), the head of the National Institute of Allergy and Infectious Diseases (NIAID), and Sonya Heath and colleagues (1995) agrees that there are many more HIVs in the lymph nodes than in the blood. Studies in his laboratory have shown there are many millions of HIV particles stuck to what are known as **follicular dendritic cells** in the lymph nodes of an infected individual.

The follicular dendritic cells, which have thousands of feathery processes emanating from them and whose **normal function is to filter and trap antigens for presentation to**

FIGURE 5-10 Anthony S. Fauci, M.D. (1989). Director, National Institute of Allergy and Infectious Diseases, National Institutes of Health; Associate Director of NIH for AIDS Research. (*Photograph courtesy of National Institute of Allergy and Infectious Diseases*)

antibody-producing B lymphocytes, serve as highly effective trapping centers for extracellular HIV particles. Virtually every lymphocyte in a lymph node is enmeshed in the processes of these cells. Follicular dendritic cells themselves are susceptible to HIV infection, and appear to place huge numbers of virus particles into intimate contact with other cells that are susceptible to HIV infection (Haase et al., 1996; Knight,1996).

During the clinically latent phase, the lymph nodes of an infected individual are slowly destroyed.

In the final phase of an infection, the follicular dendritic network completely dissolves and the architecture of the lymph node collapses to produce what Fauci called a "**burnt-out lymph node.**" With this collapse, large amounts of HIV are released into the circulatory system and an ever increasing number of peripheral cells are infected with HIV (Baum, 1992; Edgington, 1993).

In 1995, David Ho and colleagues, Xiping Wei and colleagues, and Martin Nowak and colleagues published separate reports on the rapid production of HIV which began soon after initial HIV infection. These studies, collectively demonstrated that, using early, frequent blood samples of HIV-infected persons taking the drug nevirapine, a non-nucleoside reverse transcriptase inhibitor and protease inhibitors ABT-538 and L-735.524 (protease is essential for HIV reproduction), it was possible to calculate how rapidly newly produced HIV turned over within HIV-infected persons by measuring the balance between HIV-RNA production in and HIV RNA clearance from the blood plasma. The drugs worked well, dropping the amount of HIV-RNA in blood plasma to 1% of pretreatment level. In each case the treated person's T4 cell counts significantly increased, in some by as much as 10-fold. By determining how quickly T4 cells increased once HIV RNA levels dropped, Ho and colleagues calculated that before treatment their patients had been producing over a billion T4 cells each day. That's an incredibly accelerated rate—between 25 and 75 times the rate seen in patients with less-advanced infections. But it still wasn't enough to stay ahead of the rapidly replicating virus, which was infecting and killing each new cell in less than 2 days. **The patients' immune systems were functioning at full capacity but slowly losing the battle!**

Such rapid viral replication countered by a strong immune response is not unknown; on the contrary, it is characteristic of acute, short-lived viral infections, such as measles or the flu. But in those infections, the immune system most often wins: it rids the body of the virus and builds resistance to future infection. The question is why can't the immune system do the same against HIV. Perhaps it's the length and intensity of the struggle that ultimately does in the immune system by encouraging the emergence of mutant viruses that outstrip the immune system's ability to continue making new types of antibody. During the time when the immune system is functioning well and producing new antibody against each new HIV variant, the struggle between HIV production and immune suppression of the viral load must move repeatedly from the virus to the immune system and back until the immune system is just overwhelmed with the number of different types of antibody it must rapidly produce in large quantity.

The viral load or quantitative amount of HIV RNA present in the HIV-infected persons studied by these investigators showed that about 30% of HIV and HIV-infected T4 cells in their blood plasma was turning over (produced and cleared from plasma) each day! It would appear that HIV and T4 cells are being produced at about the same rate which is why the progression of HIV disease is so gradual.

T4 Cell Generation and Destruction

Ho and colleagues calculated that about 5% of the T4 cells are being regenerated per day, indicating a destruction rate of approximately 5% per day during steady state. The 5% figure is equivalent to a daily turnover of approximately 35 million cells in the bloodstream. Since the bloodstream accounts for approximately 2% of the total lymphoid population, it can be estimated that the total population minimum production and destruction rate is approximately 1.8 billion T4 cells per day. When Ho plotted the individual rates of T4 cell repopulation against initial T4 cell counts for his 20 patients studied, an inverse correlation was found, with those subjects with the lowest T4 cell counts exhibiting the greatest production rates.

This finding was a surprise, given the expectation based on prior literature that patients with more advanced illness would have the lowest T4 cell regenerative capacity. The implication is that the immune system continues trying even in late-stage HIV disease to counter the greater T4 cell destruction rates associated with the greater quantity of actively replicating virus.

Using the analogy of a faucet and basin, Ho equated the water level in the basin to the T4 cell count, the faucet to T4 cell replenishment, and the drain to HIV destruction of T4 cells. In the analogy, the basin drains just slightly faster than replenishment occurs, with the faucet turned full on in late-stage disease. Ho maintained that the major objective of treatment should be to plug the drain rather than solely attempt to increase the amount of water going into the basin—i.e., through immune-based therapies designed to boost T4 cell production. He suggested that it is difficult to imagine how T4 cell production rate could be augmented beyond that observed in late-stage HIV disease. Ho believes that the functional capacity of regenerating T4 lymphocytes may be limited; thus, efforts to restore immune function (saving or restoring T4 cells) is essential (Ho, 1995).

Another point established by Ho and colleagues investigations is that the rate of HIV clearance from the plasma is not affected by the stage of HIV progression or disease status. Ho concluded that viral load is **not** a function of viral clearance. HIV clearance efficiency varies little among individuals yet individuals can differ in viral load by several logs. Therefore **viral load is primarily a function of viral production.**

Implications of a High HIV Replication Rate

Based on Ho and colleagues kinetic studies of viral replication, it can be estimated that the population of HIV undergoes between 3,000 and 5,000 replication cycles (**generations**) over the course of 10 years, producing a **minimum of 10^{12} (1 trillion) HIV** in an HIV-infected person. According to Ho, given the assumptions in the kinetics analysis that render the figures minimum estimates, the number of HIV produced is probably closer to 10^{13} (10 trillion) during the infected person's lifetime. This vast number of HIV replications over 3,000 to 5,000 generations creates a lot of opportunity for HIV RNA evolution, because genetic mutations occur most commonly during replication. Some mutations will weaken HIV in such a way as to expose it to attack by the immune system. But other mutations will aid the virus, speeding its replication and increasing the chances it can evade the immune system. Because HIV gets so many chances to replicate, it evolves and mutates, and those strains with the greatest replication efficiency gradually win out. **This is Darwinian evolution going on in one patient.**

HIV has been shown to have a random (any base has an equal chance of changing) mutation rate of approximately 1 mutation per 10,000 nucleotides **AND** one mutation caused by reverse transcriptase copy error (**transcription error**) per round of replication. The **combination** of events results in too many variants in the immune system to handle. For example, a single HIV-infected person, by the time he or she is diagnosed with AIDS, will have billions of HIV variants in his or her body. With so many variants present, some will be resistant to most if not all drugs currently used in therapy and to drugs that have **not yet been created!.** Thus it is the constant production of variant HIVs that produce the drug resistant mutants and not action of the drug.

In summary, the large number and excessive genetic diversity created by thousands of generations of HIV in one person allows HIV to surpass or exceed the ability of the immune system to recognize and respond to virus thus permiting resistant mutants to emerge and gain predominance. Ho and colleagues believe that because of the early rapid replication of HIV and production of such vast number of HIV variants, monotherapy (single drug) is not a rational strategy if a dramatic impact on the course of infection is to be achieved. They stated that a reasonable strategy is to use multiple potent agents simultaneously (e.g., three or four in combination), **requiring** HIV to simultaneously develop multiple mutations in order to escape the drugs. They also stated that given the dynamics of HIV in vivo, it makes sense to begin treatment as soon as infection is diagnosed, on the rationale that the more replicative cycles that are permitted, the more variants there will be when treatment is begun, stacking the deck

AN ALTERNATIVE VIEW ON THE COLLAPSE OF THE IMMUNE SYSTEM

According to David Ho and colleagues, as reported in this chapter, HIV infects, then kills T4 cells as quickly as they are produced. They are produced in the billions! The newly produced cells become infected and die. This continues until the immune system collapses from exhaustion.

The work of Marc Hellerstein and colleagues **(1999)** suggests controversial new and different findings that run counter to David Ho's group. They used a new, non-radioactive, endogenous labeling technique and directly measured circulating T cells in HIV-positive and negative people. The half-line of the cells in untreated HIV-infected people was less than one-third as long as those of the HIV negative. Furthermore, production rates of T4 cells did not compensate for the shorter half-life. Among patients receiving highly active antiretroviral therapy for 12 weeks, production rates of the T4 cells were elevated. The researchers noted that increased T4 levels resulted from greater production of the cells, not a longer half-life. Hellerstein and colleagues conclude that "T4-cell lymphopenia is due both to a shortened survival time and a failure to increase production of circulating T4 cells." In brief, results showed no T4 cell speed-up-and-collapse pattern in the infected people. What researchers found instead was that, along with **reduced cell longevity,** the virus caused **slower cell production**—the opposite of what had been assumed to occur during this critical stage of the disease.

RECAP

Old view. Multiplying HIV particles infect T4 cells. Immune system responds by cranking up production of new T4 cells in an effort to keep pace with the virus. In the end stage of disease, the immune system collapses from exhaustion.

New view. HIV infects the mature T4 cells, but not in sufficient number to explain how the disease progresses. Researchers suspect that the virus may attack the immune system's cell-making capacity. Final stage of disease caused primarily by collapse of immune system and reduced lifespan of T4 cells.

HIV virus attacks T4 cell production system, probably in bone marrow and thymus. If these new data hold up to repeat experiments, the way HIV/AIDS investigators view the pathogenesis of HIV/AIDS will be greatly altered.

Still Unknown

Just how the AIDS virus might damage T4 cell production has yet to be unraveled. There are as yet no proven therapies to address the new view of the disease, which would call for treatments that defend the immune system, insulating it from HIV or making it robust enough to withstand the virus. By comparison, today's therapies take direct aim at stopping the virus from reproducing. HIV/AIDS clinicians emphasized that nothing in the study suggest people with HIV should change their current therapy, and certainly no one should stop taking anti-viral drug combinations now in widespread use.

---

in favor of the virus' ability to avoid both immune responses and effects of drug treatment.

Thus, through early 1995, scientists had vastly **underestimated** the extent of virus activity in an HIV-infected person, particularly during the asymptomatic or clinically latent phase. These and other recent studies should satisfy the major unanswered question concerning HIV disease/AIDS, which was, "**Where is the virus?**" During the 1980s, it was difficult to find medium to high levels of HIV in persons with HIV disease or even in those persons in the later stages of AIDS. We now know that there are large amounts of the HIV present early on in the lymph nodes of HIV-diseased people. Now, however, perhaps the most troubled group of researchers may be those currently developing AIDS vaccines. (See Chapter 9 for a discussion about possible HIV vaccines.) At present, vaccine strategies propose to introduce HIV antigen repeatedly during the long asymptomatic period to bolster B cell immunity. Vaccinated antigens, however, could wind up trapped in the lymph nodes, stimulating B cells to produce the cytokines that stimulate T4 cell infection and speed lymph node architecture destruction. Thus the vaccine's effect could backfire, compressing the clinically latent period and causing a more rapid onset of AIDS.

SUMMARY

After a healthy person is HIV-infected, he or she makes antibodies against those viruses that are in the bloodstream, but not against those that have become integrated as HIV proviruses in the host immune cell DNA. Over time the immune system cells that are involved in antibody production are destroyed. Evidence is accumulating that cofactors such as nutrition, stress, and previous exposure to other sexually transmitted diseases that increase HIV expression are associated with HIV infection and HIV disease. Agents that suppress the immune system may also play a significant role in establishing HIV infection. In short, HIV infection is permanent. HIV attaches to CD4-bearing cells and enters those cells using one or more of at least four chemokine receptors—**CC-CKR-2, CC-CKR-3, CXCKR-4 (FUSIN) and CC-CKR-5.** These coreceptors allow HIV to fuse with the cell membrane, and enter the cell, after HIV attachment to the CD4 receptor. And, as seen in Chapter 3, HIV undergoes rapid genetic change, and, as far as is known, **attacks only human cells—mostly of the human immune system.** It has recently become quite clear that large amounts of the virus are produced within weeks after HIV infection. The virus remains, for the most part, in the lymph nodes until very late in the disease process.

REVIEW QUESTIONS

(Answers to the Review Questions are on page 487.)

1. Which cell type is believed to be the main target for HIV infection? Explain the biological impact of this particular infection.

2. What is CD4, where is it found, and what is its role in the HIV infection process?

3. Is there a period of latency after HIV infection—a time when few or no new HIV are being produced?

4. True or False: HIV is the cause of AIDS.

5. True or False: HIV primarily affects red blood cells.

6. True or False: Lymphocytes have a major role in the immune response to antigens.

7. True or False: All T and B lymphocytes inhibit or destroy foreign antigens.

8. True or False: CD4 and CD8 molecules are antibodies.

9. True or False: HIV belongs to the family of retroviruses.

10. True or False: Cytotoxic and suppressor T lymphocytes are the main targets of HIV.

11. True or False: The latent period is that time between initial infection with HIV and the onset of AIDS.

12. True or False: HIV can spread to infect new cells after it buds out of infected cells.

13. True or False: HIV causes the gradual destruction of cells bearing the CD4 molecule.

14. The most effective use of T4 cell counts in the clinical management of patients with HIV infection is for:
    A. Determining when to initiate therapy.
    B. Assessing risk of disease progression.
    C. Deciding on prophylaxis for opportunistic infections.
    D. Measuring the antiretroviral effect of initial therapy.

REFERENCES

AMEISEN, JEAN CLAUDE. (1994). Programmed cell death apoptosis and cell survival regulation: relavance to cancer. *AIDS*, 8:1197–1213.

AMIGORENA, SEBASTIAN, et al., (1994). Transient accumulation of new class II MHC molecules in a novel endocytic compartment in B lymphocytes. *Nature*, 369:113–120.

BAKKER, LEENDERT J., et al. (1992). Antibodies and complement enhance binding and uptake of HIV-1 by human monocytes. *AIDS*, 6:35–41.

BARR, PHILIP, et al. (1994). Apoptosis and its role in human disease. *BioTechnology*, 12:487–494.

BAXENA, S., et al. (1985). Immunosuppression by human seminal plasma. *Immunol. Invest.*, 14: 255–269.

BELSHE, R. B., et al. (1994). Neutralizing antibodies to HIV in seronegative volunteers immunized with recombinant gp120 from the MN strain of HIV. *JAMA*, 272:475–480.

BERKMAN, S. (1984). Infectious complications of blood transfusions. *Sem. Oncol.*, 11:68–75.

BLEUL, CONRAD, et al. (1996). The lymphocyte chemoattractant SDF-1 is a ligand for LESTR/fusin and blocks HIV entry. *Nature*, 382: 829–833.

BLUMBERG, N., et al. (1985). A retrospective study of transfusions. *Br. Med. J.*, 290:1037–1039.

BOBKOV, ALEKSEI, et al. (1994). Molecular epidemiology of HIV in the former Soviet Union: Analysis of ENV V3 sequences and their correlation with epidemiologic data. *AIDS*, 8: 619–624.

BOLOGNESI, DANI P. (1989). Prospects for prevention of and early intervention against HIV. *JAMA*, 261:3007–3013.

CALLEBAUT, CHRISTIAN, et al. (1993). T cell activation antigen, CD26 as a cofactor for entry of HIV in CD4+ cells. *Science*, 262:2045–2050.

CHEN, SI-YI, ET AL. (1997). Second major HIV coreceptor inactivated by intrakine gene therapy. *Proc. Natl. Acad. Sci. USA*, 94:11567–11572.

COHEN, JOEL. (1994). The HIV vaccine paradox. *Science*, 264:1072–1074.

COHEN, JON. (1993). Keystone's blunt message: "It's the virus stupid." *Science*, 260:292–293.

COHEN, JON. (1993). HIV cofactor comes in for more heavy fire. *Science*, 262:1971–1972.

COHEN, JON. (1997). Exploiting the HIV-chemokine nexus. *Science*, 275:1261–1264.

CONNER, RUTH, et al. (1994). Human immunodeficiency virus type 1 variants with increased replicative capacity develop during the asymptomatic stage before disease progression. *J. Virol.*, 68: 4400–4408.

CROWE, S.M., et al. (1991). Predictive value of CD4 lymphocyte numbers for the development of opportunistic infections and malignancies in HIV-infected persons. *AIDS*, 48:770–776.

DEAN, MICHAEL, et al. (1996). Genetic restriction of HIV infection and progression to AIDS by a deletion allele of the CKR5 structural gene. *Science*, 273:1856–1861.

DENG, HONGKUI, et al. (1996). Identification of a major co-receptor for primary isolates of HIV-1. *Nature*, 381:661–666.

DOHERTY, PETER. (1995). The keys to cell-mediated immunity. *JAMA*, 274:1067–1068.

DRAGIC, TATJANA, et al. (1996). HIV-1 entry into CD4+ cells is mediated by the chemokine receptor CC-CKR-5. *Nature*, 381:667–673.

EDGINGTON, STEPHEN M. (1993). HIV no longer latent, says NIAID's Fauci. *BioTechnology*, 11:16–17.

EMBRETSON, JANET, et al. (1993). Massive covert infection of helper T lymphocytes and macrophages by HIV during the incubation period of AIDS. *Nature*, 362:359–362.

Eugen-Olsen, Jesper, et al. (1997). Heterozygosity for a deletion in the CKR5 gene leads to prolonged AIDS-free survival and slower CD4 T-cell decline. *AIDS*, 11:305–310.

FAUCI, ANTHONY, et al. (1995). Trapped but still dangerous. *Nature*, 337:680–681.

FITZGERALD, LYNN. (1988). Exercise and the immune system. *Trends Genet.*, 2:1–12.

FOSTER, R,. et al. (1985). Adverse effects of blood transfusions in lung cancer. *Cancer*, 55:11951–12202.

FOX, CECIL H., et al. (1991). Lymphoid germinal centers for reservoirs of HIV type I RNA. *J. Infect. Dis.*, 164:1051–1057.

FOX, CECIL, (1996) How HIV causes disease. *Carolina Tips*, 59:9–11.

GELDERBLOM, H.R., et al. (1985). Loss of envelope antigens of HTLV III/LAV, a factor in AIDS pathogenesis. *Lancet*, 2:1016–1017.

GHIARA, JAYANT, et al. (1994). Crystal structure of the principal neutralization site of HIV. *Science*, 264:82–85.

GROENINK, MARTIJN, et al. (1993). Relation of phenotype evolution of HIV to envelope V2 configuration. *Science*, 260:1513–1516.

HAASE, ASHLEY, et al. (1996). Quantitative image analysis of HIV infection in lymphoid tissue. *Science*, 274:985–990.

HEATH, SONYA, et al. (1995). Follicular dendritic cells and HIV infectivity. *Nature*, 377:740–744.

HELLERSTEIN, MARC, et al. (1999). Directly measured kinetics of circulating T lymphocytes in normal and HIV-1-infected humans. *Nature Medicine* 5:83.

HENGARTNER, MICHAEL. (1995). Life and death decisions: Ced-9 programmed cell death in *C. elegans*. *Science*, 270: 931.

HILL, MARK, et al. (1996). Natural resistance to HIV. *Nature*, 382:668–669.

HO, DAVID. (1995). Pathogenesis of HIV infection. *International AIDS Society–USA*, 3:9–12.

HO, DAVID, et al. (1995). Rapid turnover of plasma virons and CD4 lymphocytes in HIV infection. *Nature*, 373:123–126.

KINGSLEY, L., et al. (1987). Risk factors for seroconversion to HIV among male homosexuals. *Lancet*, 8529:345–348.

KION, TRACY, et al. (1991). Anti-HIV and Anti-Anti-MHC antibodies in alloimmune and autoimmune mice. *Science*, 253:1138–1140.

KNIGHT, STELLA. (1996). Bone-marrow-derived dendritic cells and the pathogenesis of AIDS. *AIDS*, 10:807–817.

KWONG, PETER, et al. (1998). Structure of an HIV gp120 envelope glycoprotein in complex with the CD4 receptor and a neutralizing human antibody. *Nature*: 648–659.

LaPIERRE, A., et al. (1992). Exercise and health maintenance in AIDS. In Galantino, M.L. (ed.), *Clinical Assessment and Treatment in HIV: Rehabilitation of a Chronic Illness*, Chap. 7. Thorofare NJ: Slack, Inc.

LAURENCE, JEFFREY. (1996). Where do we go from here? *AIDS Reader*, 6:3–4, 36.

LIU, RONG, et al. (1996). Homozygous defect in HIV-1 coreceptor accounts for resistance of some multiple-exposed individuals to HIV-1 infection. *Cell*, 86:367–377.

McNicholl, Janet, et al. (1997). Host genes and HIV: The role of the chemokine receptor gene CCR5 and its allele (Δ32CCR5). *Emerg. Infect. Dis.*, 3:261–271.

Moore, John. (1997). Coreceptors: Implications for HIV pathogenesis and therapy. *Science*, 276: 51–52.

Mountain-Plains Regional HIV/AIDS Curriculum, 4th ed. (1992). Mountain-Plains Regional AIDS Office, University of Colorado Health Sciences Center, Denver, CO 80262.

Nowak, Martin, et al. (1995). HIV results in the frame: Results confirmed. *Nature*, 375:193.

O'Brien, Stephen. (1998). AIDS: A role for host genes. *Hospital Practice*, 33: 53–79.

Pantaleo, G., et al. (1993). HIV infection is active and progressive in lymphoid tissue during the clinically latent stage of disease. *Nature*, 362: 355–358.

Pennisi, Elizabeth. (1994). A room of their own. *Science News*, 145:335.

Pianz, Oliver, et al. (1996). Specific cytotoxic T Cells eliminate cells producing neutralizing antibodies. *Nature*, 382:726–729.

Rizzuto, Carlo, et al. (1998). A conserved HIV gp120 glycoprotein structure involved in chemokine receptor binding. *Science* 280: 1949–1953.

Samson, Michel, et al. (1996). Resistance to HIV-1 infection in Caucasion individuals bearing mutant alleles of the CCR-5 chemokine gene. *Nature*, 382:722–725.

Schmid, Sandra, et al. (1994). Making class II presentable. *Nature*, 369:103–104.

Schnittman, Steven M., et al. (1989). The reservoir for HIV-1 in human peripheral blood is a T cell that maintains expression of CD4. *Science*, 245:305–308.

Sinha, Animesh, et al. (1990). Autoimmune diseases: The failure of self-tolerance. *Science*, 248: 1380–1387.

Sprent, Jonathan, et al. (1994). Lymphocyte lifespan and memory. *Science*, 265:1395–1400.

Strominger, Jack, et al. (1995). The Class I and Class II proteins of the Human major histocompatibility complex. *JAMA*, 274:1074–1076.

Tulp, Abraham, et al. (1994). Isolation and characterization of the intracellular MHC class II compartment. *Nature*, 369:120–126.

Unanue, Emil. (1995). The concept of antigen processing and presentation. *JAMA*, 274:1071–1073.

Wei, Xiping, et al. (1995). Viral dynamics in HIV type I infection. *Nature*, 373:117–122.

Weiss, Robin, et al. (1996). Hot fusion of HIV. *Nature*, 381: 647–648.

Witkin, S., and Sonnabend, J. (1983). Immune responses to spermatozoa in homosexual men. *Fertil. Steril.*, 39:337–341.

Wyatt, Richard, et al. (1998). The antigenic structure of the HIV gp120 envelope glycoprotein. *Nature* 705–711.

Yeh, Edward. (1998). Life and Death of a Cell. *Hospital Practice*, 33:85–92.

Zhu, Toufu. (1993). Gentypic and phenotypic characterization of HIV-1 patients with primary infection. *Science*, 261:1179–1181.

Zinkernagel, Rolf. (1995). MHC-restricted T-cell recognition: The basis of immune surveillance. *JAMA*, 274:1069–1071.

Opportunistic Infections and Cancers Associated with HIV Disease/AIDS

WHAT IS AN OPPORTUNISTIC DISEASE?

Humans evolved in the presence of a wide range of parasites—viruses, bacteria, fungi and protozoa that do not cause disease in people with an intact immune system. But these organisms can cause a disease in someone with a weakened immune system, such as an individual with HIV disease. The infections they cause are known as **opportunistic infections** (OIs). Thus, OIs occur after a disease-causing virus or microorganism, normally held in check by a functioning immune system, gets the chance to multiply and invade host tissue because the immune system has been compromised. For most of medical history, OIs were rare and almost always appeared in patients whose immunity was impaired by cancer or genetic disease.

With improved medical technology, a steadily growing number of patients are severely immunosuppressed because of medications and radiation used in bone marrow or organ transplantation and cancer chemotherapy. **HIV disease also suppresses the immune system.** And perhaps as a corollary to their increased prevalence, or because of heightened physician awareness, OIs seem to be occurring more frequently in the elderly, who may be rendered vulnerable by age-related declines in immunity. New OIs are now being diagnosed because the pool of people who can get them is so much larger, and, in addition, new techniques for identifying the causative organisms have been developed. However, most of the infections considered opportunistic are not reportable, which interferes with a clear-cut count of their growing numbers.

Although OIs are still not commonplace, they are no longer considered rare—they occur in tens of thousands of patients. But despite this increase, physicians and their patients have reasons to be optimistic about their ability to contain these infections. The reasons are: (**1**) in a massive federal effort, driven by the HIV/AIDS epidemic, researchers are finding drugs that can prevent or treat many of the OIs; and (**2**) various anti-HIV drug therapies have shown promise for warding off OIs by boosting patients' immune systems.

THE PREVALENCE OF OPPORTUNISTIC DISEASES

The prevalence of OIs in the United States is very high. There are some 250,000 HIV-seropositive individuals with T4-cell counts below $200/\mu L$ of blood. More than 100 microorganisms—bacteria, viruses, fungi, and protozoa—can cause disease in such individuals, even though only a fraction of these (17) are included in the current surveillance definition for clinical AIDS. In a large survey from the Centers for Disease Control and Prevention (CDC), such OIs were diagnosed in 33% of individuals at 1 year and in 58% at 2 years after documentation of a T4-cell count below $200/\mu L$. In 1997, the CDC presen-

ted guidelines for prevention and treatment of OIs in HIV-infected persons (*MMWR*, 1997). The new treatments for OIs have extended the survival of AIDS patients, but they have also opened new issues. With the growing proportion of long-term AIDS survivors, new OIs have become prominent, together with concerns about cost, compliance, drug interactions, and quality of life (Laurence, 1995; *MMWR*, 1995). The six most common AIDS-related OIs were bacterial pneumonia, candidal esophagitis, pulmonary/disseminated TB, mycobacterium avium complex disease, herpes simplex reinfection, and PCP. **The year 1997 marks the first time, since the AIDS pandemic began in the United States, that the incidence of AIDS defining OIs among HIV-infected persons fell in number from the previous year's total.** The number of OIs dropped 7% in 1997 compared to 1996 (HIV/AIDS Surveillance Report 1997).

PROPHYLAXIS AGAINST OPPORTUNISTIC INFECTIONS

Drug prophylaxis against OIs has become a cornerstone of treatment for AIDS patients. For example, the prevalence of *Pneumocystis carinii* pneumonia (PCP) dropped from about 80% in 1987 to about 20% by mid-1994 because of the use of excellent drug therapy. The mortality of PCP without treatment is almost 100% (Dobkin, 1995). According to Laurence (1996), researchers at the University of California, San Francisco, found that it costs, on average, $215,000 to extend by 1 year, the life of an HIV-infected patient with *Pneumocystis carinii* pneumonia (PCP) who is treated in an intensive care unit. That is more than twice the comparable care cost for 1988. Part of the reason given was that as people with AIDS survived longer, they were presenting with second and third episodes of PCP, superimposed on other chronic infections. The downside to OI prophylaxis is that it is difficult to find drugs that work without harmful side effects. In addition, viruses and organisms that cause OIs become resistant to the drugs over time. This is one of the primary reasons researchers are looking for ways in which to boost an immunosuppressed patient's immune system (Zoler, 1991).

OPPORTUNISTIC INFECTIONS IN HIV-INFECTED PEOPLE

AIDS is a devastating human tragedy. It appears to be killing about everyone who demonstrates the symptoms. One well-known American surgeon said, **"I would rather die of any form of cancer rather than die of AIDS."** This statement was not made because of the social stigma attached to AIDS, or because it is a lethal condition. It was made in recognition of the slow, demoralizing, debilitating, painful, disfiguring, helpless, and unending struggle to stay alive.

Because of a suppressed and weakened immune system, viruses, bacteria, fungi, and protozoa that commonly and harmlessly inhabit the body become pathogenic (Figure 6-1). In addition, organisms and viruses from old infections that have lingered in the body reactivate. The suppression of the immune system presents an **opportunity** for the harmless to become harmful. About 90% of deaths related to HIV infection and AIDS are caused by OIs, compared with 7% due to cancer and 3% due to other causes.

What makes HIV disease particularly terrifying is that it leaves patients open to an endless series of infections that would not occur in people with healthy immune responses. *Pneumocystis carinii* pneumonia, toxoplasmosis, Kaposi's sarcoma, candidiasis, cytomegalovirus retinitis, cryptococcal meningitis, mycobacterium avium complex, herpes simplex, and herpes zoster are infections that sicken and disfigure, and some combinations of them eventually kill most people with AIDS.

HIV-Related Opportunistic Infections Vary Worldwide

The course of HIV infection tends to be similar for most patients: infection with the virus is followed by seroconversion and progressive destruction of T4 cells. Yet the opportunistic infections and malignancies that largely define the symptomatic or clinical history of HIV disease vary geographically. People with HIV and their physicians in different regions confront distinct problems, mainly because of differences in exposure, in access to diagnosis and care, and in general health.

Comparisons between the data about opportunistic infections in different countries must be made with care. But, most developing nations lack the facilities and trained personnel to identify opportunistic infections correctly; consequently, their prevalence may be underreported. Clinicians in developed countries can order sophisticated laboratory analyses to identify pathogens. Those in developing countries must rely on signs and symptoms to make their diagnoses. Oral candidiasis and herpes zoster are easy to diagnose without laboratory backup because the lesions are visible. While some pneumonias and types of diarrhea can be specified, others, (such as extrapulmonary tuberculosis, cytomegalovirus infections, crypto-

---

───────── **BOX 6.1** ─────────

THE DECISION

Speaking of things we don't talk about, a friend of mine was recently diagnosed with CMV. I asked her if she'd gotten a catheter and started gancyclovir. She said, "No, I haven't decided what I'm going to do. But everytime I bring up the idea of choosing not to treat this, my friends get really freaked out, even hostile. So I'm not telling anyone anymore except you." I told her that my roommate had been diagnosed with CMV and got a catheter and infused gancyclovir for four hours a day, and eventually failed on it and started foscarnet which gave him stomach problems, and started taking a new drug for his stomach. I used to watch him and wonder if I could ever do what he was able to do. I don't know if I could and I sympathize with my friend who has practiced walking around her apartment with a blindfold to experience what it would be like to be blind. I'm sure many of us wrestle with decisions about treatments and the choice not to treat at all, but this sort of consideration runs contrary to our limited notion of empowerment and our mandate to live with AIDS.

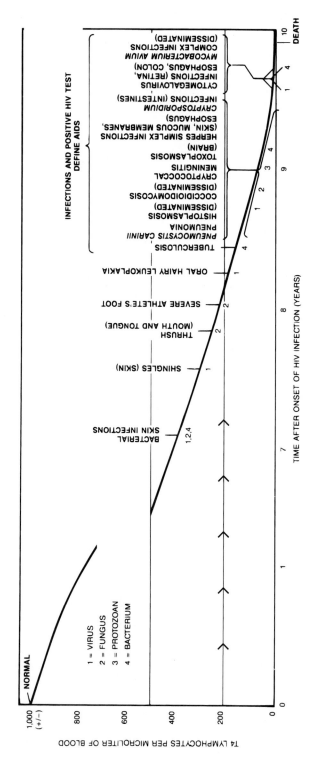

FIGURE 6-1 General Progression of Opportunistic Infections after HIV Infection. Normal T4 cell count in adolescent/adults is, on average, about 1,000/μL of blood. There is a relationship between the drop in T4 lymphocytes and the onset of opportunistic infections (OIs). The first sign of an OI begins under 500 T4 cells/μL. As the T4 cell count continues to drop, the chance of OI infection increases. Note the variety of OIs found in AIDS patients with 200 or less T4 cells/μL.

THE CHANGING SPECTRUM OF OPPORTUNISTIC INFECTIONS

Only 10 years ago there was hardly any standardized use of protective agents to block the infectious complications or opportunistic infections (OIs) associated with HIV-induced immunodeficiency. Now there is an array of drugs that can be used in strategies to prevent or delay nearly all the major OIs. With this advancement has come the need to weigh the pros and cons of various strategies. Cost, antimicrobial resistance, drug interactions, and pill overload are all important considerations.

EFFECT OF ANTI-HIV THERAPY ON OPPORTUNISTIC INFECTIONS

As presented in Chapter 4, the use of anti-HIV combination drug therapy (AIDS drug cocktails) has produced a number of unexpected results in patient response to those drugs. Soon after combination therapy began, physicians witnessed a rather **confusing** or **unusual** presentation of OIs. In some cases certain OIs improved, in others the situation deteriorated! Such changes in OIs expression are occurring now, at a time when hundreds of thousands of HIV-infected people want **Highly Active Antiretroviral Therapy (HAART).**

TREATMENT IN THE HAART ERA

HAART therapy appears to scramble the human immune system. When HIV patients take retroviral medicines that control HIV replication, their immune systems begin to recuperate in ways that are puzzling and controversial. For example, patients recover immunity to some deadly opportunistic infections but appear unable to fight diseases for which they were vaccinated as children, for example, tetanus, or to target HIV itself. Collectively, such observations indicate HAART patients can only raise successful immune responses against pathogens they see **regularly.** For example, cytomegalovirus is an organism found in everybody's blood, so the immune systems of HAART patients see the pathogen constantly and generate cells and antibodies that attack it. But tetanus is something people rarely encounter, so HAART patients, unlike their HIV-negative counterparts, fail to raise immune responses against it. And the ultimate irony is that HAART, when suc-

cessful, kills all but a few million HIVs that are forced into hiding. Therefore, the immune system doesn't see HIV any more and stops making cells and antibodies against the virus.

This information aside, when HIV is controlled with the antiretroviral drugs, immunity to infections—**other than HIV**—usually starts to return. As a result, some opportunities infections go away without specific treatment; and sometimes patients can stop prophylactic treatment for certain opportunistic diseases. However, entering year 2000, it is unclear who can stop prophylaxis safely, and who cannot. Recent studies suggest that the incidence of esophageal candidiasis and, by inference, other forms of *Candida* infection has fallen by 60% to 70% on patients treated with HAART. The use of HAART had dramatically changed the epidemiology of opportunistic infections and is clearly associated with gradual recovery of the immune system (Powderly et al., 1998).

Recent studies in 1998 and 1999, using protease inhibitor drugs in HAART have shown that virtually no patient whose T4 level rose to and stayed over 100 per microliter of blood (μl) developed and OI. This is the strongest evidence to date that immune reconstruction is occurring with protease inhibitor therapy, and suggests that it occurs early and with quite modest improvements in T4 cell levels. The implication is that the search for immunorestorative therapy other than with the current antiretroviral is somewhat less urgent than previously believed, though still a clear priority. **Clinical trails are underway to determine whether cessation of OI prophylaxis is viable in some patients and to establish guidelines if it is.**

One study indicates that patients receiving triple combination therapy may be eligible for stoppage of OI prophylaxis, but that patients on mono- or dual-therapy may not. Patients without a history of OIs who respond to antiretroviral therapy appear to be at the lowest risk for the development of the infections. Meanwhile, researchers from the University Hospital of Utrecht and Utrecht University in the Netherlands report that *Pneumocystis carinii* pneumonia prophylaxis can be safely stopped in HIV-infected patients on HAART who have T4 cell counts greater than 200 cells per microliter (Schneider et al., 1999).

In summary, William Bishai and colleagues (1999) report that Richard Moore and colleagues examined subpopulations who are benefiting from the decrease of OI incidence due to HAART by evaluating data from the Johns Hopkins HIV clinic between 1995 and 1998. Among AIDS patients who acquired HIV through homosexual contact there was an 81% reduction of OI incidence; among heterosexually transmitted HIV cases there was 60% reduction, and among injection drug users there was a 45% reduction. Hence, gay and bisexual men have benefited the most by the HAART-induced reduction in OIs, while injection drug users have benefited the least. Differences in adherence were thought to explain some of this variation. In a related study, Sullivan and colleagues found a 60% decrease in the incidence of bacterial pneumonia between 1992 and 1998, with the protease inhibitor therapy being the strongest predictor of decreasing risk.

VIRAL LOAD RELATED TO OPPORTUNISTIC INFECTIONS

HIV-clinicians have recently looked at the predictive value of plasma HIV RNA for the development of three OIs: *Pneumocystis carinii* pneumonia (PCP), CMV, and MAC. Using a database of 813 patients participating in AIDS Clinical Trial Groups (ACTG), for every 1-log increase in plasma HIV RNA level, the risk of developing one of these OIs was increased 2- to 3-fold. Plasma HIV RNA level was predictive of an increased risk of an OI independent of T4 cell count, which also predicted OI risk. This information confirms that maintaining control of viral replication may be a critical component of preventing OIs in HIV-infected patients.

coccal meningitis, and systemic infections, such as histoplasmosis, toxoplasmosis, microsporidiosis, and nocardiosis) go underreported due to the lack of laboratory facilities.

Socioeconomic Factors

Geography explains much about the varying patterns of opportunistic infections, but a decisive factor is often financial capacity. On the most fundamental level, **money** is needed to create an infrastructure that limits exposure to pathogens. Thus, while few people with HIV in wealthy countries develop certain bacterial or protozoal infections, they are a major cause of death in poor areas that cannot provide clean water and adequate food storage facilities.

Financial resources also affect clinicians' abilities to diagnose AIDS and, when appropriate, to provide the proper medicine. (Table 6-1) AIDS patients in Africa often die of severe bacterial infections because they don't have the antibiotics or the clinical care they need. They don't survive long enough to develop diseases such as PCP.

***United States, Europe, and Africa*—** The United States and Europe on one end, and Africa on the other, represent the global extremes of financial resources for health care. The most common opportunistic infections

TABLE 6-1　Cost of Care for Some Common Opportunistic Diseases

Opportunistic Infection	$ cost/month[a]
Pneumocystis carinii pneumonia	8797
Mycobacterium avium complex	3067
Toxoplasmosis	5620
Cytomegalovirus	9761
Fungal Infections	3151
Other OIs[b]	3664

[a] Costs are derived from the AIDS Cost and Services Utilization Survey. For opportunistic infections, costs include 1 month before and 2 months after AIDS diagnosis.

[b] Other OIs include lymphoma, tuberculosis, Kaposi sarcoma, and other AIDS-defining illnesses (adapted from Freedberg, et al., 1998).

each region faces reflect the overall quality of health care, sanitation, and diet. For example, Thailand and Mexico belong to the large group of nations that have intermediate incomes and correspondingly intermediate patterns of HIV complications (Harvard AIDS Institute, 1994).

AIDS patients rarely have just one infection (Table 6-2). The mix of OIs may depend on life style and where the HIV/AIDS patient lives or has lived. Thus a knowledge of the person's origins and travels may be diagnostically helpful. (Note: a number of the symptoms listed in the CDC definition of HIV/AIDS can be found associated with certain of the OIs presented.)

TABLE 6-2 Some Common Opportunistic Diseases Associated with HIV Infection and Possible Therapy

Organism/Virus	Clinical Manifestation	Possible Treatments
Protozoa		
Cryptosporidium muris	Gastroenteritis (inflammation of stomach-intestine membranes)	Investigational only
Isospora belli	Gastroenteritis	Trimethoprim-sulfamethoxazole (Bactrim)
Toxoplasma gondii	Encephalitis (brain abscess), retinitis, disseminated	Pyrimethamine and leucovorin, plus sulfadiazine, or Clindamycin, Bactrim
Fungi		
Candida sp.	Stomatitis (thrush), proctitis, vaginitis, esophagitis	Nystatin, clotrimazole, ketoconazole
Coccidioides immitis	Meningitis, dissemination	Amphotericin B, fluconazole, ketoconazole
Cryptococcus neoformans	Meningitis (membrane inflammation of spinal cord and brain), pneumonia, encephalitis, dissemination (widespread)	Amphotericin B, fluconazole, itraconazole
Histoplasma capsulatum	Pneumonia, dissemination	Amphotericin B, fluconazole, itraconazole
Pneumocystis carinii	Pneumonia	Trimethoprim-sulfamethoxazole (Bactrim, Septra), pentamidine, dapsone, clindamycin, trimetrexate
Bacteria		
Mycobacterium avium complex (MAC)	Dissemination, pneumonia, diarrhea, weight loss, lymphadenopathy, severe gastrointestinal disease	Rifampin + ethambutol + clofazimine + ciprofloxacin +/- amikacin; rifabutin, clarithromycin & azithromycin
Mycobacterium tuberculosis (TB)	Pneumonia (tuberculosis), meningitis, dissemination	Isoniazid (INH) + rifampin + ethambutol +/- pyrazinamide
Viruses		
Cytomegalovirus (CMV)	Fever, hepatitis, encephalitis, retinitis, pneumonia, colitis, esophagitis	Ganciclovir, Foscarnet
Epstein-Barr	Oral hairy leukoplakia, B cell lymphoma	Acyclovir
Herpes simplex	Mucocutaneous (mouth, genital, rectal) blisters and/or ulcers, pneumonia, esophagitis, encephalitis	Acyclovir Famciclovir
Papovavirus J-C	Progressive multifocal leukoencephalopathy	none
Varicella-zoster	Dermatomal skin lesions (shingles), encephalitis	Acyclovir, Foscarnet
Cancers		
Kaposi's sarcoma	Disseminated mucocutaneous lesions often involving skin, lymph nodes, visceral organs (especially lungs & GI tract)	Local injection, surgical excision or radiation to small, localized lesions; Chemotherapy with vincristine & Bleomycin
Primary lymphoma of the brain	Headache, palsies, seizures, hemiparesis, mental status, or personality changes	Radiation and/or chemotherapy
Systemic lymphomas	Fever, night sweats, weight loss, enlarged lymph nodes	Chemotherapy

Patients with compromised immune systems are at increased risk for all known cancers and infections (including bacterial, viral, and protozoal). Most infectious diseases in HIV-infected patients are the result of proliferation of organisms already present in the patient's body. Most of these opportunistic infections are not contagious to others. The notable exception to this is tuberculosis.

(**Disclaimer: This table was developed to provide general information only. It is not meant to be diagnostic nor to direct treatment.**)

(Adapted from Mountain-Plains Regional Education and Training Center *HIV/AIDS Curriculum*, 4th Ed., 1992 updated, 1997; and from *MMWR*, 1997.)

Fungal Diseases

In general, healthy people have a high degree of innate resistance to fungi. But a different situation prevails with opportunistic fungal infections, which often present themselves as acute, life-threatening diseases in a compromised host (Medoff et al., 1991).

Because treatment seldom results in the eradication of fungal infections in AIDS patients, there is a high probability of recurrence after treatment (DeWit et al., 1991).

Fungal diseases are among the more devastating of the OIs and are most often regional in association. AIDS patients from the Ohio River basin, the Midwest, or Puerto Rico have a higher than normal risk of histoplasmosis *(his-to-plaz-mo-sis)* infection. In the Southwest, there is increased risk for coccidioidomycosis *(kok-sid-e-o-do-mi-ko-sis)*. In the southern Gulf states, the risk is for blastomycosis. Other important OI fungi such as **Pneumocystis carinii** *(nu-mo-sis-tis car-in-e-i)*, **Candida albicans** *(kan-di-dah al-be-cans)*, and **Cryptococcus neoformans** *(krip-to-kok-us knee-o-for-mans)* are found everywhere in equal numbers. Because of their importance as OIs in AIDS patients, a brief description of **histoplasmosis, candidiasis, Pneumocystis carinii pneumonia (PCP), and cryptococcosis are presented** (Table 6-1).

Histoplasmosis (Histoplasma capsulatum)— Spores are inhaled and germinate in or on the body (Figure 6-2, A and B). This fungal pathogen is endemic in the Mississippi and Ohio river valleys. Signs of histoplasmosis include prolonged influenza-like symptoms, shortness of breath, and possible complaints of night sweats and shaking chills. In AIDS patients there is a multisystem involvement: the liver, central nervous system, lymph nodes, and gastrointestinal tract are affected, while mucosal ulcers and enlarged spleen may also occur. Histoplasmosis in an HIV-positive person is considered diagnostic of AIDS. In about two-thirds of AIDS patients with histoplasmosis, it is the initial OI. Over 90% of cases have occurred in patients with T4 cell counts below 100/µL (Wheat, 1992).

In non-HIV-infected patients, amphotericin B is almost always successful in treating histoplasmosis (Table 6-2). Only 50% to 80% of AIDS patients respond, however, and relapse is common during maintenance therapy. Nonetheless, amphotericin B followed by weekly maintenance therapy remains the recommended strategy. Amphotericin B is a toxic drug, increases one's temperature, and may also cause uncontrollable shaking; it is sometimes referred to as the "shake and bake" treatment. Earlier reports suggested ketoconazole as a potential alternative for initial or maintenance therapy, but it had an even higher failure rate than amphotericin B and is no longer recommended in HIV-infected patients.

Itraconazole (Sporanox) has recently been approved for treating histoplasmosis and some now consider it the drug of choice for HIV-related histoplasmosis (Daar et al., 1993; *Guidelines, U.S. Publich Health Service*, 1995).

Candidiasis (Candida albicans)— This fungus is usually associated with yeast infections of the vagina. It is a fungus quite common to the body and in particular inhabits the alimentary tract. It is normally kept in check by the presence of bacteria that live on the linings of the alimentary tract. However, in immunocompromised patients, especially those who have received broad spectrum antibiotics, candida multiplies rapidly. Because of its location in the upper reaches of the alimentary tract, if unchecked, it may cause mucocutaneous candidiasis or thrush, an overgrowth of candida in the esophagus and in the oral cavity (Figure 6-3, A and B). Mucosal candidiasis was associated with AIDS patients from the very beginning of the AIDS pandemic (Powderly et al., 1992). In women, overgrowths of candidiasis also occur in the vaginal area.

Gynecological conditions in women with HIV disease have been found to be more aggressive and to occur with a greater frequency than in noninfected women. Vaginal candidiasis (VC) is generally caused by *C. albicans*. VC is seen at all stages of HIV disease, with increasing frequency in conjunction with PCP and genital herpes in patients with <200 T4 cells per µL of blood. It is sometimes associated with the use of birth control pills, chemotherapeutic regimens, corticosteroids, and immune suppression related to pregnancy. Common symptoms include a thick, whitish discharge; severe

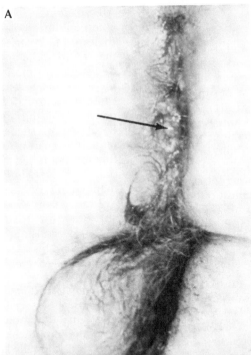

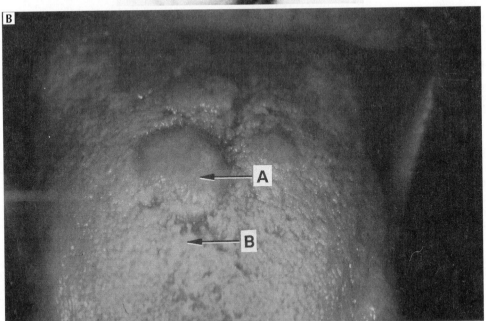

FIGURE 6-2   **A**, Anal Histoplasmosis. Histoplasmosis is caused by
Histoplasmosis capsulatum and causes infection in immunocompromised
patients. (*Courtesy CDC, Atlanta*) **B**, AIDS patient's tongue showing multiple
shiny, firm Histoplasma erythematous nodules (see arrow A) and Thrush (see
arrow B). (*Photograph courtesy of Marc E. Grossman, New York*)

A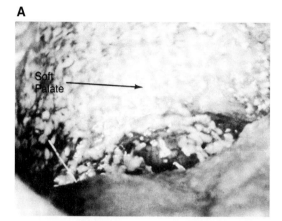

B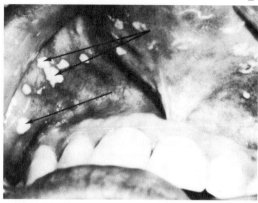

FIGURE 6-3 Thrush. **A,** An overgrowth of *Candida albicans* on the soft palate in the oral cavity of an AIDS patient. **B,** Creamy patches of candida that can be **scraped off** leaving a red and sometimes bleeding mucosa. (**A,** *Courtesy CDC, Atlanta;* **B,** *Schiodt, Greenspan, and Greenspan 1989.* American Review of Respiratory Disease, *1989, 10: 91–109. Official Journal of the American Thoracic Society. © American Lung Association)*

itching; localized, sometimes severe pain; and infrequently, the development of lesions. Vaginal manifestations of the disease are thought to precede esophageal and/or oral thrush. HIV-related vaginal candidiasis may be characterized by resistance to standard therapy. Candida infections should be monitored for frequency and response to antifungal treatments (Table 6-2). If candidiasis is limited to the mouth and oropharynx and the patient is not debilitated, treatment with topical nystatin (Mycostatin Pastilles) or clotrimazole (Mycelex Troches) is often sufficient. If the patient does not respond adequately, or esophageal involvement is possible, a systemic agent such as ketoconazole (Nizoral) or fluconazole (Diflucan) is appropriate (Daar et al., 1993; *Guidelines, U.S. Public Health Service,* 1995). Difficult to treat cases may require IV amphotericin B (Fungizone). Acidopholis supplementation appears to be beneficial, especially in women receiving PCP prophylaxis with TMP/SMX (Bactrim or Dapsone). Persons often self-administer alternative treatments, either alone or in conjunction with approved therapies. These include changes in diet, with the avoidance of foods containing yeast, sugar, and dairy products. Vitamin C supplementation and herbal products are often used (Willoughby, 1989).

Some people with oropharyngeal candidiasis may experience a sore mouth. This is particularly painful when acidic foods and juices are taken. In addition, the taste of food is often altered. Treatment is aimed at relieving the symptoms, as therapy is rarely successful at eliminating the fungus (Hay, 1991).

Impaired cell-mediated immunity, as occurs in people with HIV disease, may allow for a disseminated infection (distributed to other areas of the body). **Oral or esophageal candidiasis** causes thick white patches on the mucosal surface and may be the first manifestation of AIDS. Because other diseases can cause similar symptoms, candidiasis by itself is not sufficient for a diagnosis of AIDS.

Pneumocystis carinii— This fungus, until recent reclassification, was considered a protozoan (Edman et al., 1988). The life cycle and reproductive characteristics of *P. carinii* are not completely understood because the organism is difficult to culture for laboratory study. However, the molecular biology and molecular taxonomy are now being rapidly constructed (Walzer, 1993; Stringer et al., 1996). Virtually everyone in the United States by age 30 to 40 has been exposed to *P. carinii*. It lies dormant in the lungs, held in check by the immune system. Prior to the AIDS epidemic, *P. carinii* pneumonia was seen in children and adults who had a suppressed immune system as in leukemia or Hodgkin's disease and were receiving chemotherapy. In the AIDS patient, the onset of *P. carinii* pneumonia is insidious—patients may notice some shortness of breath and they cannot run as far. It causes extensive damage within the alveoli of the lung.

The *P. carinii* fungus develops from a small unicellular trophozoite into a cyst containing eight sporozoites (Figure 6-4A). These sporozoites are spread throughout the body, but they show a greater affinity for the lungs, where they multiply in the spaces between the lung sacs and cause pneumonia (Figure 6-4B). Prior to 1981, fewer than 100 cases of *P. carinii* infection were reported annually in the United States; yet 80% of AIDS patients develop *P. carinii* pneumonia at some time during their illness. This is one of the few AIDS-related conditions for which there is a choice of relatively effective drugs. The first of these to be made available was the intravenous and aerosolized versions of pentamidine. There are frequent recurrences of this infection. With treatment, there is a 10% to 30% mortality with each PCP episode (Montgomery, 1992). PCP accounted for a diagnosis of AIDS in over 65% of AIDS cases in 1990. In 1994 it had fallen to 20% due to available therapy (Ernst, 1990; Murphy, 1994). The triad of symptoms that almost always indicates the onset of PCP during HIV disease is fever, dry cough, and shortness of breath (Grossman et al., 1989). *P. carinii* pneumonia is unlikely to develop in people with HIV disease unless their T4 cell count drops below 200 (Phair et al., 1990).

The antifolate (blocking action to folic acid) drugs pentamidine and trimethoprim/sulfamethoxazole are currently treatments of choice for *P. carinii* pneumonia (Zackrison et al., 1991; Dobkin, 1995).

PNEUMOCYSTIS CARINII PNEUMONIA IN CHILDREN. PCP is the most common opportunistic infection in children who have AIDS. Despite the publication of guidelines for prophylaxis against PCP for children infected with HIV in 1991, ongoing AIDS surveillance has detected no substantial decrease in PCP incidence among HIV-

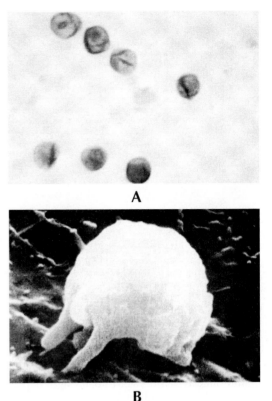

A

B

FIGURE 6-4   A, A concentration of *Pneumocystis carinii* cysts. (See text for details.) (*Photograph courtesy of Linda L. Pifer, University of Tennessee, The Health Science Center, Dept. of Clinical Laboratory Sciences*) **B,** Scanning electron microscope image of *P. carinii* attached to embryonic chick lung cells in culture. Note the tubular extensions through which it extracts nutrients from host lung tissue. (*Photograph courtesy of Linda L. Pifer.* © Pediatric Research, *1977, 11:314*)

infected infants. This continued incidence is associated with failure to identify HIV-infected children before PCP occurs and with limitations in the ability of T4 measurements to identify children at risk for PCP. In March 1994, the National Pediatric & Family HIV Resource Center, in collaboration with CDC, convened a working group to review additional data about the occurrence of PCP among HIV-infected children. This report summarizes these new data and presents revised PCP prevention guidelines that recommend: (**1**) promptly identifying children born to HIV-infected women and initiating regular diagnostic and immunologic monitoring of such children; (**2**) beginning PCP prophylaxis at 4–6 weeks of age for all children who have been perinatally exposed to HIV; (**3**) continuing prophylaxis through 12 months of age for HIV-infected children; and (**4**) making decisions regarding prophylaxis for HIV-infected children ≥12 months of age based on T4 measurements and whether PCP previously has occurred (*MMWR*, 1995).

Cryptococcosis (Cryptococcus neoformans)— Since its discovery in 1894, *C. neoformans* has been recognized as a major cause of deep-seated fungal infection in the human host. The infection can affect many sites, including skin, lung, kidney, prostate, and bone. However, symptomatic disease most often represents infection of the central nervous system. Cryptococcal meningitis is the most common form of fungal meningitis in the United States (Ennis et al., 1993). This fungus is shed in pigeon feces, the spores of which enter the lung. If the lung does not eliminate it, it gets into the bloodstream, travels to the brain, and can cause cryptococcal meningitis.

In a healthy person, *C. neoformans* may cause pulmonary problems or meningeal infection (infection of the membranes that envelop the brain and spinal cord). It is more commonly seen in people with immunosuppression, especially in HIV/AIDS cases. *C. neoformans* is a fatal OI that occurs in about 13% of AIDS patients (Brooke et al., 1990). It is acquired through the respiratory tract and most commonly causes cryptococcal meningitis. Disseminated *C. neoformans* (Figure 6-5A) infection may involve bone marrow, the central nervous system (CNS), and the lungs, causing cryptococcal pneumonia. In AIDS patients, *C. neoformans* also causes infection of the skin (Figure 6-5B), lymph nodes, and kidneys. *Cryptococcus* cannot be cured and it does recur. What drugs should be used is a subject of controversy. Eric Darr and colleagues (1993) report that fluconazole and itraconazole are effective in treating cryptococcal meningitis (Table 6-2). Since then, flucytosine and amphotericin B have been used (Dobkin, 1995; *Guidelines, U.S. Public Health Service*, 1995). For additional information read Buchanan et al., 1998).

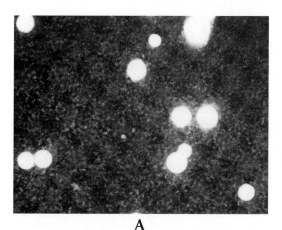

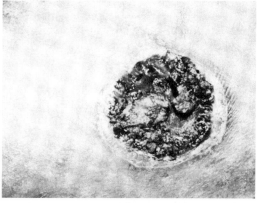

A   **B**

FIGURE 6-5 *Cryptococcus neoformans.* **A,** Large, budding, and encapsulated *C. neoformans* shows white against an India ink stain. Isolated from the cerebral spinal fluid of a patient with meningitis. **B,** Skin lesions caused by *C. neoformans* may be single or multiple. The "ulcer" is usually painless but is an early sign of infection. (**A,** *Courtesy CDC, Atlanta.* **B,** *Courtesy of Ronald P. Rapini, Texas Tech University Health Science Center and reprinted with permission from* Cutis, *1988, Vol. 42:125-128, copyright 1988 by QUADRANT HEALTHCOM, INC.)*

Viral Diseases

Because of a depleted T4 cellular component of the immune system, AIDS patients are at particularly high risk for the herpes family of viral infections: cytomegalovirus, herpes simplex virus types 1 and 2, varicella-zoster virus, and Epstein-Barr virus.

Cytomegalovirus (CMV)— This virus is a member of the human herpesvirus group of viruses. CMV is the perfect parasite. It infects most people asymptomatically. When illness does occur, it is mild and nonspecific. There have been no epidemics to call attention to the virus. Yet CMV is now considered the most common infectious cause of mental retardation and congenital deafness in the United States. It is also the most common viral pathogen found in immunocompromised people (Balfour, 1995).

A **latent state** follows initial infection, with CMV probably located in the white blood cells and involving many organs and organ systems. Later it is reactivated, usually by some form of immunosuppression such as organ or bone marrow transplantation, cancer chemotherapy, or HIV disease (Jacobson et al., 1992; Hardy, 1996).

The virus is very unstable and survives only a few hours outside a human host. It can be found in saliva, tears, blood, stool, cervical secretions, and in especially high levels in urine and semen. Transmission occurs primarily by intimate or close contact with infected secretions. The incidence of CMV infection varies from between 30% and 80% depending on the geographical community tested. However, over 90% of homosexual males have tested positive for CMV (Jacobson et al., 1988). CMV causes the more important viral infections in AIDS patients (Figure 6-6). This virus causes a broad spectrum of diseases in HIV-infected people, ranging from mild or severe gastrointestinal problems to infections of the brain, the liver (hepatitis), and the onset of fulminant (sudden and severe) pneumonia. Gastrointestinal infection may result in severe ulceration of the esophagus, stomach, small intestines, and colon. CMV pneumonia occurs in 10% to 20% of AIDS patients and can be lethal because therapy is unsuccessful.

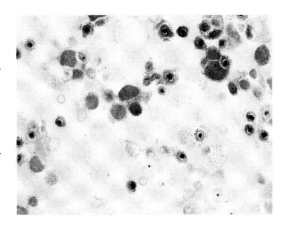

FIGURE 6-6 Cytomegalovirus. This is an electron micrograph magnified 49,200 times. (*Courtesy CDC, Atlanta*)

CMV infection of AIDS patients usually results in prolonged fever, anemia (too few red blood cells), leukopenia (too few white blood cells), and abnormal liver function. CMV also causes severe diarrhea and HIV-associated retinitis resulting in eventual blindness (*Emergency Medicine*, 1989; Lynch, 1989; Dobkin, 1995).

Perhaps 75% of AIDS patients have an eye disease, with the retina the most common site (Russell, 1990). The retina, which is a light-sensitive membrane lining the inside of the back of the eye, is also part of the brain and is nourished by blood vessels. **AIDS-related damage to these vessels produces tiny retinal hemorrhages and small cotton wool spots**—early indicators of disease that are often detected during a routine eye examination (Figure 6-7).

Symptoms of CMV retinitis may be subtle, such as blurred vision, haze or floaters, or small dim spots in side vision. In unusual circumstances, the virus can produce dramatic symptoms, such as loss of vision within 72 hours. Without treatment, CMV can destroy the entire retina in 3 to 6 months after infection.

In August 1991, the United States Food and Drug Administration (FDA) approved the use of **foscarnet** (Foscavir) and **ganciclovir** (Cytovene) for the treatment of CMV retinitis (Table 6-2). The drugs slow the progression of CMV retinitis (Danner, 1995; Hermans, 1996). Recently (1996)

——— BOX 6.2 ———

A PHYSICIAN'S AGONIZING DILEMMA

Opportunistic infections are the primary threat to patients with AIDS; they are the main causes of illness and death. The cruel irony is that although there are 26 or more FDA approved drugs to treat these infections, most cannot be used in patients receiving zidovudine (ZDV) *because the combination drug therapy is devastatingly toxic to bone marrow.*

EXAMPLE:

Cytomegalovirus (CMV) Retinitis

CMV retinitis is one of the *worst* of the opportunistic infections and it develops in 90% of AIDS patients (Gottlieb et al., 1987). Both the patient who contracts CMV retinitis and the treating physician confront an almost impossible dilemma: whether to continue the zidovudine and risk blindness or to treat the retinitis and risk death from another infection. A physician recently said, "In my experience, I have never yet been able to combine zidovudine with the experimental drug ganciclovir (DHPG), which is the only recognized treatment for CMV retinitis, because the combination destroys the bone marrow. I therefore face the virtually impossible task of asking a 25-year-old person: 'Which of these options do you prefer: to take zidovudine and keep on living but go blind, or to preserve your sight by having retinitis treatment but risk dying?' That is a truly terrible question to ask of any patient, especially a young one." He continued, "Recently, I treated just such a young patient. He had been taking zidovudine for AIDS when he contracted CMV retinitis. When I presented him with this agonizing choice, he told me, `I definitely don't want to go blind. I want to be treated for this retinitis.' So he stopped taking zidovudine and was started on DHPG treatment. He started to have seizures, which were a consequence of the toxoplasmosis brain abscess that eventually ended his life. Just before he died, he told me that the worst decision he ever made was to stop taking zidovudine. 'I should have stayed on it,' he said, 'and gone blind.'" (Robert J. Awe, M.D., 1988)

Cryptococcal Meningitis

The situation for AIDS patients with cryptococcal meningitis is similarly depressing. Because of the severe bone marrow toxicity, it is virtually impossible to use amphotericin B, the effective treatment for this meningitis, at the same time as zidovudine. Besides, the use of amphotericin B has its own side effects as can be noted from this excerpt from Paul Monette's *Borrowed Time: An AIDS Memoir:*

> Amphotericin B is administered with Benadryl in order to avoid convulsions, the most serious possible side effect. It was about nine or ten when they started the drug in his veins, and I sat by the bed as nurses streamed in and out. A half hour into the slow drip, the nurse monitoring the IV walked out, saying she'd be right back, and a couple of minutes later Roger began to shake. I gripped him by the shoulders as he was jolted by what felt like waves of electric shock, staring at me horror-struck. Though Cope [the physician] would tell me later, trying to ease the torture of my memory, that "mentation" [mental activity] is all blurred during convulsions, I saw that Roger knew the horror. (Page 336)
>
> When the nurse returned she looked at him in dismay: "How long has this been going on?" Then she ordered an emergency shot of morphine to counteract the horror. When at last he fell into a deep sleep they all told me to go home, saying they would try another dose of the ampho in a few hours. I was so ragged I could barely walk. So I left him there with no way of knowing how near it was [Roger's death], or maybe not brave enough to know.

Whichever option is chosen, the patient is bound to suffer, and perhaps die, either of cryptococcal meningitis or of some other infection.

---

eye implants of ganciclovir have shown to be a more effective way to delay CMV retinitis.

Herpes Viruses Types 1 & 2 (HSV 1 & 2)— Herpes infections are one of the most commonly diagnosed infections among the HIV/ AIDS population. Almost all HIV-infected individuals (95 percent) are co-infected with HSV 1 and/or HSV 2. Both viruses cause **severe** and progressive eruptions of the mucous membranes. HSV 1 affects the membranes of the nose and mouth. Also, when herpetic lesions

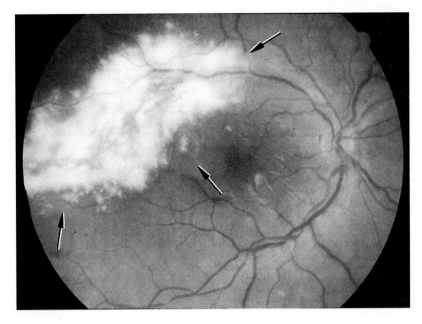

FIGURE 6-7 Cytomegalovirus retinitis. The disease, as seen in this photograph, involves the posterior pole of the right eye. Fluffy white infiltrate **(cotton wool spots)** with a small amount of retinal hemorrhage can be seen in the distribution of the superior vascular arcade. (*Courtesy of Scott M. Whitcup, M. D., National Eye Institute, National Institutes of Health, Bethesda, MD*)

involve the lips or throat, 80% to 90% of the time they either precede or occur simultaneously with **herpes-caused pneumonia** (Gottlieb et al., 1987). Bacterial or fungal superinfections occur in more than 50% of herpes-caused pneumonia cases and are a major contributory cause of death in AIDS patients.

Mortality from HSV pneumonia exceeds 80% (Lynch, 1989). Herpes may also cause blindness in AIDS patients. The following is from Paul Monette's *Borrowed Time*:

I woke up shortly thereafter, and Roger told me—without a sense of panic, almost puzzled—that his vision seemed to be losing light and detail. I called Dell Steadman and made an emergency appointment, and I remember driving down the freeway, grilling Roger about what he could see. It seemed to be less and less by the minute. He could barely see the cars going by in the adjacent lanes. Twenty minutes later we were in Dell's office, and with all the urgent haste to get there we didn't really reconnoiter till we were sitting in the examining

room. I asked the same question—what could he see?—and now Roger was getting more and more upset the more his vision darkened. I picked up the phone to call Jamiee, and by the time she answered the phone in Chicago he was blind. *Total blackness, in just two hours!*

The retina had detached. (An operation on retinal attachment was successful and sight was restored. The cause of the retinal detachment was a herpes infection of the eyes.)

HSV 2 also affects the membranes of the anus, causing severe perianal and rectal ulcers primarily in homosexual men with AIDS (Figure 6-8, A and B). Herpes of the skin can generally be managed with oral **acyclovir** (Zovirax), Foscavir or Famvir (Table 6-2).

Herpes Zoster Virus (HZV)— Like herpes simplex, this virus has the potential to cause fulminant pneumonia in AIDS patients. Untreated HZV pneumonia has a mortality rate of 15% to 35%. HZV is now monitored as

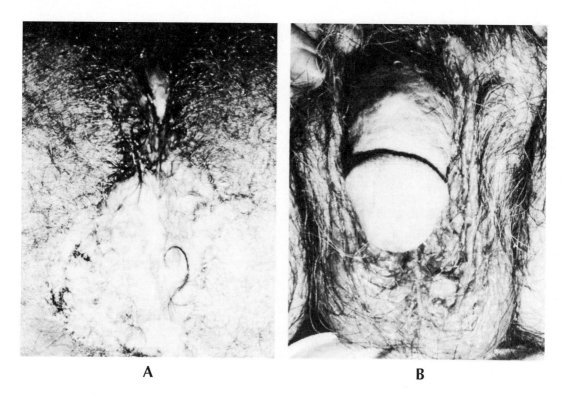

A   B

FIGURE 6-8 Perirectal Ulcer in an AIDS Patient. **A,** This ulcer was caused by the herpes type II virus. Herpes infections out of control in AIDS and other immunocompromised patients are a serious threat. **B,** The chronic expression of herpes type II virus on the scrotum of an AIDS patient. (*Courtesy of Ronald P. Rapini, Texas Tech University Health Science Center and reprinted with permission from* Cutis, *1988, 42:125-128, copyright 1988 by QUADRANT HEALTHCOM, INC.*)

an early indicator that HIV-positive people are progressing toward AIDS.

Protozoal Diseases

An increasing number of infections **which have not been observed in immunocompromised patients are being found in AIDS patients.** Three such infections are caused by the protozoans *Toxoplasma gondii, Isospora belli,* and *Cryptosporidium muris.*

Toxoplasma gondii— *T. gondii* is a small intracellular protozoan parasite that lives in vacuoles inside host macrophages and other nucleated cells. It appears that during and after entry, *T. gondii* produces secretory products that

modify vacuole membranes so that the normal *fusion* of cell vacuoles with lysosomes containing digestive enzymes is blocked. Having blocked vacuole-lysosome fusion, *T. gondii* can successfully reproduce and cause a disease called **toxoplasmosis** (Joiner et al., 1990). It can infect any warm-blooded animal, invading and multiplying within the cytoplasm of host cells. As host immunity develops, multiplication slows and tissue cysts are formed. Sexual multiplication occurs in the intestinal cells of cats (and apparently only cats); oocysts form and are shed in the stool (Sibley, 1992). Transmission may occur transplacentally, by ingestion of raw or undercooked meat and eggs containing tissue cysts or by exposure to oocysts in cat feces (Wallace et al., 1993).

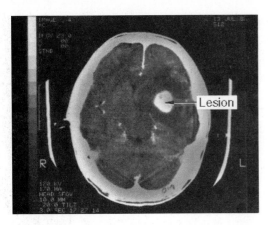

FIGURE 6-9 *Toxoplasma gondii* Lesions in the Brain. Radiographic imaging shows a deep ring-enhancing lesion located in the basal ganglia. (*By permission of Carmelita U. Tuazon, George Washington University*)

In the United States, 10% to 40% of adults are chronically infected but most are asymptomatic. *T. gondii* can enter and infect the human brain causing **encephalitis** (inflammation of the brain). Toxoplasmic encephalitis develops in over 30% of AIDS patients at some point in their illness (Figure 6-9). The signs and symptoms of cerebral toxoplasmosis in AIDS patients have been presented in detail by Rossitch and colleagues (1990). They may include fever, headache, confusion, sleepiness, weakness or numbness in one part of the body, seizure activity, and changes in vision. These symptoms can get worse and progress to coma and death unless toxoplasmosis is promptly diagnosed and treated. Toxoplasma infection affecting other parts of the body is rare, and symptoms vary depending on which organs the disease affects. Based on the projected incidence of AIDS cases in the United States in 1992, 40,000 to 70,000 cases of toxoplasmic encephalitis were expected to occur (Dannemann et al., 1989). Similar to a variety of other OIs manifested in HIV-infected people, toxoplasmosis appears to represent a reactivation of an earlier infection. In the United States, 30% of the population between the ages of 10 and 19 demonstrate serological evidence (antibody) to *T. gondii* exposure. *T. gondii* lies dormant in the reticuloendothelial system until it becomes reactivated within the immunocompromised host. Thus, for most AIDS patients, it is believed that *T. gondii* is latent within their bodies and is reactivated by the loss of immune competence. Once activated, symptoms can be as mild as chills, headaches that do not respond to common pain killers, low fevers, and delusions; or as severe as hard seizures, coma, and death. Arthur Ashe, tennis champion who died in February of 1993, was being treated for *T. gondii* infection.

Current treatment (Table 6-2) is with pyrimethamine combined with sulfadiazine or clindamycin and with trimethoprim/sulfamethoxazole (Bactrim) or dapsone (Daar et al., 1993; Hunt, 1996). Treatment keeps toxoplasma under control but it does not kill it. Therapy must be continued indefinitely.

Cryptosporidium— *Cryptosporidium* is the cause of cryptosporidiosis, and is a member of the family of organisms that includes *Toxoplasma gondii* and *Isospora*. Its life cycle is similar to that of other organisms in the class Sporozoa. Oocysts are shed in the feces of infected animals and are immediately infectious to others. In humans, the organisms can be found throughout the GI tract, including the pharynx, esophagus, stomach, duodenum, jejunum, ileum, appendix, colon, and rectum. Various case reports of patients with AIDS describe infection in the gallbladder and pulmonary dissemination which clinically resembles *Pneumocystis carinii* pneumonia. *Cryptosporidium* **causes profuse watery diarrhea of 6 to 26 bowel movements per day with a loss of 1 to 17 liters of fluid (a liter is about 1 quart).** It is an infrequent infection in AIDS patients, usually occurring late in the course of disease as immunological deterioration progresses.

Studies of transmission patterns have shown infection within families, nursery schools, and from person to person, probably by the fecal-oral route. The infection is particularly common in homosexual men, perhaps as a consequence of anilingus (oral-anal sex). Cryptosporidiosis made headlines in March and April 1993 when an outbreak of the infection in Milwaukee resulted in diarrheal illness in more than 400,000 people. Following that outbreak, testing for *Cryptosporidium* in people with diarrhea increased substantially in some areas of Wisconsin. As a result of the investigations,

Cryptosporidum contamination was found in several public swimming pools. At this time, there are no effective prophylaxes against cryptosporidiosis (Church, 1992; MMWR, 1995). The prevention of transmission rests on good hygiene, hand washing, and awareness of the risks of direct fecal-oral exposure (Wofsy, 1991).

Isospora belli— *Isospora belli* enters the body via feces-contaminated food and drink or is sexually transmitted (DeHovitz, 1988). This organism infects the bowels. Isosporiasis is characterized by an acute onset of profuse watery diarrhea (**eight to 10 stools per day**), fever, malaise, cramping, abdominal pain, and in some cases significant weight loss and anorexia. According to the 1987 revised definition for AIDS, isosporiasis persisting for longer than 1 month and a positive test for HIV is indicative of AIDS.

Treatment (Table 6-2) calls for high doses of trimethoprim-sulfamethoxazole, furazolidone, and pyrimethamine-sulfadiazine (Wofsy, 1991).

Bacterial Diseases

There is a long list of bacteria that cause infections in AIDS patients. These are the bacteria that normally cause infection or illness after the ingestion of contaminated food, such as species of *Salmonella*. Others, such as *Streptococci, Haemophilus,* and *Staphylococci are* common in advanced HIV disease. A number of other bacterial caused sexually transmitted diseases such as syphilis, chancroid, gonorrhea, and chlamydial diseases are also associated with HIV disease.

One difference between AIDS and non-AIDS individuals is that bacterial diseases in AIDS patients are of greater severity and more difficult to treat. In fact, drug treatments for HIV/AIDS patients have been associated with an increase in the incidence of bacterial infections (Rolston, 1992). Two bacterial species, *Mycobacterium avium intracellulare* and *Mycobacterium tuberculosis* are of particular importance as agents of infection in AIDS patients (Table 6-3).

Mycobacterium avium intracellulare (MAI)— Over the past 40 years, MAI has gone from a rare, reportable infection to something that is common in most large American communities.

TABLE 6-3 Categories of Organism and Viral Involvement in Opportunistic Diseases

Symptoms	Causative Agent
Generally Present	
Fever, weight loss, fatigue, malaise	*Pneumocystis carinii*
	Cytomegalovirus
	Epstein-Barr virus
	Mycobacterium avium intracellulare
	Candida albicans
Diffuse Pneumonia	
Dyspnea, chest pain, hypoxemia, abnormal chest X-ray	*Pneumocystis carinii*
	Cytomegalovirus
	Mycobacterium tuberculosis
	Mycobacterium avium intracellulare
	Candida albicans
	Cryptococcus neoformans
	Toxoplasma gondii
Gastrointestinal Involvement	
Esophagitis (sore throat, dysphagia)	*Candida albicans*
	Herpes simplex
	Cytomegalovirus (suspected)
Enteritis (diarrhea, abdominal pain, weight loss)	*Giardia lamblia*
	Entamoeba histolytica
	Isospora belli
	Cryptosporidium
	Strongyloides stercoralis
	Mycobacterium avium intracellulare
Proctocolitis[a] (diarrhea, abdominal pain, rectal pain)	*Entamoeba histolytica*
	Campylobacter
	Shigella
	Salmonella
	Chlamydia trachomatis
	Cytomegalovirus
Proctitis[a] (pain during defecation, diarrhea, itching and perianal ulcerations)	*Neisseria gonorrhoeae*
	Herpes simplex
	Chlamydia trachomatis
	Treponema pallidum
Neurological Involvement	
Meningitis, encephalitis, headaches, seizures, dementia	Cytomegalovirus
	Herpes simplex
	Toxoplasma gondii
	Cryptococcus neoformans
	Papovavirus
	Mycobacterium tuberculosis
Retinitis (diminished vision)	Cytomegalovirus
	Toxoplasma gondii
	Candida albicans

[a]Especially in those persons practicing anal sex.

Adapted from Amin, 1987

For Sexual Exposures—People should use male latex condoms during every act of sexual intercourse to reduce the risk of exposure to cytomegalovirus, herpes simplex virus, and human papillomavirus, as well as to all other sexually transmitted pathogens. Use of latex condoms will also prevent the transmission of HIV to others. People should avoid sexual practices that may result in oral exposure to feces (e.g., oral-anal contact) to reduce the risk of intestinal infections such as cryptosporidiosis, shigellosis, campylobacteriosis, amebiasis, giardiasis, and hepatitis A and B (*MMWR,* 1995).

Unlike tuberculosis, which is almost exclusively spread person-to-person, MAI is, in most instances, environmentally acquired. MAI exists in food, animals, water supplies, soil and enters people's lungs as an aerosol when they take showers.

When an elderly person develops MAI infection, it is invariably confined to his or her lungs. In contrast, MAI infections in AIDS patients run rampant and are clearly systemic (Zoler, 1991).

The fact that MAI produced disseminated disease in AIDS cases was recognized in 1982. The epidemiology of MAI continues to evolve. MAI occurs in 18% to 43% of people with HIV disease and has been implicated as the cause of a nonspecific **wasting syndrome**. AIDS patients demonstrate **anorexia** (inability to eat), weight loss, weakness, night sweats, diarrhea, and fever. Some patients also experience abdominal pain, enlarged liver or spleen, and malabsorption. In contrast to viral infections, this bacterium rarely causes pulmonary or lung problems in AIDS patients. Among persons with AIDS, the risk of developing disseminated MAI increases progressively with time. AIDS patients surviving for 30 months had a 50% risk of developing disseminated MAI. It appears most HIV-infected persons will develop disseminated MAI if they do not first die from other OIs (Chin, 1992). Some new drugs in use are clofazimine, ciprofloxacin, ethambutol, amikacin, azithromycin (Zithromax), clarithromycin (Biaxin), and rifabutin, but their effectiveness is limited— resistance to these drugs develops quickly (Table 6-2) (Kaplan et al., 1995).

Mycobacterium tuberculosis— Tuberculosis (TB) is an infectious disease caused by the bacterium *Mycobacterium tuberculosis*, which is spread almost exclusively by airborne transmission. TB has been observed in elephants, cattle, mice, and other animal species. In 1993, TB was transmitted from an infected seal to its trainer in Australia. In the United States, monkeys are the primary source of animal-to-human transmission.

The disease can affect any site in the body, but it most often affects the lungs. When persons with pulmonary TB cough, they produce tiny droplet nuclei that contain TB bacteria, which can remain suspended in the air for prolonged periods of time. (With respect to transmission, the cough to TB is similar to sex to HIV.) Anyone who breathes air that contains these droplet nuclei can become infected with TB. It has been suggested that there is a minimal chance of inhaling HIV in blood-tinged TB sputum (Harris, 1993).

A person who becomes infected with the TB bacillus remains infected for years. Usually a person with a healthy immune system does not become ill, but is usually not able to eliminate the infection without taking an antituberculosis drug. This condition is referred to as a **latent tuberculosis infection.**

About 10% of otherwise healthy persons who have latent tuberculosis infection will become ill with active TB at some time during their lives (*MMWR*, 1992). With HIV disease the risk is 10% per year (Daar et al., 1993).

Tuberculosis is not generally considered to be an OI because people with healthy immune systems contract TB. After infection with *M. tuberculosis* about 5% of immunocompetent individuals will develop TB (Daley, 1992). But, people with a depressed immune system are much more likely to develop the disease (Zoler, 1991). Tuberculosis in people with AIDS does not look like ordinary tuberculosis. In the usual presentation of the disease, TB is usually restricted to a given area in the chest. People with AIDS may have tuberculosis throughout the chest cavity.

There is a strong association between HIV disease and TB: HIV infection is the highest risk factor for progression from latent *M. tuberculosis* infection to TB (Bermejo et al., 1992).

HIV infection is now considered to be the single most important risk factor in the expression of TB. HIV disease is associated with the reactivation of a dormant or inactive TB infection (Stanford et al., 1993).

M. tuberculosis infection occurs in about 35% of HIV-infected individuals, usually as the result of *M. tuberculosis* reactivation from a latent prior infection (Brooke et al., 1990). The CDC defines extrapulmonary TB combined with an HIV-positive test as diagnostic of AIDS.

According to the World Health Organization (1995) TB is, worldwide, the leading cause of death in HIV-infected people and among adults from a single, infectious organism. TB has killed at least 200 million people since 1882, the date of the discovery of the bacterium that causes TB, and millions more die from TB each year. **March 24ᵗʰ of each year is recognized as World TB Day.**

In the United States 14% of AIDS patients are also coinfected with TB. Drugs used to treat TB are isoniazid, rifampin, pyrazinamide, streptomycin, and ethambutol. Ethambutol is used in combination with the other four drugs when the infecting organism is suspected to be drug-resistant (Bernardo, 1991; Dannenberg, 1993). However, health officials state that between 40% and 60% of those developing multidrug resistant TB will die (Ezzell, 1993).

Other Opportunistic Infections

Other opportunistic infectious organisms and viruses and the diseases they cause and possible therapies are listed in Table 6-2. Table 6-3 separates OIs into the body parts most affected by a particular organism or virus.

From diagnosis until death, the AIDS battle is not just against its cause, HIV, but against those organisms and viruses that cause OIs. Opportunistic infections are severe, tend to be disseminated (spread throughout the body), and are characterized by multiplicity. Fungal, viral, protozoal, and bacterial infections may be controlled for some time but are rarely curable.

CANCER IN AIDS PATIENTS

Because of the severe and progressive impairment of the immune system, host defense mechanisms that normally protect against certain types of cancer are lost. Cancers develop in approximately 40% of AIDS patients. Four kinds of cancer that occur with increased frequency are: **progressive multifocal leukoencephalopathy, squamous cell carcinoma** (oral and anal), **non-Hodgkin's lymphoma,** and **HIV/AIDS-associated Kaposi's sarcoma (KS).** None of these cancers, except for KS, is considered to be an opportunistic infection because they are not infections. They are cancers arising from cells that have lost control of their division processes. Of the nine types of AIDS-associated cancers, KS occurs with the greatest frequency and is discussed in some detail. Lymphomas are briefly described (Table 6-4). (For a recent review of HIV/AIDS-related cancers read Hessol, 1998.)

---------- **POINT OF INFORMATION 6.2** ----------

TB GREATEST KILLER OF WOMEN WORLDWIDE

TB accounts for 9% of deaths worldwide among women aged between 15 and 44, compared with war which accounts for 4%, HIV 3% and heart disease 3%. Women of reproductive age are most susceptible to fall sick once infected with TB than men of the same age. Women in this age group are also at greater risk from HIV infection. As a result, in some parts of Africa, young women with TB outnumber young men with TB.

Data presented by the World Health Organization, the **first International Meeting On TB and Gender** in 1998, showed unprecedented levels of infection and deaths among women and girls: Over 900 million are infected with TB worldwide, one million will die and 2.5 million will get sick this year from the disease. Most will be between the ages of 15 and 44. **This makes TB the single**

leading cause of deaths among women of reproductive age. This counters the perceptions in wealthy countries where the disease is most commonly found in elderly men. In industrialized countries, one quarter of all TB cases occur in those over 65, compared with only 10% in developing countries of Africa, Asia, and Latin America. In the developing world, TB is predominantly a disease of young adults: 60% of all cases are young men and women of reproductive age. Paul Dolin of WHO's global TB program said, "Wives, mothers, and wage earners are dying of TB in their prime and the world isn't noticing. Yet the ripple effect on families, communities and economies will be felt long after a woman has died."

TABLE 6-4 Malignancies Associated with HIV/AIDS

Kaposi's sarcoma (epidemic form)
Burkitt's lymphoma
Non-Hodgkin's lymphomas
Hodgkin's disease
Chronic lymphocytic leukemia
Carcinoma of the oropharynx
Hepatocellular carcinoma
Adenosquamous carcinoma of the lung
Cervical cancer
Squamous cell carcinoma
Progressive multifocal leukoencephalopathy

Kaposi's Sarcoma (cap-o-seas sar-comb-a)

No other HIV/AIDS-related opportunistic disease attacks and single out one segment of the population as KS does with HIV-positive gay men. Men with KS outnumber women approximately 95% to 5%; HIV-positive homosexual men with KS outnumber heterosexual men almost as significantly; and KS is extremely rare in hemophiliacs with HIV. In fact, in one major study of hemophiliacs with HIV, only one in 93 developed KD, and he happened to be a gay man.

HIV infection represents an overwhelming risk factor for the development of KS, which was a rare tumor in the United States (incidence less than 1/100,000/year) before the HIV epidemic. Today, KS remains the most frequent **neoplasm** affecting HIV-infected individuals. Approximately 25% to 30% of homosexual males with HIV infection develop clinically significant KS, and autopsy studies indicate an even higher prevalence of the disease up to approximately 40%. It is an aggressive disease, with involvement of the gut, lung, pleura, lymph nodes and the hard and soft palates.

KS, as it occurs in HIV/AIDS patients, may not be an opportunistic infection. Its cause is still unknown. It is uncertain whether KS is really a cancer because unlike cancer, which arises from one cell type, KS arises from several cell types. KS lesions are made up of an overgrowth of blood vessels.

In the United States, Kaposi's sarcoma is at least 20,000 times more common in people with HIV/AIDS than in the general population, and 300 times more common than in other immunosuppressed groups (Beral et al., 1990).

—————— **POINT OF INFORMATION 6.3** ——————

TUBERCULOSIS (TB), A MAJOR KILLER

About **2 billion** people carry the bacterium that causes tuberculosis worldwide. Tuberculosis is a killer with impressive credentials. Over the past two centuries, it has killed over a billion people. Currently it is still a leading cause of mortality from a single infectious agent—accounting for 26% of preventable adult deaths in the developing world. In the coming decade, it is slated to kill at least 30 million people.

RELATIONSHIP OF TB TO HIV

The synergy between the epidemics of TB and HIV constitutes a global public health crisis of staggering dimensions. In persons infected with HIV, infection with *Mycobacterium tuberculosis* is much more likely to progress rapidly to active disease, which is increasingly difficult to diagnose and more complicated to treat. Globally, TB is now the leading cause of death among persons with AIDS, killing 1 of every 3 people who die with AIDS. In addition, there are about 6 to 10 million people globally infected with both TB and HIV! **In 1998, worldwide, there were 8 million new TB cases and 2 million deaths from TB. About 9% of global TB cases are attributed to HIV infection.** And projected to the year 2000, 14% of TB cases will be attributed to HIV disease (DeCock et al., 1996). In the United States, about 14% of those HIV-infected are also infected with TB! To date, the dual scourge of TB and HIV has affected predominantly the African continent, but the rapid spread of HIV in Asia, home to the majority of the world's TB-infected persons, is a time bomb. While industrialized nations rejoice in the increased survival associated with the use of combination antiretroviral therapy for HIV-infected patients, it is important to remember that for most people in developing nations access to relatively inexpensive anti-TB therapy is the most significant life-prolonging medical intervention (Small, 1996).

———— BOX 6.3 ————

LIFE GOES ON!

by Wendi Alexis Modeste

What this epidemic has cost me is the complete faith I had in the medical profession. I was raised believing that doctors were second only to priests and God. They were never to be questioned. Whatever the doctor said was law. If a person didn't get well after seeing the doctor, somehow they (the patient) had done something wrong. This was pretty much standard thinking for middle-class African-Americans. For a variety of reasons (mainly no self-esteem) I became a drug addict, prostitute, convict, battered, homeless woman, in that order! With the exception of emergency room admissions (which are a joke and a whole 'nother story) I had no access to health care.

Now, as a PWA (person with AIDS) fortunately/unfortunately on SSI, Medicaid pays for my nine different AIDS medications, clinic visits, treatment, tests, etc. When I received the "exciting news" that I was eligible for "all" Medicaid benefits, I was still under the impression doctors were those super-intelligent, gifted, Christian, saint-like people. Girlfriend, I am here to tell you, AIDS has totally shot that Marcus Welby theory straight to hell!

Early in my diagnosis, I went to my physician because my tongue was almost completely covered with what looked like cottage cheese. There was also a horrible pink lesion dead center. The first time I showed it to my doctor he said, "Ugh," and made a face. He told me to wait a month. If nothing changed or got worse when I returned he'd have someone look at it. Being ignorant about the disease at that time, and still blindly believing in the medical profession, I waited a month, then returned. Again I showed the doctor my tongue. He asked me if I wanted him to write me a prescription for codeine and Valium. I totally freaked! By this time, I'd done some reading and realized I probably had thrush and some sort of herpes. This physician was aware of my serostatus. He also knew that I was a person with a 20-year history of drug abuse. I'd been in recovery less than a year and this jerk wanted to prescribe for me two of the most addictive and abused prescription drugs. I'd never mentioned being in pain or that I was experiencing any type of anxiety. I contacted the Executive Director of this health facility and asked to have a different doctor assigned to me. In an attempt to make me feel guilty about requesting a change, I was told about the problems doctors have in getting Medicaid reimbursement. I was neither intimidated nor impressed. A new doctor was assigned. My new physician was very nice. After *I* told *him* what I thought my diagnosis was, he prescribed the appropriate medications. He was not trained in AIDS/HIV. I could have dealt with his ignorance because I knew he was trying. His nurse, however, was a different story. Every time she came to do my vitals she'd say the same thing: "I always get nervous when taking the temperature of you guys." She'd then force a little chuckle and go on to say, "Oh, well, I figure we all have to die of something." (I assume this was to show me what a courageous Florence Nightingale she was.)

Let me tell you, when you are burning up with a 103 degree temp. and your bowels haven't stopped running for a week, causing your butt to feel like it's on fire, it's real hard to be the patient, understanding AIDS educator. I really get crazy when the person I'm forced to educate is someone whose been privileged to more information than myself.

But life goes on!

One day I had a toothache. I go to the dentist. After waiting half the day, I'm brought into the treatment room for an X-ray. At first I thought the dental assistant had made a mistake. Surely this room had been prepared for a paint job. Everything was draped in white towels. The entire dental unit, including where my head, arms and feet went, was completely covered. The seat of the unit was securely wrapped up, as was the metal extension arm that holds the overhead dental lamp. All surfaces of the walls were also draped. When I asked the reason for this "painter preparation," I was informed it was done because I have AIDS, and they had to protect their other patients from coming into contact with my contaminated blood. Needless to say, I saw red! I knew I had to protect myself. I mean, what kind of dentistry were they practicing if they were concerned about my blood splattering that far and wide? What was even more frightening was that they'd done all this unnecessary draping and I was only having an X-ray. I filed a complaint with the Human Rights Dept.

I became a patient at the AIDS Care Center in Syracuse. My physician, a woman (need I say more?), is a caring person and well-educated about AIDS/HIV. My nurse/social worker is excellent,

—————— **BOX 6.3** *(continued)* ——————

but as all of us living with AIDS know, shit happens. One day I awaken with enough yeast in my body to make all the baked goods in Central New York rise. My doctor isn't in. I wait a couple of days but can no longer stand the discomfort. I beg to see a doctor in the AIDS Care Unit. I'm assigned the doctor who sees the HIV-infected prisoners. (My heart and soul truly go out to those guys.) First he talks with me over the phone to find out if I can possibly wait another week when my doctor is due to return. I tell him my tongue is unrecognizable, the yeast in my esophagus is burning like a heart attack, and the Roto-Rooter service couldn't satisfy the itch caused by the yeast in my vagina. HELL NO, I can't wait another week! I go in to see him. I show him my tongue. I can't believe it, but like the first doctor, he uses that medical term, "Ugh," then says, "That does look nasty." At this point I'm ready to French kiss this idiot. But it gets worse. He won't touch me, let alone examine anything. He asks me what I think will work. I feel too badly to curse, so I tell him Mycelex, Myastatin suspension, Monistat 7. He writes the prescriptions and for good measure increases my acyclovir. For this he gets paid? He did nothing!

I'm now as educated as a lay person can be about HIV disease/AIDS. In addition to my doctor at the AIDS Care Center, I have a private primary care physician. Sometimes it's easier to get in to see this doctor. I call his office one day because there is swelling and burning on the sides of my tongue. I cannot eat. My regular doctor isn't in, but one of his associates assures me if I come in to the office he'll see me right away and give me something to ease the pain so I can at least eat. As a fat person who proudly admits a genuine fondness for food, I can tell you not being able to eat registers serious panic in my soul. Nothing stops me from eating. I was probably the only overweight homeless dope addict living on the streets of NYC. Being scared is putting it mildly. I scrape up the carfare and go to the office. After waiting an unreasonably long time, Doogie Howser's twin comes in to see me and announces he's Dr. Jones, whom I spoke to on the phone. OK, I know not to judge a book by its cover. I mean, Doogie is pretty good on TV. Dr. Jones looks at my tongue and proceeds to ask a zillion questions, all of which are answered in my chart. Finally he says he's never seen an HIV-infected person or a person with AIDS, and frankly he doesn't know what to do. He then sug-

gests I eat popsicles for a few days because the cold will soothe the pain and the sugar should help give me energy. I kid you not, this actually happened. This jerk prescribed me popsicles. I truly wished I could transmit the virus by biting at this point. He used me so he could write in his journal or resume (or somewhere) that he'd treated a person with AIDS. He could also charge Medicaid for nothing.

But this type of quackery must stop.

The last gripe I'm going to list is this patient statement I hear all the time when I get an unexplained fever or infection and no one can determine its origin. I'm told, "there hasn't been enough studies done on the paths this disease takes in women. Even less has been done on the effects of different AIDS medications on people of color." This is said to me as if it's my fault they don't know. This disease has been documented for ten years in both men and women. I know African-Americans were dying of AIDS long before the gay white community mobilized and, thank God, refused to lay down and die quietly. There is no excuse for the fact that there's no studies done on these populations.

I am thankful to Dr. Sallie Klemmens and nurse/clinician/social worker Judith Swartout at University Hospital's designated AIDS Care Center here in Syracuse. I am thankful for Dr. Barbara J. Justice who with God's help kept me alive when I lived on the streets of NYC. They are all examples of what the medical profession should be about. Dr. Justice made me feel that I counted and should be assertive about my health care. Though no longer my surgeon, she continues to be a source of inspiration and a fountain of information for me. These three are gems in a field I think is greatly overrated, overpaid, and run by capitalist male chauvinist pigs.

Though medical people wear white, that absence of color symbolizing purity and goodness, I beg you all, "Don't believe the hype!" We need a national health care system for everyone. As PWAs we must be assertive about our health care. Good health care is a right not a privilege.

As a child I cried when I learned there was no Santa Claus. When my illusion about the medical profession was shattered I got angry. I decided to fight with the only ammunition I had, education! **Knowledge gives one power**. A close friend of mine told me I shouldn't submit this article because I might offend some members of the medical profession. He also felt because I'm on Medicaid I'm not supposed to complain, I should be grateful. To his comments I respond, raised with two college-educated parents, never wanting or needing anything, I wasn't supposed to be a drug addict. Unfortunately I was. After 20 years of addiction, I wasn't supposed to be able to stop. I've been in recovery almost two years now. I'm not supposed to be living with AIDS, but with God's help I'm happy and living large (as the kids say now).

The best things I do have always been what I'm not supposed to do!
Power to all PWAs!

Wendi Alexis Modeste, a PWA who was diagnosed in 1990, died August 25, 1994.

Source: MODESTE, W.A., August 1991, Issue 68, *PWA Coalition Newsline*. Reprinted with permission.

─────── **BOX 6.4** ───────

AUTOPSY DIAGNOSES OF DISEASES IN AIDS PATIENTS

Between 1984 and 1991, autopsy diagnoses of AIDS-defining diseases were determined in 250 AIDS patients. **Forty-seven percent of diseases found at autopsy had not been diagnosed during life**. Examples of diseases found at autopsy but not in life were CMV visceral infection, mycoses, HIV-specific brain lesions, cerebral lymphomas, and progressive multifocal leukoencephalopathy. Another important finding was that only a small number of AIDS-diagnostic diseases present at some point during life were *not* observed at autopsy. This indicates that AIDS-related diseases are seldom cured (Monforte et al., 1992).

KS was first described by Moritz Kaposi in 1877 as a cancer of the muscle and skin. Characteristic signs of early KS were bruises and birthmark-like lesions on the skin, especially on the lower extremities. KS was described as a slow growing tumor found primarily in elderly Mediterranean men, Ashkenazi Jews, and equatorial Africans.

Kaposi's sarcoma as described by Moritz Kaposi is called classic KS and it differs markedly from the KS that occurs in AIDS patients (Figure 6-10, A and B). Classic KS has a variable prognosis (forecast), is usually slow to develop, and causes little pain (**indolent**). Patient survival in the United States ranges from 8 to 13 years with some reported cases of survival for up to 50 years (Gross et al., 1989). Symptoms of classic KS are ulcerative skin lesions, swelling (**edema**) of the legs, and secondary infection of the skin lesions.

The AIDS epidemic has brought a more virulent and progressive form of KS marked by painless, flat to raised, pink to purplish plaques on the skin and mucosal surfaces which may spread to the lungs, liver, spleen, lymph nodes, digestive tract, and other internal organs. In its advanced stages it may affect any area from the skull to the feet (Figure 6-11, A–C). In the mouth, the hard palate is the most common site of KS (Figure 6-12) but it may also occur on the gum line, tongue, or tonsils.

KS in AIDS patients is fulminant; it comes on swiftly and spreads aggressively. However, there have been **no reported** AIDS deaths due to just KS. Most AIDS deaths are due to opportunistic infections. But, KS can have enormous psychological impact particularly if the lesions occur on exposed areas.

Some of the most inconvenient and uncomfortable KS targets include the soles of the feet, the nose and the oral cavity. Lesions on the lower extremities or on the feet, are often associated with edema and swelling, causing not only severe pain but difficulty putting on shoes and walking. Swelling can be complicated by bacterial cellulitis, ulceration and skin breakdown, often with gram-negative bacterial infections. In addition to the obvious cosmetic damage, lesions on the face may be accompanied by swelling around the eyes that can sometimes progress to the point where patients literally cannot open their eyes. Finally, oral lesions can be painful and make eating and speaking problematic.

The prevalence of KS among gay men in 1981 was 77%; by 1987, it had fallen to 26% and by 1995 to less than 20%. This drop in KS among gay men was paralleled by a fall in CMV cases. KS continues to decline in frequency.

Two questions provide differing views of the basic nature of the KS lesion: **Is KS merely a polyclonal proliferation of blood vessel cells? Or is it a true neoplastic process?** Whatever the answer, there must be additional host factors modulating the expression of KS to explain its **male**

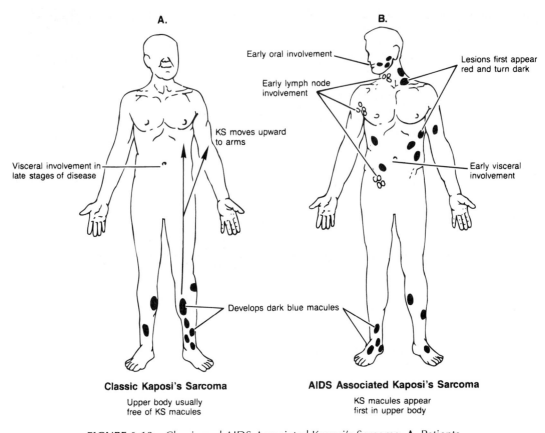

A.

Early oral involvement

Early lymph node
involvement

KS moves upward
to arms

Visceral involvement in
late stages of disease

Develops dark blue macules

Classic Kaposi's Sarcoma

Upper body usually
free of KS macules

B.

Lesions first appear
red and turn dark

Early visceral
involvement

AIDS Associated Kaposi's Sarcoma

KS macules appear
first in upper body

FIGURE 6-10   Classic and AIDS-Associated Kaposi's Sarcoma. **A,** Patients
with classic KS (non-AIDS-related) demonstrate violet to dark blue bruises,
spots, or macules on their lower legs. Gradually, the lesions enlarge into
tumors and begin to form ulcers. KS lesions may, with time, spread upward
to the trunk and arms. The movement of KS appears to follow the veins and
involves the lymph system. In the late stages of the disease, visceral organs
may become involved. **B,** For AIDS patients, initial lesions appear in greater
number and are smaller than in classic KS. They first appear on the upper
body (head and neck) and arms. The lesions first appear as pink or red oval
bruises or macules that, with time, become dark blue and spread to the oral
cavity and lower body, the legs, and feet. Visceral organs may be involved
early on and the disease is aggressive. However, death is usually caused by
opportunistic infection.

predominance both in mice and in humans and,
among AIDS patients, its preferential occurrence among gays, greater than 15:1 male: female overall, approximately 4:1 male:female in
AIDS patients when homosexual males are excluded (Looney, 1996). It has a low rate of incidence in hemophiliacs, intravenous drug users,
women with AIDS(Cooley et al., 1996), and in
pediatric AIDS cases. In summary, KS does **not**

appear to be caused by HIV, but the immunosuppression which is caused by HIV infection
may be an essential element in the evolution of
this disease.

Proposed Kaposi's Virus— Some researchers
suspect that the AIDS virus is not the primary
pathological agent for Kaposi's sarcoma. Beral
and colleagues (1990) at the Centers for Disease

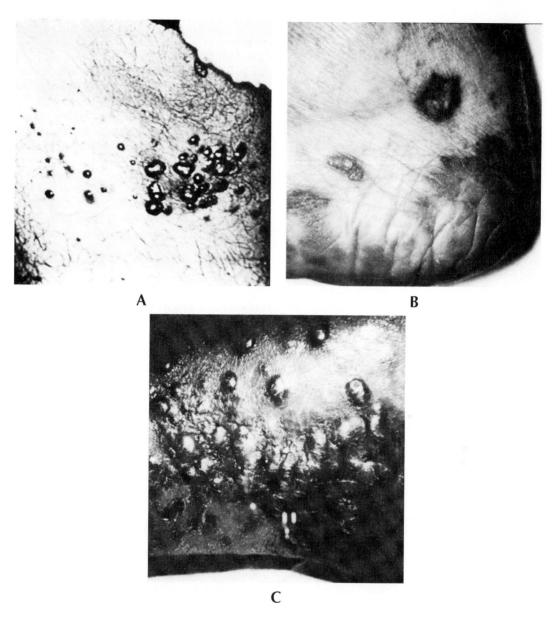

FIGURE 6-11 Kaposi's Sarcoma in AIDS Patients. **A,** KS on the right thigh. **B,** On heel and foot. **C,** On lower leg. (**A** *and* **C,** *courtesy of Nicholas J. Fiumara, M.D., Boston;* **B,** *courtesy of CDC, Atlanta*)

Control and Prevention concluded that the epidemiological data on Kaposi's distribution suggest that it is caused by a sexually transmitted pathogen other than HIV. They found, for example, that KS was more common in people infected with HIV by sexual contact (gay males) than those infected by contaminated needles or blood (Palca, 1992).

Friedman-Kien and colleagues believe that human papillomavirus-16 (HGPV-16) is a major cofactor, if not the direct cause of KS. They have detected HPV-16 DNA in 95% of KS cells tested

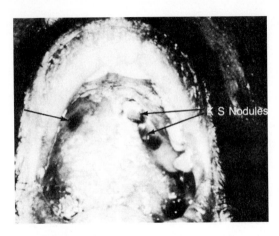

FIGURE 6-12 Oral Kaposi's Sarcoma. KS can be seen on the hard palate and down the sides of the oral cavity. (*Courtesy of Nicholas J. Fiumara, M.D., Boston*)

(Palca, 1992). Other scientists believe the cause of KS is another retrovirus. In December of 1994, Yuan Chang and colleagues reported that they found DNA sequences that appear to represent a **new human herpes virus (HHV-8)** in KS tissue. Preliminary data showed that this unique DNA sequence occurred in KS tissue of 25 out of 27 gay men who had died of AIDS, but was found in only 6 of 39 non-KS tissues from AIDS patients. Investigations are in progress (Schulz et al., 1995; Moore et al., 1996). In addition, Chang and colleague Patrick Moore reported that they have transmitted the suspect herpes virus to a cell line and identified two dozen of its genes. They also know that its DNA includes 270 kilobases (270,000 nucleotide base pairs), making it the largest known herpesvirus (Cohen, 1995). Gianluca Gaidano and colleagues (1996) report that their data confirm that HHV-8 DNA sequences are found, at high frequency, with selected types of AIDS-related KS. Evidence continues to accumulate indicating that HHV-8 is the infectious agent responsible for KS (Kledal et al., 1997; Said et al., 1997).

Update 2000. During 1998 and 1999 many research papers on whether Herpes virus 8 causes KS were published by recognized scientific/medical journals. Still there are **no** definite conclusions on whether there is a cause and effect relationship HHV-8/KS. One paper (Martin, et al., 1998) does make the claim that HHV-8 is most often sexually transmitted and that they have established a direct link between KS and HHV-8. Martin and colleagues believe infection with HHV-8 precedes the development of KS; others disagree.

KS is rare in Caucasian women, but those who acquired HIV through **heterosexual contact** were more likely to have it if their partners were bisexual men than if their partners were injection drug users (3% vs. 0.7%) (Gorin et al., 1991; Serrano et al., 1995; Cooley et al., 1996; Warren et al., 1996). For men and women who acquired HIV through heterosexual contact, Kaposi's sarcoma was more frequent among those born in the Caribbean, Mexico, Central America, or Africa than those born in the United States (6% vs. 2%).

Risk of Kaposi's sarcoma within each HIV transmission group was not consistently related to age or race and varied across the United States. Kaposi's sarcoma in AIDS patients decreased 50% between 1983 and 1988, a trend that could be due to changes in reporting, the short incubation period for Kaposi's sarcoma, and a declining exposure to the causal agent (Beral et al., 1990).

The theoretical Kaposi's virus may have entered the same population in which the AIDS virus is endemic, which would explain why the two are often transmitted together. HIV may produce the right conditions for Kaposi's development by causing growth factor production, and possibly by suppressing the body's immune defenses against cancer.

Lymphoma (lim-fo-mah)

Lymphoma is the second most common cancer in HIV and is now the seventh most common cause of death for people with AIDS. A lymphoma is a neoplastic disorder (cancer) of the lymphoid tissue (Figure 6-13). *B cell* lymphoma occurs in about 1% of HIV-infected people, but makes up about 90% to 95% of all lymphomas found in people with HIV disease (Herndier et al., 1994). Although it occurs most often in those demonstrating persistent generalized lymphadenopathy (swollen lymph glands), the usual site of lymphoma growth is in the brain, the heart, or the anorectal area (Brooke et al., 1990). The

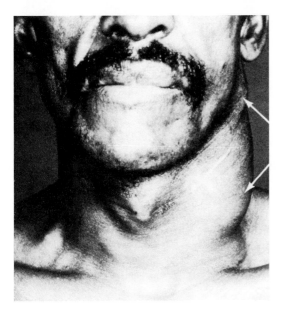

FIGURE 6-13 HIV/AIDS Patient Demonstrating a Lymphoma of the Neck. (*Courtesy CDC, Atlanta*)

and there is no treatment. In a few patients spontaneous improvement and prolonged survival have been reported. Some observations have indicated that cytosine arabinoside, a potent antiviral, may reverse the symptoms of PML. Symptoms of PML include altered mental status, speech and visual disturbances, gait difficulty and limb incoordination (Guarino et al., 1995).

HIV Provirus: A Cancer Connection

In early 1994 AIDS investigators reported that HIV, on entering lymph cell DNA, activated nearby cancer-causing genes (oncogenes). The evidence suggests that HIV itself can trigger cancer in an otherwise normal cell (Figure 6-14).

These findings may mean that a variety of retroviruses that infect humans may also cause cancer (McGrath et al., 1994). Such findings raise concerns for developing an HIV vaccine. Using a weakened strain of HIV to make the vaccine may, when used, increase the incidence of lymphoma and other cancers.

NEUROPATHIES IN HIV DISEASE/AIDS PATIENTS

Neuropathies are functional changes in the peripheral nervous system, therefore, any part of the body may be affected. Although neuropathies are not OIs, they may result from the presence of certain OIs. **Peripheral neuropathy** is caused by nerve damage and is usually characterized by a sensation of pins and needles, burning, stiffness, or numbness in the feet and toes. It is a common, sometimes painful, condition in HIV-positive patients, affecting up to 30% of people with AIDS. At autopsy, two-thirds of AIDS patients have neuropathies (Newton, 1995). Neuropathy has been a continuous problem for patients throughout the HIV/AIDS epidemic. It is most common in people with a history of multiple opportunistic infections and low T4 cell counts. There is a wide range of expression among patients with neuropathy, from a minor nuisance to a disabling weakness. The kinds of neuropathies occurring in people with HIV/AIDS are numerous and must be identified before

most common signs and symptoms are confusion, lethargy, and memory loss. Lymphomas are increasing in incidence primarily due to the extension of the life span of AIDS patients, by medical therapy (Table 6-2).

There were approximately 36,000 cases of non-Hodgkin's lymphoma (NHL) diagnosed in 1992, between 8% and 27% occurred in individuals infected with HIV. A recent large prospective observational study indicated an incidence of approximately 1.6% per year in a population with advanced HIV infection treated with zidovudine. It is clear that as HIV infection increases in the population and as individuals infected with HIV survive for longer periods because of more successful treatment, NHL cases will continue to rise (Kaplan, 1992).

Progressive Multifocal Leukencephalopathy

Progressive multifocal leucoencephalopathy (PML) is an opportunistic infection caused by a papovavirus [Jamestown Canyon virus (JCV)] affecting 4% of AIDS patients. It is usually fatal within an average of 3.5 months

Cancer: The HIV Connection

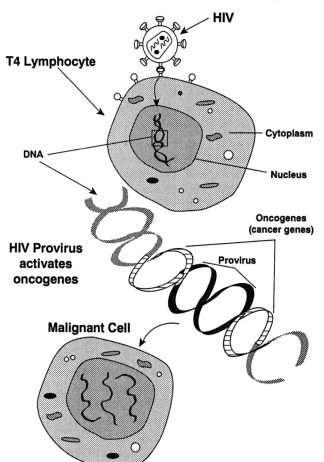

FIGURE 6-14 An HIV-AIDS cancer connection. HIV invades the lymph or other cell types. Its RNA-produced DNA enters cell DNA, becoming an HIV provirus. Sometime during or after integration into host cell DNA, dormant oncogenes, located nearby, become activated and a cancer results. In lymph cells the cancer becomes a lymphoma.

appropriate treatment can be prescribed. The underlying cause of the most common type of peripheral neuropathy remains elusive. What was a common complaint early in HIV infection of severe neuropathy—usually, burning feet, causing patients to walk on their heels—has diminished. The decrease in such complaints may be attributable to the antiviral effects of the drug ZDV. On the other hand, new varieties of drug-induced nerve damage (neuropathies) have been recognized in the use of antivirals like dideoxyinosine (didanosine) (ddI) and dideoxycytosine (zalcitabine) (ddC). Researchers have also identified cytomegalovirus as a contributing factor in some different kinds of neuropathies in HIV disease.

HIV/AIDS SURVEILLANCE CHECKLIST/REPORT

A few of the most prevalent OIs found in HIV/AIDS patients have been discussed. Figure 6-15 lists those OIs and cancers that are used in the CDC criteria for defining AIDS. The report form presented in Figure 6-15 is in use at a public health HIV/AIDS clinic in Florida.

DISCLAIMER

This chapter is designed to present information on opportunistic infections in HIV/AIDS patients. It is not intended to provide medical advice. Consult proper health care providers

HIV/AIDS SURVEILLANCE CHECK LIST/REPORT FORM

A person will have CDC defined AIDS (MMWR Vol. 36, No. 1S, Aug. 14, 1987) if they have a positive test result for Human Immunodeficiency Virus (HIV) and at least one of the following:

(If HIV test is not performed or inconclusive, please refer to the MMWR for additional criteria)

Disease   Disease

(Please indicate definitive diagnosis (D) with laboratory data or presumptive diagnosis (P) where option exists)

___Candidiasis, bronchi, trachea, or lungs	___Candidiasis, esophageal (D or P)
___Cervical cancer	___Cryptococcosis, extrapulmonary
___Coccidiodomycosis, dissem. or extrapul.	___Cytomegalovirus, other than liver, spleen, nodes
___Cryptosporidiosis, chronic intestinal	___HIV Encephalopathy (AIDS Dementia Complex)
___Cytomegalovirus retinitis with vision loss (D or P)	___Histoplasmosis, dissem. or extrapul.
___Herpes Simplex, chronic ulcers	___Kaposi's Sarcoma (D or P)
>1 month duration, or bronchitis, pneumonitis, or esophagitis	
___Isosporiasis, chronic intest.	___Lymphoma, immunoblastic or equivalent
>1 month duration	
___Lymphoma, Burkitt's or equivalent	___Mycobacterium avium, M. kansasii, dissem. or extrapul. (D or P)
___Lymphoma, primary in brain	___Mycobacterium tuberculosis, any site (pulmonary or extrapul.)
___M.Tuberculosis, dissem. or extrapul. (D or P)	___Mycobacterium, other, dissem. or extrapul. (D or P)
___Pneumocystis carinii pneumonia (PCP)(D or P)	___Pneumonia, recurrent
___Salmonella septicemia, recurrent	___Progressive multifocal leukoencephalopathy
___Wasting (>10% baseline body weight + diarr or fatigue)	___Toxoplasmosis of brain

Pediatrics (<13 years of age) include:

___Bacterial infections, multiple, recurrent   ___Lymphoid interstitial pneumonia and/or pulmonary lymphoid hyperplasia

HIV+ Test/Laboratory Date ____/____/____ Current CD4 cell count _____ Diagnosed as Inpatient _____ or Outpatient _____ ?
Risk Factor: male/male sex _____ IVDU _____ Transfusion/Hemophilia _____ male/female sex with HIV infected person _____
Mom to Baby _____ none of the above _____ other _____

Patient Name: _____ Date of Birth: ____/____/____ Male/Female: _____ Race: _____
Address: _____ Zip: _____ Country of Birth: _____

Clinic/Hospital Name: _____ Med. Record #: _____ SS Number #: ____/____/____
Physican Name: _____
Name of person notifying Surveillance Office: _____ Date sent/called: ____/____/____

Instructions: Place form in chart of each HIV+ patient. When at least one of the opportunistic infections listed is diagnosed, complete the form and either call the Surveillance Office, or mail a copy of this form in an envelope marked "CONFIDENTIAL" to:

AIDS Surveillance Office

Keep the original in each medical record for documentation of having reported to the Surveillance Office.

Thank you!

* THIS FORM WAS ALTERED TO REFLECT THE JANUARY, 1993 CDC AIDS DEFINITION. THE MAJOR CHANGE IS THAT ALL PERSONS WITH A T4 CELL COUNT OF LESS THAN 200/mm^3 HAS AIDS. THREE MORE DISEASES WERE ADDED TO THE LIST (See Table 1-3.)

FIGURE 6-15 HIV/AIDS Surveillance Report Form. The form is completed and sent to an AIDS Surveillance Office. The information is then sent on to the CDC in Atlanta.

for medical advice before undertaking any treatment discussed herein.

SUMMARY

One of the gravest consequences of HIV infection is the immunosuppression caused by the depletion of the T4 helper cell population; suppressed immune systems allow for the expression of opportunistic diseases and cancers. It is the OI that kills AIDS patients, not HIV per se. The major OIs are listed in Table 6-1. It is the cumulative effect of several OIs that creates the chills, night sweats, fever, weight loss, anorexia, pain, and neurological problems.

One tragic disease that does not result from an OI is Kaposi's sarcoma (KS), a cancer (?) that can spread to all parts of an AIDS patient's body. About 20% of AIDS patients, mostly gay men, have KS. It is not usually found in hemophiliacs, intravenous drug users, or female AIDS patients.

REVIEW QUESTIONS

(Answers to the Review Questions are on page 487.)

1. Define opportunistic infection (OI).

2. Which OI organism expresses itself in 80% of AIDS patients? Where is it located and what does it cause?

3. Which of the protozoal OI organisms causes weight loss, watery diarrhea, and severe abdominal pain?

4. Which of the bacterial OIs causes "wasting syndrome," night sweats, anorexia, and fever?

5. True or False: Kaposi's sarcoma (KS) is caused by HIV. Explain.

6. Name the two kinds of KS.

7. True or False: KS affects all AIDS patients equally. Explain.

8. True or False: Candidiasis and ulceration may be present in patients with HIV infection.

9. True or False: Oral candidiasis occurs frequently with HIV infection.

10. True or False: The use of combination anti-HIV drug therapy, especially those combinations containing a protease inhibitor have substantially **decreased** the **severity** and **number** of OIs in AIDS patients.

REFERENCES

AMIN, NAVIN M. (1987). Acquired immunodeficiency syndrome, Part 2: The spectrum of disease. *Fam. Pract. Recert.*, 9:84–118.

AWE, ROBERT J. (1988). Benefits, promises and limitations of zidovudine (AZT). *Consultant*, 28:57–72.

BALFOUR, HENRY (1995). Cytomegolovirus retinitis in persons with AIDS. *Postgrad. Med.*, 97:109–118.

BERAL, VALERIE, et al. (1990). Kaposi's sarcoma among persons with AIDS: A sexually transmitted infection? *Lancet*, 335:123–138.

BERMEJO, ALVERO, et al. (1992). Tuberculosis incidence in developing countries with high prevalence of HIV infection. *AIDS*, 6:1203–1206.

BERNARDO, JOHN. (1991). Tuberculosis: A disease of the 1990s. *Hosp. Pract.*, 26:195–220.

BISHAI, WILLIAM, et al. (1999). Opportunistic infections: down but not out. *The Hopkins Report* 11: 2, 7.

BROOKE, GRACE LEE, et al. (1990). HIV disease: A review for the family physician Part II. Secondary infections, malignancy and experimental therapy. *Am. Fam. Pract.*, 42:1299–1308.

BUCHANAN, KENT, et al. (1998) What makes *Cryptococcus neoformans* a pathogen? *Emerging Infectious Diseases*, 4:71–83.

CHANG, YUAN, et al. (1994). Identification of Herpes virus-like DNA sequences in AIDS-Associated Kaposis Sarcoma. *Science*, 266:1865–1869.

CHIN, DANIEL. (1992). Mycobacterium avium complex infection. *AIDS File: Clin. Notes*, 6:7–8.

CHURCH, DEIRDRE L. (1992). Fatal diarrhea in an AIDS patient. *Patient Care*, 26:280–283.

Coalition News. (1993). Pet guidelines for people with HIV. 2:4–5.

COHEN, JON. (1995). AIDS mood upbeat for a change. *Science*, 267:959.

COOLEY, TIMOTHY, et al. (1996). Kaposi's sarcoma in women with AIDS. *AIDS*, 10:122–125.

CURRIER, JUDITH, et al. (1997). Pathogenesis, prevention and treatment of opportunistic complications. *Improv. Manage. HIV Dis.*, 4:S17-S18.

DAAR, ERIC S., et al. (1993). The spectrum of HIV infection. *Patient Care*, 27:99–128.

DALEY, CHARLES L. (1992). Epidemiology of tuberculosis in the AIDS era. *AIDS File: Clin. Notes*, 6:1–2.

DANNEMANN, BRIAN R., et al. (1989). Toxoplasmic encephalitis in AIDS. *Hosp. Pract.*, 24:139–154.

DANNENBERG, ARTHUR M. (1993). Immunopathogenesis of pulmonary tuberculosis. *Hosp. Pract.*, 28:51–58.

DANNER, SVEN, (1995). Management of CMV disease. *AIDS*, 9:53–58.

DE COCK, KEVIN, et al.(1996).Research issues involving HIV-associated TB in research-poor countries. *JAMA*, 276:1502–1508.

DeHovitz, Jack A. (1988). Management of *Isospora belli* infections in AIDS patients. *Infect. Med.,* 5:437–440.

DeWit, Stephane, et al. (1991). Fungal infections in AIDS patients. *Clin. Adv. Treatment Fungal Infect.,* 2:1–11.

Dobkin, Jay, (1995). Opportunistic infections and AIDS. *Infect. Med.,* 125:58–70.

Edman, Jeffrey C., et al. (1988). Ribosomal RNA sequence shows *Pneumocystis carinii* to be a member of the fungi. *Nature,* 334:519–522.

Ennis, David M., et al. (1993). Cryptococcal meningitis in AIDS. *Hosp. Pract.,* 28:99–112.

Emergency Medicine. (1989). Fighting opportunistic infections in AIDS. 21:24–38.

Ernst, Jerome. (1990). Recognize the early symptoms of PCP. *Med. Asp. Hum. Sexuality,* 24:45–47.

Ezzell, Carol. (1993). Captain of the men of death. *Science News,* 143:90–92.

Freedberg, Kenneth, et al. (1998). The cost effectiveness of preventing AIDS-related opportunistic infections. *JAMA,* 279:130–136.

Ganem, Donald. (1996) Kaposi's sarcoma-associated herpesvirus. *Improv. Manage. HIV Dis.,* 4(3):8–10.

Gorin, Isabelle, et al. (1991). AIDS-associated Kaposi's sarcoma in female patients. *AIDS,* 5:877–880.

Gottlieb, Michael S., et al. (1987). Opportunistic viruses in AIDS. *Patient Care,* 23:139–154.

Gross, David J., et al. (1989). Update on AIDS. *Hosp. Pract.,* 25:19–47.

Grossman, Ronald J., et al. (1989). PCP and other protozoal infections. *Patient Care,* 23:89–116.

Guarino, M., et al. (1995). Progressive multifocal leucoencephalopathy in AIDS: Treatment with cytosine arabinoside. *AIDS,* 9:819–820.

Guidelines, U.S. Public Health Service, (1995). Preventing OIs in persons with HIV disease. *AIDS Reader,* 5:172–179.

Harden, C.L., et al. (1994). Diagnosis of central nervous system toxoplasmosis in AIDS patients confirmed by autopsy. *AIDS,* 8:1188–1189.

Hardy, David. (1996). Lessons learned from HIV pathogenesis and therapy: Implications for better management of cytomegalovirus disease. *AIDS,* 10(Suppl, 1):S31–S35.

Harris, Charles. (1993). TB and HIV: The boundaries collide. *Medical World News,* 34:63.

Harvard AIDS, Institute, (1994). *Special Report—Opportunistic Infections.* Fall issue:1–14.

Hay, R.J. (1991). Oropharyngeal candidiasis in the AIDS patient. *Clin. Adv. Treatment Fungal Infect.,* 2:10–12.

Hermans, Philippe, (1995), Haematopoietic growth factors as supportive therapy in HIV-infected patients. *AIDS,* 9:59–514.

Herndier, Brian, et al. (1994). Pathogenesis of AIDS lymphomas, *AIDS,* 8:1025–1049.

Hessol, Nancy. (1998). The changing epidemiology of HIV related cancers. *The AIDS Reader,* 8:45–49.

Hughes, Walter. (1994). Opportunistic infections in AIDS patients. *Postgrad. Med.,* 95:81-86.

HIV/AIDS Surveillance Report, Year End. (1997). 9:1–44.

Hunt, Susan, (1996). Office management of HIV: OIs. I. *Resp. Dis.,* 17:235–238.

Jacobson, Mark A., et al. (1988). Serious cytomegalovirus disease in the acquired immunodeficiency syndrome (AIDS): Clinical findings, diagnosis, and treatment. *Ann. Intern. Med.,* 108:585-594.

Jacobson, Mark A., et al. (1992). CMV disease in patients with AIDS: Introduction. *Clin. Notes,* 6(2):1–11.

Joiner, K.A., et al. (1990). Toxoplasma gondii: Fusion competence of parasitophorous vacuoles in Fe receptor-transfected fibroblasts. *J. Cell Biol.,* 109:2771.

Kaplan, Lawrence. (1992). HIV-associated lymphoma. *Clin. Notes,* 6(1):1–11.

Kaplan, Jonathan, (1995). USPHS/IDSA guidelines for the prevention of opportunistic infections in persons infected with human immunodeficiency virus an overview *Clin. Infect. Dis.,* 21 (suppl 1):S12–S31.

Kledal, Thomas, et al. (1997). A broad spectrum chemokine antagonist encoded by KS-associated herpesvirus. *Science,* 277: 1656–1659.

Laurence, Jeffrey, (1995). Evolving management of OIs. *AIDS Reader,* 5:187–188, 208.

Laurence, Jeffrey, (1996). Where do we go from here? *AIDS Reader,* 6:3–4, 36.

Looney, David, (1996). Kaposi's sarcoma. *Improv. Manage. HIV Dis.,* 4:21–24.

Lynch, Joseph P. (1989). When opportunistic viruses infiltrate the lung. *J. Resp. Dis.,* 10:25–30.

Martin, Jeffrey, et al. (1998). Sexual tranmission and the natural history of human herpesvirus 8 infection. *N. Engl. J. Med.* 338:948–954.

McGrath, Michael, et al. (1994). Identification of a common clonal human immunodeficiency virus integration site in human immunodeficiency virus-associated lymphomas. *Cancer Res.,* 54:2069.

Medoff, Gerald, et al. (1991). Systemic fungal infections: An overview. *Hosp. Pract.,* 26:41–52.

Monette, Paul. (1988). *Borrowed Time: An AIDS Memoir.* New York: Avon Books.

Monforte, Antonella d'Arminio, et al. (1992). AIDS-defining diseases in 250 HIV-infected patients: A comparative study of clinical and autopsy diagnoses. *AIDS,* 6:1159–1164.

Montgomery, Bruce A. (1992). *Pneumocystis carinii* pneumonia prophylaxis: Past, present and future. *AIDS,* 6:227–228.

Moore, Patrick, et al. (1996). Kaposi's sarcoma—Associated herpesvirus infection prior to onset of Kaposi's sarcoma. *AIDS,* 10:175–180.

Morbidity and Mortality Weekly Report. (1993). Estimates of future global TB morbidity and mortality. 4:961–964.

Morbidity and Mortality Weekly Report. (1995). USPHS/IDSA guidelines for the prevention of opportunistic infections in persons infected with HIV: A summary. 44:1–34.

Morbidity and Mortality Weekly Report. (1995). 1995 Revised guidelines for prophylaxis against PCP for children infected with or perinatally exposed to HIV. 44:1–10.

Morbidity and Mortality Weekly Report. (1997). 1997 USPHS/IDSA guidelines for the prevention of OIs in persons infected with HIV. 46:1–46.

MURPHY, ROBERT. (1994). Opportunistic infection prophylaxis. *Int. AIDS Soc.–USA*, 2:7–8.

NEWTON, HERBERT. (1995). Common neurologic complications of HIV infection and AIDS. *Am. Fam. Phys.*, 51:387–398.

PALCA, JOSEPH. (1992). Kaposi's sarcoma gives on key fronts. *Science*, 255:1352–1354.

PHAIR, JOHN, et al. (1990). The risk of *Pneumocystis carinii* among men infected with HIV-1. *N Engl. J Med.*, 322:161–165.

POWDERLY, WILLIAM G., et al. (1992). Molecular typing of Candida albicans isolated from oral lesions of HIV-infected individuals. *AIDS*, 6:81–84.

POWDERLY, WILLIAM, et al. (1998). Recovery of the immune system with antiretroviral therapy: the end of opportunism? *JAMA* 280:72–77.

ROLSTON, KENNETH. (1992). Changing pattern of bacterial and fungal infections in patients with AIDS. *Primary Care and Cancer*, 12:11–15.

ROSSITCH, EUGENE, et al. (1990). Cerebral toxoplasmosis in patients with AIDS. *Am. Fam. Pract.*, 41:867–873.

RUSSELL, JAMES. (1990). Study focuses on eyes and AIDS. *Baylor Med.*, 21:3.

SAID, JONATHAN, et al. (1997). KS-associated herpesvirus/human herpesvirus type 8 encephalitis in HIV-positive and -negative individuals. *AIDS*, 11:1119–1122.

SCHNEIDER, MARGRIET, et al. (1999). Discontinuation of prophylaxis for *Pneumocystis carinii* pneumonia in HIV infected patients treated with HAART. *Lancet*, 353:201.

SCHULZ, THOMAS, et al. (1995). Karposi's Sarcoma; A finger on the culprit. *Nature*, 373:17.

SEPKOWITZ, KENT. (1998). Effect of HAART on natural history of AIDS-related opportunistic disorders. *Lancet*, 351:228–230.

SERRAINO, DEIGO, et al. (1995). HIV transmission and Kaposi's sarcoma among European women. *AIDS*, 9:971–973.

SIBLEY, L. DAVID. (1992). Virulent strains of *Toxoplasma gondii* comprise single clonal linage. *Nature*, 359:82–85.

SMALL, PETER. (1996). Tuberculosis research: Balancing the portfolio. *JAMA*, 276:1512–1513.

SOLOWAY, BRUCE (1998) Report on the Fifth Conference on Retroviruses and Opportunistic Infections. *AIDS Clinical Care*, 10:27–29.

STANFORD, J.L., et al. (1993). Old plague, new plague, and a treatment for both? *AIDS*, 7:1275–1276.

STRINGER, JAMES, et al. (1996). Molecular biology and epidemiology of pneumocystis carinii infections in AIDS. *AIDS*, 10:561–571.

TUAZON, CARMELITA, et al. (1991). Diagnosing and treating opportunistic CNS infections in patients with AIDS. *Drug Therapy*, 21:43–53.

WALLACE, MARK R., et al. (1993). Cats and toxoplasmosis risk in HIV-infected adults. *JAMA*, 269:76–77.

WALZER, PETER D. (1993). *Pneumocystis carinii:* Recent advances in basic biology and their clinical application. *AIDS*, 7:1293–1305.

WHEAT, L. JOSEPH. (1992). Histoplasmosis in AIDS. *AIDS Clin. Care*, 4:1–4.

WILLOUGHBY, A. (1989). AIDS in women: Epidemiology. *Clin. Obstet. Gynecol.*, 32:15–27.

WOFSY, CONSTANCE. (1991). Cryptosporidiosis and isosporiasis. *AIDS Clin. Care*, 3:25–27.

ZACKRISON, LEILA H., et al. (1991). *Pneumocystis carinii:* A deadly opportunist. *Am. Fam. Pract.*, 44:528–541.

ZOLER, MITCHELL L. (1991). OI's widening realm. *Medical World News*, 32:38–44.

A Profile of Biological Indicators for HIV Disease and Progression to AIDS

CHAPTER CONCEPTS

♦ The terms **incubation** and **latency** are defined.

♦ Clinical signs and symptoms of HIV infection and AIDS are presented.

♦ Stages of HIV disease vary substantially.

♦ HIV replication is rapid and continuous in HIV-infected lymphoid cells.

♦ AIDS Dementia Complex presents as mental impairment.

♦ Viral load indicates current viral activity.

♦ T4 cell counts indicate degree of immunologic destruction.

♦ Clues to long-term survival are presented.

♦ Serological changes after HIV infection are presented.

♦ The rate of clinical HIV disease progression is variable among individuals infected with HIV.

♦ The development of AIDS over time is discussed.

♦ Classification of HIV/AIDS progression is presented.

♦ Clinical indicators to track HIV disease progression are listed.

♦ Diarrhea is the most common gastrointestinal sign and symptom of HIV/AIDS infection.

♦ Clues to pediatric AIDS diagnosis are presented.

HIV DISEASE DEFINED

By the mid-1980s the CDC had learned enough about HIV infection to call it a disease. That made sense, as the vast majority of those who became infected became ill. HIV infection leads to the loss of T4 cells, which in turn produces a variety of signs and symptoms of a **nonspecific disease** with initial acute fever associated illness or mononucleosis-like symptoms which may last up to 4 weeks or longer. After the initial symptoms most individuals enter a clinically asymptomatic phase (see Case in Point 7.1). This means the infected person feels well while

— BOX 7.1 —

— CASE IN POINT 7.1 —

DESCRIPTION OF AN AIDS PATIENT

Cecilia Worth is a registered nurse and author. In a recent edition of The New York Times Magazine she wrote:

> Clustered near the bed, framed photos show a burly athlete who placed in the triathlon, a handsome man who grins disarmingly, an arm slung around his wife's shoulders. Now, transformed into a skin and bones caricature of himself, he is ruled by fatigue. After an interminable struggle to reach the bathroom, knees buckling, leg muscles barely able to hoist his feet forward over the floor, a heroic effort of will, he collapses back in bed, exhausted, motionless, glaring from huge, haunted eyes when I speak to him.
>
> Only in his wife's presence is he calm, though no less armored. She is angry, too, and afraid of him. She cooks for him but will not touch him. His children, parents, brother visit often, struggle for words, and leave without embracing.
>
> He rejects kindness in any form. To my cordial first greeting, he responds with silence, slamming shut his eyes. To suggestions of television, music, back rubs, his response is emphatic, curt: "No!"

Worth vividly describes some of the agony of this terrible disease. She also describes the mental and emotional strain that tears at family life.

There is a point at which sickness and dying cease to offer insights into the human condition and become instead an unbearable, unredeemable absurdity. This is often how AIDS appears to those who know it.

VARIATION OF INITIAL SYMPTOMS AFTER HIV INFECTION

Case I: Male, Age 35, Los Angeles, California

One evening, for no apparent reason, John began sweating profusely. Soon after a red rash began on his arms, face, and legs and then covered his body. Simultaneously, breathing became difficult and he was rushed to an emergency room. By then he was shaking violently. After medication and a battery of tests his problem could not be defined. This brief illness passed, he felt fine but some years later, during a blood screen for insurance purposes he came up HIV-positive. He immediately reflected back on his earlier illness and its cause.

Case II: Male, Age 29, Los Angeles, California

This case is in marked contrast to case I. Feeling the pinch of a sore throat, this male went to his physician for an antibiotic. On examination, he had a yeast infection which appeared far back in his throat. This raised suspicion and he agreed to an HIV test. It came back positive. He had no other illness, he was treated, the sore throat vanished, and he is thriving in a long asymptomatic period.

HIV-infected individuals with normal T4 cell counts may predict the subsequent development of HIV-related disease, and that patients who harbor SI isolates develop immune deficiency more rapidly. It is not clear whether the appearance of more virulent strains during the chronic phase of the infection is a cause or an effect of progressive immune deficiency (Nielson et al., 1993). Several studies have demonstrated that a long period of fever around the time of **seroconversion (the presence of detectable HIV antibody in the serum)** is associated with more rapid development of immune deficiency (Pedersen et al., 1989).

Spectrum of HIV Disease

Because the immune system slowly falters, HIV disease is really a spectrum of disease. At one end of the spectrum are those infected with HIV who look, feel and are perfectly healthy. At the opposite end are those with advanced HIV disease **(symptomatic AIDS)** who are visibly sick and require significant medical and psychoso-

his or her immune system is slowly compromised. It has been shown that long-lasting symptomatic **primary** HIV infection predicts an increased risk of rapid development of HIV–related symptoms and AIDS, but it is not known whether the different responses to HIV infection are caused by viral, host factors or both. Virulent strains of HIV have been characterized by their rapid replication, **syncytium-inducing (SI)** capacity, and tropism (attraction) for various types of T cells. It is known that the biological properties of HIV strains in asymptomatic

cial support (Figure 7-1). Between these two extremes, HIV-infected people may develop illnesses that range from mild to serious. Symptoms can include persistent fevers, chronic fatigue, diarrhea, swollen lymph nodes, night sweats, skin rashes, significant weight loss, visual problems, chest pain, and fungal infections of the mouth, throat, and vagina. Illness from these conditions can be **severe** and **disabling,** and some people may die without ever being diagnosed with AIDS. Also, people with HIV disease may develop **neurologic disorders,** which can cause **forgetfulness, memory loss, loss of coordination and balance, partial paralysis, leg weakness, mood changes and demen-tia.** These symptoms may occur in the absence of any other symptoms. The interval between initial HIV infection and the presence of signs and symptoms that characterize AIDS is variable and may range from several months to a median duration of about 11 years (Figure 7-2).

Defining Incubation and Latency

Because of the long delay in determining what happened after HIV infection and progression to AIDS, the terms **incubation** and **latency** are used, in many cases interchangeably, causing some confusion. In this chapter, the two terms are used with respect

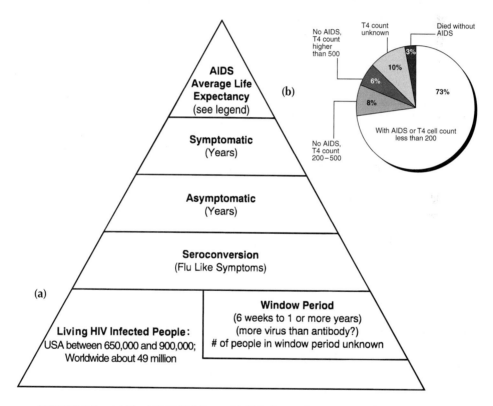

FIGURE 7-1 (**a**) The HIV/AIDS Pyramid. This figure demonstrates that current AIDS cases are coming from an existing pool of HIV-infected persons. In the United States, of those infected, about 40% do not yet know they are HIV positive. Although both the asymptomatic and symptomatic periods may last for years, once diagnosed with AIDS the average life expectancy **without AIDS drug cocktails is 2 to 3 years.** Average life expectancy for those now on combination drug cocktails has not yet been determined. (**b**) Clinical outcomes 10 to 16 years after HIV infection in a population tracked by the San Francisco City Clinic from the beginning of the AIDS outbreak.

Adult/Adolescent HIV Disease Continuum to AIDS

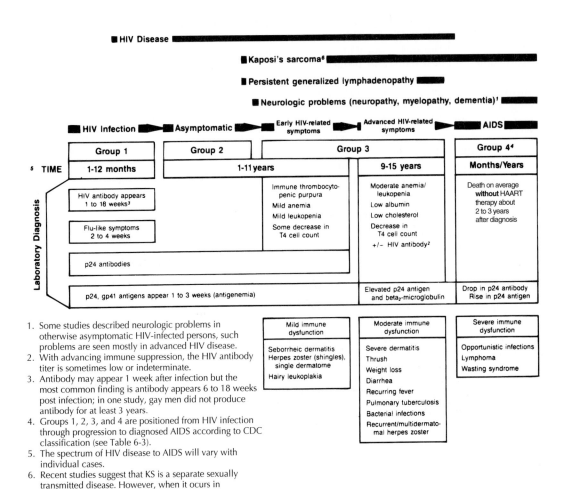

1. Some studies described neurologic problems in otherwise asymptomatic HIV-infected persons, such problems are seen mostly in advanced HIV disease.
2. With advancing immune suppression, the HIV antibody titer is sometimes low or indeterminate.
3. Antibody may appear 1 week after infection but the most common finding is antibody appears 6 to 18 weeks post infection; in one study, gay men did not produce antibody for at least 3 years.
4. Groups 1, 2, 3, and 4 are positioned from HIV infection through progression to diagnosed AIDS according to CDC classification (see Table 6-3).
5. The spectrum of HIV disease to AIDS will vary with individual cases.
6. Recent studies suggest that KS is a separate sexually transmitted disease. However, when it ocurs in HIV/AIDS patients, it occurs during this time frame.

FIGURE 7-2 Spectrum of HIV Infection, Disease, and the Expression of AIDS. Seroconversion means that HIV antibodies are measurably present in the person's serum. With continued depletion of T4 cells, signs and symptoms appear announcing the progression of HIV disease to AIDS. Although HIV antibodies have been found as early as 1 week after exposure, **most often seroconversion occurs between weeks 6 and 18; 95% within 3 months, 99% within 6 months** (see Figure 7-4). The early stage of HIV infection can be separated from the symptomatic stage by years of **clinical latency.** Early infection is characterized by a high number of infected cells and a high level of viral expression. AIDS is characterized by increased levels of viremia and p24 antigenemia, activation of HIV expression in infected cells, an increased number of infected cells, and progressive immune dysfunction. Stages of HIV disease blend into a continuum ranging

(continued on next page)

ONE MISTAKE COST HIM HIS LIFE

I held my son today while he died from AIDS. There is no pain like the pain in a mother's heart. He was 28 years old, and now, he is dead.

This wonderful young man will never have a family. He will never again have a chance to do the things he enjoyed so much — water ski, snow ski, travel. He loved *Star Trek* and music. He loved working for the airlines and traveling all over the world. He was delightful and smart — a computer whiz — could take one apart and put it back together.

He wasted away from a handsome young man to a skeleton — nothing more than skin and bones. His weight dropped from 160 pounds to 80 pounds. His hair fell out. His beautiful teeth fell out. Sores broke out all over his body. He couldn't hold down any food, and eventually, he starved to death.

No, young people, he was not gay, nor was he a drug user. He just went to bed with a girl he didn't know.

His Mom

(Source: Ann Landers, Syndicated Columnist, 1994)

to **clinical** observations as follows: **clinical incubation** is that period after infection through the window period or when anti-HIV-antibody production is measurable. **Clinical latency** is the time period from detectable anti-HIV-antibody production (seroconversion—the person now tests HIV-antibody-positive) through the asymptomatic period—a time prior to the expression of opportunistic diseases. This time period, being asymptomatic, may last from one to 20 years—

the average being 10 years. The beginning and end of these periods will vary from person to person and their susceptibility and expression of HIV disease.

STAGES OF HIV DISEASE

The course of the disease in the infected individual varies substantially. At the extremes are individuals who show either little evidence of progression (loss in T4 cells) 10 to 20 years following infection (1% to 3%) or extremely rapid progression and death within less than 2 to 3 years. In general, HIV-infected adults experience a variety of conditions, categorized into four stages: **acute infection, asymptomatic, chronic symptomatic,** and **AIDS.**

Acute or Primary HIV Disease Stage

The acute stage usually develops in 2 to 8 weeks after initial infection or exposure to the virus. Up to 70% of infected individuals develop a self-limited (brief) illness similar to influenza or mononucleosis: high spiking fever, sore throat, headaches, and swollen lymph nodes. Some may develop a rash, vomiting, diarrhea, and thrush (yeast infection in the mouth). This is referred to as the **acute retroviral syndrome.** The symptoms generally last about 1 to 4 weeks and resolve spontaneously (Table 7-1). The acute stage can be even quickly and easily missed. The acute phase is marked by high levels of HIV production. During this phase, large numbers of HIV spread throughout the body, seeding themselves in various organs, particularly lymphoid tissues such as the lymph nodes, spleen, tonsils, and adenoids (Figure 7-3). During this time frame, HIV-infected resting T4 cells become

(continued)

from the asymptomatic with apparent good health, to increasingly impaired health, to the diagnosis of AIDS. Thus the spectrum of HIV disease ranges from the silent infection to unequivocal AIDS. Clinical expression moves from one condition to another, often without a clear-cut distinction. The level of an individual's infectiousness is believed to be greatest within the first months after infection and again when the T4 cell count drops below 200 (see Figure 7-3). **However, people who are HIV-infected can transmit HIV at any time.**

TABLE 7-1 Clinical and Laboratory Analyses of Acute HIV Infection

Signs

Erythematous truncal maculopapular rash
Generalized urticaria, roseola-like exantham, palm
  and sole desquamation
Generalized lymphadenopathy, splenomegaly
Acute meningo-encephalitis
Myelopathy, Guillain-Barre syndrome
Radiculopathy (brachial or sacral plexopathy)
Peripheral neuropathy
Myopathy
Pharyngitis
Hepatitis
Mental changes
Oral or esophageal ulcerations
Weight loss

Symptoms

Fever, night sweats, chills, malaise, fatigue
Arthralgias, myalgias
Anorexia, nausea and vomiting, abdominal cramps,
  diarrhea
Headache, retro-orbital pain, photophobia, lethargy
Sore throat, dry cough

Laboratory Analysis

Mild-moderate neutropenia, relative monocytosis
Lymphopenia to lymphocytosis/atypical lymphocytes
Elevated erythrocyte sedimentation rate
Appearance of HIV antibodies
HIV in serum and/or CSF
Abnormal liver function tests
Raised levels of beta-2 microglobulin

trapped or held within the follicular dendritic cells. Within these **safe** havens, HIV can persist for years despite **Highly Active Antiretroviral Therapy (HAART).** This pool of latently infected cells is established very early after HIV enters the body—even if the person takes drug therapy immediately after exposure to HIV.

During acute primary infection, patients have extremely high levels of viral replication **but no notable antibody immune response.** However, the level of cytotoxic T lymphocytes (CTLs) targeted against HIV appears to increase significantly, an attempt by the cellular part of the immune system to contain the high rate of HIV replication. Plasma HIV-RNA levels may reach 10^5 to 10^8 coplies/ml. It is now believed that it is the increase in the presence of CTLs and chemical factors they produce that bring about a milder illness and a reduction in HIV to a **lower set point**—a

steady state of viral load. The higher this set point the more rapidly HIV disease progresses to AIDS. (The presence of CTLs and long term survival is presented in Box 7.4.) At this time in HIV infection and in some cases for weeks or months, neutralizing antibodies to HIV are not measurable—thus antibody testing, whether at clinical labs or using a home testing kit, is negative. This period of high viral replication in the absence of detectable antibody is called the **Window of infectivity before seroconversion.** During this period, patients may be highly infectious. The viral burden in genital secretion is particularly high during this time. Mathematical models suggest that 56% to 92% of all HIV infections may be transmitted during this period of acute infection (Quinn, 1997).

A true state of **biological latency,** according to the work of Xiping Wei and co-workers (1995) and David Ho and co-workers (1995), does not exist **in the lymph nodes** at any time during the course of HIV infection. The investigations of Wei and Ho show that from the time of infection HIV replication is rapid and continuous, and within 2 to 4 weeks the infecting HIV strain is replaced by drug-resistant mutants. Each day 1 billion to 10 billion HIV are produced and mostly destroyed and a billion T4 cells or more are infected, dying, and replaced. Over time the immune system fails to destroy HIV and replace its T4 cell losses and HIV disease progresses. Also, over time, many T4 cells in the lymphoid organs probably are **activated** by the increased secretion of certain cytokines such as tumor necrosis factor-alpha and interleukin-6. **Activation** allows uninfected cells to be more easily infected and causes increased replication of HIV in infected cells. Other components of the immune system also are chronically activated, with negative consequences that may include the suicide of cells by a process known as programmed cell death or apoptosis (pronounced a-po-toe-sis) and an inability of the immune system to respond to other invaders.

It is during this stage that many clinical HIV/AIDS investigators are recommending the use of **HAART** in order to stop rapid HIV reproduction and the infection of T4 lymphocyte. (Vanhems, et al. 1999)

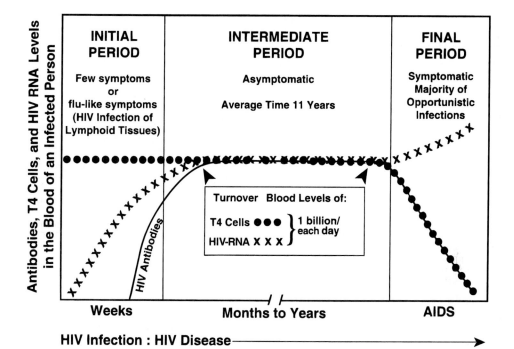

FIGURE 7-3 Relationship of T4 cells, HIV antibodies, and HIV RNA Levels (Viral Load) Beginning with HIV Infection through AIDS. Within a week to weeks after HIV infection, HIV becomes seeded throughout the body's blood and lymph system. HIV reproduction (infection of T4 cells) begins almost immediately in the lymph system. The T4 cell population also begins rapid reproduction to replace T4 cell loss. This is why the T4 graph line stays at the same level through the **asymptomatic** period. The immune system begins to turn out HIV antibodies, in general 6 to 18 weeks later (window period, see Figure 7-4). Note that during the asymptomatic period T4 cell and HIV replication and antibody production keep pace. With time, however, T4 cells fail to replace losses, HIV continues to replicate, antibody levels drop due to loss of T4 signals to B cells to produce antibodies and opportunistic infections begin—the **symptomatic** period. **Without** therapy this period lasts on average about 2 to 3 years. For people using HAART, the average number of years one can survive in the symptomatic period has yet to be determined.

Asymptomatic HIV Disease Stage

Following acute illness, an infected adult can remain free of symptoms from 6 months to a median time of about 11 years. During the asymptomatic period, measurable HIV in the blood drops to a lower level but it continues to replicate and continues to destroy T4 cells within the lymph nodes while the body continues to produce new T4 cells and antibodies to the virus (Figure 7-4). An asymptomatic in-

dividual appears to be healthy and can assume normal activities of daily living.

Chronic Symptomatic HIV Disease Stage

The chronic phase can last for months or years before a diagnosis of AIDS occurs. During this phase, as viral replication continues, T4 cells become depleted. As the number of immune system cells decline, the individual develops a variety of symptoms such as fever, weight

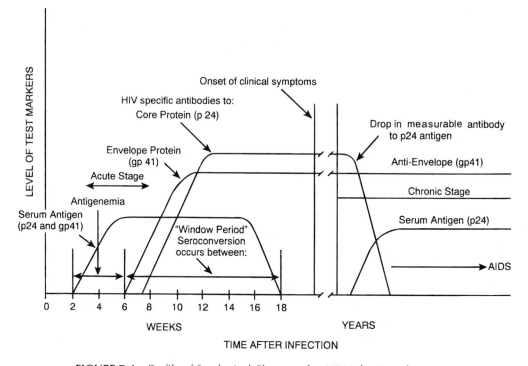

FIGURE 7-4 Profile of Serological Changes after HIV Infection. The dynamics of antibody response to HIV infection was determined by enzyme immunoassays (EIA). Note that during antigenemia, specific HIV proteins (antigens) can be detected before seroconversion occurs. Perhaps other HIV proteins will allow even earlier detection of HIV infection. Once antibodies appear, some antigens like p24 and gp41 disappear only to show up again later on. Note also that although antibody production is a sign that the immune system is working, in HIV-infected people, it is not working well enough. Although envelope and core protein antibodies are being produced as clinical illness begins, as the p24 antibody drops, the illness becomes more serious. *(Adapted from Coulis et al., 1987)*

loss, malaise, pain, fatigue, loss of appetite, abdominal discomfort, diarrhea, night sweats, headaches, and swollen lymph glands. Ultimately, HIV overwhelms the lymphoid organs. The follicular dendritic cell networks break down in late-chronic stage disease and virus trapping is impaired, allowing spillover of large quantities of virus into the bloodstream. The destruction of the lymph node structure seen late in HIV disease may stop a successful immune response against not only HIV but other pathogens as well, and heralds the onset of the opportunistic infections and cancers that characterize AIDS. Individuals at this stage, with a T4 cell count of 200 or

less/μL of blood often develop thrush, oral lesions, and other fungal, bacterial, and/or viral infections. The duration of these symptoms varies, but it is common for HIV-infected individuals to have them for months at a time. Of those persons in the chronic stage, and not using HIV drug therapy, about 30% developed AIDS-associated infections within 5 years.

AIDS: Advanced HIV Disease Stage

The diagnosis of AIDS is a marker, not an end in itself. Currently most people recover from their first, second, and third AIDS-defining

PRIMARY CASE PRESENTATIONS

Case 1. A 28-year old man visited his family physician complaining of flulike illness. The physician obtained a drug-use and sexual history, through which it was discovered that in the past year the patient had engaged in unprotected sex with several partners. HIV testing was ordered. It showed a T4 count of 550 μl and a plasma HIV RNA level of 735,000 copies/ml.

Diagnosis: Acute HIV Infection. The illness may present dramatically, with features such as painful mouth ulceration, lymphadenopathy, or a rash indistinguishable from that of mononucleosis. It may even have neurologic manifestations, such as Bell's palsy. Typically, however, the illness is hard to distinguish from influenza and an array of nonspecific viral illnesses. It may even be essentially asymptomatic. Except perhaps in retrospect, the patient usually fails to recognize the symptoms as marking acquisition of HIV infection.

Case 2. A 22-year old woman in the first trimester of her first pregnancy reported being in fair health. She underwent routine HIV screening as part of the antepartum evaluation. The T4 count was 275/μL, and the plasma HIV RNA level was 127,000 copies/ml.

Diagnosis: Asymptomatic Chronic HIV Infection. HIV infection was detected by routine screening. Here the infection has moved from its acute phase to a latent phase during which the virus is residing in sites such as the lymph nodes, its replication still-counterbalanced by host immune responses.

Today detection by routine screening occurs most often in pregnant women. Indeed, as many as two-thirds of HIV cases in women are diagnosed during pregnancy.

Case 3. A 34-year old man presented at an emergency department seeking treatment for a chest rash typical of herpes zoster. Thinking that the patient was surprisingly young to have shingles, the physician identified a history of recreational cocaine use. The patient insisted that his drug use was purely intranasal, not intravenous. He said he had never had thrush or pneumonia. The HIV anti-body test was positive.

Diagnosis: Symptomatic Chronic HIV Infection Complicated by Herpes Zoster. Although zoster can occur at any time of life, it most often occurs in patients of advanced age or in those with serious conditions such as malignancy or long-term steroid therapy. In chronic HIV infection, zoster is now recognized to have an increased chance of preceding, often by a long period, other opportunistic infections. In this instance, the patient had not had thrush of Pneumocystis carinii pneumonia. Measurement of the HIV viral RNA level was deferred because of the presence of inflammation, which may cause a misleading spike in plasma HIV RNA, regardless of the underlying condition causing the inflammation. (Reproduced with permission. John Bartlett, et al. Primary Care of HIV Infection. *Hospital Practice*, 1998; 33 (12:53–56, 61–64, 67–69) © 1998. The McGraw-Hill Companies, Inc.)

illnessess. People with AIDS are a very heterogeneous group—some feel well and continue working for several years, others are chronically ill, and some die rather quickly.

Patients with AIDS became an even more diverse group after the 1993 expansion of the Centers for Disease Control and Prevention's (CDC's) definition of AIDS. People, in excellent health, are diagnosed with AIDS because their T4 cell count is less than 200/μl of blood.

The final stage of HIV infection is called AIDS. During this time there is continued rapid viral replication which finally upsets the delicate balance of HIV production/T4 cell infection to T4 cell replacement and the virus largely depletes the cells of the immune sys-

tem. It has been suggested that during the AIDS stage, serious immunodeficiency occurs when HIV diversity exceeds some threshold beyond which the immune system is unable to control HIV replication (Nowak et al., 1990; Wei et al., 1995; Cohen, 1995).

In addition to the symptoms and conditions caused by HIV, opportunistic infections, and malignancies, HIV-infected patients often experience side effects from various drugs they are taking for primary conditions. In many instances, the toxic side effects of the drugs are life-threatening.

In the symptomatic stages of HIV disease, an individual's ability to carry on the activities of daily living is impaired. The degree of

——— BOX 7.2 ———

EVOLUTION OF HIV DURING HIV DISEASE PROGRESSION

HIV is a unique retrovirus. For example, mitosis, a form of cell division, is a requirement for the nuclear entry of most retroviral nucleic acids. In contrast, mitosis does not appear to be required for nuclear entry of HIV nucleic acids, particularly in terminally differentiated cells (e.g., macrophages and dendritic cells) (Freed et al., 1994). And second, HIV lacks any mechanism to correct errors that occur as its genetic material is being duplicated. This means that every time the virus makes a copy of itself there will be, on average, at least one genetic "mistake" incorporated in the new virus. So a few days or weeks after initial infection, there may be a large population of closely related, but not identical, viruses replicating in an infected individual. In the **quasi**—steady-state condition, there are successive generations of viral progeny, with each generation following the next by about 2.6 days. Approximately 140 generations of virus are produced over the course of a year and 1,400 generations over the course of 10 years, allowing production of an extraordinary number of genetic variants. Some variants can provide pre-existing drug-resistant forms or enable rapid development of resistance under drug pressure, and some enable the viral population to escape immune activity.

VIRAL POOL

The viral pool is estimated to be about 10 billion viruses and each is genetically different from all other HIV in the pool. It is known from experimental data that about 1 in 1,000 particles is infectious, so the infectious viral pool may be on the order of 10 million viruses. With a genome of approximately 10^4 nucleotides, and 10 billion HIV variants made daily, mutations most likely occur at every nucleotide position on a daily basis (Ho, 1996). This creates an enormous potential for viral evolution. On average, the HIV that is transmitted to another individual will be over 1,000 generations removed from the initial HIV infection. This extent of replication per transmitted infection (transmission cycle) is probably without equal among viral and perhaps bacterial infections (Coffin, 1995). Regardless of the underlying mechanism of immunodeficiency, it is becoming apparent that **the force that is driving the disease is the constant repeated cycles of HIV replication.** Simon Wain-Hobson (1995) suggests

that because of 24-hour-a-day HIV replication, as shown by Wei and co-workers (1995), an HIV-infected asymptomatic person can harbor at least 1 billion distinct HIV variants and an AIDS patient more than 100 billion HIV variants. With so many genetic variants, **some are going to be resistant to any given drug. (This would include drugs not yet used in therapy).** During the investigations of Ho and co-workers (1995) and Wei and co-workers (1995), variants of HIV, resistant to the drugs, ABT-538, **now called Ritonavir,** (a protease inhibitor), and nevirapine (a non-nucleoside agent that inhibits reverse transcriptase function), occurred within days or weeks! Alan Perelson and colleagues (1996) reported data collected from five HIV-infected people after administering Ritonavir through 7 days. Each person responded with a similar pattern of decline in plasma HIV-RNA. Their results: Infected T4 cells had an average life span of 2.2 days. Plasma HIV-RNA had an average life span of 0.3 days. The results also suggest that the minimum duration of the HIV life cycle in human T4 cells is 1.2 days on average and that the average HIV generation time—defined as the time from release of a virus until it infects another cell and causes the release of a new generation or HIV—is 2.6 days. The life time of HIV in resting or latent T4 cells may range form 6 months to perhaps an infinite amount of time. Such cells can produce HIV when activated.

Number	1 to 10 billion plus HIV produced each day.	About 2 billion T4 cells infected each day.
After two days	Half of HIV are destroyed and are replaced by about an equal amount of new HIV.	About half of the infected T4 cells die and are replaced by new T4 cells. (4% of total body T4 cell count).
After 14 days	After mutations, almost all of new HIV are drug resistant.	The healthy T4 cell count decreases and new T4 cells are infected with the drug-resistant viruses.

——— **BOX 7.2** *(continued)* ———

Effect of anti-HIV drugs	Initially, over 99% percent of HIV are destroyed using combination drugs like proteases and nucleoside and non-nucleoside reverse transcriptive inhibitors.	The healthy T4 cell count increases.

HIV CLEARANCE

Douglas Richman (1995) states that HIV clearance rate can be calculated based on the rate of reduction in viral RNA load. In cases in which viral RNA load attains a constant level, it can be assumed that there is a dynamic equilibrium resulting in a steady-state level, with production rates matching clearance rates. When viral resistance emerges, as in cases of drug resistance, the production of resistant virus may double every 2 days. The proportion of virus with resistance mutations can also be calculated. According to Richman, in the case of the nonnucleoside reverse transcriptase inhibitors, this has been found to be approximately 1 to 2 per 1,000 RNA copies circulating in the plasma; that is, a patient with 60,000 copies/mL plasma has approximately 100 copies/mL of resistant mutants prior to the **beginning of treatment.** A similar scenario probably holds for all drug-resistance mutations. Thus it is reasonable to expect the HIV-resistant mutants will emerge to almost any antiretroviral compound.

IMPLICATIONS OF HIV DYNAMICS

The rapid turnover of HIV RNA populations has several implications for pathogenesis and treatment of HIV disease. High levels of replication persist throughout HIV disease, with the rate of replication appearing to be fairly constant. Clearance of HIV RNA is also rapid and remains fairly constant throughout the disease. It is only the steady-state levels of HIV RNA that appear to change, which is affected by the production rate of HIV-RNA. Treatment can affect HIV steady-state levels by inhibiting its production. What is

not clear, however, is whether the **rate of HIV RNA clearance can be increased.**

The implications of the presence of high levels of HIV RNA and high HIV RNA turnover for the assessment of antiretroviral drug activity are clear. With new techniques for quantitating plasma HIV RNA, nearly every HIV-infected individual can be evaluated for response to antiretroviral therapy. While the immune system will recognize most members of a population of viruses, some mutants will evade the immune response for a time. Until they are brought under immune control, these so-called escape mutants will attack T4 cells. It is these cells that are key to orchestrating the overall immune response, and once they are gone the immune system collapses.

OVERWHELMING THE IMMUNE SYSTEM

As HIV multiplies and mutant forms are produced, the immune system responds to these new forms. But ultimately the sheer number of **different** viruses to which the immune system must respond becomes overwhelming. It's a bit like the juggler who tries to keep too many balls in the air: The result is disastrous. Once the immune system is overwhelmed, the latest escape mutant—which may not necessarily be the most pathogenic one to come along—will predominate and immune deficiency will progress. If, as investigators now believe, immunodeficiency or HIV disease is due to accumulated damage over the course of the clinically latent phase, then drug treatment at almost any time during this period, or even after the onset of symptoms, could have a virtually identical effect with regard to prolonging the time of progression to more serious disease or death. That is, it will make no difference when the drug is administered because the drug, whenever present, selects for HIV-resistant mutants which then replicate to increase their numbers and continue to infect T4 cells. As the number of drug-resistant mutants increase, the drug becomes less and less effective while the disease progresses (Japour, 1995).

impairment varies considerably from day to day and week to week. Many individuals are debilitated by the symptoms of the disease to the point where it becomes difficult to hold steady employment, shop for food, or do

household chores. It is also quite common for people with AIDS to experience phases of intense life-threatening illness, followed by phases of seemingly normal functioning, all in a matter of weeks. For a good review on the mechanisms of HIV disease, read "The Immunopathogenesis of HIV Infection" by Giuseppe Pantaleo et al. (1993).

HIV Can Be Transmitted During All Four Stages

A person who is HIV-infected, even while feeling healthy, may unknowingly infect others. Thus the term HIV disease, rather than AIDS, more appropriately describes the entire scope of this public health problem. **The greatest risk of HIV transmission occurs within the first several months after infection and again when the T4 cell count drops below 200.**

HIV DISEASE WITHOUT SYMPTOMS, WITH SYMPTOMS, AND AIDS

A person may have no symptoms (**asymptomatic**) but test HIV-positive. This means that the virus is present in the body. Although he or she has not developed any of the illnesses associated with HIV disease, it is possible to pass the virus on to other people.

Persons may develop some symptoms early on in HIV disease such as swollen lymph glands, night sweats, diarrhea, or fatigue. (This stage used to be called ARC for AIDS-Related Complex; the term is no longer used by the Centers for Disease Control and Prevention.)

In time, most people with HIV disease progress to AIDS. (This is sometimes referred to as full-blown AIDS. This term is not used in this text; one either is or is not diagnosed with AIDS.) A person has HIV/AIDS when the defect in his or her immune system caused by HIV disease has progressed to such a degree that an unusual infection or tumor is present or when the T4 cell count has fallen below 200/μl of blood. In AIDS patients, a number of diseases are known to take advantage of the damaged immune system. These include opportunistic infections usually caused by viruses, bacteria, fungi, or protozoa; or tumors such as Kaposi's sarcoma, a form of skin cancer (?), or lymphoma, a malignancy of the lymph glands. It is the presence of one of the opportunistic diseases, or a T4 cell count of less than 200, along with a positive HIV test that establishes the

medical diagnosis of AIDS. Thus the disease we call AIDS is actually the end stage of HIV disease. It is important to remember that AIDS itself is not transmitted—the virus is. AIDS is the most severe clinical form of HIV disease.

INCREASES IN THE NUMBER OF HIV/AIDS CASES

Prior to 1988, there were between 5 and 10 million HIV-infected people worldwide, an estimated 1.5 million of them living in the United States. Relatively few of these people were actually sick. The majority of these people were infected by HIV in the early 1980s. From 1988 on (there is an average clinical latency period of about 11 years), the number of new AIDS cases, and AIDS-related deaths began to rise rapidly; and by the beginning of year 2000, about 430,000 AIDS patients in the United States out of an accumulated 746,000 AIDS cases died. (**AIDS cases are still being underreported by 10% to 15% in the United States.**) The point is that the number of AIDS cases and deaths will continue to increase in the 1990s because those who became infected by HIV in the 1980s will progress to AIDS. In addition, the CDC estimated that there will be **at least 40,000** new HIV infections in the United States for each year from 1991 through at least the year 2000.

ASPECTS OF HIV INFECTION

HIV infection depends on a variety of events, for example, the amount and strain of HIV that enters the body (some strains of HIV are known to be more pathogenic than others), perhaps where it enters the body, the number of exposures, the time interval between exposures, the immunological status of the exposed person, and the presence of other active infections. These are referred to as cofactors that contribute to successful HIV infection.

Post Infection

Estimating the date of initial HIV infection helps predict the likely timing of disease progression. The task is easiest when there has been a known blood exposure. In other cases, the

——— BOX 7.3 ———

DEVELOPMENT OF AIDS OVER TIME

A spectrum of clinical expression of disease can occur after HIV infection. Approximately 10% of HIV-infected subjects progress to AIDS within the first 2 to 3 years of HIV infection (**rapid progressors**). People with greater than 50,000 HIV RNA copies/mL at 6 months after infection are the most likely to be rapid progressors. About 60% of adults/adolescents will progress to AIDS within 12 to 13 years after HIV infection (**slow progressors**). Approximately 5% to 10% of HIV-infected subjects are clinically asymptomatic after 8 to 15 years and have stable peripheral blood T4 cell levels (**nonprogressors**). Data from two separate studies suggest that 20 to 25 years after infection, 10% to 17% of HIV-infected individuals will still be AIDS-free (Haynes et al., 1996; Schanning et al. 1998).

In summary, true **nonprogressors**—people who have had HIV for up to 15 years or more with no symptoms and with continually normal and stable T4 cell counts—are rare. The majority of nonprogressors are, in fact, **slow progressors**—people who have had HIV for 12 years or more and who are healthy but whose T4 cell counts, although not falling below 500, have gradually declined. These people have about 10,000 or fewer HIV RNA copies/mL of plasma.

People with AIDS with a very low T4 cell count who have survived for exceptionally long periods of time (5 years or more) are usually described as long-term survivors. Thus, if one has lived with AIDS for 5 or more years, he or she is a long-term survivor. Some 6% of persons diagnosed with clinical AIDS are long-term survivors (Laurence, 1996).

Current and contained use of AIDS drug cocktails containing at least one protease inhibitor has increased the number of long-term survivors! Survival after the onset of AIDS, without HAART, has been increasing in industrialized countries from an average of less than 1 year to over 3 years at present. With therapy, survival time has been increased by 3 to 10 years, depending on the case and therapy. Survival time with AIDS in developing countries remains short and is estimated to be less than 1 year. Longer survival appears to be directly related to routine treatment with antiretroviral drugs, the use of drugs for opportunistic infections, and a better overall quality of health care.

The majority of AIDS cases occur before age 35, and over 90% of all AIDS deaths occur in people under the age of 50 worldwide.

Philip Rosenberg and colleagues (1994) reported that the length of incubation, progression from HIV infection to AIDS, varied according to the age at the time of infection. Younger ages were associated with a slower progression to AIDS. The estimated median treatment-free clinical incubation period was 12 years for those infected at age 20, 9.9 years for infection at age 30, and 8.1 years for infection at age 40.

THE CLINICAL COURSE OF AIDS AMONG MEN AND WOMEN

Andrew Phillips and colleagues (1994) compared the development of AIDS-defining diseases between 566 women and 1988 men with AIDS who were HIV-infected via similar routes, mainly by sharing IDU equipment and by heterosexual sexual contact. **They concluded that there was little if any difference between men and women in the clinical course of AIDS.**

---

patient may have experienced one or a number of high-risk sexual or drug-use exposures.

The virus may be present in the bloodstream or within cells for various lengths of time prior to antibody formation. When the antibody appears it is called **seroconversion** (sero = serum of the blood; conversion from antibody negative to antibody present or positive). Seroconversion for HIV may occur as early as 1 week, but most often is detected between 6 and 18 weeks after infection (Figure 7-4).

Immunosilent HIV Infection: Time Before Seroconversion

The period after HIV infection necessary to induce production of specific antibodies, immunological defect, and HIV disease is variable and probably depends on the characteristics of both the virus and the host. It has been estimated that between 95% and 99% of infected individuals seroconvert (produce HIV antibodies) within 6 months of infection

(Horsburg et al., 1989). However, according to some investigators, in rare cases, antibodies may not appear for up to 36 months or more (Ranki et al., 1987; Imagawa et al., 1989; Ensoli et al., 1990, 1991).

Incubation and Window Period

The time from HIV infection to the first signs and symptoms of HIV disease is called the **clinical incubation period.** Evidence suggests that the route or manner of HIV infection may influence the **incubation period.** For example, exposure through sexual intercourse has a mean incubation time of 6 months. Infection by transfusion has a mean incubation time of 2.5 years. It appears that for unknown reasons, **free viruses** (those not inside cells) present at the time of infection do not stimulate the immune system. Infected cells in the transfused blood have a **biologically latent** (inactive) proviral state and antibody will not be made until new viruses are produced which then engender HIV disease symptoms. Infected newborns have a mean incubation period of 10 months. (Newborns in general begin making their own antibody about 3 to 6 months after birth, but may not make antibody against HIV for several years.)

The time between infection and the presence of first HIV antibody or seroconversion is called the window period. Because it can be quite lengthy, HIV antibody tests performed during the window period that are negative may be **falsely negative** (Figure 7-4).

PRODUCTION OF HIV-SPECIFIC ANTIBODIES

During the 17 years (1983–2000) since the discovery of HIV, scientists have constructed a serological or antibody graph of HIV infection and HIV disease. The graph reveals how soon the body produces HIV-specific antibodies after infection and about when the virus begins its reproduction. Different parts of the graph (Figure 7-4) have been filled in by Paul Coulis and colleagues (1987), Dani Bolognesi (1989), and Susan Stramer and colleagues (1989). **The history of the HIV antibody is not yet complete, but the order of appearance and disappearance of antibodies** specific for the serologically important antigens over the course of HIV disease has been described.

The time period from seroconversion to the presentation of clinical symptoms is quite variable and may last for 10 or more years in adults and adolescents, but occurs earlier in children and older persons with HIV disease. The time period for moving from clinical symptoms of HIV disease to AIDS is also quite variable (Figure 7-2). Much depends on the individual's genetic susceptibility and his or her response to medical intervention. The **average** time from HIV infection to AIDS from 1990 through 1998 was 11 years. Note that Figure 7-4 shows that HIV plasma viremia (the presence of virus in blood plasma) and antigenemia (an-ti-je-ne-mi-ah—the persistence of antigen in the blood) can be detected as early as 2 weeks after infection. This demonstrates that viremia and antigenemia occur prior to seroconversion. Using HIV proteins produced by recombinant DNA methods (making synthetic copies of the viral proteins), antibodies specific for gp41 (a subunit of glycoprotein 160) are detectable prior to those specific for p24 (a core protein) and persist throughout the course of infection. Levels of antibody specific for p24 rise to detectable levels between 6 and 8 weeks after HIV infection but may disappear abruptly. The drop in p24 antibody has been shown to occur at the same time when there is a rise in p24 antigen in the serum. This strange phenomenon is thought to be due to the loss of available p24 antibody in immune complexes—too little p24 antibody is being made to handle the new virus being produced. It is believed that this imbalance is one of the important factors that moves the patient towards AIDS. Thus a sudden decrease in anti-p24 is considered by many scientists to be a prognostic indicator that people with HIV disease are moving towards AIDS.

It is now believed by some AIDS researchers and health care professionals that 90% to 95% of those persons infected with HIV will eventually develop AIDS. **It has been estimated that approximately 50% of people with HIV disease will progress to AIDS within 8 years after infection.** At 10 years 70% will have developed AIDS. After that, an additional 25% to 45% of the remain-

HIV EXPOSURE: FAILURE TO SEROCONVERT

A recent study has shown that 10% of 260 female prostitutes who work in Nairobi **have not developed antibodies to HIV.** They have remained seronegative for 4 years despite ongoing unprotected sexual exposure to HIV-infected men (Matheson et al., 1996 updated).

During the past few years it has become clear that apparently harmless, and possibly protective, encounters with HIV can occur. Some individuals who have been exposed to the virus and are therefore at high risk for HIV infection remain apparently uninfected; they do not have antibodies to HIV in their blood, and neither HIV nor its nucleic acids can be detected in blood samples. In one study of 97 HIV exposed individuals who were seronegative for HIV, 49% exhibit cell-mediated immunity to HIV (their T cells respond to HIV peptides in vitro), whereas only 2% of 163 individuals not known to be exposed to HIV exhibit responses to these peptides. Such HIV-specific, cell-mediated responses have been seen in gay men with known sexual exposure, injection drug users, health care workers exposed by accidental needlestick, and newborn infants of HIV-positive mothers. HIV-specific lymphoproliferation or cytotoxic T lymphocyte activity, hallmarks of cell-mediated responses, have also been observed by other investigators in some exposed, but apparently uninfected subjects (Salk et al., 1993). Mario Clerici and colleagues (1994) have also reported on HIV-specific T-helper cell activity in six of eight HIV-negative health care workers with exposure to HIV-positive body fluids. High HIV specific T-helper cell activity was detectable 4 to 8 weeks after the exposure and was lost in individuals followed up for 8 to 64 weeks. Exposure to HIV without evidence of subsequent infection appears to result in activation of cellular immunity without activation of antibody production.

Through 1997, separate cases of HIV-positive, seronegative (do not make HIV antibody) gay men have been reported on (*MMWR*, 1996). In each case the men were confirmed HIV-positive, demonstrated OIs consistent with HIV disease, e.g., PCP and low T4 cell counts, but never produced HIV antibody. To date such cases are rare—**but they do happen!** (See **Genetic Resistance to HIV Infection** in Chapter 5.)

der will develop AIDS. It is **most improbable** that 100% of those infected will develop AIDS.

AIDS Survival Time Has Nearly Doubled

A National Institute of Allergy and Infectious Diseases study followed more than 5,000 homosexual and bisexual men with HIV infection or who were at risk of infection from 1984 through 1991. For men diagnosed with AIDS during the years 1984 and 1985, 50% had a survival time of less than 11.6 months. For men diagnosed with AIDS in 1990 or 1991, a large percentage had survived for nearly 2 years. The greatest gain in survival time occurred among those whose AIDS-defining illness was *Pneumocystis carinii* pneumonia (PCP). Survival time following the diagnosis of AIDS among the participants who developed *P. carinii* increased from a median of 12.8 months in 1984 and 1985 to a median of 26.3 months in 1990 and 1991. The increase in survival time is related to effective PCP therapy. The survival estimate of persons classified with AIDS using the 1993 definition will be considerably longer than for those diagnosed with AIDS using the 1987 definition. Using the 1987 definition, the median survival for patients enrolled in the registry was 24 months; using the new definition, 53% were still alive after 57 months. Jeffrey Laurence (1995, 1996) reports that 5% to 6% of clinically defined AIDS patients now live beyond 5 years (long-term survivors). With many thousands of AIDS-diagnosed people now using the drug cocktails which include at least one protease inhibitor, the number of people with AIDS living past 5 years has greatly increased! These findings have important implications for health planners as well as for those involved in providing care and counseling for patients with AIDS.

Classification of HIV/AIDS Progression

There are several classifications that spell out the progression of signs and symptoms from HIV infection to the diagnosis of AIDS. The classifications were developed to provide a

——————— BOX 7.4 ———————

ARE THERE LONG-TERM ADULT NONPROGRESSORS OF HIV DISEASE—YES!—WHY?

Long-term nonprogressors of HIV disease are defined as those persons who are still alive 10 or more years after they tested HIV-positive, with seroconversion documented by history or stored serum samples, absence of symptoms, and normal and stable T4 cell counts (at least 600 T4 cells/microliter of blood). The long-term nonprogressors who are of particular interest to AIDS investigators, are those who after 8 years, **in the absence of antiviral therapy,** continue to maintain T4 cell counts of 500 or more. Studies through 1999 suggest that 12% to 15% of HIV-infected people remain asymptomatic with about normal T4 cell counts for at least 8 to 12 years (Conant, 1995; Levy, 1995; updated). At least one person is known to be asymptomatic now, 23 years after HIV infection! (Shernoff, 1997; updated). **The question is: How does their immune system differ from those who do not live as long?**

In 1983, at age 71, a man became HIV-positive. He received a contaminated blood transfusion while undergoing colon surgery. Unlike most long-term HIV survivors, he has suffered no symptoms and no loss of immune function. He celebrated his 81st birthday. He is one of five patients who came to the attention of an AIDS researcher as he was preparing a routine update on transfusion-related HIV infections. He has since died of natural causes. All five people were infected by the same donor. And 8 to 12 years later, none of the four has suffered any effects.

The blood donor was a gay male who had contracted the virus during the late 1970s or early 1980s and gave blood at least 26 times before learning he was infected. After locating the donor it was found that the man was just as healthy as the people who got his blood.

A 39-year-old San Francisco artist has beaten the odds by living with the virus that causes AIDS for 16 years. He has only routine medical complaints: the stuffiness of an occasional head cold or the aches and pains of a flu. He has never taken an anti-HIV drug. His own immune system seems to have held the virus at bay.

Susan Buchbinder and colleagues (1992, 1994) reviewed 588 HIV-infected gay men. Thirty-one percent were still AIDS-free 14 years after infection. They attempted to determine why these men lived while others died of HIV/AIDS. Some long-term survivors have low T4 cell counts, some have never taken antiviral therapy, and some have high T4 cell counts. The question is, what is keeping them healthy? If it

can be determined why or how their bodies have delayed the progression of HIV disease, then perhaps new approaches to treating all HIV-infected persons will follow. Understanding their defense may help in preventing HIV infection per se.

Buchbinder and colleagues have found that three aspects of the healthy survivors' immune system appear to delay HIV disease progression: Survivors have strong cytotoxic lymphocyte activity, strong suppressor or T8 cell activity, and have higher levels of antibodies against certain HIV proteins.

It is also possible that long-term nonprogressors carry a less pathogenic virus or that these men have not been reexposed to the virus through unprotected sexual activities. Nicholas Deacon and colleagues (1995 updated) have sequenced HIV DNA from a blood donor and a group of six recipients who have not shown HIV disease symptoms despite being infected for 13 to 17 years. Deletions were found in the *nef* gene. Because the lack of disease progression appears to depend on the virus instead of the host immune system, these results suggest a possible use of such HIV strains in live vaccines. However, in July of 1998, it was reported that the HIV donor and two of the 6 blood recipients experienced a drop in their T4 cell counts. This gives support to those who worry about using an attenuated (weakened) HIV vaccine—that over time, weakened HIV may still cause a problem! Physicians who treat HIV/AIDS persons do not notice any trends that would lead one to recognize the type of patient or factors that would lead to long-term survival. It was concluded that, at the moment, there is a lack of advice for longer life for persons who have become HIV-infected. No one can tell them how to live longer. AIDS investigators at the National Institute of Allergy and Infectious Diseases are studying the immune systems of 14 people who have been HIV positive for 12 or more years.

Current evidence suggests that between 5% and 10% of HIV-infected people will live up to 20 or more years. The second longest documented case of a long-term nonprogressor to date is that of a gay male who tested HIV-positive 20 years ago. His T4 count remains at 1,000 or normal. It should be mentioned that if the average time from infection to AIDS is 11 years, statistically speaking, survivors at 20+ years, are expected to occur. Time will tell! For other accounts of long-term survivors, see the articles by

Cayo (1995), Pantaleo (1995), Kirchhoff (1995), and Baltimore (1995).

VIRAL LOAD AND SURVIVAL

John Mellors studied how viral load (amount or numbers of copies of HIV RNA per milliliter of blood), regardless of treatment, affected a person's survival or nonprogression to AIDS. He found that the baseline viral load accurately predicted survival, while the person's T4 cell count did not. Mellors and his colleagues divided HIV-infected men into two groups, one including those with more than 10,190 HIV RNA copies per milliliter of blood and the other including men with fewer copies. After 10 years, he found that 70% of the people in the low viral-load group had survived compared to only 30% survival of the group with an initial high load—even though the two groups had nearly identical baseline T4 counts (Cohen, 1996). These data should transform the way physicians make decisions on when to treat and what drug(s) to use. The data strongly suggest that viral loads be determined for each HIV-infected person. In addition to viral load now being used to predict survival, the work of Laurence Meyer et al. (1997) and Michael Smith et al. (1997) shows that mutant chemokine receptors CCKR-2 and CCKR-5 can explain some cases of long-term nonprogressive HIV disease.(Refer to Chapter 4 for further explanation of viral load and chemokine receptors.)

IMMUNE CELL RECEPTORS, MOLECULES THAT BIND TO THESE RECEPTORS (LIGANDS) AND T4 AND CTL CELL INFLUENCE ON SURVIVAL

Chemokine Receptors: An Association of Coreceptors To Progression To AIDS As discussed in Chapter 4, the **chemokine receptors** (the four most familiar are CCKR5 (R5), CXCKR4 (X4), CCKR3 (R3), and CCKR2 (R2)) reside on immune system cell membranes. The receptors act as host sites for a variety of chemokines that need entry into such cells. But scientists recently learned that these chemokine receptors also act as **coreceptors** to the CD4 receptor, the receptor to which HIV first attaches. That is, HIV anchors to the cells CD4 receptor, but also needs to bind with a coreceptor, one of the chemokine receptors, in order to complete its entry into the cell. It has also been learned that **genetic mutations** that inhibit given chemokine receptor formation and/or **reduce** the numbers of such receptors on the immune cell membrane, offer such cells a complete or partial

resistance to HIV infection. A double or homozygous mutation at R5 makes a person completely resistant to HIV infection, but the R2 mutation only slows the progression of HIV disease to AIDS. Such mutations **contribute** to some people's long-term survival. Both CCKR5 and CCKR2 receptor mutations offer this protection (Balter, 1998; Collman, 1997; Cocchi et al., 1995; Feng et al., 1996). Exciting research published in 1998 reports that the genetic mutation that interferes with the X4 receptor actually interferes with the **production** of the **specific chemokine** that attaches to the R4 receptor. It appears that this mutation causes the **over production** of a chemokine called **S**tromal **D**erived **F**actor-1 (SDF-1). With an excess in the environment, SDF-1 fills available R-4 receptor sites blocking HIV attachment. So, those persons carrying an SDF-1 mutation, although HIV infected, progress through HIV disease at a much slower pace, delaying the onset of AIDS as much as 7 to 10 years later than average (Winkler, et al., 1998; Balter, 1998). Thus it appears that the progression rate to AIDS is linked to the types of coreceptors people carry on their T4 cells and macrophage.

CHILDREN

Micheline Misrashi and colleagues (1998) reported that children carrying one copy (heterozygous) of the **mutant** R5 gene demonstrate a substantial reduction in progression of HIV disease to AIDS.

T4 AND CYTOTOXIC LYMPHOCYTES (CTL)

One notable immunologic problem that occurs after HIV infection is that most people show no evidence of altered T4 cell activity. But an abnormality in immune function is present, long before a significant decline in T4 cell number is seen. In order to be able to kill off HIV-infected cells, cytotoxic T lymphocytes (CTLs) require the assistance of T4 cells—the very T4 cells that HIV infects. These T4 cells must also be able to recognize HIV antigens in order to direct the cytokine signals that will activate CTLs to kill. During the initial (acute) stage of HIV infection, even before seroconversion when antibodies are generated, the T4 cell population is devastated by HIV. Without sufficient T4 cells capable of recognizing HIV, the killer CTLs cannot in turn learn to recognize HIV antigens and kill off infected cells. Recently, it has

———— **BOX 7.4** (continued) ————

been shown that some long-term nonprogressors with low viral loads have high amounts of HIV-specific T4 cell activity, while some patients with rapidly progressive disease have no detectable HIV-specific T4 cell activity. This suggests again that T4 cell responses play an extremely important role in containing infection.

Studies in long-term nonprogressors show that they have a persistent, vigorous, virus-inhibiting CTL response, and that this response is broad and adaptable. Rapid progressors, on the other hand, appear to have a narrowly directed CTL response that is unable to adapt to changes in the virus (Hay, 1998). Thus the challenge to HIV drug therapists is to either save the T4 cells from becoming HIV infected or find a way to reconstruct the immune system for new T4 cell production.

LONG-TERM AIDS SURVIVORS

Parade Magazine, January 31, 1993, carried a review of 16 long-term AIDS survivors who date back to 1982. On April 16, 1995, the same magazine reported that 12 of the 16 had since died. Of the four survivors, all have refused zidovudine (ZDU) therapy (Figure 7-5).

On April 6, 1997, *Parade Magazine* continued its follow-up on the four survivors seen in Figure 7-5. All four remain alive and productive. They are Cristofer Shihar, 46, a Hollywood hairstylist; Michael Leonard Marshall, 51, a floral designer in Culver City; George Melton, 44, an author in San Francisco; and Niro Asistent, 51, a therapist now living in France. Of the four, only Marshall said he

was taking an AIDS cocktail: Viramune, Crixivan, and 3TC. Niro Asistent remarried in 1996. She said HIV **cannot** be detected in her body!

FIGURE 7-5 Long-Term AIDS Survivors. In 1988, *Parade Magazine* began tracking 16 long-term AIDS survivors. The longest living of these survivors was diagnosed with AIDS in 1982. Only one of the 16, a woman, was HIV-infected through a heterosexual relationship. Her sexual partner was bisexual. In June 1990, 13 were alive. By January 1993, six of the original 16 were alive. In April 1995 and April 1997, *Parade Magazine* updated interviews with the four remaining survivors. Their lives continue. (© *Blake Little*)

---

framework for the medical management of patients from the time of infection through the expression of AIDS. All classification systems are fundamentally the same—they group patients according to their stage of infection, based on signs that indicate a failing immune system (Royce et al., 1991).

The Walter Reed Army Medical Center System (WRS) classifies HIV infection through six stages. The stages are based on signs and symptoms associated with immune dysfunction. Although individual parts of the immune system appear to function independently of one another, all parts appear to depend on the function of T4 cells (see Chapter 5).

A second classification, the most widely accepted because of its greater clinical applica-

bility, comes from the CDC (Table 7-2). **The CDC classification uses four mutually exclusive groupings** (Figure 7-6). The groupings are based on the **presence** or **absence** of signs and symptoms of disease, and clinical and/or laboratory findings and the chronology of their occurrence. Group 1, acute infection, means the person is **viremic** (i.e., many virus particles are present in his or her blood or serum). There are **no measurable antibodies;** one is HIV-positive but lacks HIV antibodies and signs and symptoms of HIV disease.

The majority of people in Group 1 remain asymptomatic. Some may experience flu- or mononucleosis-like symptoms that generally disappear in a few weeks. In relatively few cases the patient moves rapidly from mild symptoms

SEROREVERSION: CHANGING FROM HIV-POSITIVE TO HIV-NEGATIVE?

INFANTS AND ADULTS WHO CLEAR HIV FROM THEIR BODIES

Infants: Update 1998

In previous yearly issues of AIDS UPDATE, studies on what appeared to be HIV clearance from HIV-infected infants were presented by Yvonne Bryson and colleagues (1995), by Pierre Roques and colleagues (1995) and by the European Collaborative Study (Newell, et al., 1996). However, recent investigations dispute their findings. In all, 42 cases of suspected transient HIV infection among 1562 perinatal (before, during or after the birthing process) infants and one mother reported in the earlier studies were re-analyzed.

HIV sequences were not found in specimens from 20; in specimens from 6, somatic genetic analysis revealed that specimens were mistakenly attributed to an infant; and in specimens from 17, phylogenetic analysis failed to demonstrate the expected linkage between the infant's and the mother's virus. These findings argue that transient HIV-1 infection, **if it exists,** will only rarely be satisfactorily documented (Frenkel and 28 co-authors, 1998).

Adults

David Schwartz of the Johns Hopkins School of Public Health reported in January 1995, a 43-year-old woman who became HIV-infected from a blood transfusion after the birth of her second child in 1981. Despite over 40 attempts, investigators have not been able to culture HIV from her blood. Although she is a nonprogressor, the donor died of AIDS and two other people who received the same virus through blood transfusions developed AIDS. Prior to learning that she received HIV-positive blood she had two children. Each was breast-fed for a year. In 1985 she read about blood transfusion HIV cases, was tested and turned up HIV-positive. At this time the mother, the children, and the husband are HIV-negative.

Schwartz's team is studying four other cases that are similar with the hope of learning what might be unusual about the virus in those people or in their immune system.

Researchers have been studying a blood donor and eight transfusion recipients in Australia who were infected prior to 1985 with an attenuated strain of HIV. Now researchers for the Sydney Blood Bank Cohort Research Group report immunologic and virologic data for seven members of that group, some 14 to 18 years after infection. The analysis describes the five surviving transfusion recipients, who remain asymptomatic and take no antiretroviral drugs; the donor, who started therapy in February; and one recipient who died in 1995 after 12 years of infection. The researchers note that three of the recipients have undetectable HIV RNA plasma concentrations. The average HIV RNA plasma concentrations of the donor and two other recipients is between 645 and 2850 copies per milliliter, and the T4 lymphocyte counts of these people has dropped by 16 to 73 cells per cubic milliliter each year. The seventh patient had a median plasma concentration of HIV-1 RNA of 1400 copies per milliliter and lost 17 T4 lymphocyte cells per cubic milliliter per year. The researchers note the signs of immunologic damage in three of the patients with detectable plasma HIV RNA but add that the T4 lymphocyte levels remain steady in the individuals who have undetectable plasma HIV RNA. (Learmont, et al., 1999) The big questions, how long can this last? And what will this teach the vaccine investigators—that attenuated vaccines are more complex than previously thought?

UPDATE 1997/1999

Through the use of combinations of AIDS drugs now available in the United States, thousands of persons with HIV disease and those with AIDS have essentially become HIV-RNA-negative. That is, **HIV has been reduced to unmeasurable levels in their blood plasma.** The AIDS drug cocktails have now suppressed HIV replication in some persons for about 4 years (1995–beginning 2000).

---

into severe opportunistic infections and is diagnosed with AIDS.

In Group 2, antibodies are present but most patients remain free of HIV disease symptoms.

Regardless of the lack of outward clinical symptoms, 90% of those who are asymptomatic experience some form of immunological deterioration within 5 years (Fauci, 1988).

TABLE 7-2 CDC Classification of HIV-Related Diseases

HIV Disease

Group 1: Acute infection—HIV antibodies absent; asymptomatic or if symptomatic mononucleosis-like symptoms which subside in most cases

Group 2: Asymptomatic infection—HIV antibodies present; eventually moves on to

Group 3: Persistent generalized lymphadenopathy; eventually moves on to group 4

AIDS
Group 4:

Subgroup A: Constitutional symptoms (previously called ARC, aids-related complex) and <200 T4 cells/µl of blood.

Subgroup B: Neurologic, symptoms (AIDS dementia complex, neuropathy)

Subgroup C
    Category C-1: Secondary infections listed in the CDC surveillance definition for AIDS: *Pneumocystis carinii* pneumonia, recurrent pneumonia, chronic cryptosporidiosis, CNS toxoplasmosis, extraintestinal strongyloidosis, isosporiasis, esophageal candidiasis, cryptococcosis, histoplasmosis, *Mycobacterium avium intracellulare* or *Mycobacterium kansasii*, cytomegalovirus, chronic or disseminated herpes simplex, progressive multifocal leukoencephalopathy
    Category C-2: Other specified secondary infections: oral hairy leukoplakia, multidermatomal herpes zoster, recurrent *Salmonella* bacterium, nocardiosis, pulmonary tuberculosis, oral candidiasis
    Subgroup D: Secondary cancers: Kaposi's sarcoma, non-Hodgkin's lymphoma (small, noncleaved lymphoma; immunoblastic sarcoma), primary brain lymphoma
    Subgroup E: Other conditions: lymphocytic interstitial pneumonitis, neoplasms, infections not previously listed

---

In Group 3, asymptomatic people from Groups 1 and 2 become symptomatic and demonstrate lymphadenopathy in the neck, armpit, and groin areas. Although a number of other diseases may cause the lymph nodes to swell, most swelling declines as the other symptoms of illness fade. However, with HIV infection, the lymph nodes remain swollen for months, with no other signs of a related infectious disease. Consequently, lymphadenopathy is sometimes called **persistent generalized lymphadenopathy** (PGL). People with PGL may experience night sweats, weight loss, fever, on and off diarrhea, fatigue, and the onset of oral candidiasis or thrush (a fungus or yeast infection of the oral cavity) (see Figure 6-3). Such signs and symptoms are prodromal (symptoms leading to) for AIDS. Studies have shown that people in group 3 appear to become more infectious as the disease progresses.

People in Group 4 have been diagnosed with AIDS. They fit the 1987 CDC criteria for AIDS diagnosis (The CDC AIDS diagnostic criteria are listed in Tables 7-2 and 1-1). Hairy leukoplakia (Figure 7-7, Table 7-2, Category C-2) is virtually diagnostic of AIDS in group 4 patients. Statistics show that about 30% of all the newly HIV-infected will progress to group 4 (AIDS) every 5 years, so that about 90% will have been diagnosed with AIDS within 15 years. Not all of the opportunistic diseases (OIs) or cancers will appear in any one AIDS patient. But some OIs, like *Pneumocystis carinii* pneumonia occur in some 80% of AIDS patients prior to their deaths.

PROGNOSTIC BIOLOGICAL MARKERS RELATED TO AIDS PROGRESSION

The **ideal marker** would be able to predict HIV disease progression, be responsive to antiretroviral therapy, and explain the variance in clinical outcome due to therapy. It is, however, unlikely that any one marker will be able to fulfill all these criteria in HIV infection completely. Therefore, individual markers used to track HIV infection to AIDS are presented.

p24 Antigen Levels

p24 is a specific protein located in the core or inner layer of HIV. Because the immune system produces antibody against foreign protein, antibody is made against p24. A positive test for p24 antigen in the blood means that HIV production is so rapid that it overcomes the available antibody (Figure 7-4). That is, there is more HIV antigen than antibody to neutralize it. During this state of HIV-antibody imbalance, HIV is believed to spread into uninfected cells. This raised p24 antigen level condition occurs at least twice; once shortly after infection and again during the AIDS period when the immune system is rapidly

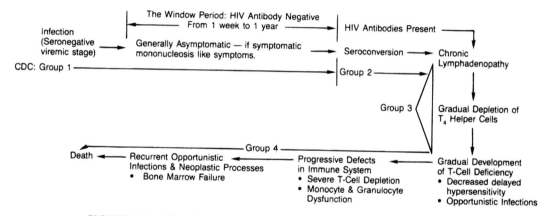

FIGURE 7-6 Clinical History of HIV Infection According to Center for Disease Control and Prevention Groupings. Seroconversion means that HIV antibodies are measurably present in the person's serum. With continued depletion of T4 cells, various signs and symptoms appear announcing the progression of HIV disease into AIDS. Although HIV antibodies have been found as early as 1 week after exposure, most often seroconversion occurs between weeks 6 and 18 but may not occur for up to 1 year or more.

deteriorating and unable to produce sufficient antibody to deal with newly produced HIV.

Those who, during the early stage of HIV infection, test p24 antigen positive are likely to progress to AIDS earlier than those who test p24 negative. Thus, a positive p24 test is an early and serious warning sign for HIV-infected people (Escaich et al., 1991; Phillips et al., 1991a).

p24 Antibody Levels

High levels of p24 antibody indicate that the immune system is functioning well and clearing

FIGURE 7-7 Oral Hairy Leukoplakia (lu-ko-pla-ki-ah) of the Tongue. An early manifestation, it is virtually diagnostic of AIDS. The white corrugated or "hairy" patches are caused by the Epstein-Barr virus (EBV).These white plaques cannot be removed. **A,** A milder form of the disease. **B,** A more severe manifestation of the disease. Note that the white plaques may cover the entire tongue. (*Schiodt, Greenspan, and Greenspan 1989.* American Review of Respiratory Disease, *1989, 10:91–109. Official Journal of the American Thoracic Society.* © American Lung Association)

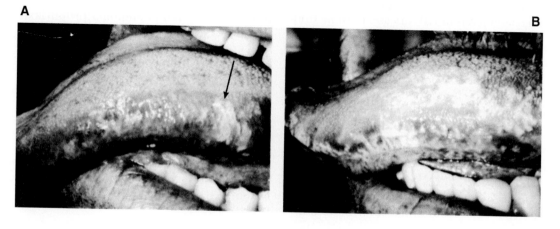

the body of free HIVs. High antibody levels appear to slow the progression toward AIDS. Typically, p24 antibody levels are high during a person's asymptomatic or latent stage (Figure 7-4). However, antibody levels begin to decrease over time. As measurable antibody falls, p24 antigen levels rise, indicating a loss of immune function. When both p24 antigen and antibody levels are high, the body may be expressing an autoimmune disorder—the immune system is attacking its own body cells. Why this occurs is not understood.

Other Markers

Beta-2 microglobulin (B-2M) is a low molecular weight protein that is present on the surface of almost all nucleated cells. As cells die, this compound is released into body fluids. Thus there is always some B-2M in the blood because of normal cell degeneration and replacement. However, in a chronic illness with increased cell destruction, as in HIV infection, B-2M increases beyond normal levels.

To date, a B-2M level of 5 mg/L or higher is the best available indicator of progression to AIDS within 3 years. Levels below 2.6 are considered normal. Similar to p24 antigen, B-2M protein increases dramatically shortly after infection occurs, then declines, and finally rises again with AIDS. B-2M can be used with T4 counts to foretell which HIV-positive individuals face the greatest immediate risk of progressing to AIDS.

Neopterin is a metabolite of guanosine triphosphate which is produced in stimulated monocytes and macrophages and is found in increased levels in HIV disease. In some studies, the presence of these two compounds have been better HIV disease progression indicators than T4 cell counts (Cohen, 1992).

Interferon is a protein that transfers chemical signals between subtypes of lymphocytes (white blood cells). Levels are generally increased in proportion to the stage of HIV disease.

T4 and T8 Lymphocyte Levels

The most extensive use of data for AIDS progression risk identification involve the number of T4 and T8 cells circulating in the blood (Anderson et al., 1991; Burcham et al., 1991; Phillips et al., 1991b).

T4 cells and the ratio of T4 cells to T8 cells found in the blood have, beginning 1997, been widely used as prognostic indicators for AIDS progression. T4 cell counts, however, are not ironclad predictors of HIV disease progression. In some cases persons with HIV disease and very low (less than 50 or 100) T4 counts remain healthy; conversely some HIV-diseased persons have relatively high counts (over 400) and are quite ill. **T4 counts are notoriously fickle,** they can vary widely between labs or because of a person's age, the time of day a measurement is taken, and even whether the person smokes (Sax et al., 1995).

CDC Insists That Performing T4 Cell Counts in HIV-Infected People Continues to Be Essential

Accurate and reliable measures of CD4+ T lymphocytes (CD4+ T cells or T4 cells) are essential to the assessment of the immune system of HIV-infected persons. The progression to AIDS is largely attributable to the decrease in T4 lymphocytes. Progressive depletion of T4 lymphocytes is associated with an increased likelihood of clinical complications. Consequently, the Public Health Service (PHS) has recommended that T4 lymphocyte levels be monitored every 3 to 6 months in all HIV-infected persons. The measurement of T4 cell levels has been used to establish decision points for initiating *Pneumocystis carinii* pneumonia prophylaxis and antiviral therapy and to monitor the efficacy of treatment. T4 lymphocyte levels are also used as prognostic indicators in people who have HIV disease and recently have been included as one of the criteria for initiating prophylaxis for several opportunistic infections that indicate an HIV infection. Moreover, T4 lymphocyte levels are a **criterion** for categorizing HIV-related clinical conditions by CDC's classification system for HIV infection and surveillance case definition for AIDS among adults and adolescents (*MMWR*, 1997).

VIRAL LOAD AND T4 CELL COUNTS: THEIR USE IN CLINICAL PRACTICE

VIRAL LOAD

Viral load assays measure only viral RNA present in the plasma. They do not account for **intact virus** in the lymph system or other tissues. Individual measurements are a good surrogate (substitute) for virus replication. Although the relationships among viral replication, T4 cell count, and disease progression are not entirely clear, viral load correlates fairly well with all of them. Viral load measurements can be effectively used to monitor therapy or the effectiveness of a given drug.

HIT EARLY, HIT HARD

The level of plasma HIV RNA at 6 months after primary infection is called the **SET POINT**, but treatment should be considered regardless of **set point** RNA values. The idea is to stop HIV replication **early rather than later.**

T4 CELL COUNTS

T4 cell counts remain helpful in determining the stage of the disease. But, T4 counts are highly variable, even from hour to hour in some HIV-infected people. They are also significantly influenced by other current viral infections. Although T4 cell counts function well as reliable markers of disease progression and antiretroviral drug efficacy in large populations, they are less reliable for prescribing therapy for the individual. Viral load determinations present a better basis on which to present therapy. Until viral load determination costs fall and until physicians become more familiar with them and insurance companies pay for them, the T4 tests will remain the more practical determinant of disease stage and when to begin therapy. (Medicaid, Medicare, and some HMOs do cover HIV RNA determinations.)

It is suggested that HIV RNA levels be used as a means to prescribe medication. RNA levels should be determined, after exposure, at 6 weeks, 12 weeks, and at 6 months—the **set point.**

SUMMARY

The level of circulating RNA in plasma is a direct reflection of ongoing HIV replication in the host. The T4 lymphocyte, being a **target** of HIV replication, is an **indirect** marker of antiretroviral therapy. The T4 count itself is a measurement of the relative production and destruction of the T4 cell population. For example, if there is a production problem, either due to nutritional, toxicity, or other factors, the T4 count may not rise proportionately to the **decrease** in viral load. Thus, the change in viral load is more precisely measuring the effect of antiretroviral therapy and, therefore, is the best marker of how well drugs are working.

Levels of HIV RNA in the Blood: VIRAL LOAD

David Baltimore, (Figure 7-8) the Nobel Prize-winning retrovirologist, and co-workers have found a useful clinical predictor of HIV disease progressors, **levels of HIV RNA in the blood. More RNA means more HIV,** and that makes patients get sicker sooner. HIV RNA is a more sensitive measure than other assays and may detect the virus earlier than it would be seen otherwise.

Since the reported work of Baltimore and others on the levels of HIV RNA in the blood, Dennis Henrard and co-workers (1995) have concluded that the stability of HIV RNA levels suggests that an equilibrium between HIV replication rate and efficacy of immunologic response is established shortly after infection and persists throughout the asymptomatic period of the disease (Figure 7-3). Thus, a defect in immunologic control of HIV infection may be as important as the viral replication rate for determining AIDS-free survival. Because individual steady-state levels of HIV RNA were established soon after infection, HIV RNA levels can, as Baltimore suggests, be useful markers for predicting clinical outcome. In that regard, when researchers analyzed HIV RNA levels in HIV-positive men they found that HIV RNA—but not T4 counts—

HIV INFECTION OF THE CENTRAL NERVOUS SYSTEM

A wide variety of central nervous system (CNS) abnormalities occur during the course of HIV infection. They result not only from the opportunistic infections and malignancies in the immunodeficient individual, but also from direct HIV infection of the CNS. It was believed that because some brain cells contain CD4-like receptors they were receptive to HIV infection. In addition, other brain cells contained a glycolipid that allowed for HIV infection (Ranki, et al., 1995). However, more recent research shows that after HIV-infected monocytes migrate to the brain, become tissue macrophage and release HIV, the virus **does not infect brain cells; rather HIV resides in brain spaces, in the cerebral spinal fluid.** HIV investigators believe that HIV may invade the brain within a few weeks to months after HIV infection. Although the precise mechanisms by which HIV produces nerve cell dysfunction are still undetermined, infection of the brain is an integral component of the biology and natural history of HIV infection; the resulting clinical manifestations are summarized in Table 7-3.

Genetic analysis and clinical studies have revealed that HIV in the cerebrospinal fluid **(CSF)** of some people with AIDS-related dementia evolves **independently** of the HIV in their blood, leading **to two genetically distinct forms of the virus.** This finding poses a new challenge for treatment of these patients, suggesting that drugs effective against HIV in their blood may not do the job in the central

FIGURE 7-8 David Baltimore, 1975 Nobel Prize molecular biologist. President of the California Institutes of Technology. (*Photograph courtesy of AP/Wide World Photos*)

tightly correlated with AIDS progression (Holden, 1994; Cohen, 1996) (see Chapter 4 for additional information on RNA-viral load). Data from many HIV RNA plasma load studies through 1998 suggest that viral load is superior to T4 cell count as a marker for HIV disease progression. **Viral load measurements indicate the amount of current HIV activity. T4 cell counts indicate the degree of immunologic destruction.** Thus, the best monitor of disease progression is the use of both, the T4 cell count and viral load (Merigan et al., 1996; Katzenstein et al., 1996; Voelker, 1995; Goldschmidt et al., 1997).

No current viral or immunological markers adequately reflect drug toxicities caused by therapy.

TABLE 7-3 Neurological Manifestations Associated with Direct HIV Infection of the Nervous System

AIDS dementia complex
Asymptomatic infection
Acute encephalitis
Aseptic meningitis
Vacuolar myelopathy
Inflammatory demyelinating polyneuropathy
Radiculopathy
Mononeuropathies
Distal sensory neuropathy

nervous system, and vice versa. Why HIV evolves independently in the CSF of the demented patients is not clear. It may reflect a separate viral subpopulation thriving in their central nervous system, or maybe those with dementia have progressed further in immunodeficiency that may allow HIV to become more virulent. Scientists at Gladstone Institute of Virology and Immunology also found preliminary evidence that in some AIDS dementia patients HIV in the CSF compartment is more resistant to antiretroviral drugs than as HIV in the blood. This can cause rebound of virus levels in the CSF while the viral load in the blood is suppressed. Ongoing studies of resistant mutations should clarify this pattern.

AIDS Dementia Complex

In addition to cancers and opportunistic diseases, a progressive dementia (mental deterioration) due to HIV infection of the central nervous system develops in 55% to 65% of AIDS patients. Pathological changes in the

CNS are observed in up to 80% of those autopsied (McGuire, 1993). But the mystery remains as to why some HIV infected people develop dementia while others do not. Some people with high viral loads were not demented, while others with low viral loads had AIDS dementia. A study by Johnson and colleagues (1996) suggests that factors other than viral load lead to dementia. Investigators have found that this dementia is solely associated with HIV infection and progression to AIDS. HIV-caused dementia has a unique set of clinical and pathological features. Some authorities estimate that 90% of AIDS patients in the terminal stages of the disease have AIDS Dementia Complex (ADC) (Hanley et al., 1988). Information presented by the National Institute of Allergy and Infectious Diseases in June 1989 indicated that **asymptomatic HIV-infected persons do not demonstrate mental impairment** (Figure 7-9). Therefore, the onset of mental impairment must begin sometime after the HIV-infected person becomes symptomatic (Update, 1989). Initially, investigators thought that OIs caused ADC,

FIGURE 7-9 Central Nervous System Events after HIV Infection. Note that aseptic meningitis (an inflammation of the membrane of the brain and spinal cord in the absence of viral or bacterial infection), when it occurs, occurs early after HIV infection. Although usually apparent later, AIDS dementia complex may begin during the early–late phase. The late phase represents the period during which major AIDS-defining opportunistic infections occur. The headings acute, latent, early-late and late refer to periods after HIV infection. It appears that OI infection of the brain occurs after the onset of ADC. (*Adapted from Price et al., 1988*)

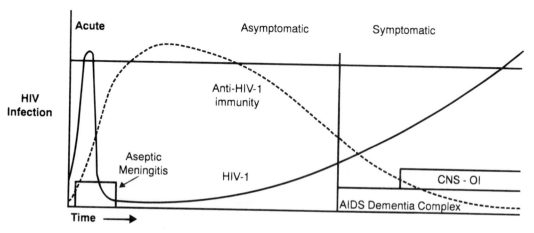

———— BOX 7.5 ————

LONG-TERM SURVIVORS AMONG HIV-INFECTED CHILDREN

Mark Kline (1995) reports that now, in the 1990s, many if not most HIV-infected children demonstrate relatively slow HIV disease progression. European studies suggest that 70% to 75% of children who truly were HIV-positive at birth survive to at least 5 years of age; 60% survive to age 10 or beyond (Italian Register for HIV Infection in Children, 1994; The European Collaborative Study, 1994). Investigators from the CDC Pediatric Spectrum of Disease Project recently reported an estimated mean survival time of 9.4 years for U.S. children born HIV-positive (Barnhart et al., 1995). The CDC study suggests that the average child is moderately or severely symptomatic for most of his or her expected life span. In a recently published study from New Jersey, 24 (57%) of 42 children over the age of 9 had a history of an AIDS-defining condition, and 33 (79%) children had moderate or severe immunologic compromise as reflected by T4 counts of 500/μL of blood or less (Grubman et al., 1995). Only 2 of 41 children over 5 years of age are asymptomatic and have no evidence of immunologic compromise. It is apparent that long-term survival with true nonprogression of disease is unusual among children born with HIV infection.

According to Susan Davis and co-workers (1995), from 1978 through 1993, approximately 14,920 children were born with vertically acquired HIV infection in the United States. Among these children, approximately 5,330 had developed AIDS and 2,680 had died. Thus, an estimated 12,240 children with vertically acquired HIV infection were alive at the beginning of 1994. More than half (56%) of these children had been born since 1988 and thus were younger than 5 years.

---

but it was later discovered that HIV is carried into the brain by HIV-infected macrophages, the Trojan Horse.

Early Symptoms— Early symptoms of cognitive (quality of knowing or reasoning) dysfunction include forgetfulness, recent memory loss, loss of concentration and slowness of thought, social withdrawal, slurring of speech, loss of balance (inability to walk a straight line), deterioration of handwriting, and impaired motor function. An early diagnosis of ADC in AIDS patients is difficult because the early signs of neurological disease are very similar to those symptoms used to identify emotional depression. However, neurological symptoms may be the first sign of illness in about 10% of adult AIDS patients.

Late Symptoms— These symptoms are characterized by loss of speech, great fatigue, muscle weakness, bladder and bowel incontinence (loss of control), headache, seizures, coma, and finally death. About 95% of patients with ADC have HIV antibodies in their cerebral spinal fluid.

PEDIATRIC CLINICAL SIGNS AND SYMPTOMS

Currently over 90% of pediatric AIDS cases are newborns and infants who received HIV from mothers who were injection drug users or the sexual partners of IDUs. For those who become HIV-infected during gestation, clinical symptoms usually develop within 6 months after birth. Few children infected as fetuses live beyond 2 years and survival past 3 years used to be rare; but with the use of anti HIV drugs and therapy for opportunistic diseases, some children born with HIV are still alive at 5, 10, and 15 years old.

The clinical course of rapid HIV disease progression in infants diagnosed with AIDS is marked by failure to thrive, persistent lymphadenopathy, chronic or recurrent oral candidiasis, persistent diarrhea, enlarged liver and spleen (hepatosplenomegaly), and chronic pneumonia (interstitial pneumonitis). Bacterial infections are common and can be life-threatening. Bacterial infection and **septicemia (the presence of a variety of bacterial species in the bloodstream)** was the lead-

HIGHLY ACTIVE ANTIRETROVIRAL THERAPY (HAART) LOWERS THE INCIDENCE OF HIV-ASSOCIATED NEUROLOGIC DISEASE

Justin McArthur reported in early 1999 on a study from the Multicenter AIDS Cohort Study (MACS) that demonstrated at least a 50% reduction in the incidence of HIV dementia, toxoplasmosis, cryptococcal meningitis, progressive multifocal leukoencephalopathy (PML), and preliminary CNS lymphoma since the introduction of HAART in 1996. This study parallels data from other investigators demonstrating dramatic declines in the incidence rates of various HIV-related illnesses and improvements in survival. While these data confirm encouraging trends in the morbidity and mortality of HIV infection, there still remains the potential that the enlarging pool of people living with AIDS may represent a group which is neurologically vulnerable for neurologic disease, with the CNS serving as a sanctuary for partially suppressed HIV replication.

tem, persistent swollen lymph nodes, and opportunistic infections. Mysteries still to be resolved are exactly why and how HIV attacks the body cells, and why some people stabilize after the initial symptoms of HIV infection while others move directly on to AIDS.

One disorder that was not immediately recognized in AIDS patients is AIDS dementia complex (ADC), a progressive mental deterioration due to HIV infection of the central nervous system. ADC develops in over 50% of adult AIDS patients prior to death. Research has shown that some of the symptoms of this dementia can be reversed with the use of the drug zidovudine.

REVIEW QUESTIONS

(*Answers to the Review Questions are on page 487.*)

1. Name the two major AIDS classification systems in use in the United States.

2. What percent of HIV-infected individuals will progress to AIDS in 5 years; in 15 years?

3. What is the neurological set of behavioral changes in AIDS patients called?

4. Name three body organs and their associated AIDS related diseases.

5. True or False: Currently the single most important laboratory parameter that is followed to monitor the progress of HIV-infection is the T4 cell count.

6. True or False: The average time from infection to seroconversion is 2 weeks. Explain.

7. True or False: Being infected with HIV and being diagnosed with AIDS are the same thing. Explain.

8. True or False: The average length of time from infection with HIV to an AIDS diagnosis is approximately 2 years.

9. Write an brief essay on: (a) In general, about how long HIV can reside in the body before one shows signs of HIV infection. (b) In general, about how long it takes the body to generate antibodies to HIV after infection.

10. The general signs and symptoms associated with HIV/AIDS include:
    a. recurrent fever
    b. weight loss for no apparent reason

ing cause of death in one group of affected infants in Florida. An excess of gamma globulin and depressed cell-mediated immunity and T cell function are frequently encountered.

Less than 25% of AIDS children express the kinds of OIs found in adult AIDS patients. Kaposi's sarcoma occurs in about 4% of them. Young AIDS children experience delayed development and poor motor function. Older AIDS children experience speech and perception problems.

SUMMARY

The clinical signs and symptoms of HIV infection and AIDS have been addressed in the previous chapters. However, there are two major classification systems used to diagnose patients as they progress from HIV infection to AIDS. The first is the Walter Reed System. It recognizes six stages of signs and symptoms which a person passes through to AIDS. The CDC uses four groupings that identify the stage of illness from infection to AIDS. Both systems revolve around the recognition of a failing immune sys-

c. white spots in the mouth

d. night sweats

e. all of the above

REFERENCES

ABOULKER, JEAN-PIERRE, et al. (1993). Preliminary analysis of the concorde trial. *Lancet*, 341: 889–890.

ANDERSON, ROBERT E., et al. (1991). CD8 T lymphocytes and progression to AIDS in HIV-infected men: Some observations. *AIDS*, 5: 213–215.

BALTER, MICHAEL. (1998). Chemokine mutation slows progression. *Science*, 279:327.

BALTIMORE, DAVID. (1995). Lessons from people with non progressive HIV infection. *N. Engl. J. Med.*, 332:259–260.

BARNHART, HUIMAN,et al. (1995). *Abstracts of the 2nd National Conference on Human Retroviruses*, Washington, D.C., p. 161, abstr. 575.

BARTLETT, JOHN, et al. (1998). Primary care of HIV infection. *Hospital Practice*, 33:53–55.

BOLOGNESI, DANI P. (1989). Prospects for prevention of and early intervention against HIV. *JAMA*, 261:3007–3013.

BRYSON, YVONNE J., et al. (1995). Clearance of HIV infection in a prenatally infected infant. *N. Engl. J. Med.*, 332:833–838.

BUCHBINDER, SUSAN P., et al. (1992). Healthy long-term positives: Men infected with HIV for more than 10 years with CD4 counts of 500 cells. *Eighth International Conference on AIDS*, Amsterdam, July 1992, abstr. TUCO572.

BUCHBINDER, SUSAN P., et al. (1994). Long-term HIV infection without immunologic progression. *AIDS*, 8:1123–1128.

BURCHAM, JOYCE, et al. (1991). CD4 % is the best predictor of development of AIDS in a cohort of HIV-infected homosexual men. *AIDS*, 5:365–372.

CAO, YUNZHEN, et al. (1995). Virologic and immunologic characterization of long-term survivors of human immunodeficiency virus type 1 infection. *N. Engl. J. Med.*, 332:201–208.

CHUN, TAE-WOOK, et al. (1998). Early establishment of a pool of latently infected, resting CD4+ T cells during primary HIV-1 infection. *PNAS*, 95:8869–8873.

CLERICI, MARIO, et al. (1994). HIV-specific T helper activity in seronegative health care workers exposed to contaminated blood. *JAMA*, 271: 42–46.

COCCHI, FIORNZA, et al. (1995). Identification of RANTES, MIP-1a and MIP-1b as the major HIV-suppressive factors produced by CD8 + T cells. *Science*, 270:1811–1815.

COFFIN, JOHN M. (1995). HIV population dynamics in vivo: Implications for genetic variation, pathogenesis and therapy. *Science*, 267:483–489.

COHEN, JON. (1992). Searching for markers on the AIDS trail. *Science*, 258:387–390.

COHEN, JON. (1993). Keystone's blunt message: It's the virus stupid. *Science*, 260:292–293.

COHEN, JON. (1995). High turnover of HIV in blood revealed by new studies. *Science*, 267:179.

COHEN, JON. (1996). Results of new drugs bring cautious optimism. *Science*, 271: 755–756.

COLLMAN, RONALD. (1997). Effect of CCR2 and CCR5 variants on HIV disease. *JAMA*, 278:2113–2114.

CONANT, MARCUS. (1995). The current face of the AIDS epidemic. *AIDS Newslink*, 6 (Fall):1–9.

COULIS, PAUL A., et al. (1987). Peptide-based immunodiagnosis of retrovirus infections. *Am. Clin. Prod. Rev.*, 6:34–43.

DAVIS, SUSAN, et al. (1995). Prevelance and incidence of vertically acquired HIV infection in the U.S.A. *JAMA*, 274:952–955.

DEACON, NICHOLAS, et al. (1995). Genomic structure of an attenuated quasi species of HIV from a blood transfusion donor and recipients. *Science*, 270:988–991.

ENSOLI, F., et al. (1990). Proviral sequences detection of human immunodeficiency virus in seronegative subjects by polymerase chain reaction. *Mol. Cell Probes*, 4:153–161.

ENSOLI, F., et al. (1991). Plasma viraemia in seronegative HIV-1 infected individuals. *AIDS*, 5:1195–1199.

ESCAICH, SONIA, et al. (1991). Plasma viraemia as a marker of viral replication in HIV-infected individuals. *AIDS*, 5:1189–1194.

FAUCI, ANTHONY S. (1988). The scientific agenda for AIDS. *Issues Sci. Technol.*, 4:33–42.

FENG, YU, et al. (1996). HIV-1 entry cofactor: Functional cDNA cloning of a seven-transmembrane, G protein-coupled receptor. *Science*, 272:872–877.

FREED, ERIC, et al. (1994). HIV infection of nondividing cells. *Nature*, 369:107–108.

FRENKEL, LISA, et al. (1998). Genetic evaluation of suspected cases of transient HIV infection of infants. *Science*, 280:1073–1077.

GOLDSCHMIDT, RONALD, et al. (1997). Treatment of AIDS and HIV-related conditions—1997. *J. Am. Board Fam. Pract.*, 10:144–167.

GRUBMAN, SAMUEL, et al. (1995). Older children and adolescents living with perinatally acquired HIV infection. *Pediatrics*, 95:657–663.

HANLEY, DANIEL F., et al. (1988). When to suspect viral encephalitis. *Patient Care*, 22:77–99.

HAY, CHRISTINE. (1998). Immunologic response to HIV. *AIDS Clinical Care*, 10:1–3.

HAYNES, BARTON, et al. (1996). Toward an understanding of the correlates of protective immunity to HIV infection. *Science*, 271:324–328.

HENRARD, DENIS, et al. (1995). Natural history of HIV cell-free viremia *JAMA*, 274:554–558.

HO, DAVID, et al. (1995). Rapid turnover of plasma virons and CD4 lymphocytes in HIV infection. *Nature*, 373:123–126.

HO, DAVID. (1996). HIV pathogenesis. *Improv. Manage. HIV Dis.*, 4:4–6.

HOLDEN, CONSTANCE. (1994). New tool for predicting AIDS onset. *Science*, 263:606.

HORSBURG, C.R., et al. (1989). Duration of HIV infection before detection of antibody. *Lancet*, ii:637–639.

IMAGAWA, D.T., et al. (1989). Human immunodeficiency virus type I infection in homosexual men who remain seronegative for prolonged periods. *N. Engl. J. Med.*, 320:1458–1462.

Italian Register for HIV Infection in Children. (1994). *Lancet*, 343:191–195.

JAPOUR, ANTHONY. (1995). Antiviral drug resistance: Clinical significance and implications for HIV pathogenesis. *AIDS Clin. Care*, 7:63–67.

JOHNSON, RICHARD, et al. (1996). Quantitation of human immunodeficiency virus in brains of demented and nondemented patients with acquired immunodeficiency syndrome. *Ann. Neurol.* 39:392–395.

KATZENSTEIN, TERESE, et al. (1996). Longitudinal Serum HIV RNA quantification: Correlation to viral phenotype at seroconversion and clinical outcome. *AIDS*, 10:167–173.

KELLY, MAUREEN, et al. (1991). Oral manifestations of human immunodeficiency virus infection. *Cutis*, 47:44–49.

KIRCHHOFF, FRANK, et al. (1995). Brief report: Absence of intact NEF sequences in a long-term survivor with nonprogressive HIV-1 infection. *N. Engl. J. Med.*, 332:228–232.

KLINE, MARK. (1995). Long-term survival in vertically acquired HIV infection. *AIDS Reader*, 5:153.

LAURENCE, JEFFREY. (1995). A primary care condition. *AIDS Reader*, 5:110–111.

LAURENCE, JEFFREY. (1996). Where do we go from here? *AIDS Reader*, 6:3–4; 36.

LEARMONT, JENNIFER, et al. (1999). Immunological and virologic status after 14 to 18 years of infection with an attenuated strain of HIV-1. *N. Engl. J. Med.*, 340:1715–1722.

LEVY, JAY. (1995). HIV and long-term survival. *Int. AIDS Soc. USA*, 3:10–12.

LIPTON, STUART. (1997). Treating AIDS dementia. *Science*, 276:1629–1630.

MATHESON, PAMELA, et al. (1996). Heterosexual behavior during pregnancy and perinatal transmission of HIV. *AIDS*, 10:1249–1256.

MCARTHUR, JUSTIN. (1999). Declining incidence of neurologic complications of HIV disease. *Hopkins HIV Report*, 11:8.

MCGUIRE, DAWN. (1993). Pathogenesis of brain injury in HIV disease. *Clin. Notes*, 7:1–11.

MERIGAN, THOMAS, et al. (1996). The prognostic significance of viral load, codon 215-reverse transcriptase mutation and CD4+ T cells on HIV Disease Progression. *AIDS*, 10:159–165.

MEYER, LAURENCE, et al. (1997). Early protective effect of CCR–5 Δ32 heterozygosity on HIV disease progression: Relationship with viral load. *AIDS*, 11:F73–F78.

MISHRAHI, MICHELINE, et al. (1998). CCR5 chemokine receptor variant in HIV-1 mother to child transmission and disease progression. *JAMA*, 279:277–280.

Morbidity and Mortality Weekly Report. (1996). Persistent lack of detectable HIV-antibody in a person with HIV-infection—Utah, 1995. 45:181–185.

Morbidity and Mortality Weekly Report. (1997). Revised guidelines for performing CD4+ T-cell determinations in persons infected with HIV. 46:1–4.

NATIONAL INSTITUTE OF ALLERGY AND INFECTION DISEASES. (1989). Tests confirm lack of mental impairment in asymptomatic HIV-infected homosexual men. June:1–2.

NEWELL, MARIE, et al. (1996). Detection of virus in vertically exposed HIV-antibody-negative children. *Lancet*, 347:213–215.

NIELSON, CLAUS, et al. (1993). Biological properties of HIV isolates in primary HIV infection: Consequences for the subsequent course of infection. *AIDS*, 7:1035–1040.

NOWAK, M.A., et al. (1990). The evolutionary dynamics of HIV-1 quasispecies and the development of immunodeficiency disease. *AIDS*, 4:1095–1103.

PANTALEO, GUISEPPE, et al. (1993). The immunopathogenesis of HIV infection. *N. Engl. J. Med.*, 328:327–335.

PANTALEO, GUISEPPE, et al. (1995). Studies in subjects with long-term nonprogressive human immunodeficiency virus infection. *N. Engl. J. Med.*, 332:209–216.

PEDERSEN, C., et al. (1989). Clinical course of primary HIV infection: Consequences for subsequent course of infection. *Br. Med. J.*, 299:154–157.

PERELSON, ALAN, et al. (1996). HIV dynamics in vivo: Viron clearance rate, infected cell lifespan and viral generation time. *Science*, 271:1582–1586.

PHILLIPS, ANDREW N., et al. (1991a). p24 Antigenaemia, CD4 lymphocyte counts and the development of AIDS. *AIDS*, 5:1217–1222.

PHILLIPS, ANDREW N., et al. (1991b). Serial CD4 lymphocyte counts and development of AIDS. *Lancet*, 337:389–392.

PHILLIPS, ANDREW, et al. (1994). A sex comparison of rates of new AIDS-defining disease and death in 2554 AIDS cases. *AIDS* 8:831–835.

PRICE, RICHARD W. (1988). The brain in AIDS: Central nervous system HIV infection and AIDS dementia complex. *Science*, 239:586–593.

QUINN, THOMAS. (1997). Acute primary HIV infection. *JAMA*, 278:58–62.

RANKI, A., et al. (1987). Long latency precedes overt seroconversion in sexually transmitted human immunodeficiency virus infection. *Lancet*, ii:589–593.

RANKI, ANNAMARI, et al. (1995). Abundant expression of HIV NEF and Rev proteins in brain astrocytes *in vivo* is associated with dementia. *AIDS*, 9:1001–1008.

RICHMAN, DOUGLAS. (1995). Antiretroviral resistance and HIV dynamics. *Int. AIDS Soc.*, 3:15–16.

ROQUES, PIERRE, et al. (1995). Clearance of HIV infection in 12 perinatally infected children: Clinical, virological and immunological data. *AIDS*, 9:F19–F26.

ROSENBERG, PHILIP, et al. (1994). Declining age at HIV infection in the United States. *N. Engl. J. Med.*, 330:789–790.

ROYCE, RACHEL A., et al. (1991). The natural history of HIV-1 infection: Staging classifications of disease. *AIDS*, 5:355–364.

SALK, JONAS, et al. (1993). A strategy for prophylactic vaccination against HIV. *Science*, 260:1270–1272.

SAX, PAUL, et al. (1995). Potential clinical implications of interlaboratory variability in CD4+ T-lymphocyte counts of patients infected with human immunodeficiency virus. *Clin. Infect. Dis.*, 21:1121–1125.

SCHONNING, KRISTIAN, et al. (1998). Chemokine receptor polymorphism and autologous neutralizing antibody response in long-term HIV infection. *J. Acquired Immune Deficiency Syndrome and Human Retrovirology*, 18:195–202.

SHERNOFF, MICHAEL. (1997). A history of hope: The HIV roller coaster. *Focus*, 12: 5–7.

SMITH, MICHAEL, et al. (1997). Contrasting genetic influence of CCR2 and CCR5 variants on HIV infection and disease progression. *Science*, 277:959–965.

STRAMER, SUSAN L., et al. (1989). Markers of HIV infection prior to IgG antibody seropositivity. *JAMA*, 262:64–69.

THE EUROPEAN COLLABORATIVE STUDY. (1994). *Pediatrics*, 94:815–819.

VANHEMS, PHILLIPPE, et al. (1999). Recognizing primary HIV infection. *Infections in Medicine*, 16:104–108, 110.

VOELKER, REBECCA. (1995). New studies say viral burden tops CD4 as a marker of HIV-disease progression. *JAMA*, 275:421–422.

WAIN-HOBSON, SIMON. (1995). Virological mayhem. *Nature*, 373:102.

WEI, XIPING, et al. (1995). Viral dynamics in HIV type 1 infection. *Nature*, 373:117–122.

WINKLER, CHERYL, et al. (1998). Genetic restriction of AIDS pathogenesis by an SDF-1 chemokine gene variant. *Science*, 279:389–393.

Epidemiology and Transmission of the Human Immunodeficiency Virus

CHAPTER CONCEPTS

♦ First evidence of HIV-1, 1959, Central Africa.

♦ In 1985, HIV-2 was isolated in West Africa.

♦ Transmission of HIV into the United States may have been via Haiti.

♦ **Lifestyle is associated with HIV transmission.**

♦ HIV is not casually transmitted.

♦ HIV is not transmitted to humans by insects.

♦ HIV transmission is being reported from 194 countries and among all age and ethnic groups.

♦ **The three basic mechanisms of HIV transmission are: sexual contact, needles and syringes and mother to child. All involve an exchange of body fluids.**

♦ The body fluids involved in HIV transmission are exchanged during sexual activities, injection drug use, blood transfusions, use of blood products and during pre- and postnatal events.

♦ No **new** routes of HIV transmission have been discovered in the last 19 years.

♦ **First documented case of HIV transmission via deep kissing.**

♦ Highest frequency of HIV transmission in the United States is among homosexual and bisexual males and among injection drug users.

♦ World wide, highest frequency of HIV transmission is among heterosexuals.

♦ Bloodbanks in several countries knowingly allowed the distribution of HIV-contaminated blood.

♦ HIV-infected athletes want to compete.

♦ Death due to HIV infection is placed in perspective.

♦ Interactions between HIV and STDs are discussed.

♦ Prenatal HIV transmission generally occurs after the 12th to 16th week of gestation, most often during childbirth.

♦ Zidovudine decreases perinatal HIV transmission.

♦ National AIDS Resources phone numbers are listed.

When a population becomes infected with a contagious disease, an epidemic results. **Epidemic** is derived from Greek and means "in one place among the people." To understand how an infectious disease can spread or remain established in a population, investigators must consider the relationship between an infectious disease agent and its host population. The study of diseases in populations is an area of medicine known as **epidemiology.**

We learned from earlier epidemics, the danger of complacency. Complacency about HIV infection is especially dangerous because the infection can remain hidden for years. Because many infected people remain symptom-free for years, it is hard to be sure just who is infected with the virus. The more sexual partners you have, the greater your chances of encountering one who is infected and subsequently becoming infected yourself.

With regard to HIV infection, it is your behavior that counts. The transmission of HIV can be prevented. HIV is relatively hard to contract and can be avoided.

The presence of HIV/AIDS is not isolated—the transmissibility of HIV between individuals and across borders and populations is what denies this global pandemic and makes it imperative that nations work together to prevent the continued transmission of HIV. On an individual level, IT IS NOT IMPORTANT HOW YOU GOT HIV, WHAT IS IMPORTANT IS HOW YOU LIVE YOUR LIFE WITH THE VIRUS.

HIV AS A COMMUNICABLE DISEASE

HIV is a communicable disease in a limited sense. A communicable disease is one in which the causative agent passes from one person to another. The modes of transmission for communicable diseases include: direct contact with body fluids, contact with inanimate objects, and contact with vectors, including flies, mosquitoes, or other insects capable of spreading disease. HIV is communicable only in the first of the three modes, that is, through direct

contact with certain body fluids. **The ways in which HIV can be transmitted have been clearly identified. However, some information that conflicts with scientific findings had been widely dispersed. Information in this chapter should correct these misperceptions. The transmission of HIV requires close contact.** HIV is transmitted in human body fluids by **three** major routes: (**1**) sexual intercourse through vaginal, rectal, or penile tissues; (**2**) direct injection with HIV-contaminated drugs, needles, syringes, blood or blood products; and (**3**) from HIV-infected mother to fetus in utero, during childbirth, from mother to infant, or during breast-feeding.

HIV and Sexual Transmission

Epidemiological data suggest that **sexual transmission, in general, is relatively inefficient, in that exposure often does not produce infection.** HIV is transmitted more efficiently through intravenous than through sexual routes. However, worldwide the predominant mode of transmission of HIV is through exposure of mucosal surfaces to infected sexual fluids (semen, cervical/vaginal, rectal) and during birth.

——————— LOOKING BACK 8.1 ———————

PUBLIC OPINION ON HOW THE DISEASE –AIDS–WAS TRANSMITTED JUNE 1983

A Newsweek poll found that nine in ten (90%) Americans over 18 had heard of AIDS, but also found that four in ten either believed it was possible to contract the disease through casual contact (25%); or were unsure whether the infection could be passed this way (16%). During this period, from 1983 to 1985, AIDS stories made regular appearances in newspapers and on television news broadcasts. Over 150 AIDS stories were broadcast by the three major TV network evening news programs in 1985—more than double the combined totals of AIDS stories broadcast in 1983 and 1984 (source: the Vanderbilt Television Archives). These initial years of press coverage of AIDS were marked by stories that often focused on the dramatic aspects of the disease (e.g., its progressive nature), the death toll among higher-risk groups, and its potential to spread to the public at large.

HIV: Other Routes of Transmission?

HIV is not communicable through contact with inanimate objects or through vectors. Thus **people do not "catch" HIV in the same way that they "catch" the cold or a flu virus.** Unlike colds and flu viruses, HIV **is not,** according to the CDC, spread by tears, sweat, coughing or sneezing. The virus **is not** transmitted via an infected person's clothes, phone or toilet seat. HIV **is not** passed on by eating utensils, drinking glasses, or other objects that HIV-infected people have used that are free of blood.

HIV **is not** transmitted through daily contact with infected people, whether at work, home, or school. Insects do not transmit the virus. Kissing is also considered very low risk: **There is only one documented case to prove that HIV is transmitted by kissing.** However, Paul Holmstrom and colleagues (1992) report that salivary HIV antibodies are detected regularly in HIV seropositive subjects. This suggests that these antibodies neutralize HIV producing the negative results of HIV cultivation or antigen detection in the saliva. The route of HIV into saliva is not fully understood. Both salivary glands and salivary leukocytes have been shown to harbor HIV. Gingival fluid, which is a transudate of serum, has been regarded as the main source of salivary HIV antibodies and infectious HIV.

In its 1990 supplemental guidelines for cardiopulmonary resuscitation (CPR) training and rescue, the Emergency Cardiac Care Committee of the American Heart Association (AHA) noted that there is an extremely small theoretical risk of HIV or hepatitis B virus (HBV) transmission via cardiopulmonary resuscitation (CPR). To date **NO** known case of seroconversion for HIV or HBV has occurred in these circumstances.

WE MUST STOP HIV TRANSMISSION NOW!

We are standing at a moment in the AIDS pandemic when we have what may be our last opportunity to stem a major new wave of HIV infection of heterosexuals and in particular minorities in the urban areas, and of gay men and drug users in the smaller cities and suburban and rural areas where infection is currently of less intensity. Stephen Joseph, Commissioner of Health of New York City from 1986 to 1990, states that 10% to 15% of reproductive age men and women in poor minority neighborhoods will be infected. This of course will lead to the birth of a steady stream of HIV-infected infants and the orphaning of a steady stream of older uninfected children as mothers and fathers die of AIDS. To get a sense of the magnitude of the AIDS orphan problem, consider the following: New York City will have 10,000 to 20,000 of these children by the year 2000 (Josephs, 1993). World wide there will be over 11 million orphans ending year 2000.

Education and advocacy for risk reduction remain important tools for preventing HIV infection, but, alone, they will never accomplish the objective of slowing the speed and extent of the virus's spread. This can only be achieved by combining educational and early intervention efforts with specific measures guiding those resources to the people who need them most: those who are already infected and to their sex and drug partners.

Beginning year 2000, there were about 746,000 AIDS cases in the United States. Figure 8-1 breaks this number down according to means of HIV infection that is, sexual behavior, drug use, medical exigencies, and undetermined causes. Figure 8–2 gives a breakdown by transmission category for 50,500 AIDS estimated cases for 1999. An estimated 48,000 AIDS cases will be reported for year 2000. There were 55,297 AIDS cases for 1998 and 60,634 for 1997.

EPIDEMIOLOGY OF HIV INFECTION

The first scientific evidence of human HIV infection came from the detection of HIV antibodies in preserved serum samples collected in **Central Africa in 1959.** The first AIDS cases appeared there in the 1960s. By the mid-1970s HIV was being spread throughout the rest of the world. The earliest places to experience the arrival of HIV were Central Europe and Haiti. Transmission into the United States may have been by tourists who had vacationed in the area of Port-au-Prince, Haiti (Swenson, 1988).

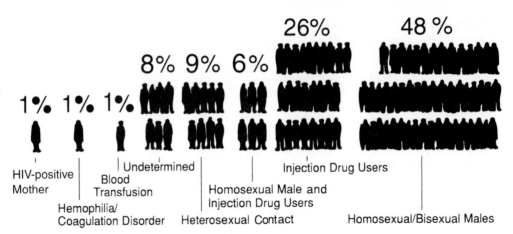

FIGURE 8-1 AIDS Cases by Route of Transmission. Beginning year 2000, there were about 746,000 AIDS cases in the United States. This diagram gives the percentage of adults and adolescents in each group through 1999. Groupings are according to sexual preference, drug use, medical conditions, and others not associated with any of these. (*Courtesy of CDC,* Atlanta— updated)

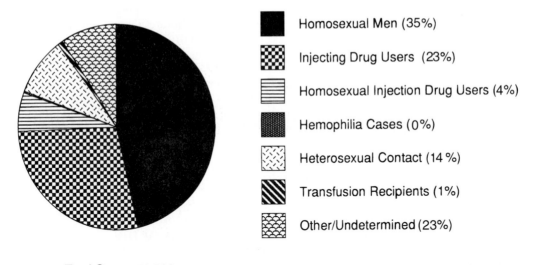

Total Cases 56,500

(Courtesy CDC, Atlanta)

FIGURE 8-2 Adult/Adolescent AIDS Cases by Transmission Category Estimated for 1999, United States. Reported for 1998, 55,298; Pediatric, 382.

On entry into the United States the virus first spread among the homosexual populations of large cities such as New York and San Francisco. The first **recorded** AIDS cases in the United States occurred in 1979 in New York. The first **CDC reported** AIDS cases were in New York, Los Angeles, and San Francisco in 1981 (Figure 8-3). In both cases, the diagnosis of AIDS was based on clinical descriptions.

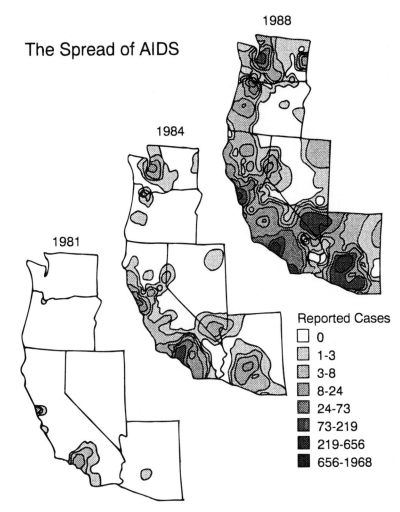

The Spread of AIDS

1988

1984

1981

Reported Cases

☐ 0
▨ 1-3
▨ 3-8
▨ 8-24
▨ 24-73
▨ 73-219
■ 219-656
■ 656-1968

FIGURE 8-3 The Spread of HIV/AIDS on the West Coast. The three separate maps show the rapid spread of AIDS along the Pacific Coast from 1981 through 1988. The maps depict a movement from urban centers into the suburbs and surrounding rural areas. The AIDS epidemic in the West first appeared in Los Angeles and San Francisco, followed by Las Vegas and Phoenix. By 1983, Seattle and Portland encountered the epidemic. The number of AIDS cases continues to increase as HIV infection continues to spread. (*Courtesy of Peter Gould and colleagues, Pennsylvania State University*)

According to the CDCs first clinical AIDS definition, at least one case of AIDS occurred in New York City in 1952 and another in 1959. Both males demonstrated opportunistic infections and *Pneumocystis carinii* pneumonia, today a hallmark of HIV infection. This early evidence of AIDS suggests that the virus might have been in the United States, Europe and Africa at about the same time (Katner et al., 1987). If HIV has been present for decades as suggested, its failure to spread may reflect a recent HIV mutation, a major change in social behaviors conducive to HIV transmission, or both. For example, the sexual revolution and the widespread use of birth control pills, which began in the 1960s, and the subsequent decrease in the use of condoms may be involved with the transmission of HIV.

TRANSMISSION OF TWO STRAINS OF HIV (HIV-1/HIV-2)

The spread of HIV-1 is global. The clinical presentation of AIDS caused by HIV-1 is similar regardless of geographical area.

HIV-2 is a genetically similar but distinct strain. HIV-2 was first discovered in 1985 in West Africa. It is believed to have been present in West Africa as early as the 1960s. Clinical data demonstrate that HIV-2 has a reduced virulence compared to HIV-1 (Marlink et al., 1994).

CONSEQUENCES OF IGNORING HIV

You may be offended by the description of how HIV is transmitted. You may feel invulnerable to HIV infection. You may think that given time, HIV infection and AIDS will go away. You may choose to ignore that it exists. **But ignoring HIV may kill you.** The cost of ignorance about any of the STDs is always high, but with AIDS the cost is a disfiguring, painful death.

What appears certain is that HIV-2, like HIV-1, will spread worldwide. HIV-2 has already spread from West Africa to other parts of Africa, Europe, and the Americas. Both HIV-1 and HIV-2 are transmitted or acquired through the same kinds of exposure.

(BECAUSE OVER 99% OF GLOBAL AIDS CASES ARE CAUSED BY THE TRANSMISSION OF HIV-1, ONLY DATA THAT PERTAIN TO HIV-1 (HIV) WILL BE PRESENTED UNLESS OTHERWISE STATED.)

IS HIV TRANSMITTED
BY INSECTS?

In spite of convincing evidence of the ways in which HIV can be transmitted, it remains difficult for the general public to believe that a virus that appears to spread as rapidly as HIV is not either highly contagious or transmitted by an environmental agent. After all, there are many viral and bacterial diseases that are highly contagious and transmitted by insects. **The question was asked: Is this virus being transmitted by insects?**

Necessary data to resolve the question were available. Epidemiological data from Africa and the United States suggested that AIDS was not transmitted by insect bites. If it were, many more cases would be expected among school-age children and elderly people, groups that are proportionally underrepresented among AIDS patients. In one study of the household contacts of AIDS patients in Kinshasa, Zaire, where insect bites are common, not a single

child over the age of 1 year had been infected with the AIDS virus, while more than 60% of spouses had become infected.

In 1987, the Office of Technology Assessment (OTA) published a detailed paper on the question of whether blood-sucking insects such as biting flies, mosquitoes, and bedbugs transmit HIV (Miike, 1987). The conclusion was that the conditions necessary for successful transmission of HIV through insect bites and the probability of their occurring rule out the possibility of insect transmission as a significant factor in the spread of AIDS. Jerome Goddard reported (1997) that blood-sucking arthropods (e.g., **mosquitoes** and **bedbugs**) for good biological reasons, **cannot** transmit HIV. The virus has to overcome many obstacles. It must avoid digestion in the gut of the insect, recognize receptors on the external surface of the gut, penetrate the gut, replicate in insect tissue, recognize and penetrate the insect salivary glands, and subsequently escape into the lumen of the salivary duct. Webb and colleagues (1989) inoculated bedbugs intra-abdominally (belly) and mosquitoes intrathoracically (chest) with HIV to enable the virus to bypass gut barriers. The virus failed to replicate in either potential carrier.

EXPOSURE TO HIV
AND SUBSEQUENT INFECTION

Early on, it was believed that certain aspects of one's lifestyle and medical status predetermined the risk of HIV infection upon HIV exposure. For example, if the HIV-exposed individual had some previous sexually transmitted disease or an open sore, used drugs, or had an already weakened immune system, he or she would be more susceptible to HIV infection. Such conditions are called **cofactors.** (See Chapter 5 for a discussion on cofactors.)

TRANSMISSION OF HIV

If most HIV-exposed people can become HIV-infected, can most infected people **transmit** HIV to others? This is a difficult question to answer. Infection with HIV appears to depend on a large number of variables that involve the donor, recipient, and portal of entry. The most important variables are mode of transmission, viral load, which subtype and variant of HIV is present, and the recipient's genetic resistance.

A report in the 1998 12th International AIDS Conference stated that for the first time strains of HIV resistant to all protease inhibitors and other widely used anti-HIV drugs are being transmitted from one person to another! Prior to this announcement, scientists did not believe that such highly mutant forms of HIV could be transmitted. It is becoming increasingly clear that, although almost every country is touched by HIV, the virus spreads very differently in different parts of the world. There are even important differences in patterns of spread in different communities and geographic areas within the same country.

CASUAL TRANSMISSION

There is overwhelming evidence that **HIV is not transmitted casually.** Assurances that HIV is not spread through casual contact are based on observation of health care workers and family members of AIDS patients. These individuals have much closer contact with AIDS patients and their body fluids than the average person would in a social, educational, or occupational setting. Thus, if HIV could be transmitted through casual contact, it would be found in much larger numbers of family members and health care workers.

HIV TRANSMISSION IN FAMILY/ HOUSEHOLD SETTINGS

Several studies of the family members of AIDS patients have failed to demonstrate the spread of the HIV through household contact. The only cases in which family members have become infected involved the sexual partners of AIDS patients or children born to mothers who were already infected with the virus. Even individuals who bathed, diapered, or slept in the same bed with AIDS patients have not become infected. In one study, 7% of the family members shared toothbrushes

with the infected person and none became infected.

Perhaps the best evidence against casual HIV transmission comes from studies of household members living with blood-transfused AIDS patients (Peterman et al., 1988). Transfusion infection cases are unique because their dates of infection are known retrospectively. Prior to the onset of AIDS symptoms, the families were unaware that they were living with HIV-infected individuals. Family life was not altered in any way, yet family members remained uninfected. In some cases, the transfusion patients were hemophiliacs who received weekly or monthly injections of blood products and became HIV-infected. From the combined studies of these households, only the sexual partners of infected hemophiliacs became infected. In some cases, the sexual partners of hemophiliacs remained HIV free after 3 to 5 years of unprotected sexual intercourse.

Although contact with blood and other body substances can occur in households, transmission of HIV is rare in this setting. Through 1999, at least nine reports have described household transmission of HIV not associated with sexual contact, injection drug use, or breast feeding (Table 8-1). Of these nine reports, five were associated with documented or probable blood contact. In one report, HIV infection was diagnosed in a boy after his younger brother had died as the result of AIDS; however, a specific mechanism of transmission was not determined.

Living in the Same Household

In December 1993, investigators found two cases in which HIV was transmitted from one child and one adolescent to others but not by the usual routes.

One case involved two New Jersey boys, whose ages were given only as between 2 and 5. The other involved two teenage brothers who are hemophiliacs.

In the New Jersey case, the older, infected child had frequent nosebleeds, and the younger child had dermatitis, a condition that can break the surface of the skin.

The brothers with hemophilia **shared** a razor on one occasion. They told investigators that they did not know who used the razor first.

TABLE 8-1 Reported Cases of HIV Infection in Which Transmission Not Associated with Sexual Contact, Injection-Drug Use, or Breast Feeding Occurred from an HIV-Infected Person to a Person Residing in the Same Household or Providing Home Care

Case-Patient	Source-Patient	Activity During Which Transmission May Have Occured	Type of Exposure	Body Substance Through Which Transmission May Have Occured	HIV DNA Sequence Match	Comment
Mother	Child	Home nursing	Cutaneous	Blood/stool	ND[a]	Mother provided extensive care without gloves (e.g., drawing blood, removing intravenous catheters, and emptying and changing ostomy bags).
Child	Child	Home intravenous therapy for hemophilia	Possible intravenous percutaneous[2]	Blood	Y	Mother administered intravenous therapy to both children in succession and placed used needles in bag within reach of case-patient.
Child	Child	Living in same household	Cutaneous[2]	Blood	Y	Source-patient had frequent bleeding; case-patient had excoriated rash.
Adolescent	Adolescent	Living in same household	Cutaneous/percutaneous	Blood	Y	Case-patient and source-patient shared a razor; each cut himself while shaving with the razor and bled as a result. Both have hemophilia.
Child	Mother	Living in same household	Cutaneous	Blood/exudate	Y	Source-patient had drained skin lesions; source-patient picked at case-patient's scabs.
Child	Child	Living in same household	Bite[b]	Not specified	ND	Source-patient bit case-patient, skin was not broken, and there was no bleeding. Details of home care not reported.
Adult	Adult	Home nursing	Cutaneous	Body secretions and excretions, including urine and saliva	ND	Case-patient wore no gloves while caring for source-patient; case-patient had eczema and small cuts on her hands.
Mother	Adult son	Home nursing	Cutaneous	Body secretions and excretions, including urine and feces	ND	Case-patient usually wore gloves.
Mother[c] (Age 61)	Adult son (Age 31)	Living in same Household	?	?	Y	?
Father[c] (Age 66)	Adult son (Age 31)	Living in same household	?	?	Y	?

[a]Not done.

[b]No definite exposure documented.

[c]Mother/father had no sexual intercourse in last 5 years and no other known risk factor. **Son had AIDS.** The three shared a genetically related strain of HIV. Two separate **casual** transmissions in a household setting is exceptionally rare and **atypical** HIV transmission is the only likely explanation for this case.

(Source: *MMWR*,1994a-updated and Laurent Belec 1998)

Each had his own injection equipment to deliver the drug needed to prevent uncontrolled bleeding.

In August of 1994, there was a single report of HIV being transmitted during a **bloody fight** (Ippolito et al., 1994). Two brothers had a severe fight. Brother (1) had been diagnosed with AIDS. Brother (2) beat (1) until blood from (1) flowed freely from both sides of his face. Brother (2) reported that blood from (1) came into contact with his mucosal surfaces of his eyes and lips. Brother (2) developed mononucleosis symptoms and tested positive 30 days after the fight. Brother(2) denied contact with needles used by his brother, or sharing of razors, toothbrushes, manicure scissors, or a common sexual partner. Tests for other bloodborne and sexually transmitted diseases were negative. HIV isolated from both brothers showed the same pattern of mutations associated with resistance to zidovudine (ZDV) even though only brother (1) had taken the drug.

Since transmission in such a manner has rarely, if ever, been documented, these cases are of special interest to scientists. But these two cases are not cause for alarm and are not cause for changing CDC recommended guidelines for allowing HIV-infected children to attend schools.

The exposures in these nine cases are not typical of the household or social contact most people would have with an HIV-infected or AIDS patient. Nevertheless, they underline the need for observance of commonsense sanitary precautions when caring for an AIDS patient.

Home Nursing Exposure

Three reports involved home nursing care of terminally ill persons with AIDS in which a blood exposure might have occurred but was not documented; in these reports, skin contact with body secretions and excretions occured.

In another reported case in the United States, a mother apparently became infected with HIV as a result of extensive, unprotected exposure to her infected child's blood, body secretions, and excretions. The child, who underwent numerous surgical procedures to correct a congenital intestinal abnormality, had become infected through multiple blood transfusions. **The mother did not wear gloves when performing procedures such as drawing blood, removing intravenous lines, emptying ostomy bags, changing diapers, and changing surgical dressings.** On numerous occasions her hands became contaminated with blood, feces, saliva, and nasal secretions. Although she reported no accidental needlesticks or open wounds on her hands, she often failed to wash her hands immediately after such exposures.

In a similar case, a woman in England developed AIDS after caring for a man who died of AIDS. Again, the care involved frequent contact with body secretions and excretions. The woman reported that she had some small cuts and eczema on her hands during the time that she cared for the man.

NONCASUAL TRANSMISSION

The routes of HIV transmission were established **before** the virus was identified. The appearance of AIDS in the United States occurred first in specific groups of people: homosexual men and injection drug users. The transmission of the disease within the two groups appeared to be closely associated with sexual behavior and the sharing of IV needles. By 1982, hemophiliacs receiving blood products, as well as the newborns of injection drug users, began demonstrating AIDS. By 1983, heterosexual female partners of AIDS patients demonstrated AIDS. **Eighteen years** of continued surveillance of the general population has failed to reveal other categories of people contracting HIV/AIDS (Table 8-2). It became apparent that the infectious agent was being transmitted within specific groups of people, who by their behavior were at increased risk for acquiring and transmitting it (Table 8-3). Clearly, an exchange of body fluids was involved in the transmission of the disease.

With the announcement that a new virus had been discovered, further research showed that this virus was present in most body fluids. Thus, even before there was a test to detect this virus, the public was told that it was transmitted through body fluids exchanged during intimate sexual contact, contaminated hypodermic needles, contaminated blood or blood products, and from mother to fetus. In addition it was concluded that the widespread dissemination of the virus was most likely the result of multiple or

TABLE 8-2   HIV Transmission and Infection

CHAIN OF HIV INFECTION

Agent causing the disease	HIV
Major reservoirs (source of HIV in the body)	Lymph nodes, blood, genitals
Replication site of HIV	Mostly inside CD4 bearing lymphocytes
Portal of exit (how does HIV leave the body)	Mucosal openings, skin breaks, bleeding or expulsion of body fluids
Transfer or transmission of HIV (from one human to another)	Via body fluids
Portal of entry (how does HIV enter the body)	Mucosal openings, skin breaks, areas of bleeding, injection

TRANSMISSION ROUTES

Blood Inoculation

Transfusion of HIV-infected blood and blood products

Needle sharing among injection drug users

Needlesticks, open cuts, and mucous membrane exposure in health care workers

Use of HIV-contaminated skin-piercing instruments (ears, acupuncture, tattoos)

Injection with unsterilized syringe and needle (mostly in undeveloped countries)

Sexual Contact: Exchange of semen, vaginal fluids, or blood

Homosexual, between men

Lesbian, between women

Heterosexual, from men to women and women to men

Bisexual men

Perinatal

Intrauterine

Peripartum (during birth)

Breast feeding

TABLE 8-3   Adult/Adolescent AIDS Cases by Sex and Exposure Categories, estimated through 1999, United States[a]

Male Exposure Category (84%)	Total No.	(%)
1. Men who have sex with men	352,957	(57)
2. Injecting drug use	136,229	(22)
3. Men who have sex with men and inject drugs	49,538	(8)
4. Hemophilia/coagulation disorder	6,192	(1)
5. Heterosexual contact:	24,769	(4)
a. Sex with injecting drug user	9,165	
b. Sex with person with hemophilia	50	
c. Sex with transfusion recipient with HIV infection	446	
d. Sex with HIV-infected person, risk not specified	15,109	
6. Receipt of blood transfusion, blood components, or tissue	6,192	(1)
7. Other/undetermined	43,346	(7)
Total male AIDS cases	619,222	(100)

Female Exposure Category (16%)		
1. Injecting drug use	51,897	(44)
2. Hemophilia/coagulation disorder		(0)
3. Heterosexual contact:	45,999	(39)
a. Sex with injecting drug user	20,699	
b. Sex with bisexual male	3,542	
c. Sex with person with hemophilia	415	
d. Sex with transfusion recipient with HIV infection	644	
e. Sex with HIV-infected person, risk not specified	20,699	
4. Receipt of blood transfusion, blood components, or tissue	4,718	(4)
5. Other/undetermined	15,333	(13)
Total female AIDS cases	117,947	(100)
Total male/female cases	737,169	

[a]Pediatric cases, 8,831.

repeated viral exposure because the data from transfusion-infected individuals indicated that they did not necessarily infect their sexual partners. In other words, it was concluded early on and later confirmed that this viral infection did not occur as easily as other blood-borne viral diseases such as hepatitis B, or viral and bacterial sexually transmitted diseases. Table 8-4 lists the means of HIV transmission worldwide. Table 8-5 lists the means of HIV transmission in the United States.

Mobility and the Spread of HIV/AIDS

Mobility is an important epidemiological factor in the spread of communicable diseases. This becomes particularly obvious when a new disease enters the scene. In the early stage of the HIV/AIDS epidemic, for example, the route of the virus could be associated with mobility.

The first HIV-infected people in some Latin American and European countries reported a

TABLE 8-4 How HIV is Transmitted Worldwide, 1999

Exposure	Efficiency, %	% of Total
Blood transfusion/blood products	>90	3
Perinatal	20–40	9
Sexual intercourse[a]	0.1–1.0	80
Injection drug use	0.5–1.0	8

[a]Heterosexual intercourse 70%
(Source: WHO/Global Programme on AIDS and UNAIDS, 1998, Updated)

TABLE 8-5 An Approximation of How an Estimated 746,000 Americans Become Infected with HIV, through 1999

75	health care workers got infected from the blood or body fluids of patients;
10,071	children infected through their mothers;
16,039	people got HIV from infected blood or blood products;
52,220	people did not know how they were infected, did not report their risk, or died before anyone could find out;
44,760	people were infected who had both unprotected sex and shared needles;
186,500	people were infected who shared needles;
436,410	people were infected through unprotected sex.

(Based on data from HIV/AIDS Surv. Rpt, CDC, End-Year Edition, 1998)

history of foreign travel. In some African countries, spread of the virus could be traced along international roads. Today, increasing numbers of HIV infection have been observed to be associated with the relaxation of travel restrictions in Central and Eastern Europe.

Few countries are **unaffected** by HIV/AIDS. This has made it clear that restrictive measures such as refusal of entry to people living with HIV/AIDS and compulsory testing of mobile populations are inappropriate and ineffective measures to stop spread of the virus. In times of increasing international interdependency, it is an illusion to think that the disease can be stopped at any border.

Number of HIVs Required for Infection

Robert Coombs and colleagues (1989) calculated the dose of HIV necessary to cause an HIV infection. They reported that **"one infective dose"** of 1,000 HIV particles is necessary to establish HIV infection in human tissue culture cells. To establish HIV infection in the body, it was reasoned that it would take **10 to 15 infective doses.** Their study indicated that a pint of blood from an AIDS patient contains 1.8 million infective doses, or about 2 billion HIV particles per pint of blood, or about 4.2 million HIV particles per milliliter.

Placing HIV Numbers in Perspective— In cases of hepatitis B there may be 100 million to 1 billion hepatitis viruses in *just 1 milliliter* of blood (25 to 250 times more virus/mL than for HIV). A pint of blood contains 473 mL. Yet only

one in three people who stick themselves with a needle contaminated with blood from a hepatitis B patient becomes infected.

The epidemiology of HIV infection is similar to that of HBV infection, and much that has been learned over the last 15 years about the risk of acquiring hepatitis B in the health care workplace can be applied

—————— BOX 8.1 ——————

RELATIVE DOSE: THE SWIMMING POOL ANALOGY

An analogy on the number of viruses it would take to cause an infection can be made by diluting the number of viruses in 1 mL of blood in a swimming pool full of water.

Basically the analogy goes like this: If 1 mL of blood carrying the hepatitis B virus were dropped into **24,000 gallons** of water and mixed well, and if 1 mL of that solution were injected into a susceptible individual, that individual would develop hepatitis B. He or she might not develop jaundice, gray stools, chocolate urine, or other clinical conditions, but would develop serological markers indicating hepatitis B infection.

In contrast, if 1 mL of blood from an AIDS patient were dropped into **a quart of water** and 1 mL of that solution were injected into a susceptible individual, there is only a 1 in 10 chance that the individual would develop HIV antibody indicating HIV infection (Cottone et al., 1990). The implication here is that HIV is not easy to acquire.

to understanding the risk of HIV transmission in similar settings. Both viruses are transmitted in the same manner.

Current evidence indicates that despite the epidemiological similarities of HBV and HIV infection, the risk of HBV transmission in health care settings far exceeds that of HIV transmission. The risk of HIV transmission from a needlestick (including probable cases) is less than 0.5% (1 in 200). By contrast, the risk of hepatitis B infection from a needlestick has been estimated at 6% to 30%, or anywhere from 12 to 60 times more likely than for HIV. Approximately 12,000 health care workers in the United States develop symptomatic hepatitis B annually. About 1,000 of them become carriers, and approximately 250 die of cirrhosis, fulminant hepatitis, or primary liver carcinoma. A health care worker's risk of dying of hepatitis B after 15 years on the job is about 27 in 10,000; the annual risk of dying in a car crash is 2.4 in 10,000 (Sadovsky, 1989).

Body Fluid Transmission

HIV has been isolated from blood, semen, saliva, serum, urine, tears, vitreous, breast milk, vaginal secretions, lung fluid, and cerebrospinal fluid (Friedland et al., 1987). HIV has not been found in sweat. Because HIV has been identified in these fluids does not mean they are important in HIV transmission. Jay Levy and colleagues (1989) isolated HIV from 10 cell-free body fluids and from the cells found in five of them (Table 8-6). The results indicated that except for cerebrospinal fluid, large quantities of cell-free HIV are not present in any of the nine other body fluids tested.

The low levels of HIV in cell-free body fluids and within the cells of these fluids does not mean that HIV cannot be transmitted via these fluids or cells—it can, but the dose (number of viruses) is so small that the risk of infection is minimal, thus the reason for the low number of health care workers contracting HIV infection after touching, being splashed by, or needlesticking themselves with blood containing HIV. Particularly interesting is the finding that there is no detectable HIV in bronchial fluid or sweat.

HIV is found in greatest numbers (100 to 10,000 infectious units per mL) **within T4 cells,**

TABLE 8-6 HIV in Body Fluids and in Cells within Body Fluids

	Estimated Number of HIV[a]
Cell-free fluid	
Plasma	10–50
Serum	10–50
Tears	<1
Ear secretions	5–10
Saliva	<1
Urine	<1
Vaginal/cervical	<1
Semen	10-50
Sweat	0
Breast milk	<1
Cerebrospinal fluid	10–1,000
Infected cells (T4 and macrophages)	
PBMC	0.001–0.1
Saliva	<0.01
Bronchial fluid	not detected
Vaginal/cervical fluid	present but not quantitated
Semen[b]	0.01–5

[a]Cell-free fluid is expressed in infectious particles per milliliter; infected cells, in percentage of infected cells observed.
[b]HIV has been detected in cell-free seminal fluid and in nonspermatazoal mononuclear cells and in the DNA of sperm from some HIV-infected men (Meriman et al., 1991; Bagasra et al., 1994).

(Source: Adapted from Levy, 1989)

macrophage, and monocytes of blood, vaginal fluids, and semen; yet these fluids carry relatively low levels of cell-free HIV. Laboratory findings, along with overwhelming empirical observations, support the scientific conclusion that the major route of HIV transmission is through human blood and sexual activities involving exchange of semen and vaginal fluids. Semen carries significantly larger numbers of HIV than vaginal fluid. It appears that of all body fluids, these three contain the largest number of infected lymphocytes (Figure 8-4), which provide the largest HIV concentration in a given area at a given time.

Cell-to-Cell Transmission

According to classic models of viral infection, which draw heavily from research on bacteriophage (viruses that attack bacteria), the process of infection starts with adherence of free virus to host cells. This view has dominated the

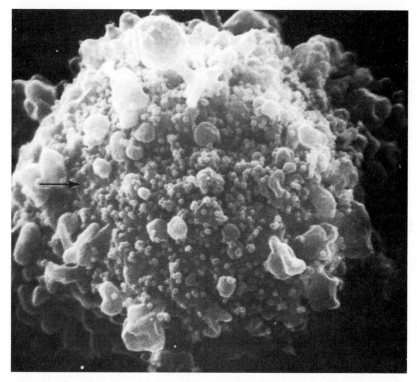

FIGURE 8-4 Electron Micrograph of an HIV-Infected T4 Lymphocyte. The T4 cell has produced a large number of HIVs that are located over the entire lymphocyte. **Each HIV leaves a hole in the cell membrane.** The photograph shows part of the convoluted surface of the lymphocyte magnified 20,000 times. (*Courtesy of The National Biological Standards Board, South Mimms, U.K. and David Hockley*)

thinking about mechanisms of infection by HIV. But now there is overwhelming evidence that HIV infection can be initiated by the direct transfer of HIV from one cell to another by syncytia (group) formation (Figure 3-8) and by cell-to-cell transmission without this formation. Cell-to-cell HIV transfer may take place either between host and donor cells within the host or between a donor HIV-infected mononuclear cell and intact host epithelium of the genital, digestive or urinary tract, or the placenta. Different mechanisms of cell-mediated transmission are illustrated in Figure 8-5.

Presence of HIV-Infected Cells in Body Fluids

Blood— Recent observations indicate that both latent and HIV-producing cells are present throughout the course of infection.

Considerable numbers of HIV-infected cells can be detected in the blood. Thus it is probable that HIV-infected people harbor substantial numbers of HIV-infected cells in their blood from soon after the initial infection through the terminal stage of the disease.

Semen— In March of 1999, Ann Kiessling of Harvard Medical School said, based on data from the World Health Organization, **every day** some 400 liters (378.4 quarts) of HIV infected semen are ejaculated worldwide. Semen is a complex mixture of fluids and cells from several organs, including the testis, epididymis, seminal vesicles, prostate, and urethra. On average, ejaculated semen specimens contain 100 million sperm. Also found in a single specimen of ejaculate are between 1 million and 10 million non-spermatozoal cells, and many leukocytes including T4 cells. But the number and type of

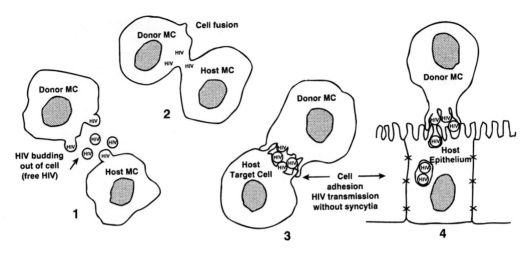

FIGURE 8-5 Schemes of Four Possible Mechanisms of Cell-Mediated Transmission of HIV. (**1**) An HIV-secreting donor mononuclear cell (MC) releases virus that infects nearby host cells; (**2**) an HIV-infected donor MC fuses with a host cell; (**3**) and (**4**) adhesion-based cell-to-cell transmission without syncytia. Cell-to-cell transfer of virus could hypothetically take place either between cells within the body of the host (**3**) or between a donor HIV-secreting MC and a host epithelial cell at the portal of entry (**4**). (*Source:* Phillips, 1994)

all mononuclear cells in the semen of a healthy man differs considerably from day to day.

Reducing the numbers of T4 cells (lymphocytes and macrophages) in semen could decrease the chances of infection. Lymphocytes and macrophages can originate from the testes, epididymis, seminal vesicles, prostate and the inflammation of the genital tract also results in an increased number of these cells in semen. In one recent study on 94 semen samples from HIV-infected men, HIV DNA was found in 35% of the mononuclear cells (Bagasra, 1996). It has been suggested that vasectomy could reduce the infectivity of HIV-infected men because it would eliminate mononuclear cells or cell-free virus in semen originating from the testes and epididymis. Although vasectomy has been reported to reduce the number of these white blood cells in semen, HIV-infected mononuclear cells have been detected in semen from seropositive vasectomized men. The work of John Krieger and colleagues (1998) shows that sperm can exist in semen for up to 3 months after vasectomy (see **sperm** below) and that there was **no significant** difference **pre** and **post-vasectomy** in

HIV RNA levels (viral load). Robert Coombs and colleagues (1998) reported that viral load in semen only weakly corresponded to the viral load in blood, and Ann Kiessling has provided evidence that HIV-infected cells found in the blood, do not co-mingle with those found in semen. She said that while it may be possible that HIV-infected cells in semen may enter the blood, infected blood cells are **not found in semen!** Thus HIV is compartmentalized in the body—and this makes it more difficult to treat HIV infection (Zhang et. al 1998). Collectively, it would appear that a vasectomy does **NOT** stop HIV transmission!

In summary, semen from HIV-infected men can contain both cell-free and cell-associated virus. From the available evidence, as in blood, it is not yet possible to determine to what extent cell-free and cell-associated virus mediate HIV infectivity (Phillips, 1994).

Sperm— Omar Bagasra (1996) reported that 33% of sperm samples, from 94 HIV-positive men, contained HIV RNA. Many investigators have now reported that the mitochondria of various tissue types are infected with HIV. The sperm midpiece contains large numbers of mitochondria.

Saliva— There has always been the fear that HIV could be transmitted by kissing, that is, by coming in contact with HIV-infected saliva. Several large-scale studies have failed to demonstrate HIV transmission to household members and health care workers by casual contact, suggesting that transmission by saliva is uncommon. Although saliva from healthy individuals may contain considerable numbers of leukocytes, very few of these are T4 mononuclear cells. Using the DNA polymerase chain reaction test (presented in Chapter 12), Goto and co-workers (1991) found that HIV provirus was present in saliva from all 20 AIDS patients they examined. Omar Bagasra (1996) found HIV DNA in nuclei of epithelial cells (cells scraped off the inner sides of the mouth oral scrapings) in 29 of 35 HIV-infected persons (83%). Saliva from the 29 people all contained HIV-positive mononuclear cells. It should also be noted that AIDS patients frequently have oral lesions that result in blood in the oral cavity. Thus, it is possible that both cell-free virus and HIV-infected mononuclear cells in saliva from AIDS patients could originate from blood (Phillips, 1994). A recent report suggests that a secretory protease inhibitor in saliva attaches to the surface of lymphocytes blocking HIV infection (see "Orogenital Sex," this chapter).

There continues to be some concern over the presence of HIV in saliva because of the exchange of saliva during **deep kissing,** the saliva residue left on eating utensils, and saliva on instruments handled by health care workers, especially in dentistry. Results of studies on hundreds of dental workers, many of whom have cared for AIDS patients, have shown no evidence of HIV infection (Friedland et al., 1987). Also, in a CDC study, none of 48 health care workers became infected after parenteral (IV) or mucous membrane exposure to the saliva of HIV-infected patients (Curran et al., 1988). Studies by Philip Fox (1991), David Archibald (1990), and Pourtois (1991) showed that human saliva contained factors that inhibit HIV infectivity. Rene Crombie and colleagues (1998) reported that the glycoprotein **thrombospondin** is concentrated in saliva and blocked

HIV infection of human cells in the **test tube!** Further saliva inhibition tests are in progress.

In 1999, Samuel Baron and colleagues at the University of Texas Medical Branch reported a provocative theory regarding a powerful protective effect from saliva. Baron's test-tube studies found in saliva, due to its **low level of salt, inactivates more than 90 percent of the HIV-infected leukocytes (blood cells) that may be the main transmitter of HIV on mucosal surfaces,** thereby blocking HIV transmission by more than 10,000-fold. This happens, Baron says, because salt is needed to keep cells alive. If you put red blood cells (that contain salt) in a glass of water, for example, they quickly pull in the liquid, swell up, and burst. The absence of salt in saliva causes HIV-infected cells to do the same thing. And that's why kissing appears so safe in many instances. Saliva, concludes Baron, "is breaking down the infected cells." When asked why saliva did not protect people during oral sex or children during breast feeding, Baron stated that "If you add HIV infected cells in the presence of a large amount of blood, semen, or breast milk and then add saliva, those three fluids protect HIV infected cells from being killed by the saliva. That coincides with the epidemiological evidence that those three fluids can transmit HIV. But in the absence of those three fluids, there appears to be only rare transmission of HIV."

First Documented Kissing Transmission Case

Although over 50 million people around the world have been HIV-infected, there are remarkably few reports of its spread by kissing, dental treatment, biting or coughing. Even among those people whose bodies are actively shedding HIV, their saliva usually contains only non-infectious components of the virus.

In July of 1997, the first documented case of HIV transmission via deep kissing was reported on by the CDC. In this **one** case, the man was HIV-positive, via IDU in 1988. Both he and his female partner had serious gum (periodontal) disease. His gums routinely bled with brushing or flossing. Investigators at the CDC believe that the HIV was transmitted via blood within the man's oral cavity due to oral

DANGER OF HIV INFECTION VIA ARTIFICIAL INSEMINATION

By early year 2000, 15 women were reported to have been HIV-infected through the use of anonymous donor sperm to initiate pregnancy; one in Germany, two in Italy, four in Australia, two in Canada, and six in the United States. Thirty recipients of semen from HIV-infected donors refused to be HIV tested. (Guinan, 1995 updated). All cases except Germany occurred **before** the availability of HIV antibody testing. About 80,000 women each year are artificially inseminated with donor sperm. But 18 years into the AIDS epidemic, the increasingly popular fertility business remains largely unregulated and unmonitored, even though it traffics in semen, long known to be one of the two main HIV transmission routes.

Only a few states (New York, California, Ohio, Illinois, and Michigan) require HIV testing of semen donors. **There are no federal regulations.**

Medical and public health experts agree that artificial insemination is an HIV risk that somehow fell through the cracks of public education and health regulations. They insist, however, that the risk is low.

In January of 1996 Washington State's top educator announced that she had AIDS. She received the virus trying to become pregnant through artificial insemination with donor sperm.

HIV announcement has once again focused national attention on the problem, the lack of federal regulations for this industry.

Omar Bagasra and colleagues (1994, 1996) state unequivocally that there is a risk of HIV transmission during artificial insemination. To quote Bagasra, "As we have shown, the sperm easily penetrates other cells— and when this occurs, the midpiece also penetrates. A single ejaculation contains [millions of] sperm. We have data to show that the mitochondria in the midpieces of HIV-infected sperm are loaded with HIV. If 1 in 1,000 of these mitochondria contains HIV, that is a significant viral load. The viral load transmitted from a man to a woman in this manner is quite likely to be high."

Lack of federal and state regulations means you must protect yourself! (1) Stay away from private physicians unless you know the doctors are using only certified sperm banks to get their products. (2) Review a doctor's or clinic's testing and record-keeping procedures and demand to see donor medical records, which can be shared even if the donor's identity is protected. (3) Accept only frozen donor semen that has been HIV-tested twice at 3- to 6-month intervals.

---

lesions onto the mucus membrane of the woman (*MMWR*, 1997).

Spitzer and colleagues (1989) reported that a male became HIV-infected during unprotected fellatio with an HIV-infected prostitute. A second similar case has also been reported. The mechanism of transmission, saliva, in these cases may be suspected, but it is not conclusive!

The precise risk of oral sex is difficult to assess because most couples engage in other sexual practices.

Mother's Milk— Several studies have shown that HIV can be transmitted by breast-feeding. Van de Perre and co-workers (1993) found that infection of babies via breast milk was most strongly correlated with the presence of HIV-infected cells in the milk, suggesting that infection might be cell-mediated. However, infection was also correlated with low lev-

els of antibodies to HIV, suggesting that infection was initiated by cell-free virus. Thus this study does not present strong evidence in favor of infection by cell-associated versus cell-free virus.

Dentist with AIDS Infects Patient During Tooth Extraction?

In July of 1990, the CDC reported on the possible transmission of HIV from a dentist with AIDS to a female patient (*MMWR*, 1990a). This case, like no other before it, sent chills through many. But why this case? Because the vast majority of people go to dentists! They don't inject drugs and are not gay. This case, however, is difficult to resolve. For example, 2 years had elapsed from the time of the dental work to when the patient, Kimberly Bergalis (Figure 8-6), was diagnosed with AIDS. Both patient and

FIGURE 8-6 Kimberly Bergalis, Age 23. Bergalis is being comforted by her mother after their train trip from their home in Florida to Washington, D.C. She made the trip to testify before a congressional committee on September 26, 1991. Bergalis favored mandatory HIV testing for health care personnel. She died December 8, 1991. (*Photograph courtesy of AP/Wide World Photos*)

dentist, David J. Acer of Stuart, Florida, were uncertain of exactly what happened. Some of the pertinent factors in this case are (**1**) review of dental records and radiographs suggest that the two tooth extractions were uncomplicated; (**2**) interviews with Bergalis and the dentist did not identify other risk factors for HIV infection. (Bergalis, age 22, stated over national TV in 1990 that she was still a virgin.) This point was disputed on June 19, 1994, during the TV program *60 Minutes* (with Mike Wallace) as well as the CDC's genetic match between Dr. Acer's HIV and his patient's strain of HIV. Yes, she did engage in sexual foreplay, but not intercourse. Yes, she was infected with the human papilloma virus, which can be sexually transmitted, but this is not at all uncommon in immune-suppressed AIDS patients with no history of sexual intercourse; (**3**) nucleotide sequence data indicated a high degree of similarity between the HIV strains infecting her and the dentist; and (**4**) the time between the

dental procedure and the development of AIDS was short (24 months), and Bergalis developed oral candidiasis 17 months after infection. Only 1% of infected homosexual/ bisexual men and 5% of infected transfusion recipients develop AIDS within 2 years of infection.

David Acer, a bisexual, was diagnosed with symptomatic HIV infection in 1986 and with AIDS in 1987. He died on September 3, 1990. Since then, 1,100 of his 2,500 patients were contacted for HIV testing. In January of 1991, the test results of 591 of these patients revealed that five were HIV-positive: a 68-year-old retired school teacher, a middle-aged father of two, a 37-year-old carnival worker, an unemployed drifter, and a 19-year-old student. As with Bergalis, infection in these patients may have come from some other source. All six patients denied having sexual contact with the dentist or with one another (*MMWR*, 1991a). **If an absolute case can be made that Acer transmitted**

DEADLY INNOCENCE: A MURDER MYSTERY?

Kimberly Bergalis had just graduated from the University of Florida. She would spend that summer, 1989, waiting tables. One day she **fainted.** This was just the beginning of her 28-month odyssey to her death from AIDS. It was the beginning of a search for her illness—an investigative medical journey into her illness proved very puzzling. It was an unending illness, chronic vaginal infections, a mouth full of yeast, dizzy spells, exhaustion, and loss of appetite. First diagnosis—could it be leukemia? (A weakened immune system allows the presentation of many look-alike symptoms.) Almost unable to breathe, she was diagnosed with Pneumocystis pneumonia. The doctor found the agent causing the pneumonia—*Pneumocystis carinii.* An HIV test was positive; **the diagnosis, AIDS!** Kim said she was a virgin, did not use drugs, had never had a blood transfusion, surgery, acupuncture, or tattoos—**but wait,** she had two teeth extracted by a dentist who was sickly but his illness was said to be cancer.

THE INVESTIGATION

The CDC and other investigators tried but failed to find evidence that Ms. Bergalis was anyone other than who she professed to be. They even asked her if she had sex with animals (Burkett, 1995). She was, after the news broke nationally, demonized by some and defended by others. Even after her funeral, her parents received hate mail—"Thank God the bitch is finally dead" or "She got what was coming to her." **Such remarks took their place among so many other even more hateful remarks that people with AIDS had to endure then and NOW.**

AUTHOR'S NOTE

The absolute truth of how Ms. Bergalis became infected will never be known—too much time has passed, the dentist is dead, and the chain of transmission was never clearly established! A recent TV documentary of this case implies that the dentist **intentionally infected** Ms. Bergalis and at least five others who were his dental patients.

the virus to Bergalis, this will be the *first documented case of a health care professional infecting a patient.* Bergalis is reported to have settled her first claim for $1 million from Acer's malpractice insurance company. She settled a second claim against his estate for an undisclosed sum.

Bergalis died of AIDS on December 8, 1991. She was 23 years old and weighed 48 pounds. In January 1993 a life-size bronze statue of Kimberly Bergalis was unveiled at the high school she attended. Her dying wish was that what happened to her should never happen to anyone else. To this end, weak and near death, she traveled by train to testify before the United States Congress urging mandatory testing for all health care workers. The 'Bergalis Bill' never got out of committee, the CDC and Congress rejected the call for mandatory testing of health care workers. On April 6, 1991, Ms. Bergalis wrote a two-page unmailed letter to the Florida State Department of Health and Rehabilitative Services. She wrote "Whom do I blame? I blame Dr. Acer and every single one of you bastards. Anyone who knew Dr. Acer was infected and had

AIDS and stood by not doing a damn thing about it. You are all just as guilty as he was."

In July 1991, a guest on the *Sally Jessy Raphael Show* stated in front of Mr. and Mrs. Bergalis, that their daughter Kimberly, knowing she was HIV-infected, went to another dentist and did not tell him that she was infected.

Acer's Dental Practice— Staff reported that by 1987 all surgical instruments were routinely autoclaved. Nonsurgical heat-tolerant instruments (e.g., dental mirrors) were autoclaved when practice conditions, such as time and instrument supply, allowed, or were immersed in a liquid germicide for varying lengths of time. Tests of the autoclave in October 1990, demonstrated that it was functioning properly (*MMWR*, 1991a).

There is no shortage of ideas as to how Acer might have infected his patients. For example, he could have used the same dental instruments on himself or his sexual partners that he used on his patients without sterilizing them. Or he could have had HIV in his sweat which could have dropped into his patients' oral cavities. However, the polymerase chain reaction (the latest in DNA detection technology) was used to detect the presence of HIV in the natural sweat from 40 HIV-infected people. All 40 tested HIV-negative. It appears that by the available methodology of detection HIV is **not** present in the sweat of HIV-positive people.

The actual route of HIV transmission in the Acer–Bergalis case will most likely never be known. There have been suggestions that the dentist did not wish to die alone and chose certain people to infect. It was suggested that he may have attempted to infect still others, but was unsuccessful. A friend of the dentist said that he believed that Acer intentionally infected his patients to call attention to the HIV/AIDS problem in the United States. Acer felt that mainstream America was ignoring the problem. In mid-1994, Continental Broadcasting Systems (CBS), during an interview with Lionel Resnick, chief of retrovirology research at Mt. Sinai Medical Center in Miami, said that, based on his research, David Acer was not the source of HIV infection for the six persons. He said the CDC "vastly overstated the reliability of the DNA tests." But, to the contrary, the CDC believes it has **understated** its case based on the evidence!

A recent CDC estimate put the theoretical risk of HIV transmission from an HIV-infected dentist to patient during a procedure with potential blood exposure at 1 chance in 263,158 to 1 chance in 2,631,579 (Friedland, 1991). **Point for Discussion: Is this last point, one of revenge against society, reasonable?**

(Read: Denis Breo (1993). The dental AIDS cases—murder or an unsolved mystery? *JAMA*, 270:2732–2734.)

Summary— The CDC states that based on the following considerations, their investigation strongly suggests that at least three patients of Acer's were infected with HIV during dental care: (**1**) the three patients had no other confirmed exposure to HIV; (**2**) they all had invasive procedures performed by an HIV-infected dentist; and (**3**) DNA sequence analyses of the HIV strains from these three patients indicated a high degree of similarity to each other and to the strain that had infected the dentist—a finding consistent with previous in-

stances in which cases have been linked epidemiologically. In addition, these strains are distinct from the HIV strain from another of Acer's patients who had a known behavioral risk for HIV infection, and from the strains of the eight HIV-infected people residing in the same geographic area and from the 21 other North American HIV isolates.

Beginning 1998, two of the six patients believed to have been infected by Acer have progressed to AIDS but are still alive: Kimberly Bergalis, Richard Driskill, John Yecs, and Barbara Webb have died. To date, the six people have received $10 million from Acer's insurance company.

See Point of Information 8.5.

Sexual Transmission

The predominant mode of HIV transmission is through sexual contact. Thus, HIV usually gains access to the immune system at mucosal surfaces. Such surfaces include the oropharynx, rectum and genital mucosa. Mucosal surfaces are rich in Langerhans cells, dendritic cells that trap antigens and virus particles. In addition, lymphoid aggregates are found throughout the tissue immediately below the mucosal surface.

Sexual transmission of HIV occurs when infected blood, semen, or vaginal secretions from an infected person enter the bloodstream of a partner. This can happen during anal, vaginal, or oral penetration, in descending order of risk. Unprotected anal sex by a male or female appears to be the most dangerous, since the rectal wall is very thin. Masturbation or self sex is the safest. In general, a person's risk of acquiring HIV infection through sexual contact depends on (**1**) the number of different partners, (**2**) the likelihood (prevalence) of HIV infection in these partners, and (**3**) the probability of virus transmission during sexual contact with an infected partner. Virus transmission, in turn, may be affected by biological factors, such as concurrent STD infections in either partner. If there is a genital ulcer caused by syphilis or chancroid or herpes, the risk of getting HIV increases 10- to 20-fold. If there is gonorrhea or chlamydial infection, which are more common, the risk increases probably 3- to 4-fold. Behavioral factors, such as type of sex practice and use of condoms, or varying levels

HIV INFECTIVITY

HIV infectivity is the average probability of transmission to another person after that person is exposed to an infected host. Infectivity plus two other parameters—the duration of infectiousness and the average rate at which susceptible people change sexual partners— determines whether the epidemic grows or slows. On a population level, host-related factors (the social, cultural, and political milieu) and agent factors (HIV) determine HIV infectivity. Host-related and environmental factors can promote the pandemic because they can accelerate infectivity (Royce et al., 1997).

BOX 8.2

HIV/AIDS ROULETTE

Case I: The Woman Executive

I am a successful executive woman. A year ago I applied for life insurance. I was required to take an HIV antibody test. It came back positive.

I am not a prostitute or promiscuous. I am not and have never been an injection drug user. I am not a member of a minority group, indigent nor homeless, and I have not slept with a bisexual male (?).

I don't fit any of the stereotypes that people have designated for those infected with HIV. I got HIV from a man I love and have been seeing for 5 years. He is not homosexual or bisexual. He has never used injection drugs. He had no idea he was carrying the virus. He believes he may have been infected about 6 years ago by a woman with whom he had a brief, meaningless relationship. For that one indiscretion we will both pay the ultimate price.

Case II: The HIV-Infected Male

In February of 1994, a 21-year-old man walked into a sexually transmitted disease clinic and told the doctor he had "the clap," or gonorrhea. But he carried HIV.

When a counselor inquired about his sex partners, he told them about several, including a 12-year-old girl. The girl had gone elsewhere to be treated for gonorrhea and tested positive for HIV.

The man admitted to sleeping with 27 women, including 13 teenagers. Ten of the partners couldn't be found. Of the 17 others, 12 tested positive for HIV.

The man has since died, and the clinic has not been able to track all of his sexual partners.

There is a message to be found in these two cases: People with HIV are much more dangerous to a community than someone with AIDS. On average, the HIV-infected are asymptomatic for 11 years. Those with AIDS and not treated, are symptomatic. They are losing weight. They are sick. They have little or no sexual appetite. Those with HIV are healthy and vigorous. That's where the sexual roulette begins.

of infectivity in the source partner (for example viral load) related to clinical stage of disease also increase the risk of HIV transmission/infection. Based on these factors, the risk for HIV infection is highest for an uninfected partner of an HIV-infected person practicing unsafe sex. Persons who have sex partners with risk factors for HIV infection or who themselves have multiple partners from urban settings with high rates of injection drug and "crack" cocaine use, prostitution, and other STDs are also at increased risk.

Personal Choice–Personal Risks

Ray Bradbury wrote that "living at risk is jumping off a cliff, and building your wings on the way down." About 90% of the HIV infections that occur within the heterosexual non-injection drug use population occur through one or more sexual activities. Some 90% of the HIV infections that occur among gay males occur through anal intercourse (Kingsley et al., 1990). In any sexual activity, HIV is transmitted via a body fluid. Transmission may occur between men, from men to women, women to men or possibly from women to women (see Chapter 11).

Antiretroviral Therapy and Personal Risk Perception

Recent studies indicate that increases in unsafe sex may be related to beliefs about the lack of severity of HIV disease. In a New York study

DISTRIBUTION OF HIV-CONTAMINATED BLOOD: FRANCE, GERMANY, THE UNITED STATES, AND OTHER COUNTRIES

Near the end he could not bring himself to visit his youngest brother, to see him dying of AIDS. He too was dying of AIDS. Their deaths would close out a family of four HIV-infected hemophiliac brothers, all diagnosed with AIDS. The first died at age 24, the second committed suicide at age 33, the last two brothers died in 1993. The four brothers became HIV-infected about 1985 from using HIV-contaminated blood-clotting factor.

FRANCE

The French National Center of Blood Transfusions (CNTS) knowingly distributed HIV-contaminated blood products to hemophiliacs in 1985. The scandal, which broke in 1991, prompted the center's director to resign and an official government investigation.

In the confidential minutes of a 1985 CNTS meeting, agency officials concluded that 100% of the concentrated blood-clotting factors used to treat French hemophiliacs were contaminated with HIV. The agency, which has a monopoly on blood for transfusions, kept its secret and ignored a 1984 recommendation from the U.S. Centers for Disease Control and Prevention that blood products be heated in order to kill the deadly virus (Dorfman, 1991).

Whatever the reasons, the secrecy and delay produced catastrophic results. Nearly half of France's 3,000 hemophiliacs were infected with HIV; subsequently over 600 have died. Bowing to public and media pressure, the French government will provide compensation for some 7,000 French citizens who have been infected with HIV through blood products and transfusions (Aldhous, 1991).

French Ministers to Face Poisoning Charges

Formal charges of "collusion in poisoning" were made in October 1994 against Laurent Fabius, the former Socialist prime minister, and four of his colleagues—Edmond Hervé, then deputy health minister, and Georgina Dufoix, then minister of social affairs, Francois Gros, scientific advisor to Fabius, and Claude Weisselberg, physician and advisor to Fabius—over their role in France's HIV-contaminated blood scandal in the mid-1980s. In 1995, Gerard Jacquin, former director of bioindustry at CNTS, was charged with poisoning and Louis Schweitzer and Patrick Baudry, advisors to

Fabius and Dufoix, respectively, were charged with delaying the screening of blood products (Reichhardt, 1995). The charge carries a maximum sentence of 30 years in prison. In addition to these three men, the charge of collusion to poison were also brought against Jean-Pierre Allain (just released from prison) and Michel Garretta (still in prison) (Butler, 1994a, 1994b).

Update 1999— Former Prime Minister Laurent Fabius, former Health Minister Edmond Herve, and former Social Affairs Minister Georgina Dufoix are accused of keeping a U. S. HIV test off the market while the Pasteur Institute completed its own test. They are charged with involuntary homocide related to the deaths of 7 hemophiliacs infected with HIV after blood transfusions. They went to trial in February 1999. Conviction could bring each a five-year jail term and a $130,000 fine.

Conclusion To The Nine Year Old Blood Transfusion Case

In March 1999, Laurent Fabius and Georgina Dufoix were acquitted of all charges in the French contaminated blood scandal. Edmund Herve was found guilty but **received no penalty.** During the case, several dozen scientists, physicians, administrators, and politicians testified on behalf of the former prime minister and two former cabinet ministers.

GERMANY

The German blood scandal began with two questions. The **first** question was how to explain several unexpected cases of HIV infection. And the **second** question, associated with the first, was how could 7,000 units of blood have been screened for HIV using 2,500 HIV blood testing kits (it takes one kit/unit of blood screened). The search for the answers led investigators to a blood supply company called UB Plasma in the city of Koblenz. The investigators concluded that either the firm had failed to test thousands of units, or it had "pooled" units from multiple donors before conducting the test, an illegal practice that reduces the chances of detecting HIV contamination.

Investigators discovered that after UB Plasma began running into financial trouble in 1991, technicians were told to pool units to save money on the $2 test kits. After German kit manufacturers stopped deliveries (UB Plasma failed to pay its bills) technicians switched to an unauthorized and less reliable test. There was also evidence that the firm may have distributed blood that was not screened at all.

Because of the extent of blood banking violations, German authorities closed UB Plasma. They arrested the manager and three employees on charges of fraud and "negligent bodily harm." The current health minister disbanded the German Federal Health Bureau. And the Director of Pharmaceuticals was relieved of his duties relating to the control of blood and blood products. In stark contrast to the HIV blood scandals in France and the United States, German investigators attribute the distribution of HIV-contaminated blood and blood products to incompetence rather than company greed—UB Plasma did not attempt to distribute known HIV-positive blood until current stocks were used up!

In 1995 the director and the head of product testing of Haemoplas, a German blood products company, have been charged with murder on the grounds that they had been motivated by greed. They are alleged to have conspired in the early 1980s to economize by not testing blood samples, and to have persisted after testing was made compulsory in 1985.

Fourteen people became infected with HIV through Haemoplas products between 1986 and 1987, and three have since died.

Both men have also been charged with 5,837 counts of attempted murder—corresponding to the number of batches distributed—infringing legislation on the control of pharmaceutical products, and fraud (Abbott, 1995).

THE UNITED STATES

In late 1993, a class action suit was filed on behalf of 9,000 hemophiliacs who became infected with HIV after using HIV-contaminated blood products, called Blood Factor 8, which is essential for proper blood clot formation following an injury. In the case of severe hemophilia, many persons spontaneously bleed into their joints. Blood Factor 8 is sometimes referred to as Anti-Hemolytic Factor or AHF. Hemophiliacs are now dying of AIDS at the rate of one a day.

The suit charges that five manufacturers of AHF and the National Hemophilia Foundation continued to pursue aggressive advertising and marketing of AHF while downplaying the risk of viral infection. In 1995, the United States Supreme court disallowed the suit.

By the mid-1970's, the dangers of viral infection from blood products was well known and methods had been provided for blood viral inactivation. Such methods were patented and available by 1977. In 1981, Donald Francis, an epidemiologist then with the CDC (Figure 9-6), and Max Essex, a retrovirologist at Harvard Cancer Biology Laboratory, were convinced that AIDS, then referred to as Gay Related Immune Deficiency or GRID, was caused by an infectious agent, most likely a virus. But others thought GRID was related to the gay male lifestyle. The disease was spreading rapidly among gay males through sexual contact. But the evidence continued to mount that the disease, or agent causing the disease, was being transmitted by blood. And in that same year, the president of an AHF-producing company stated the agent was in the blood supply. He said he thought it was a 100% fatal retrovirus. At this time, a major part of blood stocks came from paid donors, mostly from poor neighborhoods. To offset taking in contaminated blood, this single company began questioning donors face-to-face: Are you homosexual? In the first 2 weeks of new guideline operations, 308 donors said yes and their blood was not taken. Another 500 refused to answer and left.

The American Association of Blood Banks (AABB) and the American Red Cross issued a joint statement that "Direct or indirect questions about a donor's sexual preferences are inappropriate." Clearly, they kept open the door to high behavioral risk blood donors. The FDA went along with the joint statement and against the CDC recommendations for screening blood donors.

According to Milton Musin, former medical research director of Cutter Laboratories, by the end of 1983 he knew that the hemophiliacs clotting factor could transmit AIDS. And virtually all lots of the concentrate were contaminated with the AIDS virus. "You'll never convince me that profit margins and fear of the product liability and fear of losing a very lucrative business did not drive the CEOs and leaders of these companies."

In 1985, the FDA announced a blood test. "This test adds a major dimension of protection to our present safeguards. Its use will keep our blood supply safe and indeed make them even safer."

The HIV antibody test, called the ELISA test, would help make the blood supply, in large part, safe. But it came at a cost to the blood banks. Technicians had to be hired and trained. Donors had to be checked and every new unit of blood tested and logged. **But, astonishingly, the blood banks were not required to go back to the inventories on their shelves.** In hindsight, the current head of the AABB says that was a mistake.

But perhaps the worst was yet to be learned. While the AABB was stalling testing, it convinced the FDA to reduce the number of blood bank inspections at the very time HIV was entering the blood system of the country.

It wasn't until 1988 that the blood banks began recalling HIV-positive blood. FDA reports were discovered that told of errors and accidents. Between 1985 and 1987 thousands of units of potentially contaminated blood had been released and officially recalled. But recalling blood that was released for transfusion is a bit misleading because, depending on when and how you are trying to recall it, the chances of being able to get it all back are almost zero. Once it is released, it is used. Potentially contaminated blood had been used, most of it quickly, in emergency rooms. The FDA now had no choice but to take action. In September 1988, the FDA and the Red Cross entered into what is known as a voluntary agreement to comply with all FDA regulations. From now on there would be yearly inspections! But that did not work. The Red Cross kept poor records at local blood banks and the FDA constantly had to threaten license revocations.

Finally, in 1993, FDA commissioner, Dr. David Kessler, went to federal court and obtained an injunction against the Red Cross for failure to fulfill its promise under the 1988 voluntary agreement.

It is now estimated that as many as 30,000 Americans were infected with AIDS through blood. It is a tragedy that has exposed critical weaknesses in the rules and practices in blood banking. It's also a story of missed opportunities, vested interests, and lax regulations stretching back more than a decade when a mysterious virus entered America's bloodstream. In November 1993, Donna Shalala, Secretary of Health and Human Services, asked for a high-level investigation of how thousands of American hemophiliacs became HIV-infected from contaminated blood products. As a result of the investigation and other political action,

the Ricky Ray Hemophilia Relief Fund of 1998 was passed by the House on May 19, 1998, the Senate on October 21, 1998, and became Public Law 105-369 when signed by President Clinton on November 12, 1998. The law authorizes payments of $100,000 to 7,200 hemophiliacs who were infected using HIV-contaminated blood or blood extracts, to compensate them or their survivors.

In April 1997—The four major producers of blood-clotting factors made a $640 million offer to hemophiliacs of the United States. In the summer of 1997 over 6,000 hemophiliacs or their families recieved $100,000. Some refused and went to court to settle.

In March 1999, jurors in New Orleans awarded $35.3 million to the parents of a son who died of AIDS. The son was infected with the HIV from the treatment he used for his hemophilia condition. This trial, concerning "Hemophiliac AIDS Cases," began on November 4, 1998, in State court in New Orleans, Louisiana. The jurors found that two of the medication manufacturers, Cutter Biological and Alpha Therapeutics were liable for product liability, negligence and fraud. The jurors further found that after the boy was infected with the AIDS virus (which was unknown to him at the time), the continued use of the Cutter and Alpha-contaminated-product further aggravated his condition, exacerbating and accelerating the development of AIDS. The $35.3 million verdict, with prejudgement interest, comes to a total of approximately $56 million. After the jurors entered their verdict, the judge ruled that the statute of limitations expired and therefore entered a judgement for zero dollars. Plaintiff's counsel is confident that the decision by the judge will be overturned on appeal due to the finding of fraud.

Class Discussion: Do you feel this saga of delayed and controlled misuse of blood in the United States is equal to or greater than the French and/or German scandals? Explain.

USE OF HIV-CONTAMINATED BLOOD OR BLOOD PRODUCTS IN OTHER COUNTRIES

Canada and Nova Scotia

The royal Canadian Mounted Police have launched a full-scale criminal investigation into Canada's tainted blood scandal in which about 2000 people became HIV infected. To date, over 800 Canadian hemophiliacs tested positive for transfusion-associated HIV infections. About 265 have died of AIDS.

A flood of lawsuits is expected to follow the decision to award more than half a million Canadian dollars to the HIV-infected wife of a man who died of AIDS after receiving a transfusion of HIV-contaminated blood. The outcome of this case was announced on the eve of a deadline set by all except one of Canada's provinces and territories for acceptance of a compensation package by more than a thousand recipients of HIV-infected blood. To obtain the money, victims had to sign wavers promising not to sue Ottawa, the provinces or territories, hospitals, the Red Cross, and pharmaceutical and insurance companies. Compensation was set at C$515,000 per case or per person plus interest and legal fees.

The Canadian compensation is less generous than the province of Nova Scotia, which, as well as making annual payments of C$30,000 to the HIV-infected, pays for expensive AIDS drugs, as well as funeral costs and post-secondary education for their children (Spurgeon, 1994, 1996).

In late 1998 the Canadian Blood Services agency started operations, replacing the Canadian Red Cross as the country's nation blood service.

Switzerland

In mid-1994, Alfred Haessig, the former head of the Swiss Red Cross Central Laboratory in Bern was charged with causing "grievous bodily harm" by allowing HIV-tainted blood-clotting factors to be distributed to Swiss hemophiliacs in 1985 and 1986.

The Swiss case resembles the French scandal. In both cases, a national lab continued to distribute blood products that were known to be potentially infected with HIV. France stopped using such products in October 1985, but in Switzerland they were marketed through May 1986.

The suspect Swiss factors were made with blood collected some months before July, 1985, when HIV testing became routine. The trial is likely to center on his failure to stop production of the factors in July, 1985, when the New York Blood Center identified one HIV-positive sample among 3,375 units of donated Swiss blood. But the Swiss Red Cross insists that no alternative supplies were available: "If these products would not have gone out, the hemophiliacs would have bled to death." (Holden, 1994)

Results of Swiss Case: Alfred Haessig Found Guilty

In December of 1998, the former director of the Swiss central laboratory of the Red Cross, was found **guilty,** by a Geneva court, of supervising the distribution of HIV-infected blood products to hemophiliacs. The 77-year old man was given a one-year suspended sentence. The prosecutor in the case argued that Haessig should have been aware of the risk of transmitting HIV through blood products in 1982, but that he supervised the manufacture of the blood products until November 1995 and distribution through May 1996, failing to implement safer options. Haessig, who retired in 1987, asserts that he did not use pasteurization because he thought the process was less efficient in the treatment of hemophiliacs. The case was brought following complaints by eight Swiss hemophiliacs who were infected with HIV.

Japan

In July of 1993, 92 Japanese hemophiliacs filed suit against the Japanese government and blood product manufacturers. They are seeking compensation for failure to protect them from blood products contaminated with HIV. Blood product manufacturers continued to advertise untreated blood products in Japan without warning until late-1985; Japanese hemophiliacs used them into 1986. They are seeking one million dollars each in compensation.

During 1983–1985, Japan dramatically increased its imports of untreated U.S. blood products and, as a result, about 2,000 of Japan's 5,000 hemophiliacs were infected with HIV. Through 1996 over 400 have died. (Swinbank, 1993, 1994).

Update 1996— After a 7-year court battle, Japan's hemophiliacs and their families reached an out of court settlement. Each HIV-affected hemophiliac will receive about $430,000. And those with AIDS will receive an additional $1,400 a month for life (Abbott, 1996).

In August of 1996 Takeshi Abe, Japan's leading expert on hemophilia, was arrested for professional negligence—allowing the use of blood products he allegedly knew were HIV contaminated.

In September of 1996, the president and two former presidents of Green Cross Corporation (Midori Juji), Japan's leading blood product manufacturer, were arrested in connection with the infection of thousands of Japanese with HIV through the use of non-heat-treated blood products in the 1980s. They were being held on suspicion of negligence in promoting the continued sale of non-heat-treated blood coagulants in 1986, after safer heat-treated products, including those of the Midori Juji Company, were approved for sale by the government in 1985. Their trial is pending.

Update 1999— Tapes played at the trial of Akihito Matsumura, a former official of Japan's Ministry of Health and Welfare, indicate that the ministry's AIDS study panel was aware of possible HIV contamination in non-heat-treated blood products in 1983. The panel continued to suggest the use of the products, resulting in a possible 1,800 additional HIV infections. Matsumura is the only ministry official on trial, although four other individuals—including Takeshi Abe, a member of the panel and a hemophilia expert— are also on trial.

India

A blood bank in Bombay operated by the India Red Cross Society (IRCS) has been closed down following a government decision to conduct an inquiry into charges that the center had supplied blood infected with HIV to hospitals in the city between 1992 and 1994.

The medical officer in charge of the blood bank has already been removed from his post, and more officials are expected to lose their jobs. The affair has caused concern throughout India, as the IRCS is intended to set national standards for blood safety.

The affair came to light when the chairman of the IRCS subcommittee on blood transfusion services, examined the records of the blood bank. These showed that HIV-positive blood had been supplied to at least 10 city hospitals, including one that specializes in transfusion services.

In addition, there have been suspicions that employees of the blood bank, rather than discarding infected blood, have been selling it on the black market created by an overall shortage of blood supplies (Jayaraman, 1995).

CHINA

In October of 1996, China's Ministry of Public Health ordered thousands of state-run medical institutions to stop using a certain brand of serum albumin, a common blood product produced by a military-run factory.

Now, 2 years after Chinese journalists in Hong Kong and the United States first published reports that the blood product was contaminated with HIV, China's Foreign Ministry acknowledged that tests of some samples of the blood product have indicated the presence of HIV. But, at the Department of Medicine Administration and Control, a Mr. Liu said: "No, no, it has never happened. Some people outside China and some foreign journalists just made their reports on the basis of rumors."

In October of 1998, in an effort to reduce the spread of HIV, the Chinese government created a law requiring that all blood products come from volunteers. Many countries banned the practice of payment for blood donation long ago, since there were concerns that the people who would be most likely to be strapped for money—such as drug users—would also be at high risk for blood-borne diseases. However, in China it can be hard to get donors because there is a cultural aversion to blood donation. Traditionally, the donation of blood is considered disrespectful to ancestors and parents; the culture also equates blood levels with health. Because China has a chronic blood shortage, many experts believe it will take several years for the illegal blood trade to die down. The government is trying to confront this problem by actively encouraging officials, students, and soldiers to donate blood.

OTHER BLOOD SCANDALS IN BRIEF

In Columbia, a drug-addicted bisexual who knew he had AIDS sold his blood 12 times to a laboratory. Twelve patients received the blood between 1989 and 1990 and went on to infect an additional 200 people through sexual contact.

In Romania during the Communist era, HIV-contaminated blood infected 2,376 children.

Developing countries at the moment must rely on paid donors, who may include gay and bisexual men, prostitutes, and drug addicts needing money. Countries like India, Pakistan, and Russia have an open paid donor system.

Parts of Africa and Latin America use family replacement donors—a scheme that allows paid donors to pose as relatives of the patient. Paying unquestioned donors without HIV screening their blood is an invitation for the transmission of HIV. But developing countries do not have the capital nor sufficient trained personnel to do anything else at the moment.

EFFECTIVENESS OF SCREENING HEALTH CARE WORKERS FOR HIV IN THE UNITED STATES

There is considerable public concern regarding the potential transmission of the human immunodeficiency virus (HIV) from health care worker to patient. This risk is manifest in the results of a 1991 *Newsweek* poll, which found that 90% of Americans favor testing health care workers for HIV and revealing the results to patients. The same poll indicated that 49% of the public believed health care workers should be forbidden to practice if they are HIV-positive, and the majority indicated they would no longer seek treatment from an infected practitioner.

In contrast to this perceived risk, the CDC has estimated the theoretical probability of this type of transmission to be between 0.000024 and 0.0000024. Daniels (1992) points out that the risk of being infected by an HIV-positive surgeon is roughly 10% of the chance of being struck by lightning, 25% as probable as being killed by a bee, and half as likely as being hit by a falling aircraft. Part of the basis of the public's concern is that, in contrast to the other risks mentioned, transmission of HIV from a health care worker to a patient is potentially avoidable. Still, the CDC has recommended against mandatory screening because the low risk does not warrant the anticipated cost associated with testing.

William Chavey and co-workers (1994) examined a screening protocol that included a sequence of antibody tests (enzyme-linked immunoabsorbent assay and the Western blot) and culture for HIV. The incremental cost-effectiveness of applying this protocol as opposed to the status quo for the prevention of transmis-

sion of HIV from health care worker to patient was evaluated. The incremental cost-effectiveness ratio was then compared with that of other interventions. Their study showed that the expected annual cost of screening for a large hospital would be $9,177,615 per transmission prevented. (*Question:* **Is it worth over $9 million to prevent the transmission of HIV to one person?** *Discussion.*) The conclusion reached was that screening health care workers for prevention of potential HIV transmission to patients is an expensive use of health care resources.

The results of another cost-effectiveness study of HIV testing of just physicians and dentists in the United States by Kathryn Phillips and co-workers (1994) showed that, although one-time mandatory testing of surgeons and dentists with mandatory restriction of those found to be HIV-positive is more cost-effective than other policies, the cost-effectiveness varies tremendously under different scenarios. Results were highly sensitive to several data inputs, especially HIV seroprevalence of surgeons and dentists and transmission risk. For example, under a medium seroprevalence and transmission risk scenario, mandatory testing of all surgeons might avert 25 infections at a total cost of $27.9 million or $1,115,000 per infection averted and an incremental cost of $291,000 compared with current testing; however, the incremental cost-effectiveness per patient infection averted ranges from $29,807,000 under a low-risk scenario to a savings of $81,000 under a high-risk scenario. (See Point of Information 13.3)

---

of 14 serodiscordant (one person being HIV positive) male couples, 24% said they engaged in unprotected anal intercourse two months prior to the study, and half of the sample agreed that a reduction in the seropositive partner's viral load decreased the risk of transmission. Similarly, a New York study of men who have sex with men found that men who engaged in unprotected anal intercourse with partners who were seronegative or of unknown status believed that they were less infectious because they were on antiviral therapy and had negligible viral loads.

A study of 298 gay men attending a 1997 gay pride event in Atlanta found that those who had engaged in unprotected analreceptive intercourse indicated that they were less worried about unsafe sex because of treatment advances. Of 54 men surveyed in a 1997 San Francisco study, 26% were less concerned about becoming seropositive now that protease inhibitors were available: 15% were more willing to engage in unsafe sexual behaviors and risk HIV infection and 15% reported already taking such a risk (Halkitis et al., 1999). Also, a study of seronegative or untested gay and bisexual men not in long-term relationships found that 23% who had engaged in unprotected anal receptive intercourse during the previous six months believed that this was safe if an HIV-infected partner had an undetectable viral load (Kalichman et al. 1998). Canadian studies showed similar results: According to one Canadian study, 22% of HIV positive subjects believed that taking some form of antiviral therapy reduced their risk of transmitting HIV to others, and 20% believed that treatment diminished the importance of safer practices for sex and injection drug use (Kravcik et al., 1998).

Heterosexual HIV Transmission

Heterosexual HIV transmission means that the virus was transmitted during heterosexual sexual activities. As such, the **proportion** of HIV infection and AIDS cases among the heterosexual population in the United States is now increasing at a greater rate than the proportion of HIV infections and AIDS cases among homosexuals or IDUs (Friedland et al., 1987). Persons at highest risk for heterosexually transmitted HIV infection include adolescents and adults with multiple sex partners, those with sexually transmitted diseases (STDs), and heterosexually active persons residing in areas with a high prevalence of HIV infection among IDUs. In 1985, fewer than 2% of AIDS cases were from the heterosexual population; by 1989, 5% were from the heterosexual population. For the year 1993, 9% of cases were from the heterosexual population. The large increase is a reflection of the new AIDS definition of 1993. Now overall, the incidence of heterosexual AIDS cases is 9% of the total number of reported AIDS cases in the United States. Of the number of the **heterosexual AIDS cases** reported in the United States, about two-thirds (66%) of cases occur among women and one-third among men.

Heterosexual HIV Transmission Outside the United States— Studies in Africa, Haiti, and other Caribbean and Third World countries indicate that HIV transmission is most prevalent among the heterosexual population. The male-to-female ratio in Africa is 1:1. In late 1991, the World Health Organization stated that 75% of worldwide HIV transmission occurred heterosexually. By the year 2000, up to 90% will occur heterosexually. Homosexuality and injection drug use occurs in Africa but the incidence is reported to be very low. The high frequency of AIDS cases in Third World countries is thought to be due to poor hygiene, lack of medicine and medical facilities, a population that demonstrates a large variety of sexually transmitted diseases and other chronic infections, unsanitary disposal of contaminated materials, lack of refrigeration, and the reuse of hypodermic syringes and needles due to supply shortages.

Transmission from men to women in Nairobi has been shown to be facilitated by common genital ulcers, the use of oral contraceptives rather than condoms, and the presence of chlamydia, a type of bacterium. Chlamydia infection probably increases the inflammatory response in the vaginal walls and increases the likelihood of having lymphocytes there that can attach to the virus and allow transmission. The damage that sexually transmitted ulcerative diseases cause to genital skin and mucous membranes is believed to facilitate HIV transmission. Prevention and early treatment of STDs could slow HIV transmission in the United States and in other countries.

Vaginal and Anal Intercourse— Among routes of HIV transmission, there is overwhelming evidence that HIV can be transmitted via anal intercourse. In vaginal intercourse, male-to-female transmission is much more efficient than the reverse. This is believed to be due to (1) a consistently higher concentration of HIV in semen than in vaginal secretions, and (2) abrasions in the vaginal mucosa. Such abrasions in the tissue allow HIV to enter the vascular system in larger numbers than would occur otherwise, and perhaps at a single entry point.

The same reasoning explains why the **receptive** rather than the **insertive** homosexual partner is more likely to become HIV-infected during anal intercourse. It appears that the membranous linings of the rectum, rich in blood vessels, are more easily torn than are

——— BOX 8.3 ———

RE-INFECTION AMONG HIV-POSITIVE
GAY MEN

Four HIV-positive gay men were talking about sex. The conversation turned to unprotected sex. It turned out that all four had unprotected sex with their HIV-positive partners. On reflection about what they know of their other HIV-positive friends and their sexual behavior, they agreed that unprotected sex among HIV-positive partners appeared to be the norm. One of the four men said that he had unprotected sex with hundreds of men, some with and without HIV. He guessed he must have been **re-infected** many times. Another commented that **reinfection** is occurring in the gay community more times than anyone can believe—we just don't talk about it! Reinfection, beginning in 1997, has taken on new meaning because of the increasing number of reports on the transmission of **drug resistant HIV** (Hecht et al., 1998). The more commonly held belief of "we are already infected so we no longer need to practice safer sex" is wrong. If a person is infected with HIV that is responding to retroviral therapies but is reinfected with a drug resistant strain he or she may have a shortened life time.

Barebacking An Antiretroviral Drug Treatment Phenomenon?

It has recently been observed that a minority of both HIV-positive and -negative men have begun to consciously, willfully, and proudly engage in unprotected anal sex. The new phenominon, referrred to as **raw, skin-to-skin, or bareback** sex may be linked to the perceived effectiveness of antiretroviral therapy and the promise of a morning after pill exposure treatment. It is unclear how many men are engaging in such premeditated risk behavior but the number is growing. While the issue of unprotected anal sex has always been a concern, recent media attention capitalizing on the sound-bite allure of the term "barebacking," many potentially create a self-fulfilling media prophesy. Advo-cates of barebacking offer many accounts for their risky behavior. Some argue that the benefits of unsafe sex outweigh the risks. Some men even see becoming infected as a positive development in their lives. As Jesse Green relates, in his New York Times Magazine profile of Marc Ebenhock, a young ex-marine, struggling with alcoholism and his frustrated search for intimacy with other men, "In a way it's a relief. I don't have to wonder anymore. That awful waiting is gone. So now, if I do find someone, the relationship can be 100% real with nothing in the way. That's what I want: 100% natural, wholesome and real. Maybe now that I'm HIV-positive, I can finally have my life." Thoughts about HIV reinfection do little to dampen their risky behavior.

---

those of the vagina. In addition, recent studies indicate the presence of receptors for HIV in rectal mucosal tissue. A recent report by Richard Naftalin (1992) states that human semen contains at least two components, collagenase and spermine, that cause the breakdown of the membrane that supports the colonic epithelial cell layer of the rectal and colon mucosa. This leads to the loss of mucosal barrier function allowing substances to penetrate the rectal and colon mucosa.

Homosexual Anal Intercourse— Today about 60% of homosexual men in San Francisco are infected, probably the highest density of infection anywhere in the developed world. "It colors everything we do out here," says a gay activist,

"the gay community, to a large extent, is about addressing AIDS. It has to be, because it's literally a war: your entire community is under siege."

In a single year, 1982, 21% of the uninfected gay male population became infected, and for some reason not yet known, many of those infected early, died early. "Soon everyone, and I mean everyone, had a friend who was dying" (Science in California, 1993). During 1995 through mid 1997, three gay men died of AIDS each day. From mid-1997 through 1998, using aggressive anti-HIV drug therapy, daily deaths from AIDS dropped to 1 per day! (Conant, 1995 **updated**).

It appears that of all sexual activities, anal intercourse is the most efficient way to transmit HIV (DeVincenzi et al., 1989). Information

——— BOX 8.4 ———

GAY MEN PUTTING THEMSELVES AT HIGH RISK FOR HIV INFECTION

At the 11th International Conference on AIDS (July 1996), AIDS researchers said that a generation of young gay men across the industrialized world, tragically ignoring the lessons of the AIDS pandemic, risk a new wave of HIV infection by engaging in unsafe sex. In some parts of the United States, **as many as one in 10 gay men under the age of 25 carries HIV,** and the risk appears to be especially high for nonwhite minorities. Such figures are high, very high, especially considering that these young men have become sexually active in an era in which massive effort was exerted to increase awareness of HIV risk behaviors and to promote safer sex.

A variety of surveys have found that about one-third of young gay men in their 20s engage in anal sex without condoms, the riskiest form of homosexual behavior.

Gordon Mansergh reported at the 12th International AIDS Conference (June–July 1998) that of gay men in Denver, Chicago, and San Francisco, almost two-thirds engaged in **unprotected anal sex** in the previous 18 months and that 56% of gay men **under age 25 had unprotected receptive anal sex** in the same time frame.

Experts believe one reason risky behavior continues among the young is that they have not yet seen their friends die of the disease.

Another is simply the kind of risk-taking common among the young—the same impulse that prompts teenagers to drive fast or take up smoking. Indicators of increased risky sex among gay men included: gay men having **unprotected sex** increased from 35% in 1994 to 55% in early 1999. Thirty percent of 3000 gay men surveyed in four large American cities had **unprotected sex**—7% of them were HIV positive. In three large American cities, 66% of gay men surveyed had **unprotected sex** in the last 18 months. **One in five** gay men in San Francisco and New York City said they had **unprotected sex** with an HIV-negative partner or with a partner who's HIV status was unknown. In 1998 the CDC surveyed about 22,000 gay men in San Francisco. Over 39% said they have **unprotected sex**—this is a 30% increase since 1994. Much of this increase is occurring in **younger** gay men—**a group that has not watched their sex partners and friends die from AIDS.**

On January 29th 1999 the San Francisco Chronicle carried a story about an $8 admission for a night of communal sex—the rules: no clothes, no condoms, no discussion of HIV. The article also covers an Internet link offering gay men the **extreme sex party** where becoming infected or HIV-infecting another is the erotic allure of the party. Still another twist is the **Russian Roulette Party** where three non-infected men have sex with five others, one of the five being HIV-positive!

Health officials now worry that some people may be losing their fear of HIV/AIDS. With the rate of HIV infection increasing among people aged 15 to 24 years and in the rural population, the problem spreads. People are aware of HIV infection but it's not in the forefront of their minds like it used to be. Nationally, among young people, there's not the fear there used to be, and that's had a significant impact. Apathy has apparently increased in light of new high-powered anti-HIV drugs that prolongs the quality of life and significantly delays the time of death. However, the drugs are not accessible to everyone and they have a number of adverse effects.

THE HIGH RATE OF HIV INFECTION AMONG GAY MEN ON SOUTH BEACH, MIAMI, FLORIDA

Miami's South Beach—hot music, hot spots, hot sun, hot bodies. The once sagging tip of Miami Beach has been turned into the playground of the beautiful, the rich, and the chic. South Beach is also a growing mecca for young gay men. In a world where homosexuality is often confined to the closet, it is a place to be out and to be open. But behind the perfect tans, there is the dark reality! According to a new study one in every six gay men between the ages of 18 and 29 in South Beach has tested HIV-positive.

William Darrow, AIDS researcher who surveyed 87 gay men ages 18 to 29 and 70 gay men age 30 and older, said, "People seem to think they're on a holiday there from everything—including AIDS." Darrow's survey showed that about 75% of gay men in both age groups had **unprotected anal sex** in 1996.

Testimonial—One gay male living at South Beach said he fell victim to **love.** "I was in a relationship, and we both got tested together and we tested negative, so we started having unprotected sex. But during the relationship, my partner was

———— **BOX 8.4** *(continued)* ————

cheating on me and he was infected and in turn infected me."

A majority of young gay men have grown up thinking they won't reach middle age; for those who aren't positive, the specter of AIDS haunts them. For many of those who have grown up with the safer sex message, the words of warning are often lost. In addition, the media covering the new anti-AIDS drugs, protease inhibitors, are generating **hype and new hope,** but there is a downside: young gay men may be letting their

guards down because of the hype. Another contributing high-risk behavioral factor is the growing number of gay bathhouses and sex clubs cropping up across the country. People come to South Beach to have anonymous sex; they pay to use the facility and condoms are free. Safer sex posters are everywhere. But, some argue that bathhouses are centers for HIV infection; others argue that shutting down bathhouses will drive men in search of sex to public parks or other unmonitored settings.

collected from cross-sectional and longitudinal (cohort) studies has clearly implicated receptive anal intercourse as the major mode of acquiring HIV infection. The proportion of new HIV infections among gay males attributable to this single sexual practice is about 90%.

Major risk factors identified with regard to HIV transmission among gay males include anal intercourse (both receptive and insertive), active oral-anal contact, number of partners, and length of homosexual lifestyle (Kingsley et al., 1990).

Heterosexual Anal Intercourse— A number of sexuality oriented surveys of the heterosexual population indicate that between one in five and one in 10 heterosexual couples have tried or regularly practice anal intercourse. Bolling (1989) reported that 70% to 80% of women may have tried anal intercourse and that 10% to 25% of these women enjoyed anal sex on a regular basis. He also reported that 58% of women with multiple sex partners participated in anal sex. James Segars (1989) reported that the highest rates of anal sex occur among teenagers who use drugs and older married couples who are broadening their sexual experiences.

Although it may increase the risk of HIV infection, it must be emphasized that anal intercourse is not necessary for HIV transmission among heterosexuals. In fact, most HIV-infected heterosexuals say that they have never practiced anal intercourse.

Risk of HIV Infection; Number of Sexual Encounters— The risk of HIV infection to a susceptible person after one or more sexual encounters is very difficult to determine. In

some cases, people claim to have become infected after a single sexual encounter.

In some reported transfusion-associated HIV infections, the female partners of infected males remained HIV-negative after 5 or more years of unprotected sexual intercourse. Television star **Paul Michael Glaser** said that he had normal sexual relations with his wife Elizabeth for 5 years prior to her being diagnosed as HIV-infected (Figure 8-7). He was not HIV-infected. She received HIV during a blood transfusion, but was not diagnosed until after their first-born child was diagnosed with HIV. In other studies of heterosexual HIV transmission, many couples had unprotected sexual intercourse over prolonged periods of time with no more than 50% of the partners becoming HIV-infected. There are many instances of heterosexuals and homosexuals who remained HIV-negative after having repeated sexual intercourse with HIV-infected partners.

The fact that not all who are repeatedly exposed become infected suggests that biological factors may play as large a role in HIV infection as behavioral factors. For biological reasons, some people may be more efficient transmitters of HIV; while others are more susceptible to HIV infection, that is, require a smaller HIV infective dose. And, it is now known that some people carry genes that may make them resistant to HIV infection (see "Mutant CKR-2 and CKR-5 Genes" in Chapter 5).

Number of Sexual Partners and Types of Activity— One relatively large risk factor for HIV infection in both homosexuals and heterosexuals is believed to be the number of sexual part-

FIGURE 8-7 Paul Michael and Elizabeth Glaser. Elizabeth died of AIDS on Dec. 4, 1994. (*Photograph couresy of AP/Wide World Photos*)

SIDEBAR 8.2

A CAB DRIVER AND PROSTITUTES IN GHANA

In Accur, Ghana, a cab driver was asked **what is AIDS? The driver replied "I don't know but I hear if you go after a lot of women you will get it. I don't drink and I don't smoke so I don't think I will get it."** The answer to the question sums up the level of his awareness. In an interview with prostitutes on one of the FM stations they stated that although they had heard about AIDS, they were strict adherents to the customer is always right philosophy. So if a client insists on unprotected sex they had little option than to ask for double the price for sex without a condom. Among the abundant rural illiterate and (surprisingly) some urban dweller, the multiplicity of man's sexual partners is an insignia of machismo or a symbol of virility. In short, infidelity among men is a norm while the same society frowns upon a similar practice among women. This means that by their action Ghanian men are more at risk of getting infected by HIV and passing it on to others. Recent increases in prostitution and promiscuity among young girls have contributed to the transmission of HIV. Ghanaian youths, in general, consider the use of condoms as archaic and with the rising incidence of teen-age pregnancies, there is also an increase in HIV-positive teenagers.

ners. The greater the number of sexual partners, the greater the probability of HIV infection.

The scale of multiple-partnering during the late 20th century is unprecedented. With some 6 billion people on earth, an ever-increasing percentage of whom are urban residents; with air travel and mass transit available to allow people from all over the world to go to the cities of their choice; with mass youth movements advocating, among other things, sexual freedom; with a feminist spirit alive in much of the industrialized world, promoting female sexual freedom; and with **47% of the world's population made up of people between the ages of 15 and 44**—there can be no doubt that the amount of worldwide urban sexual energy is unparalleled.

The amount of protection one actually obtains from limiting one's number of partners depends mainly on who those partners are. Having one partner who is in a high-risk group may be more dangerous than having many partners who are not. An example of this is seen in prostitutes, who may be more likely to be infected by their regular injection drug using partners than by customers, who are not in a high-risk group. The risk status of a person who remains faithful to a single sexual partner depends on that partner's behavior: if the partner becomes infected, often without knowing it, the monogamous individual is likely to become infected (Cohen et al., 1989).

Sexual Activities— In addition to a high-risk partner or a number of sexual partners, the types of sexual activities that occur are also significant (Table 8-7). Any sexual activity that produces skin, anal, or vaginal membrane abrasions prior to or during intercourse increases the risk of infection.

OROGENITAL SEX. Orogenital sex may be a greater risk factor for becoming HIV-infected

TABLE 8-7 Sexual Activity According to Degree of Risk for Transmitting HIV

Lowest risk
1. Abstinence
2. Masturbating alone
3. Hugging/massage/dry kissing
4. Masturbating with another person but not touching one another
5. Deep wet kissing
6. Mutual masturbation with only external touching
7. Mutual masturbation with internal touching using finger cots or condoms
8. Frottage (rubbing a person for sexual pleasure)
9. Intercourse between the thighs
10. Mutual masturbation with orgasm *on*, not *in* partner
11. Use of sex toys (dildos) with condoms, or that are not shared by partners and that have been properly sterilized between uses
12. Cunnilingus
13. Fellatio without a condom, but never putting the head of the penis inside mouth
14. Fellatio to orgasm with a condom
15. Fellatio without a condom putting the head of the penis inside the mouth and withdrawing prior to ejaculation
16. Fellatio without a condom with ejaculation in mouth
17. Vaginal intercourse with a condom correctly used and spermicidal foam that kills HIV and withdrawing prior to orgasm
18. Anal intercourse with a condom correctly used with a lubricant that contains spermicide that kills HIV and withdrawing prior to ejaculation
19. Vaginal intercourse with internal ejaculation with a condom correctly used and with spermicidal foam that kills HIV
20. Vaginal intercourse with internal ejaculation with a condom correctly used but no spermicidal foam
21. Anal intercourse with internal ejaculation with a condom correctly used with spermicide that kills HIV
22. Brachiovaginal activities (fisting)
23. Brachioproctic activities (anal fisting)
24. Use of sex toys by more than one partner without a condom and that have not been sterilized between uses
25. Vaginal intercourse using spermicidal foam but without a condom and withdrawing prior to ejaculation
26. Vaginal intercourse without spermicidal foam and without a condom and withdrawing prior to ejaculation
27. Anal intercourse with a condom and withdrawing prior to ejaculation
28. Vaginal intercourse with internal ejaculation without a condom but with spermicidal foam
29. Vaginal intercourse with internal ejaculation without a condom and without any other form of barrier contraception

Highest risk
30. Anal intercourse with internal ejaculation without a condom

(Source: Shernoff, 1988)

than previously thought. The biological risk of HIV transmission from oral sexual contact is not known, but the risk of transmission is related to the presence of HIV at the sexual sites (oral, vaginal, penile, and anal), the amount of HIV present and whether there are physical opening such as tissue tears or open sores. (Rothenberg et al., 1998; Kahn et al., 1998) Although orogenital sex may be less efficient than needle-sharing or anal sex for HIV transmission, its increased use by gay males and in the sexual activity of **crack smokers** may help

with HIV transmission. Of 82 HIV-infected gay men in three San Francisco area studies, 14 (17%) gave orogenital sex as their only high-risk behavior (AIDS Research, 1990). A case was reported of a heterosexual male becoming HIV-infected after receiving oral sex (fellatio) from a prostitute who was HIV-infected (Fischl et al., 1987). Ireneus Keet and colleagues (1992) reported on several studies on the risk of HIV transmission among gay males. They concluded that the orogenital route of HIV transmission is difficult to assess. Information on questionnaires was frequently contradicted in follow up interviews. For example, of 20 men who denied having receptive anal intercourse, 11 later changed their statements. However, Keet and colleagues concluded that there is sufficient evidence to conclude that orogenital HIV transmission does occur.

Madalene Heng and colleagues (1994) reported that HIV-infected patients with oral herpes simplex lesions are at risk of transmitting

SPORTS AND HIV/AIDS: EARVIN "MAGIC" JOHNSON AND OTHER ATHLETES

HIV-INFECTED ATHLETES AND COMPETITION

The question regarding whether HIV-infected athletes should be allowed to compete has two facets:

1. Should these athletes be banned from competition to avoid the risk of spreading HIV infection?

2. Does the exercise that is demanded in competition accelerate progression of HIV disease?

As yet, there is no hard, fast, scientifically supported answer to either question. However, beginning 1999, there has not been a single reported case of HIV transmission in any sporting event worldwide! (Drotman, 1996, updated).

Magic Johnson: Professional Basketball Player, Los Angeles Lakers, HIV-Positive

On November 7, 1991, Magic Johnson age 32 appeared at a nationally televised press conference and said, "Because of the HIV virus I have obtained, I will have to announce my retirement from the Lakers today." He admitted having been "naive" about AIDS and added, "Here I am saying **it can happen to anybody,** even me, Magic

FIGURE 8-8 Earvin "Magic" Johnson, wife Cookie, and son Andre arrive for a special ceremony of *Hoodlum* at a Magic Johnson owned theater. Magic has become a spokesman for those struggling against HIV/AIDS. (*Photograph courtesy of AP/Wide World Photos/Michael Caulfied*)

BOX 8.5 *(continued)*

Johnson." He also assured the world that his wife, Cookie Kelly, 2 months pregnant, had tested negative for the virus.

The National AIDS Hotline lit up with 40,000 phone calls on the day of Johnson's announcement, instead of the usual 3,800. At the Centers for Disease Control and Prevention in Atlanta, AIDS-related calls, which usually average 200 per hour, jumped to 10,000 in a single hour. Prior to Magic's announcement, the black community lacked a focal point for action: no leader of national stature had claimed a prominent space from which to address HIV disease. Few black politicians, entertainers, civil rights leaders, or sports figures had tackled the issue of AIDS either through public service announcements, television talk shows, radio addresses, church pulpits, theaters, or school auditoriums. Magic's announcement—at least for a time—changed that.

Some Events Since Magic's Announced Retirement

June 4, 1992—Earvin the III is born *without* antibody to HIV. As of August 14, 1999, Magic is age 41. His wife, age 41, and baby, age 9, are HIV-negative (Figure 8-8).

Magic announced in April 1997 that his AIDS drug combination which included a protease inhibitor has reduced his viral load to unmeasurable blood levels!

June 8, 1998—Magic became the host of "The Magic Hour," television's new syndicated late-night talk show carried on Fox-owned TV stations. The program failed. Magic owns 6 companies involving real estate development, movie theaters and merchandising.

OTHER SPORTS, OTHER ATHLETES

Basketball players are not the only athletes whose behavior may place them at risk for HIV infection. In 1996 there were 6,114 professional athletes in the United States involved in Boxing (3,500), National Football League (1,590), National Hockey League (676), and the National Basketball Association (348). Of these professionals, the CDC estimated 30 to be HIV-positive.

John Elson (1991) wrote a revealing article for *Time* magazine just after Magic Johnson revealed his HIV status. Elson tells of groupies that follow athletes in all sports. They are usually college-age or older. Mainly they seek money, attention,

and the glamor of associating with celebrated and highly visible "hard bodies." According to a 31-year-old who has had affairs with athletes in two sports, "For women, many of whom don't have meaningful work, the only way to identify themselves is to say whom they have slept with. A woman who sleeps around is called a whore. But a woman who has slept with Magic Johnson is a woman who has slept with Magic Johnson. It's almost as if it gives her legitimacy."

The Girls

Baseball players call them **"Annies."** To riders on the rodeo circuit, they are **"buckle bunnies."** To most other athletes, they are **"wannabes"** or just **"the girls."** They can be found hanging out anywhere they might catch an off-duty sports hero's eye and fancy, or in the lobbies of hotels where teams on the road check in. To the athletes who care to indulge them—and many do—these readily available groupies offer pro sport's ultimate perk: free and easy recreational sex, no questions asked. Recently, an HIV-infected female stated publicly that she had had sex with at least 50 Canadian ice hockey players. She could not recall their names. **The sex may be free, but now there is a price for the lifestyle— HIV/AIDS.**

Sports/Injuries/Blood

Concerns over the transmission of HIV are shared throughout sports, particularly those sports that cause blood-letting injuries—football, hockey, and boxing. In football Jerry Smith, a former Washington Redskin, died of AIDS in 1986; no others are known at this time. In 1996, the National Football League officials estimated that there was the **possibility** of one HIV transmission from body fluid exchange in 85 million football games played. In boxing, Esteban DeJesus, WBC lightweight champion, died of AIDS in 1989. Four other boxers are known to be HIV-positive.

Tommy Morrison

In February of 1996 Tommy Morrison, age 27, a heavyweight boxing title contender (Figure 8-9) said, on announcing that he is HIV-positive, "I honestly believed I had a better chance of winning the lottery than contracting this disease. I've never been so wrong in my life. I'm here to tell

——— **BOX 8.5** *(continued)* ———

you I thought that I was bulletproof, and I'm not." Morrison went on to describe his promiscuous sexual lifestyle and ignorance about AIDS. Former heavyweight champ Floyd Paterson, the current chairman of the New York State Athletic Commission, who was asked why his organization waited until the Tommy Morrison case to institute HIV testing for boxers, said "AIDS just came out, I go back to the '50s. I fought for 23 years. There was no AIDS. **I just heard of AIDS a few weeks ago." Just heard of AIDS?** Since Morrison's announcement, 12 states have banned HIV-positive professional boxers from the ring. In November 1996, Morrison fought in Japan. He knocked his opponent out in 1:38 (1 minute, 38 seconds). The fight would have been stopped at the first sign of Morrison's blood.

FIGURE 8-10   Greg Louganis. (*Photograph courtesy of Archive Photos*)

FIGURE 8-9 Tommy Morrison. Las Vegas, NV, 7 June 1993—Tommy Morrison raises his right hand after being declared the new WBO heavyweight champion. Morrison won the 12-round bout against former heavyweight champion George Foreman at the Thomas & Mack Center. (*Photograph courtesy of Archive Photos*)

Update July 1997 POZ Magazine Interview— Tommy Morrison now believes that HIV is not a disease-causing virus! **"AIDS has been here since creation, but is doesn't do anything"** and as for AIDS, he says, **"Acquired Immune Deficiency Syndrome is something that's easily cured."** With regard to taking anti-HIV drugs, he "fired" his physician who prescribed them for him (Dr. David Ho, *Time Magazine* **Man of the Year 1996**). He points to an HIV-infected friend and says **"this kid was perfectly healthy until he started taking medication."** He says that he and his wife have **unprotected** sex because **"HIV can't hurt you. There is no way you can get HIV from sex. It's scientifically impossible."**

Morrison refers to himself as the most educated person he knows regarding HIV (he read Peter Duesberg's book, *Inventing the AIDS Virus* and Richard Wilner's *Deadly Deception*). "I don't know how I know the things I know, I just know I'm right—I have not been sick in 5 years—if it ain't broke don't fix it." Fighting weight about 240 pounds–early 1998 he weighed about 185.

By early 1998, 9 states, (Oregon, Indiana, Washington, Pennsylvania, Nevada, Arizona, Maryland, New Jersey, New York) and Puerto Rico required all professional boxers to be HIV tested. There are no known cases of HIV transmission via a boxing match.

There is a certain irony in all of this because as part of his probationary terms of a suspended sentence for weapons violation and assault

BOX 8.5 *(continued)*

charges, he gives speeches to high school and college students on HIV/AIDS!

Ice Skating

In professional ice skating, the Calgary Herald reported that, by 1992, at least 40 top United States and Canadian male skaters and coaches have died from AIDS (among them, Rob McCall, Brian Pockar, Dennis Coi, and Shawn McGill). In February 1995, Greg Louganis, the greatest diver in Olympic history, announced that he has AIDS (Figure 8-10).

Race Car Drivers

In 1996, Tim Richmond, race car driver, age 34, died of AIDS. He won 13 Winston Cup races on The NASCAR racing circuit. One report states that Richmond may have infected up to 30 women (Knight-Tribune Service, March 27, 1996, A-1). His physician estimated that he was HIV-positive for at least 8 years. During this time, according to accounts of friends and so on he was sexually promiscuous. (Richmond actually died in 1989 but his story was kept silent until 1996.)

Ice Hockey

Bill Goldsworthy, five-time NHL All-Star, age 51, died of AIDS. He played 14 seasons in the NHL. He was diagnosed with AIDS in 1994. Goldworthy said his health problem was caused by drinking and sexual promiscuity.

HIV to others through oral sex. They tested keratinocytes (live skin cells) from the oral lesions of six men with AIDS and oral herpes infection (HSV-1), and compared those biopsies from six men with HSV-1 but not HIV.

The tissue from the HIV/herpes patients was infected with both viruses, and the number of virus in that tissue was much higher than the number found in samples from the other men—736 particles per skin cell compared to 31 and 0 in the other two groups.

This shows that HIV is capable of infecting epidermal cells when herpes virus is present. Epidermal keratinocytes were thought to be resistant to HIV infection because they lack the CD4 receptor molecule.

Data from this study is similar to that from a previous study of brain cells of people infected with both HIV and cytomegalovirus. The brain tissue also lacks CD4 receptors, yet HIV was able to invade the cells and replicate.

The researchers speculated that the association of HIV envelope proteins with HSV-1 proteins may allow HIV to infect cells without the aid of CD4 molecules.

Sara Edwards and colleagues (1998), after reviewing over 100 research reports, concluded that HIV can be transmitted through oral sex. Their message is that one needs to use portection for oral sex. Also, after reading the six articles, by Jeffrey Lawrence, Alison Quayle, Gerad Ilarca, Alan Lifson, Michael Samuel, and Rebecca Young (1995), adapted from presentations at a seminar "**Oral Sex and Possible HIV Transmission: A Community Discussion;**" after reading *"The Riddle of Oral Sex"* by Robert Marks (1996); and after reading Michelle Berrey's (1997) article on "Oral Sex and HIV Transmission" the author of this text is convinced that those persons who participate in **oral receptive sex** are placing themselves at risk for HIV infection. Periodontal disease, herpes lesions, and other conditions may make an individual more susceptible to transmission. Reported within these and other articles, there are a number of cases where oral sex was the only at risk activity practiced and the oral receptive partner became HIV-positive. The articles listed are recommend reading for those who need to be convinced that orogenital sex places one at risk for HIV infection.

Additional articles are those of G.M. Liuzzi and co-workers (1995) and Omar Bagasra (1996) who determined quantitatively the cell-free HIV RNA molecules in semen, saliva, and plasma from HIV-infected people.

PROSTITUTION (SEX WORKER). There is little if any evidence that prostitutes in the United States and other developed nations play a large role in heterosexual HIV transmission (Cohen et al., 1989).

An unfortunate consequence of the attention prostitutes or sex workers have attracted in relation to AIDS in the United States and other countries, is that they tend to be seen as responsible for the spread of HIV—an attitude reflected in descriptions of prostitutes as reservoirs of infection or high-frequency transmitters. But, the sex worker is only the most visible side of a transaction that involves two people: for every sex worker who is HIV-positive there is, somewhere, the partner from whom she or he contracted HIV. Given the fact that the chance of contracting HIV during a single act of unprotected sex is not high, infection in a sex worker is likely to mean that she or he has been repeatedly exposed to HIV by clients who did not or would not wear condoms. Thus the more useful way of reading the statistics of HIV infection in prostitutes or sex workers is to view them as an indication of how strong a foothold the epidemic has gotten within a community.

Risk Estimates for HIV Infection During Sexual Intercourse in the Heterosexual Population—

Norman Hearst and Stephen Hulley of the University of California, San Francisco calculated that the odds of heterosexual HIV transmission range from **one in 500 for a single act of sexual intercourse with an HIV-infected partner when no condom is used to one in five billion if a condom is used and the sexual partner is HIV-negative at the time.**

Table 8-8 presents estimates of the risk of HIV infection from a single heterosexual encounter and after 5 years of frequent heterosexual contact for various types of partners. Risk depends on the following: (**1**) the probability that the sexual partner carries the virus; (**2**) the probability of infection given a single sexual exposure to an infected partner; and (**3**) the reduction in risk by using condoms and spermicides.

The most striking feature of the table is the large variation in the risk of HIV infection under different circumstances. The most important cause of this variation is risk status of sexual partners (Figure 8-11). Choosing a partner who is not in a high-risk group provides almost 5,000-fold protection compared with choosing a partner in the highest-risk category.

Condoms are estimated to provide about 1,250-fold protection. A negative HIV antibody test provides about 2,500-fold protection, against false negatives.

The implication of this analysis is clear: **Choose sex partners carefully and use condoms.**

In a study by Nancy Padian and colleagues (1991), only 1% of HIV-positive women passed HIV by sexual contact to their male partners. In contrast, one of every five uninfected female partners of HIV-infected men acquired the virus through sex. Overall, the study revealed that women are 17.5 times more likely to become HIV-infected from an infected male than men are to contract the disease from an infected female. The ratio of 18:1 came from a limited study (379 heterosexual couples) and as such may underestimate the relative frequency of female-to-male HIV transmission. But, Padian's findings support the Centers for Disease Control and Prevention figures showing that 90% of the more than 33,000 adults who, at the time, became HIV-positive through heterosexual contact were women.

In a follow up study, the largest epidemiological study yet conducted aimed at determining HIV infectivity via sexual intercourse, Padian and colleagues (1997) reported that it took an average of 1,000 sexual acts with an HIV positive male for an HIV negative female to become HIV positive. And 9,000 sexual acts for an HIV negative male to become HIV positive through sex with an HIV positive female. In short HIV has an extremely low infectivity rate. If a female child is raped by an HIV positive man, and the HIV seroprevalence in the country in question is 10% for instance, the odds on her becoming HIV positive due to rape would be one in 10,000.

Risk Taking in the United Kingdom (UK)

A survey of 2,000 people over age 15, by the Terrance Higgins Trust, included questions on risky or unsafe sexual behavior. It showed that although almost 3,000 people were diagnosed with AIDS in the UK during 1998, many thousands continue to have unsafe sex. Twenty percent or one in five surveyed believed there is a

Risk Category of Partner	Estimated Risk of Infection	
	1 Sexual Encounter	500 Sexual Encounters
HIV SEROSTATUS UNKNOWN		
Not in any high-risk group		
Using condoms	1 in 50,000,000	1 in 110,000
Not using condoms	1 in 5,000,000	1 in 16,000
High-risk groups[a]		
Using condoms	1 in 100,000 to 1 in 10,000	1 in 210 to 1 in 21
Not using condoms	1 in 10,000 to 1 in 1,000	1 in 32 to 1 in 3
HIV SERONEGATIVE		
No history of high-risk behavior[b]		
Using condoms	1 in 5,000,000,000	1 in 11,000,000
Not using condoms	1 in 500,000,000	1 in 1,600,000
Continuing high-risk behavior[b]		
Using condoms	1 in 500,000	1 in 1,100
Not using condoms	1 in 50,000	1 in 160
HIV SEROPOSITIVE		
Using condoms	1 in 5000	1 in 11
Not using condoms	1 in 500	2 in 3

[a]High-risk groups with prevalences of HIV infection at the higher end of the range given include homosexual or bisexual men, injection drug users from major metropolitan areas, and hemophiliacs. Groups with prevalences at the lower end of the range include homosexual or bisexual men and injection drug users from other parts of the country, female prostitutes, heterosexuals from countries where heterosexual spread of HIV is common (including Haiti and central Africa), and recipients of multiple blood transfusions between 1983 and 1985 from areas with a high prevalence of HIV infection.

[b]High-risk behavior consists of sexual intercourse or needle sharing with a member of one of the high-risk groups.

(Source: Adapted from Hearst and Hulley, 1988.)

cure for AIDS! Only **half** entering a new sexual relationship would use a condom. Eleven percent said they would ask their new sexual partner to be HIV tested. Seventy-five percent said they did nothing to alter their lifestyle because of HIV/AIDS.

Injection Drug Users and HIV Transmission

HIV entered injection users during the mid-70's and spread rapidly through 1983 largely unrecognized and unidentified. HIV transmission via IDU is the second most frequent risk behavior for becoming HIV-infected in the developed world (Table 8–9). Illicit drug injection occurs in at least 121 countries and HIV infection has been reported in IDUs in 98 of these countries. Beginning year 2000, of 746,000 cases of AIDS reported to CDC, 268,560 (36%) were directly or indirectly associated with injecting-drug use. Injecting-drug-user-associated AIDS cases include persons who are IDUs ($n = 235,259$), their heterosexual sex partners ($n = 28,199$), and children ($n = 5,102$) whose mothers were IDUs or sex partners of IDUs. About half of the females and about one third of the heterosexual males who were diagnosed with AIDS had a sex part-

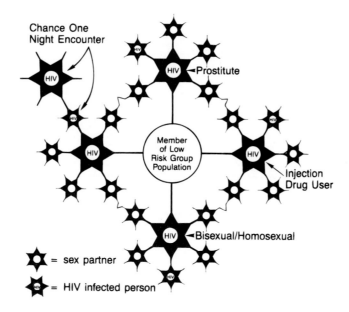

FIGURE 8-11   Risk Transmission of HIV. Sexual transmission can occur
among homosexuals or heterosexuals. Prostitutes can be either male or
female. The diagram shows possible bridges for transmission of HIV from
high-risk groups into low-risk groups. To be safe, *you* must not become
part of the chain. Be sure your sexual partner is free of HIV.

ner who was an IDU (*HIV/AIDS Surveillance
Report*, 1998). It is estimated that half of the
40,000 new HIV infections in the United States
each year, or 55 people per day through the year
2000, will be associated with IDU's. State surveillance reports for New York indicate that 60%
of IDUs are HIV-infected; in New Jersey, 80%
are HIV-infected. Because 70% to 80% of IDUs
have sex with non-drug users, IDUs are a major
source of heterosexual and perinatal HIV infection in the United States and Europe. HIV is
also spreading rapidly among IDUs in developing countries such as Brazil, Argentina, and
Thailand (Des Jarlais et al., 1989).

Andrew Ball of the WHO said that HIV
transmission through IDU has been reported
in 98 countries. It appears that IDU is the
major mode of HIV transmission in Kazakhstan,
Malaysia, Vietnam, China, North America,
Eastern Europe, The Newly Independent

States, and the Middle East. Additionally, it is
becoming more of an issue in West Africa and
Latin America. The greatest problem has been
seen in the Newly Independent states and in
Eastern Europe. In Ukraine, there were about
50 HIV infections reported annually until 1995.
However by the end of 1996, there were over
12,000 reported, with at least half of the infections due to injection drug use. Similar increases were observed in Belarus, Moldova,
and the Russian Federation.

**Over Ninety percent of injection drug users
in the United States are heterosexuals. Thirty
percent are women, of whom 90% are in their
childbearing years.** From 1988 begining year
2000, female IDUs made up 47% of all AIDS
cases in women. Of the 48% of women infected
by **heterosexual** contact 41% were infected
by having sex with a male IDU! During this
same period over 7,000 new cases of AIDS in

TABLE 8-9 Population[a] Risk of HIV Transmission Based on Type of Activity[b]

Activity	Risk	Notes and Sources
Vaginal sexual intercourse	1 infection per 1,000 acts with HIV-positive partner (0.1%)	Mean per-act risk for unprotected intercourse. Source: Isabelle de Vincenzi, European Study Group on Heterosexual Transmission of HIV, 1994
Receptive anal intercourse	5 to 10 infections per 1,000 acts with HIV-positive partner (0.5%–3%)	With no condom use. Source: Victor DeGruttola, Harvard School of Public Health, 1989
Intravenous drug injection with infected needed	10 to 20 infections per 1,000 needle uses (1%–2%)	Source: Don Des Jarlais, Beth Israel Medical Center, New York
Accidental stick in medical setting with infected needle	3 infections per 1,000 sticks (0.3%)	Source: Centers for Disease Control and Prevention
Transfusion of screened blood	1 infection per 450,000 to 650,000 donations (0.0002%–0.00015%)	Source: Centers for Disease Control and Prevention/American National Red Cross

[a]Rough estimates of the relative risks in the United States and Western Europe of the more risky activities that can transmit HIV. **THESE CALCULATIONS CANNOT BE USED AS A GUIDE TO INDIVIDUAL BEHAVIOR.** Risk to any one person depends on many factors that cannot be reduced to a single number. Recent research, for example, suggests that the infectiousness of HIV can vary greatly over the life of an infected person; infectiousness is likely to be high both at the very outset of the infection, before symptoms have appeared, and several years later. Also, women may be several times more likely than men to be infected during vaginal intercourse, a distinction that the overall risk figure obscures. (Adapted from Bennett et al., 1996)

[b]Risks vary because of differences in genital, anal, and oral mucous membranes. Sex is risk free, with or without a condom **if the partner is known to be uninfected**. Nonpenetrative sex such as mutual masturbation is nearly risk free. Sex involving penetration with fingers is low risk. Sex involving unshared sex toys is risk free. Receiving oral sex is less risky than other penetrative behaviors. Penile-receptive sexual behavior is more risky than its penile-insertive counterpart. Anal or vaginal sex without a condom is far more risky than performing oral sex without a condom. (Source: Voelker, 1996)

children occurred—37% were from IDU mothers and 18% were from mothers whose sex partners were IDUs (*MMWR*, 1992; *HIV/AIDS Surveillance Report*, 1998, updated). Women drug users and sexual partners of male drug users represent the largest part (61%) of the estimated 100,000 HIV-infected women of childbearing age. Thus, there is a direct correlation between HIV perinatal transmission and pediatric AIDS cases and injection drug use. In addition, 30% to 50% of female injection drug users have engaged in prostitution, and as such represent the largest pool of HIV-infected heterosexuals in the United States and Europe (Drucker, 1986). The use of alcohol (Avins, 1994) and the use of other noninjection drugs such as crack cocaine are believed to play an increasing role in HIV transmission among heterosexuals.

Over 80% of reported adult **heterosexual** AIDS cases are associated with people who have a history of injection drug use or have sex with IDUs. Twenty-six percent of *all* AIDS cases in the United States occur in IDUs; 21% of these cases occur where IDU is the only risk factor. Of pediatric AIDS cases, 91% occur perinatally. Of these, 60% are associated with IDU by either the mother or her sexual partner *(HIV/AIDS Surveillance Report*, 1998).

The prevalence of HIV infection in IDUs varies widely with geography. Data from more than 18,000 people tested in nearly 90 surveys consistently show high rates in eastern towns and cities with close proximity to New York City and northern New Jersey. Eighty-two percent of AIDS cases among all **reported** IDUs have been in New York City. It should be noted that most of the surveys were conducted in drug treatment facilities for heroin abuse. Since only 10% to 20% of the estimated 1,100,000 IDUs are currently in treatment, geographical conclusions based on the surveys may be mis-leading. More data are needed on HIV prevalence among injection drug users not currently in treatment.

Other Means of HIV Transmission

Other means of HIV infection have been documented. There has been a reported case of HIV transmission via **acupuncture.** It is believed that HIV-infected body fluids contaminated the acupuncture needles (Vittecoq et al., 1989). Also, **artificial insemination** can be a means of HIV transmission (Point of Information 8.2). Unlicensed and unregulated **tattoo** establishments may also present an unrecognized risk for HIV infection to patrons. If the operator does not use new needles or needles which have been autoclaved (steam sterilized), the possibility exists that infection with HIV or a number of other blood-borne pathogens may take place. In addition, single-service or individual containers of dye or ink should be used for each client.

Human Bites— According to police reports, a Florida woman with a history of arrests for prostitution and who has tested HIV-positive, bit a 93-year-old man on his arm, head, and leg while robbing him. The bites required stitches. The man initially tested HIV-negative but a test several months later was HIV-positive. A complete investigation into his personal life ruled out previous HIV infection.

In October of 1995 the CDC confirmed that a 91-year-old male became HIV-infected when, **being robbed,** he was bitten deeply, inflicting extensive tissue damage to the hand by a prostitute with bleeding gums. The man tested HIV-negative shortly after the bite but tested positive 8 weeks later. There have been other reports in the medical literature in which HIV appeared to have been transmitted by a bite. Severe trauma with extensive tissue tearing and damage and presence of blood were reported in each of these instances. Biting is not a common way of transmitting HIV. In fact, there are numerous reports of bites that did not result in HIV infection.

Sexual Assault— The subject of sexual assault, in all its forms, of children by adults or adult on adult is beyond the scope of this text. However, each time there is a date-rape or any other type of rape there is the chance that the rapist may be HIV-positive. One in five adult

——— **POINT OF INFORMATION 8.6** ———

RAPID SPREAD OF HIV THROUGH INJECTION DRUG USE

Don Des Jarlais (1995 updated) writes that the injection of illegal drugs has now been reported in 121 countries throughout the world, and HIV infection among injecting drug users (IDUs) has been reported in 98 countries. In some areas, the spread of HIV among IDUs due to the multiperson use of drug injection equipment has occurred extremely rapidly. In New York City, HIV seroprevalence among IDUs increased from less than 10% to more than 50% in 5 years; in Edinburgh, Scotland, HIV seroprevalence among IDUs increased from 0 to more than 40% in 1 year; in Bangkok, Thailand, HIV seroprevalence among IDUs increased from 2% to more than 40% in 2 years; and in the Indian state of Manipur, HIV seroprevalence increased from 0 to approximately 50% in 1 year. Clearly, HIV has spread rapidly among some populations of IDUs.

In sharp contrast, however, there are also areas where HIV was introduced into the local population of IDUs, but where HIV seroprevalence has remained low and stable for extended periods. In these areas (for example, Sydney, Australia; Toronto, Canada; Lund, Sweden; Glasgow, Scotland; Tacoma, United States), IDUs have access to sterile injection equipment.

women has been the victim of a completed rape at some time in her life (Koss et al., 1991). A **conservative estimate** of the **risk of transmission** of HIV **from sexual assault** (that invoved anal or vaginal penetration and exposure to ejaculate from an HIV-infected assailant) is greater **than two infections per 1,000 contacts,** given a variety of factors, such as clinical stage of the assailant's HIV infection, strength of the strain of HIV and repeated exposure. The per-contact risk is higher if there was violence producing trauma and blood exposure or the presence of inflammatory or ulcerative sexually transmitted diseases (Gostin, 1994). Although 61% of the women stated that physicians should routinely ask about these experiences, only 4% had been asked by their physicians (Walker et al., 1993). Without a complete sexual history that includes questions about rape and so forth,

——— BOX 8.6 ———

HIV TRANSMISSION: PREPARATION AND USE OF INJECTION DRUGS

Group use of drug paraphernalia and the type of substance used are factors in transmission of HIV from person to person.

Drugs in powdered form are usually placed in a bottle cap, or "**cooker**," an item often found on streets or in garbage cans. Water is added and heated to dissolve the powder into a solution.

The solution is withdrawn by needle through a "cotton," or a filter through which the drug solution is drawn into the syringe. Cotton swabs, lint from clothing, cigarette filters, and a variety of other materials are used.

The needle then punctures a vein wherever there is one that can be used. Blood is drawn up into the syringe to mix with the drug solution and the blood–solution mixture is injected into the vein. If the vein is missed, the drug injected subcutaneously (under the skin) hurts and may cause an abscess. Small quantities of a drug may be injected repeatedly—each time blood is drawn up into the syringe. Then, following drug injection(s), the syringe is refilled several times with blood from the vein to wash out any heroin, cocaine, or other drug left in the syringe after the injection(s). If even a tiny, invisible amount of HIV-infected blood is left on the needle or in the syringe, the virus can be transmitted to the next user.

The "**works**," as the syringe and needle are referred to, are shared and the amount of sharing depends on the number of users present. Everyone after the first user in the "shooting gallery" line receives potentially HIV-infected lymphocytes from all those with HIV infection who used the equipment before them.

It should be noted that in the injection drug culture, the sharing of needles is a sign of comradery, a sign of mutual trust. An unwillingness to share needles may cause others in the group to become suspicious of possible police connections.

Drug paraphernalia are commonly rented in **shooting galleries** by users who lack the equipment. A set of "works" is usually rented out until the needle is too dull to penetrate the skin. By then the needle has been sharpened often on whatever was available to give it a penetrating edge. Thus, not only is HIV transmitted among those in a shooting gallery but other bacteria and viruses that contaminate their surroundings are also shared.

The world of injection drug use is designed to transmit infectious agents. However, **it is not the drug that is responsible for HIV transmission, rather it is the infected blood shared by each of the users of an HIV contaminated "works."** The practice of sharing "works" appears to be equally common among homosexual and heterosexual drug users. Thus it provides a common link between the homosexual and heterosexual population placing both at risk for HIV infection.

Cocaine vs. Heroin—The use of injected cocaine is rising as is exposure to HIV in cocaine IDUs. One major difference between cocaine and heroin abuse is that IV cocaine abusers are binge users or repeat injectors, while the heroin abuser falls asleep after one injection. Studies in New York and San Francisco show that IV cocaine abusers are the more likely to test HIV-positive.

FIGURE 8-12 An Injection Drug User. Note that the arm is tied to force veins to fill with blood making them easier to reach. A candle is used to heat the drug into a solution. This is referred to as "cooking" the drug. (*Photograph courtesy of the author*)

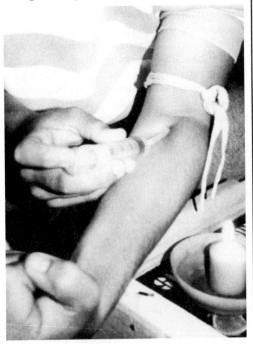

————— BOX 8.7 —————

ASSAULT WITH HIV

Reflecting a growing frustration and fear about AIDS, legislators around the country are passing an increasing number of laws intended to protect the public. This latest wave of legislation shifts the focus from earlier laws that protected the civil liberties of HIV-infected people to laws that seek to identify, notify and in some cases punish people who intentionally place others at risk of contracting the virus. At least 29 states now make it a crime to knowingly transmit or exposes others to HIV, the virus that causes AIDS, with a third of those states enacting laws within the last two years. In 1998, 16 state legislatures introduced such bills.

According to Richard Lacayo (1997), Darnell McGee, age 28, **through 1995 and 1996,** had sex with at least 61 women ranging in age from **12 to 29.** According to a Missouri public health report in February 1998, McGee had sex with at least 101 females, including four whose ages were 13 or 14. It is reported that McGee infected 18 women but Missouri officials believe he infected 30 women. Some are pregnant. At least one has given birth to an infected infant. State public health officials, who are trying to track down and notify his sex partners, expect the tally to climb as more women come forward to be tested.

McGee knew what he was doing. According to the *St. Louis Post-Dispatch*, which broke the story, state records show that he tested positive in 1992 and was told the results.

On January 15, 1997 he was shot and killed, assailant unknown!

Darnell "Bossman" McGee is just one of a number of men who recklessly and in some cases, even willfully transmitted HIV to their sex partners. Eighteen years ago it was Gatean Dugas, or patient zero, a gay male who, over three years knowingly infected an untold number (probably 50) of gays across the United States. Twelve years ago it was Fabian Bridges, a gay prostitute in Texas who was followed from city to city in the last months of his life by cameras from PBS's Frontline, as he continued to have unsafe sex.

In January 1997 James Wallace Jones, a convicted sex offender with HIV, faced charges in Michigan for failing to notify four sex partners (including a 15 year-old-girl) of his infected status. There is no evidence that he infected anyone, but he faces **non disclosure charges.**

In April 1997, an HIV-positive African American rap artist was sentenced to fourteen years in prison after having infected at least five women in Finland. And more recently, there was Nushawn Williams (Figure 8-13), age 21, who in mid-1997 admitted to having **unprotected sex** with 50 to 75 women after he was told he was HIV positive. Most of them were teenagers ages 13 and up living in New York's Chautauqua County and in New York City. To date, 13 in Chautauqua are infected, the youngest was age 13; others were age 15, 16, 18 and 21. Ten are positive in New York City but it is not known if Williams was the source of their infection. After the newspaper and TV carried this story, 625 people showed up at the county health department for HIV testing. Williams told authorities that **he did not believe health officials** when they told him he was HIV positive in September 1996. Early in 1998, **two** of the women who had unprotected sex with Williams gave birth to HIV positive babies. There are now 16 HIV positive people who are linked to Williams. In April 1999, Williams was sentenced to 4 to12 years in prison. Only two women agreed to testify against him.

In January 1998, a drifter in Michigan, who just got out of prison, kept his infection secret while having sex with 10 mostly teenage women. Two of the women directly asked him about HIV and he denied his infection. The drifter received a maximum penalty of 15 to 22 years in prison for **failing to notify a sex partner he had HIV** and for having sex with a minor. To date, **none** of the women have tested HIV positive.

In the same month, a 23-year-old HIV-infected woman in Louisiana failed to disclose her infection to a man even though he asked her prior to sexual intercourse! If convicted she can receive up to 10 years in jail.

Also, in January, in Orlando, Florida, a judge ordered a man who failed to disclose his HIV infection, **a misdemeanor in Florida** that could have resulted in a 4-year prison term, to have any women who he is going to have sex with to sign a form. It reads, "I _____, being fully informed of the fact that Jerrime Day is positive with HIV, do consent to have sex with him." Day also had to take an AIDS awareness class. However, once his two-year probation is over he will **not** have to get written permission.

In March 1998, an HIV positive army soldier received a dishonorable discharge, loss of pay,

———— **BOX 8.7** (*continued*) ————

reduction in rank and prison for having unsafe sex (no condom) with 8 women. Four of these women became pregnant; two of them are HIV positive! Also, in **Melbourne,** Australia, in a first of a kind, a man was sentenced to a **nonparole** five-year prison term for endangering the lives of three men. He knew he was HIV positive, did **not** tell his sexual partners and did **not** use a condom. Two of the three men are now HIV positive.

In July of 1998 residents of Lewisberg, Tennessee were shocked when a woman, age 29, told the police chief that she became angry and wanted revenge when a former boyfriend infected her with HIV. Knowing she was HIV positive, she had sex with up to 50 men she met at bars. She said she told the men she was HIV positive but they did not care. She faces two counts of criminal exposure to HIV and could face more. At the moment, if convicted it means up to 12 years in jail.

In July 1998, the Supreme Court of Arkansas sentenced Pierre Weaver to 30 years in prison for knowingly transmitting HIV to a woman through unprotected sex.

In September 1998, a Mississippi man age 45 was accused of having sex in 1993 with a man who was unaware he had tested positive for HIV in 1991. He denied having sex with the man. The Mississippi Department of Health met with him in January 1992 and told him to inform any potential sexual partners that he was HIV positive. The department later learned that he failed to inform his ex-wife of his condition. He was sentenced to 5 years in prison for failure to disclose and failure to use a condom with spermicide when having sex—a direct violation of a Mississippi Health Department quarantine order. Also in that month, an Ohio man, age 38, faced attempted murder for attempting to infect his sexual partner **through rape** because she introduced him to the woman that he claims gave him the virus. The prosecutor said, "He's talking to her throughout the rape and telling her she's going to die. He's telling her he's got the AIDS virus and he's going to give it to her."

In October 1998 in Lafayette, Louisiana, a jury convicted a 52-year old doctor of attempted second-degree murder for injecting his former 34-year old lover with blood from an AIDS patient, infecting her with HIV. She broke off a 10-year relationship with him. He was sentenced to 50 years of hard labor. Also in October, Swedish police reported that they were looking for an HIV-positive California man suspected of having unprotected sex with women he picked up in

Stockholm nightspots. The man, **who is aware of his infection,** has reportedly been in Sweden since 1992. Police are currently trying to contact 190 women listed in the man's address book.

In January 1999 in St. Charles, Missouri, a 32-year old father was sentenced to **life in prison** for injecting his son with HIV because he wanted to avoid paying child support. The father was convicted of injecting AIDS-tainted blood into the boy, who was then 11 months old, during a hospital visit in 1992. The child, now 7 was diagnosed with AIDS in 1996. If the boy dies, Stewart could be tried for murder. Prosecutors said the father, who worked as a hospital technician at the

FIGURE 8-13 Nushawn Williams. Chautauqua County officials issued posters like this one in October 1997, for hanging on restroom walls in Jamestown, New York. The poster was to alert the public to his sexual activities, throughout the county, that may have exposed many to HIV infection. (*Photograph courtesy of AP/Wide World Photos/Bill Sikes*)

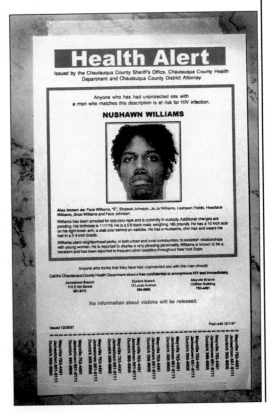

——————— **BOX 8.7** (*continued*) ———————

time, stole the HIV-infected blood from his work-place. Also **in January 1999,** a female Army private in Maryland pleaded guilty to aggravated assault charges for having unprotected intercourse, knowning that she was infected with HIV. She was sentenced to three years in military jail and will receive a bad conduct discharge from the service. The woman had been ordered by her commanding officer to use a condom and to inform her partners that she was HIV-positive; however, she disobeyed and had **unprotected** intercourse with four men and protected intercourse with five other men. None of the nine men with whom she had sex has tested positive for HIV.

In March 1999, police charged a Gainsville, Florida male with a felony for not telling his sex partners that he is HIV-positive. The man signed a statement in 1996 recognizing his legal obligation to inform potential sex partners. He admitted to having sex with 13 men without disclosing his HIV status—he said he did not tell because he was afraid that no one would have sex with him.

In May 1999, a US major in the armed forces was fired from the military after receiving a 6-year prison sentence for failure to obey an order by a superior officer. The major, who is HIV-positive was ordered to **inform** his sexual partners of his infection and to use methods, including condoms to prevent their infection. The major was convicted in 1994 of having unprotected sex with two women, willfully disobeying the "safer-sex" order. On completion of this sentence he returned to active duty and was fired by President Clinton (the case is Clinton vs. Goldsmith, 98-347).

SOME HIV/AIDS COURT CASES IN CANADA

In 1992, a man was charged in London, Ontario, with three counts of sexual assault and three charges of criminal negligence—causing bodily harm for knowingly HIV-infecting three women while under a public health order banning him from having sex. He was acquitted of the sex assault charges but died of AIDS before a verdict was reached on the other counts. The judge refused to continue with the case. Also in 1992, an HIV-infected man was found guilty of two counts of criminal negligence, causing bodily harm for having unprotected sex with two women, who later tested HIV-positive. He was sentenced to 11 years in jail by the Newfoundland Court.

In 1993, the Supreme Court of Canada upheld the conviction and the 15-month jail sentence (on the charge of being a common nuisance) of an HIV-positive Ottawa man who donated blood. He testified he hoped removing some of his infected blood would reduce the likelihood of developing AIDS and said he believed the Red Cross screening would detect the infected blood. Also **in 1993,** an HIV-infected male pleaded guilty to being a common nuisance and endangering a life for having unprotected sex with a woman, who later tested HIV-negative. He was sentenced to 12 months in jail. **In 1995,** an HIV-positive male was acquitted of being a common nuisance and endangering a life for having unprotected sex with a man, who later tested HIV-positive. The judge said it was not clear where his lover contreacted the virus. **In 1996,** a Montreal man was charged with criminal negligence causing bodily harm for having unprotected sex without informing his partner he had HIV. He died before the case came to trial.

In September 1998, in a unanimous ruling that could have far-reaching consequences, **Canada's Supreme Court** ruled that a British Columbia man who did not tell his sexual partners that he had **AIDS can be tried for aggravated assault.** The high court's decision overturns the verdicts of two lower courts and will send **Harry Currier's** case back to trial, even though the women with whom he had unprotected sex were not infected with HIV. Both women "testified at trial that if they had known that he was HIV-positive they would never have engaged in unprotected intercourse with him." The court said that "fraud" eliminated any true consent. The justices ruled, "The consent cannot apply simply to have sexual intercourse. Rather, it must be consent to have intercourse with a partner who is HIV-positive." **The ruling overturns the rulings of two lower courts that there was no assault because the women hd consented to sex with Currier and that no injury had occurred because the women were not infected with the disease.**

The case began in 1992 when Currier, then living in Squamish, British Columbia, was warned by a provincial health official that he must wear condoms and tell any sex partner that he had the AIDS virus. Currier continued to engage in unprotected sex, and in 1995, he was charged with aggravated assault against his two recent sex partners. He was acquitted at trials, and the acquittal was upheld by the B. C. Court of Appeals. Prosecutors appealed, arguing that the women could not give informed consent to sex because it was obtained through fraud in violation of Canada's Criminal Code.

The above cases of having unprotected sex after being told they were HIV positive and not informing their sexual partners raises several important issues.

Class Discussion: What is your response to these issues?

1. **Should the reckless or intentional transmission of HIV be a crime? If yes, how severe the penalty?**

2. **Do the above cases bolster arguments for more aggressive partner notification and contact tracing? Why?**

3. **Do HIV confidentiality protections help or hinder efforts to alter the course of the epidemic? Why?**

4. **Would more ready access to condoms have helped avert these tragedies? How?**

5. **Who is responsible when an HIV-infected person knowingly continues to have unprotected sexual relations with others?**

Should the infected person be warned another time, assuming that the educational message was not heard? If so, how many times should warnings go forth? Are public health officials responsible for protecting susceptible spouses or long-term lovers of those who are infected and knowingly refuse to use condoms? Should the police become involved if protective advise is not followed, or should confidentially remain in effect while educational messages go out that untold persons in the community are infected and all should use condoms? Such issues are currently being debated in the United States, and likely in many regions of developed nations.

6. **Do such incidences support calls for more sex education, or less? Or perhaps different approaches to sexuality education? What approach might work? Why?**

---

the proper care and medication can be delayed until the onset of HIV disease or AIDS.

Beginning 1999, there were at least 50 cases of purposeful HIV infection of males by HIV-infected females or vice versa. This too should be looked on as a form of **sexual assault**—one partner is being sexually deceived by the other. In some of these cases the jury found the HIV-positive persons who kept this knowledge from their sexual partner guilty of attempted murder.

Transplants— On any given day, about 20,000 Americans are waiting for a transplant. There is a small but present risk of receiving HIV along with the transplant tissue. A CDC report revealed that a bone transplant recipient became HIV-infected from an HIV-infected donor. HIV transmission has also occurred in the transplantation of kidneys, liver, heart, pancreas, and skin (*MMWR*, 1988a). In May of 1991, the CDC reported on 56 transplant patients who received organs and tissues from an HIV-infected donor in 1985. A transplantation service company supplied tissues to 30 hospitals in 16 states. All tissues came from a single young male who was shot to death during a rob-

bery. **He twice tested HIV-negative before his heart, kidneys, liver, pancreas, cornea, and other tissues were removed for transplant.** By mid-1991, three recipients of these tissues had died of AIDS and six others were HIV-positive. As of mid-1991, 32 other recipients had been located, 11 of whom tested HIV-negative. The others had not yet been tested.

In May of 1994, the CDC published guidelines for preventing HIV transmission through transplantation of human tissue (*MMWR*, 1994b).

─────────── **SIDEBAR 8.3** ───────────

MEXICAN DOCTORS TRANSPLANT HIV INFECTED KIDNEYS

In February 1999, Mexican health officials fired five physicians and warned two others for transplanting **HIV-infected kidneys** into two patients. One of the two patients has since tested HIV positive. According to the regional director of the state-run hospital, the physicians did not wait for the results of the HIV test on the kidney donor before making the transplant.

PROBABLY THE NUMBER ONE RISK COFACTOR FOR HIV IS A SEXUALLY TRANSMITTED DISEASE (STD)

TIME FOR A WAR ON STDS

According to a report, "The Hidden Epidemic," released in November 1996 by the Institute of Medicine, it is time for a national campaign against sexually transmitted diseases (STDs).

More than 12 million people in the United States, one-fourth of them teenagers, are infected with STDs such as chlamydia, genital herpes, and pelvic inflammatory disease **each year**. The situation is "a national embarrassment," says David Celentano of the Johns Hopkins University School of Hygiene and Public Health. Some of the facts:

♦ The United States has the highest rate of STDs of any developed country.

♦ Direct costs of STDs, not including AIDS, are $10 billion annually.

♦ Women and infants bear a disproportionate burden of complications from STDs, which include infertility and various types of cancers.

♦ In 1999, it was established that 40 million Americans have genital herpes but only about 20% (1 in 5 infected people) are aware that they are infected!

♦ Close to 22% of people over the age of 15 harbor the herpes virus.

♦ Teens are at an ever-increasing risk: By the 12th grade, 70% have had sexual intercourse— double the rate reported in the early 1970s— and close to 40% of them have had four or more partners.

STDs are hidden for a number of reasons, says the report by a committee headed by internist William T. Butler of the Baylor College of Medicine in Houston. They include unwillingness of parents and teachers to talk about sex, physician's ignorance, and the fact that many disorders show no symptoms in the early stages, especially in females.

The report puts special emphasis on the need for health providers to track and treat partners of people who present themselves with STDs. Butler says this recommendation is aimed in particular at easily curable bacterial infections such as gonorrhea, syphilis, and chlamydia, which otherwise will continue to cycle through the population (Figure 8-14).

Committee member Celentano says AIDS is playing a big part in bringing STDs out of the closet: **"Probably the number one risk [factor] for HIV is an STD."**

To order the $45 report, call 1-800-624-6242. (Source: *Science*, 274:1473, November 1996)

Nosocomial

Nosocomial (nos-o-ko-mi-al) refers to hospital-acquired infections. A chain of nosocomial HIV transmission has occurred in southern Russia and among children in Romanian orphanages. In Romania between 1988 and 1990, over 250 children were infected with HIV after exposure to nonsterile needles. By June 1994, 43 of these children had died of AIDS (Bobkov, 1994). Instances of nosocomial HIV transmission have also been reported from industrialized countries such as the United States, where transmission occurred from patient to patient. While nosocomial transmission accounts for a very small fraction of HIV transmission, nosocomial HIV transmission in any country at this time is unacceptable, and underscores deficiencies in present medical practices (Heymann et al., 1994).

Influence of Sexually Transmitted Diseases on HIV Transmission and Vice Versa

Sexual intercourse occurs more than 100 million times daily around the world. Results: 910,000 conceptions and over 600,000 cases of sexually transmitted disease. In the United States about 15 million new cases of sexually transmitted diseases are occurring each year. By age 21, about 1 in 5 people has received treatment for an STD. At current rates at least **one American in four** will **contract an STD at some point in his or her life.** More than **50 organisms** and syndromes are **transmitted** or occur as a **result of sexual activity** (Hooker,

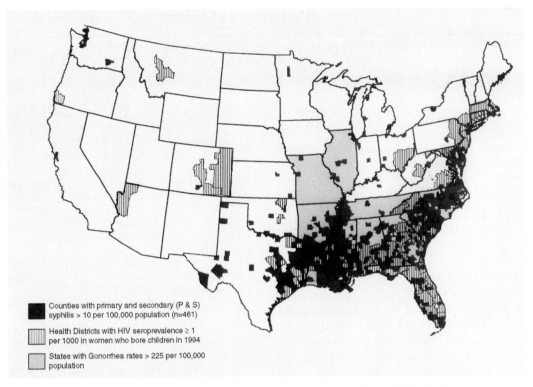

Counties with primary and secondary (P & S) syphilis > 10 per 100,000 population (n=461)

Health Districts with HIV seroprevalence ≥ 1 per 1000 in women who bore children in 1994

States with Gonorrhea rates > 225 per 100,000 population

FIGURE 8-14 HIV, Syphilis, and Gonorrhea Rates: United States. Continental U.S. health districts with the highest HIV seroprevalence among childbearing women (1994), counties reporting the highest primary and secondary syphilis rates (1993), and states reporting the highest gonorrhea rates (1993). (*Source:* St. Louis et al., 1997)

1996). Worldwide in the years 1997, 1998 and 1999 there were at least 333 million mostly curable cases of STDs. STD researchers have long recognized that the behaviors which place individuals at risk for other STDs also increase their risk of becoming infected with HIV.

Infection by sexually transmitted diseases usually occurs through the **mucosal surfaces** of the male and female genital tracts and rectum. The mucosal route also accounts for a large percentage of heterosexual and homosexual transmission of HIV. It is known that STDs increase the number of T4 cells (HIV target cells) in cervical secretions, thereby increasing the chance of HIV infection in women.

Most AIDS researchers agree that treating STDs, which cures genital sores and reduces inflammation, can raise the body's barriers against HIV infection. According to a study by Grosskurth and co-workers (1995), researchers working in rural Tanzania saw the number of new HIV infections plummet by 42% after they improved STD health care.

Because HIV can be sexually transmitted, the association between HIV and other sexually transmitted diseases can be in part attributed to the shared risk of exposure and shared modes of transmission.

For the purpose of understanding which STDs best promote HIV transmission, the sexually transmitted diseases can be divided into **genital ulcer** and **genital nonulcerative diseases**.

Genital Ulcer Disease (GUD)— Signs of genital ulcer disease appear as open sores on the penis, vagina, other genital areas and at times elsewhere on the body. The most widespread genital ulcer STDs are syphilis, genital herpes, and chancroid. In the United States, through 1999, there were about 50 million people infected with the herpes virus (about 1 in 5 people over age 12 carry the herpes virus).

In early 1997, researchers from the University of Washington showed for the first time that herpes sores contain high levels of HIV, which they believe makes the virus especially easy to spread during sexual contact. Additional research by Timothy Schacker and colleagues (1998) also showed that HIV can be consistently found in herpes genital lesions of HIV-infected people. Such data suggests that genital herpes infection likely increases the sexual transmission of HIV. The prevalence of HIV shedding in the genital tracts of both men and women increases substantially in the presence not only of accelerating immunosuppression but also of other STDs. Furthermore, the quantification of viral load in genital secretions indicates that the concentration of HIV shed may increase as much as 7-fold among men who are co-infected with HIV and STDs such as gonorrhea. Most importantly, both the prevalence and the concentration of HIV shedding can be reduced rapidly to baseline levels with appropriate STD treatment. Together with observations suggesting that HIV susceptibility is increased in women by nonulcerative STDs through recruitment of endocervical CD4 lymphocytes, these data provide a biologically plausible mechanism for the markedly increased risk of HIV transmission associated with other STDs, estimated to range between 3-fold and 50-fold depending on the STD involved (St. Louis et al., 1997).

William Cameron and colleagues (1990) reported on the bidirectional biological interactions between HIV and STDs, especially the genital ulcerative STDs. Bidirectional interaction between HIV and STDs occurs with respect to both transmission and virulence (Table 8-10). HIV transmission is a consequence of both infectivity and susceptibility, both of which can be increased by genital ulcer disease. **Virulence, the capacity of a pathogen to produce disease, is a consequence of both pathogen and host factors.** Thus, HIV infection and associated immune deficiency disease

may account for the increased prevalence of genital ulcer disease; and this in turn may further amplify HIV transmission in a network of social contacts.

Worldwide the estimated number of new STD infections for 1998 includes 12 million new cases of syphilis, 62 million new cases of gonorrhea, 89 million new cases of chlamydia, and 170 million new cases of trichomoniasis. Although trichomoniasis does not increase the risk of acquiring HIV as much as does syphilis, the huge number of trichomoniasis infections makes it at least equally important in the context of HIV transmission. **All four of these STDs are curable**—and examples from high- and low-income countries

TABLE 8-10 Bidirectional Interaction between HIV and Sexually Transmitted Genital Ulcer Disease

Types of Biological Transactions	Epidemiological Observation
GUD Increases HIV Prevalence	
Type I interaction: *Transmission*	
GUD increases susceptibility to HIV	Increased incidence of HIV in people with GUD
GUD increases infectiousness of HIV	Increased transmission of HIV with co-exposure to GUD and HIV
HIV Increases GUD Prevalence	
Type II interaction: *Virulence*	
HIV immune disease increases virulence of GUD pathogens	Increased incidence and prevalence of GUD in HIV-infected patients Decreased effectiveness of GUD therapy

Two categories of interaction between sexually transmitted genital ulcer disease (GUD) and HIV are relevant to transmission and to disease. Facilitated transmission (type I interaction), in which GUD operates to increase the prevalence of HIV, and enhanced virulence (type II interaction), in which HIV operates to increase the prevalence of GUD, may amplify the prevalence of each in a network of sexual contacts, such as a "core group" of prostitutes and clients, forming an efficient reservoir of high-frequency HIV and sexually transmitted disease transmitters.

(Source: Adapted from Cameron et al., 1990)

show that it is feasible and affordable to achieve a significant reduction of the STD burden everywhere.

Nonulcerative Disease— The nonulcerative STDs include gonorrhea, chlamydial, and trichomonal infections (also called discharge diseases), and genital warts. There are over 30 million people in the United States infected with genital wart virus. In most populations, these are much more common than genital ulcer diseases. None causes the noticeable open sores that occur in the ulcer diseases but they do cause microscopic breaks in affected tissue, and are associated with HIV transmission (Laga et al., 1993). The most common symptoms are warty growths on the genitals, discharge from the penis or vagina, and painful urination.

Collectively, worldwide there are over 250 million cases a year of just seven major STDs: syphilis, herpes, and chancroid, which cause ulcers; and trichomoniasis, chlamydia, warts, and gonorrhea, which do not (Figure 8-15).

HIV infection and other sexually transmitted diseases (STDs) share the same risk factors.

The major difference between HIV/AIDS and the other STDs is the degree of cell and tissue destruction and the mortality of HIV/AIDS.

HIV is transmitted most often during sexual contact with an infected partner. There is abundant evidence that if a sexual partner has an active STD, especially one that causes an ulcer, he or she is at greater risk of becoming HIV-infected (Laga, 1991).

The types of blood cells, lymphocytes, or macrophages most likely to become infected if exposed to HIV tend to collect in the genital tract of people with STDs. This makes an STD-infected person both more likely to transmit HIV and more vulnerable to it (Laga, 1991). The relationship of STDs to HIV can be seen in Figure 8-16.

Pediatric Transmission

Children can acquire HIV from their mothers in several ways. A pregnant HIV-infected woman can transmit the virus to her fetus in utero (**during gestation**) as the virus crosses over from the mother into the fetal bloodstream (Jovaisas et al., 1985; St. Louis et al., 1993). At least 50%

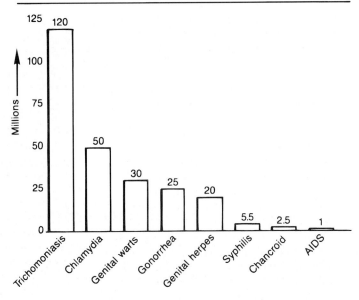

FIGURE 8-15 Global Incidence of Eight Sexually Transmitted Diseases. For 1999, Ages 15 to 49.

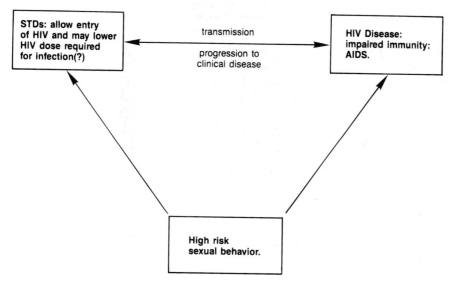

FIGURE 8-16 Bidirectional Interaction Between STDs and HIV. Medical studies support a complex bidirectional interaction between HIV and other sexually transmitted diseases with respect to transmission and virulence. In a group of sexually active, frequently HIV-exposed people with multiple sex partners (e.g., urban prostitutes), a subgroup of efficient, high-frequency HIV transmitters may occur. The epidemiology of HIV dissemination through sexual intercourse may in part be related to regional and demographic differences in the nature and size of sexually active groups and on the patterns of sexual mixing between high-risk groups and low-risk groups in the general population.

of newborn infections occur **during delivery by ingesting blood or other infected maternal fluids** (Scott et al., 1985; Boyer et al., 1994; Kuhn, et al., 1994). If breast fed, the newborn may also become **infected from breast milk** (Zigler et al., 1985; deMartino et al., 1992; Van DePerre et al., 1993). In case reports, three women who contracted HIV by blood transfusions *immediately after birth* subsequently infected their newborns via breast feeding (Curran et al., 1988). Other studies suggest that the risk of HIV transmission through breast feeding is increased if the mother becomes HIV-infected during lactation (Hu et al., 1992).

The relative efficiency of these three routes of infection is unknown. However, the data on mothers' milk add to the urgency of learning more about mucosal transmission, because the most likely explanation for HIV transmission through breast feeding is that the virus pene-trates the mucosal lining of the mouth or gastro-intestinal tract of infants. If this occurs in new-borns, then what of older children, adolescents, and adults? Does the mucosal lining change with development and become HIV-resistant?

HIV-Infected Babies— One major problem in perinatal transmission is how to determine which babies are truly HIV-infected as opposed to just carrying the mother's HIV antibodies (which would produce a false-positive test). HIV transmission can occur during pregnancy (in utero), as well as at the time of delivery (**intra-partum**) and through breast milk. HIV trans-mission is more likely if virus can be cultured from the mother's blood, or if she has later stage HIV disease, or if her T4 counts are low; and is more likely to occur in the first born than in the second born of twins. **A baby automatically ac-quires the mother's antibodies and may carry** them for 2 or more years. Usually by 18 months of age, most of the mother's antibodies will be gone. The babies may then begin to show signs of clinical AIDS-related illness. But, even at 18 months, a child cannot be unequivocally diagnosed. The most commonly used HIV antibody test is not sufficiently accurate until the child is at least 2 years old. A new antibody test in development shows promise in recognizing newborn infection by examining the **type of HIV antibodies** the infected mother is producing.

Although the rate of perinatal and breast milk HIV transmission is unknown, evidence from 1986 into year 2000 indicates that over 90% of pediatric AIDS cases acquired the virus in utero from an HIV-infected mother after the first trimester or during the birthing process (Marion et al., 1986; DeRossi et al., 1997; Backe et al., 1993). In 1990, researchers concluded that a fetus can become infected as early as the 8th week of gestation (Lewis et al., 1990). HIV has been isolated from a 20-week-old fetus after elective abortion by an HIV-positive female and from a 28-week-old newborn delivered by Caesarean section from a female who was diagnosed with AIDS (Selwyn, 1986).

Reports on the probability of a fetus becoming HIV-infected when the mother carries the virus vary widely. The most often quoted estimate is from 30% to 50%. However, the results of four in-depth follow-up studies on the frequency of HIV transmission by infected mothers to their fetuses gives a range of incidences of from 7% to 65%.

There is little documented information on maternal factors that influence vertical transmission. As with other congenital infections, only one of a pair of twins may be HIV-infected (Newell et al., 1990; Ometto, 1995). A mother's clinical status during pregnancy and the duration of her infection (stage of disease) may be important, but evidence remains circumstantial (see Chapter 11 for update information). Studies to determine mother-to-fetus transmission relative to stage of disease are in progress. One reason that all fetuses of HIV-infected mothers are not HIV-infected during gestation may be because the mothers' antibodies have a high affinity for HIV.

According to the CDC classification, children under 13 years of age are considered pediatric AIDS cases. They make up about 1.3% of all AIDS cases in the United States. Through 1999, about 5% of reported pediatric male AIDS cases occurred due to blood transfusions, 3% received HIV-contaminated blood factor VIII used in treating hemophiliacs, and in 2%, the cause was undetermined (Table 8-11).

The largest number of pediatric AIDS cases through 1999 were in New York, Florida, California and New Jersey, in that order. The highest incidence of all pediatric cases occurs in minority populations. By early 2000, there were over 8,800 pediatric AIDS cases in the

TABLE 8-11   Causes of Pediatric AIDS in the United States Estimates 1999

Pediatric[a] Exposure Category	Totals	(%)
Hemophilia/coagulation disorder	253	(3)
Mother with/at risk for HIV infection:	7,700	(91)
a. Injecting drug use	3,080	
b. Sex with injecting drug user	1,386	
c. Sex with bisexual male	154	
d. Sex with person with hemophilia	31	
e. Sex with transfusion recipient with HIV infection	30	
f. Sex with HIV-infected person, risk not specified	1010	
g. Receipt of blood tissue transfusion, blood components, or tissue	154	
h. Has HIV infection, risk not specified	1,771	
Receipt of blood transfusion, blood components, or tissue	425	(5)
Undetermined	84	(1)
Total pediatric AIDS cases (Total adult/adolescent AIDS cases, Estimated 1999, 737,169	8,461	(100)

[a]CDC classification: Pediatric cases means AIDS cases in children less than 13 years old. Different countries use other ages to define pediatric, eg. in Canada its under age 15.

(Adapted CDC *HIV/AIDS Surveillance Year End Report*, June 1998)

United States. Blacks and Hispanics make up 12% and 6% of the United States population, respectively, yet make up 55% and 20%, respectively, of all pediatric AIDS cases. Thus 75% of pediatric AIDS cases occur within two minority populations.

Over 100,000 women of childbearing age are estimated to be infected with HIV in the United States. The majority of these women may not know they are infected; they are identified as infected only after their children are diagnosed as having an HIV infection or AIDS. It is not uncommon for HIV-infected women to go through several pregnancies before expressing HIV disease. Also, there are women who become pregnant knowing they are HIV-positive. They want to have a baby regardless (see Chapter 11).

Mother-to-fetus infection or **vertical infection** could be avoided by avoiding pregnancy, but this is possible only in cases where the female is aware of her infection and takes measures to prevent pregnancy (e.g., birth control or tubal ligation). In many cases, pregnancy has occurred before the mother knew she was carrying the virus. In other cases, the mother has become infected after she has become pregnant.

Antiviral Therapy Decreases Perinatal HIV Transmission

Probably the most important step forward in the use of antiviral agents has been the discovery that zidovudine can **decrease the rate of perinatal transmission of HIV.** In a landmark study by Edward Connor and colleagues (1994) and the interim results of the AIDS Clinical Trials Group (ACTG) Protocol 076 (*MMWR*, 1994d; Goldschmidt et al., 1995; *MMWR*, 1995), the intensive use of zidovudine beginning in the 2nd trimester of pregnancy, incuding intravenous

--------- CASE IN POINT 8.1 ---------

WHEN THE WONDERFUL NEWS "YOU'RE PREGNANT" BECOMES A TRAGEDY

First Case: Amy Sloan

In 1982, AIDS was a homosexual disease. Amy Sloan became HIV-infected from a blood transfusion. She received the blood because of ulcerative colitis (ulcers in the colon). In 1985, 3 years after her transfusion, and 2 days *after* she learned she was pregnant, Amy was told she had AIDS. She was 24 years old.

By 1985, the general public was being educated to the devastating effects of AIDS; the virus had been named and it was known that blood transfusions were a major route of HIV transmission. Amy Sloan had become pregnant not knowing that she was carrying the AIDS virus. Amy delivered an uninfected son in 1986. She died in January 1987.

Second Case: Elizabeth and Paul Michael Glaser

On March 13, 1990, Elizabeth and Paul Michael Glaser, former star of *Starsky and Hutch*, testified before the House Budget Committee's Human Resource Task Force arguing for increased funding for pediatric AIDS research, education, and treatment.

Elizabeth Glaser was infected with HIV in 1981, after she was given a blood transfusion while giving birth to daughter Ariel. At that time, AIDS did not have a name, and the reporting of cases was just beginning. No one knew about the risk of contracting the virus through transfusions. Elizabeth breast-fed Ariel, unknowingly passing the virus to her daughter. Three years later, Paul and Elizabeth had a son, Jake, who later also tested positive for HIV. Paul is the only one in the family not infected.

Elizabeth testified that she watched her daughter suffer from symptoms that did not seem to affect adults. She watched her daughter's central nervous system atrophy as she became, at age 6, unable to walk, talk, or even sit up.

Ariel Glaser died in 1988, just after her 7th birthday. Elizabeth Glaser died on December 4, 1994.

In November 1996, Jake's viral load began to increase—a sign of a failing immune system. Although the FDA had not yet approved protease inhibitors (PI) for use in children, Paul Glaser took the risk; Jake received PI. Within 2 weeks his viral load dropped below measurement! Jake, now age 15 (1999), is on therapy, and is asymptomatic.

zidovudine during delivery, and 6 weeks of oral therapy in the neonate, **decreased the rate of transmission from 25% to 8%.**

UPDATE 1999

The time honored surgical maxim, **"A chance to cut is a chance to cure,"** takes on new meaning in the debate now raging about **caesarean section** as a means of reducing perinatal HIV transmission. During 1998 and 1999, clinicians have shown that the use of **cesarean section** to deliver a newborn from an HIV-infected mother is a significant help in reducing HIV infection of the newborn.

LOGIC

With about 2.3 million HIV-infected women worldwide giving birth each year, mother-to-child transmission of the virus raises serious public health concerns. (Riley et al. 1999) It is believed that most (70%) of perinatal transmission of HIV occurs during the birthing (intrapartum) period. This is the time when the fetus is exposed most directly to HIV-containing maternal genital secretions and blood. Labor contractions facilitate micro-transfusion of blood from mother to fetus. Furthermore, there have now been three recent studies suggesting that cesarean section performed before the onset of labor or ruptured membranes may reduce the risk of perinatal transmission to a greater extent than is now being accomplished with the use of zidovudine alone. Lauren Mandelbrot (1998) found that in 902 mothers receiving zidovudine (AZT) elective cesarean section was associated with lower transmission rate than emergent cesarean or vaginal delivery (0.8%, 11.4%, and 6.6% respectively). Christian Kind and colleagues (1998) noted a reduction from 17% with vaginal delivery to 0% with elective cesarean section in women receiving AZT. The European Mode of Delivery Collaboration (1999) has reported results form a randomized clinical trial of elective cesarean section at 38 weeks vs. planned vaginal delivery. Three (1.8%) of 170 infants born to women assigned cesarean section delivery were infected compared to 21 (10.5%) of 200 infants born to women assigned vaginal delivery.

Finally, an analysis of individual patient data from prospective studies in Europe and North America with at least 100 mother-infant pairs found that elective cesarean section significantly reduced perinatal transmission rates independent of antiretroviral treatment (Read et al. 1999). Although these studies sound convincing, there remain some critical unanswered: (1) use of combination therapy. None of the studies mentioned above included women traeted with antretroviral other than AZT. As in non-preganngt adults, combination antiretroviral therapy is now the standard of care in pregnant women, and although the studies were still in progress, there is reason to believe that HAART regimens will also offer greater benefit in reduction of perinatal transmission than AZT alone. (2) The question of **safety.** There is no question that in the general population cesarean section is associated with increased morbidity, particularly infectious morbidity, over vaginal delivery. And these risks are greater in some settings than others. In addition to studies already cited, a Swedish study found a 5-fold increase in maternal mortality with cesarean birth compared to vaginal delivery in a large population-based study.

In 1999, the American College of Obstetricians and Gynecologists convened a committee to review these issues and should release an opinion in the near future. At this time, researchers conclude that the subject of cesarean delivery for HIV-infected women should be discussed with every HIV-infected pregnant woman. Also, that current standards of antiretroviral therapy should be employed, and T4 cell counts and viral load should be monitored during pregnancy. Perhaps elective cesarean section may be recommended reasonably in a sub-group of women who are unable to achieve appropriate reductions in viral load.

Possible Antiretroviral Drugs to Prevent Maternal HIV Transmission

Current concepts on the prevention of vertical transmission now focus on the appropriate antiretroviral therapy of infected pregnant women. Despite the data from ACTG 076, zidovudine monotherapy alone can no longer be considered adequate for pregnant women because some newborns are still HIV positive. Some HIV/AIDS experts recommend the use of combination therapy that includes a protease inhibitor. In March 1999, five important studies were published regarding the use of combination antiretrovirals to reduce mother to infant transmission. The conclusion to be drawn from the studies is that antiretroviral prophylaxis of mother and their infant is **very effective.** With many women on HAART therapy during pregnancy, there may be less than 200 HIV-infected infants born in the United States in 1999. However, in developing countries, despite recent progress, there is still a great need for shorter, inexpensive, effective regimens. But the debate continues on when to begin therapy—first, second or third trimester?

(For an excellent review of mother-to-child transmission of HIV, see the article by Peckham and College, 1995.)

The Impact of ACTG 076 in the United States vs. Developing Countries— In late 1995, magazines and newspapers in the United States began running advertisements that show a baby lying on a quilt with these worlds superimposed over the image: **"The only thing worse than losing a child to AIDS is finding out you didn't have to."** The ad is part of a campaign launched by the Pediatric AIDS Foundation in response to the dramatic finding that the drug zidovudine can reduce the risk of a mother transmitting HIV to her baby by almost 70%. That finding has been heralded by AIDS researchers as their first real breakthrough in a decade-long effort to find a way to prevent HIV infection.

But many physicians and lay people are upset because thousands of HIV-infected pregnant women are not heeding the August 1994 recommendation of the U.S. Public Health Service (PHS) or the July 1995 PHS suggestion that all pregnant women voluntarily receive HIV testing, as there is now an effective means of preventing HIV from being transmitted to their infants.

—————— **BOX 8.8** ——————

IS BREASTFEEDING POLICY FOR WOMEN IN UNDEVELOPED NATIONS CHANGING?

During her pregnancy, a South African woman with HIV, followed a zidovudine (AZT) regimen to prevent transmission of the virus to the child she was carrying. After her daughter was born HIV-negative, she discontinued taking the AZT. No one warned her that her newborn daughter could become infected with HIV through breastfeeding. Tragically, the infant did become infected and later died. Lack of information is only one of the many problems with life-threatening consequences that confront women with HIV in resource-scarce countries.

The World Health Organization (WHO) and the Joint United Nations Program on AIDS (UNAIDS): Breastfeeding

In late 1998 WHO and UNAIDS recommended that women in undeveloped nations avoid breast feeding because it has been shown that about 10% of newborn infants become HIV-positive through breastfeeding from HIV-positive

mothers—especially if breast fed for 7 months or more.

PROBLEM—The reality is that most of the undeveloped nations cannot afford the cost of alternative feeding. Therefore, in countries like Zimbabwe, breast feeding will continue. In many of these countries there is no access to testing facilities and people could not afford them even if they existed; nor are drugs like zidovudine available to help reduce the incidence of maternal HIV transmission to the newborn. Also, many of these countries **do not have safe water** to use in making baby formula—even if free formula is available. As a result many of the children will die of diarrhoeal and respiratory diseases.

SUMMARY—Although the WHO and UNAIDS mean well, they appear to lack the economic resources or political clout to bring about the changes necessary to significantly reduce maternal HIV transmission.

But a larger problem exists in developing nations as the ACTG 076 findings are relatively meaningless to the majority of HIV-infected pregnant women in the world.

The reasons for this are as follows. **First,** poor nations can't afford the drug zidovudine nor do they have the clinics or other means by which to distribute the drug if they had it. **Second,** the ACTG 076 protocol calls for giving women zidovudine during pregnancy, and many women in developing countries won't visit medical clinics until they are in labor—if then. What's more, these women most often don't know they are infected with HIV.

Third, ACTG 076 is also out of synch with developing countries because it requires mothers to use infant formula to avoid the possibility of transmitting HIV through breast milk. For the moment, although ACTG 076 heralds a long awaited breakthrough in Pediatric HIV prevention in developed countries, it makes little if any difference to the economically impoverished nations of the world (Cohen, 1995).

HIV Infection in Older People

The latest report of the CDC (1998) on AIDS among persons aged 50 and over using 1991 through 1996 information revealed that from 1991 to 1996, the proportionate increase in incident cases of AIDS was greater among persons aged ≥ 50 years (22%) than among persons aged 13–49 years (9%). From 1991 to 1996, among men aged ≥ 50 years, the number of incident cases of AIDS among gay men remained stable (2900 cases each for 1991 and 1996), while incident cases among men whose risk was **heterosexual** contact increased 94% (from 360 cases to 700 cases) and incident cases among men reporting

---

— BOX 8.9 —

HIV AND SENIOR CITIZENS

Looking at the majority of safer sex workshops and street outreach programs, one would get the impression that only the young are at risk of contracting HIV. And it's true that most people with AIDS are under 49. But, according to the CDC, about 11% of Americans who test positive for the virus are over the age of 50. It's not just with regard ot prevention that over-50s are left behind. Older people with HIV are often misdiagnosed and typically learn that they nave the virus only later in the disease process. Medical treatment is more difficult, both because of the later diagnosis and factors related to age. Few practitioners are expert both in HIV and the health problems associated with aging. When it comes to social support services aimed at their particular needs, older HIV-infected people are all but invisible. Attitudes about HIV/AIDS and the aging reflect the beliefs built up about how people behave in the second half-century of their lives.

1. Old people are no longer interested in sex;
2. If they are interested, no one's interested in them;
3. If they do have sex, it's within a monogamous, heterosexual relationship;
4. They don't do drugs;
5. If they ever did, it' so long ago it doesn't matter.

It isn't hard to see how these misconceptions help erect barriers to effective HIV/AIDS prevention efforts, medical care, and social services for the late middle-aged and elderly. After all, if they're not doing anything risky, there's nothing to worry about, right? **WRONG,** older adults refuse to conform to the stereotypes.

Gender and Age

Older women are becoming HIV-infected at a higher rate than older men. No longer afraid of becoming pregnant, the post-menopausal woman who is uninformed of the danger of HIV transmission may become more sexually active, with more partners, and may give up a decades-old habit of using condoms. Even her biology increases her risk: after menopause, the vaginal walls thin and vaginal lubrication decreases. Thus, the vaginal membranes are more likely to tear during intercourse, providing easier access to HIV.

---

injection-drug use (IDU) increased 53% (from 850 cases to 1300 cases). Among women aged ≥50 years, cases attributed to hetero-sexual contact and IDU increased 106% (from 340 cases to 700 cases) and 75% (from 160 cases to 280 cases), respectively.

In 1998, of 55,295 persons aged 13 and over reported with AIDS, 6083 (11%) were aged ≥50 and over, this proportion has remained stable since 1991. Of those aged over 50, 48% were aged 50–54, 26% were aged 55–59, 14% were aged 60–64, and 12% were aged 65 and over. Males accounted from 84% of the cases, and blacks accounted for the highest proportion (43%) by race/ethnicity. Men who have sex with men (MSM) accounted for the highest proportion of cases by exposure category (36%). For all AIDS cases through 1999 46,252 are age 50 and above of which 5595 are age 65 and over.

HIV Transmission in the Workplace

The idea of contracting HIV from a fellow employee generates fear in many employees regardless of their jobs. It is believed that most people in the United States have been exposed to information on the routes of HIV transmission and on how to practice safer sex. But remote possibilities remain worrisome to many in the job force. In 1998, about one in six companies offered employee HIV/AIDS education. Many people still believe that HIV can be transmitted casually via handshakes, coffee cups, and food handling.

It is the anxiety of **uncertainty** that engenders suspicion about the possibility of HIV infection in the workplace—an anxiety that HIV/AIDS scientists could be wrong about the routes of HIV transmission. As Judith Wilson Ross states, **"We have spent our lives in a culture in which infectious disease does not represent a significant threat. And thus we had consigned living in fear of life-threatening contagious diseases to the pages of history books."** But today, HIV/AIDS forces us to re-examine our faith in the certainty of science. We want to believe, we want to accept—but the fear of death prevents complete surrender to education.

PUBLIC CONFIDENCE: ACCEPTANCE OF CURRENT DOGMA ON ROUTES OF HIV TRANSMISSION

Since the start of the epidemic in 1981, lawmakers have championed proposals ranging from quarantines to criminal prosecution for infecting others with HIV. In 1989, for example, eight states approved laws imposing criminal penalties for knowingly exposing people to HIV, and 30 states passed confidentiality laws to protect infected people. But many public health experts and advocates say legislatures are now far more likely to adopt laws making it a crime to intentionally expose someone to HIV. In fact, the fear associated with HIV may be growing.

A national telephone survey suggests that the public has **become more suspicious** of people who contract the virus, and **less knowledgeable about the disease.** Gregory Herek and John Capitano completed a survey in 1997 of 1,712 people randomly picked form across America. They presented their findings at the 12th International AIDS Conference, Summer 1998. They reported that 55% of Americans believed in 1997 that they could become infected by sharing drinking glass with an infected person, compared with 48% in 1991. Forty one percent believed that AIDS might be contracted from a public toilet, compared with 34% in 1991. And 54% believed that HIV might be transmitted through a cough or a sneeze compared to 47% in 1991. Twenty seven percent said they would be less likely to wear a sweater that had been worn, only one time, by a person with AIDS—even if the sweater had been cleaned and sealed in a new package. Twenty eight percent said they would be uncomfortable drinking out of a glass in a restaurant if a person with AIDS had used the same glass a few days earlier regardless that it had been washed and sterilized! Finally, compared to 1991, more respondents agreed that people who got AIDS through sex or drug use have gotten what they deserve (29%, compared to 20% in 1991). When the issues of blame and responsibility were posed in less negative terms, 55% of respondents agreed that most people with AIDS are responsible for having their illness, 51% agreed that it's their own fault if people get AIDS these days, 26% agreed that most people with AIDS don't care if they infect other people with HIV.

SUMMARY— Although no new routes of HIV transmission have surfaced over the last 19 years of this pandemic, **many people still do not believe that's all there is.** People still make the arguments that: (**1**) scientists do not yet know enough about this disease to be certain there are no other routes of transmission; and (**2**) scientists know other routes exist but are either too frightened to tell the truth, or are under political pressure not to do so for fear of creating a public panic. Many thousands of people in the United States firmly believe that in a few years they will look back and say "I told you so": You can get HIV from HIV-infected people if they breathe on you or if you touch their sweat and so on.

Question: How do you get everyone to believe what medical and research scientists say? Should we get everyone to believe scientific dogma?

NATIONAL AIDS RESOURCES

AIDS ACTION COUNCIL	1-202-547-3101
COALITION FOR LEADERSHIP ON AIDS	1-202-628-4160
GAY MEN'S HEALTH CRISIS	1-212-807-6655
MOTHERS OF AIDS PATIENTS	1-619-234-3432
NATIONAL AIDS INFORMATION CLEARINGHOUSE	1-301-762-5111
NATIONAL AIDS NETWORK	1-202-546-2424
NATIONAL ASSOCIATION OF PERSONS WITH AIDS	1-202-483-7979
PROJECT INFORM (ALTERNATIVE AIDS INFO.)	1-800-822-7422
PUBLIC HEALTH SERVICE HOTLINE	1-800-342-2437
CENTERS FOR DISEASE CONTROL AND PREVENTION TECHNICAL INFORMATION	1-404-639-2070

AMERICAN RED CROSS, NATIONAL AIDS EDUCATION	1-202-639-3223
GUIDE TO SOCAL SECURITY AND SSI DISABILITY BENEFITS FOR PEOPLE WITH HIV INFECTION	1-800-772-1213

(YOU CAN WRITE OR CALL FOR THIS SOCIAL SECURITY BROCHURE: SOCIAL SECURITY ADMINISTRATION, PUBLIC INFORMATION DISTRIBUTION CENTER, P.O. BOX 17743, BALTIMORE, MARYLAND 21235.)

SUMMARY

The World Health Organization began keeping records of AIDS cases in 1980. By the end of 1999, there were an estimated 29 million AIDS cases in 194 reporting countries and territories. About 1.5% of these cases occurred in the United States. It has been reported that the first cases of AIDS entered the United States via homosexual men who had vacationed in Haiti in the late 1970s. However, there is evidence of AIDS cases in the United States as early as 1952. While testing West Africans for HIV infection, a second strain of HIV was discovered: HIV-2. Both are transmitted in the same manner and both cause AIDS. However, HIV-2 appears to be less pathogenic than HIV-1.

There are two major variables involved in successful HIV transmission and infection. First is the individual's genetic resistance or susceptibility and second is the route of transmission. Not all modes of HIV exposure are equally apt to cause infection, even in the most susceptible individual. There have been a number of studies and empirical observations that demonstrate that HIV **is not casually acquired.** HIV is difficult to acquire even by means of the recognized routes of transmission.

HIV is transmitted mainly via sexual activities involving the exchange of semen and vaginal fluids, through the exchange of blood and blood products, and from mother to child both prenatally and postnatally (breast milk). Besides a few cases of breast milk transmission, no other body fluids have as yet been implicated in HIV infection.

The current belief is that anal receptive homosexuals have a higher risk than heterosexuals of acquiring HIV because the membrane or mucosal lining of the rectum is more easily torn during anal intercourse. This allows a more direct route for larger numbers of HIVs to enter the vascular system.

Others at high risk for acquiring and transmitting HIV are injection drug users. They infect each other when they share drug paraphernalia. Changes in sexual and injection drug use behavior can virtually stop HIV transmission among these people.

REVIEW QUESTIONS

(Answers to the Review Questions are on page 487.)

1. True or False: Africa makes up the largest percentage of **reported** AIDS cases worldwide. Explain.

2. What evidence is there that HIV may have evolved in the United States and Africa at the same time?

3. Are HIV-1 and HIV-2 related? Explain.

4. True or False: HIV-1 and HIV-2 are transmitted differently and therefore are located in geographically distinct regions of the world. Explain.

5. True or False: HIV is **not** believed to be casually transmitted. Explain.

6. Name the routes of HIV transmission.

7. True or False: Deep kissing wherein saliva is exchanged is a direct route for **efficient** HIV transmission. Explain.

8. True or False: Insects that bite or suck have been claimed to be associated with HIV transmission. Explain.

9. True or False: Among heterosexuals, HIV transmission from male to female and from female to male are equally efficient. Explain.

10. True or False: If a person has unprotected intercourse with an HIV-infected partner he or she will become HIV-infected. Explain.

11. What is the percentage of risk that a developing fetus with an HIV-positive mother in America, will be born HIV-positive, with and without zidovudine therapy? With zidovudine and "c" section?

12. Despite the warnings, groups that continue to engage in high-risk sexual activity include:

a) high school students
b) black women
c) injection drug users
d) prostitutes
e) all of the above

13. True or False: Prior to 1985, use of blood component therapy put the hemophiliac at risk for contracting HIV.

14. True or False: Relapse to risky sexual behavior can be an important source of new HIV infection in the homosexual community.

15. True or False: The body fluids shown most likely to transmit HIV are blood, semen, vaginal secretions, and breast milk.

16. True or False: Participation in risk behaviors and not identification with particular groups puts an individual at risk of acquiring HIV infection.

17. True or False: Unprotected receptive anal intercourse is the sexual activity with the greatest risk of HIV transmission.

18. True or False: Women who are HIV-infected always transmit the virus to their fetus during pregnancy or delivery.

19. True or False: A person infected with HIV can transmit the virus from the first occurrence of antigemia throughout the rest of his/her life.

20. True or False: Women constitute the fastest growing segment of the population with HIV infection.

21. True or False: The majority of HIV-infected women whose source of infection is known became infected through vaginal intercourse.

22. True or False: HIV infection in children is now a leading cause of death in children between the ages of 1 and 4.

23. True or False: Sexual contact is the major route of HIV transmission among black Americans.

24. True or False: Urine is one body fluid that remains an unproven route of HIV transmission.

25. Which of the following is **NOT** a recognized mode of HIV transmission?

a. unprotected sex with an infected partner
b. mosquito bite
c. contact with infected blood or blood products
d. perinatal transmission

REFERENCES

ABBOTT, ALISON. (1995). Murder charges brought in German HIV blood products case. *Nature,* 376:628.

ABBOTT, ALISON. (1996). Japan agrees to pay HIV-blood victims. *Nature,* 380:278.

AIDS Research. (1990). Roundup: Oral sex. *Med. Asp. Human Sexuality,* 24:52.

ALDHOUS, PETER. (1991). France will compensate. *Nature,* 353:425.

ARCHIBALD, D.W., et al. (1990). *In vitro* inhibition of HIV-1 infectivity by human salivas. *AIDS Res. Human Viruses,* 6:1425–1431.

AVINS, ANDREW., et al. (1994). HIV infection and risk behaviors among heterosexuals in alcohol treatment programs. *JAMA,* 271:515–518.

BACKE, E., et al. (1993). Fetal organs infected by HIV-1. *AIDS,* 7:896–897.

BAGASRA, OMAR, et al. (1994). Detection of HIV proviral DNA in sperm from HIV-infected men. *AIDS,* 8:1669–1674.

BAGASRA, OMAR. (1996). Use of in situ PCR for measuring viral burden. *AIDS Reader,* 6:43–47.

BARON, SAMUEL, et al. (1999). Why is HIV Rarely Transmitted by Oral Secretions? Saliva Can Disrupt Orally Shed, Infected Leukocytes. *Archives of Internal Medicine,* 159:303–310.

BELEC, LAURENT, et al. (1998). Genetically Related Human Immunodeficiency Virus Type 1 in Three Adults of a Family With No Identified Risk Factor for Intrafamilial Transmission. *Journal of Virology,* 72: 5831–5839.

BERREY, MICHELLE, et al. (1997). Oral Sex and HIV Transmission. *J. Acquired Immune Deficiency Syndromes and Human Retrovirology,* 14: 475.

BENNETT, AMANDA, et al. (1996). AIDS fight is skewed by Federal campaign exaggerating risks. *Wall Street Journal,* May 1, A1.

BOBKOV, ALEKSEI. (1994). Molecular epidemiology of HIV in the former Soviet Union: Analysis of ENV-3 sequences and their correlation with epidemologic data. *AIDS,* 8:619–624.

BOLLING, DAVID R. (1989). Anal intercourse between women and bisexual men. *Med. Asp. Human Sexuality,* 23:34.

BOYER, PAMELA J., et al. (1994). Factors predictive of maternal-fetal transmission of HIV. *JAMA,* 271:1925–1930.

BREO, DENNIS L. (1991). The two major scandals in France's AIDSGATE. *JAMA,* 266:3477–3482.

BURKETT, ELINOR. (1995). The gravest show on earth: America in the age of AIDS. Houghton Miffin: Boston.

BUTLER, DECLAN. (1994a). Allain freed to face new changes. *Nature,* 370:404.

BUTLER, DECLAN. (1994b) Blood scandal raises spectre of Dreyfus case. *Nature,* 371:548.

CAMERON, WILLIAM D., et al. (1990). Sexual transmission of HIV and the epidemiology of other STDs. *AIDS,* 4:S99–S103.

CDC (Centers for Disease Control and Prevention). (1990). *HIV/AIDS Surveillance Report,* Oct.: 1–18.

CENTERS FOR DISEASE CONTROL AND PREVENTION. AIDS among persons aged ≥50 years—US 1991–1996. *MMWR Morb Mort Wkly,* 47: 21–27.

CHU, S.Y., et al. (1990). Epidemiology of reported cases of AIDS in lesbians, United States 1980–89. *Am. J. Public Health,* 80:1380.

COHEN, J.B., et al. (1989). Heterosexual transmission of HIV. *Immunol. Ser.,* 44:135–137.

COHEN, JON. (1995). Bringing AZT to poor countries. *Science,* 269:624–626.

CONANT, MARCUS. (1995). The current face of the AIDS epidemic. *AIDS Newslink,* 6:14–8.

CONNER, EDWARD, et al. (1994). Reduction of maternal-infant transmission of HIV with zidovudine treatment. *N. Engl. J. Med.,* 331:1173–1180.

COOMBS, ROBERT, et al. (1998). Association between culturable HIV in semen and HIV RNA levels in semen and blood: Evidence for compartmentalization of HIV between semen and blood. *J. Infect. Dis.,* 117: 320–323.

COOMBS, ROBERT W., et al. (1989). Plasma viremia in HIV infection. *N. Engl. J. Med.,* 321:1526.

COTTONE, JAMES A., et al. (1990). The Kimberly Bergalis case: An analysis of the data suggesting the possible transmission of HIV infection from a dentist to his patient. *Phys. Assoc. AIDS Care,* 2:267–270.

CROMBIE, RENE, et al. (1998). Indentification of a CD36-related Thrombospondin 1-binding domain in HIV envelope glycoprotein gp120: Relationship to HIV specific inhibitory factors in human saliva. *J. Exp. Med,* 187: 25–35.

CURRAN, JAMES W., et al. (1988). Epidemiology of HIV infection and AIDS in the United States. *Science,* 239:610–616.

DANIELS, NORMAN. (1992). HIV-infected Professionals, patient rights, and the switching dilemma. *JAMA,* 267:1368–1371.

DeMARTINO, MAURIZIO, et al. (1992). HIV-1 transmission through breast-milk: Appraisal of risk according to duration of feeding. *AIDS,* 6:991–997.

DeROSSI, ANITA, et al. (1992). Vertical transmission of HIV: Lack of detectable virus in peripheral blood cells of infected children at birth. *AIDS,* 6:1117–1120.

DES JARLAIS, DON C., et al. (1989). AIDS and IV drug use. *Science,* 245:578.

DES JARLAIS, DON, et al. (1995) Maintaining low HIV-seroprevalence in populations of injecting drug users. *JAMA,* 274:1226-1231.

DeVINCENZI, I., et al. (1989). Risk factors for male to female transmission of HIV. *Br. Med. J.*, 298:411–415.

DORFMAN, ANDREA. (1991). Bad blood in France. *Time*, 138:48.

DROTMAN, PETER. (1996). Professional ing, bleeding, and HIV testing. *JAMA*, 276:193.

DRUCKER, E. (1986). AIDS and addiction in New York City. *Am. J. Drug Alcohol Abuse*, 12:165–181.

EDWARDS, SARA, et al. (1998). Oral sex and the transmission of viral STD's. *J. Infect. Dis.*, 74: 6–10.

ELSON, JOHN. (1991). The dangerous world of wannabes. *Time*, 138:77–80.

Emergency Cardiac Care Committee, American Heart Association. (1990). Risk of infection during CPR training and rescue: Supplemental guidelines. *JAMA*, 262:2714–2715.

European Mode of Delivery Collaboration, The (1999). Elective Cesarean section versus vaginal delivery in prevention of vertical HIV-1 transmission: A randomized clinical trial. *The Lancet*, 353:1035–1039.

FISCHL, M.A., et al. (1987). Evaluation of heterosexual partners, children and household contacts of adults with AIDS. *JAMA*, 257:640–644.

FOX, PHILIP. (1991). Saliva and salivary gland alterations in HIV infection. *J. Am. Dental Assoc.*, 122:46–48.

FRIEDLAND, GERALD H. (1991). HIV transmission from health care workers. *AIDS Clin. Care*, 3:29–30.

GIBBONS, MARY. (1994). Childhood sexual abuse. *Am. Fam. Phys.*, 49:125–136.

Global AIDSNEWS. (1994). A new approach to STD control and AIDS prevention. 4:13-14, 20.

GODDARD, JEROME. (1997). Why mosquitoes cannot transmit the AIDS virus. *Infect, Med.*. 14: 353–354.

GOLDSCHMIDT, RONALD, et al. (1995). Antiretroviral strategies revisited. *J. Am. Board Fam. Pract.*, 8:62–69.

GOSTIN, LAWRENCE. (1994). HIV testing, counseling and prophylaxis after sexual assault. *JAMA*, 271:1436–1444.

GOTO, Y., et al. (1991). Detection of proviral sequences in saliva of patients infected with human immunodeficiency virus type 1. *AIDS Res. Hum. Retroviruses*, 7:343–347.

GROSSKURTH, HEINER, et al. (1995). Impact of improved treatment of sexually transmitted diseases on HIV infection in rural Tanzania. *Lancet*, 346: 530–536.

GUINAN, MARY (1995). Artificial insemination by donor: Safety and secrecy. *JAMA*, 273:890–891.

HALKITIS, PERRY, et al. (1999). Beyond complacency: The effects of treatment on HIV transmissions. *FOCUS* 14: 1–4.

HEARST, NORMAN, AND HULLEY, STEPHEN B. (1988). Preventing the heterosexual spread of AIDS: Are we giving our patients the best advice? *JAMA*, 259:2428.

HECHT, FREDERICK, et. al (1998). Sexual transmission of an HIV-1 variant resistant to multiple reverse-transcriptase and protease inhibitors. *N. Eng. J. Med.* 339:307–311.

HENG, MADALENE, et al. (1994). Co-Infection and synergy of human immunodeficiency virus-I and herpes simplex virus-I. *Lancet*, 343: 255–258.

HEYMANN, DAVID, et al. (1994). The laboratory, epidemiology, nosocomial infection and HIV. *AIDS*, 8:705–706.

HIV/AIDS Surveilance Report. (1996). December 8:1–39.

HIV/AIDS Surveilance Report. (1997). December 9:1–43.

HIV/AIDS Surveilance Report. (1999). December 10:1–44.

HOLDEN, CONSTANCE. (1994). Switzerland has its own blood scandal. *Science*, 264:1254.

HOLMSTROM, PAUL, et al. (1992). HIV antigen detected in gingival fluid. *AIDS*, 6:738–739.

HOOKER, TRACEY. (1996). HIV/AIDS: Facts to consider: 1996 National Conference of State Legislators, Denver, Colorado, February. 1–64.

HU, DALE J., et al. (1992). HIV infection and breastfeeding: Policy implications through a decision analysis model. *AIDS*, 6:1505–1513.

ILARIA, GERARD, et al. (1995). Detection of HIV DNA in pre-ejaculate. *Primary Care Rev.: Update Urol.*, 29:8–9.

IPPILITO, GIUSEPPE, et al. (1994). Transmission of zidorudine-resistant HIV during a bloody fight. *JAMA*, 272:433–434.

JACKSON, BROOKS. (1999). Progress in reducing mother-to-in-infant HIV transmission. *The Hopkins Report* 11:2–3.

JAYARAMAN, KRISHNAMUNTHY. (1995). HIV scandal hits Bombay blood centre. *Nature*, 376:285.

JOSEPH, STEPHEN C. (1993). Dragon within the gates: The once and future AIDS epidemic. *Med. Doctor*, 37:92–104.

JOVAISAS, E., et al. (1985). LAV/HTLV III in 20-week fetus. *Lancet*, 2:1129.

KAHN, JAMES, et al. (1998). Acute HIV-1 Infection. *N. Eng. J. Med.*, 339: 33–40.

KAISER, JOCELYN. (1996). Pasteur implicated in blood scandal? *Science*, 272:185.

KALICHMAN, SETH, et al. (1998). Risk perception among gay men. *FOCUS* 14: 8.

KATNER, H.P., et al. (1987). Evidence for a Euro-American origin of human immunodeficiency virus. *J. Natl. Med. Assoc.*, 79:1068–1072.

KEET, IRENEUS, et al. (1992). Orogenital sex and the transmission of HIV among homosexual men. *AIDS*, 6:223–226.

KIESSLING, ANN. (1999). HIV-1 semen: Risks for transmission, disease progression and reproduction. *Physician's Research Network NYC.* NY: http//www.prn.org/prn_nb_cntnt/vol4/numl/article2_cntnt.com

KIND, CHRISTIAN, et al. (1998). Prevention of Vertical HIV Transmission: Additive Protective Effect of Elective Cesarean Section and Zidovudine Prophylaxis. *AIDS,* 12:205–210.

KINGSLEY, L.A., et al. (1990). Sexual transmission efficiency of hepatitis B virus and human immunodeficiency virus among homosexual men. *JAMA,* 264:230–234.

KOSS, M.P., et al. (1991). Deleterious effects of criminal victimization on women's health and medical utilization. *Arch. Intern. Med.,* 151: 342–347.

KRAVCIK, STEPHEN, et. al (1998). Effects of treatment on risk perception. *FOCUS* 14:8.

KRIEGER, JOHN, et al. (1998). Risk of sexual transmission of HIV unaffected by vasectomy. *J. of Urology,* 159: 820–825.

KUHN, LOUISE, et al. (1994). Maternal-infant HIV transmission and circumstances of delivery. *Am. J. Public Health,* 84:1110–1115.

LAGA, MARIE. (1991). HIV infection and sexually transmitted diseases. *Sexually Transmitted Dis. Bull.,* 10:3–10.

LACAYO, RICHARD. (1997). Assult with a deadly virus. *Time Magazine,* 149:82.

LAGA, MARIE, et al. (1993). Non-ulcerative STDs as risk factors for HIV transmission in women: Results from a cohort study. *AIDS,* 7:95–102.

LAURENCE, JEFFREY. (1995). The mechanics of HIV transmission. *Primary Care Rev: Update Urol.,* 29:4–5.

LEVY, JAY A. (1989). Human immunodeficiency viruses and the pathogenesis of AIDS. *JAMA,* 261:2997–3006.

LEWIS, S.H., et al. (1990). HIV-1 introphoblastic villous Hofbauer cells and haematological precursors in eight-week fetuses. *Lancet,* 335:565.

LICHTMAN, STUART M., et al. (1991). Greater attention urged for HIV in older patients. *Infect. Dis. Update,* 2:5.

LIFSON, ALAN. (1995). HIV transmission through specific oral-genital sexual practices. *Primary Care Rev.: Update Urol.,* 29:9–11.

LIUZZI, G.M. (1995). Quantitation of HIV genome copy number in semen and saliva. *AIDS,* 9:651–653.

MANDELBROT, LAURENT, et al. (1998). Perinatal HIV-1 Transmission: Interaction Between Zidovudine Prophylaxis and Mode of Delivery in the French Perinatal Cohort. *JAMA,* 280:55–60.

MARION, R.W., et al. (1986). Human T cell lymphotropic virus type III embryopathy: A new dysmorphic syndrome associated with intrauterine HTLV III infection. *Am. J. Dis. Child,* 140:638–640.

MARKS, ROBERT. (1996). The riddle of oral sex. *Focus,* 5:5–8.

MARLINK, RICHARD, et al.(1994). Reduced rate of disease development after HIV infection as compared to HIV-1.*Science,* 265:1587–1590.

MERIMAN, J.H., et al. (1991). Detection of HIV DNA and RNA in semen by the polymerase chain reaction. *J. Infect. Dis.,* 164:769–772.

MIIKE, LAWRENCE. (1987). Do insects transmit AIDS? Office of Technological Assessment, Sept. 1:43.

MONZON, O.T., et al. (1987). Female to female transmission of HIV. *Lancet,* 2:40–41.

Morbidity and Mortality Weekly Report. (1988a). Update: Universal precautions for prevention of transmission of human immunodeficiency virus, hepatitis B virus, and other bloodborne pathogens in health-care settings. 37:377– 382,387–388.

Morbidity and Mortality Weekly Report. (1990a). Possible transmission of HIV to a patient during an invasive dental procedure. 39:489–493.

Morbidity and Mortality Weekly Report. (1990b). HIV infection and artificial insemination with processed semen. 39:249–256.

Morbidity and Mortality Weekly Report. (1991a). Update: Transmission of HIV infection during an invasive dental procedure—Florida. 40:21–27,33.

Morbidity and Mortality Weekly Report. (1991b). Drug use and sexual behaviors among sex partners of injecting-drug users—U.S. 40:855–860.

Morbidity and Mortality Weekly Report. (1992). Childbearing and contraceptive-use plans among women at high risk for HIV infection—Selected U.S. sites, 1989–1991. 41:135–144.

Morbidity and Morality Weekley Report. (1994a). Human immunodeficiency virus transmission in household settings—United States. 43:347, 353–357.

Morbidity and Morality Weekley Report. (1994b). Guidelines for preventing transmission of HIV through transplantation of human tissue and organs. 43:1–15.

Morbidity and Morality Weekley Report. (1994c). Medical-care expenditures attributable to cigarette smoking—United States, 1993. 43:469–472.

Morbidity and Mortality Weekly Report. (1994d). Zidovudine for the prevention of HIV transmission from mother to infant. 43:285–287.

Morbidity and Mortality Weekly Report. (1995). Use of AZT to prevent perinatal transmission (ACTG 076): Workshop on implications for treatment, counseling, and HIV testing. 44:1–12.

Morbidity and Martality Weekly Report.(1997). Transmission of HIV possibly associated with exposure of mucous membrane to contaminated blood. 46:620–623.

Morbidity and Mortality Weekly Report. (1998). AIDS among people aged 50 years—United States, 1991-1996. 47: 21–27.

MUNZER, ALFRED. (1994). The threat of secondhand smoke. *Menopause Manage.,* 3:14–17.

NAFTALIN, RICHARD J. (1992). Anal sex and AIDS. *Nature,* 360:10.

NEWELL, MARIE-LOUISE, et al. (1990). HIV-1 infection in pregnancy: Implications for women and children. *AIDS,* 4:S111–S117.

NOWAK, RACHEL. (1995). Rockefeller's big prize for STD test. *Science,* 269:782.

OLESKE, JAMES M. (1994). The many needs of HIV-infected children. *Hosp Pract.* 29:81–87.

OMETTO, LUCIA et al. (1995) Viral phenotype and host-cell susceptibility to HIV infection as risk factors for mother-to-child HIV transmission. *AIDS,* 9:427-434.

PADIAN, NANCY S., et al. (1991). Female to male transmission of HIV. *JAMA,* 266:1664–1667.

PADIAN, NANCY, et al. (1997). Heterosexual transmission of HIV in northern California: Results from a ten-year study. *Am. J. Epidemiol.,* 146:350–357.

PATTERSON, JULIE, et al. (1995). Basic and clinical considerations of HIV infection in the elderly. *Infect. Dis.,* 3:21–34.

PECKHAM, CATHERINE, et al. (1995). Mother-to-child transmission of HIV. *N. Engl. J. Med.,* 333:298–302.

PETERMAN, THOMAS A., et al. (1988). Risk of human immunodeficiency virus transmission from heterosexual adults with transfusion-associated infections. *JAMA,* 259:55–58.

PETO, RICHARD. (1992). Statistics of chronic disease control. *Nature,* 356:557–558.

PHILLIPS, DAVID. (1994). The roll of cell-to-cell transmission in HIV infection. *AIDS,* 8:719–731.

PHILLIPS, KATHRYN, et al. (1994). The cost effectiveness of HIV testing of physicians and dentists in the United States. *JAMA,* 271:851–858.

POURTOIS, M., et al. (1991). Saliva can contribute in quick inhibition of HIV infectivity. *AIDS,* 5:598–599.

QUAYLE, ALISON. (1995). Mucous membrane susceptibility to HIV infection. *Primary Care Rev.: Update Urol.,* 29:6–8.

READ, JENNIFER. (1999). The Mode of Delivery and the Risk of Vertical Transmission of Human Immunodeficiency Virus Type 1—A Meta-Analysis of 15 Prospective Cohort Studies. *N. Eng. J. Med.* 340:977–987.

REICHHARDT, TONY. (1995). Top aide to face charges in French HIV blood scandal. *Nature,* 375:349.

RILEY, LAURA, et al. (1999). Elective cesarean delivery to reduce the transmission if HIV. *N. Eng. J. Med.* 13: 1032–1033.

ROGERS, DAVID, et al. (1993). AIDS policy: Two divisive issues. *JAMA,* 270:494–495.

ROTHENBERG, RICHARD, et al. (1998). Oral Transmission of HIV. *AIDS,* 12:2095–2105.

ROYCE, RACHEL, et al. (1997). Sexual transmission of HIV. *N. Eng. J. Med.* 336:1072–1078.

ROZENBAUM, W. et al. (1988). HIV transmission by oral sex. *Lancet,* 1:1395.

SADOVSKY, RICHARD. (1989). HIV-infected patients: A primary care challenge. *Am. Fam. Pract.,* 40:121–128.

SAMUEL, MICHAEL. (1995). What does risk mean? Prospective and cross-sectional studies. *Primary Care Rev.: Update Urol.* 29:11–13.

SCHACKER, TIMOTHY, et al. (1998). Frequent recovery of HIV from genital herpes simplex virus lesions in HIV infected men. *JAMA,* 280: 61–66.

Science in California. (1993). AIDS: I want a new drug. *Nature,* 362:396.

SCOTT, G.B., et al. (1985). Mothers of infants with the acquired immunodeficiency syndrome: Evidence for both symptomatic and asymptomatic carriers. *JAMA,* 253:363–366.

SEGARS, JAMES H. (1989). Heterosexual anal sex. *Med. Asp. Human Sexuality,* 23:6.

SELWYN, PETER A. (1986). AIDS: What is now known. *Hosp. Pract.,* 21:127–164.

SHERNOFF, MICHAEL. (1988). Integrating safer-sex counseling into social work practice. *Social Casework: J. Contemp. Social Work,* 69:334–339.

SPITZER, P.G., et al. (1989). Transmission of HIV infection from a woman to a man by oral sex. *N. Engl. J. Med.,* 320:251.

SPURGEON, DAVID. (1994). Canadian AIDS suit raises hope for HIV-blood victims. *Nature,* 368–281.

SPURGEON, DAVID. (1996). Canadian inquiry points the finger. *Nature,* 379–663.

ST. LOUIS, MICHAEL E., et al. (1993). Risk for perinatal HIV transmission according to maternal immunologic, virologic and placental factors. *JAMA,* 269:2853–2860.

ST. LOUIS, MICHAEL E., et al. (1997). Editorial: Janus considers the HIV pandemic—harnessing recent advances to enhance AIDS prevention. *Am. J. Public Health,* 87:10–11.

STRYKER, JEFF, et al. (1993). AIDS policy: Two divisive issues. *JAMA,* 270:2436–2437.

SWENSON, ROBERT M. (1988). Plagues, History and AIDS. *Am. Scholar,* 57:183–200.

SWINBANKS, DAVID. (1993). American witnesses: Testify in Japan about AIDS risks. *Nature,* 364:181.

UNAIDS. (1997). Inplications of HIV variability for transmission: Scientific and policy issues. *AIDS,* 11:S1–S15.

VAN DE PERRE, PHILIPPE, et al. (1993). Infective and anti-infective properties of breast milk from HIV-infected women. *Lancet,* 341:914–918.

VITTECOQ, D., et al. (1989). Acute HIV infection after acupuncture treatments. *N. Engl. J. Med.*, 320:250–251.

VOELKER, REBECCA. (1996). HIV guide for primary care physicians stresses patient-centered prevention. *JAMA*, 276:85–86.

WALKER, EDWARD A., et al. (1993). The prevalence rate of sexual trauma in a primary care clinic. *J. Am. Board Fam. Pract.*, 6:465–471.

WEBB, PATRICA, et al. (1989). Potential for insect transmission of HIV: Experimental exposure of *Cimex hemipterous* and *Toxorhynchites amboinensis* to human immunodeficiency virus. *J. Infect. Dis.*, 160:970–977.

WILL, GEORGE F. (1991). Foolish choices still jeopardize public health. *Private Pract.*, 24: 46–48.

Women's AIDS Network. (1988). Lesbians and AIDS: What's the connection? San Francisco AIDS Foundation, 333 Valencia St., 4th Floor, P.O. Box 6182, San Francisco, CA 94101-6182.

WOOLLEY, ROBERT J. (1989). The biologic possibility of HIV transmission during passionate kissing. *JAMA*, 262:2230.

YOUNG, REBECCA. (1995). The scarcity of data on cunnilingus *Primary Care Rev.: Update Urol.*, 29:13–14.

ZHANG, HUI, et al. (1998). HIV-1 in the semen of men receiving HAART. *N. End. J. Med.*, 339:1803–1810.

ZIGLER, J.B., et al. (1985). Postnatal transmission of AIDS-associated retrovirus from mother to infant. *Lancet*, 1:896–897.

Preventing the Transmission of HIV

GLOBAL PREVENTION

Because we live in a **global village,** the public health of Africa, Asia and elsewhere affects the public health of the United States. As there is one global economy, there is one global public health. **Prevention of infectious diseases in any country is prevention for all.** In 1998 and 1999, it is estimated that between 2.5 million and

3 million people died from AIDS. Worldwide ending year 2000, about 17 million people have died of AIDS. About 80% of these deaths were in Africa where AIDS is now causing one in five deaths. While waiting for an effective vaccine, how can the **out of control** spread of HIV be slowed? How can people everywhere be saved from HIV infection? In a word, **PREVENTION** is the only hope short of a vaccine.

Invest in Prevention

Spending money on prevention is a smart investment. For example, nationwide, adding $500 million to HIV prevention targeted to high-risk groups would yield total medical care savings of $1.25 billion. The potential for HIV prevention interventions to save lives and dollars emphasizes the need to spend money now rather that later, and to maintain consistent, if not increasing, funding for high-risk groups. David Holtgrave reported, at the **12th International AIDS Conference** (June/July 1998), that the current cost of lifetime treatment for an HIV-infected person is $152,402. Thus, an infection **prevented** saves this amount of money. America's prevention budget for 1998 was about $616.8 million. By **preventing 3995 HIV infections,** one theoretically saves the budget. In reality the savings would be used to prevent others from being infected. The bottom line, is prevention is paramount regardless of the means of assessment.

The Challenge

The means of preventing HIV infection exists. They must be used effectively to make an impact on this escalating pandemic. The existing methods of HIV prevention are presented in this chapter.

PREVENTION OF INFECTIOUS DISEASES: HIV/AIDS

While the hope of eradicating HIV is universal, the universe of AIDS comprises many worlds, and the goal can only be achieved if methods of prevention are designed and implemented in the context of each of these worlds. Infectious agents have persisted over the centuries through transmission from one infected person to an-

other. To combat these infections, **prevention programs have attempted to interrupt the chain of transmission.** While some methods focus on stopping transmission from person-to-person contact by using vaccines or prophylactic (preventative)therapy, other methods focus on stopping the transmission that occurs through environmental contamination by sanitary improvements.

With regard to HIV infection, there is no available vaccine against the virus, but there is a **virtual vaccine** (meaning a **procedure** as effective as a preventative vaccine): an education sufficient enough to effect prevention. Thus the leading primary preventative is education: teaching people how to adjust their behavior to reduce or eliminate HIV exposure. Because the vast majority of HIV infections are transmitted through consensual acts between adolescents and adults, the individual has a choice as to whether to risk infection or not.

Despite widely supported educational efforts at both institutional and street levels, a large number of gay males, drug abusers, and heterosexuals, continue to participate in *unsafe* sexual practices. **Unsafe sex is defined as having sex without using a condom.** This allows the exchange of potentially infectious body fluids such as blood, semen, and vaginal secretions. Unsafe sex most often occurs with injection drug users, bartering sex for drugs, and having sex with multiple partners. The sharp increase in the use of crack cocaine and its connection to trading sex for drugs has led to a dramatic rise in almost all of the sexually transmitted diseases. **The idea of *safer* sexual practices began in the early 1980s and now refers almost exclusively to the use of a latex or plastic condom with or without a spermicide.**

Among the severely drug addicted, concerns about personal safety and survival are secondary to drug procurement and use. Thus their range of acceptable unsafe behaviors leads to random sex and sex without condoms. These behaviors are in part responsible for the increased incidence of HIV and other sexually transmitted diseases (Weinstein et al., 1990).

AIDS prevention is, in a sense, more essential than, say, cancer prevention. Preventing one HIV infection now will not simply prevent one death from AIDS, as preventing one incur-

——— BOX 9.1 ———

HIV/STD PREVENTION EDUCATION—UNITED STATES

Eliminating all unsafe sex is not a reasonable goal. Preventing all future HIV and STD infections is impossible, but striving for anything less is unacceptable.

Since 1988, CDC has provided fiscal and technical assistance to state and local education agencies and national health and education organizations to assist schools in implementing effective HIV and STD prevention education for youth. These agencies and organizations develop, implement, and evaluate HIV/STD prevention policies and programs and train teachers to initiate effective prevention efforts and implement curricula in classrooms. As a result of these and other efforts, school-based HIV/STD education is widely implemented in the United States. From 1987 to 1999, the number of states requiring HIV/STD prevention education in schools increased from 13 states to 35 states plus the District of Columbia. This high level of policy support is consistent with public support; 95% of U.S. residents in a 1996 survey reported that information about AIDS and STDs should be provided in school.

The findings in this survey indicate that, despite wide implementation of HIV/STD prevention education in U.S. schools, improvements in HIV/STD prevention programs are still needed. In particular, **efforts are needed to increase the percentage of teachers who teach HIV/STD prevention in a health education setting and who receive in-service training on HIV/STD prevention.** A national health objective for the year 2000 is to increase to at least 95% the proportion of schools that provide age-appropriate HIV/STD curricula for students **in 4th through 12th grades,** preferably as part of comprehensive school health education, based on scientific information that includes the way HIV and other STDs are prevented and transmitted. Based on the findings, an 11% increase is needed in the percentage of middle/junior and senior high schools that implement HIV and STD prevention education programs.

UPDATE 1999

The following information about sexuality, STD and HIV/AIDS education is current through January 1999.

Sexuality Eduction Mandates: Only 20 states, including the District of Columbia, require schools to provide sexuality education (AL, DE, DC, GA, HI, IL, IA, KS, KY, MD, MN, NV, NJ, NC, RI, SC, TN, UT, VT, WV).

Content Requirements of Sexuality Education Mandates: Of the 20 states that require schools to provide sexuality education, four require that sexuality education teach abstinence but do not require the inclusion of information about contraception (AL, IL, KY, UT). Ten require that sexuality education teach abstinence and provide information about contraception. (DE, GA, HI, NJ, NC, RI, SC, TN, VT, WV).

STD/HIV Education Mandates: Thirty-five states, including the District of Columbia, require schools to provide STD, HIV, and/or AIDS education (AL, CA, CT, DE, DC, FL, GA, HI, IL, IN, IA, KS, KY, MD, MI, MN, MO, NV, NH, NJ, NM, NY, NC, OH, OK, OR, PA, RI, SC, TN, UT, VT, WA, WV, WI).

Content Requirements of STD/HIV Mandates: Of the 35 states that require schools to provide STD, HIV, and/or AIDS education, one requires that AIDS education teach abstinence until marriage but does not require the inclusion of other AIDS prevention methods (IN). Twenty-one require that STD, HIV, and/or AIDS education teach abstinence and other methods of prevention (AL, CA, DE, FL, GA, HI, IL, KY, MI, NJ, NM, NY, NC, OK, OR, PA, RI, TN, VT, WA, WV).

able cancer would prevent one cancer death. Preventing an HIV infection now will help break the chain of transmission, averting the risk that the infected person will knowingly or unknowingly pass the virus on to others who in turn might infect a still wider circle of people.

PREVENTING THE TRANSMISSION OF HIV

HIV/AIDS: The News Is Mostly Bad

We are now into the 19th year of an epidemic that has touched—directly or indirectly—virtually every person on the planet. We know so much about the virus, yet despite our knowledge, our only option is to **prevent** the initial infection. Prevention is foremost because there are **no vaccines, no cure** and, even using the best AIDS drug cocktails available, long-term survival remains questionable, even for those who can afford the drugs. **MOST HIV-INFECTED PEOPLE CANNOT AFFORD THESE DRUGS!** As we face this realization, alarming statistics continue to emerge about the spread of HIV infection.

In San Francisco, estimates suggest that at least 50% of homosexual African-American men are infected with HIV. Unsafe sexual practices that could lead to HIV transmission are common among adolescents and young adults, as evidenced by the epidemic of other sexually transmitted diseases in this population. And tens of millions of persons in developing countries will become HIV-infected and most likely die. Entire generations are threatened with extinction in these countries.

The Hard Questions

The political, social, cultural, economic and biological factors that have led to the HIV pandemic seem overwhelming. How can a drug user be convinced to use clean needles to prevent an infection that may kill him in 10 years, when he faces an immediate struggle in a violent environment every day? How can condom use be promoted in countries with inadequate supplies of condoms or resources to provide even basic immunizations? Why should young women on the streets of New York, San Francisco, New Delhi, or Bangkok who depend on the sex industry for daily survival care about safer sex when it might lead to rejection by their customers and an end to their livelihood?

CLASS DISCUSSION: Your answers to the hard questions are?

In many societies, there is a large power differential between men and women. Socially and culturally determined gender roles bestow control and authority on males. The subordinate status of women is reinforced by the fact that men are the main or only wage earners in the majority of families. This is compounded further by age differences: in most heterosexual relationships, the man is the older partner.

Wives in many cultures are expected to tolerate infidelity by their husbands, while remaining totally faithful themselves. **But AIDS has raised the price of such tolerance, as it puts women at great risk of infection by their husbands.** Many women feel powerless to ask their husbands to use condoms at home. Even when they can do this, their need to protect themselves may conflict with a social or personal imperative to have children.

Is There Hope?

Hopelessness threatens reason. But there is reason to believe that education may reduce the number of new HIV infections. In San Francisco, gay men organized grassroots efforts to educate themselves about HIV transmission, and the results are impressive: less than 1% of the gay male population was infected with HIV after 1985, compared to 10% to 20% in the preceding years. People **can** change their behavior when educated about the risks of transmission (Clement, 1993).

But, if these successes were known in the mid-1980s, why the continued delay in educating the general population, especially sexually active adolescents? Most likely because early efforts to describe the HIV epidemic focused on risk groups rather than the **behaviors** associated with HIV transmission. The epidemic was and still is described with labels—the gay, bisexual, injection drug user, hemophiliac, and heterosexual.

Educators Given the Job of Prevention

Former Surgeon General Joycelyn Elders (1994), Patsy Fleming, past Director of National AIDS Policy, and Sandy Thurman, current Director of National AIDS Policy (Figure 9-1) placed prevention and education at the top of their list of priorities in fighting this epidemic.

The importance of prevention is especially clear as one comes to understand the state of the art in HIV/AIDS therapy. Regardless of what can be medically done for patients with HIV disease, there is no cure. Thus officials from the Centers for Disease Control and Prevention (CDC), the World Health Organization (WHO), the United Nations Joint HIV/AIDS Program (UNAIDS), and the American Health Organization (AHO)

FIGURE 9-1 Photograph of Sandra Thurman, current Director of National AIDS Policy–United States. (*Photograph courtesy of the White House Office of National AIDS Policy, Washington, DC*)

have placed the responsibility of prevention in the hands of the educators, which include health care professionals, parents, and teachers.

Grim Reality

Steven Findlay (1991) wrote that burying those who have died from AIDS has become almost routine. With an 18-year total of 415,000 AIDS deaths in 1998, over 425,000 ending 1999 and a 20-year total estimated at 440,000 ending the year 2000, most Americans are indeed becoming accustomed to HIV/AIDS related deaths. But how many will have died, say, ending the year 2010 or 2020? Will a cure or preventive and therapeutic vaccines be produced? Will our health care system become swamped and ineffective? The best guess by scientists is that neither an effective vaccine nor cure will be found by the end of this century. Through the year 2000 it is projected that worldwide about 18 million people will have died of AIDS, about a half million of

them in the United States. San Francisco may lose 4% of its population; New York 3%; Central Africa 15% to 20%. To avoid the realization of these projections, people of the world must work on HIV/AIDS **PREVENTION.**

Based on over 18 years of intensive epidemiological surveys, scientific research, and empirical observations, it is reasonable to conclude that **HIV is not a highly contagious disease.** HIV transmission occurs mainly through an exchange of body fluids via various sexual activities, HIV-contaminated blood or blood products, prenatal events, and in relatively few cases post-natally through breast milk. **Since 1981, no new route of HIV transmission has been discovered.** The virus is fragile and, with time, self-destructs outside the human body. The most recent data show that HIV remains active for up to 5 days in dried blood, although the number of virus particles (titer) drops dramatically. But it is dangerous to assume that there are no infectious viruses remaining in the dried blood or stored body fluids from an HIV/AIDS patient. In cell-free tissue culture medium, the virus retains activity for up to 14 days at room temperature (Sattar et al., 1991). According to a recent study, HIV was found to survive between 2 and 4 days in glutaraldehyde (Table 9-1), a lubricant used to clean surgical instruments. This finding has serious implications when instruments too delicate for autoclaving: (high-pressure steam sterilization), such as endoscopes, must be used (Lewis, 1995).

Joseph Burnett (1995) reported that HIV can survive 7 days storage at room temperature, 11 days at 37°C in tissue culture extracellular fluid, and can still be infectious in refrigerated postmortem cadaver tissue for 6 to 14 days. Nadia Abdala and colleagues (1999) reported that HIV recovered in the blood from used syringes can remain active up to at least four weeks. The bottom line is that, HIV is more resistant to the environment than originally believed. Chemical agents used to destroy or inactivate the virus are listed in Table 9-1.

It appears that HIV transmission can be prevented by individual action but it will require change in social behaviors. The best way to protect against all sexually transmitted diseases is *sexual abstinence.* The next best way is a mutually monogamous sexual relationship.

TABLE 9-1 Agents Effective Against Human Immunodeficiency Virus

Agents (freshly prepared)	Recommended Concentration
Sodium hypochlorite (household bleach)[a]	Full strength (no dilution)
Chloramine-T	2%
Sodium oxychlorosene	4 mg/mL
Sodium hydroxide	30 mm
Glutaraldehyde	2%
Formalin	4%
Paraformaldehyde	1%
Hydrogen peroxide	6%
Propiolactone dilution	1:400
Nonoxynol-9	1%
Ethyl alcohol	70%
Isopropyl alcohol	30%–50%
Lysol	0.5%–1%
NP-40 detergent[b]	1%
Chlorhexidine gluconate/ethanol mix	4/25%
Chlorhexidine gluconate/isopropyl mix	0.5%/70%
Tincture of iodine/isopropyl	1/30%–70%
Betadine	0.5%
Quarternary ammonium chloride	0.1%–1%
Acetone/alcohol mix	1:1
pH of 1 or 13	
Heat[c] 56°C for 10 minutes	

[a]In 1993, the CDC and Public Health Association recommended that household bleach (e.g., Clorox) be used at full strength.

[b]To be used at 1% solution at room temperature for 2 to 10 minutes.

[c]Dried HIV is ineffective at room temperature after 3 or more days. HIV at high concentration in liquid at room temperature remained ineffective for over 1 week.

Isopropanol (35%), ethanol (50%), Lysol (0.5%), hydrogen peroxide (0.3%), paraformaldehyde (0.5%), and detergent NP-40 (1%) effectively inactivate HIV when incubated at room temperature for 2 to 10 minutes.

(Adapted from Tierno, 1988)

Following these two options is the use of a barrier method during sexual activities—male and female condoms or rubber dams.

Over 90% of **new** HIV infections now occur in the developing world. For the foreseeable future, prevention through behavioral change is the only way to slow this epidemic. In fact, **history shows that prophylaxis and immunization for most infectious diseases will only be partially effective in the absense of behavioral change.**

Sexual transmission accounts for the majority of HIV infection in the developing world; but, this is the most difficult type of transmission to prevent. The use of condoms, reducing numbers of partners and abstinence remain the mainstays of preventing sexual transmission of HIV, but they will not be enthusiastically adopted anywhere just because health authorities tell people to do so.

Behavior has changed in many populations: among gay men in San Francisco, among injecting drug users in Amsterdam and New Haven, Connecticut, and among sex workers and their clients in Nairobi, to name a few. In none of these examples is it clear how the behavioral change took place. Even so, success stories in HIV/AIDS prevention seem to have some elements in common, including consistent and persistent intervention measures over a period of time, a clear understanding of the realities of the target population and involvement of members of that population in prevention efforts. Successful interventions do far more than provide information: they teach communication and behavioral skills, change perceptions of what is **preventive behavior** and ensure that the means of prevention, such as condoms or clean needles, are readily available. (Hearst et al., 1995)

Table 9-2 provides a number of recommendations for preventing the spread of HIV. These recommendations place the responsibility for avoiding HIV infection on both adults and adolescents. **Lifestyles must be reviewed, choices made, and risky behavior stopped.** The public health service and the CDC have established guidelines that, if followed, will prevent HIV transmission while still allowing individuals to be somewhat flexible in their personal behaviors (*MMWR*, 1989).

Quarantine

With few exceptions, proposals to quarantine all individuals with HIV infection have virtually no public support in the United States. Given the civil liberties implications of quarantine, its potential cost, and the realization that alternative, less repressive strategies can be effective in limiting the spread of HIV infection, quarantine proposals in most countries have been dismissed. Despite claims that AIDS is similar to other diseases for which quarantine has been used, public health officials have insisted on distinguishing between behaviorally transmitted infections

TABLE 9-2 Guidelines for Prevention of HIV Infection

I. For the General Public:

1. Sexual abstinence
2. Have a mutual monogamous relationship with an HIV-negative partner (the greater the number of sexual partners, the greater the risk of meeting someone who is HIV-infected).
3. If the sex partner is other than a monogamous partner, use a condom.
4. Do not frequent prostitutes—too many have been found to be HIV-infected and are still 'working' the streets.
5. Do not have sex with people who you know are HIV-infected or are from a high-risk group. If you do, prevent contact with their body fluids. (Use a condom and a spermicide from start to finish.)
6. Avoid sexual practices that may result in the tearing of body tissues (e.g., penile–anal intercourse).
7. Avoid oral–penile sex unless a condom[a] is used to cover the penis.
8. If you use injection drugs, use sterile or bleach cleaned needles and syringes and *never* share them.
9. Exercise caution regarding procedures such as acupuncture, tattooing, ear piercing, and so on, in which needles or other unsterile instruments may be used repeatedly to pierce the skin and/or mucous membranes. Such procedures are safe if proper sterilization methods are employed or disposable needles are used. Ask what precautions are being taken before undergoing such procedures.
10. If you are planning to undergo artificial insemination, insist on frozen sperm obtained from a laboratory that tests all donors for infection with the AIDS virus. Donors should be tested twice before the sperm is used— once at the time of donation and again 6 months later.
11. If you know that you will be having surgery in the near future and you are able to do so, consider donating blood for your own use. This will eliminate the small but real risk of HIV infection through a blood transfusion. It will also eliminate the more substantial risk of contracting other transfusion blood-borne diseases, such as hepatitis B.
12. Don't share toothbrushes, razors, or other implements that could become contaminated with blood with anyone who is HIV-infected, demonstrates HIV disease, or has AIDS.

II. For Health Care Workers:

1. *All* sharp instruments should be considered as potentially infective and be handled with extraordinary care to prevent accidental injuries.
2. Sharp items should be placed into puncture-resistant containers located as close as practical to the area in which they are used. To prevent needlestick injuries, needles should not be recapped, purposefully bent, broken, removed from disposable syringes, or otherwise manipulated.
3. Gloves, gowns, masks, and eye-coverings should be worn when performing procedures involving extensive contact with blood or potentially infective body fluids. Hands should be washed thoroughly and immediately if they accidentally become contaminated with blood. When a patient requires a vaginal or rectal examination, gloves must always be worn. If a specimen is obtained during an examination, the nurse or individual who assists and processes the specimen must always wear gloves. Blood should be drawn from all patients—regardless of HIV status—only while wearing gloves.
4. To minimize the need for emergency mouth-to-mouth resuscitation, mouthpieces, resuscitation bags, or other ventilation devices should be strategically located and available for use where the need for resuscitation is predictable.

III. People at Risk of HIV Infection:

1. See recommendations for general public.
2. Consider taking the HIV antibody screening test.
3. Protect your partner from body fluids during sexual intercourse.
4. Do not donate any body tissues.
5. If female, have an HIV test before becoming pregnant.
6. If you are an injection drug user, seek professional help in terminating the drug habit.
7. If you cannot get off drugs, do not share drug equipment.

IV. People Who Are HIV-Positive:

The prevention of transmission of HIV by an HIV-infected person is probably lifelong, and patients must avoid infecting others. HIV seropositive persons must understand that the virus can be transmitted by intimate sexual contact, transfusion of infected blood, and sharing needles among injection drug users. They should refrain from donating blood, plasma, sperm, body organs, or other tissues. HIV-infected people should:

1. Seek continued counseling and medical examinations.
2. Do not exchange body fluids with your sex partner.

(continued)

TABLE 9-2 *Continued*

3. Notify your former and current sex partners, encourage them to be tested.
4. If an injection drug user, do not share drug equipment and enroll in a drug treatment program.
5. Do not share razors, toothbrushes, and other items that may contain traces of blood.
6. Do not donate any body tissues.
7. Clean any body fluids spilled with undiluted household bleach.
8. If female, avoid pregnancy.
9. Inform health care workers on a need-to-know basis.

V. Practice of Safer Sex:

Safer sex is body massage, hugging, mutual masturbation, and closed mouth kissing. HIV seropositive patients must protect their sexual partners from coming into contact with infected blood or bodily secretions. Although consistent use of latex condoms with spermicide containing nonoxynol-9 can decrease the chance of HIV transmission, condoms do break. If engaging in sexual intercourse with an HIV-positive person use two condoms. But even this is no guarantee. (Also see 1 through 6 under "For the General Public" in this table.)

ªTests show that HIV can sometimes pass through a latex condom. Experts believe that natural-skin condoms are more porous than latex and therefore offer less effective protection. Never use oil-based products such as Vaseline, Crisco, or baby oil with a latex condom, because they make the latex porous, cause latex deterioration and breakage thus nullifying the protection the condom provides against the virus. The use of condoms containing a spermicide is recommended.

and those that are airborne. AIDS is not tuberculosis.

Through early year 2000, HIV prevention methods in the United States and other developed nations remains education, counseling, voluntary testing and partner notification, drug abuse treatment, and syringe exchange programs. To date, the power to quarantine, for any disease, has rarely been used in the United States. In fact, only one country, Cuba, officially used the power of quarantine in 1986 to stem the spread of HIV. Data to date indicate that the use of quarantine of HIV-infected and AIDS persons in Cuba had been very effective. Cuba had 13 sanatoriums holding some 900 persons of which about 200 have AIDS. Cuba stopped the quarantine of HIV infected persons in mid-1993. Currently, there are 727 AIDS cases in Cuba, 532 people have died of AIDS.

NEW RULES TO AN OLD GAME : PROMOTING SAFER SEX

Barriers to HIV Infection

The two most effective barriers to HIV infection and other sexually transmitted diseases are (**1**) **abstinence,** which can be achieved by saying *NO* emphatically and consistently; and (**2**) **forming a no-cheating relationship with one individual, preferably for life**. These solutions

to the danger of HIV infection may not be **"cool,"** but they do work. These two completely safe approaches are endorsed by the Surgeon General as the **preferred** methods. For those who do not practice abstinence, barrier methods are necessary to prevent HIV infection/transmission.

The barrier methods used to prevent HIV infection are the same methods used to prevent other sexually transmitted diseases (STDs) and often pregnancy. They include latex condoms, plastic condoms (new in 1995), and latex dams, and diaphragms, used in conjunction with a spermicide. Barrier dams or dental dams are thin sheets of latex or similar material placed over the vagina, clitoris, and anus during oral sex. (Ask your dentist to show you a dental dam.) **Spermicides are chemicals that kill sperm.** These same chemicals have also

──────── SIDEBAR 9.1 ────────

SAFER SEX COMPLICATES SEX

The presences of HIV on the planet earth complicates and alters a great deal of human behavior—but HIV complicates sex perhaps more than anything else. Although safer sex guidelines are clear—once HIV enters a relationship, nothing else is straightforward, ever again!

EDUCATION VERSUS RISK: IT IS NOT WHAT YOU KNOW, IT IS WHAT YOU DO

Sixty-five percent of all AIDS cases in the United States have been attributed to sexual contact. HIV is transmitted from infected semen, pre-ejaculate, vaginal secretions, or blood that comes into contact with the penis, vagina, anus, or rectum of an uninfected sexual partner. Additionally, contact of the mouth with semen or pre-ejaculate has transmitted infection, but at a very low rate compared with other penetrative sexual acts (refers to penile-insertion and penile receptive anal and vaginal sex, and to oral contact with a partner's vulva or penis). The likelihood that sexually active people will become infected with HIV is determined by the probability of their having sex with an HIV-infected partner and their degree of exposure to one of these HIV-infected body fluids during sex. Any sexual act with an uninfected partner is HIV-safe (but not necessarily safe from other sexually transmissible diseases). Sex with an infected partner while correctly using a condom substantially reduces the likelihood of HIV infection. It is easy to forget, however, that unsafe sex is simply ordinary human sex.

SEXUAL RISKS

The degree to which patients may be willing to consider changing or modifying risk behaviors depends on how necessary they perceive the unchanged risk behavior to be. Although HIV transmission occurs through specific behaviors, emotional and psychological factors determine whether or not these behaviors occur. The importance that sexually active people place on sexual behavior has an influence on their willingness to take HIV risks, regardless of their current HIV status. People continue to be confused and concerned about their level of sexual risk. Consequently, people demand clear-cut answers even when factual evidence is limited. Since many people's sexual lives are structured so that the possibility of acquiring HIV infection exists, equipping them with information to make sexual decisions in the face of unavoidable uncertainties is essential.

SOME PEOPLE AT RISK

Beginning year 2000, the incidence of AIDS is six times higher among African Americans and three times higher among Latinos than among whites. About 54% of all AIDS cases in the United States have occurred among gay men and among other men who have sex with men. Sixteen percent of all AIDS cases reported in the United States occurred among women. Eighteen percent of AIDS cases in the United States occurred among 20- to 29-year olds. Vulnerability of adolescents is of particular concern because many HIV-infected 20- to 29-year olds contracted HIV in their teens. Forty two percent of AIDS cases among men of color were caused by having sex with men; 36% were caused by injecting drug use. The number of AIDS cases among heterosexuals has increased in recent years; 10% of all adult/adolescent AIDS cases in the United States are attributed to heterosexual contact.

---

been shown to kill bacteria and inactivate viruses that cause STDs (Bolch et al., 1973; Singh, 1982; Amortegui et al., 1984; Hicks et al., 1985). Spermicides are commercially available in foams, creams, jellies, suppositories, and sponges. **Use of these products may provide protection against the transmission of STDs, but the only recommended barrier protection against HIV infection is a condom with a spermicide.** National Condom Week is the week of February 14–21.

Condom—A Medical Device?

Condoms are classified as medical devices. Every condom made in the U.S. is tested for defects and must meet quality control guidelines enforced by the federal Food and Drug Administration (FDA).

Condoms are intended to provide a physical barrier that prevents contact between vaginal, anal, penile, and oral lesions and secretions and ejaculate.

At least 50 brands of condoms are manufactured in the United States. They are produced to fit every fancy. There are colored condoms—pink, yellow, and gold; flavored condoms; and condoms that are perfumed, ribbed, stippled, and phosphorescent (glow in the dark). This assortment of condoms exposes the user and his partner not only to rubber but also to a variety

A LESSON ON CONDOM USE FROM WORLD WARS I AND II

In his social history of venereal disease, *No Magic Bullet*, Allan M. Brandt describes the controversy in the U.S. military about preventing sexually transmitted disease among soldiers during World War I. Should there be an STD **prevention effort** that recognized that many young American men would succumb to the charms of French prostitutes, or should there be a more **punitive approach** to discourage sexual contact? Unlike the New Zealand Expeditionary forces, which gave condoms to their soldiers, the United States decided to give American soldiers after-the-fact, and largely ineffective, chemical prophylaxis. American soldiers were subject to court martial if they contracted a venereal disease. These measures failed. More than 383,000 soldiers were diagnosed with venereal diseases between April 1917 and December 1919 and lost 7 million days of active duty. Only the influenza epidemic was a more common illness among servicemen at that time!

This grim lesson was lost on Americans back home, and campaigns against syphilis continued to emphasize abstinence. **By the 1930s, almost 1 in 10 Americans was infected with syphilis.**

During World War II, however, the American armed forces took a more realistic approach and distributed 50 million condoms each month during the war. The military's new motto—**"If you can't say no, take a pro"**—recognized that abstinence is the best way to prevent STDs, but for those who don't say no, the next best option is to use a condom (Glantz, 1996).

HOW TELEVISION AND NEWSPAPERS VIEWED CONDOM ADVERTISING IN 1987

Television—The big three, ABC, NBC, and CBS refused to run condom advertisements. Their concerns, they said, were about public morality. They refused even after the United States **Surgeon General C. Everett Koop** encouraged such advertising to help stop the spread of AIDS. Koop, in testimony before a House subcommittee, called on the **big three** television stations to lift their ban on commercial for condoms. A number of foreign countries have been running condom commercials for several years. Representatives of ABC, NBC and CBS explained to the House panel that such ads would be offensive to viewers of many affiliated stations. All three stations, however, carry many shows with gratuitous sexual titillation (for example the soap operas), extra marital affairs, rape, incest, child abuse, and criminal behavior of all types and violence. At this time some 27,000 Americans had AIDS and over 40,000 had already died of AIDS.

Newspapers—The UNION LEADER, Manchester, New Hampshire. Publisher, Nackey S. Loeb announced that the New Hampshire Sunday News and The Union Leader will not knowingly accept any advertisements for condom sales.

THE PILOT, Archdiocese of Boston Newspaper. If the AIDS epidemic has taught us anything it is this: sexual promiscuity—that is, having sexual relations with strangers—is mortally dangerous. So condom advertisers are catering to precisely the sort of behavior that transmits AIDS. And, of course, making a profit on the deal. We know two things about automobile safety: (1) drunk drivers run a much higher risk of having an accident, and (2) when accidents happen, seat belts save lives. Does anyone seriously recommend an advertising campaign that would urge drunk drivers to buckle up? Of course not; anyone who seemed to condone drunk driving would be labeled, quite accurately, as irresponsible. By the same logic, if the media want to help stop the spread of AIDS, the only responsible approach is to campaign against sexual promiscuity. [Stop Condom Advertising.]

UPDATE 1999

A national AIDS advocacy group asked former Republican presidential candidate Bob Dole to join its efforts to persuade the nation's top six television networks, (ABC, NBC, CBS, FOX, WB, and UPN) to drop their ban on commercials for condoms.

Daniel Zingdale, aids activist, said in a letter to Dole,

"We hope you will agree that there is a dangerous contradiction in network policies that allow **Viagra ads** but ban ones for condoms. While television networks air ads promoting products to enhance sex lives, ads for products that promote safer sex lives are banned from the airwaves. Imagine if the networks aired commercials for faster motorcycles but banned ads for helmets."

The networks responded that they have no plans to lift their ban on condom commercials. Two of the networks said they allow local TV station affiliates to run condom ads in accordance with their community standards.

CLARIFYING THE ISSUES OVER CONDOM USE

Two major issues surface in the debate over advocating condom use in the prevention of HIV infection: One concerns the concept of efficacy, the condoms ability to stop the virus from passing through, and the other, the fear that making condoms available will encourage early sexual activity among adolescents and extramarital sex among adults.

EFFICACY (DO THEY WORK?)

All condoms are not 100% impermeable; they are not all of the same quality. Investigators using different testing methods have reported that latex condoms are effective physical barriers to high concentrations of *Chlamydia trachomatis, Neisseria gonorrhoeae,* the herpes and hepatitis viruses, cytomegalovirus, and HIV (Judson, 1989). But for maximum effectiveness condoms must be properly and consistently used from start to finish (Table 9-3).

To be effective, a condom must be worn on the penis during the entire time that the sex organ is in contact with the partner's genital area, anus, or mouth. Care must be taken that the condom is on before vaginal, anal, or oral penetration, and that it does not slip off. If properly used, the condom provides protection against most of the STDs that occur within the vagina, on the glans penis, within the urethra or along the penile shaft.

Because the condom covers only the head and shaft of the penis, it does not provide protection for the pubic or thigh areas, which may come in contact with body secretions during sexual activity.

Margret Fischel reported in 1987 that 17% of women whose husbands were HIV-positive became HIV-infected while using condoms properly and consistently.

Susan Weller (1993) reported on the use of condoms as a barrier to HIV transmission. She analyzed the data from 11 studies involving 593 partners of HIV-infected people. The resulting data showed that condoms are only 69% effective in preventing HIV transmission in heterosexual couples. These data surprise people because earlier contraceptive research indicated that condoms are about 90% effective in preventing pregnancy. Thus many people assume condoms prevent HIV transmission with the same degree of effectiveness. However, transmission studies by Weller do not show that to be true. Weller states that condom effectiveness in blocking HIV transmission may be as low as 46% or as high as 82%!

TABLE 9-3 Proper Placement of a Condom[a] on the Penis

1. Open the packaged condom with care; avoid making small fingernail tears or breaks in the condom.
2. Place a drop of a water based lubricant inside the condom tip before placing it on the head of the penis. Be sure none of the lubricant rolls down the penis shaft as it may cause the condom to slide off during intercourse.
3. Hold about a half an inch of the condom tip between your thumb and finger—this is to allow space for semen after ejaculation. Then place the condom against the glans penis (if uncircumcised, pull the foreskin back).
4. Unroll the condom down the penis shaft to the base of the penis. Squeeze out any air as you roll the condom toward the base.
5. After ejaculation, hold the condom at the base and withdraw the penis while it is still firm.
6. Carefully take the condom off by gently rolling and pulling so as not to leak semen.
7. Discard the condom into the trash.
8. Wash your hands after the procedure.
9. Never use the same condom twice.
10. Condoms should not be stored in extremely hot or cold environments.

[a]Males should practice putting on and removing a condom prior to engaging in sexual intercourse.

Isabelle deVincenzi (1994) reported on HIV transmission among heterosexual sexual couples in which one partner was known to be HIV-positive. Study participants were advised to use condoms during intercourse, and among the 124 (or 48%) of couples who followed that advice consistently, no seronegative partner became HIV infected during roughly 24 months of follow-up. However, despite knowledge about HIV transmission, more than 50% of the couples failed to use condoms consistently. Twelve of the seronegative individuals in that group became HIV-positive during the study.

The importance of compliance is illuminated by an analogy with pregnancy prevention programs. Although typical pregnancy rates for couples who use condoms are as high as 10% to 15%, rates are estimated to be as low as 2% for couples who use condoms correctly and consistently.

A mathematical model predicts that consistent condom use could prevent nearly half of the sexually transmitted HIV infections in persons with one sexual partner and over half of HIV infections in persons with multiple partners. A reduction of this magnitude could help interrupt the propagation of the epidemic. Therefore, promoting more widespread understanding of condom efficacy or effectiveness, and advocating their consistent use by those who choose to be sexually active, is crucial to protecting people from HIV infection and to slowing the spread of the HIV and sexually transmitted disease epidemic.

DO CONDOMS ENCOURAGE SEXUAL ACTIVITY?

Many persons assert that those who promote condom use to prevent HIV infection appear to be condoning sexual intercourse outside of marriage among adolescents as well as among adults. Recent data from Switzerland suggest that a public education campaign promoting condom use can be effective without increasing the proportion of adolescents who are sexually active. A 3-year, 10-month study showed condom use among persons aged 17 to 30 years increased from 8% to 52%. By contrast the proportion of adolescents (aged 16 to 19 years) who had sexual intercourse did not increase over that same period. A report from Deborah Sellers and colleagues (1994) and Sally Guttmacher et al. (1997) also concluded that the promotion and distribution of condoms did not increase sexual activity among adolescents. The study involved 586 adolescents who were 14 to 20 years of age. Douglas Kirby and colleagues (1998) studied 10 Seattle, Washington high schools that made condoms available to students through vending machines and school clinics. The study measured the number of condoms students took and the subsequent changes in sexual behavior. The investigators concluded that making condoms available in Seattle schools enabled students to obtain relatively large numbers of condoms but did not lead to increases in either sexual activity or condom use.

The AIDS epidemic has brought new dimensions of complexity and urgency to the debate over adolescent sexual activity. Some have urged abstinence as the only solution; while others champion condom use as the most practical public health approach. Thus a clear message about condoms may have been obscured by controversy over providing condoms for adolescents in schools while at the same time trying to discourage these same young people from initiating sexual activity.

There must be a common ground: People should be able to agree that premature initiation of sexual activity carries health risks. Therefore, young people must be encouraged to postpone sexual activity. Parents, clergy, and educators must strive for a climate supportive of young people who are not having sex. Let them know it is a very positive and intelligent decision, and so help to create a new health-oriented social norm for adolescents and teenagers about sexuality.

The message that those who initiate or continue sexual activity must reduce their risk through correct and consistent condom use needs to be delivered as strongly and persuasively as the message, **"Don't do it."** Protection of the individual and the public health will depend on our ability to combine these messages effectively (Roper et al., 1993).

ANECDOTE

Presenting facts without understanding won't work! Here is a simple story to emphasize the point: A minister following his custom, paid a monthly call on two spinster sisters. While he was standing in their parlor, holding his cup of tea, engaged in their usual chit chat, he was startled by something that caught his eye. There on the piano was a condom! 'Ladies, in all the years we've known each other I have never intruded into your private lives, and never felt the need to. But now I am forced to ask what is that thing doing there?' 'One of the ladies replied, 'Oh, that's a wonderful thing, pastor, and they really work!' The minister was agitated: 'I'm not talking about their value or effectiveness. I just want to know what that thing is doing on your piano?'

'Well, my sister and I were watching television. We heard this lovely man, the Surgeon General of the whole United States. He said that if you put one of those on your organ, you'll never get sick. Well, as you know we don't have an organ, but we bought one and put it on the piano, and we haven't had a day's sickness since!'

of different chemicals—some that can cause allergic skin reactions (**contact dermatitis**). One to two percent of people are sensitive to latex rubber a demonstrate contact dermatitis.

Condoms are also called rubbers, prophylactics, bags, skins, raincoats, sheaths, and French letters. They can be lubricated or not, have reservoir tips or not, and can contain spermicide.

Condoms: Barriers To Bad News

Over the years people have been told to use condoms to prevent STDs and pregnancy. Today people are told that an additional use of the condom is to save lives.

History of Condoms

It has been reported that early Egyptian men used animal membranes as a sheath to cover their penises (Barber, 1990). Animal intestines were flushed clean with water, sewn shut at one end and cut to the length of the erect penis. In 1504, Gabriel Fallopius designed a medicated linen sheath that was pulled on over the penis. A Japanese novel written in the 10th century refers to the uncomfortable use of a tortoise shell or horn to cover the penis.

It is interesting to note that condoms were used far more often throughout history as protection against STDs than as contraceptives. For example, an 18th century writer recommended that men protect themselves against disease by placing a linen sheath over the penis during intercourse.

The term condom came into common usage in the 1700s. According to accounts, in the early 1700s, condoms were sold and even exported from a London shop whose proprietress laundered and "recycled" them in a back room (Barber, 1990). Condoms became more widely available after 1854 when the method for making rubber was invented by Charles Goodyear (as in Goodyear tires). The latex condom was first manufactured in the 1930s.

Condoms have been available in the United States for over 130 years, but have never been as openly accepted as they are now. Their sale for contraceptive use was outlawed by many state legislatures beginning in 1868 and by Congress in 1873. Although most of these laws were eventually repealed, condom packages and dispensers until only a few years ago continued to bear the label **"Sold only for the prevention of disease,"** even though they were being used mainly for the prevention of pregnancy.

After the advent of nonbarrier methods of contraception during the 1960s (mainly the use of the birth control pill) there was an ensuing epidemic increase in most sexually transmitted infections. Condoms once again are being marketed for the prevention of disease (Judson, 1989).

In 1979, only 8% of respondents to a *Consumer Reports* survey on condoms said they used condoms for the prevention of STDs. However, in *Consumer Reports'* 1989 study, 26% said they used condoms to prevent STDs, especially HIV infection (Figure 9-2, A and B).

Safer Sex, the Choice of Condom

Although a variety of preventative behaviors have been recommended (Table 9-2), the responsibility of **safer sex,** with a condom, is a personal choice. If one decides to use a condom, then the choice is what kind, and whether or not to use a spermicide.

The Male Condom— The American-made condom most often sold is made of latex, is about 8 inches long, and in general, one size fits all. About 500 million condoms were sold in the United States in 1998. Six to nine billion are sold annually worldwide (Grimes, 1992 updated).

Intact latex condoms provide a continuous mechanical barrier to HIV, herpes virus (HSV), hepatitis B virus (HBV), *Chlamydia trachomatis,* and *Neisseria gonorrhoeae.* A recent laboratory study indicated that latex and the **new polyurethane condoms** are the most effective mechanical barrier to fluid containing HIV-sized particles (0.1 μm in diameter) available (Figure 9-3).

In a study of heterosexual couples—in which only one partner was HIV-positive—none of the 123 uninfected partners who used latex condoms consistently and correctly contracted the disease. Among a similar group who were less conscientious, 10% became infected (*MMWR,* 1993a; *Office Nurse,* 1995).

Three prospective studies in developed countries indicated that condoms are unlikely to break or slip during proper use. Reported breakage rates in the studies with latex condoms were 2% or less for vaginal or anal intercourse (*MMWR,* 1993a; Spruyt et al. 1998).

Choice— The best choice for preventing STDs and pregnancy is condoms that are made of **latex** or **polyurethane (plastic)** and contain

FIGURE 9-2 Advertisements for Condoms. **A,** A common theme centered around the use of a condom to protect against sexually transmitted diseases. (*Photograph courtesy of the Minnesota AIDS Project, Minneapolis*) **B,** A specific type of condom in a comic book style advertisement. (Photo courtesy of World Health Organization.)

a **spermicide.** The spermicide is added protection in case the condom ruptures or spills as it is taken off. During in vitro studies, condoms were artificially ruptured in a medium containing nonoxynol-9 or N-9 (*p*-diisobutylphenoxypolyethoxethanol). On examination, no virus capable of reproduction was found. Tests done without the spermicide resulted in viruses capable of reproducing (Connell, 1989). It must be mentioned, however, that there is no evidence that N-9 has any effect on HIVs that are carried within lymphocyte cells in the semen.

Nonoxynol-9— Evidence from the mid-1970s has lent support to the prophylactic use of N-9 as a spermicide against sexually transmitted diseases (Bird, 1991; Wittkowski et al.,

1998). There are reports, however, that some males and females are sensitive to it. N-9 has been shown to variably inactivate *Neisseria gonorrhoeae*, *Trichomonas vaginalis*, *Ureaplasma urealyticum*, *Treponema pallidum*, *Candida albicans*, herpes and hepatitis B viruses, and cytomegalovirus and HIV. There is some question about its effectiveness against *Chlamydia trachomatis*, and it has been reported to be ineffective against the genital wart virus (Stone et al., 1986; Rosenberg et al., 1992). More recently, concern has been raised that N-9 may serve to digest nonspecific protective mucosal coatings and induce the increased presence of lymphocytes and bleeding, which by itself might promote HIV and other viral and bacterial infections (Fisher, 1991). Ronald Moody and colleagues (1998) report that nonoxynol-9 did **not**

FIGURE 9-3 A Collection of Condoms. The condom is sealed in an aluminum foil package, sometimes along with a spermicide. On the left side is an unfolded or unrolled male condom; on the right side is a vaginal pouch (female condom) (see Figure 9-4). (*Photograph courtesy of the author*)

reduce the risk of HIV infection, nor did it reduce the risk of gonorrhea or chlamydia infection. The controversy about whether to use nonoxynol-9 continues.

Condom Advertising on American Television— On January 4, 1994, condoms danced into America's living rooms as part of the most explicit HIV prevention campaign the nation has even seen. Previously, condoms were rarely seen or even mentioned on American television. But the Centers for Disease Control and Prevention (CDC) launched a series of radio and television public service announcements targeting sexually active young adults, a group at high risk for HIV infection.

One of the television ads, entitled ***Automatic,*** features a condom making its way from the top drawer of a dresser across the room and into bed with a couple about to make love. The voice-over says, "It would be nice if latex condoms were automatic. But since they're not, using them should be. Simply because a latex con-

dom, used consistently and correctly, will prevent the spread of HIV." Another, entitled ***Turned Down,*** features a man and woman kissing, when the woman asks the man, **"Did you bring it?"** When he says he **forgot it,** she replies, "Then **forget it,**" and turns on the light. There is also a pair of abstinence ads, in which condom use is not mentioned. The ads feature a man and woman talking. She says, "There is a time for us to be lovers. We will wait until that time comes."

Language that was once forbidden on TV has now become routine. Will it prevent HIV infection? Will it save lives?

Beginning 1996, the CDC began another series of radio and TV public service condom announcements.

The spots reel along like MTV videos, interspersing personal advice from young people with scenes from dance clubs, Greenwich Village, and a drugstore. They are punctuated by headlines that scream, "Put It On," "Talk About It," and "It's O.K. to Wait to Have Sex."

The bottom line to the new announcements is: **"Respect Yourself, Protect Yourself."**

The advertising campaigns are hailed as a milestone, marking a high point of visibility for the condom on network television and introducing a new level of explicitness to the Centers for Disease Control and Prevention's public pronouncements about AIDS prevention.

Class Discussion: What is your opinion on condom advertising? When? Where? Why?

Buying Condoms— Women are taking a more active role in buying condoms. In 1985, women bought about 10% of the condoms sold. In 1989, they purchased 40% to 50%. According to surveys, most women buying condoms are single, and their concern is about HIV infection rather than pregnancy. **The fact that more women are willing to buy condoms is evidence that HIV education is working to some degree.**

Many condoms are purchased from vending machines. The FDA recommends the following guidelines when purchasing condoms from a vending machine:

1. Is the condom made of latex or polyurethane?
2. Is the condom labeled for disease prevention?
3. Is the spermicide (if any) outdated?
4. Is the machine exposed to extreme temperatures or direct sunlight?

In mid-1992, the first drive-up "Condom Hut" opened in Cranston, Rhode Island. With each purchase the customer receives a brochure on safer sex.

It is generally recommended that condoms be stored below 25°C (77°Fahrenheit; room temperature is 72°Fahrenheit or 22°Centigrade). The packaging should be impermeable to both sunlight and gas. If air, which includes ozone, enters the package, it would affect the condom very quickly—ozone is like rust to a condom. Latex is a natural product—it will go bad if you don't treat or store it properly.

Free Condoms— In the south of France, the regional tourism authority announced, in 1991, that condoms would be placed alongside soap and shampoo in all hotel rooms and would be supplied free to all people using campsites in the area. Some 8 million tourists visit the south of France every summer—the region has the highest incidence of HIV/AIDS in the country.

Condoms are now being dispensed without charge in most college and university and public health clinics, and in at least 400 high school health offices in the United States. Some cities in Canada have been providing access to free condoms in high schools since 1984.

Regardless of educational programs on safer sex and condom usage, recent studies indicate that teenagers still refuse to use condoms. **What they know is not equal to what they do!** Based on their findings, the researchers said information-oriented school- and community-based AIDS prevention programs will not succeed in getting adolescents to use condoms, because there is no association between knowledge and preventive behavior.

Condom Availability in Developing Nations

Access to condoms in developing nations, although crucial, is poor even though over **1 billion** condoms are distributed in developing nations each year. The World Health organization spends some $70 million a year for male condoms distributed in 14 countries. But, it would cost $460 million to provide just Africa with an adequate condom program. The

─────── **SIDEBAR 9.2** ───────

CHINA: CONDOM VENDING MACHINES

September 1998: According to the China Daily Newspaper, the country's first condom vending machines in the southern town of Shenzhen have been so successful that machines will be placed in other cities. In the first month of operation, each of the 50 machines sold 2,000 to 3,000 condoms at 12 cents each. The paper reported that sales have been spurred by the spread of sexually transmitted diseases.

However, overall in China about 1.5 billion condoms were used by about 12% of the population in 1998—1.2 billion of them were distributed for free. Health officials estimate that 300,000 people are HIV positive.

United Kingdom, in 1992, purchased 66 million condoms for Zimbabwe but the country needs at least 120 million condoms per year. **And it must be recognized that a number of cultures in developing nations and the Roman Catholic religion prohibit the use of condoms.**

In May 1999, researchers on sexually transmitted diseases stated to close the gap between the number of condoms now produced and the numbers needed. It will take an additional 24 billion condoms a year to slow the transmission of STDs including HIV.

―――― SIDEBAR 9.3 ――――

FAULTY CONDOMS: SOUTH AFRICA

There is only one condom manufacturer in South Africa and one "partial" manufacturer. Each produced 30 million condoms for the government in 1998, which amounts to 20% of the state's total need of 150 million. According to the South Africa Health Department's former condom logistics consultant, "to purchase condoms from another country, bids do not have specifications related to packaging, storage and shelf life. **Price is the only thing that seems to matter."** With that as a brief background, a **New York Times article** quotes experts as saying South Africa became a dumping ground for substandard products. Its procurement officers used an ineffective test regimen—inspectors would visit a factory once a year, test samples chosen by the factory and give it the South African Bureau of Standards seal of approval. Thus, imported condoms went straight into distribution without any sample testing. As a result, about 40 million **Kenzo** condoms (made in India) and **Twin Lotus** condoms (made in China) were recalled in mid-1999 because **one in four were faulty.** According to a department consultant, the recall netted only 4.7 million of the Indian-manufactured condoms, But all of this was kept from the public. The Deputy Director-General of Health, **Harm Pretorius,** said "there is no policy to inform the public. We simply took the defective condoms from the system." The Times article said that the condoms were being "dumped" in South Africa following complaints about the condoms in their own countries. The South African Health Department has since terminated these contracts for condoms.

Condom Quality

Some have touted condoms as a bulletproof vest for preventing HIV infection. This is simplistic and inaccurate because condoms may break or be used improperly. Condoms have been shown to **reduce** significantly but **not eliminate** the transmission of HIV. **Condom use is a form of safer sex but not absolutely safe sex.** The March 1989 issue of *Consumer Reports* did a rather extensive study on the quality of different brands of condoms. The reader survey revealed that one in eight people who used condoms had two condoms break in one year of sexual activity; one in four said one condom had broken. Calculated from its data, about one in 140 condoms broke, with condom breakage occurring more often during some sexual activities than others. For example, the breakage rate during anal sex was calculated at one in 105 compared to one in 165 for vaginal sex. One in 10 heterosexual men admitted to engaging in anal sex using condoms (*Consumer Reports*, 1989).

In a separate United States Public Health Service study (1994), if a couple used male condoms consistently and correctly, there was a 3% chance of unintentional pregnancy. Other studies have found that condoms reduce the risk of gonorrhea by as much as 66% in men, but by no more than 34% in women. **It doesn't appear as though they prevent papilloma virus infection—which causes genital warts—in either sex** (Stratton et al., 1993). Table 9-4 presents the results on **latex** condom efficacy in preventing HIV infection.

In 1987, FDA inspectors began making spot checks for condom quality. They fill the condom with 10 ounces (about 300 mL) of water to check for leaks. If leaks are found in more than four per 1,000 condoms, that entire batch is destroyed. Over the first 15 months of spot inspections, one lot in 10 was rejected. Import condoms failed twice as often—one lot in five.

The New Polyurethane (Plastic) Condoms— For the 1% of the general population that is sensitive to latex and for those who have a variety of other reasons not to use a latex condom, there is now a clear, thin, FDA-approved polyurethane (plastic) condom for sale in the

TABLE 9-4   Efficacy of Latex Condoms to Stop HIV Transmission

Study	Subjects	HIV-Positive
Fischl et al., 1987	Partners of HIV-positive individuals	Without barrier, 86% With barrier, 10%
Mann et al., 1987	Prostitutes, Zaire	No condom use, 26% Up to 25% use, 35% 26%–49% use, 32% 50%–74% use, 0% 75%–100% use, 0%
Ngugi et al., 1986	Prostitutes, Kenya	No condom use, 72% Less than 50% use, 56% 50% use or more, 33% 100% use, 0%
CDC, 1987	Prostitutes, United States	No or some condom use, 11% 100% use, 0%
Padian, 1987	Female partners of HIV-positive males	Relative risk, 0.6[a] (95% CI, 0.3–1.3)
Laurian et al., 1989	Female partners of HIV-positive males	No or some condom use, 17% 100% use, 0%

CDC, Centers for Disease Control and Prevention; CI, confidence interval.

[a]Extensive condom use likely to have begun after seroconversion.

(Adapted from Stratton et al., 1993.)

United States. The condoms are colorless, odorless, and **can** be used with **any** lubricant. The current cost is about $1.80 each. Manufacturers of at least five other plastic condoms are waiting FDA approval. A recent report by Ron Frezieres and colleagues (1999) reported that although polyurethane and latex condoms provide equivalent levels of contraceptive protection, the polyurethane condom's higher frequency of breakage and slippage suggests that this condom may confer less protection from sexually transmitted infections than do the latex condoms.

A Polymer Gel Condom?— In November 1997 Michael Bergeron of the Laval Infectious Disease Research Center in Canada created a liquid condom that is non-toxic, tasteless, and transparent. The condom, made out of a polymer gel, is spread over the vagina or anus using an applicator. It gelifies in response to body heat and surrounds the penis. According to the inventor: "It's a physical barrier, but it moves— that's the advantage." The polymer gel can be flushed out with water. Bergeron developed the gel due to patient complaints about tight condoms. A major benefit is that the gel condom is somewhat **invisible**—either sexual partner can use it as a protective measure. Trials to determine the contraceptive's efficacy in preventing HIV, herpes and pregnancy began in the spring of 1998 with 100 sex workers who say they do not use condoms; one-half used the polymer gel during sexual intercourse, while the other half served as a control group. Results are not available.

Why Some Men Don't Use Condoms: Male Attitudes About Condoms and Other Contraceptives— This is the name of a study published by the Henry J Kaiser Family Foundation (1997). The study shows that men, especially teenage males, don't use condoms mostly because of **embarrassment:** buying the condom, talking about the condom with their sexual partner, putting the condom on in front of their sexual partner, offending or scaring their sexual partner, losing an erection while putting the condom on, and the **reduction of sexual pleasure.** (This free report can be obtained at 1-800-656-4533 monograph #1319.)

UPDATE 1999—Recent reports reveal that HIV-infected people who have successfully suppressed HIV replication, i.e. their viral load is low or unmeasurable, may be less likely to practice safer sex. People on antiretroviral therapy, who answered a survey, said they used condoms less then half the time while having sex. With condoms being used less by gay men when compared to heterosexual men.

The Female Condom (Vaginal Pouch)— The female condom was FDA approved in May of 1993, and has become available to the general public. Before giving the Reality condom final approval, the FDA asked that two caveats be put into the labeling. First, the agency required a statement on the package label that male condoms are still the best protection against disease, and second, that the label compare the effectiveness of female condoms with that of other barrier methods of birth control. According to the FDA, in a study of 150 women who used the female condom for 6 months, 26% became pregnant. The manufacturer contends that the pregnancy rate was 21%—and only because many women did not use the condom every time they had sex. With **"perfect use,"** company officials say, the rate is 5%, in contrast to 2% for male condoms. Both conditions were met to the FDA's satisfaction. Gaston Farr and co-workers (1994) concluded that the female condom provided contraceptive efficacy in the same range as other barrier methods, particularly when used consistently and correctly, and has the added advantage of helping protect against sexually transmitted diseases.

The female condom is now called the **vaginal pouch.** It is 17 cm (about 6 3/4 inches) long and it consists of a 15-cm polyurethane sheath with rings at each end (Figure 9-4). The closed end fits into the vagina like a diaphragm. The outer portion is designed to cover the base of the penis and a large portion of the female perineum (the area of tissue between the anus and the beginning of the vaginal opening) to provide a greater surface barrier against microorganisms. Studies of acceptability, contraceptive effectiveness, and STD prevention are currently underway. Potential advantages of this product are: (**1**) it provides women with the opportunity to protect themselves from pregnancy and STDs; (**2**) it provides a broader coverage of the labia and base of the penis than a male condom; and (**3**) its polyurethane membrane is stronger than latex.

At least three versions of the female vaginal pouch are currently undergoing testing and awaiting Food and Drug Administration ap-

proval—one developed by a Wisconsin company, another by a father and son doctor team in California, and a third by a Wyoming physician. The Wisconsin version is named **Reality.** The cost for the Reality condom is about $2.50 each. The female condom has been sold in Switzerland since 1992 and in France and Britain since mid-1993.

The Wyoming pouch is folded into the crotch of a latex G-string. The three products all perform in essentially the same way. They are disposable, contain protective sheaths of rubber or plastic, and are worn on the inside of a woman instead of on the outside of a man.

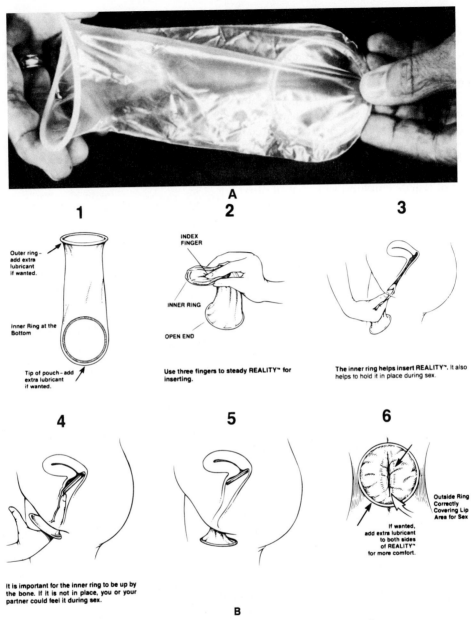

FIGURE 9-4 A, The Vaginal Pouch. **B,** The 7-inch female condom or vaginal pouch is made of lightweight, lubricated polyurethane and has two flexible rings (**1**), one at either end. It is twice the thickness of the male latex condom. The inner ring (**2**) is used to help insert the device and fits behind the pubic bone. The outer ring remains outside the body. Unlike the diaphragm, the vaginal condom protects against the transmission of HIV, which can penetrate the vaginal tissues. The pouch can be inserted anytime from several hours to minutes prior to intercourse. The vaginal pouch is inserted like a diaphragm and removed after sex. FDA approved in May 1993. (*Courtesy of Female Health Co., Chicago*)

Global Use of the Female Pouch (Female Condom)— Female Health Company (FHC) of Chicago, Illinois is the sole manufacturer of the female pouch. Under agreement between UN-AIDS and FHC the female pouch is sold for between fifty and ninety cents (US) in the developing world to encourage its use and provide greater access to women. It is also marketed in the Americas and Europe for about $2.50 (US). The pouch (condom) is in the early stages of global commercialization. Over 4 million were sold in 1997 and 14 million in 1998 within developing nations. UNAIDS hopes to get the pouch into all developing nations by the year 2000.

Redressing the Balance of Power

The symbolic importance of the female condom should not be understated: **it is the first woman-controlled barrier method officially recognized as a means for the prevention of sexually transmitted disease.** The female condom allows women to be able to deal with the twin anxieties—AIDS and unwanted pregnancy—with a method that is under their own control.

Many women become HIV infected due not to their own behavior, but to their partner's behavior. And, because of the nature of gender relations, women may have little influence over their partner's sexual behavior.

Condom Lubricants

It has been demonstrated that petroleum or vegetable oil based lubricants should not be used with latex condoms. Nick White and colleagues (1988) have reported that these

---

POINT OF INFORMATION 9.3

QUESTIONS AND ANSWERS ABOUT THE FEMALE CONDOM

1. **Does one have to be fitted for use of a female condom?** The female condom is offered in one size and is available without prescription. Unlike using a diaphragm, the female condom covers not only the cervix but also the vagina, thereby containing the man's ejaculate.

2. **Should a lubricant be used with the female condom?** A lubricant is recommended for use with the female condom to increase comfort and ease the entry and withdrawal of the penis. The female condom is prelubricated on the inside with a silicone-based, nonspermicidal lubricant. Additional water-based lubrication is included. The lubricant can be placed either inside the female condom or on the penis.

3. **Can oil-based lubricants be used with the female condom?** The female condom is made of polyurethane which is not reported to be damaged by oil-based lubricants.

4. **Can a spermicide be used with the female condom?** Use of a spermicide has not been reported to damage the female condom. It is recommended that a spermicide containing nonoxynol-9 be used with the female condom for additional protection against HIV in the event of displacement, breakage, or leakage.

5. **How far in advance of sexual intercourse can the female condom be inserted?** The female condom may be inserted up to 8 hours before sexual intercourse. Most women insert it 2 to 20 minutes before engaging in vaginal intercourse.

6. **Can the female condom be reused?** No. A new female condom must be used for each act of vaginal intercourse. After intercourse, the condom must be removed before the women stands, to ensure that semen remains inside the pouch.

7. **Should a female condom and a male condom be used at the same time?** The female condom and male condom can be used at the same time but it is not recommended because the condoms may not stay in place due to friction between the latex in the male condom and the polyurethane in the female condom.

8. **Does Medicaid cover the female condom?** Currently Medicaid does not cover this device, as it does the male condom, spermicide, and other barriers. However, Medicaid coverage is expected in 1995. (*Information provided by the New York State Department of Health AIDS Institute Division of HIV Prevention,* Info Bulletin, *Jan. 1994, Number Five.*)

lubricants weaken latex condoms. Latex condoms exposed to mineral oil for 60 seconds demonstrated a 90% decrease in strength (Anderson, 1993). There are a number of water based lubricants that do not adversely affect latex condoms; these should be the lubricants of choice (Table 9-5).

CLASS QUESTION: Reaching this point in the text and having just read the section on male and female condoms, do you think the danger for life, with respect to HIV and other life-threatening STDs, is high enough that condoms should be as familiar to everyone as are toothpaste and toilet paper and also as available? Yes or No—defend your view with credible information.

An Alternative to Condoms: Chemical Barrier Protection—Vaginal Microbicides:

While condoms, when used consistently and correctly, are effective in preventing the sexual spread of HIV, there is an urgent need for chemical barrier methods that women can use

TABLE 9-5 Water-Based[a] and Oil-Based[b] Lubricants Often Used with Latex Condoms

Lubricants Recommended	Lubricants Not Recommended
Aqualube	Petroleum jellies
Astroglide	Mineral oils
Cornhuskers Lotion	Vegetable oils
Forplay	Baby oil
H-R Jelly	Massage oil
K-Y Brand Jelly	Lard
RePair	Cold creams
Probe	Hair oils
Today Personal Lube	Hand lotions containing vegetable oils
	Shaft
	Elbow Grease
	Natural Lube

[a]Water-based lubricants can be used with latex condoms.
[b]Oil-based lubricants will chemically weaken latex, causing it to break.
The above lists are not exhaustive of all available lubricants used by consumers.
(Source: The STD Education Unit of the San Francisco Department of Public Health)

for HIV prophylaxis, such as vaginal microbicides. Vaginal microbicides are products for vaginal administration that can be used to prevent HIV infection and/or other sexually transmitted diseases (STD). An ideal vaginal microbicide would be safe and effective, and also tasteless, colorless, odorless, nontoxic, stable in most climates, and affordable.

Chemical Barriers: A Protection Method That Women Can Control

Chemical methods that can be controlled by women are likely to have a powerful effect on the spread of HIV for several reasons. They could be distributed rapidly and cheaply by well understood social marketing techniques. They have the potential to slow HIV transmission directly and to reduce other sexually transmitted diseases that are cofactors in transmission.

In the United States, HIV-infected women are among the most rapidly growing groups affected by the disease. Globally, the largest number of cases are the result of heterosexual transmission. From a public health perspective, a modest reduction in HIV transmission brought about by a vaginal microbicide made available today might save as many lives as a more effective method (e.g., a vaccine) made available in 10 years' time, when there might be 5 or 10 times as many infected people (Potts, 1994).

Chemical Barriers Used in Foam, Gel, or Sponge

In November of 1993, the WHO began a strategy to identify a safe and effective substance that can be inserted into the vagina in a **foam, gel, sponge,** or other form to destroy HIV or prevent it from infecting cells in the body. Researchers hope that it can be used by a woman without her partner's knowledge. WHO scientists believe that if there is one thing which truly can make a difference to the epidemic, it is a vaginal microbicide.

What are Microbicides?

Microbicides are chemical substances that, when applied in the vagina or the rectum, would substantially reduce transmission of STDs. They could be produced in many forms, including gels, creams, suppositories, films or as a sponge or vaginal ring. Currently, there are no products on the market, but there are some 40 microbicidal products in development, 15 of which are now being tested on humans.

In developing a vaginal microbicide, **scientists must be sure that the substance is safe, does not kill microbes naturally present in the vagina that benefit female hygiene, and does not impair a woman's ability to conceive.** Any microbicide will have to be tested to determine whether it damages spermatozoa, which could result in birth defects.

Vaginal Microbicide Research

In July of 1996 the United States government pledged $100 million to help develop virus-inactivating creams that would let women protect themselves from HIV infection without relying on their partners. The goal is to create alternatives to condoms that women can use without men's permission—especially creams that protect against HIV but would still allow them to get pregnant.

In October 1997, an American company received a U.S. patent for PRO 2000 gel, a topical microbicide that is active against a wide range of HIV strains **in the test tube.** Clinical Phase studies through 1999 are ongoing to assess the safety of the gel in women.

Current (1999) estimates are that a microbicide that works is still five years away from the market.

**INJECTION DRUG USE
AND HIV TRANSMISSION:
THE TWIN EPIDEMICS**

About 36%, almost 269,000 of the 746,000 (beginning year 2000) U. S. AIDS cases recorded since 1981 were transmitted through injection drug use. Now about half of all new HIV infections occur from IDU. About 75% of all people with IDU-related AIDS are either black (50%) or Latino (24%).

Syringes and needles used to inject drugs or steroids, or to tattoo the body, or to pierce the ears, should **never be shared.** If an individual is going to assume the risk of HIV transmission through needle sharing, the risk can be reduced by sterilizing the needle, and the

--------- BOX 9.2 ---------

THE MAGNITUDE OF THE EPIDEMICS OF DRUG USE AND HIV/AIDS

The dual epidemics of drug use and HIV/AIDS are highly destructive of public health and social life in America. **Drug use exacts an estimated annual social cost of $58.3 billion—in lost productivity, motor vehicle crashes, crime, stolen property, and drug treatment.**

IDU is the second most frequently reported risk for adolescent/adult AIDS since the beginning of the United States epidemic in 1981. **It now accounts for about half of all new HIV infections, each year** (Gostin et al., 1997). From 1994 through early year 2000, 36% of all AIDs cases occurred among IDUs, their heterosexual sex partners, and children whose mothers were IDUs or sex partners of IDUs. In contrast, in 1981, only 12% of all reported AIDS cases were associated with injection drug use. In some areas, seroprevalence among IDUs is now as high as

65%; in other areas, the rates are significantly lower. Minorities bear a disproportionately high burden. The rate of **IDU-associated AIDS** per 100,000 population is 3.5 for whites, 21.9 for Hispanics, and 50.9 for African Americans.

Transmission of HIV infection through injection drug use has a cascading effect; infections spread from IDUs to their sexual and syringe-sharing partners and from HIV-infected mothers to their children. Of the **117,000** AIDS cases among women, nearly 65% were IDUs or were sexual partners of an IDU. Further, of the **8,811** perinatally acquired AIDS cases beginning year 2000, 60% had mothers who were IDUs or had sex with an IDU. These data suggest that drug use and related behaviors are strong catalysts for spreading HIV throughout the population.

syringe, if one is used, in undiluted chlorine bleach. The needle and syringe should be flushed through twice with bleach and rinsed thoroughly with water. **(NOTE: FROM THIS POINT FORWARD, NEP STANDS FOR NEEDLE EXCHANGE PROGRAMS AND INVOLVES A NEEDLE AND A SYRINGE.)**

Both injection drug use (IDU) and HIV infection are on the increase. They are twin epidemics in the United States and Europe because the virus is readily transmitted by injection drug users and then from infected drug users to their noninfected sexual partners. Stopping injection drug-associated HIV transmission in theory is easy—just avoid injection drugs use. But, that is a difficult proposition for most of the estimated 6.5 million IDUs worldwide (2.5 million in the United States). The number of countries **reporting** IDU in 1989 was 80, in 1999, 130. IDUs will remain a major HIV connection to the homosexual, heterosexual, and pediatric populations. (Anderson et al 1998).

HIV PREVENTION FOR INJECTION DRUG USERS

What can be done and what is being done to prevent HIV transmission by this population? Available drug rehabilitation programs are far too few. It is estimated that only 15% of injection drug users in the United States are receiving treatment at any given time. Many addicts want to quit their habit but may become discouraged because of having to wait long periods of time before getting treatment because of lack of money. Even if there were a sufficient number of treatment centers, there will always be the hard core IDUs who will not enter a program.

A massive education program on how to clean drug equipment and not sharing such equipment has for the most part failed. IDUs have an economic motive to share equipment (Mandell et al., 1994). In addition, it takes time and effort to clean the syringe and it takes time to find and purchase cleansing chemicals (detergents or bleach). Beginning year 2000, several studies have shown that over 50% of IDUs continue to share drug injection equipment. Also, **less than half clean their injection drug equipment!** Perhaps the most important drawback may be that IDUs have little interest in

health care or changing their behaviors. In addition, there is always the problem of legality. **IDU is illegal throughout the United States** (see Point of Information 9.4 and Box 9.3).

IDUs know this and fear incarceration without the possibility of a "fix." A "Catch-22" situation also exists for those who want to help make injection drug use safer: **many governmental agencies and law enforcement officers interpret the intention of making drug use safer as advocating drug use.** As a result, many proponents of safer drug use have avoided becoming involved in the issue. A recent public opinion poll showed that 55% of those polled were in favor of syringe exchange programs (Marwrck, 1995). Beginning year 2000, federal money could still **not** be used to fund **NEPs.**

The Needle Exchange Program Strategy (NEPs)

The idea of syringe-needle exchange programs (NEPs) is based on the established public health policy of eliminating from any system potentially infectious agents or, where possible, carriers of infectious agents. The rationale of NEPs is similar, wherein active injection drug users exchange used, potentially contaminated syringes for new, sterile syringes. In general, these exchanges are done on a one-used needle and syringe for one-new needle and syringe basis, though some programs will add an additional number of needle and syringes on top of those already exchanged. NEPs also provide other paraphernalia and supplies, including cotton, cookers, water, and sterile alcohol prep pads. In addition, NEPs offer a variety of other services to IDUs including education, HIV testing and counseling, referrals to primary medical care, substance abuse treatment, and case management.

THE WORLD'S FIRST NEP on record began in 1984 in Amsterdam, The Netherlands. It was started by an IDU advocacy group called the **Junkie Union.**

Free Needle Exchange Programs in the United States

Jon Parker is believed to be the first person in the United States to distribute free drug injection equipment publicly in North Haven, Connecticut and in Boston, Massachusetts in November 1986.

Tacoma, Washington— Their **NEP** began in August 1988. It began as a one-man program. Dave Purchase, a 20-year drug counselor, between August 1988 and April 1989, traded out over 19,000 exchange packages. The following conversation between Purchase and an IDU offers a sense of what happens daily at the Tacoma syringe exchange site:

> Many of the 40 or so clients to use the needle exchange one day last month were regulars. They knew the drill. If they dropped five syringes into the red box marked **BIOHAZARD,** they got five clean ones in return (Figure 9-5). Soon two new clients appear. One leans close to Purchase and quietly asks for a syringe. They have none to trade.
> "You've got to give me one to get one," Purchase says loudly. "It's an absolute rule."
> "But he's going by this girl's," says the spokesman, leaning closer. "She don't know nothin' about these. I didn't know nothin' about it either."

FIGURE 9-5 Syringe Exchange. The injection drug user drops the used syringe and needle into the collection box. For each deposit, the user receives a new syringe and needle. The use of clean equipment prevents the spread of HIV. (Through 1999 federal money could not be used for syringe exchange programs.) *(Photograph courtesy of World Health Organization)*

> "It pisses me off that I can't do it. But this is an exchange program. I can't do it. I'm under too much pressure to shut this place down. Go down to Bimbo's, man, and get the ones they throw away," Purchase advises. "I don't care where you get it, just get me one."
> They leave in the direction of Bimbo's, a nearby restaurant, across the street from which junkies often shoot up, but they never come back (Perrone, 1989).

The NEP in Tacoma held the HIV infection rate to under 5% over a 5-year study period (1988–1992). During that same 5-year study period, the prevalence of HIV infection among IDUs in New York City, with few syringe exchange programs, increased from 10% to 50%!

Evaluation of Needle/Syringe Exchange Programs

Entering year 2000, there were about 130 known NEPs operating in 80 American cities exchanging about 25 million syringes annually. An IDU makes about 1000 drug injections each year (MMWR, 1997). The San Francisco AIDS Foundation operates the largest NEP in the United States.

New York City— In November 1988, after many delays, New York City began its NEP. The program was canceled in early 1990—the reason: because over 50% of NYC's 240,000 IDUs were HIV-infected, the program offered too little too late to have an impact. IDUs make up about 38% to 40% of NYC AIDS cases. The NEP was resumed in 1992. Beginning 1998 there were at least five NEPs operating in New York City.

In July 1997, and HIV prevention and NEP center in Manhattan ran a campaign featuring 1140 advertisements, on some New York City subway cars, for an NEP. The advertisements feature Michangelo's Sistine Chapel fresco, in which God's hand reaches toward Adam; in the ad, though, God is carrying needles to give to Adam. The advertisement says that "Sharing needles is the number one cause of HIV in New York City," and implores, "If you shoot drugs—please do not share needles." It also provides information on how to contact the needle-exchange program.

——————— BOX 9.3 ———————

THE CONTROVERSY: PROVIDING STERILE SYRINGES AND NEEDLES TO INJECTION DRUG USERS (IDUs)

As the HIV pandemic enters the late 1990s in the United States, the profile of those affected is beginning to change. HIV is slowly changing from a disease of men who have sex with men to a disease of injection drug users, their sexual partners and children. In 1997, 1998, and 1999, 50% of all new HIV infections were associated with IDUs. **THE SYRINGE IS THE TYPHOID MARY OF THE 1990s!** Accordingly, should IDUs be provided at public expense, with free syringe and needles to help stop the spread of HIV?

There has been no other HIV prevention activity, with the possible exception of condom distribution in public schools, that has generated as much controversy as needle exchange programs (**NEPs.**) **Pro forces** believe that it is impossible to eliminate injection drug use. Providing IDUs with free sterile equipment is an attempt at slowing the spread of HIV among drug users and their sexual partners, subsequently reducing the number of people in the general population that will become infected. In support of the Pro forces, the CDC reviewed the 37 NEPs operating in the United States in 1993 and 1997 (Beginning year 2000 there were about 130 in 80 cities) and others in Canada and Europe. **The most important conclusion is that NEPs are preventing HIV infections in drug users, their sex partners, and their children.** That finding was based, in part, on evidence collected in Tacoma and Seattle, Washington and in New Haven, Connecticut where there was a 33% reduction in the rate of new infections among NEP clients. **In another key finding, they found no evidence that distributing syringes leads to an increase in drug use.**

Besides saving lives, these NEPs deliver a huge financial payoff. Consider the case of an HIV-positive addict who infects eight others in a 1-year period (a reasonable estimate). If each turns to Medicaid to pay his or her lifetime medical costs (at an average $119,000 plus), that's about a $1 million burden for taxpayers—money that could have been saved if the one addict had been in a NEP. In addition, arguing for the availability of NEPs *is not* an argument against using police and criminal justice mechanisms to deal with drug use-related robbery, assault, or other crimes.

Con forces argue that any move to make injecting drug use safe is a move to make it attractive or socially acceptable and thus represents a step backward in the long-declared war on drugs.

Those opposing the needle exchange programs feel that to provide NEPs is to condone drug addiction and perhaps promote injection drug use among those who would not otherwise participate. Some feel that free syringe programs are just a Band-Aid that attempts to prevent HIV infection but does nothing to cope with the underlying drug addiction problem. (**These people may feel its too little, too late.**)

A side issue of course is the fact that it is **illegal to sell** needles and syringes over the counter in 8 of the 50 states and the District of Columbia. Statutes in 44 states and the District of Columbia place criminal penalties on the possession and distribution of syringes (drug paraphernalia laws) (*MMWR*, 1993b) in 28 states and Puerto Rico (*MMWR*, 1997). Laws restricting these sales were intended to discourage drug use, (Fisher, 1990). (Gostin et al., 1997). Law enforcement and community leaders (e.g., police, churches, businesses, parents, teachers, and residents) are concerned that allowing access to needles and syringes sends the wrong message, encourages initiation into drug use, and accelerates the disintegration of families. Residents and business owners fear increased street crime, lower property values, and health risks from discarded needles and syringes.

Middle Ground

There is a **third force** of people who take the middle ground. They advocate providing free bleach to IDUs to clean their drug equipment. People from many AIDS-action groups in metropolitan areas are involved in the free bleach dispensing program. With each bottle of bleach is a set of directions. But as many IDUs can not read, they are given verbal instructions, though some are too "spaced out" to listen or too uncomfortable to care. Third force people have to understand that NEPs and distribution of bleach kits have to be accessible privately, off the street and out of sight, as well as publicly. Use of a public, streetcorner NEP is a statement to the entire community that the person exchanging a syringe is an IDU. This has potentially very different consequences for women than for men. Women who have children have good reason to try to conceal their addiction as they risk losing custody or contact with their children if the state finds out they are actively using drugs—and if

——————— **BOX 9.3** *(continued)* ———————

they are using a public NEP the child welfare bureaucracy is more likely to find out than if they are not getting needles and syringes in public. Perhaps NEPs should provide some type of informal child care so that a mother doesn't have to literally exchange needle and syringes in front of her children. Addiction among women is often taken as a statement of sexual availability.

Legalize Drugs

A **fourth force** suggests that drugs be legalized so that they can be used openly, thereby reducing the threat of HIV transmission through illegal "shooting galleries." Discussion of the fourth force's position is beyond the scope of this text. However, a quasi-fourth force situation did exist in Zurich, Switzerland until 1992.

Zurich gained notoriety for its vast amounts of gold but in 1987 it gained a new image—that of an open Needle Park. The city of 250,000 people experimented in setting aside Platz Pitz park where cocaine and heroin addicts could openly buy, sell, and use injection drugs. The park attracted up to 4,000 drug users/day.

Next to the United States, Switzerland has the highest incidence of HIV/AIDS in the developed world. The city has tried to stop the spread of the disease among addicts sharing contaminated syringes by isolating them and giving them free new needles and syringes. A cart laboratory has been converted for this purpose. For every used syringe an addict turns in, he or she will get a free, sterile one. About half of the young people in the park are HIV-infected. All of them are slowly killing themselves. At the end of each day, some 7,000 dirty needles and syringes have been turned in. The needle park was not run to prevent drug use but to prevent the spread of HIV/AIDS. The program appeared to be working. The incidence of AIDS cases dropped from 50% to 5% (*Time,* 1992). An example of the people in the park are Reno, age 28, and Sophie, age 24. She has not only tested HIV-positive, she is suffering the first stages of the disease. For both of them their main concern now is not AIDS but getting the $400 or $500 a day they need to support their drug addiction.

There are no figures on how many of these addicts in the park die from AIDS. Many just drift away. It is known that more than 250 died in 1991 from addiction alone. And the rate of addiction is increasing, the population of the park is increasing, and the number overdosing is increasing. The medical staff tries to help, and the addicts, many of whom have lost hope for themselves, try to help each other. One man tries to keep the other who

has overdosed upright and moving, trying to stop death. The fight is for life. A social worker volunteer says she has watched life and death in Needle Park for 4 years. She feels that death may be the best alternative. She does not make that statement lightly. Her own 22-year-old son has been in the park for almost 5 years—one who lives from fix to fix, one of the players in the drama that is played out each day in one of the world's financial capitals. But in early 1992, the park was closed because the park became a magnet for professional dealers, especially Lebanese, Yugoslav, and Turkish gangs that overran small dealers in a violent price war. Some of the park's inhabitants clustered around the central train station, others headed off in search of methadone. Others went back to the alleys and shelters from which they came. With sales suddenly back underground, addicts complained that the price of heroin had doubled overnight to $214 a gram (454 grams/pound = $97,156). Health workers said efforts to prevent the spread of HIV/ AIDS will now be much more difficult.

UPDATE: Switzerland is currently experimenting with providing IDUs with **heroin** three times a day in 18 treatment centers across the country. So far Swiss health authorities say the program has reduced criminal activity among the participants from 59% to 10%. The program also reduced their homelessness from 12% to zero and death rate by 50%.

CLASS QUESTION: The Unites States has a zero tolerance for such activities and will not become involved in what the Swiss term, "an innovative program." The financial cost is about $20 billion annually to treat heroin addicts in the USA. Do you think the USA should, based on early Swiss data, offer heroin to the addicted in a controlled environment similar to the Swiss? Present a credible reasons/data to support your stand.

LIFTING THE BAN ON THE USE OF FEDERAL FUNDING TO SUPPORT NEEDLE-EXCHANGE PROGRAMS

Since 1990, seven national reports have reviewed the scientific evidence and recommended that the federal ban for NEP funding be lifted. Needle-exchange programs are supported by the President's Advisory Council on HIV/AIDS, the American Medical Association, the National Academy of Sciences, the Centers for Disease Control and Prevention, and the American Public Health

———————— **BOX 9.3** *(continued)* ————————

Association, as well as other prestigious medical and public health organizations. In addition, the American Bar Association and the U.S. Conference of Mayors have urged the federal government to allow states and localities to use Federal HIV-prevention funds to implement NEPs. Polls have shown that the American public supports lifting the ban on federal funding of NEPs. A recent Harris poll found that 71% of Americans believe that cities and states— and **not** the federal government—should decide whether federal HIV-prevention funds can be spent on NEPs.

On April 20, 1998, Health and Human Services Secretary Donna Shalala informed President Clinton that the scientific research has proven that NEPs effectively prevent the transmission of HIV and hepatitis, and do not lead to increased drug use. Her long-awaited action cleared the way for the President to lift the 10-year ban on federal funding for NEPs. Clinton accepted the findings, but stated that in spite of them he would continue to block the use of federal funds for NEPs. Clinton endorsed lifting the ban in his 1992 presidential campaign. HIV prevention advocates charge that the President's de-cision is politically motivated and will cause thousands of new and preventable HIV infections by the year 2000.**

Class Discussion: Do you agree with the Presidents decision? List any scientific facts or studies that could be used to support the President's action. List any scientific facts or studies that disagree with his decision.

COMMENTARY

It is hard to avoid the suspicion that the concerns about free syringe programs have less to do with science or public health than with politics: specifically, a reluctance to **detract** or **distract** the **"JUST SAY NO"** message. But there is another message that both sides would agree on— that no one should die needlessly! A leading University of California researcher estimates that nearly 10,000 lives could have been saved over the past few years by an aggressive expansion of syringe exchange programs. **Isn't the war on drugs supposed to be about saving lives? Isn't a part of government about saving lives? (Class Discussion)**

New Haven, Connecticut— Their 7-year-old program has demonstrated that NEPs dramatically slow the rate of infection without encouraging new injection drug use. Some indicators even suggest that the program has been responsible for a decrease in both crime and the amount of drugs used illegally. The city's police chief claims that crime actually dropped over 20% over the first 5 years of the program, perhaps because of the improved relationship between city workers and the community. Meanwhile, referrals to drug treatment centers increased. These results have enabled policymakers elsewhere to call for NEPs. Edward Kaplan (1993) of Yale University reported at the Eighth International AIDS Conference that the New Haven syringe program cut the rate of new HIV infections by 33% to 50%. His data come from comparing the number of HIV-contaminated syringes found on the streets versus those turned in at syringe exchange points. (There is some contraversy over this date.)

After passing a 1992 law permitting pharmacies to sell syringes without a prescription, syringe sharing has dropped 40% in the state of Connecticut. Seventy-five percent of AIDS cases in Connecticut occur among IDUs, their sex partners, and their children. Going into year 2000, Connecticut, North Dakota, Alaska, Iowa, and South Carolina allow nonprescription sales of syringes and have no drug paraphernalia law against their possession. Nine states and the District of Columbia require prescriptions both to buy syringes and to possess them. Six of the nine states (California, New York, New Jersey, Illinois, Pennsylvania, and Massachusetts) have the largest numbers of IDUs and the highest incidence of AIDS among them.

The mayor of Washington has called for syringe exchange for addicts, as well as the distribution of free condoms in city schools and jails. Movements are underway in New Jersey and California to remove legal barriers and begin officially sanctioned syringe programs. Even in the U.S. Congress, Charles Rangel, who has led the opposition to syringe exchange on the ground that it threatens blacks, has asked the General Accounting Office to reevaluate the effects of such programs.

The most important catalyst for this change has been the experiment conducted in New Haven (Thompson, 1992; *Time*, 1992).

Madison, Wisconsin— began an NEP in 1996 and now serves over 600 IDU's a month. In April 1998, the city began a **mobile mini-van** NEP for those who would not come to the clinics for free needles and condoms.

Springfield, Massachusetts— approved an NEP in July 1998. Fifty-six percent of its HIV infections are associated with IDU. Springfield became the 5th community in Massachusetts to approve a program joining Boston, Cambridge, Northhampton and Provincetown.

Hawaii— In 1990, Hawaii became the first state to legalize a statewide NEP. The state legislature felt it was necessary to stem the rate of HIV infection in women and newborns. The results of this program will have an impact on decisions to be made in other states. In Hawaii, a drug user comes in, drops a dirty syringe into a plastic bucket, and receives a fresh sterile syringe in exchange. No name is given, no questions are asked. Under the 3-year pilot project, an addict can swap a used syringe for a new one supplied by the nonprofit Life Foundation up to 5 times a day 5 days a week.

Alaska— A clinical trial NEP is underway at the University of Alaska, Anchorage. The trial, supported by the National Institute on Drug Abuse (NIDA), is a 3-year, $2.4 million study that will compare the incidence of hepatitis B virus (HBV) among participants who obtain clean syringes through syringe exchange centers and among those who purchase them at pharmacies. It will attempt to determine if one method of syringe distribution is more effective than the other at stopping the spread of disease. It will also attempt to determine if access to clean syringes encourages drug use. While the study is funded by government money, because of the federal ban, the actual NEP is supported separately by

the nonprofit Alaska Science and Technology Foundation (Benowitz, 1997).

**Needle Exchange Programs (NEPs)
in Other Countries**

Needle exchange program results from England, Austria, the Netherlands, Sweden, and Scotland, presented at the Fourth International AIDS Conference (1988) suggest that the European programs attracted IDUs who had no previous contact with drug treatment programs; and that IDUs were drawn from NEPs into treatment programs, thus the decrease in drug use. There was no indication in these studies of an increase in injection drug use in cities with exchange programs. Where HIV testing had been done, the rate of HIV infection showed a marked decline after the introduction of NEPs (Raymond, 1988; Hagen, 1991).

Some of the countries with active NEPs are:

Canada— In 1989, Montreal, Toronto, and Vancouver began NEPs. Beginning 1998, Vancouver has the highest rate of HIV infections among IDUs in the world with 40% of IDUs infected. Vancouver's NEPs gave out 2.4 million syringes and needles in 1998 just to those drug addicts who live in the city's eastside area. In Ottawa, 19% of IDUs are HIV-infected. Experts believe a shift to injecting **crack cocaine** is responsible for the rising infection rates among IDUs. While heroin users routinely use a **few** needles a day, injection cocaine users may use 30 to 40 needles a day! In **Vancouver,** 80% of IDUs preferred cocaine, it was 82% in **Ottawa** and 70% in **Montreal.**

England— Over 2 million syringes and needles were exchanged in 120 NEPs each year. Needles and syringes can be legally purchased throughout England. Free distribution and exchange remains a relatively nonpolitical and noncontroversial issue (Stimson et al., 1989). Overall, the programs in England indicate that changes in IDUs' behavior are slight (Hart et al., 1991). One problem found in England, the United States, and elsewhere is that NEPs attract IDUs who are at **lower risk** for HIV infection while **high-risk** IDUs are less likely to participate.

NEEDLE-SYRINGE LAWS—UNITED STATES

Early in this century, increasing numbers of individuals began using the more potent and easily transportable, injectable opiates instead of smoking opium. It was believed, not unreasonably, that the use of these drugs could be curtailed by limiting access to the tools necessary to inject them; therefore, all states introduced laws to govern the sale and distribution of hypodermic syringes and needles.

Lawrence Gostin and colleagues (1997) reported on the results of a national survey of laws and regulation governing the sale and possession of syringes in the United States and its territories. They reported that every state, the District of Columbia (DC), and the Virgin Islands (VI) have enacted state or local laws or regulations that restrict the sale, distribution, or possession of syringes. Drug paraphernalia laws prohibiting the sale, distribution, and/or possession of syringes known to be used to introduce illicit drugs into the body exist in 47 states, DC, and VI. Syringe prescription laws prohibiting the sale, distribution, and possession of syringes without a valid medical prescription exist in eight states and VI. Pharmacy regulations and practice guidelines restrict access to syringes in 23 states. However, ending 1997, legislatures in eight states (Connecticut, Hawaii, Maryland, Massachusetts, Rhode Island, Maine, Minnesota, New Hampshire, and the District of Columbia) either enacted or removed laws that barred the **establishment of NEPs.** For additional source material on regulations on the sale and possession of syringes on a state-by-state basis, read Scott Burris and colleagues (1996), "Legal strategies used in operating syringe-exchange programs in the United States" and Luric Drucker et al. (1997), "An opportunity lost: HIV infection associated with lack of a national needle-exchange program in the USA."

THE IMPACT—Existing laws make it difficult for injecting drug users to obtain sterile syringes.

Through 1999, the governor of New Jersey, the state with one of the highest rate of injection drug-related AIDS cases, rejected the recommendation of her Advisory Council on AIDS to distribute clean syrginges to intravenous drug users. Many NEPs in this country operate in a gray area of the law.

Question: Would changing the laws make a significant difference?

SOMETHING TO THINK ABOUT

Researchers at St. Louis University School of Public Health, Missouri, have concluded in a recent study that loosening restrictions on the sale of clean syringes may lower HIV transmission rates.

The study found that states with strict laws banning the public sale of syringes have the highest rate of HIV/AIDS transmission among intravenous drug users.

All 50 states and the District of Columbia were placed into three categories. The **first** were states that had **no laws** (no barriers) to obtaining clean needles. The **second** category included states that **had paraphernalia laws** (laws that make it a felony to have a needle/syringe). And **third,** states that **had paraphernalia and prescription laws** (only able to get a needle/syringe through prescription).

The research results came up with the following conclusions:

- **No law—HIV transmission rate: 0.83 per 100,000 people.**
- **Paraphernalia law—AIDS rate: 1.63 per 100,000.**
- **Paraphernalia and prescription law—AIDS rate: 5.78 per 100,000.**

Their conclusion is that the government should be involved in preventing drug use. But in the interim, there should be a means for reducing the potential for harm. Because the percentage increase in numbers of newly HIV- infected people will be highest among IDUs and their sexual partners by the year 2000, changing state statutes regarding NEPs may lower the rate of new HIV infections among this population.

Finally, many public health, medical, and legal organizations have supported the deregulation of syringes as a strategy to prevent HIV/AIDS and other blood-borne diseases among IDUs. Most laws, regulations, and practice guidelines that restrict the sale, possession, or distribution of syringes were promulgated before HIV/AIDS among IDUs was recognized as a pressing public health problem and without carefully contemplating the health implications. Since that time, the interconnected epidemics of drug use and HIV/AIDS have produced illness and death, particularly among poor, urban, minority communities.

NEEDLE-SYRINGE EXCHANGE SURVEY: HOW MANY ARE BEING EXCHANGED: HOW MANY ARE NEEDED?

In November 1996, the Beth Israel Medical Center in New York City, in collaboration with the North American Syringe Exchange Network (NASEN), mailed questionnaires to the directors of 101 NEPs in the United States that were members of NASEN. Although the total number of NEPs in the United States is unknown, most are believed to be members of NASEN. From November 1996 through April 1997, NEP directors were asked about when their NEP began; the number of syringes exchanged during 1995; and, for 1996, legal status, services provided, and the **number of syringes exchanged.**

Of the 73 NEPs operating in 1995, 70 reported exchanging approximately **11 million** syringes. In 1996, of the 87 NEPs that provided information about the number of syringes exchanged, 84 reported exchanging approximately **14 million syringes** (*MMWR*, 1997).

Researchers from the University of California at San Francisco, the Centers for Disease Control and Prevention, and the University of Southern California investigated injection drug use, estimating the number of injections daily in the United States and the role of pharmacies in providing IDUs with sterile syringes. Using data obtained from published articles, personal communication with experts, and national databases, the authors estimated that between 920 million and 1.7 billion injections by IDUs occur in the United States annually, with approximately 12 million injections in San Francisco and over 80 million in New York City. **The researchers note that a similar number of syringes would be needed to provide a sterile syringe for each use.** Pharmacy sale of injection kits could help increase access to sterile needles; such strategies have been used in Europe, Australia, and New Zealand. The authors note that the modification of syringe purchase and possession laws would increase the use of sterile needles (Laurie, et al. 1998).

Netherlands— NEPs are available in 40 municipalities. Data from these programs suggest that the staff operating the syringe exchange centers should be multilingual and the centers should be operational 24 hours a day.

Australia— Australia has 10% of the population of the United States but has 2,000 NEPs in five states and one territory. The remaining state and one jurisdiction may have since joined the program (Wodak, 1990 updated).

By 1994 the Sydney program was distributing around 3 million syringes, and some exchanges in the suburbs of Sydney as well as in the city center were seeing attendances rise by 1,000 clients per month from late 1992.

The Australian Broadcasting Corp. reported, in June 1999, that the Tasmanian State needle-exchange program expects to hand out over 1 million needles and syringes this year, more than doubling the breadth of the program. **Critics say** that the increased demand for clean needles is evidence of how the program promotes intravenous drug use. However, the Health Minister said that the increased demand is indicative of the program's success.

The Queensland health department says that the NEPs cost **A$**500,000/year but potentially saves **A$**279 million/year when all diseases that can be transmitted by IDU are considered.

In 1998 Ansett Australia became the world's first airline to install needle/syringe disposable bins. Ansett said the disposable bins in its 69 aircraft were installed primarily as a safety measure to avoid needle-stick injuries to staff but is intended for the use of legitimate syringe users, such as diabetics, as well as addicts.

New Mexico— A statewide NEP was launched in March 1999 to slow the spread of HIV. NEPs were first established in Roswell, Las Cruces, and Farmington.

New Zealand— A nationwide NEP began in 1988. The overall rate of HIV-infected IDUs remains low. In 1995, less than 1% of IDU men and no women tested at the syringe exchange centers were HIV-positive. There were about four to six new cases each year since 1987. Their data are too few to make statistically significant claims to support the syringe program (Lichtenstein, 1996).

Italy— Only a few limited NEPs have been reported. However, between 1983 and 1988, the sale of syringes and needles, which is legal, increased by 50%.

Summary

Through 1995, 84% of IDUs in Glasgow, 82% in Lund, 84% in Sydney, 73% in Tacoma, and 87% in Toronto reported they had changed their behavior in order to avoid HIV/AIDS. The most commonly mentioned specific behavior change: reduced sharing of injection equipment (Des Jarlais, 1995).

PREVENTION OF BLOOD AND BLOOD PRODUCT HIV TRANSMISSION

Blood Donors

There should be **no risk** in the United States or in other **developed nations** of contracting HIV **by donating blood** if blood centers use a new, sterile needle for **each** donation. Yet a 1996 survey revealed that 25% of those polled believed that they could become HIV-infected by donating blood.

Over 8 million people are donating some 18 million units of blood which are transfused into 4 million people annually in the United States. Blood can carry HIV in a cell-free state or HIV may be carried within cells of the blood. The amount of virus taken in (**dose**) and a person's biological susceptibility determines whether infection will occur after HIV exposure. **A large dose of HIV received via blood transfusion almost universally results in HIV infection. A small dose of HIV-contaminated blood, such as the blood received by a needlestick seldom results in infection** (Francis et al., 1987). The mean volume of blood injected by a needlestick has been calculated to be 1.4 µL (1.4 millionths of a liter). It is difficult to determine exactly how large a viral dose is necessary to cause infection. However, it is known that infection is more likely to occur if blood is donated close to the time the HIV-infected donor becomes symptomatic. Entering year 2000, 6% of HIV-infected adults and 10% of HIV-infected children in the United States are believed to

---

— BOX 9.4 —

UNITED STATES SUPPORTS SEX EDUCATION AND CONDOMS IN SCHOOLS AND IN TV ADS, AND NEEDLE-EXCHANGE PROGRAMS

According to the **Kaiser Family Foundation AIDS Public Information Project,** in 1996, when there were over a half million people living with AIDS and over 300,000 dead from AIDS, the Kaiser Foundation reported the following facts after reviewing the results of it's **"Survey of Americans and HIV/AIDS":** Virtually all Americans (95%) think AIDS information should be provided in the schools, including 69% who think children should start receiving AIDS education **at the latest** by age 12; seven out of ten Americans think the major television networks should accept condom advertisements; two-thirds favor providing clean needles to IV drug users; and 63% say there should be more references to condom use in movies and television. However, when asked whether they thought high schools should provide condoms or just information about AIDS to teens, Americans were almost evenly divided: 49% say just AIDS/HIV information and 46% say making condoms available should be the priority. Support for AIDS prevention measures persists across demographic groups.

Eight out of ten parents with children under 21 years of age say they are concerned about their child getting AIDS, including 53% who say they are **very** concerned. When asked about the information they want most about AIDS, five out of 10 parents say they need more information about "what to discuss with children about prevention." Getting information about where to go for help if exposed to HIV virus was the number two information needed, picked by 12% of parents.

———— **BOX 9.4** *(continued)* ————

Americans Concerned About AIDS Both Personally and for the Nation

Strong public support for specific actions may be driven by the fact that many Americans continue to report they are personally concerned about the spread of AIDS.

- Four out of ten say they are concerned about getting AIDS themselves, including 22% who are "very concerned" and 18% who are "somewhat concerned." (In a 1991 survey conducted by the Roper Organization, 27% of Americans said they were "very concerned" and 18% said they were "somewhat concerned about getting AIDS");

- 43% say AIDS is a "very serious" problem for people they know;

- 39% say they personally know or have known someone with AIDS or the HIV virus;

- 74% agree that "people who get AIDS are no different from people like me in their values and goals."

Minorities and those with low incomes and less education are most likely to be personally concerned about AIDS. A majority (55%) of African Americans and 38% of Hispanics say they are personally "very concerned" about getting AIDS, as compared to 16% of whites. More than a third (35%) of people with annual incomes under $10,000 say they are "very concerned" as compared with only 14% of those with incomes over $50,000 per year. Similarly, while 34% of those without a high school degree say they are "very concerned" about getting AIDS, only 14% of college graduate's say they are.

Americans Generally Informed about HIV/AIDS, Although Some Gaps in Knowledge Persist

Surveys generally find that Americans are ill informed about most national issues, but not when it comes to HIV/AIDS. Most Americans are fairly well informed about the disease itself—the transmission, prevention and treatment of HIV/AIDS. They know that:

- AIDS can be transmitted during sexual intercourse, 97%;

- A pregnant woman with AIDS can pass the virus to her baby, 94%;

- No vaccine is currently available to protect a person from getting AIDS, 88%; and

- Drugs are available that can "lengthen the life of a person with AIDS," 75%.

Most people (54%), however, think 100,000 or fewer Americans have died from AIDS, when in fact it's more than three times that number. While underestimating the number of lives lost to AIDS, many Americans (51%) overstate the impact here relative to the rest of the world, believing that half or more of all AIDS cases occur in America. In reality it's less that 2%. Half of the population (51%) incorrectly thinks a person can get AIDS while "giving or donating blood for use by others." (There are no known cases of HIV transmission as a result of **donating** blood. Some of the confusion may be related to perceived risk of **receiving** blood transfusions. The risk of HIV transmission in a Red Cross center is **one** for every 450,000–660,000 donations of screened blood.)

Significant Percentages of Americans Distrustful of Information About HIV/AIDS

Although most Americans believe they are receiving accurate information about the AIDS epidemic, significant percentages doubt what the government and media are telling:

- 34% do not believe the government is telling the whole truth about AIDS; and

- 25% do not believe the media is telling the whole truth about AIDS.

In addition, fewer, but still some Americans question the origins of the epidemic:

- 18% believe there is some truth to report that the AIDS virus was produced in a germ-warfare laboratory; and

- 12% believe that AIDS came from God to punish homosexual behavior.

have become infected via blood transfusions or by the use of contaminated blood products. The majority of HIV infections occurred prior to the initiation of the blood screening program in 1985. To date **only** whole blood, blood cellular components, plasma, and blood clotting factors have been involved in transmitting HIV.

——————— BOX 9.5 ———————

BLOOD TRANSFUSIONS: A GIFT OF LIFE OR THE CARRIER OF DEATH?

In 1987, a grief-stricken mother wrote an open letter to her neighbors thanking them for their compassion when her oldest son, a hemophiliac, died of AIDS. In 1989, this mother was to experience the loss of her second son to AIDS. He was also a hemophiliac. Her letter was to thank her community for caring for one of its members as a family cares for its child. It was an example of compassion to a nation demonstrating mixed emotions about those with AIDS.

In contrast, vandals burned down the home of Clifford and Cynthia Ray of Arcadia, Florida in 1987. Their three young boys, all hemophiliacs, acquired HIV from blood products used to stop life-threatening bleeding episodes. They also developed AIDS. Unlike the first family's experience, the Rays lost their friends, became the targets of hate mail, received threats of physical violence, were recipients of verbal abuse, and their children were not permitted to attend school. The Desoto County School Board, after threat of lawsuit, offered to let the boys attend class in a portable trailer in isolation from other children. Food would be served using disposable plates, cups, and utensils. The playground and library could be used only when other children were in class. Pickets again marched at the school and verbally abused the Ray family. Finally, their home was burned down and the Rays moved to Sarasota, Florida where the boys were enrolled in school without incident.

Blood Collection and Blood Screening for HIV

Testing blood for infectious diseases began with syphilis screening in the 1940's. Hepatitis-B antibody screening was added in the 1970's, HIV antibody screening in 1985, for biomedical evidence of hepatitis in 1986, the hepatitis-B-core antigen in 1986, HTL V-1 and II-antibodies in 1988, and hepatitis-C antibodies in 1990. Inclusion of the HIV antigen (p24) test in 1996 provided detection of HIV infection sooner than antibody testing. At least **8 tests** are now routinely performed on each unit of blood collected.

No Blood Purchases for Transfusion— All blood transfused in the United States comes from volunteer donors. Blood from paid donors is used for pharmaceuticals such as Rh Ig, albumin, and intravenous immuno-globulins. Under the current standards for Blood Banks and Transfusion Services of the American Association of Blood Banks, all units must be clearly labeled **"volunteer," "paid,"** or **"autologous."**

Blood Screening for HIV— **If the test indicates that the blood carries antibodies to either HIV, it is treated to destroy the HIV and then discarded.** In March of 1983, the major United States blood banking organizations instituted procedures to reduce the likelihood of HIV transmission through blood transfusions. People with signs or symptoms suggestive of HIV disease/AIDS, sexually active homosexual and bisexual men with multiple partners, recent Haitian immigrants to the United States, past or present IDUs, men and women who have engaged in prostitution since 1977 or have patronized a prostitute within the past 6 months, and sexual partners of individuals at increased risk of HIV infection **were asked to refrain from donating blood.**

A Transfusion Tragedy— In July 1989 The Boston Globe newspaper reported that in the early 1980's, the Dana-Farber Cancer Insti-tute in Boston transfused 25 patients with HIV-infected blood, and today all but two of them are dead. The incident occurred before a test for HIV became available in March 1985. The National Academy of Sciences reports that 12,000 people were infected nationwide through blood transfusions, as were additional 8,000 hemophiliacs during that same time period.

United States FDA Approves Blood Screening Test for HIV— In March of 1985, the FDA approved the ELISA test for use in commercial blood banks, plasma centers, and public health clinics to screen blood for antibodies to HIV. This test did not and still does not diagnose AIDS. It merely indicates whether or not a blood donor has antibodies against HIV (has been infected with HIV).

All blood donors who test positive for antibody to the HIV virus are rejected. This test, in conjunction with current measures to exclude blood donations by members of behavioral high-risk groups, has substantially reduced the chance of transmitting HIV through blood transfusion (ELISA and other test procedures are presented in Chapter 12). The risk of becoming HIV-infected from a blood transfusion has dropped by more than 99% from 1983 to 1991.

Blood Transfusions Worldwide— Fourteen years after the industrialized world began to screen all blood used in transfusions for HIV, about 1 in 10 people in developing countries are still being infected through this route.

A combination of the lack of screening with high levels of infected donors turns transfusion into a form of roulette. A survey carried out by WHO in 1990, for example, showed that centres in Kinshasa, Zaïre, screened fewer than a quarter of blood units for HIV, even though about 5% of donors were HIV-infected. Beginning year 2000, blood transfusions accounted for 5% to 10% of HIV infections worldwide.

Blood Safety— From 1985 through 1999, about 200 million blood or plasma donations had been screened for HIV antibody in the United States. By excluding those who test HIV-positive and by asking people from high-risk behavior groups not to donate blood, the incidence of HIV transmission from the current blood supply is relatively low. And with faster and more accurate testing procedures now in use, the risk is becoming even lower. **However, the probability or risk of receiving HIV-contaminated blood will never be zero.** The reason a small risk still exists is because some people infected with HIV may donate blood during what is known as the **"window period." During that period, a person may be infected with the HIV, but the test can not yet detect the infection. And, the test is not 100% accurate.**

In 1991, the CDC estimated the risk of receiving HIV-infected blood at 1 in 39,000 to 1 in 250,000. That is, for every 39,000 to 250,000 units of whole blood used in transfusions, one patient would receive HIV-infected blood. On average, a blood transfusion requires five or more units of whole blood (Pines, 1993).

Eve Lackritz and eight colleagues (1995) reported that, of 4.1 million blood donations from 19 regions served by The American Red Cross, one donation in every 360,000 was made during **The Window Period.** They further estimated that one in 2,600,000 donations contained HIV but was not identified (**false negative**) because of laboratory error. Their report states that collectively, of 42 Red Cross regions collecting 9 million units of blood, **the risk of HIV transmission was one transmission for every 450,000 to 660,000 available units of screened blood.** Based on their data they conclude that there are about 18 to 27 infectious units of blood available for transfusion each year. This is about half the previous estimates. (The American Red Cross collects about half the donated blood in the United States annually.)

PLAYING THE ODDS. For comparison, the odds of dying in a highway accident are 1 in 5,960 in a lifetime of driving. The CDC calculates that, **at worst,** with a risk factor of 1 HIV infection in 39,000 transfusions, there will be about 540 new HIV infections per year. (Eighteen million units of blood transfused each year means that three one thousandths of one percent (.00003%) of the 18 million units of blood are infected or 540 units.) By contrast, the CDC estimates that 10 times the number (5,400) will become infected with hepatitis C, which can be fatal.

Autologous Transfusions— These are transfusions using **"self" (your own) blood.** It has been suggested that the two groups that benefit most from autologous blood storage are: expectant mothers facing Caesarean section and certain elective surgical candidates where substantial loss of blood is anticipated. A person's blood is drawn and refrigerated for transfusion back into his or her own body when needed. **(Stored whole blood remains usable for a maximum of 42 days.)** This is the safest blood available. A healthy person taking iron supplements can donate one unit of blood a week, generally, for up to 6 weeks. In emergency situations, including some surgeries, it may be necessary to receive another person's blood through transfusion. If a transfusion is necessary to sustain life, the risk of contracting HIV from the transfusion is outweighed by the risk of refusing the transfusion and suffering further consequences from the emergency.

INFECTION CONTROL PROCEDURES

With no cure or vaccine for HIV/AIDS, prevention of infection is of paramount importance. With the advent of the HIV/AIDS epidemic, health care workers and others who are occupationally exposed to body fluids, especially blood, are understandably concerned about the risk of contracting HIV disease/AIDS. However, when precautions are observed, the risk is very small, even for those treating HIV/AIDS patients.

Two sets of infection control procedures are in use in hospitals, medical centers, physicians' offices, and units that deal with people in medical emergencies. One is called **universal precautions,** other **blood and body substance isolation.**

Universal Precautions

Universal precautions (Table 9-6) are standard practices that workers observe on the job to protect themselves from infections and injuries. These precautions or safety practices are called **universal** because they are used in all situations even if there seems to be no risk. Universal precautions had their beginnings in 1976 when barrier techniques were first recommended for the prevention of hepatitis B infection. Precautions required the use of protective eyewear, gloves, and gowns, and careful handling of needles and other sharp instruments. In 1977, hepatitis B immune globulin was recommended for those exposed to hepatitis B through needlesticks. In 1982, hepatitis B vaccine became commercially available and recommended for all health care workers exposed to human blood.

In 1983, the CDC published "Guidelines for Isolation Precautions in Hospitals." It contained a section that recommended specific blood and body fluid precautions to be followed when a patient was known or suspected to be infected with blood-borne pathogens. In 1987, the CDC published "Recommendations for Prevention of HIV Transmission in Health-

Care Settings." It recommended that blood and body fluid precautions be consistently followed for all patients regardless of their blood-borne infection status. This extension of blood and body fluid precautions to all patients is referred to as **Universal Precautions.**

Under universal precautions, the blood and certain body fluids of all patients are considered potentially infectious for HIV, hepatitis B virus (HBV), and other blood-borne pathogens.

Universal precautions are intended to prevent parenteral (introduction of a substance into the body by injection), mucous membrane, and broken skin exposure of health care workers (HCWs), teachers, or any other person who may become exposed to blood-borne pathogens. **In 1987, the CDC also published a report that got the immediate attention of most, if not all, informed health care workers.** The report stated that three health care workers who were exposed to the blood of AIDS patients tested positive for HIV. What was so startling was that until that time, needle punctures and cuts were thought to be the only dangers in a clinical setting. These three cases appeared to involve only

TABLE 9-6   Universal Precautions

DEFINITION

Universal precautions (UP) are a set of infection control practices, developed by the Centers for Disease Control and Prevention (CDC), in which health care workers (HCWs) appropriately utilize barrier protection (gloves, gowns, masks, eyewear, etc.) for anticipated contact with blood and certain body fluids of *all* patients.

1. The hands and skin must be carefully washed when contaminated with blood or certain body fluids.
2. Particular care is taken to prevent injuries caused by sharp instruments.
3. Resuscitation devices should be available where the need is predictable.
4. HCWs with exudative lesions or weeping dermatitis should refrain from patient care until the condition resolves.

BLOOD AND BODY FLUIDS TO WHICH UP APPLY

Blood is the single most important risk source of HIV, HBV, and other blood-borne pathogens in the occupational setting. Thus prevention of transmission must focus on reducing the risk of exposure to blood and other body fluids or potentially infectious materials containing visible blood.

1. UP should be used when exposure to the following body fluids may be anticipated:
   a. Blood
   b. Serum plasma
   c. Semen
   d. Vaginal secretions
   e. Amniotic fluid
   f. Cerebrospinal fluid (CSF)
   g. Synovial fluid
   h. Pleural fluid
   i. Vitneous fluid
   j. Peritoneal fluid
   k. Pericardial fluid
   l. Wound exudates
   m. Any other body fluid containing visible blood (but not feces, urine, saliva, sputum, tears, nasal secretions, or sweat, unless they contain visible blood).
2. Note: Blood, semen, and vaginal secretions have been shown to transmit HIV. The others, with the exception of fluids containing visible blood, remain a theoretical risk.

RATIONALE

1. UP reduce the risk of parenteral, mucous membrane, and skin exposure to blood-borne pathogens such as, but not limited to, HIV and HBV.
2. For several reasons, focusing precautions only on diagnosed cases misses the vast majority of persons who are infected (many of whom are asymptomatic or subclinical) and who may be as infectious as the diagnosed cases. Persons who have seen a physician and have been diagnosed with acute or active disease represent only a small proportion of all persons with infection. Infectivity always precedes the diagnosis, which often is made once symptoms develop.

(Adapted from Mountain-Plains Regional AIDS Education Training Center. *HIV/AIDS Curriculum*, Nov. 1994.)

skin exposure to HIV-contaminated blood. **One** of the three cases involved a nurse whose chapped and ungloved hands were exposed to an AIDS patient's blood.

The **second** case involved a nurse who broke a vacuum tube during a routine phlebotomy on an outpatient. The blood splashed on her face and into her mouth. A blood splash was also involved in the **third** case. The worker's ungloved hands and forearms were exposed to HIV-contaminated blood (Ezzell, 1987).

The importance of these three cases of HIV infection is that they informed health care workers in the most dramatic way that they were all **vulnerable.** In addition to hepatitis and other infections, they could now add a more lethal virus to their occupational hazards. Perhaps these three cases produced a fear among health care workers out of proportion to the actual risk of their becoming infected. However, the fear of an individual and the actual reason he or she has for it are very difficult to judge. **Although** calculations show that the risk of HIV infection after exposure to blood from an HIV/AIDS patient is about one in 200, if you are that one, probability is meaningless. Some health care workers have used, and still use, such data to support their decisions to avoid caring for HIV/AIDS patients.

The universal precautions as published by the CDC currently apply to some 5.3 million health care workers at 620,000 work sites across the United States and another 700,000 Americans who routinely come in contact with blood as part of their job, for example, people in law enforcement, education, fire fighting and rescue, corrections, laboratory research, and the funeral industry.

In July of 1991, the CDC published an additional set of recommendations. These recommendations are also for preventing the transmission of HIV and hepatitis B virus to patients during exposure-prone procedures.

In summary, the concept of universal precautions assumes that all blood is infectious, no matter from whom and no matter whether a test is negative, positive, or not done at all. Rigorous adherence to universal precautions is the surest way of preventing accidental transmission of HIV and other blood-borne pathogens.

--------- BOX 9.6 ---------

DOES THE PATIENT HAVE HIV? I CAN'T TELL

Treating disease can be difficult. Dealing with the legal system and the politics of disease can be a conundrum.

Recently, a nurse in California brought charges against a patient after the nurse accidentally cut herself with a contaminated scalpel and subsequently learned that the patient had AIDS. The prosecution based its case on the patient's failure to fully disclose information relevant to her care. The defense case centered on the failure of medical personnel to use **universal precautions** for treating all patients regardless of their health history.

I side with the nurse and so did the Los Angeles jury who convicted the patient of fraud for failing to disclose that she had AIDS.

Yes, **universal precautions** are required for all patients regardless of health history, but the practical fact is that different situations in medicine require different levels of attentiveness. Care for a young man with a simple cough will and should differ from care for a young man who has a cough and mentions he has AIDS and has had pneumocystis pneumonia in the past. In an emergency department, activity levels vary from dull and routine to harried and panicked. If medical personnel put on gloves, gowns, and masks and slowed their activities to the safest pace for every patient, regardless of known or suspected risk, gridlock would result.

Medical history is critical to proper patient care. When a patient comes to a hospital and is unable to provide a detailed history, medical records are often enormously helpful in providing proper treatment. Therefore, I was surprised recently when a new directive was tacked to the bulletin board in one of the emergency departments where I work. It said that a patient's HIV status could not be recorded on the emergency department chart. I can document syphilis, cancer, schizophrenia, or violent or suicidal behavior, which are all pertinent to a patient's history. Such documentation may help other healthcare professionals provide proper care or take "more attentive" precautions. But I cannot document HIV status. I asked why and was told that this was an effort to comply with a new part of the state of California's Health and Safety Code.

I understand the rationale for this directive. Not only do those who are HIV-positive live with the threat of a terminal disease, but they are discriminated against in the workplace and by insurance carriers, even while they are healthy. The state code is an effort to prevent paranoid discrimination. But I believe the effort to preserve a patient's privacy may put healthcare workers at greater risk.

AIDS is a terrible disease, and paranoia about it is rampant. But protecting patient privacy by putting blinders on healthcare workers is not good medicine.

(Reprinted, with permission, from Pollack B. (1994). Does the patient have HIV? I can't tell. *Postgrad Med* 96(3):19. Copyright © 1994 McGraw-Hill Inc.)

Class Discussion: Your Assessment of Dr. Pollack's concerns?

Blood and Body Substance Isolation (BBSI)

An alternative, and some believe superior, approach to the CDC's universal precautions in areas of high HIV prevalence is the system referred to as **body substance precautions or Blood Body Substance Isolation (BBSI)** (Gerberding, 1991).

In practice, these precautions are similar to universal precautions, in that prevention of needlestick injury and use of barrier methods of infection control are emphasized. Philosophically, however, the two are quite different. Whereas universal precautions place a clear emphasis on avoidance of blood-borne infection, body substance precautions take a more global view. **Body substance isolation (BSI) requires barrier precautions for all body substances (including feces, respiratory secretions, urine, vomit, etc.) and moist body surfaces (including mucous membranes and open wounds).** BSI is designed as a system to reduce the risk of transmission of all **nosocomial** (hospital-associated) pathogens, not just blood-borne pathogens. Gloves are worn for any anticipated or known contact with mucous membranes, or nonintact skin and moist body substances of all patients.

The degree of contact with the blood and body fluids or tissues of each patient is considered to determine the type of precaution (if any) required. Unlike the CDC system in which precautions are based on the premise that all blood is infectious, body substance precautions are procedure-specific, that is, based on the degree of anticipated contact. Users of the BBSI system find that this approach is actually easier to teach and implement than universal precautions, although the end result is much the same: the prevention of HIV infection.

PARTNER NOTIFICATION

One of the most controversial issues in HIV prevention is **contact tracing** or **partner notification** of sexual contacts mostly because HIV/AIDS is considered as an incurable disease with a great deal of stigma attached to the infected. Partner notification is the practice of **identifying** and treating people exposed to certain communicable diseases. The term *partner notification* is used by the CDC and some health care providers because it more comprehensively describes the process by which the physician, other health care workers such as Disease Intervention Specialists (DIS, someone who is specially trained in STD work), and the infected person may provide information to at risk partners.

There are two very different approaches to informing unsuspecting third parties about their potential exposure to medical risk.

Each approach has its own history, including a unique set of practical problems in its implementation, and provokes its own ethical dilemmas. The **first approach,** involving the moral **duty to warn,** arose out of the clinical setting in which the physician knew the identity of the person deemed to be at risk. This approach provided a warrant for disclosure to endangered persons without the consent of the patient and could involve revealing the identity of the patient. The **second approach**—that of contact tracing—emerged from sexually transmitted disease control programs in which the clinician typically did not know the identity of those who might have been exposed. This approach was founded on the voluntary cooperation of the patient in providing the names of contacts. It never involved the disclosure of the identity of the patient. The entire process of notification was kept confidential (Bayer, 1992).

History of Partner Notification

The concept of partner notification was proposed in 1937 by Surgeon General Thomas Parran for the control of syphilis (Parran, 1937). By tracing and treating all known contacts of a syphilitic patient, the chain of transmission could be interrupted. According to George Rutherford (1988), contact tracing has been successfully used in a number of STDs beginning in the 1950s. It is still used in cases of syphilis, endemic gonorrhea, chlamydia, hepatitis B, STD enteric infections, and particularly in cases of antibiotic-resistant gonorrhea.

In 1985, when the HIV antibody was first used in screening the blood supply, notification of blood donors and other HIV-infected individuals and their contacts became possible. **The strategy in HIV partner notification is the same as that used for the other STDs: to identify HIV-infected individuals, counsel them, and offer whatever treatment is available.** In asymptomatic HIV-infected people only counseling is given. But, symptomatic HIV-infected patients receive counseling and treatment, if available, for their signs and symptoms. Partner notification depends on HIV-positive people to give the names of their partners; but, they may be reluctant to do so fearing that their identification may result in physical abuse and loss of their jobs and housing. For a current review of Partner Notification read the article by Kevin Fenton, et al., 1997.

The Use of Partner Notification: An Example

In April 1993 an incarcerated male asked for an HIV test. The diagnosis was HIV-positive. Contact tracing turned up a network of 124 persons, all were linked by syringe-sharing and syringe-sharing with sex. One hundred and twenty-one were contacted and offered an HIV test; 118 accepted the test; 44 were HIV-positive. One hundred and thirteen of the 124 lived in the same county. The estimated cost for partner notification in this network was $13,969. (*MMWR*, 1995).

Opposition to Partner Notification for HIV Disease

Currently opposition to partner notification continues on the grounds that: (**1**) it is not cost effective; (**2**) there is little evidence that those who are informed of their infection will do anything about changing the high-risk behaviors that got them infected in the first place; (**3**) the threat of social discrimination undermines the intent of contact tracing; (**4**) the unintended consequences of partner notification may include violence and even death from an abusive partner (Rothenberg, 1995); and (**5**) in 24 states homosexuality is a crime, and HIV-infected homosexuals fear prosecution if they acknowledge same-sex contacts and be responsible for having their sexual partner(s) prosecuted.

In Support of Partner Notification

The following discussion is from Stephen J. Josephs' article, "The Once and Future AIDS Epidemic," *Medical Doctor*, 37:92—104. Mr. Joseph is the past Commissioner of Health of New York City, 1986–1990.

The New York City Medical Examiner looked at a large sample of dead persons who were tested postmortem and found to be HIV-positive, but who had no notification in their medical records of having been diagnosed as infected. Over 35% of those persons had a readily identifiable spouse or steady sexual partner. Josephs asks,

> How can one justify, on clinical, public health or humanitarian grounds, *not* notifying that surviving partner, who might be the source of the infection in the deceased, or the recipient of infection? Arguments against this procedure border on the absurd; one has to start with the premise that increased medical knowledge is more dangerous than helpful to the individual, and that the rights of the uninfected count for nothing against the rights of the infected.

What is your response to Stephen Josephs' point of view on partner notification?

Legality of Partner Notification

Legal Obligations to Warn Third Parties At Risk— Many state laws permit, but do not require, disclosure by physicians to third parties known to be at significant future risk of HIV transmission from patients known to be infected. Thus, if a physician reasonably believes that a patient will share drug injection equipment or have unprotected sex without informing a partner of the risk, the physician has discretion to inform the partner. Under some disclosure laws, the physician is required to first counsel the patient to refrain from the risk behavior, and, in providing the third party warning, the physician is prohibited from disclosing the patient's identity. In the absence of state laws permitting such disclosure, physicians may be held liable for breach of confidentiality for disclosing patient information to sex partners.

The **"duty to warn"** may extend to nonpatient third parties in other contexts, based on the provider's primary duty to the patient. Thus, health care professionals have a duty to inform patients that they have been transfused with HIV-contaminated blood and this duty may extend to third parties. A physician in one case failed to inform a teenager or her parents that she has been transfused with HIV-contaminated blood. When the young women's sexual partner tested positive for HIV, the court upheld his claim against the physician based on the physician's failure to inform the patient. Similarly, courts have upheld that a health care professional's duty to inform the patient of his or her HIV infection may extend to those the patient foreseeably puts at risk, such as a spouse or family member caregiver. On the other hand, courts have ruled that disclosure is wrongful in cases in which the third party, such as a family member, is not at actual risk of infection, or the physician has no knowledge that the patient has failed to disclose to the partner (Gostin, et al. 1998).

Beginning year 2000, 38 states in America have enacted partner notification laws that provide for penalties that range from a **misdemeanor** (a crime less serious than a felony, which is a serious crime) to **attempted murder** for anyone who does not reveal to a sexual partner that he or she is HIV-positive (see Box 8.7). At least 44 states passed laws requiring or permitting workers (mostly health care workers or public safety employees) to be notified of potential exposure of HIV. In some cases, the laws allow testing of the source patient. To date, Arkansas and Missouri are the only states that require patients to notify health care providers of their HIV status before receiving care (Hooker, 1996). All 50 states are now somewhere in the process of establishing the capacity for contact tracing at the request of a patient.

Class Discussion: If a law was passed that made persons who practiced high-risk behaviors and who contracted HIV/AIDS pay for their care and treatment or forgo medical help—do you think these people would continue to engage in high-risk behaviors? Would this law be an effective means of HIV transmission prevention? Present examples to support your position. (Can you relate this scenario to those who smoke and develop cancer?)

HIV VACCINE DEVELOPMENT

Vaccines work because of your immune system's ability to "remember." When you're vaccinated, dead, inactivated or weakened live forms of an infectious organism stimulate an immune response without causing the accompanying illness. Memory cells provide immunity for years or even your lifetime.

Development of a safe and effective vaccine for HIV infection remains the **"holy grail"** of HIV/AIDS research. The grim global statistics of 40 million people living with HIV disease entering year 2000 is a potent reminder of the

─────── **POINT OF VIEW 9.1** ───────

THE GOAL OF DEVELOPING AN HIV VACCINE

The ideal HIV vaccine would eradicate HIV infection. But ideals are seldom realized in biology and this ideal in particular does not seem likely. For all of the infectious agent vaccines that have been marketed, **only smallpox has been eradicated!**

For an infectious agent to be a good candidate for eradication, several conditions must exist. The most important of these are: (**1**) that only humans are affected, (**2**) that infection is easily recognized, and (**3**) that infected persons do not remain asymptomatic for long periods of time. HIV meets only the first of these conditions, and even this is questionable as it may have mutated from closely related monkey viruses. Thus a realistic HIV vaccine goal would be to develop candidate vaccines that demonstrate some percent of efficacy, with fewest side effects, and begin their distribution. Saving some lives is better than a complete loss. Because there are an increasing number of HIV strains evolving, no one vaccine is expected to be effective in everyone.

The ideal HIV vaccine has to be safe, orally administered, single dose, stable, inexpensive, confer permanent life-time immunity, and be effective against all HIV strains. This is an unrealistic expectation, at least in the next 5 to 10 years.

PRESIDENT CLINTON SETS A VACCINE GOAL

Speaking to the graduating class at Morgan State University on **May 18, 1997,** President Clinton invoked the legacy of John F. Kennedy's 1960s race to the moon and set a national target of developing an AIDS vaccine within the next 10 years (2007). This is now **The First Annual Vaccine Day in America.** The President said, "We dare not be complacent in meeting the challenge of HIV, the virus that causes AIDS." He then announced the creation of a research center at the National Institutes of Health in Bethesda, Maryland, to complete the task.

Up to 50 researchers will staff the suburban Washington facility, drawn from existing NIH programs, but no new money was earmarked for the project!

Robert Gallo stated, after the President's announcement, that "it is a serious possibility that we may never develop a vaccine for HIV."

May 18, 1999—**The second anniversary**—has passed, but only one preventive vaccine is in two phase III trials, one in America and one in Thailand. At this rate, Clinton's goal for a preventive HIV vaccine by year 2007 will not be reached.

SOUND OFF

Following President Clinton's announcement about developing an HIV vaccine in 10 years, the telephone at a large Northeastern newspaper rang nonstop that day. Here are but a few of the callers' comments concerning what the government can do to find a cure for AIDS (names are withheld).

"What is President Clinton doing with people dying of cancer who had nothing to do with getting their disease?"

"Let's use the research money on the big killers, cancer and heart disease. Stop the spread of AIDS by prohibiting blood do-

nations from drug users and homosexuals and quarantine those who are HIV positive or have AIDS."

"The government has to be really serious about it. I can compare this to the 400 blacks down south who had syphilis in the 1920s and 30s. Our government did nothing for them. AIDS is just a disease that's been made by our government to get rid of a segment of society. A form of genocide."

"Presently, there is a 100 percent cure for AIDS. It's called abstinence and that's what the government should be teaching."

"The government can't do much, but the thousands of researchers already working hard on AIDS must really have appreciated his telling them to get busy."

"The government cannot do enough. We need around-the-clock research."

"I don't know what the government can do, but in the meantime, I know what people can do. Men, keep your zippers up and women, keep your pants on and try to use some self control. This way, a million more lives will not be affected with this horrible virus."

"There will be no cure for AIDS because no viral infections has ever been cured. As far as a vaccine goes, it's possible but extremely difficult since we're dealing with a retrovirus. The best thing is abstinence and chastity, as unpopular as those words are today."

"Contact the Nation of Islam. We, as Muslims, can sit down with the government and come up with a cure for AIDS if the government has decided to rectify the difference with us so we can sit down and talk."

"They should set up a program where everyone in the United States can get an AIDS-HIV test, starting at age 15 and 16, just to know where it is heavily exposed."

pressing need for such a vaccine. In the last three years, 1997, 1998 and 1999, an estimated 17 million new HIV infections occurred worldwide and more than 90% of these new infec-

tions occurred in developing countries. In these countries, where annual per capita health care expenditures may total only a few dollars, access to potent antiretroviral therapy is essentially nonexistent. The development of an HIV/AIDS vaccine is most likely their only hope in successfully fighting this disease. Yet, 16 years have passed (1984) since the United States Secretary of Health and Human Services promised a vaccine in two years. And at this moment, no effective vaccine is in sight and none may be available for some years to come!

Goal of HIV Vaccination

The ultimate goal of an HIV vaccine is to prevent HIV infection. Realistically, this means that vaccine-induced immune responses should clear all HIV and HIV-infected cells quickly, before secondary spread occurs and persistent viral infection is established. Therefore, more than with any previous vaccine, the development of a vaccine against HIV infection must be driven by a detailed understanding of viral pathogenesis and immunity.

What the World Needs NOW Is a Vaccine for HIV Disease—Why Isn't There One?

On April 24, 1984, Margaret M. Heckler, who was then Secretary of the Department of Health and Human Services, announced the discovery of the AIDS virus. **She predicted an AIDS vaccine within 2 years.** Even though the prediction proved wrong, research was guided by the idea that finding the virus was the hard part, and vaccines could be made by simply injecting people with crucial viral proteins. Her optimism was most likely based on the success of the polio, measles, and flu vaccines. **The approach to combating these diseases was: isolate the virus, develop a vaccine, and prevent the disease!** Since then, in the rush to develop new vaccines, scientists have only belatedly understood that their technical ability to mass-produce vaccines has failed to match their knowledge about the cellular and molecular processes used by the body to protect itself from invading pathogens.

Vaccines are designed to provoke the immune system into making antibodies against a disease-causing agent. Most are made of inactivated or attenuated (genetically weakened) viruses and, in the case of some newer vaccines, extracts of viral coat proteins. In some cases, vaccination may result in worsening the disease. The distinction between protective, useless, and dangerous responses is essential for vaccine design.

Vaccines that work well are the most cost-effective medical invention known to prevent disease.

What is a Vaccine?

A vaccine is a suspension of whole microorganisms, or viruses, or a suspension of some structural component or product of them that will elicit an immune response after entering a host. In brief, vaccines mimic the organisms or virus that cause disease, alerting the immune system to be aware of certain viruses or bacteria. Because of this advance warning system, when the real organism or virus invades the body, the immune system marshals a response before the disease has time to develop.

Ideally, the body will make **antibodies** that bind to and disable the foreign invader (**humoral immunity**) and trigger white blood cells called T cells to organize **attack cells** in the body to destroy those cells that have been infected by viruses (**cellular immunity**). Once the immune system's T cells and B cells, which make antibodies, are activated, some of them turn into **memory cells.** The more memory cells the body forms, the faster its response to a future infection.

Vaccines can trigger these responses in three ways. **Some vaccines, such as those against smallpox, polio (Sabins), measles, mumps, and tuberculosis, contain genetically altered or weakened organisms or viruses that are reproduced in the body after being administered but do not generally produce disease. Yet since the virus or bacterium is still alive, there is a small risk of developing the disease.**

Whooping cough, cholera, and influenza vaccines are made of inactivated whole organisms and virus or pieces of them. Because killed organisms and inactivated virus do not replicate inside the recipient, the vaccines confer only humoral immunity, (the production of antibody) which may be short-lived.

Finally, vaccines can be made against toxic products produced by microorganisms. In diseases like tetanus, it is not the bacteria that kill, but the toxins they release into the bloodstream (Christensen, 1994a).

What Is An Effective Vaccine?

An effective viral vaccine usually blocks viral entry into a cell. **But vaccines are generally not 100% effective.** Because HIV, once inside the cell, is capable of integrating itself into the genetic material of infected cells, a vaccine would have to produce a constant state of immune protection, which not only would have to block viral entry to most cells, but also would continue to block newly produced viruses over the lifetime of the infected person. Such complete and constant protection has never before been accomplished in humans, but it has been accomplished to some degree in cats who are vaccinated against the feline leukemia virus, also a retrovirus (Voelker, 1995). But, perhaps more pertinent explanations for why there is still no HIV vaccine nor is one likely to be available soon are the facts that scientists lack sufficient understanding of HIV infection and the biology of HIV disease/AIDS is very complex.

Scientists know that the body defends itself against HIV in the early years of infection. But the great mystery has always been why it cannot neutralize HIV completely. One possibility is that the body has trouble seeing all of the variant viruses. Like a Stealth fighter plane, some HIV may have hidden parts that do not show up on the immune system scanner. As a result, the immune system may not produce the right kind of antibody to neutralize all the variant HIVs.

Vaccine Expectations

Vaccines rarely prevent infection; rather they prevent or modify disease. Most vaccines currently in use—for example those for polio, tetanus, diphtheria, measles, hepatitis B, and influenza—**prevent disease without actually preventing infection.** They reduce the number of invading microorganisms, increase the rate of clearance of the infection, prevent the secondary consequences of infection and prevent transmission. Similarly, few of the candidate HIV vaccines appear promising for preventing infection, and current expectations that HIV vaccines will prevent infection has yielded in the scientific community to the **hope** that they can produce a vaccine that can prevent disease.

Types of HIV Vaccines

Scientists are attempting to design three types of HIV vaccines: (**1**) a **preventive** or **prophylactic vaccine** to protect people from becoming HIV-infected (chance of success–doubtful); (**2**) a **therapeutic vaccine** (this is not a true vaccination, but a postinfection therapy to stimulate the immune system; the term vaccination is reserved for preventive strategies) for those who are already infected with HIV to prevent them from progressing to AIDS (chance of success–good); and (**3**) a **perinatal vaccine** for administration to pregnant HIV-infected women to prevent transmission of the virus to the fetus (chance of success–good). Stephen Straus and co-workers (1994) reported that researchers have developed a therapeutic vaccine for the herpes virus. The new, therapeutic vaccine reduces the frequency with which genital sores appear in patients infected with the herpes virus. While it fails to outperform the existing antiherpes drug, acyclovir, it sets the stage for a more effective treatment in the future.

The ability to influence the frequency of genital herpes outbreaks with this vaccine inspires optimism that similar successes may be possible with other chronic viral diseases, such as AIDS.

VACCINE PRODUCTION

To make vaccines, scientists use either **dead microorganisms** and **inactivated viruses** or **attenuated viruses** and **microorganisms**. Attenuated means that viruses and other microorganisms are modified; they are capable of reproducing and invoking the immune response but lack the ability to cause a disease.

Attenuated Viruses

Attenuated vaccine technology has its beginnings in 590 B.C. in China when smallpox matter, taken from people with mild cases of smallpox, was introduced into people who came down with a mild form of the disease and then were immune from smallpox for life. The procedure became known as **pock-sowing. In 1714 Edward Jenner created an attenuated vaccine against smallpox by using cowpox matter.** Cowpox is genetically closely related to smallpox.

HIV VACCINE THERAPIES: PREVENTION AND THERAPY

A LOOK BACK AT THE FIRST VACCINE

The year 1996 marked the 200th anniversary of the first vaccine, which was developed against smallpox. As vaccine researchers launch a new century of challenging disease, they might find inspiration in the early beginnings of immunology, Edward Jenner's discovery.

According to lore, Jenner was a country doctor who heard a rumor that the cowpox virus could provide immunity to smallpox. Investigating the theory, Jenner endured the ridicule of his colleagues before proving his point.

In a now-famous 1796 experiment, Jenner scratched (inoculated) the arm of 8-year-old James Phipps infecting the boy with cowpox pus taken from a milkmaid carrying the virus. Two months later, he again inoculated James but this time he added some smallpox residue. The rest, as they say, is history: James remained healthy! A smallpox vaccine—and the field of vaccinology—was launched.

No viral epidemic has ever been conquered by drug therapy—prevention is the key, and primary prevention via vaccine inoculation is the cornerstone. But, there are enormous problems inherent in constructing an HIV vaccine. **First,** the virus has extraordinary diversity, leaving little chance that just a few subtypes could induce broadly protective immunity. **Second,** the attenuated (weakened) virus strategy that has been so successful for many infections, including smallpox, polio, and measles, will be very difficult if not impossible to implement for HIV. **Third,** HIV attacks the very immune cells that are essential in an immunization procedure, and the immune cell activation that accompanies any immunization will activate HIV replication.

With that said, immunologists are still driven to achieve a most difficult task, the production of an effective HIV-preventive vaccine. Since the first injections of an experimental HIV vaccine were given in the United States in 1987, about 3,000 uninfected adults have received about 45 experimental vaccines; 18 people have become HIV-infected from the vaccines. (Bolognesi et al., 1998). Only a handful of HIV vaccines have made it to the second of the three stages of the trials that are needed before any vaccine can be marketed. Beginning year 2000 only one vaccine AIDSVAX, has reached the third stage, full-scale testing. Worldwide, through 1999, there are some 74 experimental vaccines in basic research or animal testing. To date most HIV vaccines being developed target surface proteins of HIV. The reason for this approach is that the immune system **first sees** the outside boundaries of HIV—the surface proteins. There is a problem, however: Different clades or strains of HIV carry surface proteins that differ from each other because their genes have changed to make the different strains. Thus a vaccine made against one strain's surface proteins, say type B (United States), may not work against type C (India).

DNA VACCINE

What is is a DNA Vaccine?

To make a DNA vaccine, a gene (or length of DNA) that is responsible for making a protein in the infectious virus or organism is inserted into a bacterial plasmid (a circular length of DNA that can replicate by itself inside a bacterial cell). Plasmids carrying the gene of choice are replicated in trillions of bacteria, the bacteria are then broken open and the trillions of plasmids, each carrying a copy of the gene, are purified. The collected plasmids are then given to a patient. **Cells of the person take up the DNA and begin to make the exact protein the gene made while it was in the virus or microorganism from which it was taken.** Such a protein is considered an antigen by the body and the immune system mounts a defense against it.

ADVANTAGES OF A DNA VACCINE

DNA vaccines have several potential advantages over the traditional methods. For one thing, there's no risk of infection, as there is for inactivated or attenuated vaccines. DNA vaccines are also superior to protein based vaccines because proteins, while being isolated to use an antigen in a vaccine, are easily destroyed. DNA vaccine produces the protein right in the host cell! Also, DNA is a very stable chemical, even at temperatures close to boiling, a decided advantage in developing countries where refrigeration is scarce.

In March of 1996, the FDA approved the **first** human testing of a vaccine made with pure DNA. Apollon Inc. and the National Institutes of Health will test the experimental AIDS vaccine on 15 (13 men, 2 women) healthy people at the NIH hospital in Bethesda, Maryland in mid-1996. However, these DNA vaccine experiments were not successful! New DNA vaccines are now in study (Calarota et al. 1998).

Potential AIDS vaccines have been given to healthy people before. But this marked the first time a healthy person has received the pure genetic material of any disease–causing agent as a vaccine.

Although scientists are mixing the old traditional ideas of vaccine production with the new, in general there remains the fear that the production of an effective HIV-preventive vaccine may not be possible. But, if it is, it won't be available before the year 2005 and even then it may only be 30% effective in preventing HIV infection! A

successful AIDS vaccine will need to induce the formation of broad, cross-reactive neutralizing antibodies against an array of HIV antigens that encompass the majority of HIV strains active in the region of interest and at the same time confer some mucosal immunity. It should be mentioned that to date, there is only one vaccine available against a sexually transmitted disease (hepatitis B vaccine) and it does not stimulate mucosal immunity.

One benefit from all of the HIV-vaccine research to date is that scientists know far less than they thought they knew about producing specific prevention vaccines. And scientists are, for the first time, learning about the mechanisms of viral-host pathologies necessary to produce preventive vaccines.

Attenuated viruses in vaccines provide a better immunity than inactivated viruses because attenuated viruses continue to reproduce, thereby acting as a constant source of antigenic stimulus to the immune system. Also, such vaccines stimulate both the production of antibodies and the movement of cytotoxic (killer) T lymphocytes that destroy already infected cells. Attenuated vaccines appear to provide lifelong immunity **without** requiring periodic boosters. One of the major concerns with using attenuated virus vaccine is the fear that first, all the virus may **not be attenuated** and second, the attenuated virus may revert to the virulent form. For example, approximately 10 cases of polio using attenuated polio virus vaccine, occur in the United States per year which represents 1 case per 2.4 million doses of vaccine administered (Stoeckle et al., 1996). There is, however, an equally great fear that has not received proper attention; that is, no one knows the long-term consequences of having an attenuated **retrovirus** (HIV) inside the body for 20 to 50 years. **Such viruses inserted into human DNA may turn genes off and on at the wrong time causing cancer and other types of disease** (see Figure 6-14).

Use of Attenuated HIV Vaccine— In 1997 data from several labs revealed that a vaccine made from weakened or attenuated SIV-HIV's simian analog can cause AIDS like symptoms in

adult monkeys. These findings that an attenuated vaccine caused AIDS like disease in monkeys has worried some investigators about attempting to use an attenuated HIV vaccine in humans. Robert Gallo, Director of The Institute for Human Virology, believes that a live HIV vaccine is too dangerous. He said that "live, low-replicating retroviruses almost always cause disease; that's been our experience in all animal systems. If those vaccinated do not get disease in three years, it will not tell you what will happen in 10 years or 30 years."

Contrasting Opinion For Use of Attenuated HIV Vaccine

Charles Farthing, chair of the Live-Attenuated HIV Vaccine Subcommittee of the International Association of Physicians in AIDS Care (IAPAC), who favors the use of an attenuated HIV vaccine, reported data from Australia and from Massachusetts on 19 people who, 10 to 14 years ago were accidentally infected with HIV that did not carry the **nef** gene. None of these people experienced immune suppression. And 7 of these people, still alive, have normal-functioning immune systems. Four of the 7 have no detectable viral load, 3 have minimal viral loads. For the 12 who have died, none of the deaths appeared to be associated with the HIV disease. Farthing believes that such data is more relevant on deciding to use an attenuated vaccine than is the data from the SIV experience (Farthing, 1998).

VOLUNTEERS TO TAKE FIRST ATTENUATED HIV VACCINE

In October 1997, 50 doctors (39 are AIDS physicians), nurses, and health advocates, members of the International Association of Physicians in AIDS Care (IAPAC), and 300 people, some outside the healthcare arena, signed up to take the attentuated vaccine of Ronald Desrosier of Harvard Medical School. The four major anti-HIV drug manufacturers have pledged lifetime free drugs for any of the people who might become HIV infected from this research. Other healthcare services will also be provided. The vaccine trial will begin with 5 volunteers **over** age 25. If all is well after 6 to 9 months, 6 more people will enter the program and on from there. At this writing, noted HIV/AIDS experts such as Anthony Fauci, David Baltimore and others do not recommend that this experiment go forward because of the risks of HIV infection to those taking the vaccine. IAPAC is seeking FDA approval for the use of Desrosier's vaccine and hope to begin the human experiment in the year 2000.

Inactivated Viruses

To inactivate viruses for use in vaccines, the viruses are treated with formalin or another chemical. There is a danger in using inactivated viruses—**they may not all be inactivated.** Inactivated virus vaccines have been made against hepatitis B, rabies, influenza, and polio (Salk vaccine). Salk's first vaccine killed a number of recipients in the late 1950s because not all the polio viruses were destroyed, that is, some could still replicate.

Jonas Salk announced at the Fifth International Conference on AIDS (1989) that he used gamma radiation to destroy the virulence of HIVs and stripped them of their outer envelope. Salk asked for and received permission to inject volunteer priests and nuns over age 65 with his vaccine. The project was suspended in 1993, as there were little if any observable beneficial effects of his vaccine.

Subunit Vaccines

Subunit vaccines are made from antigenic fragments of an organism or virus most suitable for evoking a strong immune response. Specific subunits can be mass produced and used in pure form to make a specific vaccine. Vaccine against hepatitis B is made from a subunit of the hepatitis B virus and produced in quantity in yeast.

In the United States, researchers are currently basing their vaccine strategies on the use of subunit proteins found in the envelope of HIV.

PROBLEMS IN THE SEARCH FOR HIV VACCINE

HIV poses some unique problems for making a human vaccine. **First,** scientists have not established what immune responses are crucial for protecting the body against HIV infection. Studies over the last couple of years have shown that the cell-mediated arm of the immune system may be more important than the HIV antibody response. If this turns out to be true, investigators will have to regroup with respect to producing an HIV vaccine—most vaccines in field trials are geared toward producing sustained HIV antibody responses. Without this information, they cannot tailor the vaccine to produce the most essential immune response. **Second,** it is too risky to use the entire weakened or inactive HIV.

Third, HIV undergoes a high rate of mutation as it replicates, and strains from different parts of the world vary by as much as 35% in terms of the proteins that comprise the outer coat of the virus. Even within an infected individual, over a period of years, the virus may change its proteins by as much as 10%. This degree of antigenic drift or variation means that a vaccine made from one strain of HIV may not protect against a different strain. To prove effectiveness, vaccines may have to be tested in geographic areas where the prevalent strains are the same as the strain used in the vaccine.

Fourth, the immune response raised by the vaccine may be protective for only a short period of time. In such cases, booster vaccinations would be required too frequently to be practical.

Fifth, it is possible that the vaccine may make people more susceptible to HIV infection. A vaccine induced enhancement of infection. That is, vaccine producers are concerned that if they are unable to make a completely HIV–neutralizing vaccine, the vaccine when taken might make the disease worse by **boosting the number of antibodies that might enhance entry of HIV into cells.** For example, enhancing antibodies are thought to be important in a few unusual viral infections such as dengue fever, Rift Valley fever, and yellow fever, in which the antibody binds to the virus and carries it into cells. **Thus, the more of a certain kind of antibody you have, the worse the infection can become** (Homsy et al. 1989; Levy, 1989).

Sixth, entering year 2000, no vaccine trial to date has been able to stimulate the cellular side of the immune system in the manner necessary to destroy HIV. For example, vaccine involved in current studies do not generate significant numbers of **cytotoxic lymphocytes** nor do they produce significant numbers of **memory T cells** that are necessary to recall the initial response against HIV.

Seventh, Jon Cohen, writing for the journal *Science* (1994a) reported on a *Science* survey of an international sample of over 100 of the field's leading researchers, public health officials, and manufacturers. All told, 67 people from 18 countries on six continents reponded. And when it came to describing the obstacles that hinder vaccine development, the respondents—be they from Russia, Indonesia, Egypt, Europe, India, or Brazil—had remarkably similar views. **The scientific unknowns are the highest hurdles,** they said, but they also stressed that the field lacks the strong leadership and funding to speed progress.

Predictably, money—or rather, lack of it—was one of the most frequently mentioned obstacles. Even though vaccines are among the most cost-effective medical interventions ever devised, they are not big money-makers. **Drug companies are traditionally reluctant to invest in any form of vaccine development which carries high costs, low profits, and big risks of costly legal suits should accidents occur.** Their current analysis of the state of HIV/AIDS vaccine research is particularly bleak.The National Institutes of Health (NIH) spends 8% of its AIDS budget—around $100 million a year—on vaccine work, but only about $20 million is being spent worldwide on preparing new candidate vaccines.

Eigth, economics clearly isn't the only factor that is discouraging companies from entering the search for an AIDS vaccine. Another is the

POINT OF INFORMATION 9.7

PATHOLOGICAL UNIQUENESS OF HIV

There has been much written about the uniqueness of HIV. Perhaps the most singular feature is that it directly effects the immune system and is lethal to humans. A national state of immunity has not yet been documented in humans. Once infected with HIV, there is a high probability of death within 10 to 20 years.

Most human pathological viruses, including HIV, behave in a somewhat similar fashion with one important exception. Other viruses, including all of the other sexually transmitted viruses, enter cells, cause cell lysis, and remain in the body. The body "learns" to live with viral infections, latent periods, and recurrent episodes. **But HIV is different: it infects and suppresses an immune system that holds other viral infections in check.** Perhaps it is because we survive these other viral infections that there has not been the urgency to learn more about host response to viruses: how viruses actually infect cells, become proviruses, and are activated to reproduce. There is not even a respectable spectrum of antiviral drugs available, mostly because we do not know enough about the viruses.

There are vaccines available for smallpox, polio, measles, mumps, yellow fever, rubella, rabies, and influenza; but, only one vaccine— hepatitis B—is available for a sexually transmitted viral disease. And, only one vaccine has ever eradicated a human disease, the smallpox vaccine.

fact that the science is very tough. Animal models used to test AIDS vaccines have severe limitations; the genetic diversity of HIV may require an effective vaccine to be based on many viral strains; and no researcher has successfully demonstrated which immune responses correlate with protection from HIV (Cohen, 1994b.)

Testing HIV Vaccines

Whether a vaccine will stimulate antibody production and is **safe** can be easily determined. It will not require a long time nor will it require many volunteers. But to gauge a vaccine's **efficacy** (effectiveness) will require a large number of volunteers who are free of HIV but in danger of becoming infected with it (members of behavioral high risk groups). To determine a vaccine's efficacy, two large groups of subjects are selected, one that receives the vaccine and one that receives a placebo, to see whether the vaccinated group has a lower rate of HIV infection. In carrying out efficacy trials, the number of participants, the length of the follow-up period, the rate of HIV infection, and the presumed efficacy of the vaccine are related to each other. Because the rate of HIV infection is low, even in "high-risk" populations, researchers estimate they will need to study several thousand participants to determine whether an HIV vaccine is effective. The original polio vaccine trial was completed in a year but required nearly half a million children.

An important variable in determining the parameters of the trial is whether the vaccine is prophylactic (preventative) or therapeutic. If the vaccine actually prevents infection, the trial will measure the rate at which participants develop HIV antibodies. If, however, the vaccine allows infection but prevents or greatly retards disease progression, the trial will measure the rate at which disease develops and will therefore require many more years of follow-up.

Morality of Testing HIV Vaccines

Designers of the vaccine trials are thus confronted with a paradox unique to HIV/AIDS. **If safer-sex and safe-needle practices are not taught,** volunteers who believe they are protected by the unproven vaccine (which may actually be a placebo) could take more risks and increase their chances of becoming infected. **If such practices are taught,** and by ethical standards, they must be, volunteers could cut their risk so effectively that they are never exposed to HIV leaving the vaccine with nothing to fight. **In a way, the trial depends on the failure of education.** To get around this potential conflict of interest, phase III trials of an AIDS vaccine will have to rely on populations in which the number of infections remains high regardless of education, like young high-risk gay men and injection-drug users. **There is also the additional problem of creating vaccine HIV-positive persons who can not be distinguished from those who are HIV-infected.** They, too, will be subjected to adverse social, employment, and other discrimination following a positive antibody test.

Where Will HIV Vaccines Most Likely Be Tested?

In **developed** countries, ethically, individuals in vaccine trials who are found to have acquired HIV infection must be offered **available** antiretroviral therapy, which usually reduces virus load. If vaccines cannot achieve **protection against infection,** treatment with antiretrovirals, which ethically they must be given, will compromise the ability of the trial to measure the efficacy of the vaccine in **preventing disease.** The drugs may also obscure possible secondary endpoints of vaccine efficacy, such as reduction in viral loads or immunological protection. Delaying the drug treatment until viral loads can be determined at several time points would present major ethical problems. **Because of these complications, determination of the protective efficacy of HIV vaccine candidates may only be possible in trials in developing countries where the resources are not available to provide antiretroviral drugs.** It is that circumstance, plus the fact that development of successful vaccines will be an incremental process requiring multiple trials, that presents the most challenging ethical issues (Bloom 1998).

HIV Vaccine Costs

Perhaps the most difficult moral question is the cost of the vaccine. A successful vaccine that sells for a high price will be of little use to poor and uninsured Americans and most people of developing nations, who have no more than a few dollars a year to spend on health care. Twelve years have passed since the discovery of a vaccine for hepatitis B, a viral disease that is also spread by sexual contact and the sharing of hypodermic needles. But the product has yet to reach many poor people in the United States and Third World countries largely because it costs about $120 for a series of three injections. Polio and measles vaccines are relatively cheap but are still not in universal use and they have been available for decades!

HUMAN HIV VACCINE TRIALS

The goal in producing HIV vaccines is to destroy HIV or keep it in check so that it causes no further damage. **The ideal vaccine will stop**

progressive immunodefeciency and restore the immune system. In order to determine if an HIV vaccine will meet these criteria it must ultimately be tested in human vaccine field trials.

The basic principle behind human vaccine trials has changed little since the 19th century. To test for an HIV/AIDS vaccine, several thousand people at high risk for the disease will be inoculated with the experimental agent, most likely an altered version of HIV or some portion of it. The vaccine should not be dangerous enough to cause the disease, but enough like HIV to confer immunity by triggering the production of antibodies and other virus-fighting components of the immune system. The subjects in the trial will be carefully monitored to see if they have a better rate of avoiding infection than others who were not vaccinated.

Beginning in 1999, some 40 trials using at least 25 different HIV vaccines are ongoing world wide. **Beginning year 2000, no vaccine has completed Phase III trials anywhere!**

The World Health Organization, in October 1994, recommended large-scale trials of HIV vaccines. The first candidate countries for the vaccine experiments are Uganda and Thailand. These trials began in 1996 but data from these trials will not be available for some 7 years. (HIV Vaccines Get Green Light for Third World Trials, 1994; Weniger, 1994; Cohen, 1995; Science, 1996). In February 1996 the National Institutes of AIDS and Infectious Diseases announced a joint phase II vaccine with Pasteur-Meriuex of France and Biocine Inc. of California. Early in 1997, Pasteur Merieux tested one version of the vaccine in about 100 patients. It uses the canary-pox virus to carry HIV genes into the body, where the genes produce proteins that can trigger an immune system alarm. The vaccine and a follow-up booster shot generated antibodies in all 100, a sign the immune system kicked into gear. **It also triggered the mobilization of killer T cells,** the immune system's attack force, in 40% of the people. The canary-pox vaccine is the second AIDS vaccine to make Phase II Testing in the United States. Smaller trials in Uganda and Thailand over the next few years. **The outcome is anybody's guess!**

In September 1997, the FDA approved a phase I **safety trial** of an AIDS vaccine, developed at the St. Jude Children's Research Hospital, that contains a cross section of 23 envelope proteins that have emerged worldwide. Although every viral envelope consists of protein called gp120 and gp41 (see Chapter 3), each envelope looks **different** to the immune system because the proteins vary, much like human faces differ from each other. These differences have been an obstacle to vaccine development. Just the safety evaluation of this vaccine is expected to take 1 to 2 years. The study began in October 1997 using fewer than 20 volunteers. Still waiting for the results.

On June 3, 1998 the FDA gave **VaxGen** permission to begin phase III clinical trials using their antiHIV—**AIDSVAX**—vaccine. Donald Francis, president of VaxGen (Figure 9-6) said "our primary goal at VaxGen is to develop a vaccine that can be used around the world and help put an end to an epidemic that already has caused untold suffering and taken more than 12 million lives." AIDSVAX is a preparation of recombinant gp120 (rgp120), the glycoprotein from HIV's envelope that binds to the surface of T4 cells. Injecting rgp120 into the body stimulates production of antibodies that, in any future exposure to HIV, would hopefully **prevent infection.** Initially, clinics in several cities, including Chicago, Denver, Los Angeles, Philadelphia and St. Louis began in June to inoculate volunteers with AIDSVAX. The Phase III trials will expand to an additional 10 clinics within three months, and to 50 clinics and 5000 volunteers (high risk women and gay men) in North America by early year 2000. The formulation of AIDSVAX to be used at these sites is designed to protect individuals from strains of HIV found in the Americas, Western Europe and Australia. The Phase III clinical trials of AIDSVAX in Thailand will be regulated by the Thai Health Ministry and will involve approximately 2,500 volunteers at 16 clinics. The formulation of AIDSVAX to be tested in Thailand is designed to protect against strains of HIV found in Thailand, Korea, Japan, Taiwan and Indonesia.

Vaccine Testing Confidentiality

Confidentiality must be maintained for the duration of the vaccine trials because people immunized with candidate AIDS vaccines who mount effective immune responses will appear

FIGURE 9-6 Donald Francis, Virologist and President of VaxGen, Inc. With the CDC in the early 1980's, Francis was one of the **first** people to sound the alarm about the AIDS epidemic and its possible spread through the United States' blood supply. He was instrumental in developing the newly approved **AIDSVAX**-HIV Vaccine. (*Photograph courtesy of VaxGen Inc., San Francisco, California.*)

proof identification card to help uninfected participants in vaccine trials. The seroconversion issue may play a major role in recruitment efforts and in the future welfare of vaccine trial participants (Koff, 1988).

So the question for anyone is: Should I enter an HIV vaccine trial when I know safeguards are limited? I know why I ought to do it...but! (Class discussion)

AIDS Vaccine Resources on the Internet

www.vaccineadvocates.org/avacsite/inde.htm

This address will list most of the important vaccine internet addresses and serve as linkage to others.

SUMMARY

The key to stopping HIV transmission lies with the behavior of the individual. That behavior, if the experience of the past 18 years can be used as an indicator, has proven to be very difficult to change.

Changing sexual behavior and using a condom is referred to as **safer sex.** The latex condom is the only condom believed to stop the passage of HIV; and a spermicide should be used with the condom. Oil-based lubricants must not be used because they weaken the condom, allowing it to leak or break under stress. Water-based lubricants are available and should be used. There is at least one female condom, called a vaginal pouch, approved by the FDA and is now sold worldwide. It is inserted like a diaphragm. It offers protection to both sexual partners.

Over 8000 cases of HIV infection have come from contaminated blood transfusions. A test developed in 1985 to screen all donated blood in the United States has reduced the risk of HIV transfusion infection. But blood bank screening has reduced the size of the blood donor pool. Many hospitals are encouraging people who know they might need an operation to donate their own blood for later use—autologous transfusion.

To date there is only one FDA-approved vaccine in Phase III trials, AIDSVAX: many top

positive for HIV antibody. They may be subject to the social stigma and discrimination associated with being truly HIV-positive.

Vaccine-induced seroconversion may lead to difficulties in donating blood, obtaining insurance, traveling internationally, or entering the military. Vaccine-induced antibodies may be long-lived, thus volunteers in AIDS vaccine trials must be given some form of documentation that certifies that their antibody status is due to vaccination and not HIV infection. The National Institute of Allergy and Infectious Disease has recently (1993) provided a tamper-

notch HIV/AIDS scientists have ruled out the use of an attenuated HIV vaccine before the year 2000. Inactivated whole virus vaccines are also being held back because there is no 100% guarantee that all HIV used in the vaccine will be inactivated.

Even if a vaccine does well in Phase III trials, will it be effective against all of the HIV mutants in the HIV gene pool? Can the threat of vaccine-induced enhancement of HIV infection be overcome? How are vaccine testing agencies going to handle the ethical question of vaccine seroconverting normal subjects to positive antibody status? The social repercussions may be devastating for those who, when tested, test HIV-positive even though they are HIV free.

There are some 5.3 million health care workers in the United States. It is crucial that they adhere to the Universal Protection Guidelines set down by the CDC, as a significant number of them are exposed to the AIDS virus annually. The risk of HIV infection after **exposure** to HIV-contaminated blood is about 1 in 200.

A few states have implemented HIV-partner notification; many other states are beginning to experiment with HIV-partner notification or contact tracing programs. It is too early to tell just how successful locating and testing high behavioral risk partners will be, or the cost-to-benefit ratio. One thing is certain; if these programs are to be successful, they will have to ensure confidentiality to those who are traced. Partner notification or contact tracing continues to work well for other sexually transmitted diseases.

REVIEW QUESTIONS

(Answers to the Review Questions are on page 487.)

1. Which is the better condom for protection from STDs, one made from lamb intestine or one made from latex rubber? Explain.

2. Which lubricant is best suited for condom use? Explain.

3. Briefly explain safer sex.

4. True or False: If a person has unprotected intercourse with an HIV-infected partner, he or she will become HIV-infected. Explain.

5. Yes or No: If injection drug users (IDUs) were given free equipment—no questions asked—would that stop the transmission of HIV among them? Explain.

6. What is the current risk of being transfused with HIV-contaminated blood in the United States?

7. What do you think should happen in cases where a person who knows he or she is HIV-positive, lies at a donor interview, and donates blood?

8. Why do most scientists wish to avoid using an attenuated HIV vaccine or an inactivated HIV vaccine?

9. What is the advantage of using recombinant HIV subunits in making a vaccine?

10. Explain vaccine-induced enhancement. How does it occur?

11. Why is it necessary to practice strict confidentiality with respect to volunteers for AIDS vaccine tests?

12. What are universal precautions? Who formulated them?

13. True or False: Research continues to show that AIDS prevention messages are effective in causing teens to change their sexual behaviors.

14. True or False: Latex condoms eliminate the risk of HIV transmission.

15. True or False: Partner notification is usually performed by the infected individual or a trained and authorized health department official.

16. True or False: The Centers for Disease Control and Prevention estimates that as many as 1 in 100,000 units of blood in the blood supply may be contaminated with HIV.

17. True or False: The three types of vaccines that scientists are interested in developing are preventive, therapeutic, and perinatal vaccines.

18. True or False: Used disposable needles should be recapped by hand before disposal.

19. True or False: Prompt washing of a needlestick injury with soap and water is sufficient to prevent HIV infection.

20. HIV/AIDS is not curable, but it is **preventable.** Write a short essay on the best methods of prevention.

21. True or False: The FDA approved the **first** vaccine for broad-scale testing in the United States in 1997.

REFERENCES

ABDALA, NADIA, et al. (1999). HIV-1 can survive in syringe for more than 4 weeks. *J. Acq. Imm. Def. Syndromes*, 20:73–80.

AMORTEGUI, A.J., et al. (1984). The effects of chemical intravaginal contraceptives and betadine on *Ureaplasma urealyticum. Contraception*, 30: 35–40.

ANDERSON, FRANK W.J. (1993). Condoms: A technical guide. *Female Patient*, 18:21–26.

ANDERSON, JOHN, et al. (1998). Needle hygiene and sources of needles for injection drug users: Data from a National Survey. *J. Acq. Imm. Def. Syndromes*. 18:S147

BARBER, HUGH R.K. (1990). Condoms (not diamonds) are a Girl's best friend. *Female Patient*, 15:14–16.

BAYER, RONALD et al. (1992). HIV Prevention and the two faces of partner notification. *Am. J. Public Health*, 82:1158–1164.

BENOWITZ, STEVEN. (1997). Politics polarizing issues in needle-exchange study. *The Scientist*, 11:1,6–7.

BIRD, KRISTINA D. (1991). The use of spermicide nonoxynol-9 in the prevention of HIV infection. *AIDS*, 5:791–796.

BLOOM, BARRY. (1998). The highest attainable standard ethical issues in AIDS vaccines. *Science*, 279:186–188.

BOLCH, O.H., JR, et al. (1973). In vitro effects of Emko on *Neisseria gonorrhoeae* and *Trichomonas vaginalis. Am. J. Obstet. Gynecol.*, 115:1145–1148.

BOLOGNESI, DANIEL, et al. (1998). Viral envelope fails to deliver? *Nature*, 391:638–639.

BURNETT, JOSEPH. (1995). Fundamental basic science of HIV. *Cutis*, 55:84.

BURRIS, SCOTT, et al. (1996). Legal strategies used in operating syringe-exchange programs in the United States. *Am. J. Public Health*, 86:1161–1166.

CALAROTA, SANDRA, et al. (1998). Cellular cytotoxic response induced by DNA vaccination in HIV infected patients. *Lancet*, 351:1320–1325.

CHRISTENSEN, DAMARIS. (1994a). A shot in time: The technology behind new vaccines. *Science News*, 145:344–345.

CLEMENT, MICHAEL J. (1993). HIV disease: Are we going anywhere? *Patient Care*, 27:13.

COHEN, JON. (1994a). Bumps on the vaccine road. *Science*, 265:1371–1372.

COHEN, JON. (1994b) Are researchers racing toward success, or crawling? *Science*, 265:1373–1374.

COHEN, JON. (1995). Thailand weighs AIDS vaccine tests. *Science*, 270: 904–907.

CONNELL, ELIZABETH B.(1989). Barrier contraceptives—their time has returned. *Female Patient*, 14:66–75.

Consumer Reports. (1989). Can you rely on condoms? 54:135–141.

DES JARLAIS, DON, et al. (1995). Maintaining low HIV seroprevalence in populations of IDUs. *JAMA*, 274:1226–1231.

DE VINCENZI, ISABELLE. (1994). A longitudinal study of HIV transmission by heterosexual partners. *N. Engl. J. Med.*, 331:341–347.

DRUCKER, LURIE, et al. (1997). An opportunity lost: HIV infection associated with lack of a national needle-exchange program in the USA. *Lancet*, 349:604–608.

ETZIONI, AMITAI. (1993). HIV sufferers have a responsibility. *TIME*, 142:100.

EZZELL, CAROL. (1987). Hospital workers have AIDS virus. *Nature*, 227:261.

FARR, GASTON, et al. (1994). Contraceptive efficacy and acceptability of the female condom. *Am. J. Public Health*, 84:1960–1964.

FARTHING, CHARLES. (1998). SIV vaccine for AIDS. *Science*, 279:10.

FENTON, KEVIN, et al. (1997). HIV partner notification: Taking a new look. *AIDS*, 11:1535–1546.

FINDLAY, STEVEN. (1991). AIDS: The second decade. *U.S. News World Rep.*, 110:20–22.

FISHER, ALEXANDER A. (1991). Condom conundrums: Part 1. *Cutis*, 48:359–360.

FISHER, PETER. (1990). A report from the underground. *International Working Group on AIDS and IV Drug Use*, 5:15–17.

FRANCIS, DONALD P., et al. (1987). The prevention of acquired immunodeficiency syndrome in the United States. *JAMA*, 257:1357–1366.

FREZIERES, RON, et al. (1999). Evaluation of the efficacy of a polyurethane condom: Results from a randomized, controlled clinical trial. *Family Planning Perspectives*, 31: 81–87.

GERBERDING, JULIE LOUISE. (1991). Reducing occupational risk of HIV infection. *Hosp. Pract.*, 26:103–118.

GOSTIN, LAWRENCE, et al. (1998). HIV infection and AIDS in the public health and health care systems: The role of law and litigation. *JAMA*, 279:1108–1113.

GOSTIN, LAWRENCE, et al. (1997) Prevention of HIV/AIDS and other blood-born diseases amoung injection drug users. *JAMA*, 277:53–62.

GRIMES, DAVID A.(1992). Contraception and the STD epidemic: Contraceptive methods for disease prevention. *The Contraception Report: The Role of Contraceptives in the Prevention of Sexually Transmitted Diseases*, III:1–15.

GUTTMACHER, SALLY, et al. (1997). Condom availability in New York City public high schools: Relationships to condom use and sexual behavior. *Am. J. Public Health,* 87:1427–1433.

HAGEN, HOLLY. (1991). Studies support syringe exchange. *FOCUS,* 6:5–6.

HART, GRAHAM J., et al. (1991). Prevalence of HIV, hepatitis B, and associated risk behaviors in clients of a needle exchange in central London. *AIDS,* 5:543–547.

HEARST, NORMAN, et al. (1995). Collaborative AIDS prevention research in the developing world: The CAPS experience. *AIDS,* 9 (suppl 1):51–55.

HICKS, D.R., et al. (1985). Inactivation of HTLV/LAV-infected cultures of normal human lymphocytes by nonoxynol-9 in vitro. *Lancet,* 2:1422–1423.

HIV vaccines get the green light for third world trials. (1994). *Nature,* 371:644.

HOMSY, JACQUES, et al. (1989). The Fe and not CD4 receptor mediates antibody enhancement of HIV infection in human cells. *Science,* 244: 1357–1359.

HOOKER, TRACEY. (1996). HIV/AIDS: Facts to consider, 1996. *National Conference of State Legislature.* Denver, Co., February, pp. 1–64.

JUDSON, FRANKLYN N. (1989). Condoms and spermicides for the prevention of sexually transmitted diseases. *Sexually Transmitted Dis. Bull.,* 9:3–11.

KAPLAN, EDWARD, et al. (1993). Let the needles do the talking! Evaluating the New Haven needle exchange. *Interfaces,* 23:7–26.

KIRBY, DOUGLAS, et al. (1998). The impact of condom distribution in Seattle schools on sexual behavior and condom use. *Am. J. Public Health,* 89:182–187

KOFF, WAYNE C. (1988). Development and testing of AIDS vaccines. *Science,* 241:426–432.

LACKRITZ, EVE, et al. (1995). Estimated risk of transmission of HIV by screened blood in the United States. *N. Engl, J. Med.,* 333:1721–1725.

LAURIE, PETER, et al. (1998). A sterile syringe for every drug user injection: How many injections take place annually and how might pharmacists contribute to syringe distribution. *J. Acquired Immune Deficiency Syndromes and Human Retrovirology,* 18:545–551.

LAURIE, PETER, et al. (1994). Ethical behavioral and social aspects of HIV vaccine trials in developing countries. *JAMA,* 271:295–302.

LEVY, JAY A. (1989). Human immunodeficiency viruses and the pathogenesis of AIDS. *JAMA,* 261:2997–3006.

LEWIS, DAVID. (1995). Resistance of microorganisms to disinfection in dental and medical devices *Nature Med.,* 1:956–958.

LICHTENSTEIN, BRONWEN. (1996).Needle exchange programs: New Zealand's experience. *Am. J. Public Health,* 86:1319.

MANDELL, WALLACE, et al. (1994). Correlates of needle sharing among injection drug users. *Am. J. Public Health,* 84:920–923.

MARWICK, CHARLES. (1995). Released report says needle exchanges work. *Med., News Perspect.,* 273:980–981.

Morbidity and Mortality Weekly Report. (1988). Partner notification for preventing human immunodeficiency virus (HIV) infection—Colorado, Idaho, South Carolina, Virginia. 37:393–396; 401–402.

Morbidity and Mortality Weekly Report. (1989). Guideline for prevention of transmission of HIV and hepatitis B virus to health care workers. 38:3–17.

Morbidity and Mortality Weekly Report. (1992). Sexual behavior among high school students—United States, 1990. 40:885–888.

Morbidity and Mortality Weekly Report. (1993a). Update: Barrier protection against HIV infection and other sexually transmitted diseases. 42:589–591.

Morbidity and Mortality Weekly Report. (1993). Impact of new legislation on needle and syringe purchases and possession—Connecticut 1992. 42:145–147.

Morbidity and Mortality Weekly Report. (1995). Notification of syringe-sharing and sex partners of HIV-infected persons—Pennsylvania, 1993–1994. 44:202–204.

Morbidity and Mortality Weekly Report. (1996). School-based HIV-prevention education-United States, 1994. 45:760–764.

Morbidity and Mortality Weekly Report. (1997). Update: Syringe-exchange programs—United States, 1996. 46:565–568.

Office Nurse. (1995). Contraception: how today's options stack up. 8:13–14.

PARRAN, THOMAS P. (1937). *Shadow on the Land: Syphilis.* New York: Reynal and Hitchcock.

PERRONE, JANICE. (1989). U.S. cities launch new AIDS weapon. *Am. Med. News,* March:68–71.

PINES, MAYA. (1993). Blood: Bearer of life and death. *Howard Hughes Med. Inst.,* 6:17.

POTTS, MALCOLM. (1994). Urgent need for a vaginal microbicide in the prevention of HIV transmission. *Am. J. Public Health,* 84:890–891.

RAYMOND, CHRIS ANNE. (1988). U.S. cities struggle to implement needle exchanges despite apparent success in European cities. *JAMA,* 260:2620–2621.

RODDY, RONALD, et al. (1998). A controlled trial of nonoxynol 9 film to reduce male-to-female transmission of STDs. *N. Engl. J. Med.,* 339:504–510.

ROPER, WILLIAM L., et al. (1993). Commentary: Condoms and HIV/STD prevention—clarifying the message. *Am. J. Public Health,* 83:501–503.

ROSENBERG, MICHAEL J., et al. (1992). Commentary: Methods women can use that may prevent STDs including HIV. *Am. J. Public Health,* 82: 1473–1478.

ROTHENBERG, KAREN, et al. (1995). The risk of domestic violence and women with HIV infection: Implications for partner notification, public policy, and the law. *Am. J. Public Health,* 85: 1569–1576.

RUTHERFORD, GEORGE W. (1988). Contact tracing and the control of human immunodeficiency virus infection. *JAMA,* 259:3609–3670.

SATTAR, SYED A., et al. (1991). Survival and disinfectant inactivation of HIV: A critical review. *Rev. of Infect. Dis.,* 13:430–447.

Science. (1996). Uganda may host AIDS vaccine trial. 272:657.

SELLERS, DEBORAH, et al. (1994). Does the promotion and distribution of condoms increase teen sexual activity? Evidence from an HIV prevention program for Latino youth. *Am. J. Public Health,* 84:1952–1958.

SINGH, B., et al. (1982). Demonstration of a spirocheticidal effect by chemical contraceptives on *Treponema pallidum. Bull. Pan. Am. Health Organ.,* 16:59–64.

SPRUYT, ALAN, et al. (1998). Identifying condom users at risk for breakage and slippage; Findings from three International Sites. *Am. J. Public Health,* 88:239–240.

STIMSON, GERRY V., et al. (1989). Syringe exchange. *International Working Group on AIDS and IV Drug Use,* 4:15.

STOECKLE, MARK, et al. (1996). Infectious diseases. *JAMA,* 275:1816–1817

STONE, KATHERINE M., et al. (1986). Personal protection against sexually transmitted diseases. *Am. J. Obstet. Gynecol.,* 155:180–88.

STRAUS, STEPHEN, et al. (1994). Placebo-controlled trial of vaccination with recombinant glycoprotein D of herpes simplex virus type 2 for immunotherapy of genital herpes, *Lancet,* 343:1460.

STRATTON, P., et al. (1993). Prevention of sexually transmitted infections: Physical and Chemical barrier methods. *Infect. Dis. Clin. North Am.,* 7(4):841–859.

THOMPSON, DICK. (1992). Getting the point in New Haven. *Time,* 139:55–56.

TIERNO, PHILIP M. (1988). AIDS overview: New guidelines for handling specimens. *Am. J. Cont. Ed. Nurs.,* Special Issue:1–14.

Time. (1992). Closed: Needle Park. 139:53.

U.S. Public Health Service. (1994). Counseling to prevent unintended pregnancy. *Am. Fam. Phys.,* 50(5):971.

VOELKER, REBECCA. (1995). Lessons from cat virus. *JAMA,* 273:910.

WEINSTEIN, STEPHEN P., et al. (1990). AIDS and cocaine: A deadly combination facing the primary care physician. *J. Fam. Prac.,* 31:253– 254.

WELLER, SUSAN. (1993). A meta-analysis of condom effectiveness in reducing sexually transmitted HIV. *Soc. Sci. Med.* 36:1635–1644.

WENIGER, BRUCE C. (1994). Experience from HIV incedence cohorts in Thailand: Implications for HIV vaccine efficacy trials. *AIDS,* 8:1007– 1010.

WHITE, NICK, et al. (1988). Dangers of lubricants used with condoms. *Nature,* 335:19.

WITTKOWSKI, KNUT, et al. (1998). The protective effect of condoms and nonoxynol-9 against HIV infections. *Am. J. Public Health,* 88:590–594.

WODAK, ALEX. (1990). Australia smashes international needle and syringe exchange record. *International Working Group on AIDS and IV Drug Use,* 5:28–29.

Prevalence of HIV Infections, AIDS Cases, and Deaths Among Select Groups in the United States

CHAPTER CONCEPTS

♦ **AIDS is a new plague.**

♦ Worldwide, fifty percent of new HIV infections are in people under age 25.

♦ Worldwide, about 16,000 new HIV infections occure daily.

♦ Worldwide, women represent 43% of all HIV infection.

♦ AIDS is now the world's **leading cause** of death by an infectious disease.

♦ AIDS is now **randked fourth** in causes of death worldwide.

♦ Men make up 80% of all AIDS cases; 20% are women.

♦ The majority of people with AIDS can be associated with certain **lifestyle risks.**

♦ AIDS can be associated with single or multiple exposure risks.

♦ AIDS cases can be separated by sex, age group, race, ethnicity, and sexual preference.

♦ Risk is strongly tied to **social behavior.**

♦ At risk groups include homosexual and bisexual men, injection drug users (IDUs), hemophiliacs, transfusion patients, and the sex partners of these people.

♦ HIV infection is strongly associated with injection drug use.

♦ All military personnel are tested for HIV.

♦ Two per 1,000 college students are HIV-infected.

♦ **Thirty four states report HIV-infected by name.**

♦ High rates of HIV infection have been found among prisoners.

♦ The greatest HIV threat to health care workers is needlestick (syringe) injuries.

♦ All 50 states and U.S. territories must report all AIDS cases.

♦ January 2000 **reported** AIDS cases in the United States reaches 746,000.

♦ A case can be made for national HIV reporting.

♦ People do not always tell the truth when completing questionnaires, especially with regard to sexual behavior.

♦ Ending year 2000 over 60 million people worldwide have been HIV infected.

As we leave the year 2000 and the end of the second decade of the HIV/AIDS pandemic, there does not appear to be an immediate end to either the spread of HIV infection or the devastation caused by AIDS. Looking back over the 1980s and the 1990s, it is clear that scientists and the public have consistently **underestimated** the magnitude and the potential of the HIV/AIDS pandemic. Who would have imagined in the mid-1980s that HIV would eventually spread to every country of the world, infecting over 60 million people and resulting in about 18 million deaths ending year 2000. The numbers become so large that the individual suffering and the personal, societal and economic losses become impossible to measure or to even attempt to estimate. And while there have been successes in slowing the epidemic in some communities and dramatic advances in survival in developed countries due to combination antiretroviral therapy, it is very important that people do not become complacent and focus on false beliefs that the pandemic is declining, that individuals are becoming less infectious, or that HIV control will be much better in the 21st century. If anything, the limited successes in prevention should encourage a continued effort to work harder at educating more people about how best to prevent further transmission through safer sex practices, antiretroviral therapy during pregnancy, treatment of STDs, provision of condoms, screening of the blood supply, the use of sterile needles, or many of the other avenues that can help to slow the pandemic until a vaccine can be developed.

Currently, UNAIDS reports that each day about 16,000 people become newly infected with HIV, or 11 men, women and children per minute. Ten percent of the newly infected people are under age 15. Over 50% of new infections are now occurring in people between ages 15 and 24, primarily due to sexual transmission. Women now represent 43% of all people **over 15** living with HIV infection, and there are no indications that this equalizing trend will reverse. Ending in 1998, UNAIDS reported that AIDS had become the **world's most deadly infectious disease.** It reached this level of human devastation in about 19 years (19981–1999). Of all causes of death worldwide, AIDS has moved up to **4th place.** AIDS was associated with about 2.5 million deaths in 1997, 1998, and in 1999. All of these data are a bit ironic because as Peter Piot, Executive Director of UNAIDS said, "the pandemic is out of control at the very time when we know what to do to prevent its spread."

HIV Infectious: Overwhelming Numbers in the United States

It is easy to be overwhelmed by statistics in reporting on HIV infections and AIDS cases and to lose track of the human faces of the epidemic. But certain numbers, like the **first half million documented AIDS cases reported in October 1995** and over 425,000 dead by beginning year 2000, take on a compelling quality of their own. Therefore within this chapter there are many statistics presented on all facets of the HIV/AIDS pandemic. **After reading this chapter one will have gained a deeper insight into the spread of HIV and those who are affected by HIV: those who have it, and those who don't—those who will suffer and those who won't. But in reality we are all impacted by this disease in one way or another!**

Prevalence/Incidence: How Many People Are HIV Positive?

The **prevalence** of a disease refers to the percentage of a population that is affected by it at a given time. For example, the total number of AIDS cases or number of HIV infections reported in the United States, by your state or city to the CDC, say for the year 1999. The **incidence** means the number of times an event occurs in a given time frame, for example, the number of new AIDS cases each month or new HIV infections each week (events that occur within a specified period of time). The two terms are similar. Much has been learned about the prevalence of HIV infection, HIV disease, and its terminal stage called AIDS since the 1981 CDC report that awakened the world to this new pandemic.

Although cases of AIDS appear retrospectively to have occurred in the United States as early as 1952, the **AIDS pandemic** in the United States is considered to have begun with the initial report in June 1981. Since then, there has been a rapid rise in reported AIDS cases (Table 10-1). The accelerated rise in AIDS cases means that there must be a reservoir of asymptomatic HIV-infected people who **with time** progress to AIDS.

TABLE 10-1 AIDS Cases a Percentage of Total Population: United States, 1982–2000

Year	Total Population[a]	Number of AIDS Cases[b]	Annual Percent with AIDS
1982	231,534	1,500	<0.0005
1983	233,981	2,760	<0.001
1984	236,158	4,445	0.001
1985	238,740	8,249	0.003
1986	241,078	12,932	0.005
1987	243,400	21,070	0.009
1988	245,807	31,001	0.013
1989	248,239	33,722	0.014
1990	248,710	44,755	0.018
1991	252,177	53,176	0.020
1992	254,462	49,106	0.019
1993	257,301	106,618	0.041
1994	260,507	80,901	0.031
1995	262,605	74,180	0.028
1996	265,513	69,151	0.026
1997	266,473	61,300	0.023
1998	267,521	54,937	0.021
1999[c]	269,500	50,496	0.019
2000	274,479	48,000	0.017

[a]United States total resident population × 1000.

[b]Number of AIDS cases reported, by year, to the Centers for Disease Control and Prevention, Atlanta, Georgia, USA. **AIDS cases are underreported by 10% to 20%.**

[c]1999/2000 data are estimated based on reports adapted from HIV/AIDS Surveillance Report, December 1998.

If the accumulative world AIDS deaths had occurred in a single year, this pandemic would top the list of the worst natural disasters of the 20th century. Natural disasters are often analyzed in terms of global vulnerability to events with a rapid onset such as tropical storms, earthquakes, and volcanic eruptions. The slower onset of the AIDS pandemic, with about 440,000 deaths from early 1981 early year 2000, sets it apart from natural disasters; but in terms of cost and human suffering, it is no different. If only the 35,000 to 45,000 deaths per year from 1991 through 1996 are used as reference points, the yearly numbers are still greater than the number of deaths caused by most of the natural disasters recorded in this century.

In Summary—The HIV/AIDS pandemic is the most serious pandemic to occur world-wide since the **Spanish flu** of 1918, which killed between 20 million and 50 million people but lasted less than a year!

OLD FORMULA FOR ESTIMATING HIV INFECTIONS

The estimated prevalence of HIV infection in the United States is an important measure of the extent of the nation's HIV disease/AIDS problem. Initial estimates on the number of HIV-infected people were based on the CDC formula: for every diagnosed case of AIDS, there are 50 to 100 people who are HIV-infected. The estimate was crude, but data essential for greater specificity were lacking. This was a new plague and no one was prepared for it. A workable definition for what constituted an AIDS patient had to be agreed upon. Surveillance networks also had to be set up to gather information on areas of highest incidence, who was being infected, and routes of transmission.

NEWER FORMULA FOR ESTIMATING HIV INFECTIONS

A newer formula proposed by the CDC for use in determining the number of HIV-infected persons is as follows:

National Number of Persons Living With AIDS (1997)   Number of PLWA in Your City (1997)

$$\frac{258,000}{900,000} \times \frac{(e.g.,)\,1,000}{x} = 258,000$$

(Estimated national number of HIV – infected persons

$$x = 900,000,000$$
$$x = \frac{900,000,000}{258,000}$$

$$= 3,488$$

(About 3,488 persons in this sample city are HIV-infected.)

Single or Multiple Exposure Categories

In Table 10-2, the number of AIDS cases estimated through 1999 are presented with respect to adult/adolescent, **single** or **multiple** exposure categories. For example, under *Single Mode of Exposure*, heterosexual contact accounts for 10% of all AIDS cases. Under *Multiple Modes of Exposure*, 5% of AIDS cases occurred among injection drug users who

TABLE 10-2 Adult/Adolescent AIDS Cases by Single and Multiple Exposure Categories, Estimated through 1999, United States

	AIDS Cases	
Exposure Category	No.	(%)
Single Mode of Exposure		
1. Men who have sex with men	339,098	(46)
2. Injection (IJ) drug use	151,120	(20.5)
3. Hemophilia/coagulation disorder	3,686	(.5)
4. Heterosexual contact	73,717	(10)
5. Receipt of blood transfusion	8,369	(1)
6. Receipt of transplant of tissues/organs	14	(0)
7. Other/undetermined	147	(0)
Single Mode of Exposure Subtotal	575,000	(78)
Multiple Modes of Exposure		
1. Men who have sex with men; IJ drug use	42,230	(6)
2. Men who have sex with men; hemophilia	168	(0)
3. Men who have sex with men; heterosexual contact	10,000	(1)
4. Men who have sex with men; receipt of transfusion/transplant	3,686	(.5)
5. IJ drug use; hemophilia	195	(0)
6. IJ drug use; heterosexual contact	34,858	(5)
7. IJ drug use; receipt of transfusion	1,650	(0)
8. Hemophilia; heterosexual contact	105	(0)
9. Hemophilia; receipt of transfusion/transplant	810	(0)
10. Heterosexual contact; receipt of transfusion/transplant	1,650	(0)
11. Men who have sex with men; IJ drug use; hemophilia	50	(0)
12. Men who have sex with men; IJ drug use; heterosexual contact	5,600	(68)
13. Men who have sex with men; IJ drug use; receipt of transfusion/transplant	610	(0)
14. Men who have sex with men; hemophilia; heterosexual contact	22	(0)
15. Men who have sex with men; hemophilia; receipt of transfusion/transplant	38	(0)
16. Men who have sex with men; heterosexual contact; receipt of transfusion/transplant	270	(0)
17. IJ drug use; hemophilia; heterosexual contact	82	(0)
18. IJ drug use; hemophilia; receipt of transfusion/transplant	38	(0)
19. IJ drug use; heterosexual contact; receipt of tranfusion/transplant	1,110	(0)
20. Hemophilia; heterosexual contact; receipt of transfusion/transplant	37	(0)
21. Men who have sex with men; IJ drug use; hemophilia; heterosexual contact	13	(0)
22. Men who have sex with men; IJ drug use; hemophilia; receipt of transfusion/transplant	14	(0)
23. Men who have sex with men; IJ drug use; heterosexual contact; receipt of transfusion/transplant	172	(0)
24. Men who have sex with men; hemophilia; heterosexual contact; receipt of transfusion/transplant	5	(0)
25. IJ drug use; hemophilia; heterosexual contact; receipt of transfusion/transplant	23	(0)
26. Men who have sex with men; IJ drug use; hemophilia; heterosexual contact; receipt of transfusion/transplant	7	(0)
Multiple Modes of Exposure Subtotal	103,203	(14)
Risk Not Reported or Identified	58,974	(8)
Total AIDS Cases	737,169	(100)

(Source: For exposure categories. CDC HIV/AIDS Surveillance Report, through 1998, 9:1–44, updated)

———— BOX 10.1 ————

———— BOX 10.2 ————

THE HIV/AIDS SCENARIO

HIV/AIDS is unstoppable in the short term. Because it takes an HIV infection so long to develop into AIDS, virtually all of the AIDS cases that occur during the next 5 to 15 years will be the result of existing infections. Therefore, the epidemic **cannot** be materially reduced in this time frame by any reduction in **new** HIV cases. Worldwide, millions of HIV infections will progress into AIDS into the next millenium. Is it your impression that some of your friends and associates *still* regard HIV/AIDS as someone else's problem? HIV/AIDS had invaded *all* segments of society worldwide! It is EVERYONE'S problem!

THE HIV/AIDS PANDEMIC HAS LIMITS

Epidemics or pandemics typically reach a point of saturation whereby incidence levels off under 100% of the population. This happens because some people either are naturally immune or avoid exposure to the disease. Thus AIDS will not wipe out entire populations, but the point of saturation for HIV (number of susceptible persons infected) probably varies substantially from population to population and cannot be predicted with any precision.

also had heterosexual contact. In other words, Table 10-2 lists the numbers and percentages of all reported AIDS cases broken down into seven categories of people who contracted HIV/AIDS from a single risk mode of exposure and 26 categories of people who contracted HIV/AIDS from multiple risk modes of exposure. Note that of the total number of adult/adolescent AIDS cases, 78% occurred from **single risk** modes of exposure. Of this 78%, 46% occurred among men who had sex with men and about 21% among injection drug users.

A composite representation of all AIDS cases by exposure category estimated beginning 1999 is shown in Figure 8-1.

BEHAVIORAL RISK GROUPS AND STATISTICAL EVALUATION

Behavioral Risk Groups and AIDS Cases

As the pool of AIDS patients grew in number during 1981–1983, individual case histories were separated into **behavioral risk groups.** The early case histories of AIDS patients clearly separated people according to their social behavior and medical needs. AIDS patients were placed into the following six risk behavior categories: (1) homosexual and bisexual men; (2) injection drug users; (3) hemophiliacs; (4) blood transfusion recipients; (5) heterosexuals; and (6) children whose parents are at risk. Each of these groups is considered to be at risk of HIV infection based on

some common behavioral denominator. That is, those within these groups represented a higher rate of AIDS cases than people whose needs or behaviors excluded them from these groups. However, because there is some mixing between individuals in behavioral risk groups, HIV infection has gradually spread to lower-risk behavioral groups. Over time the behavioral risk groups have been aligned and defined according to age, exposure category, and sex (see Table 10-2).

A review of AIDS cases by sex/age at diagnosis, and race/ethnicity reported through December 1998 in the United States, shows that **white, black, and Hispanic males between the ages of 20 and 44 make up 79% of all male AIDS cases. Between ages 20 and 59, they make up 97% of all male AIDS cases.** People between the ages of 25 and 44 make up over half of the Nation's 126 million workers! In 1998, in the United States, one in six worksites with over 50 employees and one in 16 small businesses have/had an employee with HIV/AIDS. Of the total number of male AIDS cases, 49% are white, 33% are black, and 18% are Hispanic. Figure 10-1 shows the distribution of **male** adult/adolescent AIDS cases estimated for 1998 in the United States.

Statistical Evaluation of Selected Risk Behavioral Group AIDS Cases

Adult/Adolescent AIDS Cases— October 31, 1995 is the time in history when the United States reached **a half million** (501,310) reported AIDS cases (*MMWR*, 1995a). Through 1998, 696,000 AIDS cases and 412,000 AIDS-related

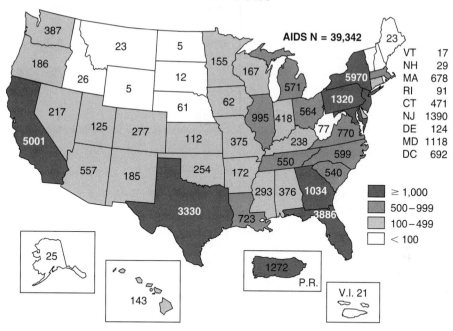

Adolescent/Adult Men AIDS Cases Reported in 1998
United States

AIDS N = 39,342

VT	17
NH	29
MA	678
RI	91
CT	471
NJ	1390
DE	124
MD	1118
DC	692

≥ 1,000
500 – 999
100 – 499
< 100

P.R.

V.I. 21

FIGURE 10-1 One Year of New AIDS Cases, Male Adults/Adolescents Across the United States, Reported for 1998. (Total adult/adolescent AIDS cases for 1998, 54,915) (*Source: Courtesy of CDC, Atlanta*)

deaths were reported to the CDC (Figure 10-2). Sixty-six percent of all AIDS cases reported occurred between 1993 through 1999. Cumulative through 1998, 10% of AIDS cases have occurred among the heterosexual population, 26% occurred among injection drug users, and 54% occurred in the male homosexual/bisexual IDU population. For 1996, the first time since the United States AIDS pandemic began, more blacks were diagnosed with AIDS (41%) then whites (38%). Table 10-3 presents the total estimated number of AIDS cases for 1998 and their distribution based on race, sex, and exposure group.

Figure 10-3 shows that the percentage of AIDS cases for ethnic related adult/adolescent groups are in striking contrast to the population percentages of each group. For example, in 1999 whites made up 72% of the population and represented 44% of adult/adolescent AIDS cases. Blacks made up 13% of the population but represented 37% of the adult/adolescent AIDS cases. Hispanics made up about 11% of the population but represented 18% of the adult/adolescent AIDS cases.

Collectively, from 1997 through 1999, blacks and Hispanics made up 22% of the population but make up 66% of adult/adolescent and 85% of pediatric AIDS cases **reported.** Thus, blacks and Hispanics contribute disproportionately high percentages to the total number of reported AIDS cases.

In 1995, 1 in 350 black men in the United States aged 25–44 years was diagnosed with AIDS, which is about **five times higher** than the annual incidence rate among white men (1 in 1,800). For black women in the same age group this rate was 1 in 1,100, for Hispanic women 1 in 2,500, whereas among white women this was 1 in 15,000 (Coutinho et al., 1996).

According to 1993 data analyzed by Philip Rosenberg (1995) of the National Cancer

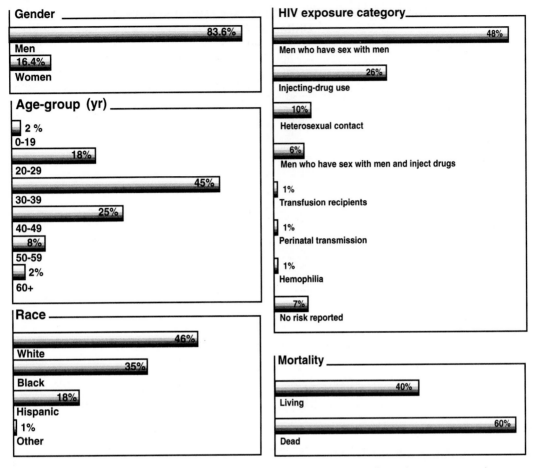

FIGURE 10-2 A Profile of the First 746,000 AIDS Cases: June 1981 Through 1999—United States. Four hundred and twentyseven thousand people died of AIDS to this point in time. Data may not add to 100% due to rounding off. (*Source: MMWR, 1995b, updated*)

Institute, 1 in every 92 American men between **ages 27 and 39 may be HIV-infected.** The findings were especially dismal for African-American men, with 1 in 33 estimated to be HIV infected. The estimate was 1 in 60 for Hispanic men. And although women were more than four times **less likely** to be infected with HIV, the statistics were equally high for women of color. One in 98 African-American women and 1 in 222 Hispanic women are estimated to be infected with HIV. By comparison, the number of white women infected with HIV was 1 in 1,667. If the trends continue, Rosenberg noted, HIV/AIDS in young people

and minorities must be considered "endemic in the United States."

The Harvard AIDS Institute reported in October 1996 that by the year 2000, more than half of all AIDS cases in the United States will be among blacks. By the same year, a black person will be nine times more likely to be diagnosed with AIDS than someone who is not black. The Institute estimated that nearly 100 blacks and Hispanic Americans are diagnosed with AIDS every day (1 every 14 minutes) in the United States.

Figure 10-4 is a United States map of estimated AIDS cases through the year 1999. Note

HIV/AIDS	No. of Cases[a]	% of Cases
Exposure Group		
Homosexual/bisexual male	22,725	45
Injection drug user (IDU)	10,605	21
Homosexual/bisexual IDU	2,525	5
Hemophiliac	50	.1
Heterosexual contact	3,535	7
Transfusion related	50	.1
None of the above	11,110	22
Total	50,500	100
Race/Ethnicity (all cases)		
White (non-Hispanic)	19,228	38
Black (non-Hispanic)	20,200	40
Hispanic	11,110	22
Other	505	1
Sex (adults only)		
Male	39,390	78
Female	1,110	22
Age Group (yrs)		
13–19	253	0.5
20–24	1,818	3.6
25–29	7,171	14.2
30–39	23,129	45.8
40–49	13,130	26.0
50–59	3,686	7.3
60 and above	1,313	2.6

[a]Estimated total in each category = 50,500 adult/adolescent plus 350 pediatric = 50,850 = 100% of cases for 1999.

(Adapted from AIDS Surveillance Report, December 1998 Updated)

that the highest incidence of AIDS cases occurs along the coastal regions. Figure 10-5 is a map showing the estimated number of people **living with AIDS by state.**

Behavioral Risk Groups and Percentages of HIV-Infected People

Beginning 1999, investigators at the CDC found that HIV infection remained largely confined to the populations at recognized behavioral risks: homosexual men, injection drug users, heterosexual partners of injection drug users, hemophiliacs, and children of HIV-infected mothers. In the general population, rates for HIV infection include 0.04% for first-time blood donors, 0.14% for military applicants, 0.33% for Job Corps entrants, 0.19% to 0.87% for child-bearing women, and 0.30% for hospital patients. Data reported by the CDC in 1990 indicated that while the number of new AIDS cases increased by 5% in cities, it had increased by 37% in rural areas. This trend continues.

Comments on a Variety of Individual Behavioral Risk Groups

It must be kept in mind that because a group of people is at risk for HIV does not mean that these people are predestined to become infected. People are placed within these groups because of their social behavior, a behavior that has been associated with a high, medium- or low-risk of becoming HIV-infected. Essentially there is no zero-risk group because a scenario can always be formulated to show that under certain circumstances one or more members of that group could become HIV-infected.

The point of placing people in behavioral risk groups is not to offend them but to provide a warning that certain behavior might make them more vulnerable to HIV infection. It is not the race or ethnic group that people belong to that places them at high or low risk for infection, it is their behavior.

The fact that AIDS was first identified in 1981 in seemingly well-defined behavioral groups (homosexual men, injection drug users, hemophiliacs, Haitian immigrants) probably contributed to a false sense of security among people who did not belong to any of these groups. However, as information about HIV and AIDS accumulated, it became clear that HIV was transmitted in body fluids. This had grave implications for all social groups. On reflection, the people in the original high-risk groups simply had the bad luck of being in the way of a newly emerging infectious agent as it first began to spread. It is highly probable that in the different behavioral risk groups there are lifestyle or medical history factors that increase the efficiency with which the virus is transmitted.

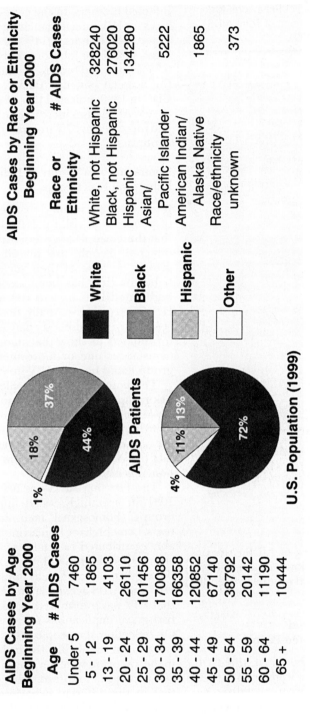

AIDS Cases by Age
Beginning Year 2000

Age	# AIDS Cases
Under 5	7460
5 - 12	1865
13 - 19	4103
20 - 24	26110
25 - 29	101456
30 - 34	170088
35 - 39	166358
40 - 44	120852
45 - 49	67140
50 - 54	38792
55 - 59	20142
60 - 64	11190
65 +	10444

AIDS Cases by Race or Ethnicity
Beginning Year 2000

Race or Ethnicity	# AIDS Cases
White, not Hispanic	328240
Black, not Hispanic	276020
Hispanic	134280
Asian/ Pacific Islander	5222
American Indian/ Alaska Native	1865
Race/ethnicity unknown	373

White ■
Black ▨
Hispanic ▨
Other □

AIDS Patients

44% 37% 18% 1%

U.S. Population (1999)

72% 13% 11% 4%

FIGURE 10-3 Aids cases by Age and Racial and Ethnic Classification. Adult AIDS cases show a disproportionate percentage among blacks and Hispanics. Fifty-five percent of reported AIDS cases occur among racial and ethnic minorities. The figures reflect higher rates of AIDS in blacks and Hispanic injection drug users and their sex partners. Percentages of the population are based on the numbers of AIDS cases in the United States estimated beginning year 2000. U.S. population data about 274 million.

FIGURE 10-4 AIDS Cases that Occurred Beginning Year 2000 and Their Approximate Location.

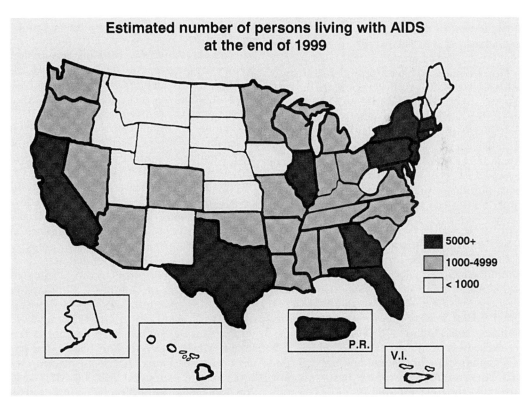

FIGURE 10-5 People Living With AIDS. Beginning the year 2000, about 313,000 men, women and children were living with HIV/AIDS in America and its territories.

Men Who Have Sex with Men (MSM)— In 1981, 100% of all AIDS cases reported to the CDC occurred in homosexual males. For 1993, 46% of the 106,618 new AIDS cases occurred in gay males: It dropped to 35% by 1997 and remained at 35% through 1998 (*MMWR*, 1995a; HIV/AIDS surveillance report, 1997, 1998). It appears that in the 1990s, the number of gay male AIDS cases, as a percentage of all AIDS cases, is decreasing. However, the *Morbidity and Mortality Weekly Report* (1995b) documents the disproportionate occurrence of AIDS among black and Hispanic MSM compared with white MSM from 1981 through 1995. This finding underscores the need for community planning groups to consider culturally appropriate prevention services when addressing the HIV-prevention needs of racial/ethnic minorities. Also, there are indicators of a second wave in the HIV/AIDS epidemic among gay males.

Studies show that a significant percentage of gay men who had originally adopted safer sex behaviors are **"relapsing"** to unsafe sex (Stall et al., 1990; Lemp et al., 1994).

Most people who have attempted to change behavior know, it is difficult to maintain new behavior patterns over a long period of time. Increasing attention and concern has been focused on the issue of **relapse.** High seroprevalence levels among homosexual men (estimated to be 50% in high incidence cities such as San Francisco) make relapse to risky behaviors a significant source of new infections and justify continued interventions to reinforce behavior change.

SIZE OF THE HOMOSEXUAL COMMUNITY. Estimating the size of the homosexual population continues to be a problem for the CDC. Because of the lack of available information on sexual practices, the CDC has relied on the 1948 Kinsey report, *Sexual Behavior in the Human Male,* for its estimate that 2.5 million (10%) American males are **exclusively** homosexual, while another 2.5 to 7.5 million have **occasional** homosexual contacts. These numbers are refuted by some recent surveys like the 1992 National Opinion Research Center that reports that among males, 2.8% are exclusively gay, and among women, 2.5% are exclusively lesbian. Judith Reisman argues in her 1990 book, *Kinsey, Sex and Fraud,* that male ho-

mosexuals make up about 1% of the population. The group, Coloradans for Family Values and the Washington-based Family Research Institute believe about 3% of the male population is exclusively gay.

Using data collected from HIV testing conducted in 1986 and 1987, the CDC estimates that 20% of exclusively homosexual men are HIV-infected. This means (depending on whose numbers are used) that between 500,000 and 625,000 gay males in the United States have been HIV-infected. For bisexuals and men with infrequent homosexual encounters, the CDC tabulates a prevalence rate of 5%, meaning that between 125,000 and 375,000 of this population are HIV-infected (Booth, 1988). Approximately 6% of all IDU-associated AIDS cases occur in the homosexual/bisexual male population. These cases may reflect HIV transmission either by contaminated syringes or sexual activity.

Injection Drug Users— According to the CDC, as many as 33% of the nation's 1.2 million injection drug users may be HIV-infected. This behavioral risk group contains the nation's **second largest group** of HIV-infected and AIDS patients. An association between injection drug use and AIDS was recognized in 1981, about 2 years before the virus was identified. **AIDS in IDUs and hemophiliacs offered the first evidence that whatever caused AIDS was being carried in and transmitted by human blood.** From the first reported IDU AIDS cases in 1981 through 1999, 28% of all adult/adolescent AIDS cases occurred in IDUs. Of IDUs, 74% listed IDU as their **only** risk factor for HIV infection; 26% were also homosexual/bisexual. The CDC estimates that through 1999 44% of women and 22% of men got AIDS through IDUs.

IDUs have been reported in all 50 states and the District of Columbia. Fifty-two percent of the accumulated 160,000 IDU-only AIDS cases occurred in New York and New Jersey. Among the adult/adolescent heterosexual AIDS cases over half had sexual partners who are/were IDUs. Beginning year 2000, of the estimated 746,000 AIDS cases, 268,560 (36%) were directly or indirectly associated with IDU (*MMWR*, 1996a, updated).

Rates for IDU-associated AIDS varied widely by area: rates in Puerto Rico, New Jersey, New York, and the District of Columbia were greater

THE LATINO COMMUNITY

" In 1983, my brother died of AIDS. It infected his wife, and a year later she died. And now we had the first set of orphans in our family. Despite everything, I saw the love and unity of my family—how we pulled together and took care of my brother and loved him. My sister left another set of orphans. My uncle, who is my godfather, lost children to HIV. His son died in a prison hospital of AIDS. His daughter died at the age of 33 and left another orphan child. My uncle wasn't around to see his daughter die because he died of suffering from watching. Last year we found out my 70-year-old uncle infected with HIV. And that's just my family. I watch the news. I watch people talk about wars in other countries and I identify with the feelings of those people. I know they have bombs thrown at them. I know they have weapons pointed at them. But we have a weapon that is killing my community. We have to silence this weapon of AIDS, which is killing us."

This was the testimony given by Marina Alvarez to 60 Latino community leaders from across the United States and Puerto Rico who had convened at Harvard University in the spring of 1998. The rate of infection among Latinos in the United States reflects an increasingly dire situation. Currently, 18% of all AIDS cases in the United States occur among Latinos, even though Latinos represent only 11 percent of the total population. The rate of HIV infection among Latino men is three times greater than the infection rate found among white, non-Hispanic men. Similarly, women and children in the Latino community have rates of infection that are seven times greater than the rates found among white women and children. If these trends continue, the Harvard AIDS Institute projects that by the year 2003 the percentage of total annual AIDS attributable to the Latino population will eclipse the percentage of such cases attributable to the white, non-Hispanic population.

The Future Threat For Latinos. The AIDS epidemic threatens not only the present generation of Latinos but also future generations. Ending the year 2000, over one-third of the members of the Latino community will be under the age of 18. With rising rates of HIV infection among all youth, the implications for the Latino community are ominous. Rafel Campo, a physician at Beth Deaconess Medical Center said, "As overwhelming as this epidemic might seem to us right now, even greater devastation faces our next generation of irreplaceable young people whose future should be to lead this country into the next millennium; to take our places in medicine, business, in government; and to share with the world the tremendous brilliance and creativity of our culture." (Kao 1999)

than 10 per 100,000; in 22 states, rates were lower than 1 per 100,000 population.

About 55% of all IDU-associated cases were reported in the Northeast, which represents about 20% of the population of the United States and its territories. The South reported 20% of IDU-associated AIDS cases, 5% from the Midwest, and the West reported the remaining 20%.

The rate of IDU-associated AIDS continues to be **higher** for blacks and Hispanics than for whites. Except for the West, where rates for whites and Hispanics were similar, this difference by race/ethnicity was observed in all regions of the country and was greatest in the Northeast. J. Peterson and colleagues (1989) reported that among **heterosexual** men, injection drug use accounts for 83% of black, 87% of Hispanic, and 45% of white AIDS cases. Through 1999, overall IDU-associated **male** AIDS cases represented 9% of all AIDS cases in whites, 36% in blacks, 37% in Hispanics, 5% in Asians/Pacific Islanders, and 15% in American Indians/Alaskan Natives. For **women** it was; white, 43%; black, 45%; hispanic, 42%; 17% Asian/Pacific Islanders; and 46% American Indian/Alaskan Natives.

Heterosexuals— The spread of HIV in the general population is relatively slow, yet potentially it is the source of the greatest numbers of HIV/AIDS cases. The CDC estimates there are 143 million Americans **without** an identified at-risk behavior, the general population.

Data from the CDC beginning year 2000 indicated that 10% of all the AIDS cases in the United States occurred through heterosexual contact. But this must be placed in perspective. Although 10% of anything may be a significant number, beginning year 2000 it stood for about 74,600 adult/adolescent/pediatric AIDS cases out of the total 746,000. Most of the heterosexual AIDS cases occurred in persons or the sexual partners of individuals with an identified behavioral risk. Relative to the general adult population, the number of heterosexual AIDS cases is only a fraction of 1%. Of the 40,000 plus new HIV infections expected to occur each year beginning 1996 through the year 2000, only 2,000 or 5% each year are expected to occur in the heterosexual population.

Global HIV/AIDS Cases and Heterosexuality

Worldwide ending the year 2000 there will be an estimated 14 million people **living with AIDS** and some 18 million will have **died of AIDS.** According to a June 1999 Worldwatch Institute report, South Africa's HIV pandemic is perhaps the worst on the globe. It has engulfed the country, and "barring a medical miracle, one of every five adults will die of AIDS over the next 10 years." This unprecedented social tragedy is also translating into an economic disaster. The working age population is being lost to AIDS. In Zimbabwe, state morgues are extending their hours to cope with the soaring death rate, mostly as a result of AIDS. An estimated 3,000 people now die every week in the southern African country, nearly 70% of them from

AIDS-related illnesses. The main hospital in Harare has opened its morgue around the clock and other hospital and mortuary facilities have extended closing time by four hours. At the University of Durban-Westville in Kwazulu-Natal, 255 of the students recently tested HIV positive. In 1997, 14% of the population was HIV positive—in 1999 it was over 22%.

Worldwide, through year 2000, **heterosexuals** will make up 75% of the estimated 43 million living HIV-infected people. About 70% of the 43 million people live in Sub-Saharan Africa (about 31 million). North America (2 million), Latin America (2.9 million), South and Southeast Asia (8 million), and Africa account for 97% of global HIV infections (Figure 10-6).

Of the 43 million living HIV-infected people worldwide, over 90% live in nonindustrial nations. It is estimated that through year 2000, 32 million people will have/had AIDS worldwide, 77% of those cases occurred in Africa, 8% in Asia, 2.4% in the United States, and 2% in Europe. The United Nations program an HIV/AIDS (UNAIDS) estimate that 18 million (58%) of the 32 million will have died (UNAIDS, 1998 updated Table 10.4).

UNAIDS estimated that almost 6 million new HIV infections occurred annually in 1997, 1998, and 1999 estimates the same number for year 2000. The AIDS death toll for these 4 years is estimated at 10 million of which 4.3 million were women and 2.1 million were under age 15. The United States will have about 800,000 AIDS cases by the end of 2000—worldwide there will be **40** times that number, (Figure 10-7). About thirty-five million people will be living with HIV disease.

TABLE 10-4   Leading Cause of Death Worldwide In 1998

Infectious Diseases	All Diseases
1. **HIV/AIDS**	1. Heart Diseases
2. Diarrheal disease	2. Cerebrovascular Diseases
3. Childhood diseases	3. Lower Respiratory Diseases
4. Tuberculosis	4. **HIV/AIDS**
5. Malaria	5. Obstructive Pulmonary Diseases
6. STDs excluding HIV/AIDS	6. Diarrheal Diseases
7. Meningitis	7. Perinatal Conditions
8. Tropical diseases	8. Tuberculosis

Data Courtesy of the World Health Organization. 1998.

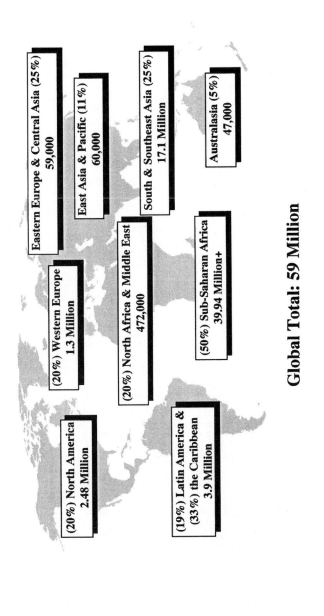

Global Total: 59 Million

(20%) North America
2.48 Million

(19%) Latin America &
(33%) the Caribbean
3.9 Million

(20%) Western Europe
1.3 Million

(20%) North Africa & Middle East
472,000

(50%) Sub-Saharan Africa
39.94 Million+

Eastern Europe & Central Asia (25%)
59,000

East Asia & Pacific (11%)
60,000

South & Southeast Asia (25%)
17.1 Million

Australasia (5%)
47,000

FIGURE 10-6 Estimated Number of Global HIV Infections Projected by United Nations AIDS Program and World Heath Organization ending year 2000. Numbers in parentheses are percent of women HIV-infected. Of the estimated 59 million HIV-infected persons, about 43 million **are living** in some state of HIV/AIDS illness. Over 4.7 million of these are children. Of 209 countries reporting to the WHO and UNAIDS, 194 have reported AIDS cases. World population over 6 billion. (Source: Pan American Health Organization Quarterly Report, September 1996 updated; UNAIDS Report on the Global HIV/AIDS Epidemic, December 1998 updated)

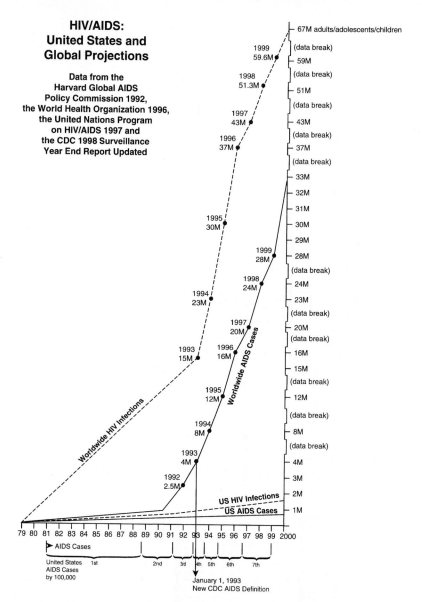

FIGURE 10-7 Estimated United States and Global Projections for Total Number of HIV and AIDS cases. Beginning year 2000, about 746,000 United States AIDS cases were reported to the CDC. Most of these cases were diagnosed according to the 1987 CDC AIDS definition guidelines. On January 1, 1993, the CDC changed the definition of AIDS to include all persons with a T4 cell count of **less than 200.** The new definition raised the number of AIDS cases in the United States from 1992 to 1993 by about 111%. The expected number of adults/adolescent cases for 1993, 49,100, became 106,618. The new definition has raised the number of persons living with daignosed and reported AIDS in the United State through 1999 to about 250,000. By the year end 2000, the United States is projected to have about 800,000 AIDS cases. Worldwide, the total number of AIDS cases is expected to reach about 33 million. Through the year 2000, about 2.2% of all HIV infections and 2.4% of total AIDS cases will be in the United States. William Hazeltine, AIDS researcher formerly at the Dana Farber Institute, and now with Human Genome Sciences, Inc., said that if prevention does not work or a cure is not found, some **1 billion** people could be HIV-positive by the year 2025. Global population through 2000, is projected to be about 6.2 billion people. (*Source: Harvard Global AIDS Policy Commission 1992 and World Health Organization 1993, 1994, National Census Bureau 1998, and projected data*)

Worldwide, one in every 100 sexually active people aged 15 to 49 is HIV infected! In sub-Saharan Africa 7 in every 100 age 15 to 49 are HIV infected. Overall, beginning in year 2000, AIDS has orphaned over 10 million children.

The World Health Organization (WHO) projected that by the end of 2000, over 90% of all HIV infections globally will be transmitted heterosexually. In 1996, the WHO estimated that there are about a half a **billion** heterosexuals who are at moderate to high risk of exposure to HIV because of multiple sexual partners. In addition, there are about 10 million homosexuals and 5 million drug users who are also at moderate to high risk because of multiple sex partners or regular sharing of injection equipment. Most HIV infections in North America have been transmitted homosexually to men who practiced receptive anal intercourse, whereas most of those occurring in Africa and parts of the Caribbean have been transmitted heterosexually to both women and men during vaginal intercourse. Heterosexual transmission accounts for a growing proportion of HIV/AIDS cases in Europe and Latin America, and is proliferating at explosive rates in parts of Thailand and India

——————— SIDEBAR 10.2 ———————

THE HIV/AIDS TIME BOMB: IT IS TICKING OUT OF CONTROL

The HIV/AIDS pandemic is like an explosion in slow motion, a slowly moving chain reaction—no sound, no blinding flash, no intense heat, no mushroom cloud, no buildings destroyed—just one silent death after another with no end in sight. Whole generations of people are in jeopardy with so little hope to go around in the developing nations.

SUB-SAHARAN AFRICA—Beginning year 2000, this region of the globe had 70% of all HIV-infected individuals. Ninety percent of all AIDS deaths have occurred in this region. Ninety percent of all children with perinatal infection and 95% of all AIDS orphans reside in sub-Saharan Africa, despite the fact that it contains only 10% of the world's population. The sheer number of Africans infected by this epidemic is overwhelming. Currently, an estimated 40 million people living in sub-Saharan Africa have been infected with HIV, 13 million people have died, about a quarter of them children. In 1998 and in 1999 AIDS was responsible for an estimated 2.5 million African deaths or about 6,800 funerals per day. Despite the scale of death, there are 30+ million Africans living with HIV. In Botswana, Namibia, Swaziland, an Zimbabwe, current estimates show that between 20% and 30% of people aged 15 to 49 are living with HIV. In Zimbabwe, 30% to 50% of all pregnant women are now found to be infected and at least one-third of these women will pass the infection on to their babies. The Republic of South Africa, which escaped much of the epidemic in the 1980s, is now being particularly hard hit. One in seven new infections on the African continent is occurring in the Republic of South Africa.

THE TIME BOMBS IN REAL TIME

The time bombs for the future are the world's most populous nations, China and India, each of which have about one billion citizens. HIV is now spreading rapidly in those countries, and because of their huge populations, even a fraction of the infection seen in Africa would be devastating. For example, if by 2010 just two percent of the Indian population is HIV-infected, that would mean 20 million people. UNAIDS thinks that about 8 million Indians are already infected, and the virus is spreading rapidly in that country, primarily through heterosexual activities.

FUNDING TO SLOW THE PANDEMIC

Peter Piot, executive director of UNAIDS said in April 1999 that, "Twenty years into this pandemic it is alarming that AIDS is expanding three times faster than the **funding** to control it. Weighed against the global catastrophe of the AIDS pandemic the level of spending on **HIV prevention is minimal.**" Between 1990 and 1997 the number of HIV-infected people **tripled** from 9.8 million to 30.3 million. During this same time frame, funding for prevention increased from $165 million to $273 million. Beginning year 2000 it is estimated that a total of 59 million people have been infected with HIV, over six times the number infected in 1990, but the funding will only have doubled.

(Aral et al., 1991). Homosexual transmission still predominates in North America, but heterosexual transmission is increasing.

Hemophiliacs— There are about 16,000 to 20,000 hemophiliacs in the United States. At least half are HIV-positive and through 1999, about 2,800 of them have died. Of those infected, over 98% received HIV in blood products that were essential to their survival. By mid-1985, an HIV blood screening test was put into effect nationally. From that point, the risk of HIV infection from the blood supply has been significantly lowered but a small risk still exists. **Scandals** in knowingly selling HIV-positive blood for transfusions and in production of the blood factor essential for hemophiliacs have surfaced in France, Germany, the United States, Canada, and Japan in the 1990s. (For discussion, see Chapter 8.)

From 1982 through 1999, 4,580 adult/adolescent hemophiliacs, 405 sexual partners of hemophiliacs, and 255 hemophilic pediatric AIDS cases have been reported. Almost all of these cases are the result of HIV infections that occurred prior to 1985.

Military— The incidence of HIV infection can be measured best in groups that undergo routine serial testing. Because active duty military personnel and civilian applicants for the service are routinely tested for HIV antibody, there is a unique opportunity to measure the incidence of HIV infection in a large, demographically varied subset of the general population.

Since October 1985, all military personnel on active duty, as well as all civilian applicants for military service, have undergone **mandatory testing for HIV antibody.** Over the next 2 years, the armed forces screened 3.96 million people for exposure to HIV. Of the 3.96 million tested, 5,890 (0.15%) were HIV-positive. This total included active duty personnel, members of the National Guard, members of reserve units, and would-be recruits. The recruits who tested positive have been barred from entering the service. Those carrying HIV who are already in uniform are allowed to stay in the service as long as they remain asymptomatic. They are, however, ineligible for overseas duty and must undergo close health monitoring.

Through 1999, there were an estimated 1,050 HIV-positive persons in the military who became HIV-positive **while serving in the military.** The Navy has the highest incidence of HIV, with 2.5 per 1,000 people. The rate in the Army is 1.4 per 1,000; in the Marines, 1 per 1,000; and in the Air Force, 0.99 per 1,000. The higher rate in the Navy may reflect that Navy personnel are based primarily on the East and West Coasts, which have the highest number of AIDS cases.

College Students— Between April 1988 and February 1989, the blood of 16,863 students enrolled at 16 large universities and three private colleges was tested for HIV antibodies.

Thirty students, 28 males and 2 females, tested HIV-positive for an overall rate of 2 per 1,000 or 0.2%. With 12.5 million students enrolled in American colleges and universities, the rate of 2 in 1,000 means there are about 25,000 HIV-infected college students.

The study was conducted by the CDC with The American College Health Association (Gayle et al., 1988). Because the samples were not identified, students who tested positive could not be informed. All blood samples from 10 of the 19 campuses were HIV-negative. At five campuses the rate of HIV-positive blood ranged from four to nine samples per 1,000. In contrast, the rate of HIV in the general heterosexual population is about 0.02% and the rate for military personnel in about the same age group tested over a similar time period was 0.14%.

The question is why college students have a rate of HIV infection 10 times higher than the general heterosexual population. Are college students less well informed than the general heterosexual population?

Surveys indicate that college students are well educated about HIV/AIDS. Why, then, the higher rate of infection? Perhaps it's the age-old dilemma of information versus behavior. **They know what to do—but they don't do it.** There is a difference between knowledge and action based on that knowledge. College students have always had information on drug use, alcoholism, pregnancies out of wedlock, and sexually transmitted diseases; but this knowledge has not appreciably reduced at-risk behaviors. STDs are at an all-time high in teenage and college students.

COLLEGE STUDENTS AND NATIONAL ISSUES. In April of 1994, the Roper College Track poll asked undergraduate students at 4-year colleges and universities, "What are the national issues you are most concerned about?" Their overwhelming response was AIDS. Of the top 10 issues listed most frequently, 50% of students listed AIDS as their main concern, followed by crime at 34%, drugs 21%, quality of education 20%, cost of health care 19%, moral values and race relations at 17% each, the budget deficit 16%, and environmental damage and poverty each at 15%.

Prisoners— Beginning year 2000, the nation's prison population was at an all-time high of about 1.9 million+ adults: contained in 1,300 state and 71 federal prisons. Ninety percent are men.

According to the United States Bureau of Justice Statistics, 2.4% of these people or 45,600 are HIV-positive and of these, 17% have AIDS (Rubel et al., 1997 updated). The total number of **prisoners, parolees,** and **probationers** in 1999 was about 5.5 million with an HIV-positive rate of 2.4% or 132,000 being HIV-positive (DeGroot et al., 1996 updated). Beginning year 2000, about 5,600 adult inmates in U.S. state and federal prisons and jails had died of AIDS. And, an additional 5,600 people with AIDS were still in prisons and jails (*MMWR*, 1996c). In 1995 through 1998, **four states**—New York, Florida, Texas, and California—had **more than half** the known cases of HIV in prison. Only six prison systems in the United States distribute or sell condoms to male inmates, Mississippi, New York City, Philadelphia, San Fransciso, Vermont and the District of Columbia (Hooker, 1996 updated) and only two of the six distribute dental dams and condoms to female inmates. **No United States correctional facility distributes bleach or syringes to inmates** (Mahon, 1996).

Blood testing of inmates at 10 selected prisons indicated that 1 in every 24 prisoners is infected with HIV. This study also found that incarcerated women under age 25 had a higher HIV seroprevalence (5.2%) than incarcerated young men (2.3%). The rate was comparable among older women (5.3%) and older men (5.6%). Non-whites were almost twice as likely to be infected as whites. Rates varied widely among the 10 institutions from 2.1% to 7.6% for men, and 2.7% to 14.7% for women (Vlahov et al., 1991). Beginning 1999, 17 states tested **all** new inmates for HIV infection. Most (about 75%) prison systems will provide an HIV test on request. The **House of Representatives** passed legislation requiring all **federal** inmates serving at least six months to be HIV-tested within four months of entering the prison system.

In general, prisoners diagnosed with AIDS were infected prior to their incarceration—most are IDUs (Francis, 1987). Infected inmates most often transmit the virus to others through homosexual and drug-related activities. For that reason 21 states segregated HIV-infected and AIDS patient prisoners from the other prisoners. In September of 1989, a court order in the state of Connecticut ended their prison segregation policy. Similar antisegregation lawsuits have been filed in Alabama and Oregon. Through 1998, 4 states had laws that allow for isolation or segregation of HIV-infected inmates, South Carolina, Alabama, Georgia and Mississippi.

A number of state prisons have instituted either mass screening programs or large-scale blind serological surveys for HIV infection. The Federal Bureau of Prisons tests a 10% sample of federal prisoners. Other jurisdictions conduct screening or testing programs for selected groups of prisoners, such as known injection drug users and homosexuals.

The incidence of AIDS is about 14 times higher in state and federal correctional systems (202/100,000 prisoners) **than in the United States population** (15/100,000 population) (Gostin et al., 1994). Anne DeGroot and colleagues (1996) report a summary of HIV testing results of prisoners in 42 state prisons and report that the rate of HIV infection is **10-to 100-fold higher** than in the general population. In 1993, the WHO reported that AIDS is responsible for 30% of all deaths in United States prisons. It is the leading cause of death in the New York and Florida prison systems. According to the *People with AIDS Coalition Newsline,* 66% of inmate deaths in New York and New Jersey resulted from AIDS. In Florida in 1999, 67% of all deaths in the state prisons were related to AIDS. Of these, 42% were due to tuberculosis or *Pneumocystis* pneumonia (24%), both are successfully treated complications of HIV infection if diagnosed early enough (De Groot et al., 1996 updated). The state of Florida spent about $19 million on HIV/AIDS inmate treatment in 1999. For year 2000 it will cost about $22 million.

TURNOVER IN PRISON POPULATIONS CREATES AN HIV TRANSMISSION PROBLEM FOR THE GENERAL POPULATION. Large numbers of at-risk individuals cycle through prisons and jails every year. The turnover rate for some prisons is as high as 50% per year. Because the prevalence of HIV is so high among the prison population, on release from prison, they can spread HIV to others. For example, the *AIDS Quarterly Report* in May 1990 stated that in one year, 15,000 HIV-infected prisoners left the New York prison system to join the general population, and because

of confidentiality laws, no one will know who they are. At Riker's Island facility in New York, of 21,000-plus prisoners, one in four are reported to be HIV-infected (National Press Release, 1989). A blinded serosurvey performed by the New York City Department of Health in 1991 determined that 26% of women and 16% of men in New York City prisons were HIV-seropositive (De Groot et al., 1996).In Florida, 3.4% of its prison population of 65,000 is HIV-positive (2,210).

The Elderly— Clearly the face of AIDS has changed dramatically from the 1980s into the 1990s. AIDS cases among men, in particular, gay men and IDUs predominated the 1980s. But in the 1990s the numbers of AIDS cases among women, their children and people of color increased significantly. What did not appear to claim; public or scientific notice is the ever increasing number of **AIDS cases among the elderly.** Thus, **the face of AIDS is aging** and as improved therapy keeps the HIV-infected alive longer, the numbers of elderly people with AIDS will continue to grow. **It should be understood that the senior citizens of today did not grow up in the age of AIDS—they did not have sex and HIV in the same thought. They have to learn a lot of unfamiliar and scary information in a hurry.**

AIDS cases among people between the ages 50 and 90 are increasing. Nationally about **11% of the total AIDS cases occur in people over age 50.** Beginning year 2000, this represented over 81,000 people. Three percent of AIDS occurs in people over age 60 or over 22,000 people and 0.7% occurs in people over 70 or over 5,000 people. One AIDS patient was 90 years old when diagnosed. He became HIV-infected from a blood transfusion during surgery at age 85. He became symptomatic about 4 years later and died 1 year later of *Pneumocystis carinii* pneumonia.

In **Florida,** a state with a large retired population, the number of recorded AIDS cases among those over age 50 has risen from 6 in 1984 to 1,341 in 1993 to over 9,000 beginning year 2000, or about 12% of the state's cases.

Age—An At Risk Factor?

In general, Americans over 50 are potentially at risk, sometimes without realizing it, when

WHERE OLDER PATIENTS WITH HIV DISEASE CAN GET HELP

The problem of AIDS in persons 50 and older has received little recognition. But a few organizations now provide literature or services aimed specifically at older patients and their physicians:

American Association of Retired Persons
601 E St. NW
Washington, D.C. 20049
(202) 434-2277
http://www.aarp.org

Publishes free fact sheet, "AIDS: A Multi-generational Crisis" (Stock No. D14942), that includes listings of referral services.

Senior Action in a Gay Environment (SAGE)
208 W. 13th St.
New York, NY 10011
(212) 741-2247

Provides a variety of services to gays in their 50s or older. Also offers training, education, counseling, and therapy and is a source of information for physicians and hospitals. Operates chiefly in the area surrounding New York City (NY, NJ, CT).

Healthcare Education Associates
70 Campton Pl.
Laguna Niguel, CA 92677
(714) 240-2179

Publishes a training manual for use in discussion groups: "AIDS and Aging: What People Over 50 Need to Know." "Leaders Guide" and "Participant's Workbook" are $13.95; workbook alone, $5.

HIV/AIDS in Aging Task Force
425 E. 25th St.

New York, NY 10010
(212) 481-7670

Arranges conferences and education seminars; participants include physicians. Provides support for establishing task forces and seminars throughout the country.

National Institute on Aging
Public Information Office
Federal Bldg. 31 (Room 5C27)
Bethesda, MD 20892
(301) 496-1752

Publishes fact sheet, "Age Page: AIDS and Older Adults," available as a FAX transmission by calling (800) 222-2225.

Other Internet Addresses:

AIDS and the Elderly:
www.caps.ucsf.edu/capsweb/elderlybib.html

Gayellow Pages—Youth, Family & Senior Resources:
http://gayellowpages.com/webfam.html

HIV/AIDS Treatment Information Service (ATIS):
(800) 448-0440; http://www.hivatis.org

Lesbian and Gay Aging Issues Network of the American Society on Aging (ASA):
http://www.asaging.org/lgain.html

National Association of HIV over 50:
www.uic.edu/depts/matec/nahof.html

National Institute on Aging: (800) 222-2225

---

they begin dating following a divorce or the death of a spouse. Others are at risk because they are IDUs or continue to have multiple sex partners or have a sex partner who practices unsafe sex. Individuals who acquire HIV infection at older ages progress to AIDS more rapidly, on average, than individuals infected at younger ages (excluding pediatric cases).

The majority of AIDS cases in the elderly once occurred due to HIV-contaminated blood transfusions; now the new elderly cases of AIDS most often include IDU or sexual activities (El Sadr et al., 1995; Gordon et al., 1995). It is believed that with the growing number of elderly people in the United States and better therapy for those who are HIV-infected in their early and late 50s, there will be an increasing number of elderly HIV-infected and AIDS patients who must be cared for. The male-female ratio in the age 50 and over is about 9:1 (Ship et al., 1991).

Health Care Workers— Health care workers are defined by the CDC as people, including students and trainees, whose activities involve contact with patients or with blood or other body fluids from patients in a health care setting. They represent 7.7% of the United States labor force.

The risk of HIV transmission from health care worker to patient during an exposure-prone invasive procedure is remote. There is a greater and well-documented risk of transmission from an infected patient to a health care worker.

As the pandemic of HIV infections continues to increase, workers in every health care field will be involved in the detection, counseling, therapy, maintenance, quantity, and quality of life for the symptomatic HIV-infected and for those expressing AIDS. Although the largest number of HIV-infected people are asymptomatic, each day many of them begin to experience signs and symptoms of HIV disease; and each day new cases of AIDS are diagnosed. As the number of patients increases, additional health care workers are required to meet their needs. As this work force enlarges to meet the demands of the estimated numbers of AIDS cases in the coming years, precautions must be practiced to prevent health care workers themselves from becoming HIV-infected.

Ruthanne Marcus and colleagues (1988) reported that, across the board, health care workers exposed to HIV-contaminated blood have about a 1 in 300 chance of becoming infected. Other more recent reports place the risk of HIV infection at 1 in 250.

Health Care Workers with HIV/AIDS

A European Commission reported that worldwide, beginning year 2000, 101 health care workers (HCW) were confirmed to have acquired HIV infection on the job **(54 cases were in America)** and an additional 200 **HCWs (132 were in America)** may have become infected on the job. In the United States, nurses account for the largest number of cases.

Twenty-five of the 54 cases have developed AIDS and eight have died. These are small numbers when compared to some 7,000 hepatitis B transmissions to health care workers every year in the United States (Osborne, 1993 Aiken et al., 1997 updated).

Of the 132 possible occupationally acquired HIV-infected workers, 60 have developed AIDS (*MMWR*, 1992b; Bell, 1996 updated). The number of persons with occupationally acquired HIV infection is probably greater than the totals presented here because not all health care workers are evaluated for HIV infection following exposures and not all persons with occupationally acquired infection are reported.

Of the estimated 737,000-plus American adult/adolescent AIDS cases entering year 2000, about 25,795 (3.5%) are health care workers, including an estimated 2,948 (0.4%) physicians, 5,061 (0.76%) nurses, and 4,053 (0.55%) dentists (Toufexis, 1991, updated). Overall, beginning year 2000, 75% of the health care workers with AIDS, including an estimated 1,300 physicians, 90 surgeons, 351 dental workers, 3,445 nurses, and 290 paramedics, have died. The Medical Expertise Retention Program (MERP) in San Francisco in November 1991 estimated that 7,000 to 10,000 physicians and 50,000 to 70,000 other HCWs in the United States are HIV-positive (Williams, 1991).

Like other adult AIDS patients, health care workers have a median age of 35 years. Males account for 91.6% of cases and 62% are white.

Ninety-five percent of the health care workers with AIDS were classified into known transmission categories. Health care workers with AIDS were significantly **less likely** to be injection drug users and **more likely** to be homosexual or bisexual men (*MMWR*, 1996b).

Needlestick Injuries— Health care workers are in a quandary about the possibility of becoming HIV-infected via needlesticks. Articles such as "Needlestick Risks Higher Than Reports Indicate" or "The Risk of HIV Transmission via Needlesticks Is Low" convey conflicting impressions.

The kinds of needlesticks most likely to occur in hospital or health care settings come from disposable syringes, IV line/needle assemblies, prefilled cartridge injection syringes, winged steel needle IV sets, vacuum tube phlebotomy assemblies, and IV catheters, in that order. Needlesticks and penetration of sharp objects account for about 80% of all health care workers' exposures to blood and blood products.

There are about 800,000 needlestick injuries from contaminated devices in health settings each year. Of these, 16,000 devices are contaminated with HIV (Miller et al., 1997).

ESTIMATES OF HIV INFECTION AND FUTURE AIDS CASES

In 1987, Otis Bowen, then Secretary of Health and Human Services, said **"AIDS would make black death pale by comparison."**

As long as the number of newly infectious people each year exceeds the number who die, the pandemic will continue to build.

Reportability

National reporting policy is developed by the Council of State Territorial Epidemiologists, in consultation with the Centers for Disease Control and Prevention (CDC). In 1997, the Council of State and Territorial Epidemiologists recommended that 52 infectious diseases be notifiable at the national level. However, which diseases are reportable vary by state, depending on individual state priorities. For instance, while syphilis is reportable in all 50 states, cocccidiodomycosis is reportable in only 11. States forward surveillance reports to the CDC. These reports include the age, sex, and race of the patients but not the names. Therefore, reporting individuals by name occurs only at local and state levels.

Who Reports AIDS Cases And To Whom? United States

Reporting systems for AIDS and HIV infection developed outside traditional programs, amid concerns about confidentiality. AIDS cases are reported to the CDC through the SOUNDEX system, which involves translating names into specific sets of numbers and letters. While the resulting codes are not unique, when they are combined with other information, such as birthdates, individual cases can be followed without revealing the names. Although SOUNDEX is used at the national level, at state and local levels the names of AIDS patients are reported in all states. In states that report HIV-infected individuals by name, reporting methods parallel those used for AIDS (Colfax, et al., 1998).

The CDC considers a case of AIDS reportable when: **(1)** an otherwise healthy person is HIV-positive and has an unusual opportunistic infection (protozoal, fungal, bacterial, or viral) or a rare malignancy; or **(2)** an individual with a positive HIV antibody test has a T4 cell count of less than 200 per cubic millimeter of blood, which is about one-fifth the normal level; or **(3)** an individual with a positive HIV antibody test has been diagnosed with pulmonary tuberculosis, invasive cervical cancer, or recurrent pneumonia.

Aids Reporting Systems

AIDS is reportable in all 50 states, the District of Columbia, and United States territories to the CDC in Alanta, Georgia. AIDS surveillance has been crucial in identifying people at risk for the disease and the modes of transmission. AIDS surveillance data together with HIV surveys are important components of public health programs directed toward controlling HIV infection, and assist in providing the most accurate picture of the HIV epidemic in the United States.

By the end of 1993, all 50 states, the District of Columbia, and four territories (Guam, Pacific Islands, Puerto Rico, and the Virgin Islands) reported adult/adolescent cases. The CDC also reported the numbers of adult/adolescent/ pediatric AIDS cases per 100,000 population by state. For the 10 leading metropolitan areas of at least 500,000 population for AIDS beginning year 2000 see Table 10-5. Beginning year 2000, the **10 states** and **territories** reporting the highest incidence of AIDS cases for adult/ adolescents can be seen in Table 10-6 .

Reporting HIV Positive Cases

Many now advocate name reporting for HIV infection from a surveillance perspective, in part because life-extending therapy is available and in part because of the success with U.S. AIDS surveillance programs. Theoretically, name reporting for HIV infection could provide more current surveillance information, since HIV infection predates AIDS. However, people with AIDS are usually reported for having a specturm of symptoms and are often ill, or are reported from death certificate summaries, thereby ensuring high reporting rates. In contrast, those with HIV infection are reported for having a positive test result, and are more likely to find ways to remain anonymous, therefore avoiding the consequences of being reported as HIV positive. At confidential test sites in Colorado, for instance, 11% of test seekers reported providing **false names.** While people with AIDS could similary provide false indentities, this would require considerably more effort, given their more frequent contact with medical professionals, and could also have negative consequences for them, since social services requires positive indentification (Col-fax, et al., 1998). Through 1999, only two states, Maryland and Washington, required that HIV infection reports carry only the name of persons who are **symptomatic.** Thirty four states and the Virgin Islands have implemented HIV case surveillance reporting HIV-infected adult/adolescents **by name** (Table 10-7). Collectively, these 34 states account for over half of the Nation's AIDS population.

Future HIV Reporting/Surveillance

In coming months, the CDC will release **HIV reporting** recommendations to states, and ultimately could tie compliance to coveted fed-

TABLE 10-5 Ten Metropolitan Areas Reporting Highest Number of AIDS Cases Beginning Year 2000

Metropolitan Area	Number of AIDS Cases
New York City	117,868
Los Angeles	42,522
San Francisco	29,094
Miami	23,126
Washington DC	21,634
Chicago	20,440
Houston	18,650
Philadelphia	17,904
Newark	16,859
Atlanta	15,442
	323,540 or 43.37% of all AIDS cases in America.

TABLE 10-6 Ten States/Territories Reporting Highest Number of AIDS Cases Beginning Year 2000

State/Territory	Number of AIDS Cases
New York	138,756
California	120,852
Flordia	75,346
Texas	52,220
New Jersey	41,776
Puerto Rico	28,372
Illinois	23,126
Pennsylvania	22,380
Georgia	21,643
Maryland	19,918
	544,389 or 73% of all AIDS cases in America.

TABLE 10-7   Status of HIV Infection Reporting—United States, 1999

Confidential HIV Reporting Required			
Name[a] (Adults/Adolescents)	Anonymous[b]	HIV Pediatric Name Reporting Only	HIV Reporting Not Required[c]
Alabama	California	Oregon (for children <6)	Delaware
Alaska	Connecticut		District of Columbia
Arizona	Georgia		Hawaii
Arkansas	Kansas		Illinois
Colorado	Kentucky		Pennsylvania
Florida	Maine		Vermont
Idaho	Massachusetts		
Indiana	Montana		
Iowa	New Hampshire		
Louisiana	Rhode Island		
Maryland[d]			
Michigan[e]			
Minnesota			
Mississippi			
Missouri			
Nebraska			
Nevada			
New Jersey			
New York			
New Mexico			
North Carolina			
North Dakota			
Ohio			
Oklahoma			
South Carolina			
South Dakota			
Tennessee[d]			
Texas			
Utah			
Virgin Islands			
Virginia			
Washington[d]			
West Virginia			
Wisconsin			
Wyoming			

All states require reporting of AIDS cases by name at the local/state level. The 1999 session of Congress expects to pass the **HIV Prevention Act of 1997.** The HIV Prevention Act adds HIV to 52 other notifiable contagious diseases such as gonorrhea, hepatitis A, B, and C, syphilis, tuberculosis, and AIDS that must be reported to the Centers for Disease Control.

[a]Names of HIV-infected people are provided to local or state health departments.

[b]Individual reports of people with HIV infection are provided to local or state health departments. Reports may contain demographic and transmission category information but do not record identifiers.

[c]Some states receive HIV reports on a voluntary basis.

[d]Requires HIV reports with names for **symptomatic** HIV-infected persons **only.**

[e]Names are reported to the local health department only.

(Source: HIV/AIDS Surveillance Report, 1998, 10:1–44 updated)

eral funding. In addition , bills are pending in Congress to make **HIV surveillance mandatory.** The California Medical Association has drafted a state HIV reporting law, and New York, Illinois, Pennsylvania, Georgia, and Massachusetts—all heavily affected states— have legislation pending or are in the process of implementation. Major advocacy organizations have come to a remarkable consensus with public health officials that AIDS-only **surveillance** puts states at risk of losing track of the changing epidemic. For example, in

JUSTIFICATION FOR HIV CASE REPORTING IN THE UNITED STATES

Reporting requirements involved an enduring conflict of values between individual rights to privacy and the collective good of society. The initial resistance to HIV case reporting may have been warranted by the sensitivity to a new and frightening disease that placed a disproportionate burden on disfavored (marginal) populations. Baffling science for years, the disease left people who had HIV infection and their partners without the benefit of effective treatment. Given the need for privacy and the near absence of effective treatments, individual rights were thought to outweigh the benefits of HIV surveillance.

With advances in **science, medicine, epidemiology,** and **law,** however, we are now at a very different point in the history of the epidemic.

Collective justifications for a national system of HIV case reporting are compelling. HIV reporting would **(1)** improve our understanding of the epidemiology of the epidemic, **(2)** prevent infections by targeting scarce resources for testing, counseling, education, and partner notification, **(3)** benefit persons with HIV or AIDS and their partners by providing a link to medical treatment and other human services, and **(4)** promote more equitable allocation of government funding (Gostin et al., 1997).

Class Discussion: There is a national standard for confidential AIDS reporting, but not for reporting HIV infection. Should there be a national standard for confidential HIV reporting? Why?; why not?

mid-1999 the **Gay Men's Health Crisis of New York** announced that the lack of information on HIV infections and their locations is endangering lives and undermining efforts to fight the spread of HIV. AIDS Project Los Angeles, the San Francisco AIDS Foundation and the Gay Men's Health Crisis have endorsed **HIV surveillance** on the condition it is done through a **"unique identifier" system** that would identify and track patients by codes. The Aids groups and the American Civil Liberties Union are on record opposing names-based reporting. These organizations argue that HIV is different from other diseases. It still has no cure. It still is associated with stigmatized groups—gay men, drugs users, prostitutes, and now ethnic minorities. Its victims, once identified, still suffer evictions, firings, hate crimes and hostility. Regardless, whether one favors name-based reporting or unique code identifiers, **all agree** that AIDS reported data is no longer sufficient to track the spread or provide other essentials for the pandemic.

Of the 10 states and territories with the highest rates of reported AIDS cases in 1999, only Florida, New Jersey, and New York conduct HIV surveillance. Alaska, and Texas will begin HIV surveillance in late 1999.

How Many People in the United States Are HIV-Infected?

How many people in the United States are HIV-infected? And just how many cases of AIDS are expected to occur and when? Although projections have been made, the numbers are in question. **Why? They come from surveys and incidence data that lack rigorous scientific documentation.** For example, publications give information such as "researchers **believe** that the current number of HIV-infected people in the United States ranges from 1% to as high as 10%" (Slack, 1991). But even if this range is to be believed, the difference between 1 in every 100 versus 10 in every 100 is very significant. This range implies that the data are questionable, and that is what raises the spectrum of concern. How many people in the United States and worldwide have HIV disease and how many have AIDS? **How large a problem is the HIV/AIDS pandemic?** The answers to such questions are crucial to the allocation of funds for research and medical programs and for medical preparedness of institutions that will be hit hard as the number of cases increases.

All health care institutions and personnel will be affected by the increase in AIDS cases as will allied health service and support occupations, insurance companies, the funeral industry, and the work force in general.

UNITED STATES ESTIMATES LEAD TO CONFUSION. WHOSE FIGURES ARE CORRECT?: ISSUES OF CREDIBILITY

In 1992, a Harvard research group of 40 HIV/AIDS experts estimated that by the year 2000 there would be between 38 and 110 million HIV-infected adult/adolescents and 10 million children world-wide. By then, there will also be 24 million adult and several million pediatric AIDS cases. The estimates of HIV infections by the Harvard group is over twice that of the World Health Organization. In retrospect they were correct. It is clear that the HIV/AIDS pandemic of the 1990s, in terms of new AIDS cases, will be much worse than for the 1980s.

The Public Health Service (PHS) in 1986 estimated that 1.5 million Americans were HIV-infected (Table 10-8). The PHS and the CDC (1987) estimated that by 1991 there would be 270,000 cases of AIDS and 179,000 people with AIDS would have died since 1981; by 1992, 365,000 Americans would demonstrate AIDS and 263,000 would have died of AIDS. These are ominous figures about the future of AIDS in the United States, **but were they responsible figures?** In fact, by 1992 there were 206,559 reported AIDS cases and 140,282 deaths—the estimates were off by 44% and 47%, respectively.

In mid-1988, New York City's Health Commissioner, who had relied on the CDC's methodology for calculating the number of future HIV-infected AIDS cases, cut the estimated number of HIV-infected gay and bisexual men from 250,000 to 50,000—an 80% reduction. The Commissioner also reduced the total estimated number of HIV-infected New Yorkers from 400,000 to 200,000—a 50% reduction. The reason for these reductions was that the numbers of AIDS cases expected from a pool of 400,000 infected people were not happening.

In mid-1989, the General Accounting Office (GAO) of the U.S. Congress reported that it analyzed 13 forecasts of the number of AIDS cases to occur by the end of 1991 and found that the range of predictions was so large—from 84,000 to 750,000—that they could not be used as a meaningful guide to health services planning. The GAO concluded that a range of 300,000 to 485,000 AIDS cases by the end of 1991 was more realistic than the PHS's and CDC's estimates of 270,000. (Recall only 206,000 were reported.)

Scott Holmberg (1996) estimated the prevalence and incidence of HIV in 96 United States,

TABLE 10-8 Estimated[a] Prevalence of HIV-Infected People in the United States

	Estimated Number in U.S.	Proportion Infected with AIDS Virus	Estimated Number Infected
Homosexual men	2.5 million	20–25%	500,000–625,000
Bisexual men and men with highly infrequent homosexual contacts	2.5–7.5 million	5%	125,000–375,000
Regular injection drug users (at least weekly)	900,000	25%	225,000
Occasional injection drug users	200,000	5%	10,000
Hemophilia A patients	12,400	70%	8,700
Hemophilia B patients	3,100	35%	1,100
Heterosexuals without specific identified risks	142 million	0.021%	30,000
Others, including heterosexual partners of people at high-risk, heterosexuals born in Haiti and Central Africa, transfusion recipients	N.A.	N.A	45,000–127,000
Total			945,000–1.4 million

[a]Public Health Service in 1986 and the CDC in 1987 made the above *rough estimates.*

(Source: CDC; Figures projected in 1986 remained unchanged through mid-1995. **In March 1995, CDC announced 50% to 59% reduction of HIV people in the United States—from 1,200,000 to between 600,000 and 700,000.**)

metropolitan areas (MSAs) with populations greater than 500,000. From his data, he estimates 565,000 HIV infections with 38,000 new infections occurring each year. This means then, there are about 700,000 total HIV-infected people in the United States.

John Karon and colleagues (1996), **based on data from three different sources and methods of calculation, estimate there are between 650,000 and 900,000 (0.3% of the population) HIV-infected people in the United States.**

According to the CDC projections in Figure 10-8, and adding in the 15% to 20% underreported AIDS cases, the estimates of the CDC fit relatively well. For 1993, before the implementation of the new AIDS definition, 65,211 cases were projected. Due to the 1993 change in the definition of AIDS, a total of 106,618 adult/adolescent cases were reported, a 37% increase over the expected and a 111% increase over the number of cases reported in 1992. The number of AIDS cases projected for 1994, 1995 and 1996 occurred. After the backlog of AIDS cases based on the new definition is reported, new AIDS cases per year is expected to decrease due to new drug therapy. Table 10-9 presents data on the time it took for the first, second, third, fourth, fifth, sixth, and seventh 100,000 AIDS cases to occur in the United States and the estimated time for the eighth 100,000 AIDS cases.

Rise in HIV/AIDS Cases Among Heterosexuals

In 1989, while the number of new AIDS cases rose by 11% among gay males, it increased by 36% or more among heterosexuals and newborns. In 1993, due to the new AIDS definition, **heterosexual contact** AIDS cases increased 130% over 1992, from 4,045 to 9,288.

The groups most affected by the expanded 1993 definition were women, blacks, heterosexual injection drug users, and hemophiliacs. The increase was greater among women (151%) than among men (105%), and greater among blacks and Hispanics than whites. Young adults ages 13 to 29 accounted for 27% of the heterosexual contact cases. On average, women accounted for about 35% of the heterosexual contact cases from 1993 through 1997.

James Chin (1990), an epidemiologist in charge of AIDS surveillance at the WHO, predicted that by the year 2000, **heterosexual transmission will predominate in most industrialized countries.** But that does not now appear likely. The growth of the epidemic may have been slower among heterosexuals than it has been among gays or injection drug users, but it will continue to increase because HIV infection is a sexually transmitted disease and most people in society are heterosexual.

TABLE 10-9 AIDS Cases Comparison—50,000 vs. 100,000

Dates in Years	Time in Months[a]	AIDS Cases by 50,000	AIDS Cases by 100,000
April 87	70	First	—
Nov. 88	19	Second	First (7.5 years)
Dec. 89	13	Third	—
Jan. 91	13	Fourth	Second (26 months)
Dec. 91	11	Fifth	—
Aug. 92	8	Sixth	Third (19 months)[b]
Dec. 92	4	Seventh	—
Aug. 93	4	Eighth	Fourth (8 months)[b]
Feb. 94	6	Ninth	—
Nov. 94	9	Tenth	Fifth (15 months)[b]
Sept. 95	9	Eleventh	—
May 96	9	Twelfth	Sixth (18 months)
Feb. 97	9	Thirteenth	—
Dec. 98	10	Fourteenth	Seventh (19 months)
Dec. 99	12	Fiftheenth	—
Feb. 2001	14	Sixteenth	Eighth (2.1 years)
Mar. 2002	14	Seventeenth	—

[a]AIDS cases reported beginning 1981 through 1999. Estimated for 2000, 2001, 2002.
[b]Rapid increase in AIDS cases due to **new** January 1, 1993, CDC AIDS definition.

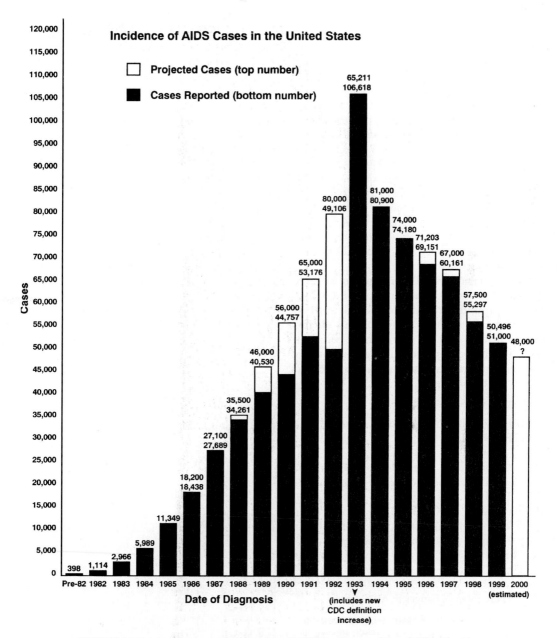

FIGURE 10-8 Incidence and Estimates of AIDS in Persons 13 and Older by Year of Diagnosis—United States, pre-1982 through 1999. The total for 1995 reflects a 9% reduction in AIDS cases from 1994 due to a drop in the backlog of persons to be identified as AIDS cases according to the 1993 definition of AIDS. Data for 1998, and estimates for 1999 and year 2000 reflect a decrease of 9%, 15% and 20% respectively from 1997.

SHAPE OF THE HIV PANDEMIC: UNITED STATES

Based on the HIV presentation of HIV infections in Figure 10-9 during the years 1981 through 1983, HIV infections reached their peak and then rapidly declined. In 1988, the number of **new** HIV infections appears to have leveled off at about 40,000 plus a year through 2000. Currently, 50% of new HIV infections are in people under 25 years old. Unless a preventive vaccine is found, the rate of new HIV infections in the United States is expected to remain at 40,000 plus annually. This may mean the new prevention campaigns, **without a vaccine, will not** significantly reduce new infections. Society has, in a sense, reduced new HIV infections to its lower limit.

Class Discussion: Is the last statement plausible—have we reached a point at which further expenditures will not significantly reduce the rate of infection—that all society can do now is wait on a preventive vaccine? What would be your solution given that we have actually reached this point and no vaccine is forthcoming?

SHAPE OF AIDS PANDEMIC: UNITED STATES

The following data reveal the recent percentage drop in numbers of AIDS cases for the years 1996 through 1998 and estimated for 1999 and year 2000. The drop in diagnosed AIDS cases is being attributed to improved drug therapy—see chapter 4. There has been a decline in new AIDS cases in every region of the United States.

Percentage Changes in AIDS Cases

Year	Men	Women	Pediatric	#Diagnosed/day adult/adolescent
1996	(-)8	(+)0.6	(-)15	189
1997	(-)14	(-)5	(-)30	164
1998	(-)10	(-)4	(-)25	150
1999	(-)10	(-)4	(-)25	138
2000	(-)10	(-)4	(-)25	131

Percentage change for $(-)$ is calculated from the previous year. For example, there was an 8% drop in the number of AIDS cases for 1996 when compared to the number of AIDS cases that occured in 1995 and so on. **1996 was the first recorded drop in new AIDS cases in the United States since the pandemic began in 1981.** Although the noticeable drop in new AIDS cases is a renewed cause for hope, the down side is the slowing progression to AIDS and to AIDS deaths (discussed below) means more HIV-infected people with better health are available to spread HIV. According to the CDC, beginning in 1999 there were 324,000 adult/adolescent people and 4,715 children **reported** to be **living** with AIDS in the United States (HIV/AIDS Surveillance Report 1998). The **estimated** number of people living with HIV was 589,000 (Laurence, 1998). The slowing of AIDS diagnosis makes tracking the epidemic harder. The ability to monitor the epidemic based on HIV infections does not, at the moment, compare with the CDC's ability to track the pandemic through the reporting of AIDS cases.

ESTIMATES OF DEATHS AND YEARS OF POTENTIAL LIFE LOST DUE TO AIDS IN THE UNITED STATES

Deaths Due to AIDS

Each year in the United States there are about 2,300,000 deaths. AIDS has, from 1991 through 1995, caused at least 40,000 of these deaths each

──────── SIDEBAR 10.3 ────────

NO AIDS-RELATED OBITUARIES: FIRST TIME IN 17 YEARS

Bay Area Reporter, a Gay Weekly San Francisco Newspaper revealed that for the **first** time (August 15, 1998) since the pandemic began, there were no AIDS deaths to report. The editor said that **no** AIDS deaths were reported over a 10 day period—it was time to rejoice. During the worst of the pandemic the paper reported as many as **35 obituaries a week.** The peak **year** in San Francisco was in 1992 with 1,816 AIDS-related deaths. A staff member at the **Metropolitan Community Church of San Francisco** said, "There have been more memorial services for people who have died of AIDS at this church than at any other church in America."

Shape of the HIV Epidemic, United States

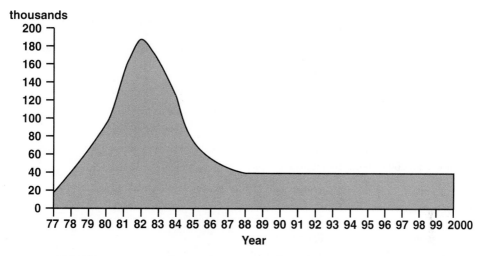

thousands

FIGURE 10-9 Rate of HIV Infection 1977 through 2000: United States.
(*Source: Dr. Don C. Des Jarlais, Beth Israel Medical Center, updated*)

year, and accounted for about 1.8% of all deaths for each of those years. That is, two people of each 100 who died, died of AIDS.

THE GOOD NEWS is that between the end of 1995 and the end of 1996, AIDS deaths dropped in the United States for the first time since the pandemic began—25% nation wide (about 12,000 fewer deaths). Between 1996 and

1997 deaths due to AIDS dropped 45% (about 76,000 fewer deaths). Between 1997 and 1998 deaths dropped again by about 25%. Compared to 1995 the drop was 67%; similar data was reported from Europe. The sudden drop in expected deaths due to AIDS is believed to be associated with the use of combination drug therapy. (See Chapter 4 for more detail on the use of combination drug therapy.) Because of this enormous change in the natural history of AIDS, mortality no longer provides an accurate picture of the extent of the epidemic.

AIDS As a Cause of Death in the United States and Worldwide Beginning year 2000

United States

◆ AIDS is the 14th leading cause of death for all ages.

◆ AIDS is the 5th leading cause of death among people ages 25 to 44.

◆ AIDS is the leading cause of death of African Americans ages 25 to 44.

◆ AIDS is currently the fourth leading cause of death among U.S. women ages 25 to 44.

◆ AIDS is the seventh leading cause of death among children ages 1 to 14.

───────**POINT OF INFORMATION 10.3**───────

UNITED STATES: 746,000 AIDS CASES BEGINNING YEAR 2000

It took nearly 6 years (June 1981–April 1987) for the first 50,000 AIDS cases to be **reported** in the United States (50,352). Over the next 5 years an additional 252,865 AIDS cases were reported (a total of 303,217 or 43% of the 696,000 AIDS cases to be reported by the beginning of 1999). During the next 6 years, between the end of 1992 and the beginning of 1999, an additional 393,000 AIDS cases were reported.

The data in Table 10-9 show that it took 7.5 years (1981 to 1988) to accumulate the first 100,000 AIDS cases. In a little over half that time (1989 to mid-1993), 4.4 years, 300,000 new AIDS cases occurred.

- In some cities in the northeastern United States, AIDS is the leading cause of death among children ages 2 to 5.

- In the United States between 4 and 5 people per hour become HIV-infected 365 days a year (40,000) and **every hour** about two people die (15,000). **Before 1996 one person died every 13 minutes.**

Worldwide

- Worldwide, 667 people become HIV infected every hour, 365 days a year (5.8 million) and **every hour 285 people die** (2.5 million).

- Beginning year 2000 about 16 million people died of AIDS.

- Ending year 2000 about 18 million people will have died of AIDS.

Former Surgeon General C. Everett Koop has said on a number of occasions that "**AIDS is virtually 100% fatal.**" Looking back over the number of AIDS cases diagnosed and comparing them to the number of AIDS patients who have died would indicate that a diagnosis of AIDS is a death sentence. Table 10-10 presents a sobering look at the numbers of AIDS patients who have died in America since those first CDC reported cases in 1981. People with HIV disease/AIDS were, through 1995, dying at the rate of about 3,000 a month. **About 95% of those diagnosed with AIDS in 1981 have now died.** And about 60% of AIDS cases diagnosed up through 1998 have died.

AIDS Deaths Postponed

Many of the deaths that would have occurred in the United States in 1996 through 1999 are being **POSTPONED** through the use of combination cocktail drug therapy and in particular the use of at least **one protease inhibitor** within the drug cocktail.

The majority of AIDS deaths have occurred among homosexual/bisexual men (59%) and among women and heterosexual men who were injection drug users (21%). Most (75%) of AIDS deaths have occurred among people 25 to 44 years of age and about 24% occurred in those over age 45. In 1992, HIV disease/AIDS became the number one cause of death among black and Hispanic men ages 25 to 44 and second among black women ages 25 to 44

TABLE 10-10   Cumulative AIDS Deaths Through Year 2000

	Pre-1981	32
	1981	128
	1982	463
	1983	1,508
	1984	3,505
12576 First 5 years	1985	6,972
	1986	12,110
	1987	16,412
	1988	21,119
	1989	27,791
108970 Total 10 years 121,546	1990	31,538
	1991	36,616
	1992	41,094
	1993	44,636
	1994	48,663
219380[a] Total 15 years 340,926	1995[b]	48,371 } 25% } 45%
	1996	36,494
	1997[c]	19,996 } 25%
	1998	15,000
	1999[d]	13,000 } 8% } 9%
96400 Total 20 years 437,326	2000[d]	12,000

Cumulative Deaths from AIDS in the United States pre-1982 to 1998 (1999 and 2000 projected). By the end of 1998 there were 60,000 deaths due to AIDS. Ending 1999 were about 425,000 AIDS deaths. About 440,000 will have died by the end of year 2000. Worldwide beginning year 2000, about 16 million have died of AIDS.

[a]From 1985 through 1995, 75% of AIDS deaths occurred; 22% occurred in the last 5 years.

[b]Drop in percent of deaths in 1995 USA. A 75% drop occurred between 1995 through year 2000. A similar or greater drop in AIDS deaths occurred in Europe over the same time frames.

[c]First time since 1987 that number of AIDS deaths/year were below 20,000 in a given year.

[d]Data for year end 1999 and 2000 estimated

(*MMWR*, 1994b). **Between January 1993 and December 1995 AIDS was the leading cause of death in all Americans aged 25 to 44** (*MMWR*, 1995a). For 1996 it dropped to second. AIDS is the **fourteenth** leading cause of death overall in the United States. (Table 10-11)

According to the CDC, deaths from HIV disease/AIDS are currently **underreported by 10% to 20%.** Although most deaths occurred among whites, proportional rates have been highest for blacks and Hispanics. During 1994, the number of deaths per 100,000 population,

Disease	% of Deaths
1. Heart Disease	31.3
2. Cancer	23.2
3. Stroke	6.9
4. Lung diseases	4.8
5. Accidents	4.0
6. Pneumonia	3.8
7. Diabetes	2.7
8. Suicide	1.3
9. Kidney disease	1.1
10. Liver disease	1.0
11. Blood poisoning	1.0
12. Alzheimer's	1.0
13. Homicide	0.8
14. **HIV and AIDS**	0.7
15. Arteries hardening	0.7
16. All other causes	15.6

*latest available data.

aged 25 to 44 years, was 178 for black men and 47 for white men. It was 9 times as high for black women (51) as for white women (6). In proportion, since 1985, the rate of death was higher for women than for men (*MMWR*, 1996d). **Through the year 1990, over 120,000 people died from AIDS. By the end of 1999, over 425,000 people died from AIDS.** To place the death rate due to AIDS in perspective, each day about 140,000 people die from assorted causes while about 95 to 100 die each day from AIDS. Worldwide, an estimated 2.5 million people died of AIDS in 1997, in 1998, and in 1999. AIDS deaths in these three years, 7.5 million, is about 45% of all people dead from AIDS; 46% were women and about 7% were children. Worldwide beginning year 2000, of an estimated 59 million HIV-infected persons, about 43 million are alive with HIV disease. About 43% are women. Although the **rate of increase in new AIDS cases has slowed,** the number of new cases continues to exceed the number who die of AIDS.

Prevalence and Impact of HIV/AIDS

However the impact of the disease is measured—by deaths, AIDS cases, or monetary losses—it is just beginning. The worldwide impact during the 1990s was 5 to 10 times that of the 1980s. As the pandemic continues, the United States will find itself progressively more involved with prevention programs and with the political changes that HIV/AIDS will bring about in countries with a high incidence of HIV/AIDS.

LIFE EXPECTANCY

In 1995, the Metropolitan Life Insurance Company published data that shows that life expectancy reached its peak in the United States in 1992 at 75.8 years. For the years 1993 and 1994 it dropped to 75.5 years. The forecast is that continued loss due to AIDS will further erode life expectancy.

A mortality analysis of homosexual males indicates that gay men in Vancouver, British Columbia, have a life expectancy 8 to 20 years shorter than non-gay men in the same area (Hog et al., 1997). The researchers note that if the trend continues, "we estimate that nearly half of gay and bisexual men currently aged 20 years will not reach their 65th birthday." The study collected vital statistics from 1987 to 1992 and assessed mortality rates based on 3%, 6%, and 9% prevalence rates of homosexual and bisexual men. "Under even the most liberal assumptions, gay and bisexual men in this urban center are now experiencing a life expectancy similar to that experienced by all men in Canada in the year 1871."

SUMMARY

In 1981, the CDC reported the first case of AIDS in the United States, and, from that time onward, has constantly tracked the prevalence of AIDS cases in different geographical areas and within different behavioral "risk" groups. In all behavioral risk groups, the common denominator is the exchange of body fluids, in particular blood or semen. The heterosexual population at large is considered to be at low risk for HIV infection in the United States. By 1993, all states and the District of Columbia, Puerto Rico, and the Virgin Islands have reported AIDS cases in people who have had heterosexual contact with an at-risk partner.

A major problem exists in attempting to determine the number of HIV-infected people. Several different approaches have been used by the CDC to estimate the total number of HIV infections. These estimates can be evaluated by examining their compatibility with available prevalence data.

With respect to race and ethnicity, the cumulative incidence of AIDS cases is disproportionately higher in blacks and Hispanics than in whites. The ratio of black to white case incidence is 3.2:1 and the Hispanic to white ratio 2.8:1. This racial/ethnic disproportion is also observed in HIV-positive blood donors and in applicants for military service. Even among homosexual and bisexual men and IDUs, where race/ethnicity-specific data are available, blacks appear to have higher seroprevalence rates than whites.

With regard to prostitution, in a large multicenter study of female prostitutes, black and Hispanic prostitutes had a higher rate of HIV infection than white and other prostitutes. This disproportion existed for both prostitutes who used injection drugs and for those who did not acknowledge injection drug use.

The risk of new HIV infections in hemophiliacs and in people who receive blood transfusions has declined dramatically from 1985 because of the screening of donated blood and heat treatment of clotting factor concentrates. Evidence also indicates an appreciable decline in the incidence of new infections in homosexual men. However, the risk of new infections appears to remain high in IDUs and in their heterosexual partners. In several metropolitan cities, the prevalence of HIV infection in IDUs has been increasing.

Studies conducted by the CDC along with the American College Health Association revealed that in 1989, two college students per 1,000 were HIV-infected.

There are some 5.3 million health care workers in the United States. Even though they are supposed to adhere to Universal Protection Guidelines set down by the CDC for their protection, a significant number are exposed to the AIDS virus annually. A relatively small number of those infected have progressed to AIDS.

Estimating the number of HIV-infected people in the United States continues to be a numbers game. Various agencies and private industries have, for different reasons, attempted to determine the number of HIV-infected people. The numbers from the different groups vary widely. However, the 1986 CDC estimated numbers of 1 to 1.5 million HIV-infected people may be too high. However, the number of new AIDS cases occurring each year are within 10% to 20% of the CDC estimated figures.

Beginning year 2000, with some 900,000 people HIV-infected in the United States and with 40,000 new infections occurring each year through the year 2000, the face of AIDS in America is changing. **It's a younger and older face than it used to be. It's more likely to be a face of color than it used to be.** And more people with HIV disease and AIDS are from areas outside of major cities. Overall, the number of new AIDS cases appears to be leveling. But the HIV epidemic should be viewed as many different epidemics in different stages that vary according to age, race, gender, and locality. Although gay and bisexual men continue to make up the largest portion of AIDS cases, the epidemic is increasing more rapidly among people who become infected though heterosexual contact and through sharing syringes to inject drugs. Every year women are composing a larger portion of AIDS cases, many who did not realize they were at risk. (See Chapter 11 for specifics on HIV/AIDS in women, children, and teenagers.)

RESOURCES

PEOPLE WITH HEMOPHILIA
Committee of Ten Thousand. (COTT) 800-488-2688

Hemophilia and AIDS/HIV Network for the Dissemination of Information (HANDI)
c/o the National Hemophilia Foundation.
800-424-2634; en español, ext. 3754

INJECTION DRUG USERS
Harm Reduction Coalition
2W. 27th St., 9th Fl., NYC, NY 10001; 212-213-6376

We the People
425 South Broad Street, Philadelphia, PA 19147; 215-545-6868; www.critpath.org/wtp

TRANSGENDER
International Foundation for Gender Education
Transgender hotline, referrals. 781-899-2212;
e-mail ifge@ifge.org

PEOPLE OF COLOR
National Asian/Pacific Islander Consortium on AIDS and STDs
c/o Asian/Pacific Islander American Health Forum.
116 New Montgomery St. Ste.531, San Francisco, CA 94105; 415-512-3408; e-mail etta@apiahf.org

National Black Leadership Commission AIDS
105 E. 22nd St. Ste. 711, NYC, NY 10010;
212-614-0023; e-mail kimblca@aol.com

Latino Commission on AIDS
80 Fifth Ave., Ste. 1501, NYC, NY 10011;
212-675-2838

National Minority AIDS Council
1931 13th St. NW, Washington, DC,
20009; 202-438-6622

National Native American AIDS Prevention Center
134 Linden St., Oakland, CA 94607; 510-444-2051;
www.nnaapc.org

LESBIANS

Lesbian AIDS Project
A division of GMHC. 212-367-1355 or 212-367-1363

PRISONERS

American Civil Liberties Union National Prison
Project
1875 Connecticut Ave. NW, Ste. 410, Washington,
DC 20009; 202-234-4830

Aids In Prison Project.
www.aidsinfonyc.org/aip/index.html

Prisoner's Rights Project of the Legal AID Society
15 Park Row, 23rd Fl., NYC, NY 10038; 212-577-3530

REVIEW QUESTIONS

(Answers to the Review Questions are on page 487.)

1. How did the CDC estimate the numbers of HIV-infected people in the United States? In what year was this done? In retrospect, how accurate are their estimates?

2. Why are people placed in potential HIV risk groups?

3. True or False: The time it takes for HIV-infected people to become AIDS patients is different for each ethnic group, risk group, and exposure route. Explain.

4. What percentage of all U.S. HIV-infected IDUs are in the New York–New Jersey region?

5. What percentage of newborns from HIV-infected mothers are HIV-infected?

6. Eventually, what percentage of HIV-infected children will result solely from HIV-infected mothers?

7. What is the rate of college students currently HIV-infected? Is this more or less than the rate for military personnel? Explain.

8. Compare the college student rate of HIV infection with the rate of HIV infection for the general U.S. population.

9. What is the risk of a health care worker converting to seropositivity after exposure to HIV-contaminated blood?

10. What single job-related event causes the greatest risk of HIV infection among health care workers?

11. What percentage of the total number of AIDS cases in the United States represents health care workers?

12. Are health care workers more apt to become infected with the hepatitis B virus or the AIDS virus?

13. In mid-1989, the government's General Accounting Office said _____ cases of AIDS will have occurred in the United States by the end of 1991. How many did the CDC predict?

14. Do you think people of the United States openly and truthfully discuss their sexual habits with survey personnel? What does the text say?

15. Worldwide how many AIDS patients are estimated to die by the end of 1999? In the United States?

16. Data on AIDS deaths indicated that of AIDS patients diagnosed between 1981 through 1999, in the United States, _____% had died.

REFERENCES

AIKEN, LINDA, et al. (1997). Hospital nurses' occupational exposure to blood: Prospective, retrospecture and institutional reports. *AM. J. Public Health*, 87:103–107.

ARAL, SEVGI O., et al. (1991). Sexually transmitted diseases in the AIDS era. *Sci. Am.*, 264:62–69.

BELL, DAVID. (1996). Occupational risk of HIV infection in health care workers. *Improv. Manage. HIV Dise.*, 4:7–9.

BOOTH, WILLIAM. (1988). CDC paints a picture of HIV infection in U.S. *Science*, 242:53.

BOOTH, WILLIAM. (1989). Asking America about its sex life. *Science*, 243:304.

CHIN, J., et al. (1990). Projections of HIV infections and AIDS cases to the year 2000. *Bull. WHO*, 68:1–11.

COLFAX, GRANT, et al. (1998). Health benfits and risks of reporting HIV-infected individuals by name. *Am J. Public Health*, 88:876–879.

COUTINHO, ROEL, et al. (1996), Summary of Track C: Epidemiology and public health. *AIDS* 10 (suppl. 3):S115–S121.

DE GROOT, ANNE, et al. (1996). Barriers to care of HIV-infected inmates: A public health concern. *AIDS Reader*, 6:78–87.

DOLAN, KATE, et al. (1995). AIDS behind bars: Preventing HIV spread among incarcerated drug infections. *AIDS*, 9:825–832.

EL-SADR, WAFFA, et al. (1994). *Managing Early HIV Infection: Quick Reference Guide for Clinicians.* Agency for Health Care Policy and Research. Publication #94-0573, Rockville, MD.

EL-SADR, WAFFA, et al. (1995). Unrecognized human HIV infection in the elderly. *Arch. Intern. Med.*, 155:184–186.

FAY, ROBERT E., et al. (1989). Prevalence and patterns of same-gender sexual contact among men. *Science*, 243:338–348.

FELDMAN, MITCHELL, et al. (1994). The growing risk of AIDS in older patients. *Patient Care*, 28:61–72.

FILLIT, HOWARD, et al. (1989). AIDS in the elderly: A case and its implications. *Geriatrics*, 44:65–70.

FRANCIS, DONALD P. (1987). The prevention of acquired immunodeficiency syndrome in the United States. *JAMA*, 257:1357–1366.

GAYLE, HELENE. (1988). Demographic and sexual transmission differences between adolescent and adult AIDS patients, U.S.A. *Fourth International Conference on AIDS.*

GORDON, STEVEN, et al. (1995). The changing epidemiology of HIV infection in older persons. *J. Am. Geriatr. Soc.*, 43:7–9.

GOSTIN, LAWRENCE, et al. (1994). HIV testing, counseling and prophylaxis after sexual assault. *JAMA*, 271:1436–1444.

GOSTIN, LAWRENCE, et al. (1997). National HIV case reporting for the United States. *N. Engl. J. Med.*, 337:1162–1167.

HENDERSON, DAVID, et al. (1990). Risk for occupational transmission of human immunodeficiency virus type 1 (HIV-1) associated with clinical exposures. *Ann. Intern. Med.*. 113:740.

HIRSCHHORN, LISA. (1995). HIV infection in women: Is it different? *AIDS Reader*, 5:99–105.

HIV/AIDS Surveillance Report. (December 1996). 8:1–36.

HIV Surveillance Report. (December 1997). 9:1–43.

HOGG, ROBERT, et al. (1997). HIV disease shortens life expectancy of gay men by up to 20 years. *Inter. J. Epidemiol.* 26:657–661.

HOLMBERG, SCOTT. (1996). The estimated prevalence and incidence of HIV in 96 large U.S. metropolitan areas. *Am J. Public Health*, 86:642–654.

HOOKER, TRACEY. (1996). HIV/AIDS: Facts to consider—1996. National Conference of State Legislatures, Denver, CO, pp. 1–64.

KAO, HELEN. (1999). Leaders unite on HIV in the Latino community. *Harvard AIDS Review Women and AIDS.* Spring: 13–15.

KARON, JOHN, et al. (1996). Prevalence of HIV infection in the United States, 1984 to 1992. *JAMA*, 276:126–130.

LAURENCE, JEFFREY. (1998). The future of HIV therapeutics. *AIDS Reader*, 8:39–40.

LEMP, GEORGE. (1991). The young men's survey: Principal findings and results. A presentation to the San Francisco Health Commission, June 4.

LEMP, GEORGE, et al. (1994). Seroprevalence of HIV and risk behaviors among young homosexual and bisexual men. *JAMA*, 272:449–454.

MAHON, NANCY. (1996). New York inmates' HIV risk behaviors: The implications for prevention policy and programs. *Am. J. Public Health*, 86:1211–1215.

MARCUS, R., et al. (1988). AIDS: Health care workers exposed to it seldom contract it. *N. Engl. J. Med.*, 319:1118–1123.

MARTORELL, REYNALDO, et al. (1995). Vitamin A supplementation and morbidity in children born to HIV-infected women. *Am. J. Public Health*, 85: 1049–1050.

MICHAELS, DAVID, et al. (1992). Estimates of the number of motherless youth orphaned by AIDS in the United States. *JAMA*, 268:3456–3461.

MILLER, PATTI, et al. (1997). Compensation for occupationally acquired HIV needs revamping. *Am. J. Public Health*, 87:1558–1562.

Morbidity and Mortality Weekly Report. (1990). HIV prevalence, projected AIDS case estimates. Workshop, Oct. 31–Nov. 1, 1989. 39:110–119.

Morbidity and Mortality Weekly Report. (1991). The HIV/AIDS epidemic: The first 10 years. 40: 357–368.

Morbidity and Mortality Weekly Report. (1992a). HIV prevention in the U.S. correctional system, 1991. 41:389–391, 397.

Morbidity and Mortality Weekly Report. (1992b). Surveillance for occupationally acquired HIV infection—United States, 1981–1992. 41:823–824.

Morbidity and Mortality Weekly Report. (1993). Update: Mortality attributable to HIV infection among persons aged 25–44 years—United States, 1991 and 1992. 42:869–873.

Morbidity and Mortality Weekly Report. (1994a). Heterosexually acquired AIDS-United States, 1993. 43:155–160.

Morbidity and Mortality Weekly Report. (1995a). Update: Trends in AIDS among men who have sex with men—United States, 1989–1994. 44: 401–404.

Morbidity and Mortality Weekly Report. (1995b). First 500,000 AIDS cases—United States, 1995. 44: 849–853.

Morbidity and Mortality Weekly Report. (1996a). AIDS associated with injection drug use—United States, 1995. 45:392–398.

Morbidity and Mortality Weekly Report. (1996b). Case central study of HIV seroconversion in health-care workers after percutaneous exposure to HIV-infected blood—France, United Kingdom, and United States, January 1988–August 1994. 44: 929–933.

Morbidity and Mortality Weekly Report. (1996c). HIV/AIDS education and prevention programs for adults in prisons and jails and juveniles in confinement facilities—United States, 1994. 45:268–271.

Morbidity and Mortality Weekly Report. (1996d). Update: Mortality attributable to HIV infection among persons aged 25–44 years—United States, 1994. 45:121–125.

Nations Health Report. (1995). Women learn of progress, share deep concerns on HIV/AIDS issues. XXV:10.

OSBORNE, JUNE E. (1993). AIDS policy advisor foresees a new age of activism. *Fam. Prac. News,* 23:1, 45.

PETERSON, J.L., et al. (1989). AIDS and IV drug use among ethnic minorities. *J. Drug Issues,* 19:27–37.

REISMAN, JUDITH. (1990). *Kinsey, Sex and Fraud: The Indoctrination of a People.* Lafayette, LA: Huntington House Press.

ROSENBERG, PHILIP. (1995). Scope of the AIDS epidemic in the United States. *Science,* 270: 1372–1376.

RUBEL, JOHN, et al. (1997). HIV-related mental health in correctional settings. *FOCUS,* 12:1–4.

RYDER, ROBERT, et al. (1994). AIDS orphans in Kinshasa, Zaire: Incidence and socioeconomic consequences. *AIDS,* 8:673–679.

SHIP, J.A., et al. (1991). Epidemiology of AIDS in persons aged 50 years or older. *J. Acquired Immune Deficiency Syndrome,* 4(1):84–88.

SLACK, JAMES D. (1991). *AIDS and the Public Workforce: Local Government Preparedness in Managing the Epidemic.* Tuscaloosa: The University of Alabama Press.

STALL, R., et al. (1990). Relapse from safer sex: The next challenge for AIDS prevention efforts. *J. Acquired Immune Deficiency Syndrome,* 3:1181.

TOUFEXIS, ANASTASIA. (1991). When the doctor gets infected. *Time,* 137:57.

UNAIDS. (1997). The HIV/AIDS situation in 1997: Global and regional highlights, Geneva 27, Switzerland, pp. 1–14.

VLAHOV, D., et al. (1991). Prevalence of antibody to HIV-1 among entrants to U.S. correctional facilities. *JAMA,* 265:1129.

WEISFUSE, ISAAC C., et al. (1991). HIV-1 infection among New York City inmates. *AIDS,* 5:1133–1138.

WILLIAMS, PATRICIA. (1991). Job fears may impede care for seropositives. *Med. World News,* 32:39.

WILSON, J., et al. (1990). Keeping your cool in a time of fear. *Emergency Medical Services,* 19:30–32.

Prevalence of HIV Infection and AIDS Cases Among Women, Children, and Teenagers in the United States

CHAPTER CONCEPTS

♦ A global estimate of HIV-positive women is presented.

♦ In proportion there are significantly more black women with AIDS than white women.

♦ Injection drug use is the major route of HIV infection for women.

♦ From 1994 through early year 2000 at least 38% of women in the United States contracted HIV from men through sexual intercourse.

♦ Injection drug use and prostitution are strongly associated with HIV infection.

♦ AIDS is the leading cause of death for women ages 25 to 34.

♦ **First 100,000 AIDS cases in women in the United States were documented in December 1997.**

♦ Between 17% and 23% of all HIV infected adult/adolescents in the USA are women.

♦ Women make up about 18% of all U.S. AIDS cases.

♦ Women make up 43% of worldwide AIDS cases.

♦ "Pediatric" means **under** age 13 in the United States and **under** age 15 in Canada.

♦ **About 95% of pediatric AIDS cases receive the virus from their HIV-infected mothers.**

♦ In proportion there are significantly more black and Hispanic pediatric AIDS cases than for whites.

♦ Perinatal HIV infection without anti-HIV drug intervention in the United States is about 25%; with drug intervention it is about 8%. With drugs and Cesarean section it's about 2%.

♦ Not all newborns that test HIV-positive are HIV-infected.

♦ Seven of 10 people are sexually active by age 19.

♦ Eighty-six percent of all sexually transmitted diseases occur in the 15 to 29 age group.

♦ The total number of HIV-infected teenagers is unknown.

♦ Black and Hispanic teenagers account for a disproportionate number of AIDS cases compared to whites.

- Teenagers are being exposed to quality HIV prevention but they choose to ignore it.
- Sex thrills but AIDS kills.
- Worldwide people ages 14 to 19 account for at least 50% of all new HIV infections.

WOMEN AND AIDS WORLDWIDE

Through 1997, women, children, and teenagers seemed to be on the periphery of the HIV/AIDS pandemic. However, they have now become the center of this pandemic. **The fear of becoming infected and an inability to ensure that they remain uninfected are common to virtually all women.**

Worldwide beginning year 2000, there were an estimated 28 million AIDS cases of 59 million people infected with HIV. About 12 million (43%) of these AIDS cases are **women**. Of the total HIV-infected, 31.5 million are men and 27 million are women (Figure 11-1). Beginning 1997, worldwide about 50% of all new HIV infections are occurring in women.

In sub-Saharan Africa, the **ratio of women to men** infected with HIV/AIDS is currently 6:5. In Rakai and Masaka, rural districts of Uganda, a 1994 national report indicated that of 1,300 young women aged 15 to 25, 70% were HIV-infected. When compared with HIV-infected young men of the same villages and age group, there was a female-to-male ratio of 6:1. In Rwanda and Tanzania, young women under age 25 accounted for 20% of female AIDS cases and men under 25 for less than 9% of male cases. The **male-to-female ratio** of AIDS cases among Ethiopian teenagers is 1:3; in Zimbabwe, it is 1:5. In Brazil, the **ratio of male-to-female** infection in Sao Paulo has changed from 42:1 in 1985 to 2:1 in 1995. In Northern Thailand, HIV prevalence among women coming to prenatal clinics reached 8% in 1993. And 72% of sex workers are HIV-infected. In the United States, women as a percentage of HIV/AIDS cases rose 8% from

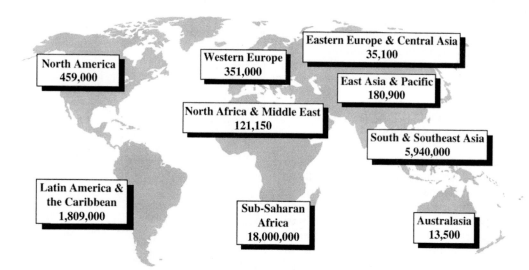

Global Total: 27 Million HIV Positive Women

FIGURE 11-1   Estimate of HIV-Positive Women Worldwide Beginning year 2000. For comparison, beginning 1993 there were **5 million** HIV-positive women worldwide; by the end of 1993, women made up an estimated **8 million** HIV cases. Beginning year 2000 there were about **27 million** HIV-positive women, or about 46% of all HIV-positives worldwide were women! About 7 million have died. There will be about 30 million HIV-positive women globally ending year 2000 and about 8 million will have died.

1981 to 1987, to 20% in 1996 through 1998. Beginning year 2000, about 80% of HIV transmission worldwide was associated with heterosexual (vaginal) intercourse. Women are biologically more vulnerable to HIV infection than men because HIV in semen is in higher concentration than in vaginal and cervical secretions and because the vaginal area has a much larger mucosal area for exposure to HIV than the penis.

Transmission of HIV from male to female is 2 to 10 times more effective than from female to male (*World Health Organization*, 1994). Paradoxically, since time immemorial women have been blamed for the spread of sexually transmitted diseases. Among certain peoples in Thailand and Uganda, STDs are known as **"women's diseases."** And in Swahili, the language of much of east Africa, the word for STD means, literally, **"disease of woman."** Also, it is not coincidental that the countries in which HIV is now spreading fastest heterosexually are generally those in which women's status is low.

Beginning 1997, every minute of every day, of the year, four women worldwide become infected by HIV, and every minute between one and two woman die from AIDS (820,000 female AIDS deaths/year). The WHO estimated that ending year 2001, about 32 million women will have been infected with HIV, about 12.5 million of whom will have progressed to AIDS.

WOMEN: HIV-POSITIVE AND AIDS CASES—UNITED STATES

In **rural** America, HIV/AIDS due to heterosexual sexual transmission is increasing faster than in any other part of the country. Women most at risk are ethnic minorities and the economically disadvantaged. Among sexually active teenagers, college students, and health care workers nationwide, nearly 60% of the heterosexual spread of HIV is among women (Pfeiffer, 1991).

In 1982, it was thought that all female AIDS cases were associated with injection drug usage. It soon became apparent that some of these women had become infected through heterosexual contact with HIV-infected males. Additional AIDS cases from heterosexual transmission appeared in 1983, and the percentage of female cases in the heterosexual category

PROFILE: WOMEN WITH HIV/AIDS IN THE UNITED STATES

It is estimated that between 17% and 23% of all HIV infected adults/adolescents in the USA are women. Beginning year 2000, about 119,000 women had/have AIDS. The majority of women with AIDS in the United States are unemployed, and 83% live in households with income less than $10,000 per year. Only 14% are currently married, compared to 50% of all women in the United States, aged 15 to 44 years. Twenty-three percent (23%) of HIV infected women live alone, 2% live in various facilities, and 1% are homeless. Approximately 50% have at least one child less than the age of 15 years. Similar to other population groups with AIDS in the United States, the majority of women with AIDS are from minority racial and ethnic groups, with 61% of all AIDS cases in 1998 diagnosed in African Americans, 19% in Latino's, and 18% in Caucasians (HIV/AIDS Surveillance 1998).

Number of Women HIV-Infected and With AIDS

Between 1991 and 1995, the number of women diagnosed as having acquired AIDS increased by 63%, more than any other group of persons reported as having AIDS, regardless of race or mode of exposure to HIV. As of January 1993, an estimated 107,000 to 160,000 women were living with HIV, and of those about 45,000 had been diagnosed as having AIDS. Beginning year 2000, an estimated 178,000 to 251,000 women were living with HIV infection and of these about 119,000 women have/had AIDS (Wortley, et al., 1997 updated).

AIDS in Older Women

Entering year 2000, about 427,000 people died from AIDS, 14.2% or about 60,600 were women. Of these, 7.6% or about 4,600 were aged 55 and older: White 1704, Black 2027, Hispanic 783, all others 92.

has continued to increase. The trend in heterosexual vaginal transmission may serve as a marker for future trends in HIV transmission.

HIV/AIDS in women is a national tragedy, but in addition women are the major source of infection in infants. From 1993 through 1998, about 96% of HIV-infected children, aged 0 to 4 years, got the virus from their mothers.

At the end of 1988, women made up 6,964 or 9% of the total adult AIDS cases in the United States. By the end of 1999, women accounted for an estimated 17% of total AIDS cases (Table 11-1). **Seventy-seven percent of all female AIDS cases have been reported in the 7 years between 1993 and the beginning of year 2000.** And they have been reported from all 50 states and territories. About 67% of these females are between the ages of 13 and 39 (HIV/AIDS Surveillance 1997). Between 1989 and the beginning of year 2000, female AIDS cases were and continue to be twice as frequent among black women than among white, and almost three times higher in black women than in Hispanic women (Figure 11-2). The number of reported female AIDS cases across the United States for 1998 is seen in Figure 11-3.

Among the most alarming HIV/AIDS statistics to emerge is that of HIV transmission through heterosexual sexual contact. Of the total reported cases acquired through heterosexual contact through 1998, in the United States, 65% are women. Of AIDS cases that occur in women ages 13 to 24, about 46% are due to heterosexual contact.

Figure 11-4 presents the cumulative **source** of United States female AIDS cases through 1998. In 1998, of the 55,000 persons aged $\geq$ 13 years reported with AIDS, 11,000 (20%) occurred

TABLE 11-1 Reported AIDS Cases for Women, United States

Year	Number	% Increase	Total
1981 (From June)	6		
1982	47		
1983	144		
1984	285		
1985	534		
1986	980		
1987	1,701		
1988	3,263		
1989	3,639		
1990	4,890		
1991	5,732	1992–1994	Through 1994
		151[b]	(58,995)
1992	6,571		
1993	16,824[a]		Through 1995
			(73,095)
1994	14,379		Through 1996
			(86,915)
1995	14,100		Through 1997
			102,383
			Through 1998
1996	13,820		(estimated)
			110,000
			Through 1999
1997	11,651		(estimated)
1998	11,000		121,000
1999	11,800		
Men	625,000		
Total:	746,000(1999)		

[a]The large increase in women's AIDS cases for 1993 over previous years was due to the January 1, 1993 implementation of the new definition of AIDS. Fifty-two percent of cases are associated with IDU. Most of the remaining cases occurred in women who are sexual partners of IDUs.

[b]Reported male AIDS cases for 1993 were up 113%, in women, 128% (Hirschhorn, 1995).

[c]Total AIDS cases include 8,811 pediatric.

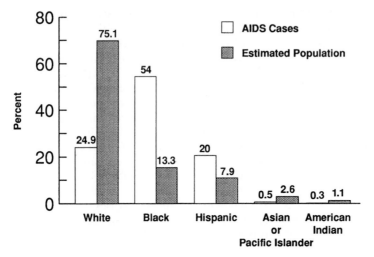

FIGURE 11-2 Incidence of AIDS Cases Among Women of Different Ethnic Groups, United States. Worldwide, every minute, four women become HIV-infected, and every minute, a woman dies of AIDS. (*Source: Global summary of HIV/AIDS Epidemic, December 1998*)

Adolescent/Adult Women AIDS Cases Reported in 1998
United States

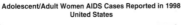

VT	2
NH	13
MA	245
RI	36
CT	194
NJ	723
DE	47
MD	509
DC	285

≥ 1,000
500 – 999
100 – 499
< 100

FIGURE 11-3 One Year of New AIDS Cases in Women Across the United States, Reported for 1998. (Total adult/adolescent AIDS cases estimated for 1999, 55,000) (*Source: Courtesy of CDC, Atlanta*)

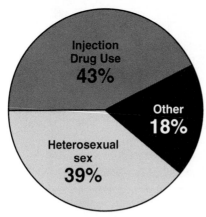

FIGURE 11-4 Cumulative Major Sources of HIV Infections in United States Women. (*Source: U.S. Centers for Disease Control and Prevention, December Surveillance Report 1998*)

among women—about 22 times more than the proportion reported in 1985. The number of women AIDS cases increased steadily through 1993 and then began to decline. The decline is attributed to the use of antiretroviral drugs. (Table 11-1). The median age of women reported with AIDS in 1998 was 35 years, and women aged 25 to 44 years accounted for 85% of female AIDS cases. Beginning year 2000, women aged 50 years and older accounted for 9% of all female AIDS cases. And most of these women became infected through heterosexual sexual activities (Schable et al., 1996, updated). More than three fourths (76%) of all AIDS cases among women occurred among blacks and Hispanics.

Beginning year 2000, the Northeast region accounted for the largest percentage of AIDS cases reported among women (44%), followed by the South (36%), West (9%), Midwest (7%), and Puerto Rico and U.S. territories (4%). In the Northeast, 1.4% of women with AIDS resided outside metropolitan areas compared with 10.2% of women who resided outside metropolitan areas in the South. **Of all AIDS cases among women, 61% were reported from five states: New York (26%), Florida (13%), New Jersey (10%), California (7%), and Texas (5%).**

Beginning year 2000, of the 39% of women reported with AIDS attributed to heterosexual contact (as opposed to 5% heterosexual contact AIDS cases for men), 44% reported contact with a male partner who was an injection drug user; 7.5%, a bisexual male; 1%, a partner who had hemophilia or had received HIV-contaminated blood or blood products; and 46%, a partner who had documented HIV infection or AIDS but whose risk was unspecified (*HIV/AIDS Surveillance Report*, 1998 updated).

[An excellent resource on HIV/AIDS in women can be obtained from the National Prevention Information Network, 1-800-458-5231 (formerly called the National AIDS Clearinghouse) and the Women's Information Service and Exchange (WISE), 125 Fifth St., N.E., Atlanta, Georgia 30308, (800) 326-3861.]

As the frequency of HIV/AIDS increases in women, the question of whether AIDS will explode in the heterosexual community of the United States becomes more a question of **when will the number of female AIDS cases equal male cases.** For the first time in 1997 American women made up over 20% (22%) of AIDS cases for the year. It was repeated in 1998 (23%) and 1999 (23%). Worldwide, women accounted for 25% of all HIV-infected adults in 1990. Ending year 2000 that percentage will have risen to about 48%. The World Health Organization predicts that by the year 2005, women will constitute the majority of HIV-infected adults.

Transmission Categories

Injection drug use (IDU) has, since 1981, been the major route of HIV transmission for women. Beginning in the 1990s, however, female AIDS cases through heterosexual contact is increasing rapidly and may equal IDU by the end of the year 2000. Blood transfusions

and use of blood products make up 4% of female AIDS cases and in 14% of cases the cause is undetermined.

HIV infection has occurred in a substantial number of women exposed to semen from a single HIV-infected male. In one case, 10 out of 17 women became HIV-infected through vaginal intercourse with one HIV-infected man (Allen et al., 1988). In October 1997, a 20-year-old male was accused of infecting 11 women (the youngest age 13) in Chautauqua County, NY, with HIV. He had unprotected sex with at least 28 women in the county, who, in turn, have admitted to having unprotected sex with another 53 people. But over 500 people turned up at the county HIV testing clinics because they thought they had come in direct contact with this male or one of his sex partners. This male told authorities that in addition to his sex partners in Chautauqua County, he had sex with **dozens** of women in New York City **after he had learned he was HIV-positive.** In many of the cases he exchanged drugs for sex. **In a precedent-setting move, the county judge, deciding "there was an imminent threat to the health and welfare of the general public," released the name of the HIV-infected male, Nushawn Williams, (Figure 8-13) to the public.** A New York State law permits such disclosure under this circumstance. A search for this male revealed that he was in jail on Rikers Island to be sentenced for selling crack cocaine.

To date there are six cases of transmission through **artificial insemination reported** in the United States, and six other cases are known to have occurred (Joseph, 1993). Four of eight Australian women who received semen from a

HIV INFECTION AMONG WOMEN

Alarming News for Women (*World News Tonight* ABC, September 7, 1993). Until now, AIDS has been perceived to be a disease primarily afflicting homosexual men and injection drug users. **In nine U.S. cities, AIDS has become the leading cause of death in women of childbearing age. For the first time, in 1992, the number of HIV women infected with HIV through heterosexual transmission approximated those women infected via virus-contaminated needles.** In 1993, the number of women infected via heterosexual transmission surpassed those infected via injection drug use.

There are 6 million women ages 18 to 40 in the United States who are unmarried and having sexual relationships. Those most at risk for HIV infection are: **(1)** those who have multiple sexual partners (defined as having more than four different partners/year), and **(2)** those women who **do not** insist on the use of a condom.

The nation's top five cities with a female population of 100,000 or more that have the highest incidence of AIDS among women are as follows:

Rate/100,000 Women (1993)

City	White	Black (non-Hispanic)
(1) New York, NY	41.5	90.9
(2) West Palm Beach, FL	38.1	295.4
(3) Ft. Lauderdale, FL	34.1	199.5
(4) Newark, NJ	29.6	105.5
(5) Miami, FL	29.1	125.2

Clearly, in all five cities, HIV disease and AIDS affects black, non-Hispanic women disproportionately.

Three of the nation's cities with the highest incidence of AIDS in women are located in Florida within a 70-mile radius of each other (West Palm Beach, Ft. Lauderdale, and Miami). This area is an epicenter of HIV infection for Florida women. Florida has about 11% of the nation's AIDS cases. The other areas of highest incidence of AIDS in women are Puerto Rico, followed by New Jersey, New York, the District of Columbia, Florida, Connecticut, Maryland, Delaware, Massachusetts, Rhode Island, Georgia, and South Carolina (Guinan, 1993). Nationally, through 1999, 16% of female AIDS cases are in women. In Florida's Palm Beach County the rate is 24%; in Broward and Dade Counties, 18%. The epidemiologist for the state of Florida stated that what is happening in Florida is happening in the inner cities nationwide. What may distinguish the AIDS epidemic in women is that it hinges on the low self-esteem and lack of personal power experienced by women in many walks of life.

Most of Florida's women with AIDS are poor and receive their medical care through the public health system. Among women in South Florida, HIV transmission is associated with crack cocaine. Pam Whittington, Director of the Boynton Community Life Center, a family support facility in southern Palm Beach County, said, "If you have 10 women on crack, probably eight of them are HIV-infected."

Crack cocaine is cheap and readily available. Its use contributes to anonymous, high-risk sex with multiple partners. Those who cannot afford crack exchange sex for it. In isolated communities of crack users, there is a high degree of sharing sex partners, many of whom are HIV-positive.

single infected donor became infected. HIV-contaminated semen had been injected into the uterus through a catheter. For other cases of men infecting women see Box 8.7.

Prostitutes

The term **prostitute** is used in preference to the more recently coined **"sex worker."** No single term can adequately encompass the range of sex for money/drugs/friendship/ accommodation transactions that undoubtedly occurs worldwide. However, the term prostitute is at least relatively clear in referring to those who are directly involved in trading sex for money or drugs. In the United States, prostitutes represent a diverse group of people with various lifestyles. About 31% of female IDUs admit to engaging in prostitution. They need money to support their drug habit, pay rent, and eat. Cities with large numbers of IDUs subsequently have large numbers of prostitutes.

———————— BOX 11.1 ————————

WOMEN + SEXUAL PARTNERS + DECEPTION = AIDS

Across the world, women in support groups or with a close friend have been telling their stories of **trusting** their sexual partners and ending up with AIDS. Their trust was violated—their lives forfeited. Here are but a few examples of the thousands of similar cases worldwide.

1. She is 48 years old with curly red hair and bags beneath her eyes. She slouches slightly in the office chair, stretching out her feet. From her eye shadow to her sneakers, everything is blue. Married to one husband for 28 years, she has children and grandchildren. She also has AIDS. She did not use drugs or have multiple sexual partners. **She did have sex with her husband without a condom!**

2. One 23-year-old had a boyfriend with hemophilia; he never used condoms and never mentioned HIV, even though he had already infected another woman.

3. A divorced man with two children did not tell his 46-year-old girlfriend he had AIDS, even when he was hospitalized with an AIDS-related infection.

4. A 7-year live-in partner of a woman denied infecting her, even though he tested positive for HIV; she did not know he was having sex outside their relationship.

5. Because she had **only two boyfriends,** because "we were perfectly ordinary," they did not use condoms.

6. This woman with a baby did not know "my man was shooting up drugs and sharing needles." Not until he died of AIDS.

7. She **never** dreamed her partner had used a needle. When the doctor said she had AIDS, she replied, **"You have made a mistake.** I cannot have AIDS. How could I have that?"

All these women discovered their HIV status only after they became seriously ill with infections that they **should not** have had. Heterosexual transmission is rising dramatically. A seldom mentioned fact is: a large percentage of the infected women are married or in committed relationships.

SPECIAL PROBLEMS FOR HETEROSEXUAL WOMEN

Heterosexual women are at far greater risk of contracting AIDS during heterosexual intercourse than heterosexual men for several reasons.

1. During vaginal contact, women are exposed to semen carrying significant quantities of lymphocytes. Conversely, men are exposed to small amounts of potentially infected vaginal fluids and cervical secretions (fewer lymphocytes).

2. The mucosal surface area available for HIV penetration is significantly smaller in men than in women. The entire vagina and cervix are of a mucosal nature. In men, the urethra is the only mucosal surface area exposed to female secretions.

3. During the act of sexual intercourse, the vaginal mucosa often suffers microscopic abrasions, drawing lymphocytes to the area. This makes vaginal tissue more susceptible to HIV-positive semen. Abrasions do not normally occur within the male urethra.

4. There are more HIV-infected men than women. In Western Europe, Australia, and North America; the ratio of HIV-infected men to HIV-infected women is about 6:1 (Aggleton et al., 1994). Because of the greater number of infected men, the sexually active woman is at increased risk. This is particularly true if the woman has multiple partners and lives in an area with a large population of HIV-infected men.

Evidence is overwhelming—IDU, prostitution, and HIV infection are strongly associated.

In a multi-city study, HIV antibody was detected in 13% of 1,378 female prostitutes; **80% of the infected prostitutes reported using injection drugs** (Darrow et al., 1989). For prostitutes **without** histories of injection drug use, HIV seroprevalence was 5%; HIV infection in this group was greater among blacks and Hispanics and among those who had had more than 200 paying sex partners. The point is that **non-injection-drug-using prostitutes in the United States play a small role in HIV transmission.** This is believed to be because of the low incidence of HIV infection among their male clients and on the insistence by many prostitutes that their clients use condoms (Felman, 1990). **In Africa and Asia, however, prostitution plays a major role in HIV transmission.**

Women Who Have Sex with Women (Lesbians)

Research on female-to-female transmission remains inconclusive. But the large numbers of HIV-positive women and women with AIDS should alert women who have sex with women that they cannot assume their partners are uninfected because they are lesbians. Daron Ferris and colleagues (1966) report that 80% of lesbian women have had sex with men during their lifetime. Also, certain sexual behaviors common among lesbians probably put them at risk for transmitting and receiving HIV through vaginal fluid, menstrual blood, sex toys, and cuts in the vagina, mouth, and on the hands.

The evidence is clear: HIV infection is present among lesbians, and lesbians engage in behaviors that put them at risk for HIV infection, in some cases, at higher risk than most others in society. Whether or not lesbians put their female sexual partners at risk is less clear. In fact, there is a great deal of controversy about this question. Some HIV/AIDS investigators believe the risk of sexual transmission increases as the number of HIV-positive lesbians increases. Others believe that lesbians are not getting infected through lesbian sex, but only through unsafe behaviors like IDU. Female-to-female transmission has been reported in one case and suggested in another (Curran et al., 1988 Chu et al., 1994).

The July 1997 CDC **Facts** sheet reported on 1,648 women who had sex with other women. Twenty percent (333) said they had sex **only** with other women but 97% of these (323) had other risks—mostly injection drug use. A separate study on over 1 million female blood donors found no HIV infected women whose **only** risk was sex with women!

George Lemp and colleagues (1995) reported on HIV infection in 498 lesbian and bisexual women. Six were HIV-positive but, there was **NO** evidence of women-to-women transmission! The Women's AIDS Network (1988) of San Francisco states in their publication *Lesbians and AIDS* that if there is a possibility that either woman is carrying the virus, she should not allow her menstrual blood, vaginal secretions, urine, feces, or breast milk to enter her partner's body through the mouth, rectum, vagina, or broken skin.

Special Concerns of HIV/AIDS Women

First, HIV/AIDS has a profound impact on women, both as an illness and as a social and economic challenge. Women play a crucial role in preventing infection by insisting on safer sexual practices and caring for people with HIV disease and people with AIDS. The stigma attached to HIV/AIDS can subject women to discrimination, social rejection, and other violations of their rights.

Women need to know that they can protect themselves against HIV infection. Women have a traditionally passive role in sexual decision making in many countries. They need knowledge about HIV and AIDS, self-confidence, the skills necessary to insist partners use safer sex methods, and good medical care.

Efforts to influence women to practice safer sex must also be joined by efforts to address men and their responsibility in practicing safer sex.

Second, women become pregnant. Women who are ill and discover they are pregnant need information about both the potential impact of pregnancy on their own health and maternal–fetal HIV transmission.

Third, women have the role of mothering. From this role come two important consequences. **First,** when a woman becomes ill with HIV disease or AIDS, her role as caretaker of the child or children or other adults in the household is immediately affected. The family is severely disrupted and each family member has to make adjustments. **Second,** the mother must cope with her own life-threatening illness while she also deals with the impact of the disease on her family. Demographic studies show that many HIV-infected or AIDS women have young children; and these women are often the sole support of these children.

Fourth, a woman's illness may be complicated further by incarceration and the threat of foster care proceedings. If the mother is healthy enough to care for her child, she must still cope with the complex issues of medical and home care, school access, friends, and family stress.

Fifth, worldwide, women make up 43% of AIDS cases and 80% of these (40%) women are mothers of children. An estimated 8.5 million women will have died of AIDS beginning the year 2001, and about 30 million will have become HIV-infected. Yet little is known about the HIV disease process in women and what is the appropriate standard for their clinical care. Initially, what became known about HIV disease was derived from studies on men, who for the most part differ from women in race, income, and risk behavior. And what female-derived information there is on HIV disease in women came mostly from studies on pregnant women which have focused primarily on perinatal issues. In 1986, only 3% of participants of federally funded HIV/AIDS clinical trials were women. By 1997, this figure rose to 28%.

To date, too little is known about the pharmacokinetics of HIV/AIDS drugs in women, nor is it clear whether gender influences specific illnesses. For example, does HIV infection increase cervical and vaginal disease and pelvic inflammatory disease? Why are certain opportunistic diseases more aggressive and damaging to females? And how are the drugs used to treat opportunistic infections metabolized in women?

Biology (male vs female) and HIV Disease

According to Arlene Bardeguez (1995) and other similar reports (Cohen, 1995; Garcia, 1995), biology does not influence the prevalence of AIDS-defining illnesses, with the exception of invasive carcinoma of the cervix and possibly Kaposi's sarcoma. Access to HIV-related care and therapies is the dominant factor influencing the prevalence of AIDS-defining illnesses among women (Table 11-2). Injection drug use in HIV-infected women leads to a higher incidence of certain diseases, particularly esophageal candidiasis, herpes simplex virus, and cytomegalovirus. Once an initial diagnosis of AIDS has been made, several major AIDS-defining illnesses appear more frequently in women: toxoplasmosis, herpes genital ulcerations, and esophageal candidiasis.

The currently proposed female-specific markers of **HIV disease** include **cervical dysplasia** and **neoplasia** (tumor), **vulvovaginal candidiasis,** and **pelvic inflammatory disease** (PID). HIV-infected women have been found to have a higher incidence of cervical abnormalities on routine screening.

─────── **POINT TO PONDER 11-1** ───────

GENDER: A BROAD DEFINITION

UNAIDS uses the following broad definition of gender. Gender is "what it means to be male or female, and how that defines a person's opportunities, roles, responsibilities, and relationships." This means **whereas sex is biological, gender is socially defined.** Gender is what it means to be male or female in a certain society as opposed to the set of chromosomes one is born with. Gender shapes the opportunities one is offered in life, the roles one may play, and the kinds of relationships one may have—social norms that strongly influence the spread of HIV. In cultures where HIV is seen as a sign of sexual promiscuity, gender norms shape the way men and women infected with HIV are perceived, in that HIV-positive women face greater stigmatization and rejection than men.

Gender norms also influence the way in which family members experience and cope with HIV and with AIDS deaths. For example, the burden of care often falls on females, while orphaned girls are more likely to be withdrawn from school than their brothers. UNAIDS states that gender roles and relations have a significant influence on the course and impact of the HIV/AIDS epidemic in every region of the world. Understanding the influence of gender roles and relations on individuals' and communities' ability to protect themselves from HIV and effectively cope with the impact of AIDS is crucial for expanding the response to the epidemic. (Adapted from UNAIDS Best Practice Materials 1998; these materials can be found at **http://www.unaids.org.**

The 14 most common AIDS-defining illnesses in HIV-infected women are listed in Table 11-2. **No biology-related differences in survival of HIV-positive persons have been documented when equal access to medical care is considered.** The shorter observed survival of women in some studies is thought to occur because of the lack of access to physicians who are knowledgeable about HIV-related care and therapies.

UPDATE—From mid-1998 through 1999, concerns have been raised with regard to the difference in viral loads in men and women and what that difference means. In current medical practice, HIV levels (viral loads) and T4 cell counts are measured, interpreted and used to help guide anti-HIV therapy without regard to sex. Recently reported studies give pause to this standard of practice (Farzadegan et. al., 1998). They suggest that there may be differences in the way HIV viral levels relate to the risk of HIV disease progression among men and women. Essentially, the studies suggest that women have progression of HIV disease (at least as measured by T4 cell counts) at lower viral levels than men. The question of how much lower, or what a lower viral level means, remains a bit unclear. The Federal Guidelines Panel—the decision-making body that creates the guidelines for the use of anti-HIV therapy— recently reviewed the new information on gender differences in viral load. It concluded that, for the time being, no changes should be made in the guidelines with regard to the use of anti-

HIV therapies in women. It concluded that these new data are not markedly different enough to warrant changing strategies for treating HIV, nor should they be cause for alarm for women living with HIV.

Female HIV/AIDS Deaths

Women's deaths in the United States, rose from 18 cases in 1981 to over 60,000 beginning year 2000. That is, 56% of all women with AIDS have died. AIDS is now the leading cause of death for all women between the ages of 25 and 34, the third leading cause of death for all women between the ages of 25 and 44, and the fifth leading cause of death in white women. It is the leading cause of death for black women between the ages of 25 and 44 (MMWR, 1996b).

Through 1994, in the United States, 48% of women who died of AIDS were diagnosed after their deaths, at autopsy. They died without knowing they had AIDS. Eleven percent of women with AIDS died within 30 days after the diagnosis.

Identifying and Preventing HIV Infection

Identification of HIV-Positive Women— At age 26, a woman and her physicians were baffled when she began suffering from a variety of strange medical conditions: fevers, throat sores, unexplained vaginal bleeding, and fatigue. **It took a variety of doctors and 7 years to find out what was wrong. She tested HIV-positive!**

TABLE 11-2 Frequency of AIDS-Defining Illnesses in HIV-Positive Women and Men

	Women (%)	Men (%)
Pneumocystis carinii pneumonia	34	36
Wasting syndrome	9	8
Esophageal candidiasis	20	12
HIV-associated encephalopathy	3	4
Herpes simplex infection	4	2
Toxoplasmosis of the brain	4	3
Mycobacterium-avium intracellulare complex	7	6
Extrapulmonary cryptococcosis	3	4
Cytomegalovirus infection	2	3
Cytomegalovirus retinitis	3	4
Extrapulmonary tuberculosis	3	2
Cryptosporidosis	2	3
Lymphoma (brain)	.3	.4
Kaposi's sarcoma	1.3	13

From Morbidity/Mortality Weekly Report 1999.

AN AIDS CLINICAL DRUG TRIAL—BY DEFINITION

A clinical drug trial is a government funded and organized study of an experimental or unproven drug to determine the drug's safety and efficacy (whether or not it works). Drug trials for HIV/AIDS, sponsored by the FDA, are organized into the AIDS Clinical Trials Group (ACTG). Guidelines are set to determine how the ACTGs may be run and who is eligible to participate. Taking part in a trial may entitle persons to receive medical examinations and checkups and to have their overall health monitored. For many people with AIDS, the ACTGs are the only means of access to certain potentially life-saving drugs, and the only form of health care they may ever receive.

Historically, FDA policy on admitting women into clinical trials has been confusing and dis-

criminatory at best. Until recently, the FDA has had an outright ban on all women of childbearing potential from participation in early stages of drug trials. In March of 1994, the Food and Drug Administration (FDA) took two important steps to ensure that new drugs are properly evaluated in women. **First,** it provided formal guidance to drug developers to emphasize its expectations that women would be appropriately represented in clinical studies and that new drug applications would include analyses capable of identifying potential differences in drug actions and value between the sexes. **Second,** the agency altered its 1977 policy to include most women with childbearing potential in the earliest phases of clinical trials.

ONE WOMAN'S COMMENTS ON HER HIV DISEASE TREATMENTS

"I was fired from my job when they found out I was HIV-positive. The boss said, 'You have a modern problem—and this is an old-fashioned business.'"

This woman, in 1995, was 29 years of age and HIV-positive since age 22. When her T4 cell count dropped to 250, her M.D. put her on zidovudine (ZDV, also called AZT). She was not given any literature or verbal explanation of how this drug would affect her. In 12 months she became anemic; she could not sleep or hold down food, and her menstrual cycle became erratic. Without explanation, she next received two other HIV repli-

cation inhibitors, ddI and ddC. The side effects were very bad: consistent premenstrual symptoms; mood swings; increased cravings for certain foods, alcohol, or drugs; breast tenderness; and bloating. But, there was no literature for her to read and her M.D. said, "I knew how the drugs affected men, but I knew nothing of what to expect when I gave these drugs to women." "I stopped taking these drugs—they were killing me faster than the virus! I have severe yeast infections, shingles, sinus infections, and a host of other infections. My M.D. is dealing with me like I'm some kind of experiment."

This woman's difficulty in getting diagnosed points out the extent to which women still are invisible when it comes to AIDS. After more than 18 years into the epidemic, the message still hasn't reached primary care physicians: Their female patients may be at risk. This young woman said, **"I went into doctors' offices and all they saw was a white, middle-class woman, not someone at risk for HIV."**

Early identification of women with HIV infection is a pressing problem. Risk-based screening at a Johns Hopkins perinatal clinic showed that 43% of HIV-positive women were

not identified as at-risk on the basis of such screening, with infection being found in 20 (9.5%) of 211 women admitting to risk behaviors and in 15 (1.6%) of 949 who were not at risk according to their response to screening questions (Garcia, 1995). Recent epidemiologic data indicate that the most common AIDS-indicator illnesses among both men and women are *Pneumocystis carinii* pneumonia, and esophageal candidiasis. (Table 11-2).

Prevention— To prevent HIV infection, women have been told to reduce their number of sexual partners, to be monogamous, and to protect themselves by using condoms. **But these goals, generally speaking, do not fit the realities of women's lives or they may not be under their control.**

Women do not wear the condom. (A female condom is now available but not yet in heavy demand. See chapter 9.) For women to protect themselves from HIV infection, they must not only rely on their own skills, attitudes, and behaviors regarding condom use, but also on their ability to convince their partner to use a condom. Gender, culture, and power may be barriers to maintaining safer sex practices.

Women who have more than one sexual partner in their lifetime often practice serial monogamy, remaining with one partner at a time. People living as couples reduce the number of their sexual partners. Still, in many phases of life, sex is practiced with new partners in new relationships. American women, on average, are single for many years before their first marriage; they might be single again after a divorce; they might marry again; and, in later phases especially, they might be widowed. **For some women, multiple partners throughout life are an economic necessity, urging them to reduce the number of partners is meaningless unless the economic situation for these women is**

improved (Ehrhardt, 1992). In addition, public health strategies not necessarily targeted to women can also play an important role for women. Syringe exchange and drug treatment are important strategies because almost half of all HIV infections in women are due to injection drug use. Because women are now more likely to be infected by men through heterosexual contact, programs that specifically target men, especially IDUs, will have a beneficial impact on women's programs. The 1999 National Conference on Women and HIV was held in Los Angeles. For information call toll free 1-877-266-3966.

Childbearing Women

Beginning year 2000, there were an estimated 100,000 HIV-infected women of childbearing age in the United States; perhaps 40% know that they are infected. Women, in general, have two children before they find out they are infected (Thomas, 1989 updated). The birth of an infected child may serve as a **miner's canary**—the first indication of HIV infection in the mother. In the United States, 0.17% of childbearing women are HIV-positive (Luzuriaga et al., 1996; 1998). Some 5,800 infants were born to HIV-infected mothers each year from 1988 through

---

——— **POINT OF INFORMATION 11.5** ———

PROPORTION OF WOMEN TAKING HIV TEST

From 1991 to 1993, the proportion of women aged 18–44 years who have ever been tested for HIV increased 60% (from 18.8% to 31.8%). Increases were similar across all sociodemographic groups (*MMWR*, 1996a). As in 1991, in the 1993 survey, higher percentages of black and Hispanic women (46.1% and 39.7%, respectively) compared with white women (27.9%) reported having been tested for HIV. Similarly, a higher proportion of women with <12 years of education reported having been tested for HIV (36.9%) compared with high school graduates (31.5%) or those with college education (30.4%). In addition, more women living in poverty reported having been tested for HIV (40.2%) than did women living at or above the poverty level (30.3%). HIV testing trends among women aged 18–44 years were similar to those in 1991 with respect to marital sta-

tus, risk perception, and region of residence; however, the proportions of women tested in all three groups increased during 1991–1993. During 1991–1993, **the proportion of women tested who had higher perceived risk for HIV did not increase; however, the proportion tested with low or no perceived risk nearly doubled.** (Data about HIV testing and other AIDS-related knowledge and attitudes were collected in 1994 and 1995.)

Class Discussion: A. How can one explain the higher level of HIV test taking among the poor and less educated than in those economically advantaged and better educated? B. Give several reasons why the percentage of those with higher perceived risk for HIV did not increase while the percentage taking an HIV test among low risk individuals doubled!

1992 and 6,500 to 7,000 such births occurred each year from 1993 through 1999.

Pregnancy and HIV Disease— Early findings in pregnant women indicated that those with T4 cell counts of less than **300**/μL of blood were more likely to experience HIV-associated illness during pregnancy. Pregnant HIV-infected women exhibit a greater T4 cell count decline during pregnancy than do women without HIV infection. T4 cell counts in the HIV-infected do not return to prepregnancy levels. However, the overall declines in HIV-infected women likely represent declines that would have occurred in the absence of pregnancy and suggest that pregnancy does not accelerate disease progression. (Newell et al., 1997; Bessinger et al., 1998)

Over the last 8 years, HIV-positive pregnant women have been attracting more attention from the medical establishment, **first,** because there are better medications for the HIV-positive mother and fetus, and **second,** because of the relatively high incidence of HIV births.

Class Discussion: Nationwide, approximately 1.7 of 1,000 pregnant women are HIV-infected, an incidence much higher than that of fetal neural tube defects, for which pregnant women are screened routinely. Should all pregnancies be screened for HIV?

Reproductive Rights— These are central to a woman's right to control her body. The choice of becoming pregnant or terminating the pregnancy are continually disputed.

However, as more women become HIV-infected and give birth to HIV-infected children, childbearing may come under the surveillance of the state. Women of childbearing age may be among the first groups to undergo mandatory testing as part of an attempt to control the birth of HIV-infected newborns.

Reproductive rights take on new meaning with HIV-infected pregnancies. The state has traditionally expressed an interest in protecting the rights of the fetus. This interest was transcended **in the 1973 *Roe v. Wade* decision when the Supreme Court recognized a woman's right to choose an abortion.** The court ruled that a woman's right to privacy must prevail against the state's interest in protecting the future life of the fetus. The state's interest in fetal survival tends to diminish, however, when the mother is infected with HIV (Franke, 1988). (See Chapter 7 for recent information on preventing HIV infection of the fetus.)

INTERNET

Among the hundreds of Web sites containing information on HIV, there are key addresses which provide comprehensive information, including links to a vast array of other resources. The key sites listed below include special areas focused on **women and HIV.**

——— **BOX 11.2** ———

HIV-INFECTED WOMEN: DIFFICULT CHOICES DURING PREGNANCY

She was 19 years old, a nursing student, pregnant, and HIV-positive. She spent 4 1/2 months of pregnancy in constant fear for herself and for her baby. She waited, her health began to falter, then she decided to have an abortion.

Several studies have reported that HIV-positive women who perceived their risk of infecting their fetus to be greater than 50% were more likely to abort than those who perceived a lower risk. HIV-positive women who chose to continue their pregnancy cited the desire to have a child, strong religious beliefs, and family pressure (Selwyn et al., 1989).

For women who are HIV-positive, pregnancy poses difficult choices. First, pregnancy may mask the presence of HIV disease symptoms and having a child poses other questions such as: Can the mother cope with a normal or infected child? and Who will care for the child if the mother becomes too ill or dies? Such questions bring up a moral issue for **Class Discussion: Do couples have a right to have children when one of the partners is known to be HIV-positive? If the woman is HIV-positive? If both are HIV-positive? Is there any stage of HIV disease/AIDS when you think a women should lose the right to become pregnant? (See Chapter 7 for the four stages of HIV disease).**

NATURAL CONCEPTION IN HIV-NEGATIVE WOMEN WITH HIV-INFECTED PARTNERS

An increasing number of couples in which the man is HIV-positive and the woman HIV-negative want children. Insemination with sperm from *seronegative donors* is the only safe option, yet most couples wish to have a child who is biologically theirs.

Between 1986 and 1996, Laurent Mandelbrot and colleagues (1997) followed 104 consecutive pregnancies in 92 HIV-negative women with HIV-positive partners. All were single pregnancies, conceived without the use of assisted reproduction techniques. Most couples had received nondirective preconceptual counseling on the risks of heterosexual HIV transmission. Prior to pregnancy, most of the men were symptom-free, 14 (13%) had HIV-related symptoms, and one died of AIDS during the pregnancy. Twenty-one men were receiving antiretroviral therapy at the time of conception.

One-third of partners reported inconsistent or no condom use; **in 68 pregnancies conception resulted from unprotected intercourse during ovulation, only after one episode of intercourse in 17 of these cases.** After conception, couples were advised to use condoms, and women were tested monthly for HIV infection. Follow-up was 3 months in all cases and more than 6 months in 95 cases. There were 92 deliveries, four abortions, six miscarriages, and two lost to follow-up in the second trimester. No seroconversions occurred within the first 3 months following conception. Seroconversion was observed in two women at 7 months of pregnancy and in two others postpartum; all such cases arose in couples reporting inconsistent condom use. The investigators suggest that their data show that male-to-female HIV transmission is infrequent during natural conception.

American Medical Association home page:
http://www.ama_assn.org

This site is an avenue for accessing several valuable resources, including the JAMA HIV/AIDS Information Center (.../special/hiv/hivhome.htm), JAMA HIV/AIDS best of the Net (.../special/hiv/bestnet/nethome.htm), and Online CME, which includes a series of Clinical Care Options for HIV. The last selection includes a female case study, "The Seropositive Patient: Initial Encounter," and a module on HIV and women.

Centers for Disease Control and Prevention (CDC) Home Page: http://www.cdc.gov

This site offers broad coverage of disease, health risk and prevention topics. Available publications include the MMWR and surveillance reports, as well as health statistics and slides. Of note is slide series L264 on women and HIV (.../nchstp/hiv_aids/graphics/women.htm).

HIV Insite: Gateway to AIDS Knowledge: http://hivinsite.ucsf.edu

This is a very comprehensive site provided by the University of California, San Francisco AIDS Program at San Francisco General Hospital and the UCSF Center for AIDS Prevention.

Sister Connect: Hotline for women with HIV
800-747-1108

Women Alive: Treatment newsletter on the Web
800-554-4876;
www.thebody.com/wa/wapage.html

WORLD: Newsletter by and for HIV positive women. P.O. Box 11535, Oakland, CA 94611

Resources for Children Whose PARENTS Have HIV/AIDS

There are workbooks tailored to the specific family situation (for example, one mother was losing her hair because of chemotherapy, so there are pictures of **mommy with and without her hair** in the workbook. Looking at pictures helps the child to accustom to the hair loss. Eventually, mommy without hair wasn't so scary.) Coloring pages, word searches, lots of fun things for parent and child to work through together. There are also a number of good children's books available such as "My Daddy Has AIDS," written by a man in Dubuque, Iowa, USA, for early elementary-age children.

Beth Wehrman
John Lewis Coffee Shop/AIDS Prevention Partnership
1202 W. Third St., P.O. Box 3733
Davenport, Iowa 52808
(319) 336-4517
email: bwehr@ix.netcom.com

Should You Tell Your Children You Are HIV Positive?
By Laura Damson, C.S.W.
BODY POSITIVE October 1997, Volume XI, Number 3
http://www.thebody.com/bp/oct97/tell.html

How to tell a 10-year-old his mom has AIDS?
Question/Answer by Michael Shernoff, *The Body*
http://www.thebody.com/shernoff/answers/quest41.html

Ruth's Poem
This poem was written by a mother living with HIV, to her daughter.
Avert UK.
http://www.avert.org/poem.html

Listening to the Children
By Julian Meldrum
Body Positive UK Issue 224, June, 1998
http://www.bodypositive.org.uk/home-page.html
No specifics, but has some useful points for discussion.

A safe place for families, by Anette Goldstein
Canadian AIDS News, Fall, 1997 (Vol. X, No. 1)
http://www.cpha.ca/cpha.docs/canews/X 1/family.html

The Twin State Women's Network (TSWN) that serves women and families in Vermont and New Hampshire (USA) who are infected/affected by HIV and/or chronic hepatitis. In addition to material that TSWN has developed (such as Positively Pregnant, TSWN's Legacy Project, and various articles on Disclosure) the website contains good links for women and kids (infected or affected) as well as a host of other things. This site also links to TSWN's home page, which contains its newsletters since 1997. Women and children's information can be reached at www.dartmouth.edu/~hivnet/women&children.htmld/index.html.

There are good books out now, and even a video about how parent, can tell their children they have HIV or AIDS. The video is entitled **"Don't Shut Me Out."** It is available for $20 by sending a check to The Center for Special Studies, B-24, New York Hospital, 525 East 68th St., NY 10021.

PEDIATRIC HIV-POSITIVE AND AIDS CASES—UNITED STATES

Pediatric AIDS in the United States affects two age groups: (1) infants and young children who became infected through perinatal (vertical) transmission, and (2) school-age children, the majority of whom acquired HIV through blood transfusions (mostly hemophiliacs).

Beginning year 2000, over 8,800 pediatric AIDS cases were **reported** and about 4,800 (55%) died from AIDS. Pediatric AIDS cases represent about 1.2% of the total number of AIDS cases to date. **It is estimated that for each pediatric AIDS case reported there are three to four other HIV-infected children.** Thus an estimated 20,000 to 28,000 children in the United States have HIV (Luzuriaga et al., 1996; Rodriguez et al., 1996). Of the pediatric AIDS cases, 3% were/are hemophilic children who received HIV-contaminated blood transfusions or blood products (pooled and concentrated blood factor VIII injections). About 5% of pediatric AIDS cases occurred in nonhemophilic children who were transfused with HIV-contaminated blood. Through 1999 over 90% of pediatric AIDS cases received the virus from HIV-infected mothers vertically. They transmitted it to them either during the fetal stage (perinatal), during labor or as the newborn passed through the birthing canal, or from breast milk soon after birth. From 1995 on virtually all HIV-infected newborns contracted HIV vertically. The Pediatric AIDS Foundation reported in April of 1995 that hospital costs for each HIV-infected newborn were $35,000 per year (*Nations Health Report*, 1995). In a study from Childrens Hospital of Wisconsin, the cost of care for children with HIV was estimated at $418,863 per patient (*AIDS Clinical Care*, 1996:8, 23). Worldwide about 1.5 million children are living with HIV disease and about 3.5 million have died from AIDS.

Reporting HIV Infected Children

Beginning 1999, 31 states and the Virgin Islands reported **HIV-infected children.** They reported a total of 1,819 HIV-positive children. **Of the 31 reporting states, only two of those states with the highest incidence of pediatric HIV, New Jersey and Texas, were on the list of the 31 reporting states.** Nine states, New York (34%), Florida (13%), New Jersey (6%), California (5%), Texas (5%), Pennsylvania (5%), Maryland (3%), Illinois (2%), and Georgia (3%) have 73% of **all reported** pediatric AIDS cases. AIDS infection is now the **sixth** leading cause of death in children between the ages of 1 and 14.

Ethnic Prevalence of Pediatric AIDS Cases

Children of color make up 14% of all children in the United States but account for 57% of pediatric AIDS cases. Whites make up 70% of children and account for 18% of pediatric AIDS cases. Hispanics make up 12% of children and account for 23% of pediatric AIDS cases (Figure 11-5).

The 1990s: Decade of the Orphans

The increasing death rate for women affects the care of their children: the estimated 90,000 HIV-infected women of childbearing age in the United States, who were alive in 1992, will leave approximately 125,000 children when they die during the 1990s. Worldwide, the former World Health Organization estimated that beginning the year 2000, over 10 million children under age 10 will be orphaned because of HIV/AIDS (*MMWR*, 1996; UNAIDS Report 1998 updated).

Vertical HIV Transmission

Vertical transmission means that HIV passes directly from the infected mother into the fetus or infant. The exact time of HIV transmission to the fetus during pregnancy is unknown. It has been shown to occur as early as the 15th week of gestation. Perinatal (at or near the time of delivery) infection occurs as does 10% to 19% of HIV transmission through breast-feeding.

Breast-feeding by mothers with HIV infection established **before** pregnancy increases the risk of vertical transmission by 14%. When a mother develops primary HIV infection **while** breast-feeding, the risk of transmission rises to 29%. In general it is believed that 50% of HIV-positive babies are infected during the last 2 months of pregnancy and about 50% are infected during the birthing process or through the early months of breast-feeding. (Miotti et al 1999).

A working definition of the timing of maternal HIV transmission has been established to differentiate infants infected **in utero (in the uterus)** from those infected near the time

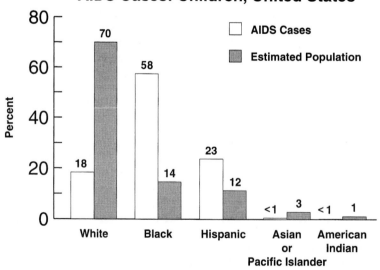

AIDS Cases: Children, United States

FIGURE 11-5 Pediatric AIDS Cases—United States through 1998. The number of pediatric AIDS cases in proportion to the ethnic pediatric population is presented in percent. Due to rounding, numbers may not add up to 100%.
(*Source: CDC, Atlanta, GA, 1996, updated*)

HIV-POSITIVE MOTHER CANNOT BREAST FEED BABY

In April 1999 a judge in Eugene, Oregon refused to give up state control of a baby born to an HIV-infected mother who wants to breast feed, saying the risk of infection to the child left the state no choice but to step in. The mother and father lost legal custody of their 4-month-old son when he was days old because of the decision to nurse him. Although the state has legal custody, he is allowed to live with his parents and his 10-year-old sister. A social worker visits the Tyson home about once a week to ensure the baby is being bottle-fed (??).

of or **during delivery (perinatally).** In utero infection occurs in approximately 20% of HIV-infected infants. Children who are infected in utero have a more rapid progression to AIDS and generally become symptomatic during the first year of life. Those infected perinatally have no detectable HIV at birth but demonstrate HIV in the blood by 4 to 6 months of age. These children constitute the majority of HIV-infected infants and have a slower progression to AIDS, about 8% per year (Diaz et al., 1996).

Rate of Transmission— The worldwide rate of HIV transmission from mother, without drug therapy, to child varies geographically. In Africa, maternal transmission is as high as 50%, producing about 1,600 infected babies **a day.** In Europe and the United States, the overall rate is 25% to 30% producing less than 500 infected babies **a year.** The U.S. Public Health Service and 16 other national health organizations have recommended that HIV testing be offered to all women at risk prior to or at the time of pregnancy. To date, late 1999, Arkansas and Tennessee are the only two states that require health care providers to HIV-test every pregnant woman as early as possible in her pregnancy unless she refuses. The state of Indiana has similar legislation pending.

Newborns: HIV Positive

Because newborns who test HIV-positive may not be HIV-infected, infected mothers in developed nations are advised not to breast-feed their children. The reason a newborn can test positive and not be infected is because the mother's HIV antibodies can enter the fetus during pregnancy (gestation). Because of the presence of material HIV antibody, the newborn may appear to be infected but is not. The HIV antibody will be lost with time and the child will revert to **seronegativity** (no HIV antibody in the serum). Mothers in **underdeveloped nations** have been advised to breast-feed because of the lack of available health care and nutrition. Thus an estimated 273,000 newborns worldwide became HIV-infected through breast feedings in 1997, 342,000 in 1998 and at least 350,000 in 1999.

Perinatal Transmission— Many factors that influence perinatal HIV transmission are not known; but, influencing factors do exist because one mother gave birth to an HIV-infected child followed by an uninfected child who was followed by an infected child (Dickinson, 1988)!

There are multiple factors involved in HIV transmission risk, including maternal immunity and viral load, placental conditions, route of delivery, duration of membrane rupture, and fetal factors (birth order, gestational age). This knowledge has led to trials of various interventions, such as drug therapy to reduce viral load (the number of HIV RNA strands present in the mother's blood at the time of birth) and cesarean section to reduce HIV exposure during delivery.

Viral RNA Load Associated with Perinatal HIV Transmission— Although the close association between stage of HIV infection in a pregnant woman and likelihood of perinatal transmission has been established, there are no precise numerical criteria for pregnancies at **high** and **low** risk of transmission. Viral load measurements are now helping to quantitate this risk.

Separate studies in 1996 and 1997 on pregnancy and viral loads showed that in one study of 30 HIV-infected women giving birth, 20 of the 22 mothers who did not transmit HIV to their children had a low plasma HIV RNA level (less

———— BOX 11.3 ————

ORPHANED CHILDREN DUE TO HIV INFECTION AND AIDS

AIDS orphans present a chilling illustration of the far reaching effects of the AIDS pandemic.

His mother was young, single, and HIV-positive. When she went to the hospital to give birth, she checked in under a false name and address and then slipped out of the hospital leaving her baby who was only a few hours old. Today the boy is 2 years old. No one has yet to offer him a home.

An increasing number of HIV-infected children are being left in hospitals because their HIV-infected mothers and fathers are unable to care for them and no one else wants them. The hospital becomes their home. James Hegarty and colleagues (1988) reported that for 37 children at Harlem Hospital Center, one third of the total in-patient days and over 20% of the cost was for social rather than medical services. By year end of 2000, there will be an estimated 30,000 AIDS-related orphan children in New York City.

As HIV continues to spread across the nation and HIV-infected women continue to become pregnant, the question is: **What will happen to their HIV-infected babies?** For one young woman who passed the HIV virus to her baby 2 years ago, the decision has been made. The baby has AIDS and is in foster care. The mother is very ill. The courts are now deciding whether her six other children should also be put in foster care.

David Michaels and Carol Levine (1994) indicate that by the end of 1995, an estimated 24,600 children, 21,000 adolescents, and 35,100 young adults in the United States will have been orphaned by the HIV/AIDS epidemic. An estimated 13% of children and 9% of adolescents whose mothers died in 1991 lost their mothers to AIDS. More youths now lose their mothers to AIDS than to motor vehicle injuries. Unless the course of the epidemic changes drastically, through the year 2000, the cumulative number of U.S. children, teens, and young adults left motherless due to AIDS will exceed 144,000. The great majority of these children, uninfected by the virus—will begin to affect already burdened social services in major American cities. Things will get immediately worse in such places, where children already spend years going from foster home to foster home and caseworkers are overwhelmed by long lists of families needing everything from housing to medical care. Using

the same projections, as many as 98,000 young people 18 and older will become motherless and in some cases be called upon to care for younger family members.

The Silent Legacy

Orphans are often referred to as the silent legacy of AIDS. It is expected that about a third of the children orphaned will be from New York City, which has the nation's largest number of AIDS cases. Other cities expected to be hit hard are Miami, Los Angeles, Washington, Newark, and San Juan, Puerto Rico. Most of these orphans will be the children of poor black or Hispanic women whose families are already dealing with stresses like drug addiction, inadequate housing, and health care. Relatives who might in other circumstances be called upon to care for the children often shun them because of the stigma attached to AIDS.

HIV-infected women who abandon their babies are often already sick themselves and face an early death. Few can bear the thought of watching their children struggle through a short life should they have passed the virus during pregnancy. Though the majority of infected mothers do not abandon their offspring, such babies are becoming an increasingly common feature of the AIDS pandemic almost everywhere. And they are part of a larger problem that is turning out to be one of the most serious social consequences of the pandemic—**THE PHENOMENON OF AIDS ORPHANS.**

By the year 2010, there will be over 40 million orphans under the age of 15 (1 in 3 under age 5) in the 23 countries (19 of them are in sub-saharan Africa) most affected by HIV. Most of these children will have lost their parents to AIDS. Children under age 15 who have lost one or both parents to AIDS will make up 33% of the population in some of these countries ending the year 2000 (Holden, 1998).

The vast majority of the children orphaned by the AIDS pandemic, will be in sub-Saharan Africa, where formerly the extended family could always be relied upon to take on the dependents of those who died. But today the continent's age-old, traditional social security system—which has proved itself resilient to so many social changes—is buckling under due to the unprecedented strain caused by the increasing number of AIDS cases.

————— BOX 11.3 (continued) —————

ADOPTION ADVERTISEMENTS FOR PARENTS WITH AIDS

"AIDS. Kids. If you have both, maybe it's time to take a closer look at adoption. Call 1-800-NCFA" (National Council For Adoption). This is the shortest of the public service announcements that began in January of 1995.

The areas targeted by the ads are those with the highest incidence of AIDS cases. The ads have prompted hundreds of calls from people wanting to adopt.

Pediatric AIDS Foundation, 1311 Colorado Ave., Santa Monica, CA 90404; 310–395–9051; 800-362-0071; www.pedaids.org; or Newborn Testing Hotlinke, Run by the HIV Law Project; 800-662-9885

————— POINT OF INFORMATION 11.6 —————

WORLDWIDE: CHILDREN WITH HIV/AIDS

Basing its HIV/AIDS estimations and projections on assumptions very similar but not identical to those applied by UNAIDS and WHO, the François-Xavier Bagnoud Center for Health and Human Rights of the Harvard University School of Public Health estimates through 1998, worldwide, about 4,000,000 children have acquired HIV infection through mother-to-child transmission since the beginning of the global epidemic. Because of the rapid progression of pediatric HIV infection to the onset of AIDS and the short survival once pediatric AIDS has set in, most of these children have already died. About one quarter of all AIDS-related deaths thus far have been in children who were infected vertically.

To date, over 85% of all children infected through mother-to-child transmission have been in sub-Saharan Africa. During 1995 through 1999, each year approximately 600,000 children were born with HIV infection (about 1,600 per day); of these children, 67% were in sub-Saharan Africa, 30% in South-East Asia, 2% in Latin America, and 1% in the Caribbean.

Beginning year 2000, about 1.8 million children are living with HIV/AIDS, of whom 65% are in sub-Saharan Africa. HIV incidence among children in sub-Saharan Africa, South-East Asia, and the Caribbean is higher than HIV incidence among adults in all other regions of the world. HIV/AIDS prevalence among children is almost 35 times higher in the developing world than in the industrialized world.

than 30,000 copies/mL) while 8 of 8 transmitting mothers had plasma HIV RNA levels above 30,000 copies/mL. In a second study (Dickover et al., 1996), 20 of 97 newborns were born HIV-positive. The 20 transmitting mothers each had over 50,000 copies of plasma HIV RNA per milliliter at delivery. An RNA copy number of **50,000** has scientists calling it a **breakpoint for defining a high-risk pregnancy for HIV transmission.** None of 63 women with less than 20,000 HIV RNA copies per milliliter produced truly HIV-positive babies. These studies are supported by the work of Donald Thea and colleagues (1997).

Nucleoside Reverse Transcriptase Inhibitor Monotherapy that Appears to Reduce HIV Transmission During Pregnancy— For many HIV investigators, the hard search for drugs to combat HIV replication has paid off. The use of **zidovudine** (ZDV), in all AIDS Clinical Trials Groups (ACTG) of pregnant women to date, has lowered the incidence of HIV-infected newborns from HIV-infected mothers.

Summary of Results of ACTG Protocol 076— On February 21, 1994, the National Institute of Allergy and Infectious Diseases and the National Institute of Child Health and Human

Development announced the interim results of a randomized, multicenter, double-blind, placebo-controlled clinical trial, ACTG Protocol 076.

Enrolled women were assigned randomly to receive a regimen of either ZDV or placebo. The estimated transmission rate was 25.5% among the 184 children in the placebo group compared with 8.3% among the 180 children in the ZDV group. This was a 60% reduction of HIV transmission into newborns within the ZDV group! Based on what has been recently learned about perinatal transmission and viral load, ZDV may be reducing HIV transmission by reducing viral load. ZDV did not appear to delay the diagnosis of HIV infection.

In short, the use of ZDV is known to reduce perinatal HIV transmission. But it is not 100%, because 8% of ZDV pregnancies will still deliver HIV-infected infants. And that is if therapy is given to all HIV-positive pregnant women! (See Chapter 8 for more information on HIV transmission and zidovudine therapy.)

The cost of medication for ACTG 076 was about US$800 per mother-and-child pair which is a prohibitive cost for most developing economies. Recent data from Thailand shows that a much shorter course of zidovudine treatment given to the mother just prior to birth cut the risk of HIV transmission by half. The cost of treatment $60. In the year 2000, over 90% of the 10 million HIV-infected children will live in developing countries.

Entering year 2000, a number of ACTG's have received a variety of combination drug therapy, which includes the use of protease inhibitors. Perhaps the best news to date is the work of Laurent Mandelbrot and colleagues (1998) presented at the 12th World AIDS Conference. The French study reported that of women who took zidovudine and had a cesarean section delivery, and did not breast feed, mother-to-newborn HIV transmission occurred **less than 1% of the time.** A study by the National Institute of Child and Human Development in Washington combined the findings of that study with several others in Europe and North America and found that Cesarean-sections when combined with zidovudine treatment, cut the transmission to about 2%.

Prenatal Therapy: Cost Effective—YES

A 1997 CDC study combined with calculations based on prevailing infection rates, costs and treatment protocols, produced the following analysis: **without intervention,** approximately 25% of HIV-infected pregnant women in the U.S. will transmit HIV to their offspring, an estimated 1,750 infants. The lifetime medical costs for care of those infants is estimated at $282 million or $161,000/child. The estimated cost of intervention, defined as counseling, testing and AZT therapy, is $67.6 million or $38,570/child. That investment will prevent 656 infections annually, representing savings of $105.7 million in medical costs, and resulting in a net savings of $38.2 million. This estimate excludes lifetime productivity savings and quality of life improvements related to prevented HIV infections. A Canadian study reported similar costs effectiveness. And the best news is that Mary Culnane and colleagues (1999) found **no** significant differences in growth, cognitive development of immunologic function in children age 5 exposed to zidovudine during pregnancy (in utero) and during their first 6 weeks of life.

Class Discussion: Present factual, medical and economic reasons for and against the use of a C-section on 100 women to prevent one HIV transmission. Also discuss the fact that neither a C-section nor abandoment of breast feeding is practical for women in poor countries.

States and Territories Most Affected by Pediatric Cases

There is evidence that HIV was present in female IDUs as early as 1977 because their babies developed AIDS (Thomas, 1988). As the number of HIV-infected women of childbearing age rises, so does the number of HIV-infected babies.

Pediatric AIDS, in the United States is most widespread among blacks, Hispanics, and **the poor of the inner cities** (Figure 11-6). Beginning year 2000, New York had the highest incidence of pediatric AIDS cases followed by Florida, New Jersey, California, Puerto Rico, and Texas. Combined, these cases accounted for 66% of all AIDS cases reported among children.

Risk Exposure for Mothers of HIV-Infected Newborns

Risk exposures for HIV infection among the mothers of children with perinatally acquired AIDS included injecting drug use (40%), sexual contact with a partner with or at risk for HIV/AIDS (35%), and receipt of contaminated blood or blood products (2%). **No risk was specified for 13% of cases.** Ten metropolitan areas account for more than 50% of all Pediatric AIDS cases: New York City, Miami, Chicago, Los Angeles, San Juan, Washington, DC, Philadelphia, Newark, New Jersey, Baltimore, and Fort Lauderdale FL. report the largest numbers. The epidemic is also spreading into smaller communities. Beginning 1999, pediatric AIDS cases have been reported in all 50 states.

AIDS Cases Among Children Declining

The findings reported in "AIDS Among Children–United States, 1996" present a documented decline in the incidence of perinatally acquired AIDS. From 1993 through year 2000, the estimated **annual** number of perinatally acquired AIDS cases declined about 75%, from 1,017 to 300.

There are many reasons to believe that, with continued HIV counseling of HIV-positive pregnant women and use of AIDS drug cocktails that lower viral loads, fewer children will become infected and this will translate into a continued decrease in children with AIDS (Figure 11-6).

This possibility was inconceivable to many clinicians, families and patients not too long ago. **The rapid decline suggests that a goal of eliminating perinatal transmission may be attainable.** Although many children have benefited from the present achievements, there are many more in the developing world awaiting the translation of progress into practice.

Contacts for Information on Use of Antiretroviral Drugs During Pregnancy

For a copy of newest "US Public Health Service Recommendations for Use of Antiretroviral Drugs in Pregnant Women Infected with HIV-1 for Maternal Health and Reduction of Perinatal HIV-1 Transmission in the United States," contact: **CDC National Prevention Information Network,** 800-458-5231, http://www.cdcnac.org or **HIV/AIDS Treatment Information Service,** 800-448-0440, http://www.hivatis.org. For information about the registration forms for the Antiretroviral Pregnancy Registration, contact: **Worldwide Epidemiology,** PO Box 13398, Research Triangle Park, NC 27709-9976, 919-315-8465 (call collect) or 800-722-9292, ext. 38465, or Fax: 919-315-8981.

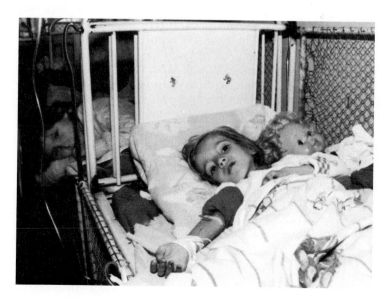

FIGURE 11-6 An HIV-Infected Orphan. This 4-year-old child lies in her hospital bed. She was left to society as her mother died of AIDS and no one has come forth to claim her. (*Photograph courtesy of Archive Photos*)

AIDS CREATING ORPHANS: EXAMPLE-ZIMBABWE

Zimbabwe offers a frightening window onto orphanhood. In this nation, where over 25% of the 5.5 million adults are HIV-infected, AIDS is already pushing hundreds of thousands of children to the brink. The government estimates that in the year 2000, 2,400 Zimbabweans a week will be dying of AIDS. Most of those deaths will be in adults, and they will be concentrated in the young adult ages when people are building up their families. Also, they may be disproportionately concentrated among single women whose death will leave a child with no parent. As early as 1992, a study in Zimbabwe's third largest city, Mutare, recorded that over 10% of children in the study area are orphaned, and that nearly one household in five had taken in orphans. By 1995, an enumeration in the same area showed that the proportion of children who were orphaned had grown to nearly 15%.

The number of children in need of care is rising just as AIDS is cutting into the number of intact families able to provide such care. Some **45% of those caring for orphans are grandparents;** often they have no income of their own, and there is a limit to how many children they can take on without outside help. One orphan-support program reports helping an 80-year-old grandmother who lives with 12 children in a single room. Another has received a request for help from a widower with 9 dependants who has just inherited another 3 grandchildren to care for. A study of households headed by adolescents and children (some as young as 11) showed that while the overwhelming majority had lost both parents, most did have surviving relatives. However, in 88% of those cases, the relatives reported that **they did not want** to care for the orphans. Children themselves are beginning to worry about orphanhood and to recognize the importance of supporting needy children. A majority of children interviewed in one study said that if orphans' needs were not met they would become delinquent. Many said the children would drift into prostitution and onto the streets. They also worried about abuse and exploitation of orphans by relatives. Reports of sexual abuse of girls have risen rapidly in recent years in Zimbabwe, prompting the establishment of a special clinic at a major Harare hospital and an initiative to promote child-friendly courts. In a single rural district of Zimbabwe one study recorded nearly 400 cases of child sexual abuse, at least a quarter of them girls under the age of 12, and at least 10% of them orphans. (UNAIDS: AIDS Epidemic Update, December 1998)

GLOBAL HIV INFECTIONS IN YOUNG ADULTS

Over the summer of 1999, especially in the United States, young people, ages 10 to 24, flocked to theaters to experience the **Dark Side Of The Force.** Many thousands saw the STAR WARS movie repeatedly. But the real **Dark Side** of their lives is the threat of HIV infection and the **Force** should be their education to prevent their infection. About half of all new HIV infections in the United States and worldwide are occuring in this age group. Hopefully, reading this section will encourage the young to stay on the **Light Side With The Force.**

In April of 1998 the Joint United Nations Program on HIV/AIDS (UNAIDS) released a report showing that 5 young people aged 10 to 24 are now infected with HIV every minute, one every 12 seconds, and drew attention to the fact that Eastern Europe has emerged as a particular trouble spot for young people and HIV.

The UNAIDS report was released to mark the launch of a year-long initiative, **"Force for Change: World AIDS Campaign with Young People",** which aims to promote the participation of young people, and strengthen support for young people in their effort to fight AIDS. This report found that overall, children and especially young people still carry the heaviest burden of new infections. Among the key findings of the report (updated) are:

♦ More than half of all new HIV infections acquired after infancy occur among young people.

♦ Of the 43 million persons living with HIV/AIDS beginning year 2000, at least a third are young people.

♦ Every day, 7,500 young people worldwide acquire the virus. This translates into 2.7 million infections each year in this group (ages 10 to 24).

- About 2 million young people are infected with HIV every year in Africa, 700,000 in Asia and the Pacific.
- **Overall, young people ages 14 to 19 account for at least 50% of all those who become HIV-infected.**

Why the High HIV-Infection Rates Among Young People?

To some extent, the high infection rates reflect the preponderance of young people in the world. The global population is young. Between 1960 and 1990, when **the world population grew by 75%,** the proportion of young people **increased by 99%.** Of the world's young people, 85% live in developing countries, and this is where over 90% of the pandemic is now concentrated. But population percentages tell only part of the story. There are special reasons why young people are exposed to infection. Remember, that above all **HIV is a sexually transmitted virus.** And that adolescence and youth are times of discovery, emerging feelings of independence and the exploration of new behavior and relationships. A time of examination, rebellion and change. By definition adolescense is about taking risks and experimentation. Sexual behavior, an important part of this, can involve risks; the same is true of experimentation with drugs. At the time, young people get mixed messages. They are often faced with double standards calling for virginity in girls but early and active sexual behavior in boys. They have been told **"Just say NO"** since early 1980's, yet abuse continues. Perhaps because they are confronted with many millions of dollars worth of media images of sex, smoking, and drinking as glamorous and risk-free. They are told to be abstinent, but exposed to a barrage of advertisements using sex to sell goods. Compounding the challenge, in the name of morality, culture, or religion young people are often denied their right to education about the heath risks of sexual behavior, and to important tools and services for protection. Among the world's young people, some are more exposed to HIV than others. Those living in what UNICEF terms "especially difficult circumstances" include young people who are out of school, who live on the streets, who share needles with other infecting drug users, engage in commercial sex, or are sexually and physically abused. Young men who have sex with men are disadvantaged by the lack of information and services available to them and directed to their needs.

TEENAGERS: HIV-POSITIVE AND AIDS CASES—UNITED STATES

Teenagers often say, **"I KNOW HOW TO BE SAFE,"** but the statistics on teenage pregnancy, smoking, drinking of alcohol, drug use and, in particular, injection drug use, sexual behavior, and the use of contraceptives indicates otherwise. **Studies on teenage risk-taking continue to show that about half of teens are NOT WORRIED ABOUT BECOMING INFECTED WITH HIV AND THAT RELATIVELY FEW TEENAGERS HAVE CHANGED THEIR SEXUAL BEHAVIORS BECAUSE OF THEIR CONCERNS ABOUT THE HIV PANDEMIC.**

Teenagers/Young Women at Greatest Risk

In the United States, 25% of all new HIV infections are estimated to occur in young people between the ages of 13 and 20. While more research needs to be done on this topic, several factors associated with teenagers and young women are clear: **(1)** they tend to be partnered with **older men** who have had more sexual partners and have a greater chance of being infected with HIV and other sexually transmitted diseases (STDs); **(2)** their risk of HIV infection is greater because of their immature cervix and relatively low vaginal mucous production presents less of a barrier to HIV; and **(3)** many lack the education, social status, economic resources, and power in sexual partner relationships to make informed choices.

Safer Sex: Sex Thrills But AIDS Kills

There was a time when safer sex meant not getting caught by your parents. With time, sexually transmitted diseases and in particular HIV/AIDS have changed the meaning of safer sex. Today, over half of teenagers (aged 13 to 19) in the United States have had sex by the time they reach 16, and 7 in 10 are sexually active by 19. Many enter the sexual arena unprepared for the responsibility of their actions (Table 11-3).

TABLE 11-3   Sexually Active Teenagers by High School Year and Age (%)

High School Year	Age Group	Female	Male	Combined
9th Grade	14–15	31.9	48.7	40.3
10th Grade	15–16	42.9	52.5	47.7
11th Grade	16–17	52.7	62.6	57.7
12th Grade	17–19	66.6	76.3	71.5

Source: CDC HIV/AIDS Surveillance Report, November 1991:1–18.

--- **POINT OF VIEW 11.2** ---

PERILS OF UNSAFE SEX

As one teenager put it, "When you're a teenager, your hormones are raging and you think you're indestructible. But sex is how I got AIDS."

About 1 million teenage women become pregnant, outside of marriage, each year.

Whether or not society openly discusses it, teens are having sex. Many women—and most men—have their first sexual relations prior to marriage, usually during their teens, and most often those first encounters are unprotected. Research in family planning has revealed that the quality of reproductive health information is generally low among adolescents. This is a reflection in part of the lack of social acceptance of providing sex education and contraceptive services to teens in many countries. In the developing world, contraceptive services are often available only to married women, and in some situations, only to women who have already borne one or more children.

The guiding philosophy in dealing with sexuality in many cultures is **"if you don't talk about sex they won't do it."** This logic, however, is critically flawed. Teenagers are sexual beings at varying stages of self-awareness and understanding. Many teenagers continue to engage in sexual intercourse despite lack of access to any accurate information about sex and, in most cases, engage in "unsafe sex." **Safer sex requires an ability to distinguish between risky and non-risky sexual activities and the emotional security to choose safer sex.**

In a 1990 nationwide survey, 19% of high school students have had four or more sex partners by their junior year and 29% had four or more by their senior year (*MMWR*, 1992). Teenagers are experiencing skyrocketing rates of sexually transmitted diseases (STDs). **Every 10 seconds a teenager somewhere in the United States is infected with an STD.** (That's equal to 3 million teenagers aged 13 to 19 years.) People under age 25 account for 66% of all new STDs every year. One in four teens will contract an STD before finishing high school. In California, for example, 15-to 19-year-olds have the highest rates of gonorrhea and chlamydia of any age group in the state. Experts fear that if these diseases are being transmitted, then HIV is too (California Department of Health Services, 1995).

In April of 1992, the *MacNeil/Lehrer Report* estimated that 40,000 to 80,000 teenagers will become HIV-infected during the 1990s. Yet AIDS cases are relatively rare among **13- to 19-year-olds**. This is because of the 9- to 15-year time average from HIV infection to AIDS diagnosis. In 1981, there was **one** reported teenage AIDS case, by 1991 there were **789** reported cases. Beginning 1999, there were an estimated **5,222** (0.7% of the total AIDS cases). From 1991 to 1999 the number of AIDS cases in 13- to 19-year-olds increased by 84%.

--- **POINT OF INFORMATION 11.7** ---

A NEW ADOLESCENT SEXUALLY TRANS-MITTED DISEASE CENTER

In late 1998 the National Institute of Health awarded a $7 million grant for the creation of a sexually transmitted disease center at the Indiana University School of Medicine. The center will focus on adolescents, and will be the only one of its kind in the United States. The National Institute of Medicine recently identified STDs as a hidden epidemic, with 14 million reported STD cases in the United States in 1995—3 million of which occurred among adolescents. The grant will also fund a study on chlamydia infection in up to 600 girls aged 14 to 16 and their mothers.

HIV/AIDS Won't Affect Us!

In 1996, it was estimated that at least two teenagers per hour were becoming HIV-infected (8,760 per year) and this is expected to increase markedly through and beyond the year 2000. (Office of National AIDS Policy 1996). This means that regardless of available information on prevention, and, according to several large-scale studies, **teenagers do know** how HIV is transmitted, there has been a continuing increase in HIV infections within the preteen and early teen population! They continue to engage in sexual intercourse **without condoms!**

Two groups, teenage (adolescent) gay men and teenage women infected via heterosexual sex, account for about 75% of teenage HIV infections. Race is an important factor with regard to who becomes infected. Sixty-one percent of AIDS cases that occur in people ages 20 to 24, occur in blacks and Latino's but, they were HIV-infected in their teens! (Collins et al. 1997).

How Large Is the Teenage and Young Adult Population?

In the United States there are about 28 million teenagers between the ages of 13 and 19. This number is expected to increase to 43 million over the next 24 years (Sells et al., 1996). There are 18 million people ages 20 to 24, and about 22 million between the ages of 25 and 29. That's 68 million people between the ages of 13 and 29. Eighty-six percent of all STDs occur in the age group of 15 to 29. It has been

--------- POINT OF INFORMATION 11.8 ---------

FEDERAL PUBLIC HEALTH POLICY SUPPORTS ABSTINENCE-ONLY PROGRAM

Will it work? Years of heated debate on the failure vs. benefits of **sexual abstinence-only** education continues. However, in August of 1996, with the passage of the **Welfare Reform Act** the United States Congress allocated $50 million annnually for 5 years to states that would institute abstinence-only educational programs (no discussion of contraception). To qualify, states had to match $4 of federal money with $3 of state money. Thus the yearly total for abstinence-only education is $87 million. This legislation specifically requires funded programs to teach the social, psychological and health gains to be realized by abstaining from sex; that abstaining from sexual activity outside marriage is the expected standard for all school-age adolescents; that a mutually faithful monogamous relationship in the context of marriage is the expected standard of human sexual activity; and that **abstaining from sexual activity in the only certain way to avoid pregnancy, STD's, and other associated health problems.**

Clearly there are numerous health, economic and social benfits in delaying sexual onset in teenagers but disease prevention scientists say they can **not** identify a long-term advantage to abstinence programs relative to safer sex programs. Perhaps young people should receive two messages: one, promoting abstinence and the delay of sexual activity, the other, warning against high-risk behaviors and teaching teens how to protect themselves. These messages are not contradictory, but they are complex. "Don't drink, but if you drink, don't drive" is a similar complex message which has saved many people from death on highway. Prevention scientists offer significant evidence that safer-sex interventions work. But there is no clear and compelling evidence that abstinence-only programs work (DiClemente, 1998). The work of John Jemmott and colleagues (1998) and Douglas Kirby and colleagues (1998) demonstrate that for those adolescents that were sexually experienced at the beginning of their study, there was **no difference** in the proportion of the adolescents in the abstinence program relative to those in a safer-sex program with regard to having sexual intercourse. For those **not** sexually experienced, abstinence education was effective for a short time (3 months). After 12 months the adolescents in the two groups were each as likely to become sexually active and were similar to their pregnancy and sexually transmitted disease rates. At press time two states, California and New Hampshire, decided against the program and returned the Federal money.

Class Dsicussion: Is it good science, poor politics or poor science and good politics or some other combination of events that made it logical for the federal government to earmark tax dollars specifically for abstinence-only educational programs? Do you sense a religious involvement in the government policy? Explain.

estimated that 34% of all heterosexual adults with AIDS were infected with HIV as teenagers.

Estimate of HIV-Infected and AIDS Cases Among Teenagers and Young Adults

The total number of HIV-infected teenagers is unknown. **Federal health agencies estimate that teen-agers make up about 20% of the HIV-infected population.** They are a silent pool for eventual cases of AIDS. About 20% of the total number of AIDS cases in the United States, or 1 in 5 occur in people ages 20 to 29 (about 149,000 through 1999). Most of these people, given the 9- to 15-year period before being diagnosed with AIDS, were infected as teenagers! About 50% of HIV-infected teenagers come from six locations: New York, New Jersey, Texas, California, Florida, and Puerto Rico. The overall male-to-female ratio of AIDS cases in the United States is now about 7:1, but among 13- to 19-year-olds, the ratio in 1997 became about 1:1.

In New York City, teenagers account for 29% of all New York AIDS cases. AIDS is the sixth leading cause of death among persons aged 15 to 24 (*MMWR*, 1995a,b). Sixty percent of new HIV infections in women now occur during their adolescent years. **About 500,000 people in the United States under the age of 29 are HIV-positive.**

Teenagers and Incidence of AIDS Cases by Gender and Color

Teenagers of color account for 47%, whites 32% and Hispanics 19% of reported teen AIDS cases. Among males, white teenagers account for 41% of reported teen AIDS cases, followed by blacks (36%) and Hispanics (21%). Among females, black teenagers account for 66%, white for 17%, and Hispanic for 17%. **Teenage females, unlike their adult counterparts, are more likely to become infected with HIV through sexual exposure than through injection drug use.** A program that followed a large group of HIV-infected adolescents found that although 85% of females contracted HIV infection through heterosexual intercourse, very few were aware that their male partners had HIV infection at the time of their exposure (Futterman et al., 1992).

Categories of Teenage HIV Infection

From 1996 through 1999, the breakdown for HIV infection among teenagers was estimated as follows: males who have sex with males account for about 29%, patients with hemophilia and other coagulation disorders account for about 31%, heterosexual contact represents about 14%, intravenous drug use about 17%, and nonhemophilia-related transfusions and organ donations about 9%.

David Rogers, vice chairman of the National Commission on AIDS (which dissolved in 1993), said, "We have let issues of taste and morality interfere with the delivery of potentially life-saving information to young people."

Class Discussion: Do you think he is correct? What is the down side of his statement?

Runaway Teenagers

Each year since 1993, an estimated 3 million adolescents dropped out of high school. Youth drop-outs have higher frequencies of behaviors that put them at risk for HIV/STDs, and are less accessible to prevention efforts. Many teenagers choose to run away from home.

─────── BOX 11.4 ───────

HIV SPREADING RAPIDLY AMONG TEENAGE GAY MALES

Results of the first national survey of young gay and bisexual men (teens and early twenties) showed that 7% in the survey were HIV-infected (HEALTH, 1996). Researchers interviewed 1,781 men ages 15 to 22 in six urban counties: Miami's Dade County, Dallas County, and Los Angeles, San Francisco, and Santa Clara and Alameda counties in California. Results of the interviews demonstrated that 5% of the young men ages 15 to 19 and 9% of those ages 20 to 22 were HIV-infected. Four percent of whites, 7% of Hispanics, and 11% of blacks were infected. And, 38% reported unprotected anal sex within the previous 6 months. These 1996/1997 data, when compared to those data of a similar 1994/1995 survey, were about the same! Education and prevention control measures do not appear to be changing teenage gay sexual behaviors with regard to preventing HIV infection.

THOUGHTS AND COMMENTS FROM A GENERATION AT RISK

"If you're going to educate kids about AIDS, you have to educate them about drugs as well. If you're a youth, you're going to experiment with drugs, especially if you live in a metropolitan area. Even though you get stupid with drugs, you still think about things you don't want to do, but you do it anyhow."—**16-year-old HIV-positive youth from San Francisco**

"We grow up hating ourselves like society teaches us to. If someone had been 'out' about their sexuality. If the teachers hadn't been afraid to stop the 'fag' and 'dyke' jokes. If my human sexuality class had even mentioned homosexuality. If the school counselors would have been open to a discussion of gay and lesbian issues. If any of those possibilities had existed, perhaps I would not have grown up hating what I was. And, just perhaps, I wouldn't have attempted suicide."—**Kyallee, 19**

"People say HIV is this or that group's problem, not mine. But for HIV, it's a matter of risk behaviors, not risk groups. Because if you say it's a risk group thing, I don't identify with that group, so I'm not at risk. That makes people feel invincible to HIV."—**HIV-positive youth**

"I was infected with HIV by my first partner when I was 16 years old. Now at 20 I have this virus that's taking my life because everything I heard when I was younger was sugar-coated. We need more complete information than what we are being given. Even the pamphlets concerning HIV/AIDS prevention are too basic and bland. We need to know real stuff."—**Ryan, age 20**

"We, the young people of this country, need a place where we can go to ask our questions, where we won't be teased or ridiculed. We need a place where we can ask about our mixed up feelings, about sex, and about AIDS."—**15-year-old high school student from Concord, NH**

"If I could talk to the President, or a Senator, or anyone in the Federal Government who can make a difference, I'd tell them to take a look, learn a lesson from the youth that are currently dealing with the disease. Listen to them, hear their stories and then see that they have a future. If they don't have that future, then we don't have an America."—**Allan, San Francisco**

(Adapted from a *Report to the President*, March 1996)

The Homeless

An estimated 1 to 2 million runaway teenagers are homeless each year. These youth may engage in behaviors such as injection drug use, having multiple sexual partners, exchange of sex for money or drugs, and unprotected sexual intercourse that place them at risk for HIV infection. HIV seroprevalence studies among homeless youth have shown rates that are higher than adolescents in other settings (Rotheram-Borus et al., 1991).

In nine of 16 sites, HIV seroprevalence rates among homeless youth were higher than the median rates of youth attending STD clinics within the same city. These studies can only hint at how far HIV has infiltrated specific groups of teenagers.

If you are a teenager or know of one who needs help or has HIV/AIDS questions, call: **NATIONAL TEENAGERS AIDS HOTLINE** 1-800-234-8336.

Adolescent AIDS Program: Montefiore Medical Center, 111 E. 210th St., Bronx, NY 10467; (718) 882-0023

AIDS Community Alliance: Works with HIV postitive and HIV-affected individuals. 44 North Queens St., Lancaser, PA 17603; (717) 394-3380

Bay Area Young Positive: Youth-run, offers couseling, resources, newsletter. 518 Waller St., San Francisco, CA 94117; (414) 487-1616; e-mail: **BAYPOZ@aol.com.**

INTERNET

1. The Coalition for Positive Sexuality **(CPS)** Website **(http://www.positive.org/cps),** which provides information and advice on sexuality, is produced by and targets adolescents. The Coalition is a grassroots volunteer group based in Chicago. Their self-described mission is: "to

CAN LIFE GET WORSE THAN BEING YOUNG, HOMELESS AND DYING OF AIDS?

Regardless of cause, to have AIDS and nowhere to go, no one to help, to have no one who cares about you means that one has hit the bottom rung of life. **Many young people are there now!** There are about 15,000 to 20,000 young homeless people just in New York City. About 40% of them sell themselves in order to survive. Too many are becoming HIV-infected because of the demands of their lifestyle. A few examples should make the point.

Case 1. In the 10th grade he was kicked out of his home—he meandered over time to San Francisco's tenderloin area. He began working as a street prostitute and injecting drugs. **He wanted to become HIV infected.** He said he actually cried when he tested HIV negative. At age 19 he tested positive. This time he cried because he now recognized the mistakes he made along the way. He had no money, no job skills and no friends. Once, near death he said he had an epiphany (a sudden perception—the essential nature of a thing). He went to a clinic, received job training, gave up drugs, got a job and said, "I really want to live!"

Case 2. Her history begins in a North Carolina group home. She left and drifted to Georgia where she believes she became infected from a man who gave her $300 to have unprotected sex. She said she was a night child—she went to night clubs and adult bookstores to pick up her tricks. By her 21st birthday she said she was ready to jump off the Golden Gate Bridge—if she could find it. After finding help at a clinic she is now preparing to take her GED (graduate equivalence degree).

Case 3. He was raped by the son of his foster parents. His life was spent in a series of Milwaukee foster homes. He became HIV-positive at age 13. He found his way to San Francisco. He said "I felt I would be accepted there, it's a city that understood AIDS and the people that have it." He goes to a clinic that services 3,000 other young homeless people. He believes he has been sentenced to death. He has not been able to become comfortable with his HIV status. (Adapted from Russell Sabin, San Francisco Chronicle, Oct. 18, 1998.)

give teens the information they need to take care of themselves and in doing so, affirm their decisions about sex, sexuality, and reproductive control; second, to facilitate dialogue, in and out of the public schools, on condom availability and sex education." Included among the topics is information about safe sex, birth control, STD's, pregnancy, and being gay. Homosexual relations are discussed pretty much in the same manner as heterosexual relationships.

2. Although CPS is aimed at youth in general, **Oasis (http://www.oasismag.com)** targets and is written primarily by gay youth. Most of the columns written by contributors, who range in age from 14 to 22, read a lot like personal high school journals, and approach that undoubtedly makes readers feel comfortable—like hearing from a friend. A monthly advice column on sexual health is written by a physician and an epidemiologist, who are based in San Francisco area.

SUMMARY

AIDS surveillance and HIV seroprevalence studies indicate that a significant proportion of HIV infection among women in the United States is acquired through heterosexual contact. Because more men than women are HIV carriers, a woman is more likely than a man to have an infected heterosexual partner. The predominance of heterosexually acquired HIV infection in women of reproductive age has important implications for vertical HIV transmission to their offspring: nearly 30% of children with AIDS were infected by mothers who acquired infection through heterosexual contact. One of the greatest tragedies of the AIDS pandemic is orphaned children. They are left in hospitals because **(1)** their parents have died of AIDS or cannot care for them, or **(2)** no one wants them. AIDS at present is relatively rare among 13- to 19-year-olds, but 20% of AIDS cases, due to a 9- to 15-year period before AIDS diagnosis, had to begin with HIV-infected teenagers.

REVIEW QUESTIONS

(Answers to the Review Questions are on page 487.)

1. Beginning year 2000, how many women are expected to be HIV-positive worldwide?

2. Globally, what percent of **new** HIV infections occur in women?

3. What are the major routes of HIV transmission into women?

4. What is the most likely way a female prostitute in the United States becomes HIV-infected?

5. By the year 2001, how many women worldwide will have become HIV-positive and how many will have died from AIDS?

6. AIDS is now the _____ cause of death for all women between ages _____ and _____. It is the _____ leading cause of death in _____ women between the ages of _____ and _____ and the _____ cause of death for black women ages _____ to _____.

7. Beginning year 2000, in the United States how many HIV-infected women are of childbearing age?

8. Beginning year 2000, how many states have **not** reported a pediatric AIDS case.?

9. Since 1995, what percentage/year of HIV-infected newborns received HIV from their mothers?

10. List three major factors that are associated with perinatal HIV transmission.

11. Where do most of the orphaned AIDS children come from? Why are they called AIDS orphans?

12. What percentage of teenagers have had sexual intercourse by the age of 16?

13. Currently, _____ teens/hour are being HIV-infected in the United States? Worldwide _____/hour?

14. How many teenagers are there between the ages of 13/19 in the United States?

15. What two groups of teenagers in the United States (adolescents) account for the most HIV infections _____, _____? What is the percentage _____?

16. What percentage of all new HIV infections in the United States occur between ages 13 and 20?

17. About how many people in the United States under the age of 29 are estimated to be HIV-positive?

18. Is race an important factor with regard to who becomes HIV infected _____? What race or races have the largest number of HIV-infected young adults _____?

19. Of the 43 million people living with HIV disease, how many are young adults _____?

20. What is the National Teenagers AIDS Hotline telephone number?

REFERENCES

AGGLETON, PETER, et al. (1994). Risking everything? Risk behavior, behavior change, and AIDS. *Science*, 265:341–345.

ALLEN, J. R., , et al. (1988). Prevention of AIDS and HIV infection: Needs and priorities for epidemiologic research. *Am. J. Public Health*, 78:381–386.

BARDEGUEZ, ARLENE. (1995). Managing HIV infection in women. *AIDS Reader, Suppl.*, Nov/Dec, pp. 2–3.

BESSINGER, RUTH, et al. (1997). Pregnancy is not associated with the progression of HIV disease in women attending an HIV outpatient program. *Am J. Epidemiol.*, 147:434–440.

BUTLER, DECLAN. (1993). Who side is focus of AIDS research? *Nature*, 366:293.

California Department of Health Services, STD Control Branch. (1995). Sexually transmitted diseases in California. *Surveillance Report.*

CHU, SUSAN, et al. (1994). Female-to-female sexual contact and HIV transmission. *JAMA*, 272:433.

COHEN, JON. (1995). Women: Absent term in the AIDS research equation. *Science*, 269:777–780.

COLLINS, CHRIS, et al. (1997). Outside the prevention vaccum: Issues in HIV prevention for youth in the next decade. *AIDS Reader*, 7:149–154.

COTTON, PAUL. (1994). U.S. sticks head in sand on AIDS prevention. *JAMA*, 272:756–757.

CULNANE, MARY, et al. (1999). Lack of long-term effects of in utero exposure to Zidovudine among uninfected children born to HIV-infected women. *JAMA*, 281:151–157.

CURRAN, JAMES W., et al. (1988). Epidemiology of HIV infection and AIDS in the United States. *Science*, 239:610–616.

DARROW, W.W., et al. (1989). HIV antibody in 640 U.S. prostitutes with no evidence of intravenous (IV) drug abuse. *Fourth International Conference on AIDS*. Bio-Data Publishers: Washington, DC.

DIAZ, LESLIE, et al. (1996). Factors influencing mother-child transmission of HIV. *J. Fla. Med. Assoc.*, 83:244–248.

DICKINSON, GORDON M. (1988). Epidemiology of AIDS. *Int. Ped.*, 3:30–32.

DICKOVER, RUTH, et al. (1996). Identification of levels of maternal HIV-RNA associated risk of perinatal transmission: Effect of maternal Zidovudine treatment on viral load. *JAMA*, 275:599–610.

DICLEMENTE, RALPH. (1998). Preventing sexually transmitted infections among adolescents: A clash of ideology and science. *JAMA*, 279:1574–1575.

EHRHARDT, ANKE A. (1992). Trends in sexual behavior and the HIV pandemic. *Am. J. Public Health*, 82:1459–1464.

FARZADEGAN, HOMAYOON, et al. (1998). Sex differences in HIV-1 viral load and progression to AIDS. *Lancet* 352:1510–1514.

FELMAN, YEHUDI M. (1990). Recent developments in sexually transmitted diseases: Is heterosexual transmission of HIV a major epidemiologic in the spread of AIDS? III: AIDS in Sub-Saharan Africa. *Cutis*, 46:204–206.

FERRIS, DARON, et al. (1966). A neglected lesbian health concern: Cervical neoplasia. *J. Fam. Pract.*, 43:581–584.

FRANKE, K. (1988). Turning issues upside down. In: *AIDS: The Women*, I. Rieder and P. Ruppelt, eds., San Francisco: Cleis Press.

FUTTERMAN, DONNA, et al. (1992). Medical care of HIV-infected adolescents. *AIDS Clin. Care*, 4:95–98.

GWINN, MARTA, et al. (1991). Prevalence of HIV infection in childbearing women in the U.S.: Surveillance using newborn blood samples. *JAMA*, 265:1704–1708.

HEALTH. (1966). AIDS still spreading rapidly among young gay men. *Am. Med. News*, 39:30 (March issue).

HIV/AIDS Surveillance Report. (1995). Year end, 7:1–36.

HIV/AIDS Surveillance Report. (1996). Year end, 8:1–39.

HIV/AIDS Surveillance Report. (1997). December, 9:1–34.

HOLDEN, CONSTANCE. (1998). World-AIDS The worst is yet to come. *Science* 278:1715.

JEMMOT, JOHN, et al. (1998). Abstinence and safer sex high risk-reduction interventions for African American adolescents: A randomized controlled trial. *JAMA*, 279:1529–1536.

JOHNSON, TIMOTHY, et al. (1995). Current issues in the primary care of women with HIV. *Female Patient*, 20:51–58.

JOSEPH, STEPHEN C. (1993). The once and future AIDS epidemic. *Med. Doctor*, 37:92–104.

KIRBY, DOUGLAS, et al. (1997). The impact of the Postponing Sexual Involvement Curriculum among youths in California. *Fam. Plann. Perspect.*, 29:100–108.

LEMP, GEORGE, et al. (1995). HIV seroprevalence and risk behaviors among lesbians and bisexual women in San Francisco and Berkley, California. *Am. J. Public Health*, 85:1549–1552.

LUZURIAGA, KATHERINE, et al. (1996). DNA polymorase chain reaction for diagnoses of vertical HIV infection. *JAMA*, 275:1360–1361.

LUZURIAGA, KATHERINE, et al. (1998). Prevention and treatment of pediatric HIV infection. *JAMA*, 280:17–18.

MANDELBROT, LAURENT, et al. (1997). Natural conception in HIV-negative women with HIV-infected partners. *Lancet*, 349:850–851.

MANDELBROT, LAURENT, et al. (1998). Perinatal HIV-1 Transmission. *J. Am Med Assoc*, 280:55–60.

MIOTTI, PAOLO, et al. (1999). HIV transmission through breastfeeding. *JAMA*, 282:744–749.

Morbidity and Mortality Weekly Report. (1992). Selected behaviors that increase risk for HIV infection among high school students–United States, 1990. 41:236–240.

Morbidity and Mortality Weekly Report. (1993). Update: Acquired immunodeficiency syndrome–United States, 1992. 42:547–557.

Morbidity and Mortality Weekly Report. (1994). Update: Impact of the expanded AIDS surveillance case definition for adolescents/adults on case reporting–United States, 1993. 43:160–170.

Morbidity and Mortality Weekly Report. (1995a). Update: AIDS–United States, 1994. 44:64–67.

Morbidity and Mortality Weekly Report. (1995b). Update: AIDS among women–United States, 1994. 44:81–84.

Morbidity and Mortality Weekly Report. (1996a). HIV testing among women aged 18–44years–United States, 1991 and 1993. 46:733–736.

Morbidity and Mortality Weekly Report. (1996b). Update: Mortality attributable to HIV infection among persons aged 25–44 years–United States, 1994. 45:121–125.

Morbidity and Mortality Report. (1999). Surveillance for AIDS-defining opportunistic illnesses, 1992–1997. 48:1–20.

NEWELL, MICHAEL, et al. (1997). Immunological markers in HIV-infected pregnant women: The European Collaborative Study and the Swiss HIV Pregnancy Cohort. *AIDS*, 11:1859–1865.

Office of National AIDS Policy. (1996). Youth and HIV/AIDS: An American agenda. *Report to the President*, pp. 1–14.

PFEIFFER, NAOMI. (1991). AIDS risk high for women; care is poor. *Infect. Dis. News*, 4:1, 18.

RODRIGUEZ, ZOE, et al. (1996). Medical care and management of infants born to women infected with HIV. *J. Fla. Med. Assoc.*, 83:255–261.

ROTHERAM-BORUS, M., et al. (1991). Sexual risk behaviors, AIDS knowledge, and beliefs about AIDS among runaways. *Am. J. Public Health*, 81:208–210.

SCHABLE, BARBARA, et al. (1996). Characteristics of women 50 years of age or older with heterosexually aquired AIDS. *Am. J. Public Health*, 86:1616–1618.

SELLS, WAYNE, et al. (1996). Morbidity and mortality among US adolescents: An overview of data and trends. *Am. J. Public Health*, 86:513–519.

SELWYN, PETER A., et al. (1989). Knowledge of HIV antibody status and decisions to continue or terminate pregnancy among intravenous drug users. *JAMA*, 261:3567–3571.

THEA, DONALD, et al. (1977). The effect of maternal viral load on the risk of perinatal transmission of HIV. *AIDS*, 11:437–444.

THOMAS, PATRICIA. (1988). Official estimates of epidemic's scope are grist for political mill. *Med. World News*, 29:12–13.

THOMAS, PATRICIA. (1989). The epidemic. *Med. World News*, 30:41–49.

UNAIDS, (1997). Global summary of the HIV/AIDS epidemic. *Report on the global HIV/AIDS epidemic*. December 1997:1–25.

Women's AIDS Network. (1988). Lesbians and AIDS: What's the connection? *San Francisco AIDS Foundation*, 333 Valencia St., 4th Floor, P.O. Box 6182, San Francisco, CA 94101-6182.

World Health Organization, Geneva. (1994). *Women's Health*, p. 18.

WORTLEY, PASCALE, et al. (1997). AIDS in women in the United States. *JAMA*, 278:911–916.

YAO, FAUSTIN K. (1992). Youth and AIDS: A priority for prevention education. *AIDS Health Promotion Exchange No. 2*, Royal Tropical Institute, The Netherlands: 1–3.

Testing for Human Immunodeficiency Virus

CHAPTER CONCEPTS

♦ ELISA means enzyme linked immunosorbent assay; it is a screening test for HIV infection.

♦ Western blot is a confirmatory HIV test.

♦ The ELISA test has been used to screen all blood supplies in the United States since March 1985.

♦ HIV screening tests can produce both false positives and false negatives.

♦ A positive ELISA test only predicts that a confirmatory test will also be positive.

♦ False-positive readings result from a test's lack of specificity.

♦ There is a relationship between the incidence of HIV in the population being tested and the number of false positives reported. The higher the incidence, the fewer the false positives.

♦ Levels of test sensitivity and specificity must be determined.

♦ Several new screening and confirmatory HIV tests are now available.

♦ June 27th is national HIV Testing Day.

♦ Improved screening of nation's blood supply.

♦ Saliva and urine HIV tests are FDA approved.

♦ The polymerase chain reaction test is the most sensitive HIV RNA test currently available.

♦ Other HIV RNA tests available are Amplicor and the branched-DNA test.

♦ Rapid HIV test kits used in other countries now to be used in the USA.

♦ AIDS cases have been reported in all 50 states.

♦ Forty three states have some form of HIV reporting.

♦ Competency and informed consent are necessary for most HIV testing.

♦ Mandatory HIV testing does not mean people can be forced to undergo testing.

♦ HIV testing for the most part is on a voluntary basis.

♦ Compulsory HIV testing is used in the military, prisons, and in certain federal agencies.

♦ FDA approves two home HIV Test Kits.

- U.S Public Health Service guidelines for annual prenatal HIV counseling and **voluntary** HIV testing of all pregnant American women.
- New York is first state to legislate mandatory HIV testing and disclosure of newborn HIV status to mother and physicians.
- American Medical Association endorses mandatory HIV testing of all pregnant women and newborns.

Let's begin this chapter by asking, **WHY DOES TESTING MATTER? ANSWER:** The most basic epidemiology holds that early knowledge of where a virus is moving—into which populations—is essential to slowing its spread. Even if a disease cannot be cured, knowing who the infected people are may help prevent the transmission of the disease to other people.

Beginning year 2000, of the estimated **43 million** living HIV-infected people worldwide, at least **half** became infected **before age 25.** About 10% know they are HIV-positive. HIV testing is not readily available in many places in developing nations, and in developed nations, HIV testing is a controversial component of the HIV/AIDS spectrum of services. This chapter presents HIV testing information and some of the important problems connected with **whom, how,** and **where** to test.

DETERMINING THE PRESENCE OF ANTIBODY PRODUCED WHEN HIV IS PRESENT

HIV-antibody testing is the most readily available, inexpensive, reliable, and accurate method to identify whether a person is infected with HIV. **HIV antibodies are found in the blood and in other body fluids.** When properly performed, HIV-antibody testing is highly sensitive and specific.

REQUESTS FOR HIV TESTING

Public HIV/AIDS clinics in the United States are becoming overwhelmed with requests for HIV testing. Most of the requests are repeats. Beginning year 2000 about 65% of the CDC-estimated 650,000 to 900,000 living HIV-infected persons in the United States have been tested. However, a large percentage of

people **at risk** for HIV infection have **not** been tested, including most **hardcore IDU's and sexual partners of IDU's.** Currently, about 200 million blood and plasma samples are HIV tested annually worldwide. In the United States about 50 million blood and plasma samples are HIV tested annually.

In the developing world, where the greatest number of HIV-infected people are concentrated, the picture is very different. About 90% of the infected people do not know they are infected! HIV testing is done mostly for purposes of surveillance, which involves very small population samples and is done anonymously. **Few people have any hope of treatment, so they feel little incentive to get tested.** But even those who would want to know may not be able to find out. In many countries, there are no voluntary testing and counseling facilities; people have no acceptable way of learning if they are HIV-infected. An ongoing study at a rural hospital in South Africa suggests that only 2% of people who are HIV-positive know their status. The situation in urban Kenya is equally poor. Of 63 randomly chosen women who tested HIV-positive in one study, only one was aware that she was infected.

REASONS FOR HIV TESTING

HIV testing is done to monitor the pandemic; to determine how many people are infected, how many are being infected in a given time

————— BOX 12.1 —————

AN ASSUMPTION OF AIDS WITHOUT THE HIV TEST: IT SHATTERS LIVES

Case I

San Francisco— For six years, a 53-year-old gay male lived in the nether world of AIDS. He stopped working, suffered the painful side effects of experimental drugs, and waited to die.

Now his doctors say he never had the disease.

His health shattered by AIDS treatment, his livelihood lost, he filed a $2 million claim against Kaiser Permanente health maintenance organization. He claims he underwent sustained treatment for full-fledged AIDS without receiving an HIV test.

His attorney said, "For six years he thought that the most he had was six months to live. So everyday he'd wake up and think 'Is this the last day of my life?'"

To begin, this male says he checked into a San Jose hospital affiliated with Kaiser in 1986 with respiratory problems and doctors told him he had pneumocystis pneumonia, considered a sure sign of AIDS at the time.

He underwent tests but **was not** given one to determine the presence of HIV, the virus associated with AIDS.

In 1986, he began taking the drug zidovudine in high doses, which gave him a chronic headache, high blood pressure, and peripheral neuropathy— permanent pins and needles pains from his calves to his feet. He is battling an addiction to Darvon and other prescription drugs.

Under doctors' orders, he quit his job as a skin care technician and lives on government welfare and disability benefits of $600 a month.

(Associated Press, 1992)

Case II

In 1980, she had received a blood transfusion during surgery at a hospital in this Southeast Georgia town. During a checkup for a thyroid problem a decade later at a clinic in Hialeah, Florida, her blood was taken for testing.

On November 13, 1990, her telephone rang. She was asked to come down to the local health clinic where she was told she had AIDS. They were not sure how long she had to live. She was 45 years old. Her three sons were then teenagers; their father had died.

She kept the television on continually in a usually unsuccessful effort to block out the thought of AIDS.

The nights were the worst.

"I'd go to bed every night thinking about dying. What color do you want the casket to be? What dress do you want to be buried in? How are your kids going to take it? How will people treat them? I was afraid to go to sleep."

In 1992, her doctor put her on didanosine, (ddl), which brought on side effects that included vomiting and fatigue.

"I had put my kids through hell. They were scared for me."

When she joined a local hospice group for AIDS patients, counselors heard her story and noted that her T-cell counts had remained consistently high. At their suggestion, she was retested.

In November 1992, nearly 2 years to the day she was told she was HIV-positive, another call came. She was greeted at the clinic with these words: **"Guess what? Your HIV test came out negative!"**

She sued the Florida Department of Health and Rehabilitative Services—the agency that performed the test—and the clinic and doctor who treated her.

A jury awarded her $600,000 for pain and suffering but cleared the clinic and said the bulk must be paid by the agency.

Case III

Everyday, four times each day, for six years Mark swallowed his antiretroviral drugs. Mark was told by a "fine physician" that he was HIV positive in July of 1990. Regardless of the drugs, Mark felt sick and suffered further physical effects and depression. For reasons not given, Mark moved from Chicago to Ohio. His new physician was puzzled that Mark did not demonstrate signs or symptoms of HIV infection. His tests on Mark came back **HIV-negative.** On investigation, the Chicago clinic could not produce any documents showing that Mark was ever HIV tested! Mark is now suing the physician who put him on antiretroviral drugs without first testing him for HIV. Case pending.

---

period (incidence), and their location; to determine the impact of prevention efforts to slow the spread of HIV; to prompt behavior change, to provide entry into clinical care, if necessary, to provide a starting point for partner notification and education, and to protect

THE MIAMI HERALD, SEPTEMBER 1985

The **Metro Commissions** of Dade County Florida voted to require **all** food service workers in Dade County by HIV tested. The commissioners' motivation is not in question. Aware of how the AIDS panic has hurt restaurant patronage in several other cities, they properly want to calm fears and protect food service jobs. Inadvertently, though, the Metro commissioners may have sent the wrong signal. Medical experts say that AIDS is not spread by the kinds of contact that food handlers have with a restaurant's patrons or with their food. By ignoring this advice, commissioners implicitly seem to be questioning it. This may heighten the public's fears, not calm them. That is especially evident when one notes the test's inability to do what is expected of it. Richard Morgan, Dade's Health Director, says that testing all 80,000 of Dade's food service workers would take a year. So even if AIDS were spread through food—which it is not—the test would be of dubious value. A food service worker might test clean one day, become infected the next, continue working, and escape detection for months. Moreover, the high worker turnover in this seasonal business means that large numbers of these costly tests would continue to be given indefinitely. Thus the costs would not necessarily diminish much after all current workers had been tested. The AIDS-test requirement is part of a broader health-care ordinance that commissioners tentatively approved.

Chicago Tribune, 1985

In early 1985 Chicago's Health Commissioner protested federal guidelines requiring that blood donors be informed if their blood was HIV positive. In response the following article appeared on April 10, 1985: "Chicago Health Commissioner Lonnie Edwards is wrong to protest federal guidelines requiring that blood donors be informed if tests show that their blood contains antibodies to the virus linked to AIDS (acquired immune deficiency syndrome). The new test for AIDS antibodies is being phased into use in the nation's blood banks, after its approval early in March by the Department of Health and Human Services. Blood found to have AIDS antibodies will be discarded. By late April, all blood banks in the Chicago area will be doing such screening. But the antibody test is still controversial. It cannot show whether a blood donor currently has AIDS. A positive result only indicates that a person has been exposed at some time and in some way to the virus believed to cause the usually fatal disease—not that he is immune to it or that he will eventually become ill and die of it. Another problem is that the test is not totally accurate; both false negatives and false positives can occur. Some homosexual groups oppose the use of the AIDS screening test on the grounds it could violate individual privacy or that a third party might obtain test results and use them to discriminate against donors. But the presence of AIDS virus in the nation's blood supply is becoming a serious problem. Dozens of recipients of blood or blood-based products have already died of AIDS presumably acquired from blood donors. In a few instances, these victims have passed the disease on to their spouses and young children.

the nation's blood supply. **There is an immediate need to change perception about being HIV-positive so that people feel good about taking the test to protect themselves and others, rather than the discrimination that now exists against those who have taken the test.**

AIDS is diagnosed by evaluating the results of clinical findings during a physical examination, and by laboratory analysis of blood samples for the presence of antibodies to HIV. Until early 1983, the causative agent of AIDS was unknown. Up to that time, infection could only be determined after the person began to express the signs and symptoms associated with AIDS. Once the virus was identified, a means of detecting viral exposure was developed. There is no single diagnostic test for AIDS. To date all **screening** tests, in the United States, measure the absence or presence of antibodies against **HIV subtype B,** a strain of HIV found primarily in North America and Europe (Hu et al., 1996). HIV counseling and testing are important components of HIV prevention programs.

IMMUNODIAGNOSTIC TECHNIQUES FOR DETECTING ANTIBODIES TO HIV

Refinements in the field of immunological testing, serology, and the study of antigen–antibody reactions have produced test names that reflect the component parts of the test being used. In most cases, tests are based on the detection of antibodies present in the serum, in this case antibodies to HIV. One immunological test uses antibodies, which if present in the person's serum, form a complex with a given antigen. An enzyme is then connected to the antibody. The presence of the antibody can be determined by adding a reagent that will form a colored solution if antibody to HIV is present. This is called the **enzyme linked immunosorbent assay** (ELISA, el-i-sa). The ELISA test was first used in 1983 to detect anti-bodies against HIV.

Recently, a technique has been developed that uses fluorescently labeled antibodies against HIV-specific antigens. This technique, called the fluorescent antibody technique, and others are described.

ELISA HIV-ANTIBODY TEST

The initial application of the ELISA test outside the research laboratory, was used primarily in large-scale screening of the nation's blood supply. ELISA **testing of the existing blood supply and all newly donated blood began in the United States in March 1985.** The ELISA test is used as a screening test because of its low cost, standardized procedures, high reproducibility, and rapid results.

The ELISA test is now semi-automated. It uses an antigen derived from HIV cultivated in human cell lines. One such line is the human leukemia cell line **H9.** Whole viruses isolated from H9 **are disrupted** into subunit antigens for use. The subunits of HIV are then bound to a solid support system.

Two different solid support systems are used in the seven ELISA screening test kits licensed in the United States. Some attach the antigens

onto small glass beads (Figure 12-1A), while others fix the antigens onto the sides and bottoms of small wells (microwells) in a glass or plastic microtiter plate (Figure 12-1B). The serum to be tested is separated from the blood and is diluted and applied to the HIV-coated solid support systems (Figure 12-2). The ELISA test takes from 2.5 to 4 hours to perform (Carlson et al., 1989) and costs between about $8 in state sponsored virology laboratories and about $60 to $75 in private laboratories.

In accordance with FDA recommendations, effective June 1992, blood collection centers in the United States began HIV-2 testing on all donated blood and blood components. The CDC does not recommend routine testing for HIV-2 other than at blood collection centers (*MMWR*, 1992).

Understanding the ELISA Test

The ELISA test determines if a person's serum contains antibodies to one or more HIV antigens. Although there are some minor differences among the FDA licensed kits, test procedures are similar.

Problems with the ELISA Test

Any HIV screening test must be able to distinguish those individuals who are infected from those who are not. **The underlying assumption of an ELISA test is that all HIV-infected people will produce detectable HIV antibodies.** However, the HIV-infected population in general does not produce detectable antibodies for 6 weeks to 1 or more years after HIV infection, the **window period. Most often, HIV antibody is detectable within 6 to 18 weeks.** Thus, HIV-infected people can test HIV-negative. **This is**

Solid Supports

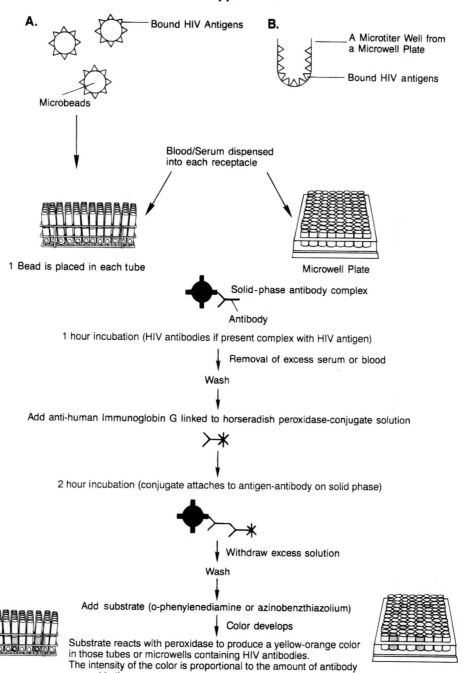

A. — Bound HIV Antigens

Microbeads

B. — A Microtiter Well from a Microwell Plate

— Bound HIV antigens

Blood/Serum dispensed into each receptacle

1 Bead is placed in each tube

Microwell Plate

Solid-phase antibody complex

Antibody

1 hour incubation (HIV antibodies if present complex with HIV antigen)

Removal of excess serum or blood

Wash

Add anti-human Immunoglobin G linked to horseradish peroxidase-conjugate solution

2 hour incubation (conjugate attaches to antigen-antibody on solid phase)

Withdraw excess solution

Wash

Add substrate (o-phenylenediamine or azinobenzthiazolium)

Color develops

Substrate reacts with peroxidase to produce a yellow-orange color in those tubes or microwells containing HIV antibodies. The intensity of the color is proportional to the amount of antibody present in the serum.

FIGURE 12-1   The ELISA Test. **A,** Microbeads with attached antigen in test tubes. **B,** Antigens bound to walls and bottom of microtiter wells.

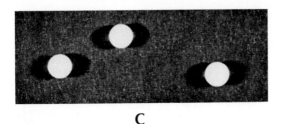

C

D

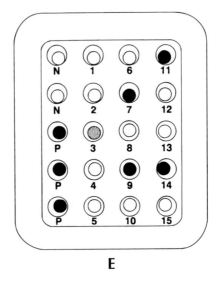

E

FIGURE 12-1 *(continued)* **C**, Microbeads are 7 mm in diameter. The test takes between 2.5 and 4 hours to perform. **D**, Microwell plates showing positive (yellow) and negative (clear) test results. The microtiter dispenser handles eight microwells at a time. (**C** *and* **D**: *Photographs courtesy of the author*) **E**, Specimens positive for HIV antibody have a deeper color in this microwell tray. Serum specimens from 15 patients were tested for antibodies to HIV. Two negative and three positive control specimens are provided in the first column. In wells 7, 9, 11, and 14, the dark yellow color change, matching the color in the three positive control wells, indicates that the specimens are positive. Well 3 shows a weakly reactive result. The remaining specimens showed no color change and were interpreted as negative for HIV antibodies. (Adapted from Fang et al., 1989)

a false negative result. In some HIV-infected persons, as their disease progresses, the virus ties up the available antibody. Testing at this time may also produce false negative results. Other reasons for false negative results are: immunosuppressive therapy, blood transfusions, some forms of cancer, B-cell dysfunction, bone marrow transplantation, and starch powder from laboratory gloves (Cordes et al., 1995).

An Unusual Case— At the VA Medical Center, Salt Lake City, Utah, in 1997, a man tested **HIV-negative 35 times over a 4-year period.** Because his wife was HIV-positive and because he demonstrated symptoms of HIV disease, p24 and PCR assays were performed—he was HIV-positive. [p24 antigen (see Point of Information 12.2) and PCR test are explained later in this chapter.] This case is unusual because (1) he is falsely negative almost 4 years beyond the window period; (2) the strain of HIV is typical of that found in the United States; and (3) the strain of HIV is closely related to the strain infecting his wife (Reimer et al., 1997).

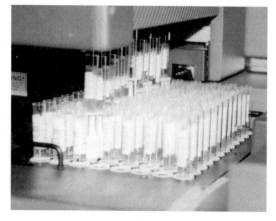

A

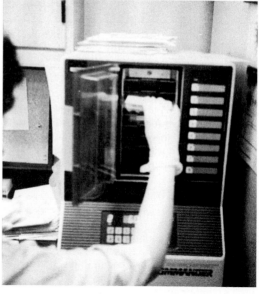

C

B

FIGURE 12-2   Semi-Automated ELISA Test. **A,** Serum samples are individually machine diluted with a special solution 1 to 400 to achieve a balance between the amount of antibody to the amount of antigen. Antibodies, if present in the individual sera, attach to the HIV antigens in each well. **B,** Excess serum is withdrawn from each sample and the beads are washed. The antihuman immunoglobulin-horseradish enzyme conjugate is added to each prepared sample. **C,** The samples are incubated. **D,** Each sample receives the chromogen or substrate (o-phenylenediamine or azinobenzthiazolium). A yellow-orange color appears in samples that contain antibodies to HIV. (*Courtesy of Florida Department of Health, Retrovirology Unit, Jacksonville*)

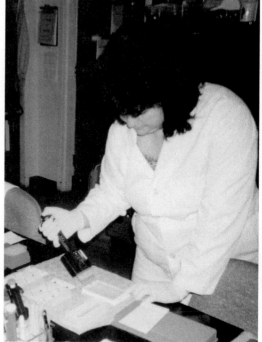

D

——— BOX 12.2 ———

PRE-TEST PATIENT INFORMATION

It is important that you read and understand the following information before having the HIV-antibody test.

The HIV-antibody test is a blood test. It was first used in 1985 to screen donated blood. **This test is not a test for AIDS. But a positive result does mean that there is a high probability an HIV-infected person will eventually develop AIDS.**

Only anonymous test results remain absolutely confidential.

What Does the Test Reveal?

The test reveals whether a person has been exposed to HIV. The test detects the presence of HIV antibodies, an indication that infection has occurred. It does not detect the virus itself nor does it indicate the level of viral infection, **only that one has been infected.**

It is important to find out whether you are HIV-positive or negative so that you can prevent spreading the virus to others and seek early medical intervention.

What Tests Will Be Done?

Wherever the test is done, the procedures are similar. A blood sample is taken from the arm and analyzed in a laboratory using a test called *ELISA* (*enzyme linked immunosorbent assay*). If the first ELISA test is positive, the laboratory will run a second ELISA test on the same blood sample. If positive, another test, called Western blot, is run on the same blood sample to confirm the ELISA result.

What Do the Test Results Mean?

See Table 12-1.

Positive Result: A positive result suggests that a person has HIV antibodies and may have been infected with the virus at some time. People with

TABLE 12-1 The Meaning of Antibody Test Results

A Positive Results	B Negative Results
If you test positive, it does mean: 1. Your blood sample has been tested more than once and the tests indicate that it contained antibodies to HIV. 2. You have been infected with HIV and your body has produced antibodies. **If you test positive, it does not mean:** 1. That you have AIDS. 2. That you necessarily will get AIDS, but the probability is high. You can reduce your chance of progressing to AIDS by avoiding further contact with the virus and living a healthy lifestyle. 3. That you are immune to the virus. Therefore, if you test positive, you should do the following: 1. Protect yourself from any further infection. 2. Protect others from the virus by following AIDS precautions in sex, drug use, and general hygiene. 3. Consider seeing a physician for a complete evaluation and advice on health maintenance. 4. Avoid drugs and heavy alcohol use, maintain good nutrition, and avoid fatigue and stress. Such action may improve your chances of staying healthy.	**If you test negative, it does mean:** 1. No antibodies to HIV have been found in your serum at the time of the test. Two possible explanations for a negative test result exist: 1. You have not been infected with HIV. 2. You have been infected with HIV but have not yet produced antibodies. Research indicates the most people will produce antibodies within 6 to 18 weeks after infection. Some people will not produce antibodies for at least 3 years. A very small number of people may never produce antibodies. **If you test negative, it does not mean:** 1. That you have nothing to worry about. You may become infected, be careful. 2. That you are immune to HIV. It has not yet been shown that anyone is immune to HIV infection. 3. That you have not been infected with the virus. You may have been infected and not yet produced antibodies.

(Adapted from the San Francisco AIDS Foundation)

a positive result should assume that they have the virus and could therefore transmit it:

- by sex (anal, vaginal, or oral) where body fluids, especially semen or blood, get inside the partner's body;
- by sharing injection drug "works" (needles, syringes, etc.);
- by donating blood, sperm, or body organs; or
- to an unborn baby during pregnancy or to a newborn by breast feeding.

Negative Result: If it has been 4 to 6 months since the last possible exposure to HIV, then a negative result suggests that a person is probably not infected with the virus. However, **false negatives** can occur. For example, if the test was too soon after HIV exposure, the body may not have had time to produce HIV antibodies. **Collectively, studies show that 50% of HIV-infected persons seroconvert (demonstrate measurable HIV antibodies) by 3 months after HIV infection and 90% seroconvert by 6 months. A small percentage seroconverted at 1 year or later. Most important, a negative result does not mean that you are protected from getting the virus in the future—your future depends on your behavior.**

Inconclusive Result: A small percentage of results are inconclusive. This means that the result is neither positive nor negative. This may be due to a number of factors that have nothing to do with HIV infection, or it can occur early in an infection when

there are not enough HIV antibodies present to give a positive result. If this happens, another blood sample will be taken at a later time for a retest.

How Accurate Is the Test Result?

The result is accurate. However, as with any laboratory test, there can be false positives and false negatives.

False Positive: A small percentage of all people tested may be told they have the HIV antibody when, in fact, they do not. This can be due to laboratory error or certain medical conditions which have nothing to do with HIV infection.

False Negative: Some people are told that they are not HIV-infected when, in fact, they are. This can happen when the test is taken too soon after being infected—the body has not had time to produce HIV antibodies. This is called the **window period.**

Should the Test Be Performed?

Whether to have the test done is a personal decision. However, knowing your antibody status can help you address some very difficult questions. For example, if the test is negative, what lifestyle changes should be made to minimize risk of HIV infection? If the test is positive, how can others be protected from infection? What can be done to improve your chances of staying healthy?

(Source: Roche Biomedical Laboratories, Inc.)

False-positive reactions may also occur. This means that the person's serum does not contain antibodies to HIV but the test results indicate that it does.

Purpose of the ELISA Test

Because the original purpose of the ELISA test was to screen blood, the sensitivity (ability to detect low-level color formation; see Figure 12-1) of the test was purposely set high. It was reasoned that it was better to have some false positives and throw away good blood rather than to take in any HIV contaminated blood. Thus the ELISA test is a **positive predictive value** test. It only predicts that the serum tested will continue to test positive when a test with greater specificity, called a **confirmatory test**, is done.

In 1985, during the first month of donor screening, 1% of all blood tested HIV-antibody-positive. On ELISA retesting of these samples, only 0.17% (17/10,000) were HIV-antibody-positive. On subjecting these samples to a confirmatory test, only 0.038% (4/10,000) were actually HIV-positive. These early tests produced about 24 false positives for every true positive result. The main reason for such a high false positive rate or **lack of specificity** was that something other than HIV produced an antibody or other substance that reacted with HIV antigen causing the HIV test to appear positive.

People may test **false positive** who have an underlying liver disease, have received a blood transfusion or gamma globulin within 6 weeks of the test, have had several children, have had

IMPROVED ELISA TESTING, SCREENING FOR p24 ANTIGEN, AND BLOOD DONOR INTERVIEWS LOWER HIV IN NATION'S BLOOD SUPPLY

The increased sensitivity of current HIV-antibody ELISA tests, improved donor interviewing about behaviors associated with risk for HIV infection, and deferral of donors who test positive for HIV, hepatitis, human T-cell leukemia virus type 1 (HTLV-1), or syphilis have considerably improved the safety of the U.S. blood supply. In 1993, approximately six per 100,000 blood donations collected by the American Red Cross tested positive for HIV antibody. In addition, an estimated one in 450,000 to one in 660,000 donations per year (i.e., 18–27 donations) were infectious for HIV but were not detected by current screening tests.

Screening for p24 Antigen— In August 1995, the FDA mandated that all blood and plasma collection centers screen all blood for p24 antigen. The FDA recommended p24 screening as an additional safety measure because recent studies indicated that p24 screening reduces the infectious window period (the FDA approved Coulter p24 antigen blood test detects HIV as early as 16 days after infection). Among the 12 million-plus annual blood donations in the United States, p24-antigen screening is expected to detect four to six infectious donations that would not be identified by other screening tests. FDA regards donor screening for p24 antigen as an interim measure pending the availability of technology that would further reduce the risk for HIV transmission from blood donated during the infectious window period (*MMWR*, 1996 updated).

rheumatological diseases, malaria, alcoholic hepatitis, autoimmune disorders, various cancers, **accute cytomegalovirus infection,** or DNA viral infections; are injection drug users; or have received vaccines for influenza or hepatitis B (Fang et al., 1989; MacKenzie et al., 1992). In each case, the person may have antibodies that will cross-react with the HIV antigens giving a false-positive reaction. Other reasons for false positives are laboratory errors and mistakes made in reagent preparations for use in the test kits.

Although high sensitivity tests eliminate HIV-contaminated blood from the blood supply, there is a downside to high sensitivity testing when proper procedure is **not** used. People told that they have tested positive have become emotionally distraught. Former Senator Lawton Chiles of Florida, at an AIDS conference in 1987, told of a tragic example from the early days of blood screening in Florida. Of 22 blood donors who were told they were HIV-positive by the ELISA test, seven committed suicide.

There continue to be false positive reactions among blood donors and low-level risk populations because of a **low prevalence** of HIV infection in such populations. The American Red Cross Blood Services laboratories report that using current ELISA methodol-

ogy, a specificity of 99.8% can be achieved (*MMWR*, 1988a).

Positive Predictive Value— The positive predictive value of the ELISA test indicates the percentage of true positives among total positives in a given population. To determine a positive predictive value:

Number of true positives ÷ Number of positives or true positives + false positives × 100%

There is also a negative predictive test. A negative predictive value refers to the percentage of individuals who test truly negative; they **do not** have HIV. It is determined by:

Number of true negatives ÷ Number of true negatives + false negatives × 100%

To safeguard against false-positive tests, the CDC recommends that serum that tests positive be retested twice (in duplicate). If both tests are negative, the serum is considered HIV-antibody-negative and further tests will only be done should signs or symptoms of HIV infection occur (Figures 12-3A and 12-3B). If one or both of the tests is positive, the serum is subjected to a confirmatory test, usually a Western blot (WB) (Figure 12-4). At blood banks, if the

A

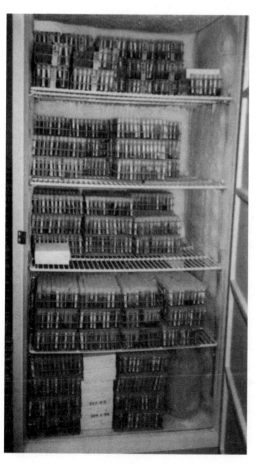

B

FIGURE 12-3 **A.** Incoming specimens. About 900 to 1000 blood specimens are received each day and are prepared for testing. From those samples, those testing negative are reported out in 8 to 10 hours. Positive samples are reported out in 3 to 4 days. **B.** A collection of frozen serum samples. The 7,500 samples in this freezer are part of over 20,000 samples collected each month in the Jacksonville Retrovirology Unit (one of two such units in Florida). From 1992 through 1999 an average of 8,000+ samples of 245,000+ tests each year were HIV-positive (3.5%/year). **C.** Typical algorithm followed by public health laboratories, blood banks and clinical settings. Some ELISA testing uses recombinant protein and whole viral lysates instead or synthetic peptide HIV. Either blood, serum or saliva can be screened for antibodies to HIV. If the ELISA test is positive, it is repeated in duplicate. If duplicate tests are positive, a Western blot (WB) test is performed. Borderline reactions indicate that antibodies to HIV may be present but in small numbers (low titer) or that other antibodies present in the serum are cross-reacting with HIV antigens. To determine the situation, the more specific WB test is required. For information on antibody screening procedures, see *MMWR*, 1988. (*Courtesy of Florida Department of Health, Retrovirology Unit, Jacksonville.*)

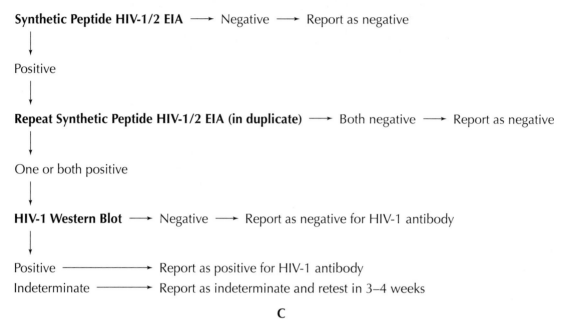

Florida Department of Health
HIV Antibody Testing Algorithm (Effective 12/10/97)

Synthetic Peptide HIV-1/2 EIA ⟶ Negative ⟶ Report as negative

↓

Positive

↓

Repeat Synthetic Peptide HIV-1/2 EIA (in duplicate) ⟶ Both negative ⟶ Report as negative

↓

One or both positive

↓

HIV-1 Western Blot ⟶ Negative ⟶ Report as negative for HIV-1 antibody

↓

Positive ⟶ Report as positive for HIV-1 antibody

Indeterminate ⟶ Report as indeterminate and retest in 3–4 weeks

C

FIGURE 12-3 *(continued)*

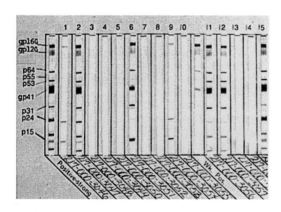

FIGURE 12-4   Western Blot Strips. Each WB strip contains nine separate antigenic proteins of HIV. Human serum or blood is applied directly to the strips. See text for details on reactions. Because false positives can sometimes occur with the ELISA test, additional testing is needed to evaluate specimens which are repeatedly reactive by ELISA. The Western blot is more specific but less sensitive than the ELISA and is recommended for blood banks and organ donor centers. Its clinical usefulness in trials to aid in evaluating specimens which are questionably positive by other methods has been proven. It is not a screening test because it lacks a high level of sensitivity. (*Courtesy of Roche Biomedical Laboratories, AIDS Testing Brochure*)

initial ELISA test is positive, the blood is discarded. If an individual's serum subjected to a confirmatory test is positive, the person is considered to be HIV-infected.

Although confirmatory tests can be used to determine true-positive results, they are too labor intensive and expensive to be used in screening a large population. Thus the positive predictive value of an ELISA test is an important first step in large-scale screening. Recall, however, that **the predictive value depends on the prevalence of HIV infection in the population tested.** The higher the prevalence of HIV infection in a given population, the more likely a positive ELISA test is to be a true positive; and conversely, the lower the prevalence of infec-

———— BOX 12.3 ————

THE PERFORMANCE RATE FOR THE COMBINED ELISA AND WESTERN BLOT HIV TEST—IS 99% ACCURACY GOOD ENOUGH? THE ANSWER: NO!

The Centers for Disease Control and Prevention (CDC) states that the two tests used to identify HIV— the ELISA and the Western blot (WB)—used in combination, have a better than 99% overall accuracy rate, but only if they are performed repeatedly. (The exact rate is unknown and the CDC states that it has no data on just how many false positives versus false negatives occur!) **The rate is the percentage of correct test results in all specimens tested.** With a 99% rate, if a population of 10,000 were tested, 9,900 would receive correct results, but 100 would receive erroneous results—either false positives or false negatives—including indeterminates.

If 99% accuracy is used as an example, then false positives would have to be less than 6/10ths (0.6%) of the erroneous results, because the CDC estimates that 0.6% of Americans are HIV-positive. That is, if false positives accounted for fully 0.6% of the errors, then the 0.6% of people who are HIV-positive would all be false positives, and that is not the case. However, if one assumes that only 0.2% are false positives, this leaves 0.8% as false negatives. So, of those same 100 people with erroneous results, 20% would be false positives and 80% would be false negatives. **False-negative people are an unwitting threat to sex partners.** But there is still another ramification: using the CDC estimate that 0.6% of Americans are HIV-positive, in a population of 10,000, 60 Americans that would test positive! This 60 must include all the false positives, 30, leaving only 30 people actually infected. This leads to the following conclusion: using a 99% accuracy, one finds as many false positives as true positives.

Even if the results of both AIDS tests, the ELISA and WB, are positive, the chances are only 50-50 that the individual is infected. This is why people with HIV-positive results must be tested repeatedly over the following 6 months to 1 year. The error rate at 99% accuracy is high with only two tests. (The CDC's *Morbidity and Mortality Weekly Report* shows an overall performance rate of only 98.4% on the Western blot alone— a lower accuracy than that used in this example). Even with repeated HIV-positive tests, the rare person may just be a false-positive tester.

The implications resulting from a false-positive test are broad for people tested at random. For example, a person was recently HIV-tested for a routine insurance examination. Because he or she had no behavioral risk factors and was in excellent health there was no concern about testing HIV-positive. The major concern is testing falsely positive— that risk, with a 99% accuracy testing procedure, is 30 out of 10,000 (0.3%). **This may appear to be a low risk, but it isn't if you are one of the 30—after all, some 30 people out of 10,000 will be false positive.** The results can destroy one's personal and professional life; other people believe your test results even if they are later found to be in error. It's like the newspaper scenario: retractions are found in small print on the back page — somewhere.

There is also the danger that false-positive people will not feel the need to avoid sex with truly infected people—a good route to infection.

But there is also room for optimism in these statistics. An individual who is a random false positive can find hope in them. This does not mean that he or she can take chances with other people's lives, so each person must behave as though he or she is actually infected. But, inwardly, the random false HIV-positive individual can be cautiously optimistic.

The occurrence of even a small number of false-positive HIV tests can have profound implications. This is especially true when testing blood donors, since false-positive results waste resources in discarded blood units and require verification of positive results using more expensive tests. A false negative result, indicating that an individual is not infected, can have serious consequences for the blood recipient. Therefore, attempts to improve tests are a continuous challenge.

---

tion, the less likely a positive ELISA test is to be a true positive.

Levels of Sensitivity and Specificity in Testing for HIV— A test's **sensitivity** is its capacity to identify **all** specimens that **have** HIV antibodies in them. A test's **specificity** is its capacity to identify **all** specimens that **do not have** HIV anti-bodies in them.

Sensitivity is determined as follows:

Number of true positives ÷ Number of true positives + the false negatives × 100%

If 100 persons are actually HIV-infected and the test identifies only 90 of them as such, then we can say that the test has 90% sensitivity.

Specificity is determined as follows:

Number of true negatives ÷ Number of true negatives + false positives × 100%

Assume that, in a group of 500 people being tested for HIV antibodies, 100 individuals are actually not infected. If test results show that only 90, out of the 100, are identified as not having the virus, then the test has 90% specificity.

WESTERN BLOT ASSAY

The gold standard for determining a true-positive HIV-antibody test is still the **Western blot** (WB). This test is a method in which individual HIV proteins are used to react with HIV antibody in a person's serum. It should be understood that the WB test is not a **true gold standard** because it is not 100% certain, but it can come close to 100% if properly used.

Human cells in which HIV is being cultured are lysed or broken open, and the mixture of cell components and HIV components (proteins) are separated from each other. The viral proteins are placed on a polyacrylamide gel which then gets an electrical charge. The electrical current separates the viral proteins within the gel. **This is called gel electrophoresis.** The smallest HIV proteins will move quickly through the gel, separating from the next larger size, and so on.

Each different protein will arrive at a separate position on the gel. After proteins of similar molecular weight collect at a given site, they form a band; and these bands are identified based on the distance they have run in the gel. Because each band is a protein produced as a product of a different HIV gene, the gel band patterns give a picture of the HIV genes that were functioning and the location of each gene's products on the gel (Figure 12-4 and Table 12-2). The protein or antigen bands within the gel are "blotted," that is, transferred directly, band for band and position for position, onto strips of nitrocellulose paper (Figure 12-4).

Once the antigen bands have been formed, serum that is believed to carry HIV antibodies is placed directly on them. That is, a test

TABLE 12-2 Description of Major Gene Products of HIV and Capable of Inducing Antibody Response

Gene Product[a]	Description
p17/18[b]	GAG[c] protein
p24/25	GAG protein
p31/32	Endonuclease component of POL[d] translate
gp41	Transmembrane ENV[e] glyco-protein
p51/52/53	Reverse transcriptase component of POL translate
p55	Precursor of GAG proteins cleaved to p18 and p24
p64/65 p66	Reverse transcriptase component of POL translate
gp110/120 gp120	Outer ENV glycoprotein
gp160	Precursor of ENV glycoprotein

[a]Number refers to molecular weight of the protein in kilodaltons; measurement of molecular weight may vary slightly in different laboratories.

[b]Where two bands on acrylamide gel are too close for easy identification, they are presented as such, p51/53, etc.

[c]GAG = core.

[d]POL = polymerase.

[e]ENV = envelope.

(Adapted from *MMWR*, 1988a)

serum is added directly to antigen bands located on the nitrocellulose strip. If antibodies are present in the serum, they will form an antigen–antibody complex directly on the antigen band areas. After the strip is washed, the HIV antibody–antigen bands are visualized by adding enzyme-conjugated antihuman immunoglobulin G to the strip. Then the substrate or color agent is added. If antibody has complexed with any of the banded HIV proteins on the nitrocellulose strip, a color reaction will occur at the band site(s) (Figure 12-4). Positive test strips are then compared to two control test strips, one that has been reacted with known positive serum and one that has been reacted with known negative serum.

In contrast to the ELISA test, which indicates only the presence or absence of HIV antibodies, the WB strip qualitatively identifies which of the HIV antigens the antibodies are directed against. **The greatest disadvantage of the WB test is that reagents, testing methods, and test-interpretation criteria are not standardized.** The National Institutes of Health (NIH), the American Red

Cross, DuPont company, the Association of State and Territorial Health Officers (ASTHO), and the Department of Defense (DoD) each define a positive WB differently.

It can be concluded from the criteria set up by different organizations to define a positive result that although the WB may be the gold standard of confirmatory testing, there is **no** agreement on what constitutes a positive WB test (Miike, 1987).

The WB procedure is labor-intensive, takes longer to run (12 to 24 hours), and is therefore more costly than the ELISA test. **The WB is less sensitive than the ELISA but more specific.**

Because the WB lacks the sensitivity of the ELISA test, it is not used as a screening test. Despite the high specificity of the WB, false positives do occur, but they occur less frequently than with ELISA tests because the WB is only run on serum, blood, oral fluid, and urine already **suspected** of containing HIV antibodies.

Indeterminate WB

Western Blots may also turn out to be **indeterminate** in HIV infections. Meaning, a person can be infected but the blot is not conclusive—it looks positive but it may not be or is too poor to tell. The indeterminate WB results can occur either during the window period for HIV seroconversion, or during end-stage HIV disease. Indeterminate WBs have **occurred in uninfected individuals because of cross-reacting autoantibodies** related to recent immunization, prior blood transfusion, organ transplantation, autoimmune disorders, malignancy, infection with other retroviruses (e.g., HIV-2), or pregnancy. Some patients have a persistent pattern of indeterminate reactivity that remains stable over several years in the absence of true HIV infection.

Relative Costs of ELISA and Western Blot Tests

The cost incurred to identify each true-positive HIV individual in a population with a 10% pre-valence for HIV infection is presented in Table 12-3. Cost estimates are for tests performed on a contractual basis, that is, for screening large numbers of people versus testing of individuals. Note that the difference is quite significant.

OTHER SCREENING AND CONFIRMATORY TESTS

There are a variety of HIV-antibody and HIV-antigen detection tests now on the market and others are on their way. A few of these tests have been singled out because they are currently in use or because of their potential to make a contribution in the field of HIV antibody–antigen testing methodology.

Immunofluorescent Antibody Assay

The **immunofluorescent antibody assay** uses a known preparation of antibodies labeled with a fluorescent dye such as fluorescein isothiocyanate (FITC) to detect antigen or antibody.

TABLE 12-3   Incurred Costs to Identify True-Positive HIV-Antibody-Containing Sera in a Population with a 10% Prevalence of HIV-Antibody-Infected People

Price of testing under negotiated contracts:
 Low estimate: $4.41 per specimen tested
 High estimate: $7 for each ELISA, $60 for each Western blot
Price for individual testing: $47.50 for each ELISA, $121 for each Western blot

Best case:
 Number of true positives: 9,920
 Number of ELISAs performed: 100,000
 Number of Western blots performed: 9,960 + 900 = 10,860
 Low estimate: $4.41 × 100,000 divided by 9,920 = $44
 High estimate: $7 × 100,000 plus $60 × 10,860 divided by 9,920 = $136.25
 Individual testing: $47.50 × 100,000 plus $121 × 10,860 divided by 9,920 = $611.30

(Adapted from Mike, 1987; adjustment reflects charges through 1999)

SOME RELATIVE DRAWBACKS TO THE CURRENT HIV SCREENING TEST— ELISA/WESTERN BLOT

Regardless of the high sensitivity and high specificity or overall test performance, there are a few test-related problems. For example, although only 24 hours are required to complete the ELISA/Western blot testing procedures, most labs batch specimens for processing, causing a 1- to 2-week wait for definitive results. One study has shown that 40% of those tested never returned for their test results. Perhaps because of concern about maintaining anonymity or other factors, many individuals with HIV are not tested until they develop symptoms. Up to one-third of patients receive their HIV diagnosis within 2 months of an AIDS diagnosis.

Cost and time investment of the two-step process (pre- and post-test counseling) have also impeded testing. Even when tested without charge at a publicly funded clinic, clients must take time out of work or caring for children for waiting time, counseling, or travel. The true cost of this process has been estimated at $41 per test. And, individuals tested in private offices may incur not only the cost of the test by commercial or hospital labs, but also the physician's office fee (Sax et al., 1997).

In the direct fluorescent antibody test, fluorescent antibodies detect specific antigens in cultures or smears. In the indirect fluorescent antibody test, specific antibody from serum is bound to antigen on a glass slide.

The indirect procedure is modified for use in detecting antibodies to HIV. Cells that are HIV-infected will have HIV antigens on their cell membranes and will later fluoresce when the antihuman fluorescent conjugate is added.

A diluted sample of a person's serum is placed on a slide prepared with HIV antigens and incubated to allow antibody–antigen reaction if there are HIV antibodies in the serum. The slides are then rinsed to get rid of excess serum and other materials. Fluorescent antihuman antibody is then placed on the slide and incubation allows antihuman HIV antibody complexing to occur. The slide is again rinsed and dried. Fluorescence, if any, is observed using a special fluorescence microscope. The

indirect fluorescent antibody (IFA) test is now used in many laboratories as a screening procedure. In as much as the sensitivity and specificity of the IFA test are similar to the Western blot, this method has been proposed as an alternative confirmatory test for positive ELISA results. The IFA method is relatively simple and is one of the quickest tests available.

In late 1992, the FDA approved Fluorognost for marketing, the first assay for HIV-IFA confirmation and screening.

The assay allows doctors to do in-office tests for antibodies to HIV in human serum or plasma. As opposed to the Western blot test, the current standard confirmation test, Fluorognost posts almost no indeterminate test results. In addition, the test takes only 90 minutes to complete, while the Western blot takes from 12 to 24 hours to process. This FDA-approved test allows smaller health care facilities, emergency rooms, and doctors' offices to conduct in-office HIV screening and confirmation with accuracy, ease, and low overhead.

Polymerase Chain Reaction

Interactions between HIV and its host cell extend across a wide spectrum, from latent to productive infection. The virus can persist in cells as unintegrated DNA, as integrated DNA with alternative states of viral gene expression, or as a defective DNA molecule. Determining the fraction of cells in the blood that are latently or productively infected is important for the understanding of viral pathogenesis and in the design and testing of effective therapies. Determining the number of infected cells in a heterogeneous cell population and the proportion of those cells that are carrying the virus but not producing new viruses requires the identification of the proviral DNA and viral mRNA in single cells.

The **polymerase chain reaction (PCR) is a technique by which any DNA fragment from a single cell can be exponentially multiplied to an amount large enough to be measured.** This technique indirectly measures **viral load** and thus enables an assessment of viral load and viral expression in different parts of the body of HIV-infected patients. Thus PCR could be an ideal diagnostic test for HIV infection, since it directly amplifies proviral HIV DNA and does not require antibody formation by the host. It is already used in settings where antibody production is unpredictable or difficult to interpret, such as in acute HIV infection or in the perinatal/postnatal period. For HIV testing, the technique used is to copy a segment of proviral DNA found in cells such as T4 lymphocytes and macrophages that carry the virus. **The PCR is so sensitive that it can detect and amplify as few as six molecules of proviral DNA in 150,000 cells or one molecule of viral DNA in 10 μL of blood.**

Finding these few molecules of DNA to copy and amplify is, as the saying goes, like finding a needle in a haystack. The needle in this case is the proviral DNA molecule. It is only a miniscule fraction of the total DNA content of a given cell and far less when mixed in with the DNA of over one million cells as used in some HIV PCRs.

One of the first uses of the PCR in HIV/AIDS research was to show that HIV was present in people who were suspected of being infected but did not produce HIV antibodies. The HIV provirus was detectable in their cells. The PCR's most recent use has been in the detection of HIV provirus in newborns of HIV-infected mothers (Rogers et al., 1989).

In 1993, researchers reported that, in HIV-infected persons, 4% to 15% of peripheral blood lymphocytes were infected with HIV. The percentage of these cells that contained HIV mRNA, an indicator of viral replication, ranged from less than 1% to 8%. The data indicate that, in HIV-positive individuals, a significant proportion of peripheral blood lymphocytes are infected with HIV, but that the virus is in a latent state in the majority of these cells.

Now that there are some good anti-HIV therapies available to help slow the onset of AIDS, the diagnosis of individuals who carry the provirus is critical because they may benefit from early treatment. The PCR test will become even more important with the advent of an HIV vaccine. Vaccinated people will become HIV-antibody-positive. The PCR test will be used to identify those who are truly HIV-infected.

How PCR Works— The PCR (Figure 12-5) was developed by Kary Mullis and colleagues at Cetus Corporation (Mullis et al., 1987; Keller et al., 1988). The procedure requires the synthesis of **oligomer primers** (short 16 to 25 nucleotide sequences of DNA) that will hybridize or bind to the segment of DNA to be copied. The primer is required because the polymerase enzyme used to copy the desired DNA sequence needs an initial DNA sequence of nucleotides to add on to.

In the case of the HIV provirus, the primers need to hybridize to a DNA sequence adjacent to the gene area to be copied—for example, the GAG gene DNA sequence that produces the p24 protein. Because DNA is double-stranded, a primer molecule is attached to each strand (Figure 12-5). After primer attachment, a polymerase enzyme (usually Taq 1) is added. The polymerase enzymes, using the primers as starting points, add one nucleotide at a time as they copy their single DNA strands.

After each strand of the DNA segment has been copied at a temperature of 70°C, the reaction is heated to 93°C to **separate** the newly synthesized strands from the original DNA strands. This is called **denaturation.** The temperature is then cooled to 37°C to permit the additional primer molecules to attach (anneal) to the newly synthesized DNA sequences and the original DNA sequences (Figure 12-5). After **primer attachment,** the temperature is raised to 70°C and the polymerase enzyme molecules, starting at the primer sites, **copy** each strand of DNA. At the end of this second cycle there are four copies of each of the original DNA single strand sequences. By repeating the cycles of **denaturation, annealing,** and **synthesis,** the original DNA sequence can be amplified exponentially according to the formula 2^n where n is the number of cycles.

RAPID HIV TESTING

Until now, HIV testing required two visits. During the first visit, a person receives pretest counseling and blood is drawn for HIV testing. During the second visit, test results are communicated and additional counseling is provided. For people who need them, referrals are given for additional services.

Polymerase Chain Reaction

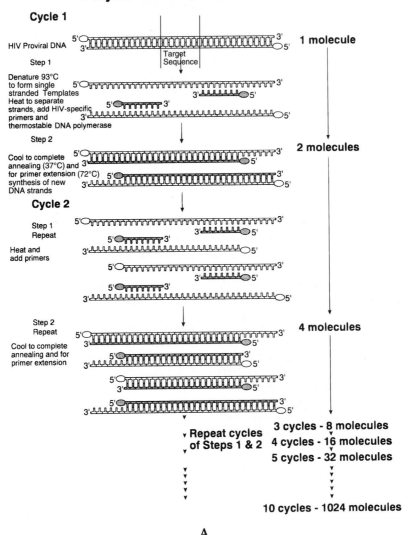

Cycle 1

HIV Proviral DNA — 5' ... 3'
3' ... 5'
Target Sequence

1 molecule

Step 1

Denature 93°C to form single stranded Templates
Heat to separate strands, add HIV-specific primers and thermostable DNA polymerase

Step 2

Cool to complete annealing (37°C) and for primer extension (72°C) synthesis of new DNA strands

2 molecules

Cycle 2

Step 1 Repeat

Heat and add primers

Step 2 Repeat

Cool to complete annealing and for primer extension

4 molecules

Repeat cycles of Steps 1 & 2

3 cycles - 8 molecules
4 cycles - 16 molecules
5 cycles - 32 molecules

10 cycles - 1024 molecules

A

FIGURE 12-5 Polymerase Chain Reaction. A source of DNA containing the sequence to be amplified is mixed with two primers, nucleotide triphosphates and the DNA polymerase Taq 1 (Taq 1 is a DNA polymerase derived from the bacterium *Thermus aquaticus*). The primers are synthetic oligonucleotides of *known sequence*. Therefore, the boundaries of the DNA sequence to be copied are already known. *Cycle 1:* Primers are added in excess (1 million-fold). Using DNA polymerase, the primers are extended by DNA synthesis. The first cycle takes about 5 minutes. *Cycle 2:* The product from cycle 1 is denatured, reannealed, and primer extension occurs once again. This cycle takes place as many as 50 times so that the primer-extended sequence increases thousands of times. The original DNA sequence can be amplified exponentially according to the

(continued on next page)

What Is Rapid HIV Testing?

A rapid test for detecting antibody to HIV is a screening test that produces very quick results, usually in 5 to 30 minutes. The **only** rapid HIV test licensed (1992) by the FDA for use in the United States is the Single Use Diagnostic System for HIV-1 (SUDS). Rapid assays for detecting HIV were developed in the late 1980's. A number of rapid testing kits are being used in other countries but they do not have FDA approval for use in the United States. The **HEMA-STRIP** and **SERO-STRIP** are used to test for HIV antibody in over 20 countries. Those and other rapid HIV tests used outside the USA require no additional equipment. These "one step" assays have all reagents contained in a tube-like device that has a strip containing antigens. Whole blood, oral fluid, or serum is placed at the tip of the device and allowed to diffuse along the strip with impregnated reagents where reaction with the antigen occurs. These test can be completed in less than 10 minutes, require no addition of reagents, and contain a built-in quality control reagent. These assays offer attractive features and may be the tests of the future in the USA.

What Is the Difference Between a Rapid HIV Test and an Enzyme Immunoassay (EIA)?

The rapid HIV test is easier to use and produces results more quickly than the EIA does. The sensitivity and specificity of the rapid HIV test are just as good as those of the EIA. Persons whose rapid HIV tests are **negative** can be

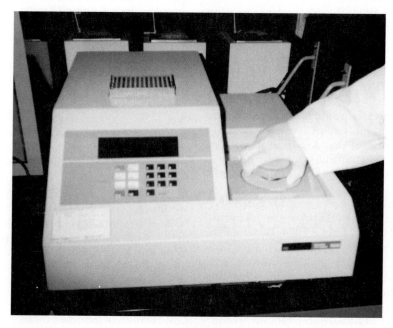

B

FIGURE 12–5   *(continued)*

formula 2^n, where *n* is the number of cycles. In theory, 25 PCR cycles would result in a 34 million-fold amplification. However, since the efficiency of each cycle is less than 100%, the actual amplification after 25 cycles is about 1 to 3 million-fold. The size of the amplified region is generally 100 to 400 base pairs, although stretches of up to 2,000 bases can be efficiently amplified. In HIV-infected cells, the DNA template is a specified region within the provirus (Keller et al., 1988). **B.** Photograph of a thermocycler— the machine used to run the PCR. (*Photograph courtesy of the author*)

given a **definitive** negative result without a return visit. Persons whose tests are **positive** can be counseled about the likelihood of their infection and to return for confirmation testing.

In March 1998, the CDC changed its policy on the use of rapid HIV screening tests in the USA. Based on it's studies, about **700 thousand** out of **2 million** people annually **do not** return to anonymous test centers to find out their **ELISA** test results. The CDC now feels that early disclosure of results (10 minutes rather than 2 weeks) will cut down on the number of people who do not return for their results. Therefore the CDC is recommending that physicians use the SUDS test, and the FDA will most likely approve other rapid test kits before the year 2000 (MMWR, 1998b updated).

Saliva and Urine Tests

By the end of 1994, there were at least nine **serum** HIV-antibody **screening** tests and four confirmatory tests available in the marketplace. Because HIV antibodies are present in body fluids other than blood (e.g., urine and saliva), HIV-antibody assays have been developed for their use. **Collecting samples of urine and saliva is noninvasive, easier, less dangerous, and less expensive.** Such tests are particularly useful in developing nations where there is a shortage of refrigeration and sterile equipment (Constantine, 1993); Frerichs et al., 1994).

At least one HIV saliva test, **OraSure** (Figure 12–6), had received FDA approval for use in the United States in late 1994. But the FDA withheld permission to market OraSure until June of 1996. Previously, people whose OraSure pad tested positive had to give a blood sample to confirm the results. Now both the initial and confirming (Western blot) tests can be performed with the OraSure pad. OraSure is comparable to blood tests in cost and speed. But the FDA says it is slightly less accurate than blood tests. Although the OraSure kit performs well, at this point its niche in the clinical lab is not entirely clear. The manufacturer suggests it may be useful for needle-phobic patients and in mass screening programs in which phlebotomy (drawing blood) is not readily available. A home version now being

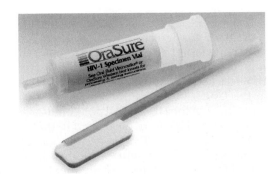

FIGURE 12–6 HIV-antibody testing and OraSure technology. OraSure is a simple lollipop-like device that is an easy-to-use oral HIV-antibody testing technology now available to consumers through health professionals. The product does not require the use of needles or blood and is about as effective as traditional HIV blood tests. A specially treated pad attached to the handle is placed by the patient between the lower cheek and gum for 2 minutes. The pad is then placed in a vial with preservative and sent to a clinical laboratory for testing for the presence of HIV antibodies, the same way blood samples are tested. *(Photograph courtesy of Epitope, Inc.)*

developed for marketing here should have a wide appeal (Gallo et al., 1997). Entering year 2000, one urine test, **Sentinel,** (Seradyn) a screen HIV test, was FDA approved on August 6, 1996. In May of 1998, the FDA approved a **urine Western blot** test. Now the confirmatory test can be done in urine. It is **less** sensitive and specific than blood testing. **The test is available only to professional laboratories** (Sax et al., 1997 updated, 1999).

HIV Gene Probes

Gene probes or genetic probes are an idea borrowed from methodologies used in recombinant DNA research. The idea is to isolate a DNA segment, make many copies of it, and label these copies with a radioisotope or other tag compound. If the DNA sequence copied is contained in any of the HIV genes, then the labeled copies of this DNA sequence can be used to hybridize or attach to DNA of cells that contain HIV DNA. This method of DNA probe analysis

BLOOD VS. URINE: OUTCOME OF HIV TEST MAY DIFFER (DISCORDANT RESULTS)

In December 1998 **Clinical Reference Laboratory (CRL),** one of the country's leading clinical reference laboratories, announced the results of the largest study to date showing a significant proportion of individuals—approximately one of every 1,000—within the low-risk population test **positive** for HIV antibodies **in their urine** and **negative in their blood.** The results suggest that people being tested for HIV antibody should have both their blood and urine screened. And the presence of antibody against HIV in the urine indicates, at the very least, a prior exposure to the virus. Scientists believe that the reason some persons show antibodies in their saliva, but not in their blood is based on their ability to compartmentalize the virus, the virus is contained to specific body tissues.

eliminates the need of searching for HIV gene products or antibodies to these products to prove that a person is HIV-infected.

At least two HIV-specific probes are on the market. One uses a radioactive sulfur label on the DNA for detection (^{35}S). This probe hybridizes to about 50% of the entire HIV genome and most specifically hybridizes to the HIV polymerase region. A second probe, also using ^{35}S, is an **RNA probe.** It is being used to detect HIV RNA in peripheral blood or tissue samples. **RNA probe hybridization allows detection of one HIV-infected cell out of 400,000 uninfected cells.** Specifically, the assay detects the presence of HIV in whole white blood cells as soon as the virus begins to replicate. The ^{35}S-labeled probes enter the white blood cells and combine with HIV; the procedure does not require DNA extraction and results are obtained in just over 1 day (Kramer et al., 1989).

Passive Hemagglutination Assay

There continues to be an urgent need for an inexpensive, accurate assay for anti-HIV screening in the developing world. Scheffel's results (1990) indicate that the **passive hemagglutination assay (PHA)** is a good candidate. The assay is simple to perform and requires no expensive equipment or precision pipettes. The reagents appear very stable even in adverse conditions.

SALIVA HIV TEST?

According to Wesley Emmons and co-workers (1995), a blinded study to determine the accuracy of detecting HIV antibody in human saliva was performed at the United States National Naval Medical Center. Naval medical center personnel tested commercially available (OraSure) saliva test kits. Test kit performance was as follows:

An absorbent pad, mounted on a lollipop-like plastic stick, was placed between the lower cheek and gum, rubbed along the tooth-gum margin 20 times, and held there for 2 minutes; the collected fluid was then centrifuged. A 1:4 dilution of saliva was used for ELISA tests (1:400 is usual for serum testing) and 200-μL samples for Western blot confirmation (as compared with 20 μL for serum testing).

All 195 HIV-seropositive adults were positive for HIV antibodies on ELISA tests of saliva samples, and 190 of them (97.4%) had strongly positive Western blot results (the other five blots were indeterminate). Saliva samples from all 198 HIV-seronegative controls were negative on ELISA (del Rio et el., 1996).

Oral disease, mouth sores, and tobacco use did not seem to affect its accuracy. The OraSure test is the first saliva test to received FDA approval in December of 1994. **The OraSure kit is only available to professional health care workers as of year 2000.**

However, the assay requires a minimum of 3 hours before a reading can be made.

The accuracy of the assay is reportedly excellent, with 100% sensitivity on some 890 HIV-positive serum samples and sensitivity equal to or better than the second-generation ELISA.

VIRAL LOAD: MEASURING HIV RNA

Recent studies have shown that a very rapid turnover of HIV RNA occurs in the plasma of infected patients, with approximately 30% of the total virus population in the plasma being replenished daily. Continual viral replication and rapid T4 cell turnover play a central role in the pathogenesis of HIV infection. **High plasma HIV RNA levels have been shown to be a strong predictor of rapid progression to AIDS after HIV seroconversion, independent of T4 cell count.** These findings have led to increased interest in quantitating HIV RNA in patients for prognostic purposes. HIV RNA viral load measurements are used in the management of HIV-infected patients, both in predicting rate of progression and monitoring response to antiretroviral therapy (see Chapters 4, 7, and 11 for additional information on HIV RNA load). Until June of 1996, reliable measures of HIV RNA were available only in research laboratories. And, the quantitation of HIV from a clinical specimen required very expensive, labor-intensive, and difficult-to-reproduce culture techniques. Recently, however, quantitation of HIV viral load in plasma specimens has been accomplished with a variety of techniques that measure HIV RNA and are less expensive and easier to perform. Currently, three commercial assays are available: (**1**) FDA Approved Amplicor HIV Monitor Test, which couples reverse transcription to quantitative polymerase chain reaction (PCR) (Roche Molecular Systems); (**2**) Quantiplex HIV RNA assay, a branched DNA (bDNA) technology (Chiron Corporation); and (**3**) Nucleic Acid Sequence-Based Amplification (NASBA) (Organon) (Technika). Because in June of 1996 the FDA approved Amplicors' first generation test and in March 1999 the **ULTRASENSITIVE** test, they are briefly reviewed. In addition, the branched-DNA test will also be presented because it is near FDA approval.

Amplicor HIV Monitor Test

This multistep test includes specimen preparation, reverse transcription and PCR, and nonradioactive detection. Plasma samples (200 µL) are treated to lyse the viral particles and release RNA; isopropanol is then added to precipitate the RNA. This first step renders the sample noninfectious and safe for laboratory workers.

The new Amplicor HIV Monitor Ultra-Sensitive Test has a three-log dynamic range; its lower limit of sensitivity is about 50 copies of HIV RNA/mL. The assay is reproducible and fivefold differences in HIV RNA copy number are easily discernible. Using stored samples for comparison, it is possible to detect the high levels of HIV RNA that occur in acute infection and the suppressed levels of circulating virus that follow seroconversion.

Branched-DNA Testing

Another new assay for measuring HIV levels directly, developed by Chiron, uses a branched-DNA (bDNA) technology to amplify a signal to indicate the presence of HIV RNA. The bDNA assay was first designed in 1989 (Horn et al., 1989) and improved upon for use in HIV detection by Urdea (1993) and Dewar (1994).

———— SIDEBAR 12.3 ————

VIRAL LOAD FALSE POSITIVES

Josiah Rich and colleagues (1992) and Brown University of Medicine reported on three cases of false HIV-infection diagnosis by plasma viral load testing. Two of the false positive cases were detected by branched-chain DNA plasma viral load assays, with the other case detected by reverse transcriptase-polymerase chain reaction plasma viral load assay. All of the tests yielded the positive results with low values. All three of the patients subsequently tested negative by HIV ENZYME-LINKED IMMUNOSORBENT ASSAY AND REPEATED PLASMA VIRAL LOAD TESTING. The authors suggest that doctors use care when implementing viral load assays for HIV detection, especially when the pretest possibility of infection is low.

Using the bDNA assay, nucleic acids can be detected **directly in clinical samples** by means of signal amplification. Signal amplification of HIV RNA occurs when branched oligodeoxyribonucleotides (bDNAs) hybridize to the target HIV RNA and incorporate many alkaline phosphate molecules onto the target HIV RNA. The complex of HIV RNA plus bDNA plus alkaline phosphate is then exposed to dioxetane which is then triggered by an enzyme to luminesce (give off a light that can be detected with a measuring device, a luminometer) (see Figure 12-7). The luminescence means that HIV RNA is present and the brighter the light (or signal) the more HIV

RNA is present in the test sample. The current assay, Chiron's Quantiflex 340 can measure down to 50 copies of HIV RNA/ml of sample.

DETECTION OF HIV INFECTION IN NEWBORNS

Early detection of HIV infection in newborns and children is important because it may prevent unwarranted toxicity from the use of antiretroviral agents in children who are not infected, and it can allay the fears of parents with potentially afflicted but uninfected newborns. However, early diagnosis of HIV

— SIDEBAR 12.4 —

DISCUSSION OF HIV DIAGNOSTIC TESTS

The standard procedure of diagnosing HIV infection is clear. When HIV infection is possible, the first test is an ELISA. A positive antibody assay should be repeated. If the second ELISA is positive, a Western Blot test or immunofluorescence assay is done to confirm the diagnosis. Both of these serologic tests are highly sensitive and specific. Their main limitation is their inability to detect acute HIV infection, the presence of HIV in the first couple of weeks after infection. It takes an average of 25 days after infection for detectable antibodies to develop. (MMWR 1999) If a person in the very early stages of HIV infection, the most sensitive test is the viral load test, or HIV PCR, although this is not usually recommended as a diagnostic test because of the possibility of false-positive results. (Merriman et.al 1999)

New Possibilities

The latest HIV test possibility is the availability of the rapid HIV test, which people can use at home. It detects antibody to HIV in a blood sample and is as sensitive and specific as the standard ELISA. Like the ELISA, the rapid test is only a screening test; a positive result requires a confirmation by another test. Also, if the test subject has not yet developed antibodies, the result will be a false-negative one. Although the rapid test was designed for home use by people who wished for greater anonymity than that afforded

by a public health clinic or physician's office, clinics and individual physicians also can use the test. However, obtaining test results within half an hour, while the patient remains in the office, is a two-edged sword. First, a person learns immediately whether the test is positive or negative. But second, it may make counseling harder, especially if the person is shocked by a positive result. People must be told about the necessity of a confirmatory testing to establish the diagnosis and the importance of precautions to prevent the spread of a possible HIV infection. This information should be communicated both before and after testing to assure that the results do not affect a person's understanding of the test results. A rapid test may use a blood sample, urine specimen or saliva. Although these tests may be easier for people to use at home, confirmatory testing requires an extra step.

HIV Monitoring Test

Another important development in HIV testing is the ability to monitor the effectiveness of treatment. The ELISA and Western Blot tests cannot do this; they simply detect antibodies to HIV. But repeated viral load testing with HIV RNA and determination of the T4 lymphocyte count can be used to assess how the patient is responding to therapy. If the viral load is increasing, the therapeutic regimen can be altered.

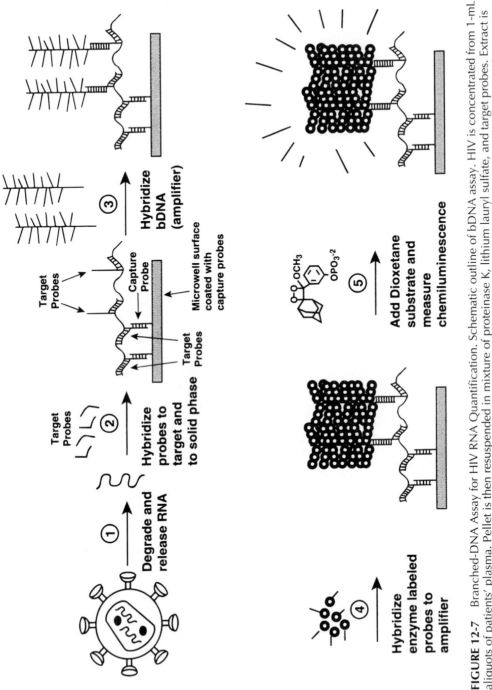

FIGURE 12-7 Branched-DNA Assay for HIV RNA Quantification. Schematic outline of bDNA assay. HIV is concentrated from 1-mL aliquots of patients' plasma. Pellet is then resuspended in mixture of proteinase K, lithium lauryl sulfate, and target probes. Extract is added to wells of 96-well microplate, incubated overnight at 53°C, and washed. bDNA amplifier is then added, incubated at 53°C, and washed, followed by addition of alkaline phosphatase probes, another wash, and addition of the chemiluminescent substrate. (*Adapted from Dewar et al., 1994*)

infection in children born to HIV-positive women is difficult owing to the presence of maternal IgG antibody in the newborn which may persist until the child is about 2 years old. Thus serological detection of HIV infection in neonates is complicated by the presence of immune complexes, consisting of passively transferred maternal antibodies and HIV antigens. Steven Miles and colleagues (1993) have used a rapid assay designed to disrupt these immune complexes in order to permit the detection of a specific HIV antigen. Their preliminary work correctly identified p24 antigen in the blood of 29 of 29 children. These children tested HIV-positive, but when the maternal antibodies were treated to separate the p24 antigen present (the antigen would be present because the virus was present), they found p24 antigen. Children who were HIV-positive but lacked the presence of the p24 antigen were falsely positive—they contained only the mother's HIV antibodies. Although this technique is simple to perform and accurate when compared to other methods of HIV detection in newborns, additional testing of this technique is required before its adoption. If a mother is known to be HIV-infected, cells from the newborn can be subjected to HIV blood culture or polymerase chain reaction tests. These two techniques, although very accurate, are not yet ready for the screening of all newborns. They can detect the presence of HIV in the first few days of life.

FDA APPROVES TWO HOME HIV-ANTIBODY TEST KITS

On May 14, 1996, the FDA, which for years opposed home-based HIV testing kits because of the lack of face-to-face counseling, reversed its stance by saying the benefits of early detection of HIV infection outweigh any risks posed by the test. FDA Commissioner David Kessler said, "We are confident that this new home system can provide accurate results while assuring patient anonymity and appropriate counseling."

The FDA approved the home test **because** in a 1994 study by the Federal Centers for Disease Control and Prevention of people at increased risk of infection, like intravenous drug users and sexually active homosexual men, 42% indicated that they would use a home test. To 1994, over 60% of Americans **at risk** for HIV infection had not been tested.

Test Kit Operation

A person who buys the kit uses an enclosed lancet to prick his or her finger and places three drops of blood on a test card with an identification number. The card is mailed to a laboratory for HIV testing, and samples that test positive are retested to ensure reliability. People who use the home system do not submit names, addresses, or phone numbers with the specimen sent in on filter paper. Therefore, **the HIV test results are anonymous.** To get results, the individual calls a week later and punches into the phone his or her identification number.

If the caller's test results are positive or inconclusive, he or she will be connected to a counselor who will explain the results, urge medical treatment, and, if necessary, make a referral to a local doctor or health clinic. If the person's results are negative, he or she will be connected to a recording that will note that it's possible to be infected with HIV and still test negative, if the antibodies to HIV haven't yet developed. A counselor will be available for anyone who tests negative and wants to discuss the results.

The FDA said that the kit is as reliable as tests conducted in doctors offices and clinics.

Test Kit Availability

The first FDA-approved HIV test kit, called Confide HIV Testing Service was made available in June 1996. It was withdrawn from the marketplace in June 1997 due to poor sales. A second FDA-approved HIV home test kit went on sale nationwide in July 1996. This kit (Figure 12-8), called *Home Access Express-HIV Test* (1-800-448-8378), lets people take a blood sample at home, mail it to a laboratory, and, 3 business days later, learn by phone their results. The two tests are very similar to each other with regard to use and performance.

Pro and Con of Home-Use HIV Test (Collection) Kits

Pro— Those who favor HIV home testing point to the discouraging statistics using current

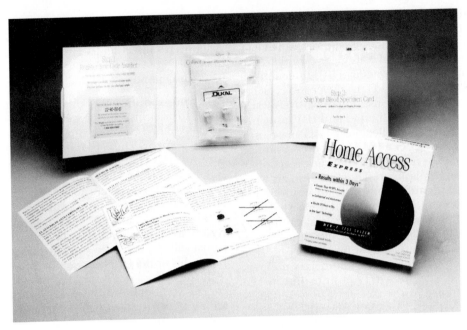

FIGURE 12-8   Home Access Anonymous HIV Test Kit. This kit was FDA approved in July 1996 and provides access to professional counseling and medical/social service referrals 24 hours a day, 7 days a week. (*Photograph courtesy of Home Access Health Corp., Illinois*)

HIV testing methods stating that 80% of people in high-risk populations are unwilling to use standard testing services. They note frequent instances of substandard HIV testing even in physician's offices, with test results delivered in brief phone calls or left on telephone answering machines. Also offered **in support of home-based testing is that the existing system of testing is inefficient and wasteful because 98% of tests are HIV-negative.** Thus, there is an enormous amount of money spent on HIV-negative results that should be spent elsewhere. With respect to the need for face-to-face counseling, they cite the success of suicide hotlines and crisis counseling services in support of using home telephone HIV counseling procedures.

Con— Those against the use of home testing say that this is just one more instance of the telephone substituting for face-to-face contact in medical practice. They suggest that **bad health news of this magnitude and complexity should be given in person.** They also point out that these test kits will probably be used mostly by affluent people who are well but worried, while the poor at-risk people continue to remain untested. In addition, there is the risk of the tests being used coercively by employers or the police. The potential for human rights violations are great.

Possible use of the tests by adolescents and young adults is especially controversial. Currently, there are no age restrictions on the purchase of home kits. Although the box states that the test is "recommended for persons 13 and over," anyone who can afford a kit can buy one. If youth use the tests, how can they be served effectively by telephone counselors? In addition, many advocates for youth have expressed concern about possible coercion of youth to be tested by parents, guardians, or older adults.

For a more thorough review of the issues on HIV home testing, read **"Testing for HIV at home: What are the issues?"** (Schopper et al., 1966).

Remarks— It remains to be seen just how many of the undiagnosed thousands of HIV-infected persons home-based kits manage to capture—and at what cost—and whether the system will be more helpful than abused. **CLASS QUESTION: WHAT DO YOU THINK WILL HAPPEN, GIVE SOME EXAMPLES OF WHAT YOU THINK WILL HAPPEN.**

REPORTING HIV INFECTIONS

In May of 1998 a report from the CDC advised all states not currently requiring name-based, laboratory reporting of positive HIV tests to implement **"integrated HIV and AIDS surveillance"** (MMWR 1998a). The CDC now believes that surveillance AIDS data is sufficient to track the AIDS pandemic in the United States. The CDC believes that HIV diagnosis and reporting will more accurately reflect trends and incidence of the national pandemic. The issue **now** is **not** whether HIV infections should be reportable at the **federal level,** but rather how to proceed with the reporting! Because of the enormous change in the natural history of AIDS, with significant reductions in the diagnosis of new AIDS cases and AIDS-related deaths associated with the use of newer and more effective treatment options.

All 50 states and the District of Columbia require health care providers to report new cases of AIDS to their state health departments. Beginning year 2000, 34 states required some form of **name** reporting of HIV-infected people (Table 10-4). The 43 states that require HIV reporting have over 60% of the population and accounted for over 60% of reported AIDS cases through 1999. Based on the number of HIV infections reported to the CDC, about 200,000–300,000 people out of the estimated 900,000 HIV-infected in the United States **do not know** they are infected. Beginning 1999, only 12 states classified HIV/AIDS as a sexually transmitted disease. Sixteen states classify HIV/AIDS as a communicable disease and 23 states classify HIV/AIDS as a **separate** category of disease. Only 12 states authorized physicians to inform partners of HIV/AIDS patients (*Medical Tribune*, 1991, updated).

Does the United States Need a National System for HIV Reporting?

Lawrence O. Gostin of Georgetown University Law Center in Washington, D.C., and colleagues state that the United States now needs a national system of HIV case reporting. Up to now, the AIDS surveillance system has ". . . formed the cornerstone of the nation's efforts to monitor and characterize the HIV pandemic." Because this system reports **advanced** cases of HIV infection, [AIDS] it only ". . . provides a snapshot of a decade-old epidemic." Gostin and colleagues believe that the compelling need for accurate monitoring of HIV infection and for effective medical and public health interventions **mandates a fundamental reevaluation of AIDS surveillance.** They argue that there is now a **new era** of more sophisticated

――――――― **SIDEBAR 12.5** ―――――――

FEDERAL TRADE COMMISSION WARNS OF POOR HIV HOME TEST KITS

In mid-June 1999, the U.S. Federal Trade Commission issued a warning about a dozen home HIV tests sold over the Internet. In a consumer alert, the agency cautioned, "Using one of these kits could give a person who might be infected with HIV the false impression that he or she is not infected." FTC tests showed that when a known HIV-infected sample was used, the tests indicated that the sample was not infected—**a false negative.** One HIV home test has already led to criminal prosecution against the Lei-Home Access Care of Sunnyville, California. The FDA said the tests were **"medically useless"** and the distributor was sentenced to five years in prison. The Internet advertisements have falsely stated or suggested that the kits were cleared by various well known health groups, the FDA and the WHO. However, the U.S. Food and Drug Administration has approved only The Home Access Express home HIV test kits.

treatments for HIV infection, that current treatment of HIV using reverse transcriptase inhibitors and protease inhibitors reduces mortality and delays progression of disease. Thus anti-HIV therapy has provided a **defining moment** in this pandemic and, unless our current HIV/AIDS surveillance system is revised to include reporting of all HIV infections, authorities will not have reliable information about the prevalence, incidence, and future directions of HIV infection. And they will not be able to provide therapy intervention when it will do the infected the most good.

WHO SHOULD BE TESTED FOR HIV INFECTION?

Testing for antibodies to HIV is an important first step in establishing a diagnosis of HIV infection. Testing every person may be counterproductive. Attempts to isolate or publicly identify people with HIV/AIDS can actually fuel the spread of HIV. People outside the "tested" group may feel invulnerable, and then fail to make necessary changes in their behavior.

The effects of stimulating a false sense of security are well illustrated in Germany where, in certain towns, prostitutes are required to be checked for certain STDs every week and are given health inspection cards. Many customers think these cards guarantee against disease and so refuse to use condoms.

The problem is not confined to Germany. **A European shipbroker whose work frequently takes him to south-east Asia, where he has a regular sex partner, said: "I tell her when I'm due to arrive and she has an HIV test just before. If she has an up-to-date health card I know I'm safe."**

In fact, a negative HIV test is no guarantee the tested person is **truly** HIV-negative—he or she may be in the **window period.** The decision to test must be based on people's risk behaviors and/or symptoms. As to who **needs** to be HIV tested; a complete history and physical examination will give the best answer to this question. Decisions based on individual indications are often more appropriate than decisions based on one's classification (i.e., all pregnant women or all single men between 20 and 49 years of age). Initial assessment

for current or past behaviors (within the past 10 years) should include:

1. Persons with **risk behaviors** such as:
    a. Anal sexual activity, male or female
    b. Injection drug use
    c. Frequent casual heterosexual activity
    d. Encounters with prostitutes
    e. Previous treatment for sexually transmitted diseases (*Condyloma acuminata* (genital warts), herpes simplex virus, gonorrhea, syphilis, *Chlamydia*)

f. Blood transfusions, especially before 1985

g. Sexual activity with partners having any of the above

h. Infants born to women involved in any of the above

2. Persons with symptoms such as:

a. Fever, weight loss (unexplained)

b. Night sweats

c. Severe fatigue

d. Recent infections, especially thrush and shingles (varicella-zoster)

3. Persons with **signs** (based on physical exam) such as:

a. Weight loss

b. Enlarged lymph nodes and/or tonsilar enlargement

c. Oral exam (candidiasis, oral hairy leukoplakia)

d. Skin lesions (e.g., Kaposi's sarcoma, varicella-zoster, psoriasis)

e. Hepatosplenomegaly (enlarged liver)

f. Mental status examination showing changes

4. Pregnant women who have demonstrated high-risk behavior.

WHY IS HIV TEST INFORMATION NECESSARY?

A recent CDC random-digit-dial telephone survey of American ages 18 to 65 revealed that 42% of those surveyed said they had been HIV tested. Geographic differences were great. For example, 26% of people in South Dakota said they had been tested vs. 60% in Washington, D.C. (MMWR 1999). The CDC also reported that about 25 million Americans are HIV tested each year. Publicly funded counseling and testing programs conduct about 2.5 million of these tests annually.

Thirty percent of adults who **seek HIV testing** do so to **find out their HIV status;** 12% are tested because of hospitalization or surgery; 16% for application for insurance; and 7% to enter the military. Another 1% are referred by their doctor, the health department, or sexual partner, and 4% are tested for HIV for immigration reasons (Hooker, 1996).

Knowledge of HIV infection status allows infected persons and their infected partners to seek treatment with retroviral agents, prophylaxis against *Pneumocystis carinii* pneumonia,

tuberculosis skin testing and tuberculosis prophylaxis (if appropriate), and other types of therapy and vaccines that may delay or prevent the opportunistic infections associated with HIV infection. Such measures have been shown to delay the onset of AIDS in infected persons and to prolong the lives of persons with AIDS. Counseling and testing may help some persons change high-risk sexual and drug-use behaviors thereby preventing HIV transmission to others.

Entry into Foreign Countries

An increasing number of foreign countries require that foreigners be tested for HIV prior to entry. This is particularly true for students or long-term visitors. Information available through 1999, reveals that 30 of 45 foreign countries queried require an HIV test prior to entry, on arrival, or on application for residency. **Before traveling abroad, check with the embassy of the country to be visited to learn entry requirements and specifically whether or not HIV testing is a requirement.** If the foreign country indicates that U.S. test results are acceptable "under certain conditions," prospective travelers should inquire at the embassy of that country for details (i.e., which laboratories in the United States may perform tests and where to have results certified and authenticated) before departing the United States. **For a copy of HIV Testing Requirements for Entry into Foreign Countries,** send a self-addressed, stamped, business-size envelope to: Bureau of Consular Affairs, Room 5807, Department of State, Washington, DC 20520.

Testing for HIV Infection

With an estimated 43 million HIV infected people **living** in almost all the countries of the world, and with about 650,000 to 900,000 of them living in the United States beginning year 2000, the question of HIV testing on either a **voluntary** or a **mandatory** basis cannot be ignored. Perhaps the decision on **who** is to be tested should be decided by society since most everyone is at risk.

The case in which a Florida dentist, David Acer, is believed to have infected six of his patients has led members of Congress, the Senate, and others to demand changes in HIV testing and confidentiality procedures. Under current

UNITED STATES PUBLIC HEALTH SERVICE (USPHS) RECOMMENDATIONS FOR HIV COUNSELING AND VOLUNTARY TESTING FOR ALL PREGNANT WOMEN

Knowledge of HIV status is important for several reasons. **First,** women who know their serostatus can gain access to HIV-related care and therapies (such as *Pneumocystis carinii* pneumonia prophylaxis, TB screening, and antiretroviral therapy) during pregnancy and post-partum as needed. **Second,** HIV-infected women can be offered zidovudine to potentially block maternal–fetal transmission of HIV. **Third,** an obstetrician would postpone rupture of amniotic membranes and avoid scalp electrodes or other potentially invasive procedures, all of which may be cofactors for enhanced transmission of HIV. **Fourth,** zidovudine can be offered to infants of HIV-seropositive women who have recently delivered.

This report (*MMWR,* 1995) contains the recommendations of a 10-member USPHS task force on the use of zidovudine to reduce perinatal transmission of HIV. In summary, the recommendations are:

1. Health-care providers should encourage all pregnant women to be tested for HIV infection, both for their own health and to reduce the risk for perinatal HIV transmission. Four million women, in the United States, each year become pregnant.

2. HIV testing of pregnant women and their infants should be voluntary. In voluntary testing, the reason for the test, how it is administered, and the person's right to privacy and confidentiality must be explained. This allows the person the choice of taking or refusing the test and giving or not giving demographic data.

3. Three states require health-care providers to counsel and offer HIV testing to women as early in pregnancy as possible so that informed and timely therapeutic and reproductive decisions can be made.

4. Uninfected pregnant women who continue to practice high-risk behaviors (e.g., IV-drug use and unprotected sexual contact with an HIV-infected or high-risk partner) should be encouraged to avoid further exposure to HIV and to be retested in the third trimester of pregnancy.

5. For women who are first identified as being HIV-infected during labor and delivery, health-care providers should consider offering intrapartum and neonatal zidovudine.

In Mid-1995 the CDC reported that routine HIV counseling and voluntary testing for all pregnant women has already proved effective in several communities nationwide. In one inner-city hospital in Atlanta, for example, 96% of women chose to be tested after being provided HIV counseling.

According to the Pediatric AIDS Foundation, about $350 million a year could be saved by testing pregnant women for the AIDS virus. The average hospital bill for a baby born infected with HIV is $35,000 annually for the 8 to 10 years the child lives.

---

policy, people are tested only with their **informed consent.** Although the **number** of people who test positive is reported to the CDC, their names are not. No attempts are made to track down contacts of infected individuals.

Marcia Angell, executive editor of the *New England Journal of Medicine,* wrote in 1991 that it was time to adopt a traditional public health approach to HIV/AIDS. She said, "Tracing and notification of the sexual partners of HIV-infected persons, and screening of pregnant women, newborns and hospitalized patients and health care professionals are warranted. This should be adopted only if steps are taken to protect HIV-infected people from discrimination and hysteria. Jobs, housing and insurance benefits, for example, should be protected by statute."

Regardless of protection, it has been estimated that 10% to 28% of persons in the United States choose not to be HIV-tested. These percentages most likely are in keeping with personal attitudes toward HIV testing in the other developed countries (Clezy et al., 1992).

Benefits of HIV Testing— Testing for the presence of HIV antibodies or antigens as early as possible when HIV is suspected can provide substantial health benefits to the individual, for example, early combination anti-HIV drug therapy reduces viral load and extends the quality of life for the infected, and reduce the risk of transmitting the virus to others. The infected person can be better treated for other infections if his or her HIV status is known. For example, 2% of HIV/AIDS patients in the United States contract tuberculosis. Unless otherwise indicated, HIV-infected patients with a past or present positive TB test should receive therapy as early as possible because active TB may be the first sign of AIDS. If other diseases occur prior to treatment, treatment for those diseases may interfere with the therapy for TB and vice versa.

The presence of HIV in people with syphilis may also alter the recommended therapy and follow-up. Influenza and pneumococcal polysaccharide vaccines are recommended for *all* people infected with HIV. The recommendation for pneumococcal immunization explicitly states that the vaccine be given as early in the course of HIV infection as possible to maximize antibody response (Rhame et al., 1989). In addition, the early detection of asymptomatic HIV infection provides an even more important health benefit: a chance for a lifestyle change to reduce stress. This would lessen the chance of acquiring other microbial infections that may stimulate the immune system, activate HIV to reproduce and destroy T4 cells, and

begin AIDS progression. However, too few of the HIV-infected are aware of their infections, and so, do not take these precautions.

TESTING, COMPETENCY AND INFORMED CONSENT

Competency is often used interchangeably with capacity; it refers to a person's ability to make an informed decision. For example, to consent to medical treatment, a person must be mentally capable of comprehending the risks and benefits of a proposed procedure and its alternatives. While a health care provider can assess competence, a legal finding of competency is often required based on the testimony of a mental health professional. Mental illness by itself does not indicate that a person is incompetent to make medical decisions. Various degrees of mental incapacity may occur with HIV infection, requiring an assessment of competency. AIDS dementia complex (ADC) occurs in approximately 70% of HIV-infected patients at some point in HIV disease/AIDS and may interfere with the patient's capacity to provide an informed consent.

Informed Consent

Informed consent is not just signing a form but is a process of education and the opportunity to have questions answered. The concept of informed consent includes the following components: full disclosure of information, patient competency, patient understanding, voluntariness, and decision making. The process of obtaining informed consent involves appropriate facts being provided to a competent patient who understands the information and voluntarily makes a choice to accept or refuse the recommended procedure or treatment.

When the concept of informed consent is applied clinically, complexities arise regarding both the content and the process. The concept contains ambiguous requisites such as "appropriate" facts, "full" disclosure, and "substantial" understanding. The process is affected by many variables including the communication skill and range of practice style of the physician; the maturity, intelligence, and coping strategies of the patient; and the interaction

――――― POINT OF INFORMATION 12.5 ―――――

A WEAK LINK IN HIV TESTING

The movement from a **positive** test result to treatment is a weak link in the overall care of the HIV-infected. Jeffrey Samet and colleagues (1998) reported that although a majority of patients (61%) sought medical care in the first year after their diagnosis, 39% delayed treatment for longer than one year, 32% for longer than 2 years, and 18% for longer than 5 years. Their report did not include those persons using home testing kits. It is believed that these people are even further removed from testing to medical care follow up.

———— BOX 12.5 ————

BOXER STRIPPED OF FEATHERWEIGHT TITLE AFTER POSITIVE HIV TEST

It took Ruben Palacio 12 years to win a world title. On the eve of his first defense, he became the first champion to test positive for HIV. The British Boxing Board of Control said, "We can't risk the life of another boxer by letting him fight. It's a kind of disease that can be spread via blood contact, and boxing is a sport where that is likely to happen."

Palacio is the first active world title holder known to have tested positive for the AIDS-causing virus. Esteban DeJesus, who held the WBC lightweight boxing title in the 1970s, contracted AIDS after his retirement and died in 1989.

HIV testing has been a routine part of the pre-fight medical examination in Britain for several years. In February 1990, African heavyweight champion Proud Kilimanjaro of Zimbabwe was barred from a fight with Britain's Lennox Lewis because he refused to give details of an HIV test to the British Boxing Board of Control.

His manager said, "This brings the HIV thing into perspective. Instead of going home with the largest paycheck of his life, he is going home with an HIV test result that means he will die."

In March of 1996 Tommy Morrison, a former United States heavyweight contender disclosed that he is HIV-positive (see Figure 8-10). Beginning year 2000 nine states require professional fighters and kickboxers, licensed in those states, to be HIV-tested.

4. Test is needed to protect the health of other patients, health workers, emergency or law enforcement personnel.
5. Several states require post-test counseling.
6. In 27 states, teenagers must have signed parental consent to be HIV tested. In 23 states (Alabama, Arizona, California, Colorado, Connecticut, Delaware, Florida, Georgia, Hawaii, Illinois, Iowa, Michigan, Nebraska, Nevada, New Mexico, North Carolina, North Dakota, Ohio, Rhode Island, Tennessee, Utah, Washington, Wyoming) **minors can consent** to HIV testing and treatment.

Generally, HIV antibody testing without consent is legally considered **battery.** Legal liability for "unlawful touching" may result from performing an HIV antibody test without consent. Such a procedure may also constitute an illegal search.

Question: **Are federal and state governments overly stressing personal privacy at the expense of prevention? (Defend your answer with examples/situations.)**

Voluntary Named HIV Testing

In **voluntary named** HIV testing, the individual freely provides his or her name. In this type of testing, an individual voluntarily seeks to learn his or her HIV status and receives a result which is known both to the individual and the test provider/testing agency. An advantage to named testing is that health care providers can contact the person tested if he or she does not return for the testing result.

Voluntary Unnamed HIV Testing

HIV testing is voluntary, but the identity of the person being tested is not placed on the blood sample or the testing form. As a result, the only person who can link the test result with an individual is the person being tested. This form of testing may encourage people concerned about the HIV infection status to obtain testing as it eliminates risk of discrimination or stigmatization. However, it places the exclusive responsibility for seeking counseling, support, and preventive measures on the individual who is infected. **Unnamed testing permits reporting of test data to public health authorities without the risk of breaching confidentiality.** Many pub-

between the physician and the patient (Hartlaub et al., 1993).

Testing Without Consent

Beginning 1998, 29 states had laws that allowed HIV testing without informed consent under certain conditions. The required conditions vary and include:

1. Patient or other authorized person is unable to give or withhold consent.
2. Test result will help determine treatment.
3. Patient is unable to give consent, and physician can document that a medical emergency exists and that the test is needed for diagnosis and treatment.

lic health authorities currently prefer voluntary unnamed testing to named testing.

Testing that is voluntary may miss populations that disproportionately need to be reached. **The people least likely to have the virus, it appears, are the most likely to say yes to a test, and the people most likely to be infected are the most likely to say no. In one study, infection rates were 5.3 times as high among people who refused HIV testing as among people who consented to it. In voluntary anonymous, the downside is that such testing reduces the probability that they will return for post-test counseling and linkage to follow-up services and a substantial reduction in partner notifications (Moser 1998).**

Mandatory HIV Testing

HIV testing is **mandatory** if it is required to participate in a process or activity which is not itself required. For example, if an HIV test is required for travel to some foreign countries, or to donate blood, this is considered mandatory testing because, while the test is required, one is not required to travel or donate blood. In mandatory testing, care must be taken to ensure that people are not in fact **forced** to undergo testing. At least **in theory,** mandatory testing is a form of voluntary testing: people can decide not to participate in the process or activity for which testing is required. **In practice,** however, the degree of voluntary consent is in some cases questionable. For example, in a situation where employment is not possible unless one agrees to be tested and one needs that job, the **voluntary** nature of the test appears to have vanished (*AIDS, Health and Human Rights,* 1995).

Many people advocate mandatory HIV screening for everyone. They believe that this is the best way to stop the transmission of HIV. But others oppose such testing. For example, some Christian groups that are morally opposed to homosexuality oppose mandatory testing because if pregnant women knew they were infected with HIV, this would increase abortion, which these groups also oppose on moral grounds. Others oppose testing because of civil rights concerns and the fear that individuals identified as HIV positive would be subjected to social stigma.

Mandatory Screening Has Been Tried

Mandatory premarital syphilis screening has often been cited as a precedent for premarital HIV screening. The logic appears to be: we screen for syphilis, but HIV infection and AIDS are a lot worse; therefore we should screen for HIV infection to prevent the spread of AIDS.

It can be answered that **syphilis is curable;** yet history reveals that syphilis screening has turned out to be ineffective and unnecessary. For example, in 1978, premarital syphilis screening found only 1% of the total number of new syphilis cases in the United States at a cost of 80 million dollars. Over the past dozen years, many of the 50 states have dropped this screening program and others have indicated that they will follow. Still, in 1988, legislation was pending in 35 states that would require premarital HIV testing. Louisiana, Illinois, and Maryland passed mandatory testing legislation; by 1989, all three states repealed that legislation.

Why Mandatory Testing?

Mandatory testing is for the protection of a certain group or the public at large. Although it is not anonymous, results are kept confidential on a need-to-know basis. Mandatory testing for HIV continues to be angrily debated primarily because of the possibility of error when running large numbers of test samples, inadvertent loss of confidentiality, and lack of overall benefit to those who are found to be HIV-positive. Mandatory HIV testing is routine for blood donors, and military and Job Corps personnel.

Compulsory HIV Testing

In compulsory testing, a person cannot refuse to be tested. Compulsory testing may be forced onto an individual, groups, communities, or even entire populations. A court may order an individual to be tested, or a government may decree or legislate that, for example, commercial sex workers, homosexuals, prisoners, hospital patients, or persons seeking immigration must be tested.

In Colorado, Florida, Georgia, Kentucky, Illinois, Michigan, Nevada, Rhode Island, Utah, and West Virginia, HIV **testing is compulsory for people convicted of prostitution.** However, many prostitutes are back on the streets before their test results are in. In many cases, the prostitutes could not be found for follow-up counseling. In Duval County Florida, county judges agreed to impose a 30-day jail term for convicted prostitutes, a time period long enough to get their test results and provide counseling. Pros-

titutes have to sign the test results sheet. They are released as soon as they do.

Under Florida law, a prostitute who knows he or she is carrying the AIDS virus but continues to offer sexual favors can be jailed for 1 year.

In June 1999, the state of Oregon passed legislation that allows a judge to order a person accused of a crime to be tested for HIV. The person

---

POINT OF VIEW 12.2

MANDATORY HIV TESTING

Historically, the idea of mandatory HIV testing is nothing new. Mandatory reporting already occurs for 60 infectious diseases, including herpes, syphilis, gonorrhea, and tuberculosis. But what makes mandatory testing attractive? To many it looks like a quick fix. And if you are a politician, it makes constituents think legislators are doing something about the problem. Two classic examples that showed why mandatory testing is ineffective at preventing HIV transmission were the premarital screening programs in Illinois and Louisiana in the late 1980s. Both states required people seeking marriage licenses to be tested for HIV, then disclosed test results to both partners. The programs' advocates argued that knowledge of test results would prevent transmission by allowing partners to reconsider marriage, adopt safer sex practices, and make informed child-bearing decisions. These programs, however, caused many people to avoid testing, identified few cases of HIV infection, and incurred great social and financial costs.

It is uncertain whether the programs were effective in preventing people from being exposed to HIV. Many couples avoided testing altogether by marrying in another state or deciding not to marry. Since testing a population with low seroprevalence causes a greater proportion of false positives, the premarital screening programs falsely identified some people as being infected with HIV, resulting in broken engagements, aborted pregnancies, and psychologic distress. The programs had financial costs as well: Illinois spent $243,000 per positive test result.

TESTING AS A PUBLIC HEALTH TOOL

HIV testing can be a very effective public health tool, but it 's only effective when deployed

in ways that are socially, politically, and medically appropriate. If it's not, it can actually be a detriment to public health. Being identified as having HIV has caused people to lose their jobs, health and life insurance, housing, family members, and friends, and to be denied employment, medical treatment, and education. Legal safeguards such as the Americans with Disabilities Act and the Fair Housing Act have helped people with HIV fight discrimination, but discrimination remains for those diagnosed with HIV disease or AIDS.

WHEN DOES MANDATORY TESTING FOR A DISEASE MAKE SENSE?

Many people argue that routine HIV testing will drive high-risk behavior people underground and make people avoid the health-care system altogether. **There is no empirical proof that this will or won't occur to a greater extent than it already does, under a voluntary testing program.** Ultimately one must ask whether people who would go underground because of self-interest should dictate policy and also whether such people would cooperate in disease-prevention efforts under any circumstances.

There is a set of guidelines that were developed in the mid-to-late 1980s and incorporated into the HIV testing policies of the Centers for Disease Control and Prevention (CDC) and the World Health Organization. These guidelines require, roughly, that the level of seroprevalence in the targeted population be high enough to keep the number of uninfected people subjected to testing small; the testing be conducted in a context in which the risk of HIV transmission is high; the test results enable policymakers or health officials to take actions that could not be taken otherwise; the harm or potential harm of testing not

---

be so disproportionate to whatever benefit that testing is unfair; and there be no means of achieving the same information in a less restrictive manner. Under these guidelines the case for man- datory testing in most situations is very weak. One exception is the screening of all blood and tissue. Actually, blood screening differs significantly from mandatory testing because screening, which allows the discarding of infected blood, is clearly effective in preventing HIV transmission. Also, blood donors are told prior to donation that their blood will be screened. So, **although donors do not necessarily provide informed consent, they retain the option of not donating, thereby avoiding testing.**

OTHER USES FOR MANDATORY HIV TESTING

Another instance in which mandatory testing could be used is rape, where it could be appropriate to test the assailant. There is evidence from survivors' groups that rape victims have considerable anxiety about being infected with HIV. Therefore, it is possible that the advantages to the rape survivor override the concerns about nonconsensual testing.

In many states there is a waive of consent requirement for HIV testing under specific circumstances. In a recent 50-state survey, people tested without informed consent variously include inmates of many state and federal prison systems; health care workers; hospital patients; those arrested for or convicted of sexual assault; people charged with other sex-related crimes, such as prostitution; people charged with certain drug-related crimes; and applicants for health and life insurance. Other calls for mandatory testing have targeted children in foster care, children awaiting adoption, and professional athletes

(Harvard AIDS Institute, 1995).

MANDATORY HIV TESTING OF NEWBORNS AND DISCLOSURE OF TEST RESULTS TO MOTHERS AND PHYSICIANS

Perhaps no call for mandatory HIV testing has caused as much recent controversy as those requiring all pregnant women or all newborns to be tested. Entering year 2000 only the states of Arkansas, Tennessee and Indiana (pending) require health care providers to HIV test **every** pregnant woman as early in her pregnancy as possible—**unless she refuses the test.** The results of AIDS Clinical Trials Group 076 (ACTG 076), which showed that Zidovudine (ZDV) could reduce perinatal transmission by two-thirds, have again promoted debate and prompted policymakers to take action. Most recently, on June 26, 1996, New York became the **first** state in the nation to mandate and disclose the HIV status of newborns to mothers and physicians. Governor George E. Pataki signed into law legislation known as the **"Baby AIDS" bill,** which authorizes the State Health Commissioner to establish a comprehensive program of HIV testing of newborns. The Governor said, "This law extends our efforts to do everything we can to reduce the number of babies born with this devastating virus. This information can be used to help mothers make important health care decisions for themselves and their children. We will no longer allow infants to be used as statistical tools in some scientific study. Today, we recognize the HIV infant as a living, breathing human being whose right to medical treatment must be respected."

The passage of this bill was immediately followed by the American Medical Association announcement endorsing mandatory testing of all pregnant women and newborns for the AIDS virus.

also would be tested for other communicable diseases if he or she transmitted bodily fluids to a victim. The results would not become public records.

At least 45 states and the District of Columbia authorize HIV testing for charged or convicted sex offenders (Hooker, 1996 updated).

Fear of Mandatory Testing— If a massive mandatory screening test program was im-

plemented, would it be possible to keep results confidential? (See Box 12.6) What would be done with the information? For example, would the state prevent an uninfected person from marrying an infected one? Officials fear that mandatory testing will drive many people who might have volunteered for anonymous testing underground and away from the health care system. These people will be lost to the counseling and education that would benefit them and others. The reason for going underground would be fear of discrimination and social ostracism if found to be HIV-infected.

A case can be made that a compulsory program could maintain strict confidentiality even with large numbers of people being tested. But it would appear that the political powers and public in general are not ready for broad-scale compulsory screen testing in the United States.

Confidential, Anonymous, and Blinded Testing

Both confidential and anonymous testing involve the use of informed consent forms which are, to date, with exception of the U.S. military, Job Corps workers, and certain criminals, done on a voluntary basis. Blinded testing **does not,** because of procedure, require informed consent.

Confidential HIV Testing— A consent to HIV testing must be given freely and without coercion. The volunteer does not have to provide any information unless he or she wants to. Individuals to be tested in a state laboratory complete a HIV Antibody Form. The consent form explains in simple terms what the presence of HIV antibody does and does not signify. It explains the uncertain medical outcome and potential social and legal implications of a positive test result. The consent form assures confidentiality but **warns that the result is part of the patient's medical record, to which others may have access under certain conditions.** Some centers provide completely anonymous

---

————— SIDEBAR 12.6 —————

TYPE OF TESTING CAN FOSTER MEDICAL CARE

At University of California, San Francisco a study evaluated whether anonymous HIV testing was associated with earlier HIV testing and HIV-related medical care than confidential HIV testing among people diagnosed with AIDS. The study found that HIV-infected individuals who used anonymous testing services got tested and entered care earlier than those who used confidential services. Many people tested in confidential settings did not seek testing prior to becoming ill. The findings underscore the critical need to educate individuals at risk about the need to learn their HIV status, and if infected, to seek early care. (Bindman et al., 1998)

————— BOX 12.6 —————

FLORIDA: HEALTH CARE WORKER LEAKS CONFIDENTIAL HIV INFORMATION

As of October 1, 1996, the state of Florida mandated there be at least one anonymous and confidential HIV counseling and testing site in each of its 67 counties. As of January 1, 1997, all persons testing HIV-positive who request health services must give up their identity.

Penalties: Any person who violates the confidentiality of the new laws commits a misdemeanor of the first degree punishable by imprisonment not to exceed 1 year.

Update 1997: The state of Florida delayed the identification of HIV-positive people until they are more certain that the state can assure patient confidentiality. This concern resulted from a Pinellas County health care worker who left the office, in September 1996, with a computer disk containing the names of about 4,000 HIV-infected people. The disk included names of people with AIDS and other personal information, including telephone numbers, addresses, date of birth, information about how they got AIDS, and other health information. The disk was sent to the *Tampa Tribune* and the *St. Petersburg Times* newspapers! **Florida began mandatory HIV reporting in July 1997.**

LEGISLATURE SURVEY OF STATE CONFIDENTIALITY LAWS

1. At least 39 states have laws providing for the confidentiality of HIV/AIDS-related information. At least 28 states have laws that specifically regulate medical records. The remaining states may protect confidentiality of HIV information under other statutes.

2. Almost every state allows for disclosure of HIV-related information in certain circumstances, according to a recent report on state confidentiality laws. The most frequently cited permission to disclose is given to a health care provider involved with a patient's care, to blood banks and under a subpoena or court order. Almost all states allow disclosure of data to epidemiologists and researchers, with provisions for removal of in-formation that may identify the patient. Most states have penalties for unauthorized disclosure of information.

3. Most states impose a duty on physicians and health care institutions to maintain the confidentiality of medical records. About half the states extend this duty to other health care providers. Only four states have specific legislation imposing the duty on insurance companies and few states impose a similar duty on employers or other nonhealth care institutions. Fewer than half the states have specific laws imposing a duty to maintain confidentiality of electronic or computerized medical records (Hooker, 1996).

testing; this option is available to the patient and his or her physician.

The following example demonstrates one of the problems with confidential testing. A young homosexual male with signs of oral thrush agreed to an HIV test. Later that day, he called and asked that his blood **not** be sent to the lab. He was a teacher in a parochial school and feared the results would be revealed. His sample was set aside, but the laboratory courier mistakenly took it for testing. The result was positive, yet no one could tell the patient. A malpractice attorney said to make certain all records of the test were deleted and to send the patient a letter urging him to return for a blood test. He never appeared (Wake, 1989).

Many national public health agencies and committees favor a confidential screening and counseling program that includes all individuals whose behavior places them at high risk of HIV exposure. These agencies recommend that the following eight groups seriously consider volunteering for periodic HIV-antibody testing:

1. Homosexual and bisexual men
2. Present or past injection drug users
3. People with signs or symptoms of HIV infection
4. Male and female prostitutes
5. Sexual partners of people either known to be HIV-infected or at increased risk of HIV infection

6. Hemophiliacs who received blood clotting products prior to 1985
7. Newborn children of HIV-infected mothers
8. Emigrants from Haiti and Central Africa since 1977

The Public's Question— The situation surrounding confidentiality for patients infected with HIV is comparable to circumstances that dictate our reporting of sexually transmitted diseases and cancer. The reason for reporting these cases to the CDC is that valuable epidemiological data may be gathered. These data can then be used to document geographical location, prevalence, routes of transmission, and those groups of people that may be at highest risk for these diseases. **If these diseases are reportable for the ultimate benefit of individuals susceptible to sexually transmitted diseases and cancer, shouldn't HIV-postive people be reportable for the same reasons?** It is questions such as this that are bringing about a change in public opinion concerning confidential HIV testing and use of test results.

Anonymous HIV Testing— This is also a form of voluntary testing. It differs from confidential testing only in that those who request anonymity receive a bar coded identification number. They provide no personal information and they come back at a predetermined time to find out if their test number is positive or negative. No follow-up occurs. Forty states have anonymous test sites providing over 2 million tests a year (Nash et al. 1998).

Blinded HIV Testing— This occurs when blood or serum is available for HIV testing as a result of another medical procedure wherein the patient's blood has been drawn for analysis. In this case, the demographic data have been recorded and can be used for epidemiological studies even if the name of the individual is withheld and a bar code is used. In 1988, the CDC asked for a blinded study of all 1989 newborn blood samples taken in certain cities in 45 states for metabolic studies. The name and other demographics of each newborn were recorded on the label of each tube. After the metabolic tests were completed, the name was changed into a bar code and leftover blood was sent to a state HIV testing center.

New York began a study of this type in 1987. Since then, New York has blind-tested about 250,000 newborns. The statewide incidence rate for HIV-infected newborns was 0.7%, or seven per 1,000. (Recent anti HIV therapy before and following birth has cut the incidence by half.)

In May of 1995, and for a period of several months, the Department of Health and Human Services suspended all blinded HIV testing of newborns. The reason was that Representative Gary Ackerman, D-N.Y., introduced legislation, the Newborn Infant Notification Act (HR 1289), that would require that mothers of HIV-exposed children be told of the test results. According to Ackermen and supporters, the bill was to protect the health and rights of the newborn/infants exposed to HIV. The pro and con arguments were many but in the end, the bill was withdrawn.

CLASS DISCUSSION: Present reasons for and against the Ackerman bill; one of each is provided to get you started.

Pro: The state health department reports 17% of pregnant women agree to be tested, 24% know their HIV status, and the rest have unknown HIV status.

Con: Some people have argued that only mandatory testing will give mothers the information to do what's best for themselves and their babies. But, mandatory testing would not prevent HIV in children. This would be too little too late.

PERSON-TO-PERSON: DISCLOSURE OF HIV INFECTION

Michael Stein (1998) surveyed 129 patients 989 men, 40 women) at two eastern urban hospitals, seeking their first primary care for HIV infection, that had sex in the past 6 months. They were asked if they had disclosed their HIV infection to their sex partners. **Forty percent said they did not inform their sexual partners.** And of these 40%, over half (58%) did **not** use a condom all the time. The survey subjects were mostly poor, often illegal drug users, and commonly lacked high school education, but researchers believe withholding HIV information is widespread. Among those surveyed, 46% were black, 23% were Hispanic and 27% were white. Whites and Hispanics were three times more likely to tell partners than blacks. Subjects with only one sexual partner were three times more likely to have told their partners than subjects with multiple partners. "The public health message is that if you are not absolutely sure of the HIV status of your partner, you should be having safer sex."

Reasons for Nondisclosure: risk of rejection, stigmatization, revealing sexual orientation, revealing IDU, breach of privacy that could lead to loss of employment, housing and insurance.

Class Discussion: Provide other reasons for nondisclosure?

Rapid HIV Test Kits

Rapid HIV test kits require no special equipment and can be performed in less than 10 minutes. These tests are a major contribution in Third World countries that do not have the equipment or trained personnel to run ELISA and WB tests. And, the CDC recently (mid 1998) encouraged physicians to use such tests in the USA.

The polymerase chain reaction (PCR) is a process wherein a few molecules of HIV proviral DNA can be amplified into a sufficient mass of DNA to be detected by current testing methods. It can determine if newborns of HIV-infected mothers are truly HIV-positive. The branched-DNA assay offers a means of detecting the presence of HIV RNA directly within the clinical sample.

DECIDING CONFIDENTIALITY—IS THE MAJORITY RIGHT?

In Lagos, a chief port in Nigeria, a company doctor said during a conference that she had been withholding the HIV test result of a client from her employer. The employer had sent its worker to the doctor for an HIV test to be done. The test was done (apparently without the knowledge of the patient) and it was positive. She kept the result both from the patient and from the company because she did not know what to do. One of the participants at the conference said that it would be correct if the doctor had sent the result of the test to the company since it was the company that sent the patient in the first place and also retained the doctor on its payroll. "He who pays for the piper dictates the tune." His position received acclamation from the majority of the participants who were also doctors. The doctor who raised the issue said she would now feel comfortable in sending the result to the company, having received massive support from her colleagues. Her decision received a general approval which was reinforced by another resounding round of applause.

CLASS QUESTION: Is this physician violating one of the basic principles of her profession—confidentiality? Present a list of reasons for or against her sending the HIV report to the company? In your presentation incorporate the idea of informed consent, who requested the test be done, who paid for the test, and given the discrimination associated with being HIV positive in Africa, comment on the fairness of the test to the person who took the test and the likely outcome because the person was HIV positive.

CONFIDENTIALITY AND SEXUAL PARTNER BETRAYAL

This story took place in an HIV/AIDS clinic in the South. A husband and wife came into the clinic for an HIV test. They said the **only** reason for requesting the test was that they wanted to begin a family and hoped that nothing in their past would have led to either of them being HIV-positive. The tests were completed.

The husband came in on a Monday; the wife came in that Friday.

Monday A.M.

Counselor: Mr. X, your test came back HIV-positive.

Reactions and counseling were similar to those presented in this chapter. Then MR. X said he wanted to be the one to tell his wife; **he insisted on it**. The counselor agreed, Mr. X left the clinic agreeing to come back for a follow-up counseling session.

Friday A.M.

Counselor: Mrs. X, your HIV test was negative.

Mrs. X: That's wonderful news. I can't wait to tell my husband. We've been waiting for my results. I want to get pregnant immediately.

Mrs. X received HIV-negative counseling and left the clinic very happy.

Clearly, the husband did not tell his wife the truth about his test results. A follow-up phone call to the husband went unanswered; so did a letter from the clinic. Several months later, Mrs. X called the counselor to tell her that she was pregnant!

Question: What do you think the counselor should do now?

1. Inform the woman about her husband.
2. Take no action.
3. Call the husband and discuss the situation.
4. Threaten the husband with legal action if he does not tell his wife.
5. Your position!

Discuss the moral, ethical, and legal responsibilities of each participant.

Gene probes are also being used to detect small HIV proviral DNA sequences in cells of people who are HIV-infected but not yet making antibodies.

The CDC recommends that pregnant women and people in high-risk groups volunteer for HIV testing if there is any reason to suspect they may have been exposed. Broad-scale testing in low-risk populations is not advocated because the rate of false positives rises with decreasing rates of HIV infections.

In June 1996, New York became the first state to pass legislation allowing for mandatory HIV testing of all newborns and disclosure of their HIV status to mothers and physicians. This action was quickly followed by the American Medical Association's endorsement of mandatory HIV testing of all pregnant women and newborns.

Many people have voiced opinions that HIV testing should be mandatory for all people in high-risk groups. They believe that identifying HIV carriers is the way to stop HIV transmission. However, mandatory premarital testing has failed to stop the spread of syphilis. But, nearly everyone agrees that voluntary HIV testing and counseling should be available to anyone who wants it. For those who want to be tested, there are at least two routes to consider: voluntary confidential and voluntary anonymous. In the first, people give their names and addresses and the test results become a confidential part of their medical records. In the latter, no personal identification information is asked for; a bar code (number) is used to label the blood sample.

SUMMARY

HIV infection can be detected in two ways: first, by HIV-antibody testing prior to the signs and symptoms of AIDS; and second, by physical examination after symptoms occur.

The test most often used to screen donor blood at blood banks and individuals referred to testing centers is the ELISA test. ELISA (enzyme-linked immunosorbent assay) is a highly sensitive and specific test that determines the presence of HIV antibodies in a person's blood or serum. The ELISA test was first used in 1985 to reduce the number of HIV-infected blood units for blood transfusions.

Because the ELISA test is only a predictive test which gives the percentage chance that a person is truly positive or truly negative, serum from those who test positive is retested in duplicate. If still positive, the serum is then subjected to a Western blot (WB) test. The WB is a confirmatory test. If it is also positive, the person is said to be HIV-infected.

Other screening and confirmatory tests are available. The indirect immunofluorescent antibody assay (IFA) is relatively quick and easy to perform. Although it can be used as a screening test, it is generally used as a confirmatory test. The test is similar to the ELISA test except that the analysis is made by looking for a fluorescent color, indicating the presence of HIV antibodies, with a dark field light microscope.

REVIEW QUESTIONS

(*Answers to the Review Questions are on page 487.*)

1. What is the acronym for the most commonly used HIV-antibody test and what does each letter stand for?

2. What basic immunological assumption is this test based on?

3. Does a single positive HIV antibody result mean the person is HIV-infected? Explain.

4. Is there a specific test for AIDS? Explain.

5. What is currently the most frequently used HIV confirmatory test in the United States?

6. What is the name of one additional confirmatory test in use in the United States?

7. How is HIV antibody detected in the ELISA test?

8. True or False: All newborns who are antibody positive are HIV-infected and all go on to develop AIDS. Explain.

9. What is the greatest shortcoming of the ELISA and WB tests?

10. What are the two major problems in interpreting ELISA test results?

11. What two factors may account for false-positive and false-negative results?

12. What is the relationship between false-positive results and prevalence of HIV in the population?

13. In an HIV screening test, what is a positive predictive value? Why is it called a predictive value?

14. What is the current gold standard of confirmatory tests in the United States?

15. What is the major problem in using this test?

16. Why is the polymerase chain reaction (PCR) considered so useful in HIV testing? Name two situations when PCR can be significant in HIV testing.

17. True or False: Two home use HIV-antibody test kits are now available in the United States.

18. Using the ELISA test, when are HIV antibodies first detectable?

19. How early are HIV antigens detectable in human serum?

20. What are three benefits of early identification of HIV-infected people?

21. Name the four kinds of testing privacy available to people who want to take an HIV test.

22. What is the major difference between an anonymous and a blind HIV test?

23. Why would someone want an anonymous test?

24. True or False: The ELISA serological test is adequate to confirm HIV infection.

25. True or False: Pre- and post-HIV-antibody test counseling is recommended any time an HIV-antibody test is performed.

REFERENCES

AIDS, Health and Human Rights. (1995). Francois-Xavier Bagnoud Center for Health and Human Rights–Harvard School of Public Health, pp. 1–162.

ALLEN, BRADY. (1991). The role of the primary care physician in HIV testing and early stage disease management. *Fam. Pract. Recert.*, 13:30–49.

ANDERSON JOHN, et al. (1992). HIV antibody testing and post-test counseling in the United States: Data from the 1989 National Health Interview Study. *Am J Public Health*, 82:1533–1535.

ANGELL, MARCIA. (1991). A dual approach to the AIDS epidemic. *N. Engl. J. Med.*, 324:1498–1500.

BAYER RONALD et al. (1995).Testing for HIV infection at home (Sounding Board). *N. Engl.J. Med.*, 332:1296–1299.

BELONGIA, EDWARD A., et al. (1989). Premarital HIV screening *JAMA,* 261:2198.

BINDMAN, ANDREW, et al.(1998). Multistate evaluation of anonymous HIV testing and access to medical care. *JAMA*, 280:1416–1420.

CARLSON, DESIREE A., et al. (1989). Testing for HIV risk from therapeutic blood products. In:

Pathology and Pathophysiology of AIDS and HIV Related Diseases (Eds. Jami J. Harawi and Carl J. O'Hara), St. Louis: C.V. Mosby Co.

CLEZY, K., et al. (1992). AIDS-related secondary infections in patients with unknown HIV status. *AIDS*, 6:879–893.

CONSTANTINE, NIEL T. (1993). Serologic tests for the retroviruses: Approaching a decade of evolution. *AIDS*, 7:1–13.

CORDES, ROBERT, et al. (1995). Pitfalls in HIV testing. *Postgrad. Med.*, 98:177–189.

DEL RIO, CARLOS, et al. (1996). The use of oral fluid to determine HIV prevalence rates among men in Mexico City. *AIDS*, 10:233–234.

DEWAR, ROBIN, et al. (1994). Application of branched DNA signal amplification to monitor HIV-I type burden in human plasma. *J. Infect. Dis.*, 170: 1172–1179.

EL-SADR, WAFAA, et al. (1994). Managing early HIV infection: Agency for Health Care Policy and Research. *Clinical Practice Guideline on Evaluation and Management of Early HIV Infection.* January, 7:1–37.

EL-SAHR, W., et al. (1994). *Managing Early HIV Infection: Quick Reference Guide for Clinicians.* AHCPR Publication No. 94-0573. Rockville, MD.

EMMONS, WESLEY, et al. (1995). A modified Elisa and Western Blot accurately determine anti-human HIV-I antibodies in oral fluids obtained with a special collecting device. *J. Infect. Dis.*, 171: 1406–1410.

FANG, CHYANG T., et al. (1989). HIV testing and patient counseling. *Patient Care*, 23:19–44.

FRERICHS, RALPH R., et al. (1994). Saliva-based HIV-antibody testing in Thailand. *AIDS*, 8:885–894.

GALLO, DANA, et al. (1997). Testing oral secretions for HIV. *AIDS Clin.Care*, 9:26.

GOSTIN, LAWRENCE, et al. (1997). National HIV case surveillance is urged. *N. Engl. J. Med.*, 337: 1162–1167.

HARTLAUB, PAUL, et al. (1993). Obtaining informed consent: It is not simply asking "do you understand?" *J. Fam. Pract.*, 36:383–384.

HARVARD AIDS INSTITUTE, (1995). Mandatory HIV testing: the search for a quick fix. *Harvard AIDS Letter*, May/June:4–8.

HEGARTY, J.D., et al. (1988). The medical care costs of human immunodeficiency virsu infected children in Harlem. *JAMA*, 260:1901–1905.

HOOKER, TRACEY. (1996). HIV/AIDS: Facts to consider—1996. Natural conference of State Legislatures, February, pp. 1–64.

HORN, THOMAS, et al. (1989). Forks and combs and DNA: The synthesis of branched oligodeoxyribonucleotides. *Nucleic Acids Res.* 17:6959–67.

HU, DALE, et al. (1996). The emerging genetic diversity of HIV. *JAMA*, 275:210–216.

Intergovernmental AIDS Report. (1989). Illinois court overrules mandatory HIV testing for prostitutes and sex offenders, 2:1–18.

KELLER, G.H., et al. (1988). Identification of HIV sequences using nucleic acid probes. *Am. Clin. Lab.*, 7:10–15.

KRAMER, F.R., et al. (1989). Replicatable RNA reporters. *Nature*, 339:401–402.

MACKENZIE, WILLIAM R., et al. (1992). Multiple false positive serologic tests for HIV, HTLV-1 and hepatitis C following influenza vaccination, 1991. *JAMA*, 268:1015–1017.

Medical Tribune. (1991). Beware of blanket AIDS solutions. June:14.

MERCOLA, JOSEPH M. (1989). Premarital HIV screening. *JAMA*, 261:2198.

MIIKE, LAWRENCE. (1987). *AIDS Antibody Testing.* Office of Technological Assessment Testimony To The U.S. Congress. Oct.:1–21.

MILES, STEVEN A., et al. (1993). Rapid serologic testing with immune-complex-dissociated HIV p24 antigen for early detection of HIV infection in neonates. *N. Engl. J. Med.*, 328:297–302.

Morbidity and Mortality Weekly Report. (1988a). Update: Serologic testing for antibody to human immunodeficiency virus. 36:833–840.

Morbidity and Mortality Weekly Report. (1992). Testing for antibodies to HIV-2 in the United States. 41:1–9.

Morbidity and Mortality Weekly Report. (1995).U.S. Public Health Service recommendations for HIV counseling and voluntary testing for pregnant women. 44:1–12.

Morbidity and Mortality Weekly Report. (1996). U.S. Public Health Service Guidelines for testing and counseling blood and plasma donors for HIV type I antigen. 45:1–9.

Morbidity and Mortality Weekly Report. (1998a). Diagnosis and reporting of HIV and AIDS in the Unites States with integrated HIV and AIDS surveillance—United States January 1994–June 1997, 47:309–314.

Morbidity and Mortality Weekly Report. (1998b). Update: Counseling and testing using rapid tests—United States, 1995, 47:211–215.

Morbidity and Mortality Weekly Report (1999). HIV Testing—United States, 1996. 48:52—55.

MOSER, MICHAEL, (1998). Anonymous HIV testing. *Am. J. Pub. Health*, 88:683.

MULLIS, KARY B., et al. (1987). Process for amplifying, detecting, and/or cloning nucleic acid sequences. (U.S. Patent No. 4,683,195). *Official Gazette of the U.S. Patient and Trademark Office,* Volume 1080, Issue 4, July.

NASH, GRANT, et al. (1998). Health benefits and risks of reporting HIV-infected individuals by name. *Am. J. Pub. Health*, 88:876–879.

PASSANNANTE, MARIAN R., et al. (1993). Responses of health care professionals to proposed mandatory HIV testing. *Arch. Fam. Med.*, 2:38–44.

PHILLIPS KATHRYN, et al. (1995a).Potential use of home HIV testing. *N. Engl. J. Med.*, 332:1308–1310.

PHILLIPS KATHRYN, et al. (1995b).Who plans to be tested for HIV or would get tested if no one could find out the results? *Am. J. Prevent. Med.*, 11(3):156.

REIMER, LARRY, et al. (1997). Undetectable antibody reported in a patient with typical HIV. *Clin. Infect. Dis.* 25: 98–103.

RHAME, FRANK S., et al. (1989). The case for wider use of testing for HIV infection. *N. Engl. J. Med.*, 320:1242–1254.

RICH, JOSIAH, et al. (1999). Misdiagnosis of HIV infection by HIV-1 Plasma Viral load testing: A case series. *Annals of Internal Medicine*, 130:37—39.

ROGERS, MARTHA F., et al. (1989). Use of the polymerase chain reaction for early detection of the proviral sequences of human immunodeficiency virus in infants born to seropositive mothers. *N. Engl. J. Med.*, 320:1649–1654.

SAMET, JEFFREY, et al. (1998). Trillion Niron delay: Time from testing positive for HIV to presentation for primary care. *Arch. Int. Med.*, 158:734–740.

SAX, PAUL, et al. (1997). Novel approaches to HIV antibody testing. *AIDS Clin. Care*, 9:1–5.

SCHEFFEL, J.W. (1990). Retrocell HIV-1 passive haemagglutination assay for HIV-1 antibody screening. *J. Acquired Immune Deficiency Syndromes*, 3:540–545.

SCHOPPER, DORIS, et al. (1996). Testing for HIV at home: What are the issues? *AIDS*, 10:1455–1465.

STEIN, MICHAEL, et al. (1998). Sexual ethics: Disclosure of HIV-positive status to partners. *Arch. Int. Med.*, 158:253–257.

URDEA, MICKEY. (1993). Synthesis and characterization of branched DNA (bDNA) for the direct and quantitative detection of CMV, HBV, HCV, and HIV. *Clin. Chem.*, 39:725–726.

WAKE, WILLIAM T. (1989). How many patients will die because we fear AIDS? *Med. Econ.*, 66:24–30.

WOFOY, C. B. (1987). HIV infection in women. *JAMA*, 257:2074–2076.

AIDS and Society: Knowledge, Attitudes, and Behavior

CHAPTER CONCEPTS

- AIDS is here to stay
- The HIV/AIDS **"devastation"** is **now.**
- Inaccurate journalism leads to public hysteria.
- Vignettes on AIDS.
- It's 19 years later, and what do we know about HIV/AIDS?
- Use of explicit sexual language on TV and in journalism.
- Goal of sex education: To interrupt HIV transmission.
- Education, **Just Say Know.**
- **The Red Ribbon.**
- Education is not stopping HIV transmission.
- Students still have **misconceptions** about HIV transmission.
- The general public and homophobia.
- Employees are not well informed and fear working with HIV/AIDS-infected co-workers.

- Teenagers are not changing sexual behaviors that place them at risk for HIV infection.
- First World AIDS Day observed in 1988.
- Orphaned children due to AIDS-related deaths.
- Physician–patient relationships in the HIV/AIDS era.
- United States Supreme Court renders its FIRST ever ruling in the 19 years of the HIV/AIDS pandemic.
- Educating employees about AIDS.
- Placing the risk of HIV and infection in perspective.

AIDS IS HERE TO STAY

The past 19 years of the global AIDS pandemic has taught us that HIV disease is a **permanent part of life** on planet earth. **HIV is simultaneously a virus and a phenomenon.** When it is viewed only as a virus, it is hard to see why HIV prevention is a problem. People normally want to avoid harming themselves and others. But when viewed as a phenomenon, HIV points to the many personal and societal causes of disease transmission. It points to the difficulties people have **in making their intentions match their behavior.** It points to the inequalities in relationships, which help to spread HIV. It points to societies' reluctance to prepare young people to manage their intimate relations, to admit to sexual diversity, to responsibly manage complex health and social problems such as drug abuse, and to provide access to health care. It points to the world's failure to care about improving living conditions in poor countries.

AIDS

Acquired Immune Deficiency Syndrome (AIDS) is an illness characterized, according to CDC criteria, as the presence of antibody to HIV and a T4 cell count of less than $200/\mu L$ of blood, or by the presence of HIV and certain opportunistic infections and diseases that affect both the body and brain.

AIDS was first described in the United States in 1981. Who would have thought then that 19 years later that about 60 million people or 1 in every 100 people on planet earth would be infected with the virus that causes AIDS? And, that this virus, regardless of the involvement of the world's governments, the best scientists, and the expenditure of about 200 billion of dollars, continues to spread out of control in many nations of the world. There is no way at the present time to prevent the 5 to 6 million new HIV infections each year. No drugs offer a cure, and most of the 43 million living with HIV-infected (beginning year 2000) can not afford and will not receive those drugs that offer **some** temporary improved quality of life. An HIV preventative vaccine **now** is wishful thinking. But, there is one commodity in plentiful supply— BLAME—**enough for everyone, everywhere.**

BLAME SOMEONE, DÉJÀ VU

The greater **hostility** and greatest **stigma** tends to be assigned to diseases in which individuals are seen as responsible for having the disease; in which the disease's course is fatal; in which fear of transmission is a major issue; and in which the disease leads to highly visible and frightening physical expressions. All the above conditions are associated with HIV disease and AIDS. With AIDS more than any other disease in history, people have found verbal mechanisms for distancing themselves from thoughts of personal infection. Worldwide, from the onset of this pandemic, people have learned in a relatively short time to **categorize, rationalize, stigmatize,** and **persecutize** those with HIV disease and AIDS. AIDS statistics are published in categories to identify how many gay or bisexual men, injection drug users, persons with hemophilia, and so on, have developed AIDS. Also listed are the countries, states, and cities with the highest incidence of the disease, along with which racial and ethnic groups are highest among reported AIDS cases. **By focusing on categories of people, have we made it possible for society to rationalize that AIDS belongs to somebody else? Have we made the thought of HIV/AIDS somewhat impersonal? HAVE WE FOUND A WAY TO BLAME SOMEONE ELSE?**

Placing blame does not always require reason and tends to focus on people who are not considered normal by the majority. Thus minorities and foreigners are often singled out to blame for something, sometimes anything. Epidemics of plague, smallpox, leprosy, syphilis, cholera, tuberculosis, and influenza have historically focused social blame onto specific groups of people for spreading the diseases by their "deviant" behavior. Blaming others leads to their stigmatization and persecution.

While the Black Death, an epidemic of bubonic plague, swept across Europe in the 14th century, blame was variously attached to Jews and witches, followed by the massacre and burning of the alleged culprits. In Massachusetts between 1692 and 1693, some 20 people were hanged or burned at the stake after being accused of having the powers of the devil. Eighty percent of those accused were women. And when Hitler blamed Jews, communists, homosexuals, and other **undesirables** for the economic stagnation of Germany in the 1930s, the result was death camps and ultimately the second World War. Now there is a new plague— HIV/AIDS. **What blame comes packaged with this new disease?**

Jonathan Mann, former head of the World Health Organization's Global Program on AIDS, said in 1998 that there are really three HIV/

AIDS epidemics, which are in fact phases in the invasion of a community by the AIDS virus.

First is the epidemic of silent infection by HIV, often completely unnoticed. **Second,** after a period of incubation/clinical latency that may last for years, is the epidemic of the disease itself, with an estimated total of 24.7 million AIDS cases worldwide beginning 1999. (ending 1999 about 28 million)

Third, and perhaps equally important as the disease itself, is the epidemic of social, cultural, economic, and political reaction to HIV/AIDS. **The willingness of each generation to place blame on others when believable explanations are not readily available simply recycles history. We have been there before; we have placed blame on others and it will continue.** With respect to the HIV/AIDS pandemic, blame has been disseminated among nations. And, there is no shortage of political, economic, social, or ethical issues associated with this **new** disease.

——————— **POINT OF VIEW 13.1** ———————

DO PEOPLE REALLY TRUST SCIENCE?

Earlier in this century, scientists achieved enormous prestige following their discovery of antibiotic "miracle drugs" and the polio vaccine. But since then, the public has witnessed few miracles. Moreover, few of us, whether patient or health professional, thought we would experience an epidemic with the ferocity of AIDS that has so far stymied the biomedical community. To make matters worse, medical journals, followed quickly by terrifying paperbacks in airport bookstores, carry word of new or newly virulent killer microbes. Why can't scientists sweep away the present and future dangers posed by such organisms? Why does the **war on cancer** remain a skirmish? Public confidence in biomedical science has fallen, and alternative medicine, with its primal message hope, is currently filling a vacuum.

The essence of science is to find **TRUTHS.** To attack **IGNORANCE!** Sounds good—but most science as presented in scholarly works is difficult for the lay person to understand. And scientists themselves do little to make their work more easily understood. They produce **conflicting information,** confusing both doctors and patients. Contradictory results emerge every week from epidemiologic data and clinical trials. When people ask for advice about risk prevention the answers depend on **who** gives the answer. Information in the medical journals judged best for managing the interplay of obesity, blood pressure, lipids and family history may change monthly. Is salt in or out this year? When is wine healthy, and when is it not? What's best for back pain? How do coronary stents (a device to keep an artery open) or the drugs, phentermine, or interferons fare this week? When so much appears uncertain or contradictory, taking the doctor's counsel seriously becomes difficult. Then there is the news media's constant clamor to expose scientific fraud and to overplay scientific reports to sell copy. Note the 1997 explosion of media events following the announcement of **cloning sheep.** But what the news media didn't say was **more important,** that the technique for cloning sheep had been used some 45 years ago to clone frogs. But suddenly, our society was at risk? Earlier there were the announcements of **electromagnetic fields** causing cancer and a **"tremendous breakthrough"** at Fusion Research Laboratories that would virtually solve the **world's energy problems.** The media hype, then too, was out of proportion to the available known facts and this turned out to be a fraud. And remember, President Nixon promised a cure for cancer in the 1970s! These and many, many other accounts of media-hyped scientific promises may be the cause of public **distrust** (?) or at the very least the public's hesitancy to believe in science. How do we know lay persons are hesitant about science? **Here we are in 2000,** and we live in a country in which half of the people believe that our earth is no more than 10,000 years old, and where many people still believe the earth is **flat,** that humans never landed on the moon, or that Sojourner rolling across the martian surface is really a series of photographs of a battery-operated toy moving over rocky earth terrain! Yet, the majority of people believe in paranormal phenomena! The fact that most newspapers carry a daily column on astrology but not on astronomy gives us a strong indication of where the public's faith lies. To add to this confusion, with respect to HIV/AIDS, are the headlines and TV programs that continue to provide a wide array of opinions, by some reputable scientists, physicians, and politicians that declare HIV does not cause AIDS. (See Chapter 2 for a discussion of this material.) **Is it any wonder that millions of people, worldwide, do not believe what scientists are saying about the HIV/AIDS pandemic?**

FEAR: PANIC AND HYSTERIA OVER THE SPREAD OF HIV/AIDS IN THE UNITED STATES

With the 1981 announcement by the U.S. Public Health Service and the CDC that there was a new disease, AIDS quickly became a symbol for our darkest fears. Responsible public officials gave out conflicting messages: **reassurance** on one hand and **alarm** on the other. **Public panic and hysteria began.**

People with HIV disease and AIDS are still abused, ridiculed, and maligned. Some people believe that AIDS is divine retribution for immoral lifestyles. People who have not indulged in high-risk lifestyles (e.g., newborns and recipients of blood products) continue to be labeled as **innocent victims,** implying perhaps that other HIV-infected individuals are guilty of the behavior that led to their infection and therefore deserve their illness.

Families and communities continue to be divided on their beliefs and acceptance of HIV/AIDS patients. Federal and state agencies stand accused of a lack of commitment and compassion in the war against AIDS. The bottom line is that **value judgments** are associated with HIV/AIDS because the disease involves the most private areas of people's lives—**sex, pregnancy, drug use, and finances.**

The Fear Factor

Worldwide, the political, medical, and legal communities used the media or vice versa, to scare people about a new disease called AIDS. **The result has been to scare people into fearing other people rather than the disease.** For example, a 1990 survey of 1,000 black American church members in five cities found that more than one third of them believed the AIDS virus was produced in a germ warfare laboratory as a form of genocide against blacks.

Another third said they were **unsure** whether the virus was created to kill blacks. That left only one third who disputed the theory.

These findings held firm even among educated individuals, said one of the authors of the 1990 survey. Rumors that AIDS was created to kill blacks have circulated in the black community for years, and the belief is still endorsed by some black leaders.

A poster in the Swiss STOP AIDS Campaign focuses on how we think about people with AIDS:

> For the doctors, I am HIV-positive; for some neighbors, I am AIDS-contaminated; for my friends, I am Claude-Eric.

Soon after young homosexual men began dying in large numbers, a barrage of frightening rhetoric began filling the airwaves, television, the popular press, and even the most reputable scientific journals. The AIDS disaster was here. One health care administrator stated **"We have not seen anything of this magnitude that we can't control except nuclear bombs."**

In 1986, Myron Essex of the Department of Cancer Biology at the Harvard School of Public Health noted,

> The Centers for Disease Control and Prevention (CDC) has been trying to inform the public without overly alarming them, but we outside the government are freer to speak. The fact is that the dire predictions of those who have cried doom ever since AIDS appeared haven't been far off the mark ... The effects of the virus are far wider than most people realize. It has shown up not just in blood and semen but in brain tissue, vaginal secretions, and even saliva and tears, although there's no evidence that it's transmitted by the last two.

In 1987, columnist Jack Anderson reported that the Central Intelligence Agency (CIA) concluded that in just a few years heterosexual AIDS cases would *outnumber* homosexual cases in the United States. Also in 1987, Otis Bowen, former Secretary of Health and Human Services, said AIDS would make the Black Death that wiped out one-third of Europe's population in the Middle Ages pale by comparison. In 1988, sex researchers William Masters, Virginia Johnson, and Robert Kolodny stated in their book, *New Directions In the AIDS Crisis: The Heterosexual Community*, that there was a possibility of HIV infection via casual transmission—from toilet seats, handling of contact lenses from an AIDS patient, eating a salad in a restaurant prepared by a person with AIDS, or from instruments in a physician's office used to examine AIDS patients.

———— BOX 13.1 ————

AN EXAMPLE OF UNCONTROLLED FEAR AND HOSTILITY

In September 1985, in Stamford, Connecticut, a murder suspect with AIDS is escorted into court by police wearing rubber gloves. Fourteen prospective jurors promptly ask to be removed from the case for fear of catching AIDS from being in the same room. In Atlanta, November 1985, a man develops a nosebleed while being driven to a jail in a squad car. After he tells police that he has AIDS he is rushed to the hospital and his car is quarantined for 21 days.

In 1986, a very ill AIDS patient on an airline flight spilled some urine from a drainage bag on the seats and carpet. After the patient deplaned, several rows of seats around the area were taped off; the contaminated area would be removed. Crew members and airline union leaders expressed outrage at the exposure of passengers and crew to this threat. The clean-up crew refused to remove the carpet, even using rubber gloves.

BUT WE HAVE LEARNED SINCE THEN, RIGHT? **WRONG!**

In April 1993, an American Airlines flight crew requested that all pillows and blankets be changed on a plane that carried a group described by an internal communication as "gay rights activists." A representative for American Airlines apologized saying that such action was completely and totally inappropriate and that American Airlines has a strict policy against discrimination of any individual or group. **But it happened.**

Well, that is behind us, not to worry, **it will not happen again—certainly not in 1995** when we have all been exposed to so much AIDS education- RIGHT? **WRONG AGAIN! In June of 1995** a delegation of 50 gay and lesbian elected officials was met at the White House by uniformed Secret Service officers wearing rubber gloves. The officers were apparently concerned about the risk of infection from HIV.

CAN IT HAPPEN AGAIN? In September 1997 — His sister told ambulance personnel that her sick brother had AIDS and had to go to a specific county hospital. They said they could **not** go outside their county line. She called a multi-county ambulance service. After arrival, neither ambulance service personnel would go near her brother. Her young daughter held an oxygen mask on him while she drove him to the hospital. He died later that evening. The sister filed an AIDS discrimination suit against the ambulance services. (Case pending)

SURELY NOT IN 1998! In September 1998 — a federal judge barred the eviction of an HIV-positive man who was to be evicted from public housing. The infected male is suing the manager of a federally subsidized housing complex. The manager is said to have told people that she did not want the complex contaminated with AIDS.

In Cleveland, November 1998, a black American sex worker was forced as part of his plea bargain to appear on an NBC-affiliated TV station WKYC to identify himself as a person living with HIV. The judge said her goal with the unusual sentencing was to protect the public. In December 1998, **Gugu Dlamini of South Africa was beaten to death the day after she revealed that she was HIV positive during a commemoration of World AIDS Day, December 1, 1998.** She was one of many unsung heroes of the daily struggle against HIV. Her death reminds us how stigmatizing a disease AIDS still is, and how much courage it takes for people with HIV to be open about their condition because fear still controls people's behavior.

In **Colonial Heights, Virginia, February 1999,** the 4th U.S. Circuit Court of Appeals found that although the Americans With Disabilities Act prohibits discrimination against people with AIDS, the law does not require **U.S.A. Bushidokan,** in Colonial Heights, to accept **Michael Montalvo,** now 14, for its Japanese-style sparring classes. Michael was 12 when he attempted to join the karate school. The owners barred him from class after learning about his HIV infection. The decision drew criticism from AIDS groups, which said it reflects fear rather than legitimate health concerns. An AIDS Action leader said, "Communities are right to become concerned about HIV, but cases like this district from giving young people the information they need and give a false sense of how it is transmitted." The Montalvos, who have since **moved out of Virginia** because they believe public disclosure of Michael's infection led to discrimination, plan to appeal to the U.S. Supreme Court.

The way people with HIV are treated is likely to continue to be based more on fear, prejudice and judgmentalism than on the facts of medical reason. Supreme Court rulings are not going to change that. What the courts do, however, is make it absolutely clear that such prejudicial behavior will not be legally tolerated.

In contrast to these reports is the 1988 article by Robert Gould in *Cosmopolitan* reassuring women that there is practically no risk of becoming HIV-infected through ordinary vaginal or oral sex even with an HIV-infected male. The vaginal secretions produced during sexual arousal keep the virus from penetrating the vaginal walls. His explanation was: **"Nature has arranged this so that sex will feel good and be good for you."**

In December 1997, a male entered a bar and held a syringe full of blood against a female patron's throat. He said the blood contained HIV. He demanded money. He was caught and charged with robbery and attempted murder, pending the outcome of HIV tests of the blood from the syringe. **The man felt the fear of AIDS would be sufficient to rob the bar.**

Fear of AIDS is understandable, given that AIDS is fatal and is communicable. AIDS appeared suddenly and spread quickly—it took a number of years to identify the virus that causes it and the mechanisms for spread. Yet, today the routes of transmission are well established and widely known, as are the precautionary measures that can be taken to prevent its spread. Early on, fear of AIDS took an unhealthy turn, anxieties are projected on to those who are hit the hardest by the disease, and the fear of AIDS became an irrational fear of people with AIDS.

ACTIONS OF COURAGE

There are far too many heroes of this pandemic to list here. And the heroes come from every walk of life and profession. So many have answered the call for help, for compassion and human kindness. But there are a few examples that might be considered either sufficiently different or unusual enough to mention. And although unusual, even these actions have been demonstrated by many.

First, in 1979, when gay pride was rising, a teenage male took a boy to the prom in Sioux Falls, South Dakota. The national media were there to record the moment; the camera lights glared as he and his date danced; gay activists were elated. It was the first time a same-sex couple had been allowed to attend an American prom. This male couple faded into obscurity—

until January 1993. **His death from AIDS was reported to the CDC.**

The next examples of courage are about two courageous teenagers. He was a husky eighth-grader from Missouri who wore a peace symbol around his neck and was diagnosed with HIV in the summer of 1991. He recalled the day his mother called him into the house to give him the news. "I was getting ready to play baseball. I said, 'Does this mean I'm going to die?' She said, 'Not today.' I said, 'Can I go play baseball now?'".

He had only one problem with prejudice. A few parents in neighboring towns didn't want their children to play against his baseball team. **"I quit the team because I didn't think it was fair for them to not get to play because of me."**

A 13-year-old girl in Detroit who goes by the nickname of Slim has very grown-up responsibilities at home.

"It's hard, but I won't show you or let you see my pain," she said. "I was always taught to be strong."

She's had to be. Her mother, a single parent, is infected with HIV. At the same time, she learned that her sisters were also infected.

One sister died in April of 1993 at age 9. Then in November, her 6-month-old sister died. Now, another sister is battling HIV.

Through it all, she has been responsible for giving medications and intravenous feedings to everybody, around the clock.

"I want people to know what's going on. But I don't want anyone's pity."

WHOM IS THE GENERAL PUBLIC TO BELIEVE?

Because of the complexity of HIV disease, a great deal of press coverage of AIDS issues reflects what scientists say to journalists. A journalist's responsibility is to check that the facts are accurate, but not necessarily to judge their overall merit. Why should a good story be spiked just because other scientists disagree with the data interpretation? **When scientists say contradictory things to the public, how can the public assess whom to believe?** Science has a duty to inform and educate the public, but it must neither frighten people unnecessarily nor give them unjustified expectations. Claims

PATSY CLARKE'S CORRESPONDENCE WITH UNITED STATES SENATOR JESSE HELMS ABOUT AIDS

This letter to Senator Jesse Helms—running for reelection to the US Senate November 1995—was written by Patsy M. Clarke, of Raleigh, North Carolina. Clarke, whose husband had once worked as a consultant for the Senator, wrote to ask Jesse Helms to reconsider his harsh views about the AIDS epidemic and homosexuality. Mrs. Clarke's son Mark had died from AIDS the previous year.

Senator Helms replied. Patsy Clarke found Helms' response so unsatisfactory that she founded an organization called **MAJIC—Mothers Against Jesse In Congress**—to work against the Senator's reelection.

RALEIGH, NORTH CAROLINA, JUNE 5, 1995

Dear Jesse,

This is a letter I have wanted to write for a long time. I do it now because its time has come.

When my husband (and your strong friend), Harry Clarke, died in a plane crash at the Asheville airport on March 9, 1987, you called me in the night. You told me of your sorrow at our loss and of what Harry had meant to you as a friend. You placed your praise for him and his principles in the Congressional Record. You sent me the flag flown over the capitol in his memory. You did all of these things and I am grateful.

Harry and I had another son, Mark. He was almost the image of his father, though much taller. He was blessed with great charm and intelligence and we loved him. He was gay. On March 9, 1994, exactly seven years to the day that his father died, Mark followed him—a victim of AIDS. I sat by his bed, held his dear hand and sang through that long last night the baby song that I had sung to all of our children. "Rock-a-bye and don't you cry, rock-a-bye little Mark. I'll buy you a pretty gold horse to ride all around your pasture...."

A few days before he died, Mark said these words: "This disease is not beating me. When I draw my last breath I will have defeated this disease—and I will be free." I watched him take his last breath and claim his freedom. He was 31.

As I write these words I re-live the most difficult time of my life. The tears will smudge this if I don't take care. No matter, I will type it so it is legible. My reason for writing to you is not to plead for funds, although I'd like to ask your support for AIDS research; it is not to ask you to accept a lifestyle which is abhorrent to you; it is rather to ask you not to pass judgment on other human beings as "deserving what they get." No one deserves that. AIDS is not a disgrace, it is a TRAGEDY. Nor is homosexuality a disgrace; we so-called normal people make it a tragedy because of our own lack of understanding.

Mark gave me a great gift. A quote returns to me from long ago: "I have no lamp to light my feet save the lamp of experience...." I think Patrick Henry said it. Mark's life and death have illuminated my own and I am grateful for him.

So, that's what this letter is about, and I hope I have written it well. I wish you had known Mark. His life was so much more eloquent than any words which I might put on paper. I ask you to share his memory with me in compassion.

Gratefully,
Patsy M. Clarke

WASHINGTON, D.C., JUNE 19, 1995

Dear Patsy,

I hope you will forgive my first-naming you. Having known Harry as I did and having read your poignant letter, I just don't feel like being formal in this response.

I know that Mark's death was devastating to you. As for homosexuality, the Bible judges it, I do not. I do take the position that there must be some reasonableness in allocation of federal funds for research, treatment, etc. There is no justification for AIDS funding far exceeding that for other killer diseases such as cancer, heart trouble, etc.

And, by the way, the news media have engaged in their usual careless selves by reporting that I am "holding up" the authorizing legislation that includes AIDS funding. One of the homosexual activists sent out a totally erroneous press release (and he knew what he was saying was not true) hoping to cause me problems. He failed. I did file a "notify" request because I have two or three amendments that I intend to offer to restore balance to the spending of taxpayers' money for research and treatment of various diseases.

I understand the militant homosexuals and they understand me. They climbed onto the roof of Dot's and my home and hoisted a giant canvas condom.

As for Mark, I wish he had not played Russian roulette with his sexual activity. He obviously had a great deal to offer to the uplifting of his generation. He did not live to do all of the wonderful things that he might otherwise have done.

I have sympathy for him—and for you. But there is no escaping the reality of what happened.

I wish you well always.
Sincerely,
Jesse H.

An Open Letter to Senator Jesse Helms

Dear Jesse,

I'm not sure that it is proper to call you "Jesse" anymore. When last I wrote it was as one who felt you to be a friend. And, of course, you were a friend to my husband, Harry. No question about it. And he to you. And now here I am, his widow, in a position of opposition. I guess that's what this letter is all about.

It has been almost exactly a year since last we exchanged letters. I wonder if you even remember my writing you about my son, Mark, and his death from AIDS. For some reason, I feel compelled to tell you about what has happened to my thinking as a result of your answer to me.

I believe that if you had just said, "I'm

so sorry" I would have retained my belief that Senator Helms was basically a kind human being. When I read your response saying that "the Bible judges homosexuality, I do not" and that you were sorry that Mark had chosen to play Russian Roulette with his sexuality, I sat down in my chair and let the tears flow. The tears were for what I perceived as my ineptitude in reaching your heart. I kept saying to myself over and over, "He didn't understand. He didn't understand."

The sense of helplessness that I felt at that time finally turned to a reexamination of beliefs and opinions which I had long held, beliefs about political stands, about human beings who are different from myself, about many things which I had simply taken for granted. Being a part of mainstream, conservative American life for so many years had seemed comfortable and right until my very being was forced to consider differently.

In a way, Senator, I wish that you had simply written, "I'm sorry" in response to my letter. I could have remained sad, but secure in my comfortable belief that you were beneficient in your view of other human beings; however, in another way I am grateful that you wrote to me as you did. It caused me to question the source and validity of the words you used in referencing the gay community with their "disgusting, revolting behavior."

I remembered my son. There was nothing disgusting or revolting in his behavior. He was honorable, kind, and accepting of differences in other human beings. The young man who was his partner has the highest of life principles. As I journeyed back through these memories I began to question myself for accepting such opinions as you expressed in your letter. I searched out biblical writings and opinions on many of these concerns, but mostly, I searched out my own heart. Believe me, this has not been easy, but it has been enlightening.

So, now, I find myself as a founding mother of MAJIC (Mothers Against Jesse In Congress) actively opposing your reelection. I can hardly believe this. I have always voted for Jesse Helms. I have always been a Republican. To change my life at age 67 is shocking to me but, so also, is the lesson you have taught me. It would have been infinitely easier to have grieved for Mark, put the loss behind me, and gotten on with my life as so many suggested that I do. But I can't. I offered you a quote from Patrick Henry in my last letter. This time I tender one from J. R. Lowell: "They are slaves who fear to speak for the fallen and the weak." I am trying to unslave my mind and heart from the rigid thought patterns which I accepted for so long. Frankly, the freedom I have gained is heavy at times.

I would have preferred to cook Sunday dinners for my grandchildren, to travel a bit, do my needlework, and rest on the past. But the strongest principle that Harry and I taught at the kitchen table was that with every ounce of freedom comes a pound of responsibility.

This letter has wandered some. Forgive that, if you can.
Patsy Clarke

Thanks to Patsy M. Clarke and MAJIC for their permission to reprint these letters.

of "**AIDS cures**" in the popular press need to be based on much more than just test tube data. **Whatever the need to attract research funding, is 5 minutes of fame ever worth a day of fear or weeks of false hopes for many?** The popular press has provided HIV-infected persons with a roller coaster ride between hopelessness and fantasies of imminent cure.

As a result of journalistic promises, there was and still is a range of emotions that run from real hope of a cure to public panic and hysteria. In at least five states, children with AIDS were barred from attending local public schools. The case of 12-year-old Ryan White of Kokomo, Indiana was made into a TV movie, *The Ryan White Story*, in 1989. Many parents fail to place HIV infection in perspective. In reality, automobile accidents, voluntary and involuntary exposure to tobacco smoke, drugs, and alcohol present greater risks to their children than does casual contact with an HIV-infected person.

In some localities, police officers and health care workers put rubber gloves on before apprehending a drug user or wear full cover protective suits when called to the scene of an accident (Figure 13-1). In other communities, church members, out of fear of HIV infection, have declined communion wine from the common cup.

Since the epic announcement in 1981, HIV/ AIDS has refashioned America. It is not the first epidemic to alter history; measles, smallpox, bubonic plague, leprosy, polio, cholera, and

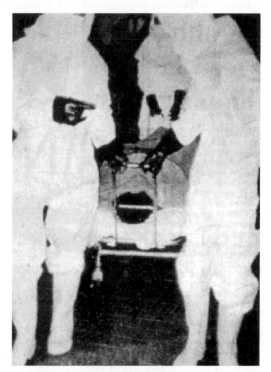

FIGURE 13-1 Ambulance Workers Protecting Themselves from the AIDS Virus at the Scene of an Accident. (*Photograph courtesy of AP/Wide World Photos*)

other diseases have ravaged their eras. But HIV/AIDS is a disease molded to the times, one that has wedged apart the thinly concealed fault lines of American society by striking hardest at the outcasts—gay men, injection drug users, prostitutes, and impoverished whites, blacks, and Hispanics. **HIV/AIDS has brought forth uncomfortable questions about sex, sex education, homosexuality, the poor, and minorities.** The disease has inevitably polarized the people, accentuating both the best and worst worldwide. Many churches, schools, and communities have responded to the new disease with compassion and tolerance; others have displayed hate and reprisals of the worst kind.

As Camus wrote in *The Plague,* **"The first thing (the epidemic) brought . . . was exile." Anyone who carried the disease could inspire terror. They became pariahs in society.**

People with hate in their hearts torched the house of the Ray family and their three HIV-positive hemophilic children in Arcadia, Florida. And someone shot a bullet through the window of Ryan White's home to let the teenager know he should not attend the local high school. After he died, his 6 foot 8 inch gravestone was overturned four times and a car ran over his grave!

Early on, the federal government and its public health apparatus showed little interest in the HIV/AIDS epidemic. **Former president Ronald Reagan never once met with former Surgeon General C. Everett Koop to talk about AIDS despite Koop's pleas.** Koop said, "If AIDS had struck legionnaires or Boy Scouts, there's no question the response would have been very different."

By the time former president of the United States Ronald Reagan delivered his first speech on the AIDS crisis in 1987 over 40,000 men, women and children had been diagnosed with AIDS and over 28,000 Americans had died of AIDS. It took 9 years and over 115,000 AIDS deaths before the U. S. Congress and former president George Bush enacted the nation's first comprehensive AIDS-care funding package— the Ryan White CARE Act (1990).

The primary purpose of the act is to provide emergency financial assistance to localities that are disproportionately affected by the human immunodeficiency virus epidemic. An eligible metropolitan area (EMA) is any metropolitan area for which there have been reported to the CDC a cumulative total of more than 2,000 acquired immune deficiency syndrome (AIDS) cases for the most recent five years for which data

are available. The amended act requires formula grants based on the estimated number of persons living with AIDS in the EMA. Estimates were derived by using methods specified in the act. The amount of funds received by each EMA (under Title I) or state (under Title II) is determined by the locality's proportion of the total estimated number of living persons with AIDS.

Misconceptions About AIDS Linger

Despite widespread reports that casual contact does not spread the virus, families have walked out of restaurants that employed gay waiters and hospital workers have quit rather than treat HIV/AIDS patients.

In 1989, a man was barred in Anderson, Indiana from coaching his daughter's intramural basketball team because he had been diagnosed with AIDS. The 37-year-old father received the virus in a blood transfusion during open heart surgery in 1984.

Each example points out that regardless of education, the public still assumes the virus can be casually transmitted. **IT IS FEAR THAT IS BEING TRANSMITTED BY CASUAL CONTACT—NOT THE VIRUS.** How would you react if a good friend, classmate, or coworker told you he or she was HIV-positive? What if you found out that your child's schoolmate, a hemophiliac, had AIDS? What if you were told this child had emotional problems or a biting habit? What if your work put you in direct physical contact with people who might be HIV-positive?

An AIDS diagnosis for one person resulted in his physician's refusal to treat him, his roommate left him, his friends no longer visited him, his attorney advised him to find another attorney and his clergyman failed to be there for him. They were all afraid of **"catching" AIDS.** In another case, a mother whose young son has AIDS sent cupcakes to his classmates on his birthday. School officials would not permit the children to eat the cupcakes. The elementary school principal said **the school had a policy against homemade food because it could spread diseases such as AIDS.**

In Jacksonville, Florida, Leanza Cornett (Miss America 1993) using her reign as a national platform to teach about AIDS, was told by public school officials not to use the word

A SOUTHERN UNIVERSITY STUDENT'S EXPERIENCE

A southern university student taking a senior semester course on HIV/AIDS in 1996, experienced first-hand the fear-related ignorance that still exists.

This student had accompanied her friend to the office of a physician in a nearby small town. While the student was in the waiting room, she was using the time to read the given class assignment in her textbook, *AIDS Update 1996.* People coming into the office would **look over** at the empty chairs on each side of her—then look at her and decide to sit elsewhere. Even as the office filled and fewer and fewer seats were available, no one chose to sit near her. She said, "People took one look at the book and chose another seat some distance from me." She said, **"So much for education on nontransmission of HIV by casual contact!!"**

ally transmitted via pens, pencils, toilet seats, toothbrushes, towels, bedding, medical procedures, and kissing. Immediate concern about casual transmission indicates the depth of human fears about disease and sexuality. Concerns about hygiene, contamination, contagion, pain, and death are expressions of anxieties that reveal much about contemporary society.

Fallout from AIDS: The Spector of Discrimination— The biggest difference between HIV/AIDS and other diseases is the larger amount of social discrimination. Society does not reject those with cancer, diabetes, heart disease, or **any** other health problems to the degree that it rejects people with HIV disease or AIDS.

The AIDS pandemic has taught people about risk behavioral groups, homosexuals in particular. In some, this has promoted tolerance and understanding; in others, it has reinforced feelings of hatred. Information on HIV disease and AIDS, how it is spread and how to prevent becoming infected has, over the past 18 years, become a part of TV talk shows, movies, TV advertisements, and newspaper and magazine articles.

Phil Donohue, host of a popular TV show, said **"On *Donohue*, we're discussing body cavities and membranes and anal sex and vaginal lesions. We've discussed the consequences of a woman's swallowing her partner's semen. No way would we have brought that up five years ago. It's the kind of thing that makes a lot of people gag."**

The language, photography, and art work used by the media are explicit and have upset certain religious groups. They believe that open use of language about condoms, homosexuality, anal sex, oral sex, vaginal sex and so on promotes promiscuity.

Class Question: How can people learn to prevent HIV infection and AIDS without talking about sexual behavior and injection drug use? Does it seem at times as if opponents of sex education would rather have people suffer with AIDS than have them learn about sex?

Regardless of who is correct, few could have predicted in 1980 the casualness with which these topics are now presented in the media. If the AIDS pandemic has done nothing else, it surely has affected the nature of public discourse. In

condom while addressing student groups. In Bradford County, Florida she was told she could not mention the name of the disease (AIDS) in three elementary schools she planned to visit.

In Hinton, West Virginia, one woman was killed by three bullets and her body dumped along a remote road. Another was beaten to death, run over by a car, and left in the gutter. Each woman had AIDS and told people so. And each, authorities say, was killed because she had AIDS. Lawyers and advocates for AIDS patients say the similar slayings, two counties and 6 months apart, illustrate AIDS' arrival in the American countryside and the fear and ignorance it can unearth.

What Do We Know?— Beginning year 2000, and nearly 19 years of the AIDS epidemic it is clear that the scare headlines and tactics lack substance. From what has been learned about the biology of HIV, it appears that the virus is not casually nor easily spread but it has reached the magnitude of the great plagues and a vaccine has not yet been found!

AIDS Epidemic Parallels the Syphilis Epidemic in the United States—The social and medical history of the AIDS epidemic parallels the syphilis epidemic. It was feared that syphilis was casu-

————— BOX 13.2 —————

HOW SOME PEOPLE RESPONDED AFTER LEARNING THAT SOMEONE HAD AIDS

VIGNETTES ON COMMUNITY BEHAVIOR AND AIDS

In **Colorado Springs, Colorado,** Scott Allen's wife Lydia had contracted HIV from a blood transfusion hours before son Matthew was born. A second son Bryan was also born before Lydia learned of her HIV infection. Scott wasn't infected, but was **dismissed** as minister of education at First Christian Church in Colorado Springs, when he sought his pastor's consolation. Matt was **kicked out** of the church's day-care center and the family was told to find another church.

When the family moved to Dallas and moved in with Scott's father Allen and his wife, church after church refused to enroll Matt in Sunday school. **Allen, a former president of the Southern Baptist Convention** wrote in his book, *Burden of a Secret: A Story of Truth and Mercy in the Face of AIDS,* "Good churches. Great churches. Wonderful People. Churches pastored by fine men of God, many of whom I had mentored. **Nobody had room for a boy with AIDS."**

Bryan, an infant, died in 1986, Lydia died in 1992, and Matt died in 1995.

In **Ohio**, a man was erroneously diagnosed as being HIV-infected. Within 12 days of learning the test results, he lost his job, his home, and he almost lost his wife. **The error almost cost him his life:** He had planned to commit suicide on the very day he received notice that he had received the wrong test results!

In **Maryland,** a court of appeals upheld a lower court's ruling permitting courtroom personnel to wear gloves to prevent picking up the AIDS virus. The court did suggest that the gloves were unnecessary.

In **New York,** a minister said that he was in a religious "Catch-22": He wanted to show concern and compassion for AIDS patients, but there are definite biblical injunctions against homosexuality. "How," he asked, "can I support these people without supporting homosexuality?"

In **Florida,** 1987 Mrs. Ray, the mother of three hemophilic HIV-infected sons (Ricky, 14; Robert, 13; and Randy, 12) turned to her pastor for confidential counseling. He responded by expelling the family from the congregation and announcing that the boys were infected. As a result, the boys were not allowed to go to church, school, stores, or restaurants. Barbers refused to cut their hair. Some townspeople interviewed said they were terrified at having the boys in the community. They had to move to another town! The Rays sued the DeSoto County School District. They agreed to pay a $1.1 million out of court settlement in 1988. Ricky Ray died of AIDS on December 13, 1992 at age 15. Robert, age 15, was diagnosed with AIDS in 1990; Randy, age 14, was diagnosed with AIDS in May of 1993. Both are living at this writing.

In **Florida, Broward County** parents packed school meetings and teachers filed a class action grievance saying a student's presence endangered their health. District officials determined that the 17-year-old mentally handicapped boy with AIDS would be educated by a teacher in an isolated classroom in the school. The student received 3 1/2 years of isolated education. For the 1 1/2 years , maintenance people would not walk into the portable classroom. Instead, they would leave packages or supplies in the doorway. The teacher and an assistant got used to doing their own cleaning. The one student cost over $50,000 per year to educate.

In **Duval County, Florida** (1990), the foster parents of a 3-year-old AIDS child who was infected by his mother, were forced to leave their church because other parents insisted that the child not be allowed to attend the church nursery. The pastor went along with the majority. When presented with CDC findings that HIV is not casually transmitted, one parent scoffed. **"I called the CDC for information and they asked me what I was going to use it for."** He then asked the congregation, "How can you believe anyone like that?"

In **California,** a young man arrived home one evening to find that the locks had been changed. A few days later he discovered that everything he had ever touched had been thrown out—clothes, books, bed sheets, toothbrush, curtains, and carpeting. Even the wallpaper had been stripped from the walls and trashed. The day before, he had told his friends he had AIDS. "Overnight, I had no friends. I slept on park benches. I stole food. I passed bad checks. No one would come near me. I was told that I had 14 weeks to live."

In a second California incident, volunteer fire fighters refused to help a 1-year-old baby with AIDS at a monastery that cares for unwanted infants. The baby was reported to be choking. Although the fire department has agreed to respond to such calls in the future, one fire fighter quit, saying he was frightened because he had not been trained to deal with victims of acquired immune deficiency syndrome.

———— BOX 13.2 (continued) ————

most from the indignities, lies, and meanness of his classmates and his classmates' parents in Kokomo, Indiana. They accused him of being a "fag," of spitting on them to infect them with the virus, and other fabrications. Ryan said he understood that this discrimination was a response of fear and ignorance. Ryan got the virus from blood and blood products essential to his survival. Ryan's wish was to be treated like any other boy, to attend school, to study, to play, to laugh, to cry, and to live each day as fully as possible. But AIDS was an integral part of his life. AIDS may not have compromised the quality of his life, as much as the residents of his community. One day, at age 16, as Ryan talked about AIDS to students

in Nebraska, another boy asked Ryan how it felt knowing he was going to die. Showing the maturity that endeared him to all, Ryan replied "It's how you live your life that counts." Ryan White died, a hero of the AIDS pandemic, at 7:11 A.M. on April 8, 1990. He was 18 years old.

Gregory Herek and colleagues (1993) determined via telephone interviews that HIV/AIDS stigma is still pervasive in the United States. They interviewed 538 white adults and 607 African-Americans. Nearly all respondents indicated some stigma. Regardless of race, men were more likely than women to support preventive policies such as quarantine and said they would **avoid** persons with AIDS.

HATRED EXPRESSED

CASE I.

In April 1996, a youth traffic safety program coordinator with the Florida Department of Transportation received a pledge form requesting a donation for the Florida AIDS Ride, an Orlando to Miami Beach bike race to raise money for AIDS programs. The person responded to the pledge form by writing to the race organizers that AIDS "was created as a punishment to the gay and lesbian communities across the world. As far as the gay and lesbians of this world . . . **let them suffer the consequences!"**

CASE II.

In May of 1998 a St. Charles, Missouri, county man was arrested and charged with assault for **allegedly** injecting his 11 month old son, in 1992, with blood containing HIV that police believe he stole from a hospital where he worked as a phlebotomist (a person who draws blood). **He allegedly injected his son with HIV because he hated paying child support.** The boy is now age 7 (at 5/98) and was diagnosed with AIDS in 1996. The trial was held in late 1998. In January 1999 he was sentenced to life in prison.

NEWSPAPER HEADLINE SHOCKS READERS

On June 7, 1996, a large bold print headline of the Jacksonville, *Florida Times Union* newspaper, a **very conservative** Southern newspaper, read:

STUDY: ORAL SEX POSES HIV INFECTION RISK

This headline in this newspaper would **NEVER** have happened in this author's lifetime without the 15-plus years of HIV/AIDS history. During this time, declarations involving all types of sexual behavior and drug use appeared during prime time TV programming and voluminous publication of such topics in all types of regional, state, and federal pamphlets, magazines, journals, and public reading materials. Yet, this newspaper received 124 phone calls from people angry about this headline in their newspaper. Their main objection concerned the use of the words "ORAL SEX." Many others objected but did not call—some dropped their subscription! **Class Discussion: How would you feel about this headline appearing in your hometown newspaper?**

1987, prior to the TV broadcast of the **National IDS Awareness Test,** viewers were warned of objectionable material. By 1990, few if any such fewer warnings were given. It is as if to say that o one can afford to be ignorant of this information because it may save your life.

How Can Information Help?— There is great hope that information will lead the nation past its social prejudice and forward to compassion for those who are HIV-infected or have AIDS. More than any disease before, AIDS has proved that ignorance leads to fear and in-

——————— **BOX 13.2** (continued) ———————

In **New Jersey,** a bartender could not tell his parents or friends he had AIDS. It meant confessing that he was gay. He feared it might also mean the loss of family, friends, lovers and insurance. He was expressing signs and symptoms and paying out of pocket for medical bills rather than file an insurance claim.

In **Charlotte, North Carolina,** 1993, a bride wore an ankle-length chiffon dress, white above and flowered black below. She'll wear the dress again, at her husband's funeral. The groom was diagnosed with progressive multifocal leukoencephalopathy (PML), a relentless viral infection that attacks the nervous system and causes brain lesions. About 4% of people with AIDS contract the disease. No cure exists.

A legal marriage would make the groom ineligible for Medicare and Medicaid, which he can't afford to lose. So their wedding had to be symbolic and legally nonbinding.

Both wanted a wedding as a sign of commitment. He also wanted it for psychological support: It tells him that his wife will be there as things get worse.

PML has already begun to separate the newlyweds. He has developed speech problems. Sometimes he doesn't remember what he's doing.

From the wreckage, they have clutched love.

"Love is — period," she said "We don't qualify it. We don't distinguish between heterosexual and homosexual. When love is there, the physical form it takes is just a detail."

In **Charleston, West Virginia** in July of 1995, a 10-year postal mail carrier who refused to deliver mail to a couple with AIDS was indefinitely suspended with pay after an educational class failed to change his mind. The mailman said he was afraid of cutting himself on the home's metal mail slot and becoming infected from envelopes or stamps the couple had licked. Postal sources said the mailman would eventually be fired.

In **Texas,** a father with AIDS cried while praying that his three children would not be treated as cruelly. The man, in his mid-20s, was mugged because he looked too weak to fight, hit with rocks by people who found out he had AIDS, and in one incident a man broke a bottle over his head screaming that he was out to kill AIDS. During one beating, an attacker said, "After we kill you, we will kill your wife and children in case they have AIDS."

Almost daily, similar senseless acts of violence and cruelty occur across the United States as a response to AIDS. Such episodes of panic, hysteria, and prejudice are perpetuated by the very people society uses as role models: clergy, physicians,

teachers, lawyers, dentists, and so on. F Jonathan Moreno said, **"Plagues and epi AIDS bring out the best and worst of so to face with disaster and death, people a down to their basic human character, t evil. AIDS can be a litmus test of huma**

THE LIFE OF RYAN WHITE

In Kokomo, Indiana, Ryan White w unacceptable. He was not gay, a drug or Hispanic. He was a hemophiliac; he His fight to become socially acceptabl school, and to have the freedom to leav for a walk without ridicule made Ryar hero (Figure 13-2).

Ryan's short life was a profile in c understanding. Like many other p AIDS, Ryan tried to change the public ception of how HIV is transmitted. Ry

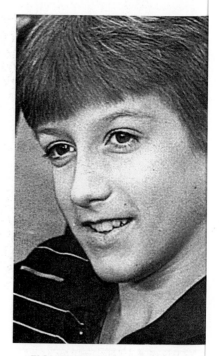

FIGURE 13-2 Ryan White was di with AIDS in 1984 and died on Ap 1990. This young male became an teenage AIDS tragedy. He gained t spect of millions across the United before he died of an AIDS-related I fection. (*Courtesy AP/Wide World*

formation can lead to compassion. **The need for compassion is great.**

Admiral James Watkins, Chairman of the 1988 Presidential Commission on the HIV Epidemic, reported that "33% to 50% of physicians in some of our major hospitals would not touch an AIDS patient with a ten-foot pole."

A friend of a dying AIDS patient who was in the hospital with pneumocystic pneumonia went to visit him. As he was leaving, his dying friend said, **"Thank you for coming—thank you for touching me."** He said, "I can't even imagine being at a point in my life that I would be so grateful for someone touching me, that I would have to say thank you."

It would appear that although biotechnology has provided methods of HIV detection, new drugs, and hope for a vaccine, human emotional responses have not changed much from those demonstrated during previous epidemics like the plague, cholera, influenza, and syphilis.

AIDS EDUCATION AND BEHAVIOR

I said education was our **"basic weapon."** Actually it's our **only** weapon. We've got to educate everyone about the disease so that each person can take responsibility for seeing that it is spread no further.

C. Everett Koop
Former U.S. Surgeon General

For some, the occurrence of an estimated 40,000 new HIV infections in the United States

---

———— **BOX 13.3** ————

WHEN ONE WITH AIDS COMES FORTH

Father Paul made the decision to preach on AIDS because of a phone call he had received informing him that a former parishioner was coming to Jacksonville. "George has AIDS. He will be in Church on Sunday. With your permission, he will be receiving Communion."

Father Paul granted permission and welcomed George's attendance. He sensed, however, that some might not agree to have George in church or receive Communion. In his parish a number of people refuse to believe that HIV is not transmitted by saliva from the lavitha (Communion cup).

By Sunday morning's sermon, over 60% of the congregation had learned of George.

Father Paul began his sermon, "Today's Gospel lesson, Luke 10:25-37, tells us the Parable of the Good Samaritan . . . The Parable challenges us to take stock of who our neighbors are who have needs that we can meet. . . . is it not also true that our neighbors are being harmed by AIDS? . . . Many Orthodox Christians are good about reaching out to the needy and indigent. But we are not so willing to reach out to those with AIDS."

Father Paul said that he would be dishonest if he did not admit having concerns about people receiving Communion after George. "I did not fear for myself, I feared for them, and especially for my two daughters who receive Communion regularly."

Father Paul reminded everyone about the faith of The Church. He addressed the question, "Can AIDS be contracted through casual contact and specifically from the Eucharistic Chalice?" He said, "Communion is the Body and Blood of our Lord. In the Gospel of John, chapter 6, Jesus speaks to us of His Body and Blood as being agents of life, NOT of sickness and death. Similarly, St. Ignatius of Antioch alludes to Communion as the 'medicine of immortality' which allows one to eternally abide with God. To believe, therefore, that one can contract sickness and AIDS from Holy Communion is blasphemy against the Holy Spirit. It is also to render everything that the Bible and the Church teaches about Communion meaningless."

As he spoke, he saw George near the back of the Church. Though only 47 years old, George looks 60. George was weak, abnormally thin, spoke with a rasp, and walked with a cane.

Before Sunday's Liturgy, Father Paul had discussed with George his pastoral concerns about the people's anxieties. Without asking, he proposed a solution, he would receive Communion last.

Father Paul introduced George to the congregation and announced that he would be receiving last.

At Communion time, several of the congregants assisted George to the front of the church. Then, the same individuals and several others lined up behind him.

About 10 people received Communion after George. Father Paul asked one why he did it. "Father, did you not tell us in your sermon that we had to be Good Samaritans. It would have been a very unloving and discriminatory act to

————— **BOX 13.3** (*continued*) —————

allow George to go last."

George died several weeks later in New York City. The news was received with sadness.

George, thank you for coming to Jacksonville. God brought you to us to help us grow. May God remember you in his Kingdom.

(Adapted with permission from Father Paul Costopoulos, Jacksonville, Florida)

FREEDOM FOR COMPASSION: CHILDREN AND AIDS

Charlie the doll was pressed against the antique glass case. Jeff, a blue-eyed, blond-haired boy of 10, looked at him closely. "Someday Charlie will leave this cage," he pondered, "and someday he will be free."

A few visits later, Jeff devised a plan to buy Charlie. He began his task by seeking employment as a leaf raker, a car washer, and the best of panhandlers among friends. After a while his hope faded and his energy waned, but he did not despair. He had met adversity before, in fact, for most of his life. When Jeff was 4 years old, he had contracted AIDS.

The family of John Calvin Presbyterian Church met Jeff because other congregations had turned him away, telling his family that Jeff's illness was a punishment from God. Jeff planned his own memorial service, but five different congregations ignored him by making excuses that the songs he had chosen from the play, "Peter Pan," and the balloons he had requested would not be appropriate.

One Sunday morning as I was beginning my sermon, Jeff's mother wheeled him down the main aisle to a front pew. I wondered what would happen if this church rejected him, too? How would the other children treat him? Could this congregational family risk enough to love Jeff and his family in the same way in which God loved them? But my fears were relieved after church at the coffee fellowship. Parents introduced themselves to Jeff's mother and the children included Jeff within their circle of games. People earnestly gathered to accompany Jeff and his family on their special journey: members ran errands, provided transportation, and brought in food. This outpouring of help came at a crucial time: Jeff's mother had given up her own business to take care of Jeff, emotional pressures contributed to Jeff's parents' divorce, the family lost their home to bankruptcy court, and Jeff's brother and sister suffered from the prejudice of schoolmates and others.

A few weeks prior to Jeff's death, I bought Charlie the doll. As Jeff's fragile hands began to untie the shiny silver ribbon that secured the purple box, large tears began to trickle down his sunken cheeks. When he discovered what was inside, Jeff smiled and said, **"Now he's free . . . and someday I will be, too."** Jeff's freedom arrived on March 2, 1988.

Those of us who knew Jeff have gained freedom as well. Jeff, and others like him, have introduced us to a new appreciation of life. Through them we have been reminded of how fragile we are and of our precious responsibility to live each moment fully. Together we have discovered the gift that no person, no circumstance, no condition—even the AIDS virus—can take from us. We have discovered God's gift to us in Jesus Christ.

John Calvin Presbyterian Church is now a better-educated congregation. Jeff helped to teach us the importance of risking and reaching out. Our church has adopted a resolution stating that we are committed to minister intentionally with people with AIDS and their families. We provide office space to the Tampa AIDS Network, an advocacy, support, and fund-raising group serving the Tampa Bay area. A number of members and I are part of a growing coalition of volunteers serving on task forces, care-giving programs, and support groups.

In the midst of our congregation's activity we are still keenly aware of the continuing apathy, ignorance, and prejudice permeating much of the religious community. Many underestimate the possibility of this disease intruding into their lives and communities. Much work remains to be done with few to accomplish it.

Jeff's memorial service was just as he had planned it. The sanctuary of the church was filled with the nurses and doctors who had worked with him, the many hospice volunteers who had given him solace, his buddies from the Tampa AIDS Network who had held his hand, the many other friends he had made on his journey. Hundreds of brightly colored helium balloons were released into the sky at the end of the service. We celebrated Jeff's life—and ours as well. Our celebration continues as we minister with others who are traveling this very difficult path. Our strength and hope are renewed with each encounter, for we have been given the freedom for compassion.

(Adapted from Rev. Jim Hedges, pastor of John Calvin Presbyterian Church, Tampa, Florida, in *Church and Society*, Vol. 79, No. 3(January/ February 1989). Reprinted with permission.)

each year is evidence that HIV education and prevention efforts have failed. If HIV prevention programs are held to a standard of perfection and are expected to protect 100% of people from disease 100% of the time, the efforts are by definition doomed to failure. No intervention aimed at changing behaviors to promote health has been or can be 100% successful, whether for smoking, diet, exercise, or drinking and driving. For example, even though warnings regarding the health effects of smoking were issued in 1964, warning labels on cigarettes were not mandated until 1984, and smoking-related illness still remains a major cause of death.

Because some of the behaviors and activities that need to change in order to avert HIV infection are **pleasurable,** it should be no surprise if short-term interventions do not lead to immediate and permanent behavior

changes. An important difference between HIV infection and other life-threatening diseases is that HIV can be contracted by a single episode of risk-taking behavior. **Once HIV-infected there is no second chance—no giving up the behavior, like drinking alcohol or smoking, that will make any difference;** HIV disease progresses to AIDS. The adverse effect of smoking, alcohol or diet on health usually manifests after years of smoking thousands of cigarettes or after years of drinking or poor dietary choices.

More than a dozen years of experience with HIV has demonstrated that lasting changes in behavior needed to avoid infection can occur as a result of carefully tailored, targeted, credible, and persistent HIV risk-education efforts. Given experience in other health behavior change endeavors, no interventions are likely

––––––––– POINT OF INFORMATION 13.1 –––––––––

AIDS STIGMA: CHANGING ATTITUDES

A January 1996 opinion poll on public attitudes concerning HIV/AIDS revealed that to some degree people, through education, are changing their beliefs about HIV/AIDS.

In 1985, 31% of adults believed children with HIV disease or AIDS should be barred from school. In 1996, only 6% held that belief.

In 1985, 6% of adults knew someone with AIDS. In 1996, 40% of adults know someone with AIDS.

In 1985, 70% of adults believed that AIDS was the nation's greatest health problem. In 1996, 37% held that belief.

In 1985, 33% believed it was unsafe to associate with someone who had AIDS. In 1996, 15% still held that belief!

In 1990, 30% of randomly picked adults feared they would become HIV-infected. In 1996, 15% had the same fear.

In 1996, 73% of adults polled favored spending tax dollars to treat people with AIDS.

UPDATE: PREVALENCE OF AIDS STIGMA IN AMERICA 1997 VS. 1990–1991 SURVEY

(Results from a national telephone survey by Gregory Herek and John Capitano, Department of Psychology, University of California, Davis presented at the 12th International AIDS Conference, Geneva, Switzerland, July 1, 1998)

Method

List-assisted random digit dialing was used to generate a probability sample of telephone numbers, representing private households in the 48 contiguous states. Telephone interviews were conducted with 1712 English-speaking adults, with each respondent randomly selected from among the members of his or her household. The analyses reported were conducted with respondents who self-identified as heterosexual and who reported that they were HIV negative. The margin of error due to sampling is approximately plus-or-minus two percentage points.

Key Findings

1. In 1997, the majority, 77%, believed that people with AIDS are unfairly persecuted in our society. Also, the majority supported mandatory HIV testing for immigrants (77%), people considered to be at high risk for AIDS (74%), and pregnant women (84%).

2. The proportion of the population supporting extreme coercive policies against PWA (People With AIDS) has declined since 1991. In the 1997 survey, 17% supported quarantine for PWA and 19% supported public disclosure of the names of PWAs. In a 1991 survey, these

policies were supported by, respectively, 36% and 30% of the US public.

3. Compared to 1991, fewer people said that they would avoid a PWA in various hypothetical situations. Ten percent would have their child avoid another schoolchild with AIDS (compared to 16% in 1991), 12% would avoid a coworker with AIDS (compared to 20% in 1991), and 33% would avoid shopping at a neighborhood grocery store whose owner has AIDS (compared to 47% in 1991). However, many of those who would not actively avoid a PWA in these situations would nevertheless feel uncomfortable about being around the person with AIDS: 27% of respondents would feel uncomfortable about having their child interact with a schoolchild with AIDS, 25% would feel uncomfortable working the same office as the coworker with AIDS, and 30% would feel uncomfortable about shopping in the store whose owner has AIDS.

4. The proportion of the public that believes casual social contact might spread HIV has **increased since 1991.** Fifty-five percent believed that it was possible to contract AIDS from using the same drinking glass as a PWA (compared to 48% in 1991), 41% believed that AIDS might be contracted from a public toilet (compared to 34% in 1991), and 54% believed that AIDS

might be transmitted through a cough or sneeze (compared to 45% in 1991). Most of the shift in beliefs since 1991 has occurred as a result of fewer people believing that AIDS definitely cannot be transmitted through these routes, and more believing that these routes are merely somewhat unlikely to spread AIDS. The proportion believing that these forms of casual contact are somewhat likely or very likely to transmit AIDS has remained fairly stable.

5. Consistent with widespread inaccurate beliefs about casual contact, 27% of respondents reported that they would be less likely to wear a sweater that had been worn one time by a PWA than if it had been worn once by another person—even if the sweater had been cleaned and sealed in a new package so that it looked like it was new. Twenty-eight percent of respondents said that they would feel uncomfortable drinking out of a glass in a restaurant if a PWA had used the same glass a few days earlier, even if it had been washed and sterilized.

Class Discussion: Does it appear that the belief that HIV CANNOT be transmitted casually is declining? Explain why or why not. How do these data impact the push for mandatory names reporting of those who test HIV positive?

to reduce the incidence of HIV infection to zero; indeed, insisting on too high a standard for HIV risk-reduction programs may actually undermine their effectiveness. A number of social, cultural, and attitudinal barriers continue to prevent the implementation of promising HIV risk-reduction programs. The remote prospects for a successful prophylactic vaccine for HIV and the difficulty in finding long lasting effective drug treatments have underscored the importance of sustained attention to HIV prevention and education.

The goals of educating people about HIV infection and AIDS are to promote compassion, social understanding and to prevent HIV transmission. To achieve these goals, accurate information must be provided that makes people aware of their risk status. People at minimum risk can continue their safer sexual lifestyles. People at high risk should determine their HIV status, alter behaviors, and practice safer sex.

This sounds so easy: Educate people and they will do the right thing. **Wrong.** Knowledge does not guarantee sufficient motivation to change sexual behavior or stop the biological urge to have sex. Education has not stopped teenage pregnancy, nor has the knowledge about cigarettes causing lung cancer stopped people from smoking.

Perhaps the reason education is not as effective as it could be is because the public receives its education by daily doses from the mass media. With so much going on in the world, people have become more or less dependent on the media for information essential to their well being. Gordon Nary (1990) said, **"The public wants to know what's right or wrong in five three-second images or 25 words or less. It wants simple problems with simple solutions. It wants *Star Wars* with good and evil absolutely defined. The media often responds to these demands."**

———— BOX 13.4 ————

THE WAY WE ARE

The young man worked alongside a 35-year-old woman helping to teach the disabled to function. She was very attractive with a wonderful sense of humor. He watched her every movement - he was in love from a distance; she inspired him to new heights in his work and thoughts about his future. She understood his emotions so she was not surprised when, after work one evening, he offered dinner and a moonlight walk around the lake. During the walk, she could tell from the conversation that his young hormones were flowing—as were her own. In the awkwardness of saying goodnight she said, "I know you want me—I would like that very much." As his broad face glowed she said, **"I must share something with you. I have been diagnosed with AIDS but there are no outward signs yet."** The love that moments before surged through his body crashed down around him; he felt ill yet sympathetic - his urge for sex vanished. He could not bring himself to look at her as he said, "Please forgive me, I just can't." She nodded her understanding and explained it was either a blood transfusion in 1984 following an auto accident or death. **"I made the choice for life and whatever comes with it."** The young man cried as he walked home. He left the job the next morning.

Mathew Lefkowitz (1990) says that the word AIDS has been infused with an irrational fear that has nothing to do with the illness. He states that the word AIDS has been politicized in such a way that it can and has been used as a weapon. Lefkowitz relates the parallel between today's use of the word AIDS with Eugene Ionosco's classic absurdist play *The Lesson*. In the play, a professor stabs a girl to death with the word *knife*—not with a knife but with the **word knife**. It appears that today the word *AIDS* is being used to stab those whom we fear; namely the HIV-infected. How long will it take the educational process to work? **THE VIRUS IS NOT AS MUCH OUR ENEMY AS OURSELVES.** For example, in Washington, D. C., 1995, because a national public radio commentator was furious at U. S. Senator Jesse Helms of South Carolina for having the audacity to suggest that the government spends too much money on

AIDS research, said, "I think he ought to be worried about what's going on in the good Lord's mind because if there is retributive justice, he'll get AIDS from a transfusion—or **one of his grandchildren will get it."**

Perhaps zealots need to realize that the purpose of civil exchange is to arrive at wisdom through reasoned debate, not to verbally intimidate those who differ into silence. **Class Discussion: WHAT IS YOUR OPINION?**

Public AIDS Education Programs

The 1988 Presidential Commission on the Human Immunodeficiency Virus Epidemic stated:

> No citizen of our nation is exempt from the need to be educated about the HIV epidemic. **The real challenge lies in matching the right educational approach with the right people.** During the last year, there has been a great deal of debate over the content of HIV/AIDS education. The Commission is concerned that, in the promotion of the personal, moral and political values of those from both ends of the political spectrum, the consistent distribution of clear, factual information about HIV transmission has suffered. HIV/AIDS education programs, for example, should not encourage promiscuous sexual activity; however, they need to be explicit in nature so that there is no confusion about how to avoid acquiring or transmitting the virus. The Commission firmly believes that it is possible to develop educational materials and programs that clearly convey an explicit message without promoting high-risk behaviors.

West Coast health officials told this same Presidential Commission that the censorship of sexually explicit materials is hampering the fight against AIDS. Many of those who testified before the Commission expressed frustration about the barriers posed by government bureaucracy. Several representatives of AIDS organizations told commissioners of the difficulties they have experienced in trying to produce effective educational materials and programs. They said that state officials prohibit the use of certain language when they fund educational programs. It is sometimes necessary, however, to use slang or sexually explicit language in order for readers to understand the material. As an example, they cited women who had given birth but claimed they had never had "vaginal intercourse." They did not know what

vaginal meant, but they did know what *cunt* meant. Educational material must use recognizable language to be effective.

Over the years billions of dollars have been spent by federal and state health departments to inform the public about cardiovascular risks, health risks associated with sexually transmitted diseases (STDs), smoking and lung cancer, chewing tobacco and oral cancer, drug addiction, alcohol consumption, and seat belt use to name just a few. In some cases these campaigns were eventually supported by specific state and federal legislation. Tobacco advertisements were outlawed on TV and drivers in some states who are not buckled up must pay a fine. But even with laws to support these educational programs, many adults have failed to change established behavior patterns. In the larger cities, educators must combat the fear that AIDS is a government conspiracy to eliminate society's "undesirables"—minorities, drug addicts, and homosexuals. They must overcome cultural and religious barriers that prevent people from using condoms to protect themselves.

Another problem is that educational programs are not preventing individuals from acquiring one or more STDs, using drugs, drinking to excess, driving drunk, failing to use seat belts or motorcycle helmets, and smoking. It might also be added that there have always been educational programs against crime, but in 1989 through 1998 more new jails were built in the United States than ever before. And this expanding jail-building program continues! In short, educational programs on TV, radio, in newspapers, and in the popular press have achieved only limited success in changing peoples' behavior.

It is not that education is unimportant; it is essential for those who will use it! But that is the catch. Although education must be available for those who will use it, too few, relatively speaking, are using the available education for their maximum benefit. In general, people, especially young adults, **do not do what they know. They sometimes do what they see, but most often do what they feel.** In short, knowledge in itself may be necessary but it is insuffi-

---

POINT OF VIEW 13.5

THE RED RIBBON

We were sitting in a small Italian restaurant. I had just come back to town from a presentation on AIDS. The jacket I wore still had the red ribbon on the lapel. As we enjoyed our meal I noticed a woman at the next table who appeared to be glaring at me and making statements to her companion. At one point her voice became loud enough for us to hear her say, "I am sick and tired of those people trying to push the lifestyle of homosexuals down our throats" as she was looking right at me. She then said, "That red ribbon is a sign of a sick person trying to make all of us sick too. That ribbon and all that it stands for ruins my day." With that she and her companion left the restaurant.

That outburst left my family and me embarrassed and confused. My children deserved an explanation. I don't think I have ever explained the idea of the red ribbon to anyone before. Like so many things we observe in life, after a while they become understood by each in his or her own way. This woman expressed her way rather forcefully. To my children I said that the ribbon is a symbol to call attention to a social problem that needs a solution. I went on to say, "Do you recall the song 'Tie A Yellow Ribbon Round the Old Oak Tree' in 1973 and what that meant? And, do you recall the ribbons tied around trees, on car antennas, mailboxes and so on while our 56 service men were held captives in Iran in 1980 and again for out captives in the 1991 Persian Gulf War? Remember the first lady Nancy Reagan's campaign using red ribbons for 'Just say no to drugs' and more recently the pink ribbons for women against breast cancer and most recently the purple ribbons for stoppng violence in our schools? These are all symbolic gestures to show support for those enduring suffering and pain. All of the ribbons then and now serve to connect people emotionally, to help unite people in a common cause, to help people feel less isolated in a crisis."

I explained to my children that the problem with the red ribbon now is similar to what occurred over the long time period our soldiers were in captivity—people begin to wear the ribbon as an accessory.

The Author

---

cient for behavioral change. A variety of studies have failed to show a consistent link between knowledge and preventive behaviors (Fisher, 1992; Phillips, 1993)

Costs Related to Prevention— The assertion that spending more money on educational programs will ensure disease prevention for the masses, as the examples given above suggest, may not be the case. In particular, peoples' behaviors regarding the prevention of HIV infection do not appear to be changing significantly despite the billions of dollars used to produce, distribute, and promote HIV/AIDS education. The major educational thrust is directed at how not to become HIV-infected. Most of this information is being given out to people from age 13 and up.

The problem with AIDS education is that communicating the information is relatively easy but changing behavior, particularly addictive and/or pleasurable behavior, is quite difficult. Most cigarette abusers know that smoking causes lung cancer and would like to quit or have tried to quit. Likewise, many alcoholics have tried to quit. But both cigarettes and alcohol are pleasurable to those who use them. The mass media have provided near saturation coverage of key AIDS issues and **it is very unlikely that significant numbers of future HIV infections in the United States will occur in individuals who did not know the virus was transmitted through sexual contact and IV drug use. Yet new infections continue at an incidence of 40,000/year.**

Evidence accumulated between 1983 and 1987 indicates that homosexual and bisexual males modified their sexual behavior and this resulted in a drop in AIDS cases among them. There is also evidence that some IDUs have begun wearing condoms. But recent evidence suggests that some of these people are being drawn back to their former practice of unprotected sex and needle sharing. The disease is showing signs of a resurgence in the gay community as health workers struggle to get their message to a new generation and to older gays who seem to have lost the motivation to protect themselves.

In 1982, there were 18 new HIV infections for every 100 uninfected gay men in San Francisco. That rate dropped to 1 per 100 in 1987 but rose to 2 per 100 in 1993. For gay men younger than 25, it is 4 per 100.

Although humans are capable of dramatic behavioral changes, it is not known what really initiates the change or how to speed up the process.

Public School AIDS Education: JUST SAY KNOW

Some information relevant to AIDS education can be learned from educational programs that have been designed to reduce pregnancy and the spread of STDs among teenagers. However, data from a variety of high school sex education classes offered across the country indicate that teenagers are learning the essential facts but they are not practicing what they learn. **They do not do what they know.**

Risky sexual behavior is widespread among teenagers and has resulted in high rates of STDs. Over 25% of the 12 million STD cases per year occur among teenagers. One teenager is infected with an STD every 10 seconds in the United States. One in six teens has been infected with an STD. Over 50% of sexually active teenagers (11 million) report having had two or more sexual partners; and fewer than half say they used a condom the first time they had intercourse (Kirby, 1988). A National Research Council panel (1989) stated that 75% of all female teenagers had sexual experience and that 15% had four or more partners.

Teenage Perceptions About AIDS and HIV Infection in the United States— A recent survey run by People magazine indicated that 96% of high school students and 99% of college students knew that HIV is spreading through the heterosexual population; but the majority of these students stated that they continued to practice unsafe sex. Combined data from surveys performed in 1988-1989 indicate that among sexually active teenagers, only 25% used a condom. Peter Jennings stated in a February, 1991, AIDS Update TV program that 26% of American teenagers practice anal intercourse. Data such as these have prompted a number of medical and research people to express concern for the next generation. If HIV becomes widespread among today's teen-agers, there is a real danger of losing tomorrow's adults.

Available data suggest that teenagers have not appreciably changed their sexual behaviors in response to HIV/AIDS information presented in their schools or from other sources (Kegeles et al., 1988).

Teenagers at high risk include some 200,000 who become prostitutes each year and others who become IDUs. About 1% of high school seniors have used heroin and many from junior high on up have tried cocaine (Kirby, 1988). A large number of children from age 10 up consume alcohol. **Is it possible that too much hope is being placed on education to prevent the spread of HIV?** Beginning year 2000, teenagers made up about 0.5% of 746,000 AIDS cases, or about 3,800 cases. Teenagers must be convinced that they are vulnerable to HIV infection and death. Until then, it only happens to someone else. The World Health Organization estimates that worldwide, through 1999, there were between 6 and 7 million HIV-infected teenagers.

Teenagers, like adults, must be convinced of their risk of infection but not with scare tactics. Behavior modification as a result of a scare is short lived. However the information is given, it must be internalized if it is to be of long-term benefit.

College Students— Everyone must know and act on the fact that a wrong decision about having sexual intercourse can take away the future. For example, a young college student had a 3-year nonsexual friendship with a local bartender. She was bright, well educated, and acutely aware of AIDS. After drinks one evening, as their friendship progressed towards sexual intercourse, she asked him if he was "straight" (a true heterosexual) and he said yes. **But he was a bisexual! It was a single sexual encounter.** She graduated and left town. She found out that the bartender died of AIDS 3 years later. She did not think much of it until she was diagnosed with AIDS 5 years after their affair. **This young, talented, bright, and personable girl lost her future.** This brings home the point that it is difficult to change something as complex as personal sexual behavior regardless of knowledge. It also brings up at least one other important point in personal relationships: **telling the truth.**

During the 1988 Psychological Association Convention, the following facts were presented with respect to telling the truth or lying in order to have sex. The data come from a survey of 482 sexually experienced southern California college students:

1. 35% of the men and 10% of the women said they had lied in order to have sex.

2. 47% of the men and 60% of the women reported they had been told a lie in order to have sex.

3. 20% of the men and 4% of the women said they would say they had a negative HIV test in order to have sex.

4. 42% of the men and 33% of the women said they would never admit a one-time sexual affair to their long-term partner.

Although these data come from a 1988 study, data on telling the truth, when it comes to acquiring sex, is most likely the same today.

Adult Perceptions About HIV Infection and AIDS in the United States— R.J. Blendon and colleague (1988) summarized the results of telephone surveys and personal interviews with a

---

BOX 13.5

ANECDOTE: FATHER AND SON EXCHANGE SEX INFORMATION

"Son, I think it's time we had a talk about sex." "O.K., Dad, what do you want to know?"

SIDEBAR 13.2

RECOMMENDED FILM: DAWN'S GIFT

She relates her story of becoming HIV-infected **during one sexual encounter while on vacation.** Running time, 38 minutes. Cost $5.50 plus about $2.00 for shipping.

Make check out to L. Schwitters-Dawn's Gift and send to

L. Schwitters-Dawn's Gift
1745 Brookside Dr. SE
Issaquah, WA 98027
(425) 392-9161 fax (425) 837-9971
email dawnsgift@hotmail.com
Purchase orders accepted

large number of adults. They provide an insight into adult knowledge and emotions regarding transmission of the AIDS virus. The data may also serve as a potential indicator for discrimination against HIV-infected people. Of the adults surveyed:

General Public—

1. Over 50% felt that the AIDS pandemic has lead to increased homophobia (fear of homosexuals) and discrimination.
2. 10% actively avoided social contact with homosexuals.
3. 20% said AIDS was punishment against immoral homosexual behavior and that the punishment was deserved.
4. 29% said HIV-infected people should wear a tattoo or other means of identification to warn the public.
5. 30% supported the idea of quarantine for HIV-infected and AIDS patients.
6. 40% believed that HIV-infected school employees should be fired.
7. 33% said they would keep their child out of school if a student there were HIV-infected or had AIDS.

Although it would be comforting to believe these attitudes have changed significantly since 1988, most likely they have not. Perhaps there have been a reasonable (5% to 10%) percentage reduction in items 4, 6 and 7. The others probably remain unchanged. A 1996 survey by the **Kaiser Foundation: Survey on Americans and AIDS/HIV** found that with respect to having confidence in AIDS sources, 6% **did not** believe the public health officials, 9% **did not** believe the US Surgeon General, 11% **did not** believe the newspapers, 16% **did not** believe Magic Johnson, 17% **did not** believe church/religious organizations, 25% **did not** believe the media and 34% **did not** believe the US government. Thirty-three percent of people **did not** know there are drugs and other treatments for HIV disease or AIDS and 29% **did not** know there were ways to **prevent** or reduce the spread of AIDS. And, less than 5% could name a single person that they thought was a **national leader** on AIDS.

The Workplace— If a person is **not** working near or beside someone who is HIV-positive, they will be relatively soon! But they may not know it because a person's right to privacy prevails over an employee's right to know.

Recently, 2,000 adult employees were phone interviewed during a national survey concerning their attitudes about AIDS. The survey (Hooker, 1996) revealed that:

--------------- SIDEBAR 13.3 ---------------

EXAMPLES OF ADULT REACTIONS TO HIV/AIDS VS. THE COURTS

Case 1. In September 1998, a mother in the state of Maine, who feared powerful AIDS drugs would kill her HIV-infected 4-year-old son rather than extend his life won the right Monday to refuse treatment for the boy. A state judge refused to give custody of her son to the state Department of Human Services, which had argued his mother was jeopardizing his health. The mother said that she saw her 3-year-old daughter go through an agonizing death while on the drug, AZT, and she did not want her son to suffer the same way. The mother is infected with HIV and has stopped taking medication as well. She passed the virus on to her son at birth. She has two older children who do not have the virus. The judge found that there was insufficient evidence that the boy's health was in jeopardy from being denied medical treatment. He also found that evidence suggested the combination of drugs was still experimental and carried the risk of side effects while not guaranteeing a positive outcome.

Case 2. In April 1999, a juvenile court judge in the state of Oregon **refused** to give up state custody of an infant born to an **HIV positive woman** who wants to breast feed. **The judge said,** "The parents may choose to run that risk with the child, but the court may second-guess that decision." The four-month-old boy will be allowed to live with his family, but a caseworker from Oregon's Office for Services to Children will visit periodically to ensure that the order is being fulfilled. The mother's case rested on the contention that HIV does not cause AIDS and cannot be transmitted through breast milk, however, witnesses for the state testified that breast milk from an infected mother has a very high risk of virus transmission.

Class Discussion: What do you think, can this work for the child?

- AIDS is the **chief health concern** among 20% of U. S. employees (cancer, was the primary concern of 32%; heart disease of 7%).
- 67% of employees predicted that their co-workers would be uncomfortable working with someone with HIV or AIDS.
- 32% thought an HIV-positive employee would be fired or put on disability at the first sign of illness.
- 24% said that an HIV-positive employee should be fired or put on disability at the first sign of illness.
- 75% of employees said they wanted their employers to offer a formal AIDS education program.

AIDS can have a variety of impacts in the workplace. The obvious one, of course, is on the individual employee who is diagnosed with HIV. The probability of sickness and death obviously affects the individual and his or her ability to continue to contribute to the organization's activities and goals.

Employees fear AIDS can create a widespread loss of teamwork and productivity, and create an environment that is inhumane and insensitive toward the infected employee.

As the incidence of HIV and AIDS increases, the impact on organizations will obviously increase as well. While there are important logical and moral reasons for ensuring that infected people are not discriminated against, there are also practical reasons for addressing the employee HIV/AIDS problem.

An educated work force, aware of the facts regarding diagnosis, testing, treatment, and transmission of HIV, can have a positive impact on the overall health of all employees. People are more inclined to openly acknowledge their HIV status and to seek treatment when assured of a supportive workplace environment. This increases productivity.

The results of a survey of people in the workplace by Blendon and Donelan (1988) revealed that:

1. 25% believed HIV was transmitted by coughing, spitting and sneezing.
2. 20% believed HIV infection could occur from a drinking fountain or toilet seat.
3. 10% believed touching an HIV-infected or AIDS patient was dangerous.
4. 25% refused to work with an HIV-infected or AIDS person.
5. 25% believed employers should be able to fire

people who were HIV-infected or had AIDS. In May of 1998, a TV news broadcast stated that a survey of United States workers revealed that **21%** favored **firing** or restricting activities of HIV-infected employees.

These 1988 data are, as shown for item 5, about what would be found today if the same survey was repeated.

Whether additional public education will change these attitudes over time is unknown. The Surgeon General's brochure, "Understanding AIDS" (Figure 13-3), was sent to 107,000,000 homes in 1988 at a cost of 22 million dollars. Fifty-one percent of those who received the pamphlet did not read it because they did not remember getting it or they chose to ignore it (Eickhoff, 1989).

A recent Georgia Tech survey found that workers who received educational brochures or pamphlets about the disease were more likely to have negative feelings toward HIV-infected co-workers than employees who received no information. The results suggest companies should be more concerned about the kind of messages workers receive. By focusing only on practices that can transmit the disease, many educational materials neglect to deal with the social, emotional, and humanitarian aspects of the problem. Education may be the key, but we need to look closely at what kind of education that should be.

THE CHARACTER OF SOCIETY

Rumors of Destruction

Today, friends are asked on street corners, at social gatherings, or over telephones: "Did you hear that he/she has AIDS?" Or: "Do you believe that he/she might be infected? You never know with the life they lead!" Some of the famous people **rumored** to have HIV/AIDS are Madonna, Elizabeth Taylor, Burt Reynolds, Richard Pryor, and Joe Penny from the TV show *Jake and the Fatman*.

Rumors ruin lives. People suddenly subtly lose services; the lawn boy quits, no reason given. Quietly job applications are turned down or car and homeowner insurance policies are canceled, and so on. In one case, after rumors of HIV infection spread in a small town, a man, if he was served in local bars at all, re-

What You Should Know About AIDS

Facts about the disease
How to protect yourself and your family
What to tell others

An Important Message from the U.S. Public Health Service
Centers for Disease Control

FIGURE 13-3 Understanding AIDS. This public health pamphlet contains eight pages of HIV/AIDS information. It was sent to 107 million homes in the United States in 1988, at a cost of $22 million. Fifty-one percent of those who received the pamphlet did not read it! (*Courtesy CDC, Atlanta*)

ceived his drinks in plastic cups. A health club refunded his membership dues. His apartment manager asked him to leave and when his toilet backed up the maintenance man came in wearing a hat over a World War II gas mask, deep water fishing boots, raincoat, and rubber gloves. In frustration, he had an HIV test. The results were negative and he gave copies of the test to every "joint" in town, his physician, dentist, theater manager, grocery store manager. . . . **He felt this approach was better than running. Do you agree?**

In another case in Brantly County, Georgia, population 11,077, the 22-year-old mother of a 2-year-old son was the subject of a rumor that she was HIV-infected. The rumor also stated that she had had intercourse with 200 men in the past year. To convince the townspeople, she

THE RIGHT TO PRIVACY

Arthur Ashe died of AIDS in February of 1993; he was *forced* to go public about having AIDS. Did that affect his life? Yes. He had to alter a lifestyle he was trying to live with his family. He could no longer live a normal day with his family.

The immediate defense from the press, which forced Arthur Ashe to go public, is that it serves the public interest to identify prominent people with AIDS. The news media says that it increases awareness—and therefore, theoretically, action—to fight back against AIDS. **DO YOU AGREE?**

Organizations can write and broadcast about HIV/AIDS any time, about anybody they want to. **IS THIS MORALLY RIGHT?**

Some defenders of the news media say that a public figure cannot have it both ways—cannot deliberately keep himself in the limelight with sports-equipment endorsements, and so forth, and then turn around and say, "I want to be a private person." **But what is public and what is private?**

As Ellen Hume asks, "Is there no moment of a public figure's life that is not open to prurient exposure? Does being a political officeholder or, as in Ashe's case, a sports champion, mean that the public owns all of your life, including your life in the bedroom, the doctor's office, the church confessional or the psychiatrist's couch?" **WHAT IS YOUR REPLY TO HER?**

Ms. Hume asks, "Should journalists end their scrutiny of public figures?" She thinks not. But she says, "**Not every revelation serves the public interest.** No American president and few athletes, astronauts, journalists or other heroes could have survived this Spanish Inquisition. Isn't it time for the press to develop a more sophisticated sense of priorities and ethics to go along with its extraordinary new power?" **WHAT IS YOUR REPLY?**

tients. Rumor quickly spread that the entire neighborhood was in danger, especially after the mail carrier refused to deliver mail and was ordered to wear rubber gloves and return to the post office for disinfection! To help the neighborhood understand AIDS and stop unfounded fears, a seminar was held at the AIDS home, **but no one would enter the house.**

A young person with AIDS reluctantly returned home—it meant revealing that he was gay to his family—and to the community. He said to his parents, **"I have good news and bad. The bad is that I'm gay. The good is I have AIDS. I won't be around long enough to interfere with anyone."** Once the word got out a catering service refused to do the annual family Christmas party They could not hire a practical nurse. Family and friends who used to drop in stayed away. People whispered that the son had gay cancer . . . that it was lethal . . . that it could be caught from dishes, linens, a handshake and breathing in the same air he breathed out.

In Athens, Alabama, the headline of the town's newspaper read, **"Athens doctor: 'I don't have AIDS.'"** This doctor, a prominent pediatrician in this town for 18 years, had to produce a public defense to dispel the gossip that he had AIDS. Because in a small town everybody knows everybody, the rumor spread with lightning speed to the town hospital, grocery store, hair salon, school, newspaper, and so on. The doctor offered a $2000 reward for any information about who began the rumor. To date no one has collected the money, but the townspeople have gathered to support him.

In Nebraska, a man sued a prominent woman for starting a **rumor** that he had AIDS. The Nebraska Supreme Court upheld a lower court ruling that the man was **slandered** and he received $25,350 in damage.

In Atlanta, in March 1999, the CDC said it had received many inquiries about reports that drug users infected with HIV had left contaminated needles in public places. Some reports have falsely indicated that CDC confirmed the presence of HIV in the needles. But the CDC said in a statement, "CDC has not tested such needles. Nor has the CDC confirmed the presence or absence of HIV in any samples related to these rumors. The majority of these reports and warnings appear to have no foundation in fact. CDC is not aware of any cases where HIV

took the HIV test and was not HIV-infected. This young woman had to circulate the results of her blood test around town, but it was still not enough to stop the rumor. A newspaper in nearby Waycross quoted an unnamed source saying that this woman was HIV-positive. **The rumor goes on!**

There was the case of a compassionate person who opened a home for helping AIDS pa-has been transmitted by a needle-stick injury outside a health-care setting."

Mathilde Krim of the American Foundation of AIDS Research said, "From the beginning, the people with AIDS have been a disposable group—gays, blacks and drug users. People don't care about them. They'd rather they wouldn't be here anyway." When the National Gay Rights Advocates asked the nation's 1,000 biggest companies whether employee medical plans covered AIDS-related expenses, one anonymous answer read, "Just enough to defray the cost of the bullet." However, HIV has spread to infect most other groups of society and attitudes are slowly changing.

Good News, Bad News, and Late News

Good News— Good news is thinking and believing you're HIV-infected and you're not. There must be a reason to think you're infected, so not being infected is, as some would say, a new lease on life. All too often that feeling is soon forgotten and many people continue lifestyles that place them at risk for infection.

Bad News— Bad news is thinking you're HIV-infected and you are. It is difficult to predict what a person sees, hears, or does after being told he or she is HIV-positive. For example, some have contemplated suicide, some have committed suicide, others have become completely fatalistic and proceeded to live a reckless and careless lifestyle that endangered others. Some have said it's like death before you're dead.

You are never prepared to hear the bad news regardless of how sure you are that you're infected. For example, one man who had suffered from night sweats, fevers, weight loss, and other classic symptoms just knew he was infected. Yet when told of the positive test results he said "I got so angry, I ripped a shower out of the wall." Being told you are infected is totally devastating, said another infected person. "You feel that everyone is looking at you, everyone can tell you're dirty." Another person said, "The fact that I'm HIV-positive completely dominates my life. There is not a waking hour that I do not think about it. I feel like a leper. I live between hope and despair." Still another said, "I did not leave the house for two days after being told I

was HIV-positive. The initial shock was that I was contaminated—unhealthy, soiled, unclean. I carried this burden in isolation for over two years. After all, I had met people who had AIDS but never a person who said—I'm HIV-infected."

The bad news is not confined to those hearing they are HIV-infected; it touches everyone they know—lovers, family, friends, and employers. Nothing remains the same. The more symptomatic one becomes, the greater the social and human loss. One symptomatic mother said, "Whenever I tell my four-year-old I am going to the doctor, he screams because he knows I could be gone for weeks." I try to put him in another room playing with his sister when they come for me." This woman has died. Relatives care for her children.

Late News— Late news is remaining **asymptomatic** after infection. Asymptomatic can be defined as when no clinically recognizable symptoms appear that would indicate HIV infection. During this time period, the virus can be transmitted to sexual partners. By the time either antibodies and/or clinical symptoms appear, **the news is too late** for those who might have been spared infection had their sexual partner tested positive or demonstrated clinical symptoms early on.

Munchausen's Syndrome: When People Pretend to Be HIV-Positive or to Have AIDS

As Hector Gavin wrote in 1843 in his 400-page history *Of Feigned and Factitious Diseases*, "The monarch, the mendicant, the unhappy slave, the proud warrior, the lofty statesman, even the minister of religion . . . have sought to disguise their purposes, or to obtain their desires, by feigning mental or bodily infirmities."

Donald Craven (1994) reported that a growing number of people may pretend to have AIDS either because of emotional disorders or because they want to gain access to free housing, medical care, and disability income. In one case, seven patients with self-reported HIV infection were treated for an average of 9.2 months in a clinical AIDS program before their sero-negative status was discovered at the hospital (in general, hospitals in the United States do not require a written copy of the HIV test results proving that someone is HIV-positive). Craven noted that "because patients with AIDS often have preferred access to drug treatment, prescription drugs, social security disability insurance, housing and comprehensive medical care, the rate of malingering may increase and reach extremes."

Another doctor who treats many AIDS patients said that he and his colleagues have seen many patients who repeatedly come to their offices fearing that they have AIDS, even though multiple tests have shown that they do not (Zuger, 1995).

An Oklahoma physician has written about the AIDS **Munchausen's syndrome**—an emotional disorder in which people pretend to have the disease simply to get attention from doctors. Confidentiality requirements make AIDS a perfect illness for people suffering from Munchausen syndrome, because the laws shield them from being discovered.

Physicians' Public Duty: A Historical Perspective of Professional Obligation

Physicians enjoy a virtual monopoly on medical care, social status, and generous financial remuneration. Thus the medical profession is uniquely entrusted with the knowledge to care for those with HIV disease/AIDS or any other contagious disease and has a clear responsibility to do so, absent compelling considerations to the contrary. A fair and reasonable share of medical risk just naturally goes with the professional territory.

Robert Fulton and Greg Owen (1988) stated that throughout history, plagues and pestilences have challenged humankind. In his book, *Plagues and People,* William McNeill (1976) cited the many death-dealing epidemics in Europe. He wrote that one advantage the West had over the East in the face of deadly epidemics was that caring for the sick was a recognized duty among Christians. The effect of a prolonged epidemic more often than not strengthened the Church when other social institutions were discredited for not providing needed services. McNeill further observed that the teachings of Christianity made life meaningful, even in the immediate face of death: Not only would survivors find spiritual consolation in the vision of heavenly reunion with their dead relatives or friends, but

God's hand was also seen in the work of the life-risking caregivers.

The United States has also had its share of plagues and epidemics; one of the most notable was the outbreak of yellow fever in Philadelphia in 1793. Thousands of citizens of the capital city perished. William Powell (1965), in his book *Bring Out Your Dead,* describes Philadelphia at the time of the yellow fever plague: the dying were abandoned, the dead left unburied, orphaned children and the elderly wandered the streets in search of food and shelter. Nearly all who could fled the city, including the President, leaving the victims of the fever to their fate. Among those who remained, however, were Benjamin Rush, M.D., the mayor, a handful of medical colleagues and their assistants, and a number of clergy. With the help of a small group of laborers and craftsmen, they undertook the enormous tasks of maintaining law and order, providing medical care, food and shelter to the sick and helpless, as well as gathering up and burying the dead.

Dr. Rush and the others remained at their posts because of their overriding sense of professional obligation along with a conviction inspired by the precept of the New Testament **"Blessed are the merciful, for they shall obtain mercy"** (Matthew 5:9).

But this vision, shared by Christians for centuries, along with a sense of professional commitment, may not be sufficient to persuade contemporary health care workers to stay at their posts.

The baby boom generation, well-educated and self-oriented, has learned to blame AIDS on groups whom society defines as deviant: homosexuals, prostitutes, and drug abusers. There is a significant probability, therefore, that today's young health care practitioner may turn away from HIV/AIDS patients despite the fact that a 1981 Gallup Poll of the religious beliefs and practices of 14 countries showed that the United States leads the world not only in church membership but also in charitable services.

According to an article in the November 13, 1987, issue of *The Wall Street Journal,* Arthur Caplan, a director for the Center of Biomedical Ethics at the University of Minnesota in Minneapolis, believes that doctors and nurses who refuse to treat AIDS patients do so out of disapproval. "They know how to deal with violent patients and infectious diseases like hepatitis." So why not treat someone with AIDS? "It's more than fear. They're making a value . . . judgment about AIDS victims. They're saying they won't treat people (they find) disgusting." Others refuse to place their lives at risk for someone who became HIV-infected through immoral behavior. **C. Everett Koop said the refusal by some medical personnel to treat AIDS patients "threatens the very fabric of health care in this country." DO YOU AGREE?**

Medical Moral Issues

HIV/AIDS represents a new era in medicine, one in which physicians are faced with complex moral issues. When the American Medical Association (AMA) issued a statement to the effect that it is unethical to refuse to treat HIV/AIDS patients, that statement reflected a deep concern in the medical community about the possibility of their becoming infected by treating patients with HIV/AIDS.

The AMA statement for an ethical call to arms is unprecedented in this century. It is the result of a spreading fear that HIV/AIDS is too contagious to tolerate in spite of the knowledge that the virus is not transmitted via casual contact. **Emotions, not education, are in control of those whose fears exceed reality.** But these emotions are real and they are having an impact on the medical community.

There is an ongoing dilemma concerning the rights of the physician and other health care

DEALING WITH DISCRIMINATION

Although incidents of discrimination are disheartening, they are only a part of the story. Another part of the story is how discrimination has been fought and how courts, legislatures and other social institutions have responded with attempts to reduce HIV/AIDS discrimination and to minimize its impact. To this end a brief synopsis of the Americans with Disabilities Act is presented along with the first United States Supreme Court ruling based on this act.

AMERICANS WITH DISABILITIES ACT

The primary federal nondiscrimination statute that prohibits discrimination on the basis of a person's disability or health status is the **Americans with Disabilities Act (ADA) of 1990.** The ADA provides that no individual **"shall be discriminated against on the basis of disability in the full and equal enjoyment of the goods, services, facilities, privileges, advantages or accommodations of any place of public accommodation."** The ADA's definition of "public accommodation" specifically includes hospitals and professional offices of health care providers. A critically important issue under the ADA is whether persons with **asymptomatic** HIV infection have a disability and thus are protected under the ADA. **Disability is defined as a physical or mental impairment that substantially limits one or more of the major life activities of the individual, a record of such impairment, or being regarded as having an impairment.** In the past, many courts have ruled or assumed as undisputed that HIV infection, as the underlying cause of a life-threatening illness, is a disability. However, several recent court decisions have held that HIV does not automatically qualify as a disability, and in each case there must be an individualized determination as to whether the infection actually limits, in a substantial way, a major life activity. The ADA's legislative history, however, indicates that Congress intended to include HIV infection within the definition of disability, and the Equal Employment Opportunity Commission's regulations embody that view. **In its first AIDS case ever, the Supreme Court had to decide whether and to what extent persons with HIV infection are protected under the ADA.**

THE CASE OF *BRAGDON V. ABBOTT*: UNITED STATES SUPREME COURT

1998

In the seventeen year history of the HIV/AIDS pandemic, the U.S. Supreme Court has never considered a case directly involving HIV or AIDS. That changed on Monday, March 30, 1998 when oral arguments began in the case of *Bragdon v. Abbott.* The *Bragdon* case is also the first time the court has ever heard a case involving the ADA. On September 16, 1994, Sidney Abbott, age 37, went to her dentist in Bangor, Maine to get a cavity filled. Dr. Randon Bragdon refused to fill a gum line cavity in his office when he read on her medical form that she was HIV positive. He told her he could do the procedure in a hospital, a change of venue which would have added approximately $150 to the bill. According to a **cover story** about the case in the *American Bar Association Journal*, Dr. Bragdon did not have privileges to practice in any area hospitals, nor had he applied for them. Dr. Bragdon maintains that he could have sought and received permission to perform occasional procedures without having been granted full privileges.

Abbott's Lawsuit

Her lawsuit argues that in refusing to treat her in his office, Bragdon violated the ADA and the Maine Human Rights Act. Federal district and appeals courts both agreed with her.

United States Supreme Court: Decision June 25, 1998

The Supreme Court ruled 5 to 4, upholding the District Court and the First Circuit Court of Appeals, finding that **Bragdon** violated Abbott's rights to treatment under the provisions of the ADA of 1990. Abbott's HIV infection constituted a **disability** under the ADA in that her HIV infection **"substantially limits"** a major life activity—her ability to reproduce and bear children (to have a child she places her husband at risk for HIV infection and risks infecting her child). **Justice Kennedy** in delivering the opinion of the court held that from the moment of infection and throughout every stage of the disease, HIV infection satisfies the statutory and regulatory definition of a "physical impairment." Applicable Rehabilitation Act regulations define "physical or mental impairment" to mean "any physiological disorder or condition affecting the body['s] hemic and lymphatic [systems]." HIV infection falls well within that definition. The medical literature reveals that the disease follows a predictable and unalterable course from infection to inevitable death. It causes immediate abnormalities in a person's blood, and the infected person's white cell count continues to drop throughout the course of the disease, even during the intermediate stage when its attack is concentrated in the lymph nodes. Thus, HIV infection must be regarded as a physiological disorder with an immediate, constant, and detrimental effect on the hemic and lymphatic systems.

DISSENT IN PART

Justice O'Conner stated that Abbott's claim of a disability should be evaluated on an individual basis and that she has not proven that her symptomatic HIV status substantially limited one or more of her major life activities. "In my view, the act of giving birth to a child, while a very important part of the lives of many women, is not generally the same as the representative major life activities of all persons—caring for one's self, performing manual tasks, walking, seeing, hearing, speaking, breathing, learning, and working"— listed in regulations relevant to the Americans with Disabilities Act of 1990. Based on that conclusion, there is no need to address whether other aspects of intimate or family relationships not raised in this case could constitute major life activities; nor is there reason to consider whether HIV status would impose a substantial limitation on one's ability to reproduce if reproduction were a major life activity.

NATIONAL SURVEY OF DENTISTS IN CANADA: REFUSAL TO TREAT HIV-INFECTED PATIENTS

According to the report by Gillian McCarthy and colleagues (1999) there are about 15,232 dentists in Canada. A random sample of 6,444 answered a survey on whether they would refuse to treat HIV-infected patients. Some 4,281 dentists responded. The conclusion presented by the investigators was that 1 in 6 (or 17%) would **refuse to treat** an HIV-infected patient. The refusal was associated with the respondent's lack of belief in an ethical responsibility to treat patients with HIV and their fears of becoming infected from their patients.

Class Discussion: Clearly a 5 to 4 ruling is not an overwhelming mandate to support Abbott's lawsuit. Do you feel that a simple majority, 55% in this case is sufficient or because this case has vast implications, should it require a two-thirds majority, 6 in favor-3 against (67%). What are the legal and moral issues in accepting a simple majority vs. a two-third ruling?

Bragdon raised a question for you to consider as you research the question above. Looking at **MAGIC JOHNSON**—he asks if it really makes sense to consider someone "disabled" who can play professional basketball, or go to work, or otherwise perform the tasks of daily living. YOUR RESPONSE IS. . . .

workers to practice medicine in a safe environment as opposed to the rights of HIV-infected and AIDS patients to receive care and medical support. Although the risk of HIV transmission through medical occupational exposure appears to be quite low, the fact that it is possible at all, coupled with the uniformly fatal prognosis associated with AIDS, suggests that physicians, nurses, and other health care workers have legitimate concerns about risk to their health.

In a recent report by C.E. Lewis and colleagues (1992), almost half of the primary care physicians in the Los Angeles area had refused to treat HIV-infected patients or had planned not to accept them as regular patients.

In a more recent survey of American doctors in residency training, 39% said that a surgeon or other specialist had refused to treat a patient with AIDS in the resident's care. In Canada, only 13% reported a specialist had refused to treat a patient with AIDS; in France, only 8% said a specialist had rejected care.

Further, 23% of American doctors would not care for AIDS patients if they had a choice as compared with 14% of Canadian physicians and 4% of French doctors.

The survey results may be a disturbing indicator of how American physicians view their work and a reflection of cultural and political attitudes here that view with veiled hostility those with AIDS.

To combat that fear, medical schools are now providing their students with disability insurance that covers AIDS. Hospitals have adopted policies that require physicians to treat AIDS patients or face dismissal.

HIV-Infected Physicians

The fact that medical students and residents fear contracting HIV infection and AIDS can be tied to the reality that by mid-1991, 720 physicians were reported to the CDC as having AIDS. This number is probably underreported by 20% to 50%. At least 5,000 physicians were HIV-infected by 1990 (see Chapter 10 for additional information on HIV-infected health care workers) (Breo, 1990). The question residents and medical students are asking is: How did so many become infected?

Along with the recognition of infected physicians come many disturbing questions

on whether they should be permitted to continue their medical practice. In 1987, a Gallup Poll (as reported in January 1989 in *The Wall Street Journal*) asked that very question, and 57% of the general public said **not under any circumstances.**

Medical experts say the public should not be worried about patients contracting the virus during a routine physical examination or through casual contact with an infected doctor or hospital worker.

But the public does worry about what might happen during surgery and procedures involving contact with mucous membranes (Table 13-1). For example, some medical experts fear that if an infected surgeon is accidently cut in the operating room, the surgeon's blood could cause an infection. Because of the controversial dentist-to-patient HIV transmission that occurred in Florida, the CDC has convened its advisors to help revise CDC guidelines for infection control. A major question under review is whether any HIV/AIDS professional who performs invasive procedures should be allowed to continue. Restriction of HIV-infected surgeons and dentists would greatly curtail or even destroy their practices. It was suggested by one consultant to the CDC that any restriction would lead to mandatory screening of all health care workers. At the moment, the CDC says that the question of whether an infected physician or dentist should be allowed to perform invasive procedures (those

TABLE 13-1 Poll on Revelation of HIV Status

A. Should the following health care workers who are HIV-infected be forbidden to practice?

	Yes	No
Surgeons	63%	28%
All physicians	51%	42%
Dentists	60%	33%
All health care workers	49%	43%

B. Should patients be required to inform their physicians, dentists and other health care workers whether they are HIV-infected?

Yes	No
97%	2%

(Source: *Newsweek*, July 1, 1991)

in which bleeding may occur) can only be answered on a case-by-case basis.

Risk Protection of the Patient: Political Dimensions— The American Medical Association's position is that HIV/AIDS doctors "should consult colleagues as to which activities the physician can pursue without creating a risk to patients."

There must be a rational relationship between the degree of risk, the morbidity and mortality of the disease, and the consequences of policies and procedures implemented to reduce the risk. Invasive surgery involves many risks. There is an overriding ethical obligation to reduce these risks in ways that do not create more harm than good. The justification of policies to prevent life-threatening illnesses with a risk factor of 1 in 10,000 is significantly greater than those with a risk factor of one in 100,000. More lives can be saved by innovative approaches to preventing the kind of injuries that have accounted for 75% of all cases of occupationally related HIV infection than by preventing HIV-infected health care workers from performing invasive surgery.

The risk of HIV infection by an HIV-infected surgeon is only one-tenth the chance of being killed by lightning, one-fourth the chance of being killed by a bee, and half the chance of being hit by a falling aircraft. The 1 in 100,000 risk of transmission from an infected surgeon equals the probability of death we face bicycling 2 miles each way to school for 1 month, or commuting 15 miles round-trip by car for a year.

Similarly, our chance of dying from anesthesia during an operation is roughly 10 times our chance of being infected by a surgeon known to be infected with HIV. A mother who approves a penicillin injection for her toddler with pharyngitis accepts a 1 in 100,000 risk of death from anaphylactic shock. In one study, the most successful coronary artery bypass graft surgeon surveyed had a 1.9% mortality rate, while the least successful had a 9.2% mortality rate. Patients selecting the least successful surgeon thus face 7,300 times the extra risk of death posed by an HIV-infected surgeon. The risk to a person undergoing invasive surgery by a surgeon of **unknown** HIV status is about 1 in 20 million of becoming HIV-infected (Daniels, 1992). The CDC estimates the risk to a patient

HIV TRANSMITTED BY AUSTRALIAN AND FRENCH SURGEONS

A breakdown in infection control procedures is being blamed for the transmission of HIV to five patients of an Australian surgeon. It's believed the virus was transmitted from one patient to four others during minor skin surgery on a single day in November, 1989. Health officials say one patient, a gay man, is believed to have been the infection source. **The CDC calls the case the first known patient-to-patient transmissions of HIV in a health care setting** (*American Medical News.* 1994 37:2.) In October 1995, after the HIV seropositive status of an orthopedic surgeon in Saint Germain en Laye (west suburb of Paris) was announced in the medical press, the French Director of General Health decided to inform and offer testing to the patients operated on by this surgeon. Review of the medical history of the surgeon suggests that he most probably became infected with HIV in May 1983. The diagnosis of HIV infection and AIDS were made simultaneously in May 1994. This investigation identified 3,004 patients who had undergone at least one invasive procedure by the surgeon; 2,458 patients were able to be contacted by mail. The serologic status of 968 patients was ascertained; 967 are negative. Only one patient, who was negative before prolonged operation performed by the surgeon in 1992 (10 hours) is HIV positive. Typing of the viral strains of the patient and the surgeon comparing nucleotide sequences of 2 viruses showed they are closely related. Similar to the more recent dentist-Bergalis case, the source of HIV in these cases will most likely never be documented to everyone's satisfaction.

by an **HIV-infected surgeon** to be between 1 in 42,000 and 1 in 420,000 (Lo et al., 1992).

Patients' Right to Know if Their Physician Has HIV/AIDS— In a 1987 Gallup Poll, 86% of those polled felt that they had the right to know if a health care worker treating them was HIV-infected. Many lawyers also take this position. The courts appear to be moving toward an interpretation of the doctrine of informed patient consent as **"what a reasonable patient would want to know,"** rather than **"what a reasonable physician would disclose."** Because it is so difficult for surgeons to avoid occasionally cutting themselves during surgery, it has been suggested that the best solution is not to have HIV-infected surgeons perform surgery at all.

Public anxiety on HIV/AIDS and medical care is becoming increasingly tinged with hysteria. A national Gallup Poll undertaken for *Newsweek* (June 20, 1991) asked a representative sample of 618 adults, "Which of the following kinds of health care workers should be required to tell patients if they are infected with the AIDS virus?"

The answers were: surgeons 95%; all physicians 94%; dentists 94%; all health care workers 90%.

Clearly, people do not differentiate between doctors who perform invasive procedures and those who do not. However, the patient could ask what the probability is of a single dentist (Acer) infecting six of his patients (Bergalis, Web, and four others). Extremely low, **yet it did happen!** The lowest of probabilities and best of guidelines and precautions do not stop the fire of fear. It must also be mentioned that the same *Newsweek* poll found that 97% of those interviewed felt that HIV-infected patients should tell their health care workers that they are infected (Table 13-1B).

There is one important aspect related to this poll that needs to be addressed. That is, many of the people interviewed stated that if they **knew** their surgeon was HIV-infected, they would "get another surgeon." This *switching dilemma* may, at some point, have the majority of the population needing surgery standing in line for the uninfected surgeons. Services provided by the reduced number of surgeons, it could be claimed, at some point, result in increased costs and diminished quality.

HIV-infected Health Care Professional's Duty to Disclose— Several courts have held that health care professionals have a duty to disclose their HIV status to patients or health authorities, assuming that their professional activities pose a risk of transmission to patients. The Maryland Court of Appeals ruled that a surgeon has a duty to inform his patients of his infection; even if the patient has not actually been exposed and tests HIV negative, the contact with the surgeon may subsequently give rise to a claim for their infliction of mental distress due to fear of transmission. Courts justify orders to disclose

based on a duty to protect patients and on the doctrine of informed consent. Requiring disclosure to patients, of course, can severely jeopardize a health care professional's career. To avoid this result, some states allow the professional to continue practicing, with appropriate restrictions and supervision, but without disclosing his or her HIV status (Gostin, et al. 1998).

Physician–Patient Relationships

In most states, physicians may **not test** a patient for HIV antibodies without written permission from the patient. Physicians can run tests for any other infectious diseases without written permission.

Class Question: Do you think this is fair and equal medical practice?

Because the majority of HIV-infected patients are asymptomatic, there is no way of telling **who is** or **is not** infected. Therefore, for the protection of health care workers, everyone must be treated as though they were HIV-infected.

HIV is the silent medical threat of the 1990s. Wherever blood is drawn, a wound is examined, a dressing is changed, or anything that involves blood, needles, or surgery is done, there is an unspoken fear that HIV might be present. The patient sees this preventive attitude of physicians and other health care workers by the new look of the 1980s and 1990s: wraparound smocks, gloved hands, and masks. The patient wonders whether his or her physician is an HIV carrier and the physician assumes

--- **BOX 13.6** ---

MY FIRST AIDS EXPERIENCE

"A man was waiting for his mother in my office while she consulted me about a back problem," reports a family practitioner in a Gulf state city. "He had a grand mal seizure, bit his tongue and bled quite a bit. I put a tape-wrapped tongue depressor into his mouth to keep him from further injuring his tongue and then just supported him until the seizure was over. Later the physician who had been caring for him at a hospital clinic called to ask that I confirm the seizure, adding, **"By the way, this man has AIDS."** I never gave a thought to AIDS in this situation," continues the family practitioner. "It was my first experience in dealing with an AIDS patient" (Polder et al., 1989).

that the patient may be a carrier. A recent Associated Press article presented some examples of how the AIDS pandemic has changed doctor–patient relationships:

1. In many operating rooms, doctors and nurses wear wraparound glasses in case of blood splashes.

2. In many emergency rooms, health care workers cover themselves with caps, goggles, masks, gowns, gloves, shoe covers, and blood-proof aprons.

3. Infection control specialists must spend time convincing hospital workers it is safe to enter an HIV/AIDS patient's room to perform normal duties ranging from picking up dinner trays to fixing the plumbing.

4. At some hospitals, all emergency room trash, no matter how innocuous, is treated as hazardous waste.

5. Many doctors and nurses routinely pull on gloves whenever they give an injection or draw blood.

6. Mouth-to-mouth resuscitation is simply not done at many hospitals. Instead a mask and valve device is used to avoid direct contact with the patient's mouth.

7. A surgeon at San Francisco General Hospital takes the AIDS drug zidovudine whenever he operates on people he thinks are infected.

8. Physicians are increasing life insurance policies to provide for their families in case they become HIV-infected.

The health care professionals' fear of getting AIDS will persist as long as there is a risk that HIV can be transmitted in the workplace. **The goal is not to eradicate that fear, but to prevent it from compromising the quality of patient care and from threatening the health professional's own well-being** (Gerbert et al., 1988).

In early 1990, Lorraine Day quit her post as chairperson of the Orthopedics Department at San Francisco General Hospital and abandoned her surgical practice because of her fear of exposure to HIV. "I have two children to think about and operating was too dangerous. If I was a skydiver, people would say I was an irresponsible mother, but to me, surgery now, it was just as risky."

Dr. Day said during a TV interview, "Our risk is one in 200 per single (needle) stick with AIDS blood and it can be the first one—it doesn't take 200. And I ask you, if you came to work every **day and flipped the light switch on in your office and only one out of 200 times you were electrocuted would you consider that low-risk?"**

While Dr. Day has been called a scaremonger, many surgeons in private conversations call her a "hero" for raising the risk issue.

Physicians with AIDS

According to Mandelbaum-Schmid (1990), one overriding issue facing physicians who become HIV-infected is that there is no support system to compensate them once they become too sick to work (Figure 13-4). For example, a 32-year-old gastroenterologist, claimed she was infected with HIV in 1983 when she accidently pricked herself with a needle that an intern had used to draw blood from an AIDS patient. She was working at the time at Brooklyn's Kings County Hospital.

She tested HIV-positive in 1985. On subsequent consultations with physicians at the

FIGURE 13-4. Together for Now: Judy (39), Andy (41), Robbie (12), and Amy (8) Lipschitz. In 1986, during his residency, Andy was asked to work with an AIDS patient. He inserted an arterial line into the patient's arm. After withdrawing the needle from the catheter, he accidentally punctured his thumb. Several weeks later Andy had severe flu-like symptoms. He demonstrated *Pneumocystis carinii* pneumonia in mid-1992—he had AIDS. His ever-changing regimen of ZDV, ddI, ddC, d4T and protease inhibitors all cause side effects. He is hoping a more successful drug will soon be available. ©Shonna Valeska 1995.

CURRENT VIEW OF HEALTH CARE WORKERS WHO ARE HIV-INFECTED

The legal system traditionally has been reluctant to challenge the professional autonomy of physicians by removing or altering their choice of medical decisions. In this light, neither the courts, legislatures, nor public health agencies have produced clear, binding rules for the treatment of HIV-positive health care workers (HCWs). The legal developments of the last several years return the question of HIV-infected HCWs to the HCWs themselves, to the institutions that employ them, and therefore to the colleagues and professional organizations with whom HIV-infected HCWs are associated.

What Is the Medical Profession Doing with This Authority Relative to HIV Infection/AIDS?

Individual HCWs have evidently concluded that they should keep their infections secret for as long as possible, while **many institutions seem to have adopted a "don't ask, don't tell" policy.** When a HCW's HIV infection is discovered, most institutions evaluate the worker under applicable guidelines and negotiate a mutually acceptable course of action, all in strict confidence. Few institutions go public; in some cases the HCW has no option but to sue (Burris, 1996).

who died of AIDS in June, 1989, at the age of 46, set an unwelcome precedent. Before his illness, he had a thriving practice in Princeton, New Jersey. He believed he was infected in 1984 after being splashed in the face with blood while performing an emergency tracheotomy. He became ill with an undifferentiated pneumonia in 1987. Physicians performed a bronchoscopy and gave him an HIV test without telling him.

Word of his positive HIV test soon spread through the hospital and soon after into his community. By the time he was released from the hospital, his telephone lines were jammed with calls from hysterical patients. He eventually lost about half his practice. When he tried to schedule operating room time, the hospital's president convened an emergency meeting of trustees. They made the unprecedented decision that he may operate but only with the informed consent of his patients.

He filed a lawsuit based on breach of ethics and doctor–patient confidentiality—his diagnosis had been leaked by the hospital laboratory—and violation of antidiscrimination laws. In April, 1991, New Jersey Superior Court Judge Philip S. Carchman ruled that the Medical Center of Princeton, after learning the doctor had AIDS, made a "reasoned and informed response" in barring him from performing surgery without informed consent of his patients.

The last example of the plight of a physician with AIDS involves a physician at Johns Hopkins Medical Center. He contracted an HIV infection from a patient in 1983 while working as a resident in the bone marrow transplantation unit, he had an accident with the blood of a young leukemia patient who had been transfused many times. The accident was followed 3 weeks later by an acute febrile illness with cough, sore throat, rash, and lymphadenopathy. After this, he was in good health for the next 3 years. He completed his residency and became a chief resident and a fellow in cardiology. He married, and had a daughter. In November of 1986, unexplained weight loss led him to be HIV tested. He tested positive for HIV.

The hospital's reaction was unexpected. He had to sue the institution and some members of the faculty in an effort to defend his reputation and obtain appropriate benefits. After endless

National Institutes of Health (NIH), she was told that she was a healthy carrier of the virus who could become ill. She continued training at Kings County, completing a residency in internal medicine and a fellowship in gastroenterology.

She became symptomatic in 1987. A year later, she filed a 175 million dollar lawsuit charging that the hospital's negligence had caused her illness. At the time of the trial, her life expectancy was estimated at less than a year. Following a stormy 2-month trial that challenged the integrity of all the physicians and expert witnesses who testified, she reportedly settled for just under 2 million dollars.

Another pressing issue pointed out by Mandelbaum-Schmid is that HIV-infected doctors are not protected against discrimination. The experience of an otolaryngologist

months of legal battle, Johns Hopkins proposed a settlement 3 weeks before the trial date.

He said, "By actions like this, medical institutions send an awful message to the health care worker: we ask you to be in the front lines, but if something happens to you, we will not stand behind you, you will be abandoned, and you will be deprived of the privilege of practicing medicine" (Aoun, 1990). He died of AIDS in February of 1992, 9 years after becoming infected with HIV.

SUMMARY

In 1981, the CDC announced a new disease affecting the homosexual population. This disease was later called AIDS. Many religious people believed this was a sign that homosexuality should be punished. The few facts available at that time gave rise to a great deal of fantasy and fear. Affected people were either innocent victims or they deserved the disease. Contracting AIDS labeled a person as less than desirable, a homosexual or one who practiced deviant forms of sexual behavior. But even the so-called innocent victims, the children, the hemophiliacs, and the recipients of blood transfusions were not spared social ostracism. If you had AIDS, you were twice the victim—first of the virus and second of the social behavior.

Children were barred from attending school, adults from their jobs, and both from adequate medical care. For example, there are still relatively few dentists who will treat AIDS patients and a significant number of surgeons refuse to operate on AIDS patients. Fourteen years have passed, but many misconceptions about HIV/AIDS linger on.

Fear is being casually transmitted rather than the virus. A significant number of people, after 16 years of broad scale education, still believe that the AIDS virus can be casually transmitted from toilet seats, drinking glasses, and even by donating blood.

The fallout from the fear of the AIDS pandemic has been a major change in sexual language in TV advertisements, magazines, and radio. Condoms once spoken about only in hushed tones in conversations and kept under the counter in most drug stores, are now spoken of everywhere as a means of safer sex. AIDS, perhaps more than any other disease, has demonstrated that ignorance leads to fear and knowledge can lead to compassion.

To achieve understanding and compassion, people must be educated as to their HIV risk status and how they can keep it low. Many hundreds of millions of dollars have been spent to inform the public of the kinds of behavior that either place them at risk or reduce their risk for HIV infection. The problem is that although people are getting the information, too many refuse to act on it. Former Surgeon General C. Everett Koop's office mailed 107 million copies of the brochure "Understanding AIDS" to households in the United States. Fifty-one percent of those who received it said they never read it. But even among those who read the brochure are those who refuse to change their sexual behavior. Old habits are difficult to break.

To date, the hard evidence shows that only the homosexual population has significantly modified their sexual behavior as evidenced by the drop in the number of new cases of AIDS among them from 1988 through 1999.

A major problem looming on the horizon is the prospect of HIV being spread in the teenage population. Large numbers of teens use drugs and alcohol, have multiple sex partners, and believe they are invulnerable to infection.

The AMA stated in 1988 that physicians may not refuse to care for patients with AIDS because of actual risk or fear of contracting the disease. Some physicians get around this through referral to other physicians who will treat AIDS patients. There is one area of medicine that takes issue at having to treat AIDS patients. That area is surgery. Because it is difficult not to accidently get cut during surgery, surgeons have been the leading advocates for HIV testing of all surgical patients so they will know their risks before performing surgery.

On the other hand, patients feel they have a right to know if their physician, especially a surgeon, is HIV-infected. Surveys indicate that most people would not want to be treated by an HIV-infected physician.

On June 25, 1998, the United States Supreme Court ruled that the Americans with Disabilities Act protected HIV-infected people. Even though they experienced **no symptoms,** they are to be treated as handicapped.

REVIEW QUESTIONS

(Answers to the Review Questions are on page 487.)

1. Name three major sources of information that contributed to the early panic and hysteria about the spread of AIDS.

2. Give three examples of unfounded public fears to AIDS infection.

3. Fear of the casual transmission of AIDS parallels what other earlier STD epidemic?

4. What evidence is there that it is difficult to get people to change their behavior even though they know it is harmful to their well being?

5. What is the major thrust of AIDS education in the United States?

6. If education is the key to preventing HIV infection and new cases of AIDS; and most people interviewed say they have been 'educated,' why is it not working?

7. Why are today's teenagers in danger of contracting and spreading HIV?

8. Yes or No: Do physicians have a right to refuse to treat AIDS patients? Support your answer.

9. Do patients have a right to know if their physician is HIV-infected or has AIDS?

10. What is the primary means of offsetting the bias toward people with AIDS in the workplace?

REFERENCES

American Medical Association News. (1991). Ruling fuels debate over HIV-infected doctors. May:1,41–43.

AOUN, HACIB. (1990). A handful helped us. *Medical Doctor*, 34:31–32.

BLENDON, R.J., et al. (1988). Discrimination against people with AIDS: The public's perspective.*N. Engl. J. Med.*, 319:1022–1026.

BREO, DENNIS L. (1990). The slippery slope—handling HIV-infected health care workers. *JAMA*, 264:1464–1466.

BURRIS, SCOTT. (1996). Human immunodeficiency virus-infected health care workers. *Arch. Fam. Med.*, 5:102–106.

CRAVEN, DONALD, et al. (1994). Factitious HIV Infection. *Ann. Intern. Med.*, 121:763–766.

DANIELS, NORMAN. (1992). HIV-infected professionals, patient rights and the 'switching dilemma'. *JAMA*, 267:1368–1371.

EICKHOFF, THEODORE C. (1989). Public perceptions about AIDS and HIV infection. *Infect. Dis. News*, 2:6.

FISHER, J.D., et al. (1992). Changing AIDS risk behavior. *Psychol. Bull.*, 111:455–474.

FOURNIER, A.M., et al. (1989). Preoperative screening for HIV infection. *Arch. Surg.*, 124: 1038–1040.

FULTON, ROBERT, et al. (1988). AIDS: Seventh rank absolute. In *AIDS: Principles, Practices and Politics.*, Inge B. Corliss, et al., eds. Bristol, PA: Hemisphere.

GERBERT, BARBARA, et al. (1988). Why fear persists: Health care professionals and AIDS. *JAMA*, 260:3481–3483.

GOSTIN, LAWRENCE, et al. (1998). HIV infection and AIDS in the public health and health care systems: The role of law and litigation. *JAMA*, 279:1108–1113.

HAGEN, M.D., et al. (1988). Routine preoperative screening for HIV: Does the risk to the surgeon outweigh the risk to the patient? *JAMA*, 259: 1357–1359.

HEGARTY, JAMES D., et al. (1988). The medical care costs of HIV-infected children in Harlem. *JAMA*, 260:1901–1909.

HEREK, GREGORY M., et al. (1993). Public reaction to AIDS in the United States: A second decade of stigma. *Am. J. Public Health*, 83:574–577.

JAPENGA, ANN. (1992). The secret. *Health*, 6:43–52.

KEGELES, S.M., et al. (1988). Sexually active adolescents and condoms: Changes over one year in knowledge, attitudes and use. *Am. J. Public Health*, 78:460–461.

KIRBY, D. (1988). The effectiveness of educational programs to help prevent school-age youth from contracting AIDS: A review of relevant research. United States Congress.

LEFKOWITZ, MATHEW. (1990). A health care system in crisis: The possible restriction against HIV-infected health care workers. *PAACNOTES*, 2:175–176.

LEWIS, C.E., et al. (1992). Primary care physicians' refusal to care for patients infected with HIV. *West. J. Med.*, 156:36–38.

LO, BERNARD et al. (1992). Health care workers infected with HIV. *JAMA*, 267:1100–1105.

MANDELBAUM-SCHMID, JUDITH. (1990). AIDS and MDs. *Medical Doctor*, 34:33–40.

McCARTHY, GILLIAN, et al. (1999). Factors associated with refusal to treat HIV infected patients: The results of a National Survey of Dentists in Canada. *Am. J. of Public Health*, 89:541–545

McNEILL, WILLIAM H. (1976). *Plagues and People*, Garden City: Anchor Press.

MICHAELS, DAVID, et al. (1992). Estimates of the number of youth orphaned by AIDS in the United States. *JAMA*, 268:3456–3461.

Morbidity and Mortality Weekly Report. (1990). HIV-related knowledge and behavior among high school students—Selected U.S. cities, 1989. 39: 385–396.

NARY, GORDON. (1990). An editorial. *PAACNOTES*, 2:170.

PHILLIPS, KATHRYN A. (1993). Subjective knowledge of AIDS and Use of HIV testing. *Am. J. Public Health*, 83:1460–1462.

POLDER, JACQUELYN A., et al. (1989). AIDS precautions for your office. *Patient Care*, 23:161–171.

POWELL, JOHN H. (1965). *Bring Out Your Dead*. New York: Time-Life Inc.

RHAME, FRANK S., et al. (1989). The case for wider use of testing for HIV infection. *N. Engl. J. Med.*, 320:1242–1254.

ROWE, MONA, et al. (1987). *A Public Health Challenge: State Issues, Policies and Programs, Volume 2.* Intergrovernmental Health Policy Project, George Washington University.

STRYKER, JEFF, et al. (1995). Prevention of HIV infection *JAMA*, 273:1143–1148.

VOELKER, REBECCA. (1989). No uniform policy among states on HIV/AIDS education. *Am. Med. News*, Sept.:3,28–29.

ZUGER, ABIGAIL. (1995). The high cost of living. *Sci. Am.* 273:108.

Epilogue

While the achievements of the global response to date should not be underestimated, neither should the challenge ahead. HIV disease/AIDS is a catastrophe in slow motion, and it is essential that the world community pace itself for the long haul. The task ahead calls for clear vision, renewed will, and greatly increased resources. **But it also calls for greater determination to use the resources in the interest of everyone: HIV disease or AIDS must not be allowed to join the list of problems, like poverty and hunger, that this world has learned to live with because the powerful have lost interest, and the powerless have no choice.**

Whatever we do now, the die is already cast for many millions of people. The HIV/AIDS epidemic will exact an enormous toll. What is at stake is the possibility of shaping the future course of the pandemic, of moderating or lessening the suffering that inevitably will occur as a result of the infections that already exist. **An epidemiological catastrophe is unavoidable. What can be limited are the dimensions of the catastrophe.**

In the United States, as of year 2000, through 19 years of the HIV/AIDS epidemic, we can reflect on the fact that we have learned a lot about this new disease but there is a long way to go. While it is easy to criticize, condemn, ridicule, and find mistakes in the way the HIV/AIDS pandemic has been handled, society should not lose sight of the outstanding scientific and social achievements made since HIV/AIDS was discovered.

Excellent progress has been made in understanding the biology of HIV, its transmission, therapy, and its relationship to the progression of AIDS. If the 1980s are to be remembered as the period of AIDS violence, hatred, and irrational fears, the 1990s will hopefully be remembered as the time of optimism. HIV disease has become a more treatable illness wherein the quality and duration of life are being extended. Promising antiviral treatments are now in use and vaccines for the prevention of HIV infection are in field trials. HIV/AIDS education will soon reach every grade school, high school, and workplace in the United States. Courts have affirmed the rights of HIV-infected/AIDS patients to a public education and to remain at their jobs.

WHAT A DIFFERENCE A DECADE MAKES:

We have come a long way.

In October of **1988** protesters demanded access to new drugs and an end to fear and discrimination because someone was infected with HIV. They swarmed the doors of the Food and Drug Administration, were held back for nine hours by police wearing rubber gloves and carrying riot shields, the protesters smashed windows, chanted slogans, staged "die-ins" and burned President Ronald Reagan in effigy. By **1998** the climate of fear surrounding AIDS has largely dissipated. Large numbers of people now know how human HIV is contracted and how to avoid it. The fire-bombings and firings have largely stopped. When they still happen lawsuits are filed under the ADA. And, the kinds of treatments the protesters were demanding a decade ago are now on sale at the local drugstore. Death rates from AIDS have plummeted in the United States and other industrialized countries. The protesters who once smashed windows at the FDA now sit in government committees that oversee AIDS research. Special programs permitting early access to promising drugs have been in place for years. **But, insurance coverage still remains an issue**.

To the Future

Despite the gains against HIV, we are somewhere between the end of the beginning and the beginning of the end of AIDS. Exactly where depends on our willingness to make the most of our opportunities. The things that we still need to do about HIV also illustrate a fundamental truth: fighting HIV isn't only about HIV. It's about our vision of a better society where we are clearer about our responsibilities to ourselves and one another, and act accordingly.

There is great hope that in the early years of the next decade, the diagnosis of HIV infection will *not* indicate the almost certain destruction of the immune system, the onset of AIDS, and death. It is hoped that HIV will be just another controllable viral infection.

The divisiveness that has marked the 1980s and the early 1900s has been too costly, the human toll too great. We can spend the present debating the past but we must look to the future—we must go forward from where we are. Let history be history—learn from it and move on! The way for us to get through this crisis is hand in hand.

Humans shall overcome HIV!

Answers to Review
Questions

CHAPTER 1

1. Acquired Immune Deficiency Syndrome
2. No. AIDS is a syndrome. A syndrome is made up of a collection of signs and symptoms of one or more diseases. AIDS patients have a collection of opportunistic infections and cancers. Collectively they are mistakenly referred to as the AIDS disease.
3. In 1983 by Luc Montagnier
4. 1981
5. LAV
6. Five; 1982, 1983, 1985, 1987, and 1993
7. The ARC or AIDS Related Complex definition was a middle ground used before AIDS was better understood. It became meaningless after the 1985 expanded definition listed organisms and symptoms which indicated that two states existed: HIV infection and progression to AIDS.
8. It allows HIV-infected persons earlier access into federal and state medical and social programs.

CHAPTER 2

1. The unbroken transmission of infection from an infected person to an uninfected person.
2. The answer to both questions is unknown at this time.

CHAPTER 3

1. Because it contains RNA as its genetic message and a reverse transcriptase enzyme to make DNA from RNA
2. GAG-POL-ENV; at least six
3. Because HIV has demonstrated an unusually high rate of genetic mutations; (1) the reverse transcriptase enzyme in HIV is highly error prone (makes transcription errors), and (2) a variety of HIV mutants have been found within a single HIV-infected individual.

4. The reverse transcriptase enzyme is highly error prone, making at least one, and in many cases more than one, deletion, addition, or substitution per round of proviral replication.

CHAPTER 4

1. Not really, because there is no way as yet to remove the provirus from the cell's DNA.
2. A physiological measurement that serves as a substitute for a major clinical event.
3. March; November; 12
4. 5
5. Zidovudine, didanosine, zalcitabine, stavudine, lamivudine.
6. Becoming incorporated into DNA as it is being synthesized, thereby stopping reverse transcriptase from attaching the next nucleotide.
7. (a) Clinical biological side effects; (b) the selection of drug-resistant HIV mutants.
8. (a) The number of copies of HIV RNA present in the plasma; (b) this number indicates the reproductive activity of HIV at the time and, if therapy is being used, the effect of the therapy on the reproductive ability of the virus.
9. Saquinavir, mesylate, saquinavir (Fortovase) ritonavir, indinavir, nelfinavir, amprenavir.
10. Indinavir, because either as a monotherapy or in combination, it is the most successful at dropping the viral load.
11. They physically interact with the reverse transcriptase enzyme and interfere with its function.
12. To suppress HIV replication, thereby reducing the number of mutant RNA strands produced.
13. A
14. B
15. To be determined by the instructor.
16. Answer depends on the credible facts the student presents.

CHAPTER 5

1. T4 helper cells; because T4 cells are crucial for the production of antibodies, a depletion of T4 cells results in immunosuppression which results in OIs.

2. CD4 is a receptor protein (antigen) secreted by certain cells of the immune system, for example, monocytes, macrophages, and T4 helper cells. It becomes located on the exterior of the cellular membrane and happens to be a compatible receptor for the HIV to attach and infect the CD4-carrying cell.

3. The question of true latency after HIV infection has not been settled. Most HIV/AIDS investigators currently believe there is a latent period, a time of few if any clinical symptoms and low levels of HIV in the blood. Other scientists, currently the minority, believe there is no true latency. The virus hides out in the lymph nodes slowly reproducing, and slowly killing off the T4 cells. The virus is always present, increasing slowly in numbers over time.

4. True
5. False
6. True
7. False
8. False
9. True
10. False
11. True
12. True
13. True
14. B assessing risk of disease progression.

CHAPTER 6

1. OI is caused by organisms that are normally within the body and held in check by an active immune system. When the immune system becomes suppressed, for whatever reason, these agents can multiply and produce disease.

2. *Pneumocystis carinii*; lungs, pneumonia

3. *Isospora belli*

4. *Mycobacterium avium intracellulare*

5. False. HIV has not been found in KS tissue. KS is believed to develop as a result of a suppressed immune system and not the virus *per se*.

6. Classic KS, as described by Moritz Kaposi; and KS associated with AIDS

7. False. KS normally affects gay males. It is highly unusual to find KS in hemophiliacs, intravenous drug users and female AIDS patients.

8. True

9. True

10. True (Answer provided in POI 6.1)

CHAPTER 7

1. The 6-stage Walter Reed System and the 4 group CDC system

2. About 30%; about 90%

3. AIDS dementia complex

4. Skin—Kaposi's sarcoma

   Eyes—CMV retinitis

   Mouth—thrush or hairy leukoplakia

   Lungs—pneumocystis pneumonia

   Intestines—diarrhea

5. True

6. False. The average time is 6 to 18 weeks.

7. False. HIV infection leads to HIV disease. AIDS is the result of a weakened immune system that allows opportunistic infections to occur.

8. False. The average length of time is about 10 to 11 years.

9. Instructors evaluation

10. Answer is E. (all of the above)

CHAPTER 8

1. False. The United States currently **reports** most of the world's AIDS cases.

2. Cases of AIDS-related death, according to the CDC definition, can be traced back to 1952 in the United States and to the mid-1950s in Africa.

3. HIV-1 and HIV-2 show a 40% to 50% genetic relationship to each other.

4. False. HIV-1 and HIV-2 are both transmitted via the same routes. HIV-2 is spreading globally in similar fashion to HIV-1.

5. True. All scientific and empirical evidence to date indicates that HIV is **not** casually transmitted.

6. Through sexual activities: exchange of certain body fluids—blood and blood products, semen and vaginal secretions; and from mother to fetus or newborn by breast milk.

7. False. There is only one documented case of HIV infection caused by deep kissing. HIV has been found in the saliva of infected people in very low concentration, and saliva has been shown to have anti-HIV properties.

8. True; but this assertion has been proven to be untrue. Insects, in particular mosquitoes, have not been shown to transmit HIV successfully.

9. False. According to studies involving the sexual partners of injection drug users and hemophiliacs, HIV transmission from male to female is the more efficient route. This is believed to be due to a greater concentration of HIV found in semen than in vaginal fluid.

10. The answer may be True or False. There have been cases in which a single act of intercourse has resulted in HIV infection. However, the majority of surveys on the sexual partners of injection drug users and hemophiliacs indicate that the number of sexual encounters may increase the risk of HIV infection but does not guarantee infection. Sexual partners of infected people have remained HIV-free after years of unprotected penis–vagina or penis-anus intercourse.

11. The percentage of fetal risk varies widely in a number of hospital studies. At the moment, the risk as reported without zidovudine therapy, varies from less than 30%. For Africa the figures most commonly used are 30% to 50%. With the use of zidovudine therapy, the risk has been cut to about 8%. Using zidovudine and a cesarean section reduces HIV transmission to about 2%.

12. E, all of the above

13. True

14. True

15. True

16. True

17. True

18. False

19. True

20. True

21. False

22. True

23. True

24. True

25. b. mosquito bites

CHAPTER 9

1. Latex condoms. They are known to stop the transmission of viruses. This may not be true for animal intestine condoms.

2. Water-based lubricants. Oil-based lubricants weaken the latex rubber causing them to leak or break under stress.

3. Safer sex is having sexual intercourse with an *uninfected* partner while using a condom.

4. The answer may be True or False. There have been cases where a single act of intercourse resulted in HIV infection. However, the majority of surveys completed by sexual partners of injection drug users and hemophiliacs indicate that the number of sexual encounters may increase the risk of HIV infection but does not guarantee infection. Sexual partners of infected persons have remained HIV-free after years of unprotected penis-vagina intercourse.

5. No. IDUs exist between "fixes." They lose things, they may not care to pick up new equipment—they need the "fix" now, it may be easier to share. Circumstances vary considerably among the IDU. Just giving them free equipment is no assurance that they will use it.

6. Between 1 in 39,000 and 1 in 200,000.

7. Have several students read their answers for promoting class discussion. Compare their response to that given in the text (that they should be punished).

8. Because attenuated HIV may mutate to a virulent form causing an HIV infection; there is no absolute guarantee that 100% of HIV are inactivated.

9. Because at no time will a whole HIV be present in the vaccine. Only a specific subunit of the HIV will be present in pure form so the vaccine should be free of any contaminating proteins that might prove toxic to one or more persons receiving the vaccine.

10. It is a situation wherein HIV antibodies might predispose the host to become more easily HIV-infected. For whatever reason, it appears that the HIV antibody complex enters the cell more easily than HIV alone.

11. Because of the severe forms of social ostracism that occur when it is learned that someone is HIV-positive or belongs to a high-risk group (gay, injection drug user, bisexual).

12. Universal precautions are a list of rules and regulations provided by the CDC to help prevent HIV infection in health care workers.

13. False

14. False

15. True

16. True

17. True

18. False

19. False

20. Instructor evaluation

21. False, 1998

CHAPTER 10

1. The CDC said for each AIDS case there are from 50 to 100 HIV-infected people in the population. They got the 1 to 1.5 million figure using the 50 HIV-infected/AIDS cases; 1986; they are believed to be within plus or minus 10%.

2. Because their social and sexual behaviors and medical needs place these people at a greater risk for HIV exposure than those not practicing these social and sexual behaviors or who do not need blood or blood products.

3. False. Studies show that the time for progression from HIV infection to AIDS is the same regardless of parameters.

4. 52%

5. 30% to 50%

6. 99%

7. Two per 1,000 students; more: the rate for military personnel is 1.4 per 1,000.

8. College students 2/1,000, general population 0.02/1,000; this means the rate of HIV infection on college campuses is about 10 times higher than in the general population.

9. One in 20 to 300

10. Needlestick injuries

11. 3.5%

12. Hepatitis B virus

13. GAO—300,000 to 485,000
     CDC—270,000

14. Student's answer; text says no. People most often do not tell the truth—much depends on where, when, why, and who is doing the survey. There are just too many variables involved to believe sexual surveys.

15. Worldwide about 16 million, United States about 425,000.

16. 57% $\dfrac{425,000}{746,000}$

CHAPTER 11

1. Approximately 235 million

2. About 50%

3. Injection drug use, being a sexual partner of an IDU, and through heterosexual contact.

4. IDU

5. 32 million; 8.5 million

6. Leading; 25 and 34; fifth; white; 25 and 44; leading; 25 and 44

7. An estimated 100,000

8. None, all states now have reported pediatric cases.

9. Virtually all - 100%

10. (1) maternal viral load; (2) route of delivery, and (3) duration of early membrane rupture.

11. Most orphaned AIDS children have mothers who are IDUs and are themselves HIV-infected. They are AIDS orphans because (1) their parents abandon them due to illness or death; and (2) these children are HIV-infected or demonstrate AIDS and therefore no one wants them.

12. Over 50%

13. Two; 300

14. 28 million

15. Teenage gay men and teenage women; 75%

16. 25%

17. About 500,000

18. Yes; black and Latinos

19. One third or about 14 million

20. 1-800-234-8336

CHAPTER 12

1. ELISA; enzyme linked immunosorbent assay

2. That the body will produce antibody against antigenic components of the HIV virus after infection occurs

3. No; a positive antibody result must be repeated in duplicate and if still positive, a confirmatory test is performed prior to telling people they are HIV-infected.

4. No; AIDS is medically diagnosed after certain signs and symptoms of specific diseases occur.

5. Western blot

6. Indirect immunofluorescent assay

7. By a color change in the reaction tube; the peroxidase enzyme oxidizes a clear chromogen into color formation. This occurs if the HIV antibody–antigen enzyme complex is present in the reaction tube.

8. False. Some newborns receive the HIV antibody passively during pregnancy. About 30% to 50% of HIV-positive newborns are truly HIV-positive; it is unknown whether all HIV-positive newborns go on to develop AIDS. Not all have been discovered and it has not been determined whether 100% of HIV-infected adults or babies will develop AIDS.

9. They are not 100% accurate.

10. Determining that positive and negative tests are truly positive and negative and not falsely positive or negative.

11. Using either too high or too low cutoff points in the spectrophotometer and the presence of cross-reacting antibodies

12. The percentage of false positives will increase as the prevalence of HIV-infected people in a population decreases.

13. It is a screening test value that represents the probability that a positive HIV test is truly positive; because screening tests are not 100% accurate.

14. Western blot

15. There is no standardized WB test interpretation. Different agencies use different WB results (reactive bands) to determine that the test sample is positive.

16. Because the PCR allows for the detection of proviral DNA in cells before the body produces detectable HIV antibody; PCR reactions can be used to determine if high-risk (or anybody) antibody-negative people are HIV-infected but not producing antibodies and whether newborns are truly HIV-positive or passively HIV-positive.

17. False. The FDA approved two home-use HIV antibody test kits in 1996 but one was withdrawn from the market.

18. Between 6 and 18 weeks after HIV infection.

19. As early as 2 weeks after infection.

20. (1) Changes in their lifestyles that reduce stress on their immune systems may delay the onset of illness.

    (2) They can practice safer sex and hopefully not transmit the virus to others.

    (3) The earlier the detection, the earlier they can enter into preventive therapy.

21. Mandatory with confidentiality; voluntary with confidentiality, anonymous and blinded.

22. For anonymous testing no personal information is given; in blind tests, the name is deleted but the demographic data remain.

23. Because there are many example of breaches of confidence, which destroys trust and subjects people to social stigma.

24. False

25. True

CHAPTER 13

1. Newspapers, TV, radio, magazines, etc.

2. Barring children from public schools, police wearing rubber gloves during arrests, not going to a restaurant because someone who works there has AIDS, firing AIDS employees, etc.

3. Syphilis

4. Use of tobacco products, alcohol, drugs; nonuse of seat belts and motorcycle helmets, etc.

5. The ways by which one can become HIV-infected and how not to become HIV-infected.

6. Because most of the new cases of HIV infection and AIDS occur in high-risk groups that will not or cannot change sexual and drug practices.

7. Because a larger percentage of teenagers are sexually active with more than one partner, use drugs, use alcohol, and think that they are invulnerable to infection and death.

8. According to the AMA, No. Physicians may not refuse to care for patients with AIDS because of actual risk or fear of contracting the disease.

9. The CDC and AMA feel that a patient's right to that information should be determined on a case-by-case basis where surgery will be performed. There is no legal requirement for physicians to tell their patients of their HIV status.

10. Worker information sessions that explain how the virus can and cannot be transmitted.

Glossary

ACRONYMS

ACTG AIDS clinical trial group

AIDS acquired immunodeficiency syndrome

AZT azathioprine (a misnomer for zidovudine or azidothymidine)

CD4 a protein imbedded on the surface of a T lymphocyte to which HIV most often binds—a T4 cell.

CD8 a protein imbedded on the surface of a T lymphocyte suppressor cell—a T8 cell.

CDC Centers for Disease Control and Prevention (part of PHS)

3TC Lamivudine; nucleoside analog

DHHS Department of Health and Human Services

DNA deoxyribonucleic acid

d4T stavudine; nucleoside analog

ddC dideoxycytosine; nucleoside analog

ddI dideoxyinosine; nucleoside analog

FDA Food and Drug Administration (part of PHS)

IDU Injection Drug User

LAV lymphadenopathy-associated virus

NCI National Cancer Institute (part of NIH)

NIAID National Institute of Allergy and Infectious Diseases (part of NIH)

NIH National Institutes of Health (part of PHS)

PCR (polymerase chain reaction) a very sensitive test used to detect the presence of HIV

PHS Public Health Service (part of DHHS)

RNA ribonucleic acid

ZDV Zidovudine major drug in treating HIV/AIDS persons; nucleoside analog

For the newest anti-HIV drugs, names and use, see Chapter 4.

TERMS

Acquired immunodeficiency syndrome (AIDS): A life-threatening disease caused by a virus and characterized by the breakdown of the body's immune defenses. (See AIDS.)

Acute: Sudden onset, Short-term with severe symptoms.

Acyclovir (Zovirax): Antiviral drug for herpes 1 and 2 and herpes zoster.

AIDS (acquired immunodeficiency syndrome): A disease caused by a retrovirus called HIV and characterized by a deficiency of the immune system. The primary defect in AIDS is an acquired, persistent, quantitative functional depression within the T4 subset of lymphocytes. This depression often leads to infections caused by opportunistic microorganisms in HIV-infected individuals. A rare type of cancer (Kaposi's sarcoma) usually seen in elderly men or in individuals who are severely immunocompromised may also occur.

AIDS dementia: Neurological complications affecting thinking and behavior; intellectual impairment.

Analog (analogue): A chemical molecule that closely resembles another one but which may function differently, thus altering a natural process.

Anemia: Low number of red blood cells.

Antibiotic: A chemical substance capable of destroying bacteria and other microorganisms.

Antibody: A blood protein produced by mammals in response to a specific antigen.

Antigen: A large molecule, usually a protein or carbohydrate, which when introduced into the body stimulates the production of an antibody that will react specifically with that antigen.

Antigen-presenting cells: B cells, cells of the monocyte lineage (including macrophages and dentritic cells) and various other body cells that 'present' antigen in a form that T cells can recognize.

Antiserum: Serum portion of the blood that carries the antibodies.

Antiviral: Means against virus; drugs that destroy or weaken virus.

Asymptomatic carrier: A host which is infected by an organism but does not demonstrate clinical signs or symptoms of the disease.

Asymptomatic seropositive: HIV-positive without signs or symptoms of HIV disease.

Attenuated: Weakened.

Atypical: Irregular; not of typical character.

Autoimmunity: Antibodies made against self tissues.

B lymphocytes or B cells: Lymphocytes that produce antibodies. B lymphocytes proliferate under stimulation from factors released by T lymphocytes.

B- and T-cell lymphomas: Cancers caused by proliferation of the two principal types of white blood cells—B and T lymphocytes.

Bacterium: A microscopic organism composed of a single cell. Many but not all bacteria cause disease.

Blood count: A count of the number of red and white blood cells and platelets.

Bone marrow: Soft tissue located in the cavities of the bones. The bone marrow is the source of all blood cells.

Cancer: A large group of diseases characterized by uncontrolled growth and spread of abnormal cells.

Candida albicans: A fungus; the causative agent of vulvovaginal candidiasis or yeast infection.

Candidiasis: A fungal infection of the mucous membranes (commonly occurring in the mouth, where it is known as Thrush) characterized by whitish spots and/or a burning or painful sensation. It may also occur in the esophagus. It can also cause a red and itchy rash in moist areas, e.g., the vagina.

Capsid: The protein coat of a virus particle.

CC-CKR-5 (CKR-5): Receptor for human chemokines and a necessary receptor for HIV entrance into macrophage.

CD: Cluster differentiating type antigens found on T lymphocytes. Each CD is assigned a number: CD1, CD2, etc.

CD4(T4 cell): White blood cell with type 4 protein embedded in the cell surface—target cell for HIV infection.

CD8 cell: Suppressor white blood cell with type 8 protein embedded in the cell surface.

Cell-mediated immunity: The reaction to antigenic material by specific defensive cells (macrophages) rather than antibodies.

Cellular immunity: A collection of cell types that provide protection against various antigens.

Chain of infection: A series of infections that are directly or immediately connected to a particular source.

Chemokines: Chemicals released by T cell lymphocytes and other cells of the immune system to attract a variety of cell types to sites of inflammation.

Chemotherapy: The use of chemicals that have a specific and toxic effect upon a disease-causing pathogen.

Chlamydia: A species of bacterium, the causative organism of *Lymphogranuloma venereum*, chlamydial urethritis and most cases of newborn conjunctivitis.

Chromosomes: Physical structures in the cell's nucleus that house the genes. Each human cell has 22 pairs of autosomes and two sex chromosomes.

Chronic: Having a long and relatively mild course.

Clade: Related HIV variants classified by degree of genetic similarity; nine are known for HIV.

Cleavage Site: One of nine sites (peptide bond) within the gag-pol polyprotein (peptide precursor) that is cleaved by HIV-1 protease to form functional subunits of gag (p 17, p7, p24) and pol (protease, reverse transcriptase, integrase).

Clinical latency: Infectious agent developing in a host without producing clinical symptoms.

Clinical manifestations: The signs of a disease as they pertain to or are observed in patients.

CMV: See cytomegalovirus

Cofactor: Factors or agents which are necessary or which increase the probability of the development of disease in the presence of the basic etiologic agent of that disease.

Cohort: A group of individuals with some characteristics in common.

Communicable: Able to spread from one diseased person or animal to another, either directly or indirectly.

Condylomata acuminatum (venereal warts): Viral warts of the genital and anogenital area.

Congenital: Acquired by the newborn before or at the time of birth.

Core proteins: Proteins that make up the internal structure or core of a virus.

Cross-resistance: Development of resistance to one agent as an antibotic, that results in resistance to other, usually similar agents.

Cryptococcal meningitis: A fungal infection that affects the three membranes (meninges) surrounding the brain and spinal cord. Symptoms include severe headache, vertigo, nausea, anorexia, sight disorders and mental deterioration.

Cryptococcosis: A fungal infectious disease often found in the lungs of AIDS patients. It characteristically spreads to the meninges and may also spread to the kidneys and skin. It is due to the fungus *Cryptococcus neoformans*.

Cryptosporidiosis: An infection caused by a protozoan parasite found in the intestines of animals. Acquired in some people by direct contact with the infected animal, it lodges in the intestines and causes severe diarrhea. It may be transmitted from person to person. This infection seems to be occurring more frequently in immunosuppressed people and can lead to prolonged symptoms which do not respond to medication.

Cutaneous: Having to do with the skin.

CXCR4 (FUSIN): Receptor for human chemokines and a necessary receptor for HIV entrance into T4 cells.

Cytokines: Powerful chemical substances secreted by cells. Cytokines include lymphokines produced by lymphocytes and monokines produced by monocytes and macrophages.

Cytomegalovirus (CMV): One of a group of highly host-specific herpes viruses that affect humans and other animals. Generally produces mild flu-like symptoms but can be more severe. In the immunosuppressed, it may cause pneumonia.

Cytopathic: Pertaining to or characterized by abnormal changes in cells.

Cytotoxic: Poisonous to cells.

Cytotoxic T cells: A subset of T lymphocytes that carry the T8 marker and can kill body cells infected by viruses or transformed by cancer.

Dementia: Chronic mental deterioration sufficient to significantly impair social and/or occupational function. Usually patients have memory and abstract thinking loss.

Dendritic cells: White blood cells found in the spleen and other lymphoid organs. Dendritic cells typically use threadlike tentacles to "hold" the antigen, which they present to T cells.

Didanosine: Also known as videx; see ddI-inhibits HIV Replication.

Dissemination: Spread of disease throughout the body.

DNA (deoxyribonucleic acid): A linear polymer, made up of deoxyribonucleotide repeating units. It is the carrier of genetic information in living organisms and some viruses.

DNA viruses: Contain DNA as their genetic material.

Dysentery: Inflammation of the intestines, especially the colon, producing pain in the abdomen and diarrhea containing blood and mucus.

Efficacy: Effectiveness.

ELISA test: A blood test which indicates the presence of antibodies to a given antigen. Various ELISA tests are used to detect a variety of infections. The HIV ELISA test does not detect AIDS but only indicates if viral infection has occurred.

Endemic: Prevalent in or peculiar to community or group of people.

Enteric infections: Infections of the intestine.

ENV: HIV gene that codes for protein gp160

Envelope proteins: Proteins that comprise the envelope or surface of a virus, gp120 and gp41.

Enzyme: A catalytic protein that is produced by living cells and promotes the chemical processes of life without itself being altered or destroyed.

Epidemic: When the incidence of a disease surpasses the expected rate in any well-defined geographical area.

Epidemiology: The study of the factors that impact on the spread of disease in an area.

Epivir: See 3TC.

Epstein-Barr virus (EBV): A virus that causes infectious mononucleosis. It is spread by saliva. EBV lies dormant in the lymph glands and has been associated with Burkitt's lymphoma, a cancer of the lymph tissue.

Etiologic agent: The organism which causes a disease.

Etiology: The study of the cause of disease.

Extracellular: Found outside the cell wall.

Factor VIII: A naturally occurring protein in plasma that aids in the coagulation of blood. A congenital deficiency of Factor VIII results in the bleeding disorder known as hemophilia A.

Factor VIII concentrate: A concentrated preparation of Factor VIII that is used in the treatment of individuals with hemophilia A.

False negative: Failure of a test to demonstrate the disease or condition when present.

False positive: A positive test result caused by a disease or condition other than the disease for which the test is designed.

Fellatio: Oral sex involving the penis.

Fitness: The ability of an individual virus to replicate successfully under defined conditions.

Follicular dendritic cells: Found in germinal centers of lymphoid organs.

Fomite: An inanimate object that can hold infectious agents and transfers them from one individual to another.

Fortovase: A more easily assimilated form of saquinavir.

Fulminant: Rapid onset, severe.

Fungus: Member of a class of relatively primitive organisms. Fungi include mushrooms, yeasts, rusts, molds and smuts.

Fusin: See CXCR4

Gammaglobulin: The antibody component of the serum.

Ganciclovir (DHPG): An experimental antiviral drug used in the treatment of CMV retinitis.

Gene: The basic unit of heredity; an ordered sequence of nucleotides. A gene contains the information for the synthesis of one polypeptide chain (protein).

Gene expression: The production of RNA and cellular proteins.

Genitourinary: Pertaining to the urinary and reproductive structures; sometimes called the GU tract or system.

Genome: The genetic endowment of an organism.

Genotype: The sequence of nucleotide bases that constitutes a gene.

GP41: Glycoprotein found in envelope of HIV.

GP120: Glycoprotein found in outer level of HIV envelope.

GP160: Precusor glycoprotein to forming gp41 and gp120.

Globulin: That portion of serum which contains the antibodies.

Glycoproteins: Proteins with carbohydrate groups attached at specific locations.

Gonococcus: The specific etiologic agent of gonorrhea discovered by Neisser and named *Neisseria gonorrhoeae.*

Granulocytes: Phagocytic white blood cells filled with granules containing potent chemicals that allow the cells to digest microorganisms. Neutrophils, eosinophils, basophils and mast cells are examples of granulocytes.

Hemoglobin: The oxygen-carrying portion of red blood cells which gives them a red color.

Hemophilia: A hereditary bleeding disorder caused by a deficiency in the ability to synthesize one or more of the blood coagulation proteins, e.g., Factor VIII (hemophilia A) or Factor IX (hemophilia B).

Hepatitis: Inflammation of the liver; due to many causes including viruses, several of which are transmissible through blood transfusions and sexual activities.

Hepatosplenomegaly: Enlargement of the liver and spleen.

Herpes simplex virus I (HSV-I): A virus that results in cold sores or fever blisters, most often on the mouth or around the eyes. Like all herpes viruses, it may lie dormant for months or years in nerve tissues and flare up in times of stress, trauma, infection or immunosuppression. There is no cure for any of the herpes viruses.

Herpes simplex virus II (HSV-II): Causes painful sores on the genitals or anus. It is one of the most common sexually transmitted diseases in the United States.

Herpes varicella zoster virus (HVZ): The varicella virus causes chicken pox in children and may reappear in adulthood as herpes zoster. Herpes zoster, also called shingles, is characterized by small, painful blisters on the skin along nerve pathways.

Histoplasmosis: A disease caused by a fungal infection that can affect all the organs of the body. Symptoms usually include fever, shortness of breath, cough, weight loss and physical exhaustion.

HIV (human immunodeficiency virus): A newly discovered retrovirus that is said to cause AIDS. The target organ of HIV is the T4 or CD4 subset of T lymphocytes, which regulate the immune system.

HIV-positive: Presence of the human immunodeficiency virus in the body.

Homophobia: Negative bias towards or fear of individuals who are homosexual.

Human leukocyte antigens (HLA): Protein markers of self used in histocompatibility testing. Some HLA types also correlate with certain auto-immune diseases.

Humoral immunity: The production of antibodies for defense against infection or disease.

Immunity: Resistance to a disease because of a functioning immune system.

Immune complex: A cluster of interlocking antigens and antibodies.

Immune response: The reaction of the immune system to foreign substances.

Immune status: The state of the body's natural defense to diseases. It is influenced by heredity, age, past illness history, diet and physical and mental health. It includes production of circulating and local antibodies and their mechanism of action.

Immunoassay: The use of antibodies to identify and quantify substances. Often the antibody is linked to a marker such as a fluorescent molecule, a radioactive molecule or an enzyme.

Immunocompetent: Capable of developing an immune response

Immunoglobulins: A family of large protein molecules, also known as antibodies.

Immunostimulant: Any agent that will trigger a body's defenses.

Immunosuppression: When the immune system is not working normally. This can be the result of illness or certain drugs (commonly those used to fight cancer).

Incidence: The total number of new cases of a disease in a given area within a specified time, usually 1 year.

Incubation period: The time between the actual entry of an infectious agent into the body and the onset of disease symptoms.

Indinavir: Crixivan, a protease inhibitor drug.

Infection: Invasion of the body by viruses or other organisms.

Infectious disease: A disease which is caused by microorganisms or viruses living in or on the body as parasites.

Inflammatory response: Redness, warmth and swelling in response to infection; the result of increased blood flow and a gathering of immune cells and secretions.

Injection drug use: Use of drugs injected by needle into a vein or muscle tissue.

Innate immunity: Inborn or hereditary immunity.

Inoculation: The entry of an infectious organism or virus into the body.

Integrase: HIV enzyme used to insert HIV DNA into host cell DNA.

Interferon: A class of glycoproteins important in immune function and thought to inhibit viral infection.

Interleukins: Chemical messengers that travel from leukocytes to other white blood cells. Some promote cell development, others promote rapid cell division.

Intracellular: Found within the cell wall.

In utero: In the uterus

In vitro: "In glass"—pertains to a biological reaction in an artificial medium.

In vivo: "In the living"—pertains to a biological reaction in a living organism.

IV: Intravenous.

Kaposi's sarcoma: A multifocal, spreading cancer of connective tissue, principally involving the skin; it usually begins on the toes or the feet as reddish blue or brownish soft nodules and tumors.

Lamivudine: Nucleoside analog inhibits HIV replication.

Langerhans cells: Dendritic cells in the skin that pick up antigen and transport it to lymph nodes.

Latency: A period when a virus or other organism is still in the body but in an inactive state.

Latent viral infection: The virion becomes part of the host cell's DNA.

Lentiviruses: Viruses that cause disease very slowly. HIV is believed to be this type of virus.

Lesion: Any abnormal change in tissue due to disease or injury.

Leukocyte: A white blood cell.

Leukopenia: A decrease in the white blood cell count.

Log: 10-fold difference.

Lymph: A transparent, slightly yellow fluid that carries lymphocytes, bathes the body tissues and drains into the lymphatic vessels.

Lymph nodes: Gland-like structures in the lymphatic system which help to prevent spread of infection.

Lymphadenopathy: Enlargement of the lymph nodes.

Lymphadenopathy syndrome (LAS): A condition characterized by persistent, generalized, enlarged lymph nodes, sometimes with signs of minor illness such as fever and weight loss, which apparently represents a milder reaction to HIV infection.

Lymphatic system: A fluid system of vessels and glands which is important in controlling infections and limiting their spread.

Lymphocytes: Specialized white blood cells involved in the immune response.

Lymphoid organs: The organs of the immune system where lymphocytes develop and congregate. They include the bone marrow, thymus, lymph nodes, spleen and other clusters of lymphoid tissue.

Lymphokines: Chemical messengers produced by T and B lymphocytes. They have a variety of protective functions.

Lymphoma: Malignant growth of lymph nodes.

Lymphosarcoma: A general term applied to malignant neoplastic disorders of lymphoid tissue, not including Hodgkin's disease.

Lytic infection: When a virus infects the cell, the cell produces new viruses and breaks open (lyse) releasing the viruses.

Macrophage: A large and versatile immune cell that acts as a microbe-devouring phagocyte, an antigen-presenting cell and an important source of immune secretions.

Major histocompatibility complex (MHC): A group of genes that controls several aspects of the immune response. MHC genes code for self markers on all body cells.

Malaise: A general feeling of discomfort or fatigue.

Malignant tumor: A tumor made up of cancerous cells. The tumors grow and invade surrounding tissue, then the cells break away and grow elsewhere.

Messenger RNA (mRNA): RNA that serves as the template for protein synthesis; it carries the information from the DNA to the protein synthesizing complex to direct protein synthesis.

Microbes: Minute living organisms including bacteria, viruses, fungi and protozoa.

Microorganisms: Microscopic plants or animals.

Molecule: The smallest amount of a specific chemical substance that can exist alone. To break a molecule down into its constituent atoms is to change its character. A molecule of water, for instance, reverts to oxygen and hydrogen.

Monocyte: A large phagocytic white blood cell which, when it enters tissue, develops into a macrophage.

Monokines: Powerful chemical substances secreted by monocytes and macrophages. They help direct and regulate the immune response.

Morbidity: The proportion of people with a disease in a community.

Morphology: The study of the form and structure of organisms.

Mortality: The number of people who die as a result of a specific cause.

Mucosal immunity: Resistance to infection across, mucous membranes.

Mucous membrane: The lining of the canals and cavities of the body which communicate with external air, such as the intestinal tract, respiratory tract and the genitourinary tract.

Mucous patches: White, patchy growths, usually found in the mouth, that are symptoms of secondary syphilis and are highly infectious.

Mucus: A fluid secreted by membranes

Mutant: A new strain of a virus or microorganism that arises as a result of change in the genes of an existing strain.

Natural killer cells (also called NK cells): Immune cells that kill infected cells directly within 4 hours of contact. NK cells differ from other killer cells, such as cytotoxic T lymphocytes, in that they do not require contact with antigen before they are activated.

Neisseria gonorrhoeae: The bacterium that causes gonorrhea.

Neonatal: Pertaining to the first 4 weeks of life.

Neoplasm: A new abnormal growth, such as a tumor.

Neuropathy: Group of nerve disorders—symptoms range from tingling sensation, numbness to paralysis.

Nevirapine: Non-nucleostide analog inhibits HIV replication.

Notifiable disease: A notifiable disease is one that, when diagnosed, health providers are required, usually by law, to report to state or local public health officials. Notifiable diseases are those of public interest by reason of their contagiousness, severity, or frequency.

Nucleic acids: Large, naturally occurring molecules composed of chemical building blocks known as nucleotides. There are two kinds of nucleic acid, DNA and RNA.

Nucleoside analog: Synthetic compounds generally similar to one of the bases of DNA.

Nucleotide of DNA: Made up of one of four nitrogen-containing bases (adenine, cytosine, guanine or thymine), a sugar and a phosphate molecule.

Oncogenic: Anything that may give rise to tumors, especially malignant ones.

Opportunistic disease: Disease caused by normally benign microorganisms or viruses that become pathogenic when the immune system is impaired.

p24 antigen: A protein fragment of HIV. The p24 antigen test measures this fragment. A positive test result suggests active HIV replication and may mean the individual has a chance of developing AIDS in the near future.

Parenteral: Not taken in through the digestive system.

Parasite: A plant or animal that lives, grows and feeds on another living organism.

Pathogen: Any disease-producing microorganism or substance.

Pathogenic: Giving rise to disease or causing symptoms of an illness.

Pathology: The science of the essential nature of diseases, especially of the structural and functional changes in tissues and organs caused by disease.

Perianal glands: Glands located around the anus.

Perinatal: Occurring in the period during or just after birth.

Phagocytes: Large white blood cells that contribute to the immune defense by ingesting microbes or other cells and foreign particles.

Phenotype: A defined behavior; specifically drug susceptibility with regard to HIV drug resistance.

PID (pelvic inflammatory disease): Inflammation of the female pelvic organs; often the result of gonococcal or chlamydial infection.

Placebo: An inactive substance against which investigational treatments are compared to see how well the treatment worked.

Plasma: The fluid portion of the blood which contains all the chemical constituents of whole blood except the cells.

Plasma cells: Derived from B cells, they produce antibodies.

Platelets: Small oval discs in blood that are necessary for blood to clot.

PLWA: Persons Living With AIDS.

Polymerase chain reaction: Method to detect and amplify very small amounts of DNA in a sample.

Positive HIV test: A sample of blood that is reactive on an initial ELISA test, reactive on a second

ELISA run of the same specimen and reactive on Western blot, if available.

***Pneumocystis carinii* pneumonia (PCP):** A rare type of pneumonia primarily found in infants and now common in patients with AIDS.

Prenatal: During pregnancy.

Prevalence: The total number of cases of a disease existing at any time in a given area.

Primary immune response: Production of antibodies about 7 to 10 days after an infection.

Prophylactic treatment: Medical treatment of patients exposed to a disease before the appearance of disease symptoms.

Protease: Enzyme that cuts proteins into peptides (breaks down proteins).

Protease inhibitors: Compounds that inhibit the action of protease.

Proteins: Organic compounds made up of amino acids. Proteins are one of the major constituents of plant and animal cells.

Protocol: Standardization of procedures so that results of treatment or experiments can be compared.

Protozoa: A group of one-celled animals, some of which cause human disease including malaria, sleeping sickness and diarrhea.

Provirus: The genome of an animal virus integrated into the chromosome of the host cell, and thereby replicated in all of the host's daughter cells.

Quasispecies: A complex mixture of genetic variants of an RNA virus.

Race: Beginning in 1976 the federal government's data systems classified individuals into the following racial groups: American Indian or Alaskan Native, Asian or Pacific Islander, black, and white.

Rate: A rate is a measure of some event, disease, or condition in relation to a unit of population, along with some specification of time.

Receptors: Special molecules located on the surface membranes of cells that attract other molecules to attach to them. (For example CD4, CD8, and CC-CKR-5)

Recombinant DNA: DNA produced by joining pieces of DNA from different sources.

Recombinant DNA techniques: Techniques that allow specific segments of DNA to be isolated and inserted into a bacterium or other host (e.g., yeast, mammalian cells) in a form that will allow the DNA segment to be replicated and expressed as the cellular host multiplies.

Remission: The lessening of the severity of disease or the absence of symptoms over a period of time.

Retroviruses: Viruses that contain RNA and produce a DNA analog of their RNA using an enzyme known as reverse transcriptase.

Reverse transcriptase: An enzyme produced by retroviruses that allows them to produce a DNA analog of their RNA, which may then incorporate into the host cell.

Ritonavir: Novir, a protease inhibitor drug.

RNA (ribonucleic acid): Any of various nucleic acids that contain ribose and uracil as structural components and are associated with the control of cellular chemical activities.

RNA viruses: Contain RNA as their genetic material.

Sarcoma: A form of cancer that occurs in connective tissue, muscle, bone and cartilage.

Saquinavir: Invirase, a protease inhibitor drug.

Secondary immune response: On repeat exposure to an antigen, there is an accelerated production of antibodies.

Sensitivity: The probability that a test will be positive when the infection is present.

Septicemia: A disease condition in which the infectious agent has spread throughout the lymphatic and blood systems causing a general body infection.

Seroconversion: The point at which an individual exposed to the AIDS virus becomes serologically positive.

Serologic test: Laboratory test made on serum.

Serum: The clear portion of any animal liquid separated from its more solid elements, especially the clear liquid which separates in the clotting of blood (blood serum).

Shigella: A bacterium that can cause dysentery.

Specificity: The probability that a test will be negative when the infection is not present.

Spirochete: A corkscrew-shaped bacterium; e.g., *Treponema pallidum.*

Spleen: A lymphoid organ in the abdominal cavity that is an important center for immune system activities.

Squamous: Scaly or plate-like; a type of cell.

Stavudine: Also known as Zerit; See d4T—inhibits HIV replication.

STD (sexually transmitted disease): Any disease which is transmitted primarily through sexual practices.

Subclinical infections: Infections with minimal or no apparent symptoms.

Subunit vaccine: A vaccine that uses only one component of an infectious agent rather than the whole to stimulate an immune response.

Suppressor T cells: A subset of T cells that carry the T8 marker and turn off antibody production and other immune responses.

Surrogate marker: A substitute; a person or agent that replaces another, an alternate.

Surveillance: The process of accumulating information about the incidence and prevalence of disease in an area.

Susceptible: Inability to resist an infection or disease.

Syndrome: A set of symptoms which occur together.

Systemic: Affecting the body as a whole.

T cell growth factor (TCGF, also known as interleukin-2): A glycoprotein that is released by T lymphocytes on stimulation by antigens and which functions as a T cell growth factor by inducing proliferation of activated T cells.

T Helper cells (also called T4 or CD4 cells): A subset of T cells that carry the CD4 marker and are essential for turning on antibody production, activating cytotoxic T cells and initiating many other immune responses.

T lymphocytes or T cells: Lymphocytes that mature in the thymus and which mediate cellular immune reactions. T lymphocytes also release factors that induce proliferation of T lymphocytes and B lymphocytes.

T8 cells: A subset of T cells that may kill virus-infected cells and suppress immune function when the infection is over.

Thrush: A disease characterized by the formation of whitish spots in the mouth. It is caused by the fungus *Candida albicans* during times of immunosuppression.

Thymus: A primary lymphoid organ high in the chest where T lymphocytes proliferate and mature.

Titer: Level or amount.

Tolerance: A state of nonresponsiveness to a particular antigen or group of antigens.

Toxic reaction: A harmful side effect from a drug; it is dose dependent, i.e., becomes more frequent and severe as the drug dose is increased. All drugs have toxic effects if given in a sufficiently large dose.

Toxoplasmosis: An infection with the protozoan *Taxoplasma gondii*, frequently causing focal encephalitis (inflammation of the brain). It may also involve the heart, lungs, adrenal glands, pancreas and testes.

Transcription: The synthesis of messenger RNA on a DNA template; the resulting RNA sequence is complementary to the DNA sequence. This is the first step in gene expression.

Translation: The process by which the genetic code contained in a nucleotide sequence of messenger RNA directs the synthesis of a specific order of amino acids to produce a protein.

Treponema pallidum: The bacterial spirochete that causes syphilis.

Tropism: Involuntary turning, curving, or attraction to a source of stimulation.

Tumor: A swelling or enlargement; an abnormal mass that can be malignant or benign. It has no useful body function.

V3 loop: Section of the gp120 protein on the surface of HIV; appears to be important in stimulating neutralizing antibodies.

Vaccine: A preparation of dead organisms, attenuated live organisms, live virulent organisms, or parts of microorganisms that is administered to artificially increase immunity to a particular disease.

Vector: The means by which a disease is carried from one human to another.

Venereal warts: Viral *Condylomata acuminata* on or near the anus or genitals.

Viral load: The total amount of virus in a person's body.

Viremia: The presence of virus in the blood.

Virulence: The ability on the part of an infectious agent to induce, incite or produce pathogenic changes in the host.

Viïrus: Any of a large group of submicroscopic agents capable of infecting plants, animals and bacteria; characterized by a total dependence on living cells for reproduction and by a lack of independent metabolism.

Western Blot: A blood test used to detect antibodies to a given antigen. Compared to the ELISA test, the Western Blot is more specific and more expensive. It can be used to confirm the results of the ELISA test.

Wild Type: A genotype or phenotype circulating prior to selection of drug resistance.

X-ray: Radiant energy of extremely short wavelength used to diagnose and treat cancer.

Zalcitabine: Also known as HIVID; see ddC—inhibits HIV replication.

Zidovudine: Also known as Retrovir; see ZDV—inhibits HIV replication. Mistakenly referred to as AZT.

INDEX

Note: Italicized letters *f* and *t* following page numbers indicate figures and tables, respectively.